Quantum Aspects of Light Propagation

Antonín Lukš · Vlasta Peřinová

Quantum Aspects of Light Propagation

Springer

Antonín Lukš
Joint Laboratory of Optics
Palacký University and
Institute of Physics of the Czech
Academy of Sciences
772 07, Olomouc
Czech Republic
luks@prfnw.upol.cz

Vlasta Peřinová
Joint Laboratory of Optics
Palacký University and
Institute of Physics of the Czech
Academy of Sciences
772 07, Olomouc
Czech Republic
perinova@prfnw.upol.cz

Consulting Editor

D.R. Vij
Kurukshetra University
E-5 University Campus
Kurukshetra 136119
India

ISBN 978-1-4899-8480-7 ISBN 978-0-387-85590-5 (eBook)
DOI 10.1007/b101766
Springer Dordrecht Heidelberg London New York

Printed on acid-free paper

Springer is part of Springer Science+Business Media (www.springer.com)

Preface

Quantum descriptions of light propagation frequently exhibit a replacement of time by propagation distance. It seems to be natural since a propagation lasts some amount of time. The primary intention was to inform more fundamentally inclined, open-minded readers on this approach by this book. We have included also spatio-temporal descriptions of the electromagnetic field in linear and nonlinear optical media. We call some of these formalisms one dimensional (more exactly 1 + 1-dimensional), even though they comprise the time variable along with the position coordinate. These descriptions, however, are 3 + 1-dimensional in principle. The rapid development of applications of photonic band-gap structures and experiments on lasing in a disordered medium has directed us to pay attention even to these topics, which has influenced the style of the book, which becomes a very review of these streams.

This book has the following features. It reviews both macroscopic and microscopic theories of the electromagnetic field in dielectrics. It takes into account parametric down-conversion experiments. It covers results on nonlinear optical couplers. It includes optical imaging with nonclassical light. It expounds basics of quasimode theory. It respects success of the Green-function approach in describing optical field at dielectric devices, left-handed materials and the Casimir effect for some geometries. It refers to quantization in waveguides, photonic crystals, disordered media, and propagation in strongly scattering media, incoherent and coherent random lasers, and important problems in optical resonators including chaotic cavities. In our opinion it is appropriate to do something more than only formal comparison of various approaches in the future, even though the reader will already have formed an idea of their scope.

The simplest approach with one variable (time or propagation distance) and with several frequencies has proven its vitality in the development of the quantum information theory and the quantum computation. At present there exist even books devoted to these fields: Alber, G., Beth, T., Horodecki, M., Horodecki, P., Horodecki, R., Rötteler, M., Weinfurter, H., Werner, R., and Zeilinger, A. (2001), *Quantum Information: An Introduction to Basic Theoretical Concepts and Experiments*, Springer-Verlag, Berlin; Nielsen, Michael A. and Chuang, Isaac L. (2000), *Quantum Computation and Quantum Information*, Cambridge University Press, Cambridge.

The fundamental problem of light propagation in dielectric media is connected with the role of nonclassical light in applications and has been pursued intensively in quantum optics since about 1984. In the present book we review spatio-temporal descriptions of the electromagnetic field in linear and nonlinear dielectric media applying macroscopic and microscopic theories. We mainly pay attention to canonical quantum descriptions of light propagation in a nonlinear dispersionless dielectric medium and linear and nonlinear dispersive dielectric media. These descriptions are regularly simplified by a transition to the one-dimensional propagation, which is illustrated also by descriptions of some optical processes.

Quantum theories of light propagation in optical media are generalized from dielectric media to magnetodielectrics. Classical and nonclassical properties of radiation propagating through left-handed media will be presented. The theory is utilized for the quantum electrodynamical effects to be determined in periodic dielectric structures which are known to be a basis of new schemes for lasing and a control of light field state. Quantum descriptions of random lasers are provided.

It is an interesting question, to what extent the topic of this book overlaps with the condensed-matter theory. Restricting ourselves to optical devices, we cannot exclude such overlap in principle, because many of them are made of condensed matters. The condensed-matter theory, however, is devoted mainly to problems of conductors and semi-conductors. Photonic crystals can be studied similarly as ordinary electronic crystals, even though for instance the conductivity is replaced by the transmissivity. This does not mean any thematic overlap.

Texts on quantum optics have so far based the spatio-temporal description on the quantization of the electromagnetic field in a free space in the hope that differences from the field in a medium are negligible or can be easily included in other ways. A rare exception was for instance the text Vogel, W. and Welsch, D.-G. (1994), *Lectures on Quantum Optics*, Akademie Verlag, Berlin, where a choice of a suitable approach, albeit a selection of one of possibilities, was declared.

The book will be useful to research workers in the field of general optics, quantum optics and electronics, optoelectronics, and nonlinear optics, as well as to students of physics, optics, optoelectronics, photonics, and optical engineering.

Olomouc Vlasta Peřinová
Olomouc Antonín Lukš

Acknowledgments

We have pleasure in thanking Dr. J. Peřina, Jr., Ph.D., for communicating files to the publisher, graphics, and word processing and Ing. J. Křepelka, Ph.D., for the careful preparation of figures. This book has arisen under the financial support by the Ministry of Education of the Czech Republic in the framework of the project No. 1M06002 "Optical structures, detection systems, and related technologies for few-photon applications".

Contents

Chapter 1
Introduction

The importance of quantum optics has been recognized by both specialists and public since Roy J. Glauber was awarded the Nobel Prize in Physics 2005. Quantum informatics is closely connected with this field. Ingenious, but simple solutions are preferred to intricacies of the quantized field theory with the hope that experimenters realize the simple proposals with appropriate means.

From the historical viewpoint, the problem of quantization of the electromagnetic field in vacuo was solved by Dirac (1927) long ago and the quantization of a nonlinear theory is due to Born and Infeld (1934, 1935). With respect to the propagation in linear dielectric media it is appropriate to refer first to Jauch and Watson (1948). A revived interest in this problem can be perceived since the 1990s. First it resembled some dissatisfaction with the situation following the advent of laser in 1958. The new optical effects are analyzed both by the methods of nonlinear optics which belong to classical physics and by those of quantum optics (Shen 1969). In quantum optics, the normal-mode expansion approach is used that is well suited for systems in optical cavities, such as an optical parametric oscillator, but is not appropriate for open systems such as a parametric amplifier. In nonlinear optics (Bloembergen 1965, Shen 1984), the Maxwell equations completed by the constitutive relations are solved with the method of slowly varying envelope approximation and the resultant equations are sometimes simplified on the assumption of parametric approximation. It has become standard that the phenomenological Hamiltonians of quantum optics are frequently introduced without a quantitative connection to the classical equations describing the nonlinear optical effects.

The quantization of the electromagnetic field in the presence of a dielectric is possible. This can be done in two ways which are called the macroscopic and microscopic approaches. In the first, the macroscopic approach, the medium is completely described by its linear and nonlinear susceptibilities. No matter degrees of freedom appear explicitly in this treatment. After a Lagrangian which produces the macroscopic Maxwell equations for the field in a nonlinear medium is found, the canonical momenta and the Hamiltonian are derived. Quantization is accomplished by imposing the standard equal-time commutation relations. In the second approach, the microscopic, a model for the medium is constructed and both the field and the matter degrees of freedom appear in the theory. Both are quantized. The result is a

A. Lukš, V. Peřinová, *Quantum Aspects of Light Propagation*,
DOI 10.1007/b101766_1,

theory of mixed matter-field (polariton) modes, which are coupled by a nonlinear interaction.

Hillery and Mlodinow (1984) have used the electric displacement field as the canonical variable for nonlinear quantization and they have explored the macroscopic approach to the quantization of homogeneous nondispersive media. They have pointed out that there is a difficulty in including the dispersion in the quantized macroscopic theory.

In the past, many authors that dealt with macroscopic quantum theories of light propagation wrote also on space displacements, shifts, and translations of the electromagnetic field along with the time displacements, shifts, and translations, or simply on the (time) evolution. Accordingly, they used the term "space evolution" in the former case. In the following, we will use the term space progression instead of space evolution. Abram (1987) intended to overcome the difficulties of the conventional quantum optics by reformulating its assumptions. He has based the formalism on the momentum operator for the radiation field and investigated in this way not only the spatial progression of the electromagnetic wave but also refraction and reflection. The importance of a proper space–time description of squeezing has been recognized (Bialynicka-Birula and Bialynicki-Birula 1987). The problem of a proper quantum mechanical description of the operation of optical devices has been addressed (Knöll et al. 1986, 1987). Besides this, an attempt at a formulation of quantum theory of propagation of the optical wave in a lossless dispersive dielectric material has been made (Blow et al. 1990). The vacuum propagation and low-order perturbation theory have sufficed for spatio-temporal descriptions of parametric down-conversion experiments (Casado et al. 1997a, Casado et al. 1997b). The experiment on the "induced coherence without induced emission" has been described on restriction to spatial behaviour of fields and the multimode description has been restored too (Peřinová et al. 2003).

The applications have used the fact that the nonlinear processes of quantum optics are described quantum optically in the parametric approximation with linear mathematical tools so that quantization procedures and solutions of the dynamics need not face immense difficulties as for the really nonlinear formalism (Huttner et al. 1990). The formalism of the macroscopic approach to the quantization has been developed (Abram and Cohen 1991). The space–time displacement operators have been related to the elements of the energy–momentum tensor (Serulnik and Ben-Aryeh 1991). The macroscopic quantization of the electromagnetic field was applied to inhomogeneous media (Glauber and Lewenstein 1991). The theoretical methods for investigating propagation in quantum optics, in which the momentum operator is used along with the Hamiltonian, have been developed (Toren and Ben-Aryeh 1994). The excellent review of linear and nonlinear couplers (Peřina, Jr. and Peřina 2000), where the restriction to merely spatial behaviour of interesting optical fields is accepted, has used a similar approach.

The optical solitons have been studied in the nonlinear optics and their quantum properties have been calculated using spatio-temporal descriptions by erudite authors. The dispersion has been treated on the assumption of a narrow-frequency interval (Drummond 1990). Drummond (1994) has presented a review of his theory and its applications. In addition to previous work (Lang et al. 1973) devoted to the concept of quasinormal modes, the modes of the universe have been used in the treatment of the spectrum of squeezing (Gea-Banacloche et al. 1990a,b). An original approach to the description of a degenerate parametric amplifier (Deutsch and Garrison 1991a) has been related to the theory of paraxial quantum propagation (Deutsch and Garrison 1991b). One-dimensional description of beam propagation has been completed with transverse position coordinates. It has taken into account the existence of small photodetectors or pixels (Kolobov 1999). Abram and Cohen (1994) have developed a travelling-wave formulation of the theory of quantum optics and have applied it to quantum propagation of light in a Kerr medium. A quantum scattering theory approach to quantum-optical measurements has been expounded (Dalton et al. 1999a). In addition to (Lang et al. 1973) and along with an independent work (Ho et al. 1998) devoted to the concept of quasinormal modes, quasimode theory of macroscopic canonical quantization has been invented and applied (Dalton et al. 1999b,c). A macroscopic canonical quantization of the electromagnetic field and radiating atom system involving classical, linear optical devices, based on expanding the vector potential in terms of quasimode functions has been carried out (Dalton et al. 1999b). The relationship between the pure mode and quasimode annihilation and creation operators is determined (Dalton et al. 1999c). A quantum theory of the lossless beam splitter is given in terms of the quasimode theory of macroscopic canonical quantization. The input and output operators that are related via scattering operator are directly linked to multi-time quantum correlation functions (Dalton et al. 1999d). Brown and Dalton (2001a) have generalized the quasimode theory of macroscopic quantization in quantum optics and cavity quantum electrodynamics developed by Dalton, Barnett, and Knight (1999a,b). This generalization admits the case where two or more quasipermittivities are introduced. The generalized form of quasimode theory has beeen applied to provide a fully quantum-theoretical derivation of the laws of reflection and refraction at a boundary (Brown and Dalton 2001b).

Huttner and Barnett (1992a,b) have presented a fully canonical quantization scheme for the electromagnetic field in dispersive and lossy linear dielectrics. This scheme is based on the Hopfield model of such a dielectric, where the matter is represented by a harmonic polarization field (Hopfield 1958). Following (Huttner and Barnett 1992a,b), Gruner and Welsch (1995) have calculated the ground-state correlation of the quantum-mechanical fluctuations of the intensity. Gruner and Welsch (1996a) have realized the expansion of the field operators which is based on the Green function of the classical Maxwell equations and preserves the equal-time canonical commutation relations of the field. They have found that the spatial progression can be derived on the assumption of weak absorption. In (Schmidt et al. 1998), the microscopic approach to the quantum theory of light propagation has been extended to nonlinear media and the generalized nonlinear Schrödinger equation well known from the description of quantum solitons has been derived for a dielectric with a Kerr nonlinearity. Dung et al. (1998) have developed a quantization scheme for the electromagnetic field in a spatially varying three-dimensional linear dielectric which causes both dispersion and absorption. In the case of a homogeneous dielectric, the well-known Green function has been used and it has been shown that the indicated quantization scheme exactly preserves the fundamental equal-time commutation relations of quantum electrodynamics. The Green function has also been used in the more complicated case of two dielectric bodies with a common planar interface. Spontaneous decay of an excited atom in the presence of dispersing and absorbing bodies has been investigated using an extension of this formalism (Dung et al. 2000). A microscopic theory of an optical field in a lossy linear optical medium has been developed (Knöll and Leonhardt 1992). Dutra and Furuya (1997) have considered a single-mode cavity filled with a medium consisting of two-level atoms that are approximated by harmonic oscillators. They have shown that macroscopic averaging of the dynamical variables can lead to a macroscopic description. Dutra and Furuya (1998a,b) have observed that the (full) Huttner–Barnett model of a dielectric medium does not comprise all the dielectric permittivities of the medium which can be expected from the classical electrodynamics, although the field theory in linear dielectrics should have such a property. Dalton et al. (1996) have dealt with the quantization of a field in dielectrics and have applied it to the theory of atomic radiation in one-dimensional Fabry–Pérot resonator.

Yablonovitch (1987) suggested that three-dimensional periodic dielectric structures could have a photonic band gap in analogy to electronic band gaps in semiconductor crystals, namely, a band of frequencies for which an electromagnetic wave cannot propagate in any direction. This idea and its subsequent experimental proof in the macrowave domain have led to extensive activity aimed at the optimization of photonic band-gap structures for the visible domain and the exploration of their potential applications (Journal of Modern Optics 1994, Journal of the Optical Society of America B 2002, etc.).

As soon as quantization in nonhomogeneous dielectric media is solved, not only the case of a finite dielectric medium is worth a treatment but also the case of an infinite periodic medium (Caticha and Caticha 1992, Kweon and Lawandy 1995, Tip 1997). The idea of one-dimensional propagation may be compared with results concerning a mirror waveguide. Nonlinear optics in a photonic band-gap structure has been studied (Tricca et al. 2004, Peřina, Jr. et al. 2004, 2005, Peřina, Jr. et al. 2007). Sakoda (2002) has formulated quantization of the electromagnetic field in photonic crystals. Spontaneous parametric down conversion in a finite-length multilayer structure has been considered (Centini et al. 2005, Peřina, Jr. et al. 2006).

Both localization and laser theory, which were developed in the 1960s, were jointly applied in the study of random laser. They were used in strongly scattering gain media. Lasing in disordered media has been a subject of intense theoretical and experimental studies. Random lasers have been classified into incoherent and coherent random lasers. Research works on both types of random lasers have been summarized in the monographic chapter (Cao 2003). In order to understand quantum-statistical properties of random lasers, quantum theory is needed. Standard quantum theory for lasers applies only to quasidiscrete modes and cannot account for lasing in the presence of overlapping modes. In a random medium, the character of lasing modes depends on the amount of disorder. Weak disorder leads to a poor confinement of light and to strongly overlapping modes. Statistics naturally belongs to the theory of amplifying random media (Beenakker 1998, Patra and Beenakker 1999, 2000, Mishchenko et al. 2001), which is restricted to linear media.

Hackenbroich et al. (2002) have developed a quantization scheme for optical resonators with overlapping (nonorthogonal) modes. Cheng and Siegman (2003) have derived a generalized formalism of radiation-field quantization, which need not rely on a set of orthogonal eigenmodes. True eigenmodes of a system will be non-orthogonal and the method is intended for quantization of an open system, in which a gain or loss medium is involved.

We will use units following original papers and although the system of international (SI) units prevails, there are exceptions, so some of the relations (2.25)–(2.69) and (3.276)–(3.322) are in the Gaussian units, the relations (3.323)–(3.392) are in the rationalized cgs units, and the relations (2.1)–(2.2), (3.14)–(3.108), (3.494)–(3.578), (3.109)–(3.125), and (2.15)–(2.24) are in the Heaviside–Lorentz units.

Chapter 2
Origin of Macroscopic Approach

With the birth of quantum optics in the 1960s it became clear that it would be easy to describe the interaction between the electromagnetic field and the matter in a cavity even on elimination of matter degrees of freedom. A similar travelling-wave description for the electromagnetic field–matter interaction was considered to be possible in terms of a virtual cavity and a momentum operator of the field.

This approach to quantization was rather distant from the quantum theory of the electromagnetic field. On a fundamental level the theory of the electromagnetic field in the free space does not differ from the theory of this field in the matter. Macroscopic approaches to quantization of the electromagnetic field are not fundamental theories and modify the free-space electromagnetic-field theory. Especially, quantization of the field power has been assumed.

Although the virtual cavity has been beaten, the momentum operator has still enabled one to study quantum aspects of nonlinear optical processes. Quantization restrictions of any kind such as the frequency dispersion of the refractive index were apparent on published work.

Efforts emerged to formulate so simple a quantum theory of the electromagnetic field that it allows one to recognize the role of the momentum operator. Formalisms were presented which, to the contrary, did not consider the momentum operator. With the progress in (classical) optics interest in the quantization of the field power in quantum optics has increased.

Not always is it necessary to utilize the formalism of the electromagnetic field in the matter. For description of experiments with correlated photons it suffices to describe the electromagnetic field between optical devices and to know the input–output relations for the optical elements, both passive and active, with which the radiation is transformed.

2.1 Lossless Nonlinear Dielectric

An approach to the quantum theory of light propagation was considered standard until the critique by Hillery and Mlodinow (1984) and is still. Concerning this approach, let us consider papers by Shen (1967, 1969). Shen (1967) studied

A. Lukš, V. Peřinová, *Quantum Aspects of Light Propagation*,
DOI 10.1007/b101766_2, © Springer Science+Business Media, LLC 2009

quantum statistics of nonlinear optics. He contributed to the contemporary research (Glauber 1965). Quantum theory of radiation had long been formulated (Heitler 1954). For investigation of properties of a medium, incoherent scattering has been a useful tool. For nonlinear optics, coherent scattering has been interesting as well or more.

Weak nonlinearity has a significant effect on light only after a longer interaction distance. Light can cover a longer distance easily when contained in a cavity resonator. Quantum statistics has been determined using descriptions suited to the case of a cavity.

In principle, the same treatment can be applied to problems of light propagation in media (Shen 1967). But for coherent scattering, it becomes difficult. This case should be treated by the method of many-body transport theory (Ter Haar 1961). In quantum optics, the cavity treatment of the problems of light propagation in media seems to be valid on the following assumption. Photon fields may be quantized in a box of finite volume, which moves in the z direction with a light velocity c ($\frac{c}{\sqrt{\epsilon}}$ in Shen 1969). One is advised to imagine a box of length cT, where T is the counting time of photodetectors. A partial interaction of the light with the medium can be approximated with no interaction and a complete interaction, which lasts for a time t. The finite medium can be extended to infinity. The resultant change of statistical properties of fields in the box can now be calculated using the cavity treatment (Shen 1967).

In nonlinear optics a number of classical descriptions have been developed both as a cavity problem and as a steady-state propagation problem. Then a cavity problem can be converted to a corresponding steady-state propagation problem by replacing t by $-\frac{z}{c}$ in the field amplitudes and the latter problem can be changed to the former one by replacing z by $-ct$ when $\mathbf{e}_z$ is the direction of propagation. It raises expectations that the same is true in the quantum treatment. Shen (1969) pays attention to replacing t by $-\frac{z\sqrt{\epsilon}}{c}$ and to replacing z by $\frac{ct}{\sqrt{\epsilon}}$.

Here the dependence on the time seems to be more fundamental. It is evident that one is interested in a conversion of a cavity problem to a corresponding steady-state propagation problem. The operators will be space dependent (localized) instead of time dependent. Transformations will be generated with a localized momentum operator instead of the Hamiltonian operator.

On quantizing in a volume L^3 and assuming that the field does not vary appreciably over a distance d large compared with the wavelength, and associating the discrete values of the wave vector k with d (instead of L), the localized annihilation and creation operators $\hat{b}_k(z)$ and $\hat{b}_k^\dagger(z)$ have been proposed. An appropriate component of the vector-potential operator has the expansion of the form

$$\hat{A}(z,t) = c\sum_k \sqrt{\frac{\hbar}{2\omega_k \epsilon_k L^3}} \left\{\hat{b}_k(z)\exp[-\mathrm{i}(\omega_k t - kz)] + \text{H.c.}\right\}, \qquad (2.1)$$

where ω_k is the frequency, $\hbar$ is the Planck constant divided by 2π, $\epsilon_k = \epsilon(\omega_k)$ is the value of the dielectric function ϵ at ω_k, and H.c. denotes the term Hermitian conjugate to the previous one. The annihilation and creation operators $\hat{b}_k(z)$ and $\hat{b}_k^\dagger(z)$, respectively, satisfy the equal-space commutation relation

$$\left[\hat{b}_k(z), \hat{b}_{k'}^\dagger(z)\right] = \delta_{kk'}\hat{1}. \tag{2.2}$$

The small variation of the field has been formulated as that of the normally ordered moments $\left\langle \hat{b}_k^{\dagger m}(z)\hat{b}_k^n(z)\right\rangle$. It is also specified that $k = \frac{2\pi n}{d}$, where n is an integer.

There is a difficulty. The above picture of a moving box requires a light velocity c independent of the frequency ω_k. Shen (1969) utilizes the notation $\frac{c}{\sqrt{\epsilon}}$ for this velocity. Here it is replaced by the phase velocity $\frac{c}{\sqrt{\epsilon_k}}$, with $c \equiv c_0$, the free-space speed of light. There is another difficulty in view of this picture that d has been used instead of cT. The localized photon-number operator is realized as a configuration-space photon-number operator (Mandel 1966)

$$\hat{n}(z) = \frac{Ad}{L^3}\sum_k \hat{b}_k^\dagger(z)\hat{b}_k(z), \tag{2.3}$$

where $\mathcal{A}$ is the cross-sectional area of the beam.

A Hamiltonian density $\hat{H}(z,t)$ is considered. The Hamiltonian is

$$\hat{\mathcal{H}}(t) = L^2 \int_L \hat{H}(z',t)\,\mathrm{d}z'. \tag{2.4}$$

A third difficulty is that the localized momentum operator is defined as $\frac{\hat{H}(z,t)}{c}$ essentially, not by using an integration with respect to time. It has been assumed that $k = \frac{2\pi n}{d}$, not that $\omega_k = \frac{2\pi n}{T}$. For free fields, the localized momentum operator is

$$\hat{\mathcal{P}}(z) = \sum_k \hbar k\left[\hat{b}_k^\dagger(z)\hat{b}_k(z) + \tfrac{1}{2}\hat{1}\right]. \tag{2.5}$$

For interacting fields, the localized momentum operator has the form of a Hamiltonian, but with $\hat{b}_k(z)$ and $\hat{b}_k^\dagger(z)$ replacing $\hat{a}_k(t)$ and $\hat{a}_k^\dagger(t)$.

A momentum operator should have the form $\mathbf{e}_z\hat{\mathcal{P}}(z)$, be a vector, but in fact one does not utilize this. The momentum operator generates translations

$$\frac{\mathrm{d}}{\mathrm{d}z}\hat{b}_k(z) = \frac{\mathrm{i}}{\hbar}\left[\hat{b}_k(z), \hat{\mathcal{P}}(z)\right]. \tag{2.6}$$

The electric strength vector is derived from the vector potential according to the relation

$$\hat{E}(z,t) = -\frac{1}{c}\frac{\partial}{\partial t}\hat{A}(z,t). \tag{2.7}$$

We decompose this operator as

$$\hat{E}(z,t) = \hat{E}^{(+)}(z,t) + \hat{E}^{(-)}(z,t), \tag{2.8}$$

where $\hat{E}^{(+)}(z,t)$ $(\hat{E}^{(-)}(z,t))$ contains the functions $\exp(-\mathrm{i}\omega_k t)$ $(\exp(\mathrm{i}\omega_k t))$. In Shen (1967) the opposite convention is used. Then

$$\frac{\mathrm{d}}{\mathrm{d}z}\hat{E}^{(+)}(z,t) = \frac{\mathrm{i}}{\hbar}\left[\hat{E}^{(+)}(z,t), \hat{\mathcal{P}}(z)\right]. \tag{2.9}$$

Something is more suitable for propagation problem: We define all the quantities at a given plane $z = z_0$ for all times and try to obtain the propagation towards $z \geq z_0$. According to equations (2.6) and (2.9), the unitary translation operator is

$$\hat{U}(z, z_0) = \mathcal{S}\exp\left[\frac{\mathrm{i}}{\hbar}\int_{z_0}^{z}\hat{\mathcal{P}}(z')\,\mathrm{d}z'\right], \tag{2.10}$$

where $\mathcal{S}$ is the space-ordering operation. The space-ordered product has a similar definition as the time-ordered product. Field operators at different spatial points z, z_0 are connected by this unitary operator:

$$\hat{E}(z,t) = \hat{U}^{\dagger}(z,z_0)\hat{E}(z_0,t)\hat{U}(z,z_0). \tag{2.11}$$

There are indications that any "alternative" quantum theory is avoided. Such an indication is the fact that the localized momentum operator has been derived from the Hamiltonian density. With this in mind, we pass from the "spatial Heisenberg picture" to a spatial Schrödinger picture. In the latter picture, a localized density matrix (statistical operator) progresses:

$$\hat{\rho}(z) = \hat{U}(z,0)\hat{\rho}(0)\hat{U}^{\dagger}(z,0). \tag{2.12}$$

Here $\hat{\rho}(0)$ is a given statistical operator. Then the correlation function of fields at different times is expressed in two forms:

$$\begin{aligned}
&\left\langle\hat{E}^{(-)}(z,t_1)\ldots\hat{E}^{(-)}(z,t_n)\hat{E}^{(+)}(z,t_n)\ldots\hat{E}^{(+)}(z,t_1)\right\rangle \\
&= \mathrm{Tr}\left\{\hat{\rho}(0)\hat{E}^{(-)}(z,t_1)\ldots\hat{E}^{(-)}(z,t_n)\hat{E}^{(+)}(z,t_n)\ldots\hat{E}^{(+)}(z,t_1)\right\} \\
&= \mathrm{Tr}\left\{\hat{\rho}(z)\hat{E}^{(-)}(0,t_1)\ldots\hat{E}^{(-)}(0,t_n)\hat{E}^{(+)}(0,t_n)\ldots\hat{E}^{(+)}(0,t_1)\right\}.
\end{aligned} \tag{2.13}$$

The equation of motion for a statistical operator $\hat{\rho}(z)$ is

$$\frac{\partial}{\partial z}\hat{\rho}(z) = \frac{\mathrm{i}}{\hbar}\left[\hat{\mathcal{P}}(z), \hat{\rho}(z)\right]. \tag{2.14}$$

With the help of these localized operators, the calculations for steady-state propagation in a medium become the same as the corresponding calculations for a cavity with t replaced by $-\frac{z}{c}$ (Shen 1967) and by $\frac{z\sqrt{\epsilon}}{c}$ (Shen 1969). The problem of beam splitting was mentioned. Essentially, the same proposal has been included in Shen (1969).

2.2 Nondispersive Lossless Linear Dielectric

The study of nonlinear optical phenomena and their inclusion in an effective nonlinear theory of the electromagnetic field has utilized the asymmetry of most optical media, which are nonlinear with respect to the electric-field, but linear relative to the magnetic field. The canonical momentum should be the magnetic induction in place of the more usual electric-field strength. Such a theory may not be capable of describing the Bohm–Aharonov effect. Besides such a theory we expound a simple quantization connected to considerations of the role of the Poynting vector operator and the momentum operator.

A description of the field distribution in space must be completed with a quantum state of the field in quantum physics. A renewed interest in the spatio-temporal description leads to the study of the wave functional of the electromagnetic field despite the doubts of the pioneers of theoretical physics of the photonic wave function.

On neglecting dispersion and nonlinearity, a macroscopic theory of the quantized electromagnetic field in a medium can be very close to the usual theory of this field in free space. In contrast to this, solutions have been disseminated, which include the dispersion and the nonlinearity at least approximately.

2.2.1 Quantization in Terms of a Dual Potential

According to a pioneering paper of Hillery and Mlodinow (1984), the standard macroscopic quantum theory of electrodynamics in a nonlinear medium is due to Shen (1967) and has been elaborated upon by Tucker and Walls (1969). Hillery and Mlodinow (1984) have pointed out some problems with the standard theory, above all that it is not consistent with the macroscopic Maxwell equations.

One approach to the derivation of a macroscopic quantum theory would be to begin from a quantum microscopic theory as explored in the linear case by Hopfield (1958). The other approach is to take the expression for the energy of the radiation in nonlinear medium, which differs from the free-field Hamiltonian in part, and to keep interpreting the electric-field (up to the sign) as the canonically conjugated variable to the vector potential. Then, this macroscopic classical theory is quantized. (Let us note that it differs from Shen (1969)). The Hamiltonian formulation of the theory consists in the noncanonical Hamiltonian

$$\hat{H}_{\mathrm{noncan}} = \hat{H}_{\mathrm{EM}} + \hat{H}_{\mathrm{Inoncan}}, \tag{2.15}$$

where

$$\hat{H}_{\mathrm{EM}} = \frac{1}{2}\int (\hat{\mathbf{E}}^2 + \hat{\mathbf{B}}^2)\,\mathrm{d}^3\mathbf{x}, \tag{2.16}$$

$$\hat{H}_{\mathrm{Inoncan}} = \frac{1}{2}\int \hat{\mathbf{E}}\cdot\hat{\mathbf{P}}\,\mathrm{d}^3\mathbf{x}, \tag{2.17}$$

with $\hat{\mathbf{E}}$ being the electric field strength operator and $\hat{\mathbf{P}}$ being the polarization of the medium, and the Heaviside–Lorentz units having been used. The polarization is a function of the electric field which may be written as a power series. This theory may be called standard. It can easily be seen that, as an undesirable "quantum effect", we obtain an improper expression for the time derivative of the magnetic-induction field $\hat{\mathbf{B}}$.

It is assumed that the medium is lossless, nondispersive, and homogeneous. A Lagrangian is considered which gives proper equations of motion. The electric and magnetic fields are expressed in terms of the vector potential $\mathbf{A}$ and the scalar potential A_0:

$$\mathbf{E} = -\frac{\partial\mathbf{A}}{\partial t} - \nabla A_0, \mathbf{B} = \nabla\times\mathbf{A}. \tag{2.18}$$

The appropriate Lagrangian density depends on the first partial derivatives of the four-vector $A = (A_0, \mathbf{A})$. The momentum canonical to A is $\Pi = (\Pi_0, \mathbf{\Pi})$, where $\Pi_0 = 0$. The vanishing of Π_0 indicates that the system is constrained. It has been shown how to utilize the Dirac quantization procedure for constrained Hamiltonian systems (Dirac 1964). It can be derived that the canonical momentum is $\mathbf{\Pi} = -\mathbf{D}$. The canonical Hamiltonian has the form

$$H = H_{\mathrm{EM}} + H_{\mathrm{I}}, \tag{2.19}$$

where

$$H_{\mathrm{I}} = \int \mathbf{E}\cdot\left[\mathbf{P} - \int_0^1 \mathbf{P}(\lambda\mathbf{E})\,\mathrm{d}\lambda\right]\mathrm{d}^3\mathbf{x}. \tag{2.20}$$

In order to simplify the quantization of the macroscopic Maxwell theory, the dual potential Λ has been introduced along with $\mathbf{\Lambda}$ and Λ_0, which we call the dual vector and scalar potentials. The relation (2.18) is replaced by

$$\mathbf{D} = \nabla\times\mathbf{\Lambda}, \quad \mathbf{B} = \frac{\partial\mathbf{\Lambda}}{\partial t} + \nabla\Lambda_0. \tag{2.21}$$

It can be shown that the canonical momentum is $\mathbf{\Pi}^{\times} = \mathbf{B}$. Upon expressing the canonical Hamiltonian functional in terms of the electric-displacement and magnetic-induction fields, the results are the same:

$$H = H^{\times}. \tag{2.22}$$

Then, the usual Hamiltonian theory for the electromagnetic field in a nonlinear dielectric medium and the alternative have been quantized in the ordinary way. We can compare

$$\left[\hat{A}_i(\mathbf{x}, t), \hat{\Pi}_j(\mathbf{x}', t)\right] = \mathrm{i}\delta_{ij}^{\perp}(\mathbf{x} - \mathbf{x}')\hat{1}, \tag{2.23}$$

with

$$\left[\hat{\Lambda}_i(\mathbf{x}, t), \hat{\Pi}_j^{\times}(\mathbf{x}', t)\right] = \mathrm{i}\delta_{ij}^{\perp}(\mathbf{x} - \mathbf{x}')\hat{1}. \tag{2.24}$$

The transverse δ function has been used and made a reference to Bjorken and Drell (1965). Hillery and Mlodinow (1984) do not mention propagation except a paragraph on the interpretation problems, where they recommend to confine the medium to part of the quantization volume and to place the field source and the detector outside of the medium, being aware that they require the consideration of propagation.

It is added that different diagonalizations indicated by the quadratic part of the total Hamiltonian generate different kinds of normal ordering. A doubt is expressed that there is an appropriate kind and the microscopic approach is propounded. Dispersion is also considered a reason for a microscopic theory to be contemplated.

2.2.2 *Momentum Operator as Translation Operator*

In the late 1980s, the problem of propagation did not seem to be typical of quantum optics. Abram addressed the problem of light propagation through a linear nondispersive lossless medium (Abram 1987). Although this model can be an appropriate limit of the Huttner–Barnett model, we expound the main ideas of Abram (1987). Abram criticized the modal Hamiltonian formalism, especially the inclusion of the linear polarization term in the Hamiltonian:

$$\mathcal{H} \stackrel{?}{=} \frac{1}{8\pi}\int_V (E^2 + H^2 + 4\pi\chi E^2)\,\mathrm{d}V, \tag{2.25}$$

where E (H) is the magnitude of the electrical (magnetic) field strength, χ is the (linear) susceptibility of the material, and V is a quantization volume. This would lead to an incorrect result, mainly to a change of the frequencies of the modes which does not occur. He decided to extend the traditional theory of quantum optics to describe propagation phenomena without invoking the modal Hamiltonian. According to him one of the propagation phenomena, refraction, suggests the momentum as the concept appropriate for the description of these phenomena. Quantum mechanically, space and momentum are canonically conjugate variables. Let us remark that microscopic models demonstrate that a Hamiltonian including light–matter interaction can be considered. These are a good antidote against the idea that "space and momentum are canonically conjugate variables like time and energy".

Propagation of the electromagnetic field is described by the Maxwell equations:

$$\nabla \times \mathbf{H} = \frac{1}{c}\frac{\partial \mathbf{D}}{\partial t}, \tag{2.26}$$

$$\nabla \times \mathbf{E} = -\frac{1}{c}\frac{\partial \mathbf{B}}{\partial t}, \tag{2.27}$$

$$\nabla \cdot \mathbf{B} = 0, \tag{2.28}$$

$$\nabla \cdot \mathbf{D} = 0, \tag{2.29}$$

where $\mathbf{D} = \mathbf{E} + 4\pi \mathbf{P}$ is the electric displacement, $\mathbf{B}$ is the magnetic induction, $\mathbf{E}$ and $\mathbf{H}$ are the electric and magnetic field strengths, respectively, $\mathbf{P}$ is the (linear and nonlinear) polarization induced in the medium, and c is the speed of light. We assume that there are no free charges or currents and that we are dealing with nonmagnetic materials, so that $\mathbf{B} = \mathbf{H}$.

For simplicity, we shall consider only the case of plane waves propagating along the z-axis, with the electric field polarized along the x-axis and the magnetic field along the y-axis. This reduces the Maxwell equations to scalar differential equations, the directions of all vectors being implicit. We shall further assume that light is propagating in a linear dielectric, where the induced polarization is at all times proportional to the incident electric field:

$$P = \chi E, \tag{2.30}$$

where we assume the susceptibility of the material for simplicity to be a scalar (neglecting its tensorial properties), independent of frequency (no dispersion). It is convenient to define also the dielectric function ϵ of the material

$$\epsilon = 1 + 4\pi \chi, \tag{2.31}$$

and the refractive index n,

$$n = \sqrt{\epsilon}. \tag{2.32}$$

The change in the total energy which is given by the integrated energy flux (the Poynting vector) over the surface of a body or volume is proposed in Abram (1987) as the proper quantum-mechanical Hamiltonian. The change in the total momentum is given as the integrated flux of the Maxwell stress tensor. The momentum is treated on the same footing as the Hamiltonian. However, the enigma of the Hamiltonian (2.25) is solved. We may consider a square pulse which enters a dielectric. The total energy is conserved, but the energy density is increased by a factor of n, because the volume V reduces to $V' = \frac{V}{n}$. In volume V' the wavelengths of the modes become $\lambda' = \frac{\lambda}{n}$, but the oscillator frequencies remain unchanged. It is interesting that in the absence of reflection, the electric and magnetic fields of the transmitted (T) waves in the dielectric are related to the corresponding incident (I) fields in free space by

$$E_T = \frac{1}{\sqrt{n}} E_I, \tag{2.33}$$

$$H_T = \sqrt{n} H_I. \tag{2.34}$$

This change in the energy density implies a similar increase for the total momentum of the pulse, the components of which are always proportional to the wave vectors of the excited modes. In propagation along the z-axis the Maxwell stress tensor is replaced by the energy density.

When the propagation along the $\pm z$-axis in free space is considered with the electric field polarized along the x-axis and the magnetic field along the y-axis ($\chi = 0$, $\epsilon = 1$), the electromagnetic vector-potential operator $\hat{A} \equiv \hat{A}(z, t)$ is usually written as ($\hbar = 1$)

$$\hat{A}(z, t) = c \sum_j \left(\frac{2\pi}{V\omega_j} \right)^{\frac{1}{2}} \left(\hat{a}_j^\dagger e^{\mathrm{i}\omega_j t - \mathrm{i}k_j z} + \hat{a}_j e^{-\mathrm{i}\omega_j t + \mathrm{i}k_j z} \right), \tag{2.35}$$

where $\hat{a}_j^\dagger$, $\hat{a}_j$ are the creation, annihilation operators, respectively, for a photon in the jth mode of the wave vector k_j (with $k_{-j} = -k_j$) and the frequency $\omega_j = c|k_j|$ fulfilling the Bose commutation relations. To simplify the notation, we omit unit vectors. It is convenient to rearrange equation (2.35) in a manner that is familiar to solid-state physicists,

$$\hat{A}(z, t) = c \sum_j \left(\frac{2\pi}{V\omega_j} \right)^{\frac{1}{2}} \left(\hat{a}_j^\dagger e^{\mathrm{i}\omega_j t} + \hat{a}_{-j} e^{-\mathrm{i}\omega_j t} \right) e^{-\mathrm{i}k_j z}. \tag{2.36}$$

The electric and magnetic field operators may be obtained as

$$\begin{aligned} \hat{E}(z, t) &= -\frac{1}{c} \frac{\partial}{\partial t} \hat{A}(z, t) = \sum_j \hat{e}_j \\ &= -\mathrm{i} \sum_j \left(\frac{2\pi\omega_j}{V} \right)^{\frac{1}{2}} (\hat{b}_j^\dagger - \hat{b}_{-j}), \end{aligned} \tag{2.37}$$

and

$$\hat{H}(z,t) = \frac{\partial}{\partial z}\hat{A}(z,t) = \sum_j \hat{h}_j$$

$$= -\mathrm{i}\sum_j s_j \left(\frac{2\pi\omega_j}{V}\right)^{\frac{1}{2}} (\hat{b}_j^\dagger + \hat{b}_{-j}), \tag{2.38}$$

where $s_j \equiv \operatorname{sgn} j$ and

$$\hat{b}_j = \hat{a}_j e^{-\mathrm{i}\omega_j t + \mathrm{i}k_j z}. \tag{2.39}$$

When products of these operators are encountered, we suppose that they are symmetrized. The Hermiticity of the operators $\hat{E} \equiv \hat{E}(z,t)$ and $\hat{H} \equiv \hat{H}(z,t)$ can be verified using the relations

$$\hat{e}_j^\dagger = \hat{e}_{-j}, \tag{2.40}$$

$$\hat{h}_j^\dagger = \hat{h}_{-j}. \tag{2.41}$$

The energy density operator can be written as

$$\hat{u} = \frac{1}{8\pi}\left(\hat{E}^2 + \hat{H}^2\right) \tag{2.42}$$

$$= \sum_j \hat{u}_j \tag{2.43}$$

$$= \frac{1}{8\pi}\sum_j \left(\hat{e}_j\hat{e}_{-j} + \hat{h}_j\hat{h}_{-j}\right)$$

$$= \frac{1}{2V}\sum_j \omega_j \left(\hat{b}_j^\dagger\hat{b}_j + \hat{b}_{-j}^\dagger\hat{b}_{-j} + \hat{1}\right) \tag{2.44}$$

$$= \frac{1}{V}\sum_j \omega_j \left(\hat{b}_j^\dagger\hat{b}_j + \frac{1}{2}\hat{1}\right). \tag{2.45}$$

The energy fluxes due to the forward (backward) waves alone can be expressed uniquely:

$$\hat{u}_+ = \sum_{j(>0)} \frac{\omega_j}{V}\left(\hat{b}_j^\dagger\hat{b}_j + \frac{1}{2}\hat{1}\right), \; \hat{u}_- = \sum_{j(<0)} \frac{\omega_j}{V}\left(\hat{b}_j^\dagger\hat{b}_j + \frac{1}{2}\hat{1}\right). \tag{2.46}$$

The total momentum operator $\hat{G}$ is then

$$\hat{G} = \frac{V}{c}(\hat{u}_+ - \hat{u}_-) = \sum_j k_j \left(\hat{b}_j^\dagger \hat{b}_j + \frac{1}{2}\hat{1} \right). \tag{2.47}$$

It is important to understand the relations (3.9) and (3.10) in Abram (1987) well. We interpret (3.9) concerning elementary quantum mechanics as

$$\langle z|\hat{p}_z|\psi\rangle = -i\hbar \frac{\partial}{\partial z}\langle z|\psi\rangle, \tag{2.48}$$

where $|z\rangle$ are position coordinate states and $|\psi\rangle$ is an arbitrary pure state. The similarity with equation (3.10) from Abram (1987)

$$\frac{\partial \hat{Q}}{\partial z} = -i[\hat{G}, \hat{Q}], \tag{2.49}$$

where $\hat{Q}$ is any operator, fades. We would prefer a definition of the operator $\hat{Q}$. Let us consider

$$\hat{Q} \equiv \hat{Q}(z,t) = Q[\hat{E}(z,t), \hat{H}(z,t)], \tag{2.50}$$

where $Q[\bullet, \bullet]$ is a formal series in $\hat{E}$ and $\hat{H}$. Since the differential operator $\frac{\partial}{\partial z}$ is just as differentiation as the superoperator $-i[\hat{G}, \bullet]$, it suffices to verify the relation (2.49) for $\hat{Q} = \hat{E}, \hat{H}$. It is true at least in the situations treated in Abram (1987).

Although the operators $\hat{b}_j \equiv \hat{b}_j(z,t)$ are studied using (2.49), the Heisenberg equation of motion, and the initial condition

$$\hat{b}_j(0,0) = \hat{a}_j \tag{2.51}$$

as appropriate for any operator $\hat{Q}(z,t)$, we perceive that the operators do not obey our definition of the operator $\hat{Q}$.

We may calculate the Poynting vector operator as

$$\hat{S} = \frac{c}{4\pi}\hat{E}\hat{H} = \frac{c}{4\pi}\sum_j \hat{e}_j \hat{h}_{-j} \tag{2.52}$$

$$= \sum_j s_j \frac{c\omega_j}{V}\left(\hat{b}_j^\dagger \hat{b}_j + \frac{1}{2}\hat{1} \right). \tag{2.53}$$

The Poynting vector operators due to the forward (backward) waves alone can be expressed uniquely:

$$\hat{S}_+ = \sum_{j(>0)} \frac{c\omega_j}{V}\left(\hat{b}_j^\dagger \hat{b}_j + \frac{1}{2}\hat{1} \right), \quad \hat{S}_- = -\sum_{j(<0)} \frac{c\omega_j}{V}\left(\hat{b}_j^\dagger \hat{b}_j + \frac{1}{2}\hat{1} \right). \tag{2.54}$$

The total energy operator of the free field inside the volume of quantization is thus

$$\hat{\mathcal{H}} = \hat{U} = \frac{V}{c}\left(\hat{S}_+ - \hat{S}_-\right) = \sum_j \omega_j \left(\hat{b}_j^\dagger \hat{b}_j + \frac{1}{2}\hat{1}\right). \tag{2.55}$$

The investigation of the case $\chi \neq 0$, $\epsilon \neq 1$ does not lead to any new expansions of the field operators $\hat{E}$ and $\hat{H}$. The individual components of the rearranged electric and magnetic field operators according to (2.37) and (2.38) satisfy a modified operator algebra with respect to that of the harmonic oscillator:

$$[\hat{e}_j, \hat{e}_l] = [\hat{h}_j, \hat{h}_l] = \hat{0}, \tag{2.56}$$

$$[\hat{e}_j, \hat{h}_l] = s_{-j}\left(\frac{4\pi\omega_j}{V}\right)\delta_{-j,l}\hat{1}, \tag{2.57}$$

where $\delta_{j,l}$ is the Kronecker δ function. The knowledge of these commutators and of the generalized total momentum operator $\hat{G}$, the derivation of (2.42) through (2.47), should have been generalized accordingly, e.g. the relation (2.42) becoming

$$\begin{aligned}\hat{u} &= \frac{1}{8\pi}\left(\epsilon\hat{E}^2 + \hat{H}^2\right) \\ &= \frac{1}{8\pi}\sum_j \left(\epsilon\hat{e}_j\hat{e}_{-j} + \hat{h}_j\hat{h}_{-j}\right),\end{aligned} \tag{2.58}$$

which enables us to derive the Maxwell equations both via the temporal derivatives and via the spatial derivatives.

The energy density operator (2.58) can be generalized. In the expansion (2.43) we can set $\hat{u}_j = \hat{u}_{j\,\mathrm{refr}}$,

$$\hat{u}_{j\,\mathrm{refr}} = \frac{\omega_j}{2V}\left\{\hat{b}_j^\dagger\hat{b}_j + \hat{b}_{-j}\hat{b}_{-j}^\dagger - 2\pi\chi\left(\hat{b}_j^\dagger - \hat{b}_{-j}\right)\left(\hat{b}_{-j}^\dagger - \hat{b}_j\right)\right\}. \tag{2.59}$$

The energy density operator $\hat{u}_{\mathrm{refr}}$ may be diagonalized through a Bogoliubov transformation. To this end we introduce an anti-Hermitian operator $\hat{R}$ of the form

$$\hat{R} = \sum_j \left(\hat{b}_j\hat{b}_{-j} - \hat{b}_j^\dagger\hat{b}_{-j}^\dagger\right) \tag{2.60}$$

and introduce the operators

$$\hat{B}_j = e^{-\gamma\hat{R}}\hat{b}_j e^{\gamma\hat{R}} = (\cosh\gamma)\hat{b}_j - (\sinh\gamma)\hat{b}_{-j}^\dagger, \tag{2.61}$$

where

$$\gamma = \frac{1}{4}\ln\epsilon = \frac{1}{2}\ln n. \tag{2.62}$$

On substitution

$$\hat{b}_j = e^{\gamma \hat{R}} \hat{B}_j e^{-\gamma \hat{R}} = (\cosh \gamma)\hat{B}_j + (\sinh \gamma)\hat{B}^{\dagger}_{-j}, \tag{2.63}$$

the operator $\hat{R}$ takes the form

$$\hat{R} = \sum_j \left(\hat{B}_j \hat{B}_{-j} - \hat{B}^{\dagger}_j \hat{B}^{\dagger}_{-j} \right), \tag{2.64}$$

and the energy density operator has the diagonal form

$$\hat{u}_{j\,\mathrm{refr}} = \frac{n\omega_j}{2V} \left(\hat{B}^{\dagger}_j \hat{B}_j + \hat{B}_{-j} \hat{B}^{\dagger}_{-j} \right). \tag{2.65}$$

The momentum operator is then given by

$$\hat{G}_{\mathrm{refr}} = \frac{V}{c} (\hat{u}_+ - \hat{u}_-) = \sum_j K_j \left(\hat{B}^{\dagger}_j \hat{B}_j + \frac{1}{2} \hat{1} \right), \tag{2.66}$$

with $K_j = nk_j$ and the Hamiltonian can be calculated as

$$\hat{\mathcal{H}}_{\mathrm{refr}} = \sum_j \omega_j \left(\hat{B}^{\dagger}_j \hat{B}_j + \frac{1}{2} \hat{1} \right). \tag{2.67}$$

By inserting (2.63) into (2.37) and (2.38), respectively, we can obtain the electric and magnetic field operators inside the dielectric:

$$\hat{E}(z,t) = -\mathrm{i} \sum_j \left(\frac{2\pi \omega_j}{nV} \right)^{\frac{1}{2}} (\hat{B}^{\dagger}_j - \hat{B}_{-j}) \tag{2.68}$$

and

$$\hat{H}(z,t) = -\mathrm{i} \sum_j s_j \left(\frac{2\pi n \omega_j}{V} \right)^{\frac{1}{2}} (\hat{B}^{\dagger}_j + \hat{B}_{-j}). \tag{2.69}$$

Similarly as above, this relation can be interpreted as a result of the replacement $\hat{b}_j \mapsto \hat{B}_j$ and a consequence of the quantized classical equations (2.33) and (2.34).

For normal incidence on a sharp vacuum–dielectric interface, both reflection and diffraction occur. We will not treat this more general case according to Abram (1987).

2.2.3 *Wave Functional Description of Gaussian States*

Białynicka-Birula and Białynicki-Birula (1987) have tried first to define the squeezing that is a generalization of the standard definition for one mode of radiation. This definition can be reformulated with respect to Białynicki-Birula (2000). The Riemann–Silberstein–Kramers complex vector has been introduced

$$\mathbf{F}(\mathbf{r},t) = \frac{1}{\sqrt{2}}\left[\frac{\mathbf{D}(\mathbf{r},t)}{\sqrt{\epsilon_0}} + \mathrm{i}\frac{\mathbf{B}(\mathbf{r},t)}{\sqrt{\mu_0}}\right], \tag{2.70}$$

where we have divided by $\sqrt{\epsilon_0}$, $\sqrt{\mu_0}$, as is appropriate with SI units. It has been shown how the Green function method can be used for solving linear equations for the field operator $\hat{\mathbf{F}}(\mathbf{r},t)$. This approach allows that the medium under investigation is inhomogeneous and time dependent. It is not clear whether the complex vector (2.70) is then useful. It has been suggested that the periodicity of the electric permittivity tensor $\epsilon(\mathbf{r},t)$ or the magnetic permeability $\mu(\mathbf{r},t)$ can be important for the generation of squeezed states. Only the dispersion of the medium has not been considered. It has been derived that photon pair production is a necessary condition for squeezing.

It is tempting to generalize the concept of a Gaussian state of the finite-dimensional harmonic oscillator to the case of an infinite oscillator. Białynicka-Birula and Białynicki-Birula (1987) treat the time development of the Gaussian states in the free-field case. There the Schrödinger picture is adopted and an analogue of the Schrödinger representation in quantum mechanics has been introduced. Let us recall the quadrature representation in quantum optics. This representation is a wave functional $\Psi[\mathbf{A},t]$. Let us observe that contrary to the operator $\hat{\mathbf{A}}(\mathbf{r},t)$, the argument $\mathbf{A}(\mathbf{r})$ of the wave functional does not depend on t, but the wave functional does depend on t. The Hamiltonian in this representation has the form

$$H = \frac{1}{2}\int\left\{-\frac{\hbar^2}{\epsilon_0}\frac{\delta^2}{\delta\mathbf{A}(\mathbf{r})^2} + \frac{1}{\mu_0}[\nabla\times\mathbf{A}(\mathbf{r})]^2\right\}\mathrm{d}^3\mathbf{r}. \tag{2.71}$$

In Białynicka-Birula and Białynicki-Birula (1987), the wave functional of the vacuum state, i.e. the simplest Gaussian state of the electromagnetic field, can be found, as well as that of the "most general". Thus, the exposition is confined to pure Gaussian states while it is possible to generalize it also to mixed Gaussian states of the electromagnetic field. The pure Gaussian state is determined by a complex matrix kernel, i.e. by two real matrix kernels. It is shown that the expectation values $\langle\hat{\mathbf{B}}\rangle = \mathcal{B}$ and $\langle\hat{\mathbf{D}}\rangle = \mathcal{D}$ (equivalently, $\langle\hat{\mathbf{E}}\rangle = \mathcal{E}$) evolve according to the free-field Maxwell equations and also the equations which the complex matrix kernel obeys can be found there.

The whole electromagnetic field is treated as a huge infinite-dimensional harmonic oscillator. The wave function and the corresponding Wigner function become then functionals of the field variables. Mrówczyński and Müller (1994) have considered only the scalar field. Białynicki-Birula (2000) starts from the wave functional of the vacuum state (Misner et al. 1970)

$$\Psi_0[\mathbf{A}] = C \exp\left[-\frac{1}{4\pi^2\hbar}\sqrt{\frac{\epsilon_0}{\mu_0}}\int\int \mathbf{B}(\mathbf{r})\frac{1}{|\mathbf{r}-\mathbf{r}'|^2}\cdot\mathbf{B}(\mathbf{r}')\,\mathrm{d}^3\mathbf{r}\,\mathrm{d}^3\mathbf{r}'\right] \tag{2.72}$$

and from the wave functional (we change $\tilde{\mathbf{A}} \to -\mathbf{D}$)

$$\tilde{\Psi}_0[-\mathbf{D}] = C \exp\left[-\frac{1}{4\pi^2\hbar}\sqrt{\frac{\mu_0}{\epsilon_0}}\int\int \mathbf{D}(\mathbf{r})\frac{1}{|\mathbf{r}-\mathbf{r}'|^2}\cdot\mathbf{D}(\mathbf{r}')\,\mathrm{d}^3\mathbf{r}\,\mathrm{d}^3\mathbf{r}'\right]. \tag{2.73}$$

The normalization constant C is an issue and it has not been completely solved in Białynicki-Birula (2000). The analogy with the one-dimensional harmonic oscillator leads to other notions. The Wigner functional of the electromagnetic field in the ground state is

$$W_0[\mathbf{A}, -\mathbf{D}] = \exp\{-2N[\mathbf{A}, -\mathbf{D}]\}, \tag{2.74}$$

where

$$N[\mathbf{A}, -\mathbf{D}] = \frac{1}{4\pi^2\hbar}\int\int\left[\sqrt{\frac{\epsilon_0}{\mu_0}}\mathbf{B}(\mathbf{r})\frac{1}{|\mathbf{r}-\mathbf{r}'|^2}\cdot\mathbf{B}(\mathbf{r}') + \sqrt{\frac{\mu_0}{\epsilon_0}}\mathbf{D}(\mathbf{r})\frac{1}{|\mathbf{r}-\mathbf{r}'|^2}\cdot\mathbf{D}(\mathbf{r}')\right]\mathrm{d}^3\mathbf{r}\,\mathrm{d}^3\mathbf{r}'. \tag{2.75}$$

The expression (2.75) also plays the role of a norm for the photon wave function (Białynicki-Birula 1996a,b). The Wigner functional for the thermal state of the electromagnetic field has been presented. This state is mixed and it even has infinitely many photons in the whole field. In each of the subsequent cases, the wave functional and the Wigner functional have been introduced. The exception, the mixed state, has no wave functional. Let us remark that for (the statistical operator of) such a state the matrix element can be considered which is a functional of two arguments, $\mathbf{A}$ and $\mathbf{A}'$.

In particular, the Wigner functional for the coherent state of the electromagnetic field $|\mathcal{A}, -\mathcal{D}\rangle$ has been presented, where $\mathcal{A}(\mathbf{r})$, $\mathcal{D}(\mathbf{r})$ are the vector potential and the electric displacement vector, respectively, which characterize the state. The exposition is related to the hot topic of the superpositions of coherent states of the electromagnetic field. The exposition continues with the Wigner functionals for the states of the electromagnetic field that describe a definite number of photons. An example of the functional for the one-photon state with the photon mode function $\mathbf{f}(\mathbf{r})$ has been included.

The norm (2.75) has not been related to any inner product of the photon wave functions, but these notions are connected. In contrast to Białynicka-Birula and

Białynicki-Birula (1987), we introduce quadrature operators as

$$\hat{X}_1[\mathcal{D}] = \int \hat{\mathbf{D}}(\mathbf{r}, 0) \cdot \mathbf{f}(\mathbf{r})\, \mathrm{d}^3\mathbf{r}, \tag{2.76}$$

$$\hat{X}_2[\mathcal{B}] = \int \hat{\mathbf{B}}(\mathbf{r}, 0) \cdot \mathbf{g}(\mathbf{r})\, \mathrm{d}^3\mathbf{r}, \tag{2.77}$$

where

$$\mathbf{f}(\mathbf{r}) = \frac{1}{4\pi^2\hbar} \int \sqrt{\frac{\mu_0}{\epsilon_0}} \frac{1}{|\mathbf{r} - \mathbf{r}'|^2} \mathcal{D}(\mathbf{r}')\, \mathrm{d}^3\mathbf{r}', \tag{2.78}$$

$$\mathbf{g}(\mathbf{r}) = \frac{1}{4\pi^2\hbar} \int \sqrt{\frac{\epsilon_0}{\mu_0}} \frac{1}{|\mathbf{r} - \mathbf{r}'|^2} \mathcal{B}(\mathbf{r}')\, \mathrm{d}^3\mathbf{r}'. \tag{2.79}$$

The commutator of the $\hat{X}_1$ and $\hat{X}_2$ operators is

$$[\hat{X}_1[\mathcal{D}], \hat{X}_2[\mathcal{B}]] = \mathrm{i}\hbar \int \mathbf{f}(\mathbf{r}) \cdot [\nabla \times \mathbf{g}(\mathbf{r})]\, \mathrm{d}^3\mathbf{r}. \tag{2.80}$$

Let us note that the right-hand sides of (2.78) and (2.79) comprise the operator $|\nabla|^{-1}$ up to a certain factor (cf. Milburn et al. 1984). Without resorting to this notation, we obtain that

$$[\hat{X}_1[\mathcal{D}], \hat{X}_2[\mathcal{B}]] = \frac{\mathrm{i}}{4\hbar} \int \mathcal{D}(\mathbf{r}_1) \cdot \mathcal{A}(\mathbf{r}_1)\, \mathrm{d}^3\mathbf{r}_1\, \hat{1}. \tag{2.81}$$

We see easily that the usual commutator $-\frac{1}{2}\mathrm{i}\hat{1}$ is yielded by the field $(\mathcal{D}, \mathcal{B})$ (or $(\mathcal{A}, -\mathcal{D})$) with the property

$$\int [-\mathcal{D}(\mathbf{r}_1)] \cdot \mathcal{A}(\mathbf{r}_1)\, \mathrm{d}^3\mathbf{r}_1 = 2\hbar. \tag{2.82}$$

We have not deepened the contrast by introducing the notation $\hat{X}_1[-\mathcal{D}]$ and $\hat{X}_2[\mathcal{A}]$ on the left-hand sides of (2.76) and (2.77).

Białynicki-Birula (2000) presents the Wigner functional for the squeezed vacuum state:

$$W_{\mathrm{sq}}[\mathbf{A}, -\mathbf{D}] = \exp\left[-\frac{1}{\hbar} \int\int \left(\sqrt{\frac{\epsilon_0}{\mu_0}} \mathbf{B} \cdot \mathbf{K}_{\mathrm{BB}} \cdot \mathbf{B}' \right.\right.$$
$$\left.\left. + \sqrt{\frac{\mu_0}{\epsilon_0}} \mathbf{D} \cdot \mathbf{K}_{\mathrm{DD}} \cdot \mathbf{D}' + \mathbf{B} \cdot \mathbf{K}_{\mathrm{BD}} \cdot \mathbf{D}' \right) \mathrm{d}^3\mathbf{r}\, \mathrm{d}^3\mathbf{r}' \right], \tag{2.83}$$

where $\mathbf{K}_{\mathrm{BB}}$, $\mathbf{K}_{\mathrm{DD}}$, and $\mathbf{K}_{\mathrm{BD}}$ are real matrix kernels. The kernel $\mathbf{K}_{\mathrm{BD}}$ is not independent of $\mathbf{K}_{\mathrm{BB}}$ and $\mathbf{K}_{\mathrm{DD}}$, but it must obey the condition that is reminiscent of the Schrödinger–Robertson uncertainty relation (Białynicki-Birula 1998). The problem of the time evolution is also discussed. It has been conceded that the Wigner function is not a very powerful tool for making detailed calculations. Just as in the field theory, the symmetric ordering is vexed. Another open question is how the projection of this Wigner functional onto Wigner functions of any orthogonal (complete or incomplete) modal system looks out. It is appropriate to mention here work concerning the photon wave function (Inagaki 1998, Hawton 1999, Kobe 1999), although it is relevant mainly to the electromagnetic field in vacuo.

Using a straightforward procedure, Mendonça et al. (2000) have quantized the linearized equations for an electromagnetic field in a plasma. They have determined an effective mass for the transverse photons. An extension of the quantization procedure leads to the definition of a photon charge operator. Zaleśny (2001) has found that the influence of a medium on a photon can be described by some scalar and vector potentials. He has extended the concept of the vector potential to relativistic velocities of the medium. He has derived formulae for the mass of photon in resting and moving dielectric and the velocity of the photon as a particle.

2.2.4 Source-Field Operator

Knöll et al. (1987) have compared the problem of quantum-mechanical treatment of action of optical devices with the input–output formalism (Collett and Gardiner 1984, Gardiner and Collett 1985, Yamamoto and Imoto 1986, Nilsson et al. 1986, cf. also Gea-Banacloche et al. 1990a,b). Apart from the fact that only a very particular setup is considered in the input–output formalism, the theory does not take into account the full space–time structure of the field.

Knöll et al. (1987) have elaborated on the approach developed on the basis of quantum field theory and applied to the problem of spectral filtering of light (Knöll et al. 1986). The only assumptions are that the interaction between sources and light is linear in the vector potential and the optical system is lossless and that the condition of sufficiently small dispersion is fulfilled. First, the classical Maxwell equations with sources and optical devices are formulated and solved by the procedure of mode expansion and the quantized version is derived. The classical Maxwell equations comprise the relative permittivity

$$\epsilon(\mathbf{r}) = n^2(\mathbf{r}), \tag{2.84}$$

where $n(\mathbf{r})$ is the space-dependent refractive index. The mode functions $\mathbf{A}_\lambda(\mathbf{r})$ are introduced as the solutions of equation

$$\nabla \times (\nabla \times \mathbf{A}_\lambda(\mathbf{r})) - \epsilon(\mathbf{r})\frac{\omega_\lambda^2}{c^2}\mathbf{A}_\lambda(\mathbf{r}) = \mathbf{0}, \tag{2.85}$$

where ω_λ^2 is the separation constant for each λ, from which the gauge condition can be derived

$$\nabla \cdot [\epsilon(\mathbf{r})\mathbf{A}_\lambda(\mathbf{r})] = 0. \tag{2.86}$$

It is assumed that these solutions are normalized and orthogonal in the sense of equation

$$\int \epsilon(\mathbf{r})\mathbf{A}_\lambda(\mathbf{r}) \cdot \mathbf{A}_{\lambda'}(\mathbf{r})\,\mathrm{d}^3\mathbf{r} = \delta_{\lambda\lambda'}\hat{1}. \tag{2.87}$$

In terms of these functions, the vector potential can be decomposed. In the standard manner the destruction and creation operators $\hat{a}_\lambda$ and $\hat{a}_\lambda^\dagger$ are defined, which have the properties

$$\begin{aligned}
[\hat{a}_\lambda, \hat{a}_{\lambda'}^\dagger] &= \delta_{\lambda\lambda'}\hat{1}, \\
[\hat{a}_\lambda, \hat{a}_{\lambda'}] &= \hat{0} = [\hat{a}_\lambda^\dagger, \hat{a}_{\lambda'}^\dagger].
\end{aligned} \tag{2.88}$$

On inserting the operators $\hat{a}_\lambda$ and $\hat{a}_\lambda^\dagger$ into the decomposition of the vector potential, the operator of the vector potential $\hat{\mathbf{A}}(\mathbf{r}, t)$ is defined:

$$\hat{\mathbf{A}}(\mathbf{r}, t) = \sum_\lambda \mathbf{A}_\lambda(\mathbf{r}) \left[\hat{a}_\lambda(t) + \hat{a}_\lambda^\dagger(t)\right]. \tag{2.89}$$

The source quantities $\mathbf{r}_a$ and $\mathbf{p}_a$ are considered as the operators $\hat{\mathbf{r}}_a$ and $\hat{\mathbf{p}}_a$, which obey the standard commutation relations

$$\begin{aligned}
[\hat{r}_{ka}, \hat{p}_{k'a'}] &= \mathrm{i}\hbar\delta_{aa'}\delta_{kk'}\hat{1}, \\
[\hat{r}_{ka}, \hat{r}_{k'a'}] &= \hat{0} = [\hat{p}_{ka}, \hat{p}_{k'a'}],
\end{aligned} \tag{2.90}$$

and the commutation relations

$$\begin{aligned}
[\hat{r}_{ka}, \hat{a}_\lambda] &= \hat{0} = [\hat{r}_{ka}, \hat{a}_\lambda^\dagger], \\
[\hat{p}_{ka}, \hat{a}_\lambda] &= \hat{0} = [\hat{p}_{ka}, \hat{a}_\lambda^\dagger].
\end{aligned} \tag{2.91}$$

The operator $\hat{\mathbf{A}}(\mathbf{r}, t)$ can be used for the derivation of the electric-field strength operator which is associated with the radiation field by the relation

$$\hat{\mathbf{E}}(\mathbf{r}, t) = -\frac{\partial}{\partial t}\hat{\mathbf{A}}(\mathbf{r}, t) \tag{2.92}$$

and for the derivation of the magnetic field strength operator

$$\hat{\mathbf{B}}(\mathbf{r}, t) = \nabla \times \hat{\mathbf{A}}(\mathbf{r}, t). \tag{2.93}$$

Nevertheless, the mode functions are redefined so that they obey the normalization condition

$$\int \epsilon(\mathbf{r})\mathbf{A}_\lambda(\mathbf{r}) \cdot \mathbf{A}_{\lambda'}(\mathbf{r})\, \mathrm{d}^3\mathbf{r} = \frac{\hbar}{2\epsilon_0\omega_\lambda}\delta_{\lambda\lambda'}. \tag{2.94}$$

The form of the normalization conditions (2.87) and (2.94) is tailored to real-mode functions and the necessity of modification of some fundamental relations is commented on by Knöll et al. (1987). All of these field operators may be written in the form

$$\hat{\mathbf{F}}(\mathrm{r}, t) = \sum_\lambda \left[\mathbf{F}_\lambda(\mathbf{r})\hat{a}_\lambda(t) + \mathbf{F}^*_\lambda(\mathbf{r})\hat{a}^\dagger_\lambda(t)\right]. \tag{2.95}$$

In dependence on the choice of the operator $\hat{\mathbf{F}}(\mathbf{r}, t)$, the functions $\mathbf{F}_\lambda(\mathbf{r})$ can be derived from the mode functions of the vector potential $\mathbf{A}_\lambda(\mathbf{r})$.

It is often convenient to decompose a given field operator $\hat{\mathbf{F}}(\mathbf{r}, t)$ into two parts by the relation

$$\hat{\mathbf{F}}(\mathbf{r}, t) = \hat{\mathbf{F}}^{(+)}(\mathbf{r}, t) + \hat{\mathbf{F}}^{(-)}(\mathbf{r}, t), \tag{2.96}$$

where

$$\hat{\mathbf{F}}^{(+)}(\mathbf{r}, t) = \sum_\lambda \mathbf{F}_\lambda(\mathbf{r})\hat{a}_\lambda(t), \tag{2.97}$$

$$\hat{\mathbf{F}}^{(-)}(\mathbf{r}, t) = [\hat{\mathbf{F}}^{(+)}(\mathbf{r}, t)]^\dagger. \tag{2.98}$$

Further, the Heisenberg equations of motion for the field operators are derived, so that the field operators can be expressed in terms of free-field and source-field operators. It is typical of the approach of Knöll et al. (1987) that any field operator $\hat{F}_k^{(+)}$ is decomposed into a free-field operator and a source-field operator as follows:

$$\hat{F}_k^{(+)}(\mathbf{r}, t) = \hat{F}_{k\mathrm{free}}^{(+)}(\mathbf{r}, t) + \hat{F}_{k\mathrm{s}}(\mathbf{r}, t), \tag{2.99}$$

where

$$\hat{F}_{k\mathrm{free}}^{(+)}(\mathbf{r}, t) = \sum_\lambda F_{k\lambda}(\mathbf{r})\hat{a}_{\lambda\mathrm{free}}(t), \tag{2.100}$$

$$\hat{F}_{k\mathrm{s}}(\mathbf{r}, t) = \int\int \theta(t - t')K_{kk'}(\mathbf{r}, t; \mathbf{r}', t')\hat{J}_{k'}(\mathbf{r}', t')\, \mathrm{d}^3\mathbf{r}'\, \mathrm{d}t'. \tag{2.101}$$

Here vector components are labelled by the index k and repeated indices k' mean summation.

Unfortunately, the operator $\hat{a}_{\lambda\mathrm{free}}(t)$ was not defined, so that we may only guess that $\hat{a}_{\lambda\mathrm{free}}(t)|_{t=t_0} = \hat{a}_\lambda(t_0)$ for $t = t_0$, and the dynamics for $t \geq t_0$ can be found in Knöll et al. (1987). In equation (2.101), the kernel $K_{kk'}$ is defined by the relation

$$K_{kk'}(\mathbf{r}, t; \mathbf{r}', t') = -\frac{1}{\mathrm{i}\hbar} \sum_\lambda F_{k\lambda}(\mathbf{r}) A^*_{k'\lambda}(\mathbf{r}') \exp[-\mathrm{i}\omega_\lambda (t - t')]. \tag{2.102}$$

Inserting equation (2.99) yields the following representation of $\hat{F}_k^{(+)}$:

$$\hat{F}_k^{(+)}(\mathbf{r}, t)$$
$$= \int \int \theta(t - t') K_{kk'}(\mathbf{r}, t; \mathbf{r}', t') \hat{J}_{k'}(\mathbf{r}', t') \, \mathrm{d}^3\mathbf{r}' \, \mathrm{d}t' + \hat{F}^{(+)}_{k\mathrm{free}}(\mathbf{r}, t). \tag{2.103}$$

In particular, if $\hat{F}_k^{(+)}$ is identified with the vector potential $\hat{A}_k^{(+)}$, it holds that $F_{k\lambda} = A_{k\lambda}$ and the kernel $K_{kk'}$ takes the form

$$K_{kk'}(\mathbf{r}, t; \mathbf{r}', t') = -\frac{1}{\mathrm{i}\hbar} \sum_\lambda A_{k\lambda}(\mathbf{r}) A^*_{k'\lambda}(\mathbf{r}') \exp[-\mathrm{i}\omega_\lambda (t - t')]. \tag{2.104}$$

Analogously, if one is interested in the electric-field strength of the radiation $\hat{E}_k^{(+)}$, the appropriate form of the kernel $K_{kk'}$ is

$$K_{kk'}(\mathbf{r}, t; \mathbf{r}', t') = -\frac{1}{\hbar} \sum_\lambda \omega_\lambda A_{k\lambda}(\mathbf{r}) A^*_{k'\lambda}(\mathbf{r}') \exp[-\mathrm{i}\omega_\lambda (t - t')]. \tag{2.105}$$

So the symmetry relations

$$K^*_{kk'}(\mathbf{r}, t; \mathbf{r}', t') = \mp K_{k'k}(\mathbf{r}', t'; \mathbf{r}, t) \tag{2.106}$$

are valid for $\hat{A}_k^{(+)}$ and $\hat{E}_k^{(+)}$, respectively. The information on the action of the optical instruments on the source field is contained in the space–time structure of the kernel $K_{kk'}$, which may be regarded as the apparatus function also used in classical optics.

Further, the commutation relations for various combinations of field operators at different times are studied and relationships between field commutators and source-quantity commutators are derived. The following abbreviations of the notation are used:

$$x = \{\mathbf{r}, t\} \tag{2.107}$$

and others, by which the superscripts +, − are introduced also for $\hat{J}_k(x)$ and $K_{kk'}(x, x')$. With these generalizations, it holds that

$$\hat{F}_k^{(j)}(x) = \hat{F}_{k\text{free}}^{(j)}(x) + \hat{F}_{ks}^{(j)}(x),\ j = +, -, \tag{2.108}$$

$$\hat{F}_{ks}^{(j)}(x) = \int \theta(t - t')K_{kk'}^{(j)}(x, x')\hat{J}_{k'}^{(j)}(x')\,\mathrm{d}x'. \tag{2.109}$$

When appropriate, the time ordering symbols $\mathcal{T}_+$ and $\mathcal{T}_-$ are used. Let us consider any operator product $\hat{A}_1(t_1)\ \hat{A}_2(t_2)...\hat{A}_n(t_n)$. The symbol $\mathcal{T}_+$ introduces the time ordering of the operators $\hat{A}_i(t_i)$ with the latest time to the far left:

$$\begin{aligned}&\mathcal{T}_+\hat{A}_1(t_1)\hat{A}_2(t_2)...\hat{A}_n(t_n)\\ &= \hat{A}_{i_1}(t_{i_1})\hat{A}_{i_2}(t_{i_2})...\hat{A}_{i_n}(t_{i_n}) \text{ with } t_{i_1} > t_{i_2} > \cdots > t_{i_n},\end{aligned} \tag{2.110}$$

and the symbol $\mathcal{T}_-$ introduces time ordering of the operators $\hat{A}_i(t_i)$ with the latest time to the far right:

$$\begin{aligned}&\mathcal{T}_-\hat{A}_1(t_1)\hat{A}_2(t_2)...\hat{A}_n(t_n)\\ &= \hat{A}_{i_1}(t_{i_1})\hat{A}_{i_2}(t_{i_2})...\hat{A}_{i_n}(t_{i_n}) \text{ with } t_{i_1} < t_{i_2} < \cdots < t_{i_n}.\end{aligned} \tag{2.111}$$

From (2.91) it follows that

$$\begin{aligned}[\hat{F}_{k_1}^{(j_1)}(x_1), \hat{F}_{k_2}^{(j_2)}(x_2)] &= [\hat{F}_{k_1\text{free}}^{(j_1)}(x_1), \hat{F}_{k_2\text{free}}^{(j_2)}(x_2)]\\ &\quad + \hat{D}_{k_1k_2}^{(j_1,j_2)}(x_1, x_2) - \hat{D}_{k_2k_1}^{(j_2,j_1)}(x_2, x_1),\end{aligned} \tag{2.112}$$

where

$$\begin{aligned}\hat{D}_{k_1k_2}^{(j_1,j_2)}(x_1, x_2) &= -\int\int \theta(t_2 - t_2')\theta(t_2' - t_1')\theta(t_1' - t_1)\\ &\otimes K_{k_1k_1'}^{(j_1)}(x_1, x_1')K_{k_2k_2'}^{(j_2)}(x_2, x_2')[\hat{J}_{k_1'}^{(j_1)}(x_1'), \hat{J}_{k_2'}^{(j_2)}(x_2')]\,\mathrm{d}x_1'\,\mathrm{d}x_2'.\end{aligned} \tag{2.113}$$

From an inspection of equation (2.113), we readily learn that

$$\hat{D}_{k_1k_2}^{(j_1,j_2)}(x_1, x_2) = \hat{0} \text{ if } t_1 > t_2. \tag{2.114}$$

The commutators in (2.112) are

$$[\hat{F}_{k_1\text{free}}^{(j)}(x_1), \hat{F}_{k_2\text{free}}^{(j)}(x_2)] = \hat{0},\ \ j = +, -, \tag{2.115}$$

$$[\hat{F}_{k_1\text{free}}^{(+)}(x_1), \hat{F}_{k_2\text{free}}^{(-)}(x_2)] = F_{k_1k_2}(x_1, x_2)\hat{1}, \tag{2.116}$$

where

$$F_{k_1k_2}(x_1, x_2) = \sum_{\lambda} F_{k_1\lambda}(\mathbf{r}_1)F^*_{k_2\lambda}(\mathbf{r}_2)\exp[-\mathrm{i}\omega_\lambda(t_1 - t_2)]. \tag{2.117}$$

It would be interesting to find the particular forms of the commutators between $\hat{A}$ and $\hat{E}$ or between $\hat{A}^{(+)}$ and $\hat{E}^{(-)}$.

Further, these commutation relations are used to express field correlation functions of free-field operators and source-field operators and to describe the effect of the optical system on the quantum properties of light fields.

The method of transformation of normal and time orderings is demonstrated for the following important class of correlation functions:

$$G^{(m,n)}_{k_1\ldots k_{m+n}}(x_1, \ldots, x_{m+n}) = \left\langle \left[\mathcal{T}_- \prod_{j=1}^{m} \hat{F}^{(-)}_{k_j}(x_j)\right] \left[\mathcal{T}_+ \prod_{j=m+1}^{m+n} \hat{F}^{(+)}_{k_j}(x_j)\right] \right\rangle. \tag{2.118}$$

This transformation is understood in the relation

$$\mathcal{T}_+ \hat{F}^{(+)}_{k_1}(x_1)\hat{F}^{(+)}_{k_2}(x_2) = \mathcal{O}_+ \left[\hat{F}^{(+)}_{k_1\mathrm{free}}(x_1) + \hat{F}^{(+)}_{k_1\mathrm{s}}(x_1)\right] \left[\hat{F}^{(+)}_{k_2\mathrm{free}}(x_2) + \hat{F}^{(+)}_{k_2\mathrm{s}}(x_2)\right]. \tag{2.119}$$

In equation (2.119) and the following ones the ordering symbols $\mathcal{O}_+$ and $\mathcal{O}_-$ are used. The symbol $\mathcal{O}_+$ introduces the following ordering of operators $\hat{F}^{(+)}_{k_i\mathrm{s}}(x_i)$, $\hat{F}^{(+)}_{k_j\mathrm{free}}(x_j)$:

(i) Ordering of the operators $\hat{F}^{(+)}_{k_i\mathrm{s}}(x_i)$, $\hat{F}^{(+)}_{k_j\mathrm{free}}(x_j)$ with the operators $\hat{F}^{(+)}_{k_j\mathrm{free}}(x_j)$ to the right of the operators $\hat{F}^{(+)}_{k_i\mathrm{s}}(x_i)$.
(ii) Substitution of equation (2.109) for the operators $\hat{F}^{(+)}_{k_i\mathrm{s}}(x_i)$ and $\mathcal{T}_+$ time ordering of the source-quantity operators $\hat{J}_{k_i'}(x_i')$ in the resulting source-quantity operator products before performing the integrations with respect to t_i'.

The symbol $\mathcal{O}_-$ introduces the following operator ordering in products of operators $\hat{F}^{(-)}_{k_i\mathrm{s}}(x_i)$, $\hat{F}^{(-)}_{k_j\mathrm{free}}(x_j)$:

(i) Ordering of the operators $\hat{F}^{(-)}_{k_i\mathrm{s}}(x_i)$, $\hat{F}^{(-)}_{k_j\mathrm{free}}(x_j)$ with the operators $\hat{F}^{(-)}_{k_j\mathrm{free}}(x_j)$ to the left of the operators $\hat{F}^{(-)}_{k_i\mathrm{s}}(x_i)$.
(ii) Substitution of equation (2.109) for the operators $\hat{F}^{(-)}_{k_i\mathrm{s}}(x_i)$ and $\mathcal{T}_-$ time ordering of the source-quantity operators $\hat{J}^\dagger_{k_i'}(x_i')$ in the resulting source-quantity operator products before performing the integrations with respect to t_i'.

Equation (2.119) may now be generalized:

$$\mathcal{T}_+ \prod_{j=1}^{n} \hat{F}_{k_j}^{(+)}(x_j) = \mathcal{O}_+ \prod_{j=1}^{n} \left[\hat{F}_{k_j \mathrm{free}}^{(+)}(x_j) + \hat{F}_{k_j \mathrm{s}}^{(+)}(x_j) \right], \tag{2.120}$$

$$\mathcal{T}_- \prod_{j=1}^{n} \hat{F}_{k_j}^{(-)}(x_j) = \mathcal{O}_- \prod_{j=1}^{n} \left[\hat{F}_{k_j \mathrm{free}}^{(-)}(x_j) + \hat{F}_{k_j \mathrm{s}}^{(-)}(x_j) \right]. \tag{2.121}$$

Using relations (2.118), (2.120), and (2.121), we may represent the correlation functions as

$$G_{k_1 \ldots k_{m+n}}^{(m,n)}(x_1, \ldots, x_{m+n}) = \left\langle \left\{ \mathcal{O}_- \prod_{j=1}^{m} \left[\hat{F}_{k_j \mathrm{free}}^{(-)}(x_j) + \hat{F}_{k_j \mathrm{s}}^{(-)}(x_j) \right] \right\} \otimes \left\{ \mathcal{O}_+ \prod_{j=m+1}^{m+n} \left[\hat{F}_{k_j \mathrm{free}}^{(+)}(x_j) + \hat{F}_{k_j \mathrm{s}}^{(+)}(x_j) \right] \right\} \right\rangle. \tag{2.122}$$

When at the points of observation the following conditions are fulfilled

$$\left\langle \ldots \hat{F}_{k\mathrm{free}}^{(+)} \right\rangle = 0 = \left\langle \hat{F}_{k\mathrm{free}}^{(-)} \ldots \right\rangle, \tag{2.123}$$

then the relation (2.122) can be simplified:

$$G_{k_1 \ldots k_{m+n}}^{(m,n)}(x_1, \ldots, x_{m+n}) = \left\langle \left[\mathcal{O}_- \prod_{j=1}^{m} \hat{F}_{k_j \mathrm{s}}^{(-)}(x_j) \right] \left[\mathcal{O}_+ \prod_{j=m+1}^{m+n} \hat{F}_{k_j \mathrm{s}}^{(+)}(x_j) \right] \right\rangle. \tag{2.124}$$

When written in more detail, into the relation (2.124), the complex kernels $K_{k_j k'_j}(x_j, x'_j)$ are introduced.

It is noted that the effect of the beam splitter that is used for mixing of source light with the reference beam in the case of homodyne detection is described by the assumption that the reference light beam is a free field. In Knöll et al. (1987) the relation (2.122) is specialized to a multimode coherent free field $|\{\alpha_\lambda\}\rangle$,

$$\hat{F}_{k\mathrm{free}}^{(+)}(x)|\{\alpha_\lambda\}\rangle = \mathcal{F}_k(x)|\{\alpha_\lambda\}\rangle, \tag{2.125}$$

that is to say

$$G^{(m,n)}_{k_1\ldots k_{m+n}}(x_1, \ldots, x_{m+n})$$
$$= \left\langle \left\{ \mathcal{O}_- \prod_{j=1}^{m} \left[\mathcal{F}^*_{k_j}(x_j)\hat{1} + \hat{F}^{(-)}_{k_j s}(x_j) \right] \right\} \right.$$
$$\left. \otimes \left\{ \mathcal{O}_+ \prod_{j=m+1}^{m+n} \left[\mathcal{F}_{k_j}(x_j)\hat{1} + \hat{F}^{(+)}_{k_j s}(x_j) \right] \right\} \right\rangle . \qquad (2.126)$$

Finally, the theory is applied to the photocount statistics. Following Glauber's theory of photon detection (Glauber 1965, Kelley and Kleiner 1964), the probability of observing precisely n events in a counting time interval $[t, t + \Delta t)$ is given by the relation

$$p_n(t, \Delta t) = \left\langle \mathbf{\Omega} \frac{1}{n!} \left[\hat{\Gamma}(t, \Delta t) \right]^n \exp\left[-\hat{\Gamma}(t, \Delta t) \right] \right\rangle , \qquad (2.127)$$

where

$$\hat{\Gamma}(t, \Delta t) = \sum_i \int_t^{t+\Delta t} \int_t^{t+\Delta t} S(t_1 - t_2) \hat{E}^{(-)}_k(\mathbf{r}_i, t_1) \hat{E}^{(+)}_k(\mathbf{r}_i, t_2)\, \mathrm{d}t_1\, \mathrm{d}t_2 \qquad (2.128)$$

may be interpreted as the operator of the integrated intensity. Here $\mathbf{r}_i$ are position vectors of the detector atoms and $S(t)$ is a response function. Let us note that one usually assumes that

$$S(t_1 - t_2) = \eta\delta(t_1 - t_2), \qquad (2.129)$$

with some η. In relation (2.127), the ordering symbol $\mathbf{\Omega}$ introduces the following operator ordering:

(i) The normal ordering of the operators $\hat{E}^{(-)}_k(x)$, $\hat{E}^{(+)}_k(x)$ with the operators $\hat{E}^{(-)}_k(x)$ to the left of the operators $\hat{E}^{(+)}_k(x)$.
(ii) $\mathcal{T}_+$ ordering of the operators $\hat{E}^{(+)}_k(x)$ and $\mathcal{T}_-$ ordering of the operators $\hat{E}^{(-)}_k(x)$.

In analogy with (2.122), relation (2.127) becomes

$$p_n(t, \Delta t) = \left\langle \mathcal{O} \frac{1}{n!} \left[\hat{\Gamma}(t, \Delta t) \right]^n \exp[-\hat{\Gamma}(t, \Delta t)] \right\rangle , \qquad (2.130)$$

where the $\boldsymbol{\Omega}$ ordering is simply replaced by the $\mathcal{O}$ ordering defined as follows:

(i) The normal ordering of the operators $\hat{E}_{ks}^{(-)}(x)$, $\hat{E}_{k\text{free}}^{(-)}(x)$, $\hat{E}_{ks}^{(+)}(x)$, $\hat{E}_{k\text{free}}^{(+)}(x)$ with the operators $\hat{E}_{ks}^{(-)}(x)$, $\hat{E}_{k\text{free}}^{(-)}(x)$ to the left of the operators $\hat{E}_{ks}^{(+)}(x)$, $\hat{E}_{k\text{free}}^{(+)}(x)$.
(ii) $\mathcal{O}_+$ ordering of the operators $\hat{E}_{ks}^{(+)}(x)$, $\hat{E}_{k\text{free}}^{(+)}(x)$ and $\mathcal{O}_-$ ordering of the operators $\hat{E}_{ks}^{(-)}(x)$, $\hat{E}_{k\text{free}}^{(-)}(x)$.

The fulfilling of the conditions (2.123) causes a modification of relation (2.128) as follows:

$$\hat{\Gamma}(t, \Delta t) = \sum_i \int_t^{t+\Delta t} \int_t^{t+\Delta t} S(t_1 - t_2)\hat{E}_{ks}^{(-)}(\mathbf{r}_i, t_1)\hat{E}_{ks}^{(+)}(\mathbf{r}_i, t_2)\,\mathrm{d}t_1\,\mathrm{d}t_2. \quad (2.131)$$

In the case of mixing the source field light with a coherent free-field reference beam, there is an analogy with the relation (2.126):

$$\hat{\Gamma}(t, \Delta t) = \sum_i \int_t^{t+\Delta t} \int_t^{t+\Delta t} S(t_1 - t_2)$$
$$\times \left[\mathcal{E}_k^*(\mathbf{r}_i, t_1)\hat{1} + \hat{E}_{ks}^{(-)}(\mathbf{r}_i, t_1)\right]\left[\mathcal{E}_k(\mathbf{r}_i, t_2)\hat{1} + \hat{E}_{ks}^{(+)}(\mathbf{r}_i, t_2)\right]\mathrm{d}t_1\,\mathrm{d}t_2. \quad (2.132)$$

A generalization of the Wick theorem on transforming a time-ordered product onto a sum of normally ordered terms was performed by Agarwal and Wolf (1970).

The quantum theory of the radiation field interacting with atomic sources in the presence of a linear, dispersionless, and absorptionless dielectric with space-dependent refractive index has been applied to the description of the action of a resonator-like cavity with input–output coupling and filled with an active medium (Knöll and Welsch 1992).

2.2.5 Continuum Frequency-Space Description

Blow et al. (1990) have formulated the quantum theory of optical wave propagation without recourse to cavity quantization. This approach avoids the introduction of a box-related mode spacing and enables one to use a continuum frequency-space description. In this chapter and in that by Blow et al. (1991) a continuous-mode quantum theory of electromagnetic field has been developed. As usual in the quantum field theory, the box-related modes are considered whose creation and destruction operators satisfy the usual independent boson commutation relations:

$$[\hat{a}_i, \hat{a}_j^\dagger] = \delta_{ij}\hat{1}. \quad (2.133)$$

Different modes of the cavity, labelled by i and j, have frequencies given by different integer multiples of the mode spacing $\Delta\omega$. The mode spectrum becomes continuous as $\Delta\omega \to 0$ and in this limit the transformation to continuous-mode operators is convenient:

$$\hat{a}_i \to \sqrt{\Delta\omega}\,\hat{a}(\omega). \tag{2.134}$$

A complete orthonormal set of functions was considered which may describe states of finite energy. The set is numerable infinite and to each function in it a destruction operator is assigned. Such operators have all the usual properties of the operators of the monochromatic mode.

Further specific states of the field have been treated such as coherent states, number states, noise and squeezed states. With the use of noncontinuous operators, a generalization of the single-mode normal ordering theorem was proved. Field quantization in a dielectric has been treated including the material dispersion and the theory has been applied to the pulse propagation in an optical fibre. A comparison with results by Drummond (1990, 1994) would be in order.

Let us consider the fields in a lossless dielectric material with the real relative permittivity $\epsilon(\omega)$ and the refractive index $n(\omega)$ related by

$$\epsilon(\omega) = [n(\omega)]^2. \tag{2.135}$$

Let us recall the definition of the phase velocity

$$v_{\mathrm{F}}(\omega) = \frac{\omega}{k} = \frac{c}{n(\omega)} \tag{2.136}$$

and that of the group velocity $v_{\mathrm{G}}(\omega)$

$$\frac{1}{v_{\mathrm{G}}(\omega)} = \frac{\partial k}{\partial\omega} = \frac{1}{c}\frac{\partial}{\partial\omega}[\omega n(\omega)]. \tag{2.137}$$

The normalization of the field operators is fixed by requirement that the normally ordered total energy density operator $\hat{U}(z,t)$ has the diagonal form:

$$\hat{H}_{\text{free}} = \mathcal{A}\int_{-\infty}^{\infty} \hat{U}(z,t)\,\mathrm{d}z = \int \hbar\omega\hat{a}^\dagger(\omega)\hat{a}(\omega)\,\mathrm{d}\omega. \tag{2.138}$$

The field operators are obtained in accordance with the relation

$$\hat{E}^{(+)}(z,t) = -\frac{\partial}{\partial t}\hat{A}^{(+)}(z,t),\; \hat{B}^{(+)}(z,t) = \frac{\partial}{\partial z}\hat{A}^{(+)}(z,t) \tag{2.139}$$

and with the expansion of the vector-potential operator

$$\hat{A}^{(+)}(z,t) = \int_{-\infty}^{\infty}\sqrt{\frac{\hbar v_{\mathrm{G}}(\omega)}{4\pi\epsilon_0 c\omega n(\omega)\mathcal{A}}} \times \sum_{\lambda=1,2} \boldsymbol{\epsilon}(k,\lambda)\hat{a}(k,\lambda)\exp[-\mathrm{i}(\omega t - kz)]\,\mathrm{d}k. \tag{2.140}$$

Noting that

$$dk = \frac{d\omega}{v_G(\omega)}, \hat{a}(k, \lambda) = \sqrt{v_G(\omega)}\hat{a}(\omega), \tag{2.141}$$

and taking the polarization to be parallel to the x-axis, it follows from (2.139) that the field operators are

$$\hat{E}^{(+)}(z, t) = i \int \sqrt{\frac{\hbar\omega}{4\pi\epsilon_0 c \mathcal{A} n(\omega)}} \times \hat{a}(\omega) \exp\left\{-i\omega\left[t - \frac{n(\omega)z}{c}\right]\right\} d\omega \tag{2.142}$$

and

$$\hat{B}^{(+)}(z, t) = i \int \sqrt{\frac{\hbar\omega n(\omega)}{4\pi\epsilon_0 c^3 \mathcal{A}}} \times \hat{a}(\omega) \exp\left\{-i\omega\left[t - \frac{n(\omega)z}{c}\right]\right\} d\omega. \tag{2.143}$$

Alternatively, the propagation constant can be expanded to the second order in frequency and a partial differential equation can be obtained (cf. Drummond 1990). Assuming a narrow bandwidth, the slowly varying field envelope can be represented by the operator $\hat{a}(z, t)$, which obeys the equation

$$i\frac{\partial}{\partial z}\hat{a}(z, t) + \frac{k''}{2}\frac{\partial^2}{\partial t^2}\hat{a}(z, t) = \hat{0}, \tag{2.144}$$

where k'' is the second derivative with respect to the frequency of the propagation constant, evaluated at the central frequency. The equation has been simplified using the transformation of envelope into a frame moving with the group velocity. This is necessary for the envelope to be slowly varying. In the classical nonlinear optics the stationary fields have also envelopes, but they seem to be defined otherwise.

The treatment of this problem in the noncontinuous basis proceeds from the replacement

$$\hat{a}(z, t) = \sum_j \phi_j(z, t)\hat{c}_j, \tag{2.145}$$

where $\phi_j(z, t)$ are a complete orthonormal set of functions on z and $\hat{c}_j$ are destruction operators obeying the usual commutation relations. The advantage of this treatment is that the functional dependence on z and t is contained in the c-number functions rather than the operators $\hat{a}(z, t)$ as in the propagation equation (2.144) for example. It is not emphasized by Blow et al. (1990) that the solution of equation (2.144) preserves the equal-space, not equal-time, commutators. Similarly, the set of functions $\phi_j(z,t)$ enjoys the orthonormality and completeness only as the equal-space, but not equal-time, properties. The propagation equation (2.144) now yields the following equations for the noncontinuous basis functions:

$$\mathrm{i}\frac{\partial}{\partial z}\phi_j(z,t) + \frac{k''}{2}\frac{\partial^2}{\partial t^2}\phi_j(z,t) = 0. \tag{2.146}$$

Finally, the process of photodetection in free space is considered and the results applied to homodyne detection with both local oscillator and signal fields pulsed. The results of sets of measurements in which the photocurrent is integrated over periods T can be predicted by the use of an operator

$$\hat{M} = \int_{\tau}^{\tau+T} \hat{a}^{\dagger}(t)\hat{a}(t)\,\mathrm{d}t. \tag{2.147}$$

Here τ is the start time of the measurements, the detector is placed at $z = 0$, and

$$\hat{a}(t) = \frac{1}{\sqrt{2\pi}}\int \hat{a}(\omega)\exp(-\mathrm{i}\omega t)\,\mathrm{d}\omega. \tag{2.148}$$

Let us further consider a balanced homodyne detector in which the light beam under study is superposed on a local oscillator by combining them at a 50:50 beam splitter. The measured quantity is the difference in the photocurrents of two detectors placed in the output arms of the beam splitter and it can be represented by the operator (Collett et al. 1987)

$$\hat{O} = \mathrm{i}\int_{\tau}^{\tau+T} [\hat{a}^{\dagger}(t)\hat{a}_{\mathrm{L}}(t) - \hat{a}_{\mathrm{L}}^{\dagger}(t)\hat{a}(t)]\,\mathrm{d}t, \tag{2.149}$$

where $\hat{a}_{\mathrm{L}}^{\dagger}(t)$ and $\hat{a}_{\mathrm{L}}(t)$ are the continuum creation and destruction operators of the local oscillator field and $\hat{a}^{\dagger}(t)$ and $\hat{a}(t)$ correspondingly for the signal field.

For homodyne detection of pulsed signals it is advantageous to use a pulsed local oscillator. The pulsed signal is described by the noncontinuous basis function $\phi_0(t)$ and the local oscillator is described by a normalized function $\phi_{\mathrm{L}}(t)$, the field of the local oscillator being in the coherent state $|\{\alpha_{\mathrm{L}}(t)\}\rangle$, where

$$\alpha_{\mathrm{L}}(t) = \sqrt{N_{\mathrm{L}}}\exp(\mathrm{i}\theta_{\mathrm{L}})\phi_{\mathrm{L}}(t), \tag{2.150}$$

with N_{L} the mean total number of photons in the pulse and θ_{L} the externally controlled local oscillator phase. Let us recall the definition of a coherent state:

$$|\{\alpha(t)\}\rangle = \hat{D}(\{\alpha(t)\})|0\rangle, \tag{2.151}$$

with

$$\hat{D}(\{\alpha(t)\}) = \exp\left\{\int [\alpha(t)\hat{a}^{\dagger}(t) - \alpha^{*}(t)\hat{a}(t)]\right\}, \tag{2.152}$$

which is close enough to that by Blow et al. (1990) except for the exchange of space for time. It is assumed that the signal field is described by a set of noncontinuous operators $\hat{d}_i$ at the output of a nonlinear system and the signal field at the input to the system is described by a similar set of operators $\hat{c}_i$. The action of the nonlinear system is defined by the relations

$$\hat{d}_0 = \mu\hat{c}_0 + \nu\hat{c}_0^\dagger,$$
$$\hat{d}_i = \hat{c}_i, i > 0. \tag{2.153}$$

In the relation (2.149) it is necessary to substitute

$$\hat{a}(t) = \sum_i \phi_i(t)\hat{d}_i. \tag{2.154}$$

In analogy, we consider

$$\hat{a}_{\mathrm{L}}(t) = \sum_i \phi_{i\mathrm{L}}(t)\hat{c}_{i\mathrm{L}}, \tag{2.155}$$

where the subscript L only modifies the familiar meanings and $\phi_{0\mathrm{L}}(t) = \phi_{\mathrm{L}}(t)$.

It is shown how the formulation of the quantum field theory is modified for the one-dimensional optical system. The fields are defined in an infinite waveguide parallel to the z-axis, but of finite cross-sectional area $\mathcal{A}$ of the rectangular form with sides parallel to the x- and y-axes. The x and y wave-vector components are thus restricted to discrete values and any three-dimensional integral over this spatial region is converted according to

$$\int \mathrm{d}^3k \rightarrow \frac{(2\pi)^2}{\mathcal{A}} \sum_{k_x,k_y} \int \mathrm{d}k_z. \tag{2.156}$$

On the assumption that the modes with $k_x \neq 0$ or $k_y \neq 0$ are vacuum ones, a reduced Hilbert (namely Fock) space can be exploited. The summation in (2.156) can, therefore, be removed and putting $k_z = k$, the other conversions are

$$\delta^{(3)}(\mathbf{k} - \mathbf{k}') \rightarrow \frac{\mathcal{A}}{(2\pi)^2}\delta(k - k'), \tag{2.157}$$

$$\hat{a}(\mathbf{k}, \lambda) \rightarrow \frac{\sqrt{\mathcal{A}}}{2\pi}\hat{a}(k, \lambda). \tag{2.158}$$

The vector-potential operator has been modified for the dispersive lossless medium and compared with Drummond (1990) and Loudon (1963), the positive-frequency part is

$$\hat{A}^{(+)}(\mathbf{r}, t) = \int \sqrt{\frac{\hbar v_{\rm G}(\omega)}{16\pi^3 \epsilon_0 c\omega n(\omega)}}$$

$$\times \sum_{\lambda=1,2} \boldsymbol{\epsilon}(\mathbf{k}, \lambda)\hat{a}(\mathbf{k}, \lambda) \exp[-\mathrm{i}(\omega t - \mathbf{k} \cdot \mathbf{r})] \, \mathrm{d}^3\mathbf{k}, \tag{2.159}$$

$$\omega = \frac{c}{n(\omega)}|\mathbf{k}|. \tag{2.160}$$

The expression (2.159) can be converted to the one-dimensional form easily (as indicated above, cf. (2.140)).

McDonald (2001) has considered a variation of the physical situation of "slow light" to show that the group velocity can be negative at central frequency. A Gaussian pulse can emerge from the far side of a slab earlier than it hits the near side and the pulse emission at the far side is accompanied by an antipulse emission, the antipulse propagating within the slab so as to annihilate the incident pulse at the near side.

2.3 Quantum Description of Experiments with Stationary Fields

Burnham and Weinberg (1970) found that the measured value of the correlation time between the two optical photons produced in a parametric process was very small, an effect of a practical interest. Laboratory techniques for doing experiments with single photons also have advanced. Since 1985, such photon pairs have become familiar for the study of nonclassical aspects of light (Horne et al. 1990).

The process of optical parametric three-wave mixing in a second-order nonlinear medium consists of the coherent interaction between pump, signal, and idler waves. This process may occur as frequency down conversions, specifically as an optical parametric oscillation and an optical parametric amplification. In a travelling-wave setting, the optical parametric generation is called a spontaneous parametric down-conversion. The photon pairs (biphotons) produced in parametric down-conversion are useful in experiments concerning fundamental questions of quantum theory. The description of experiments has been facilitated by studies of Campos et al. (1990). The autocorrelation and cross-correlation properties of the signal and idler beams produced in the parametric down-conversion have been studied, e.g. in Joobeur et al. (1994). A unified treatment of the experiment on the interference of a "biphoton with itself" and of other three experiments has been provided by Casado et al. (1997a). A fourth-order interference has been obtained in the four cases, and the uniformity has been achieved also by the use of the Wigner (or Weyl) representation of the field operators. A similar treatment of the famous experiment and of another one has been presented by Casado et al. (1997b). A second-order interference has been treated in the two cases and the stochastic properties of the pump beam have been respected.

The studies of the fundamentals of quantum mechanics underlie such interesting applications as quantum cryptography and quantum computing (Bowmeester et al. 2000). The experiments have become very popular (Shih 2003). Design of experiments for undergraduate students has become feasible (Galvez et al. 2005).

Here we return from the Wigner to the Hilbert-space formalism as in Peřinová and Lukš (2003). First we consider the three-dimensional expansion of the operator of a chosen component of the electric vector after Casado et al. (1997a). As in the schematics of the experiments the field is restricted to paths leading to detectors, we introduce one-dimensional expansions of the electric-field operator. We attempt to consider orthogonal modal functions, although we cannot define them everywhere, but only on the paths. We are aware of the dangerous position, where one cannot evaluate the orthogonality property for the lack of a complete definition. In this approach we do not start with the description of the process of parametric down-conversion from a Hamiltonian, but with the response of the output fields of a nonlinear crystal to the input fields (Casado et al. 1997a) when two paths cross such a crystal. Such a response depends also on stochastic properties of the pump beam, which is assumed to be monochromatic however. The experiment on the interference of signal and idler photons (Ghosh et al. 1986) can do with the simple description, when the lack of the second-order interference is derived. In the use of two detectors we consider four paths and modify (double) the description. Nevertheless, we do not reproduce the well-known result.

Similarly we proceed in the case of the experiment of Rarity and Tapster (1990), which was also used to test Bell's inequality using phase and momentum. In contrast, in the case of the experiment of Franson (1989) we are allowed to return to the simple description, as essentially two paths are involved, even though the schematic is more complicated. This experiment was proposed in order to test a Bell inequality for energy and time.

Next we deal with induced coherence and indistinguishability in two-photon interference (Zou et al. 1991). In this case the schematic comprises two nonlinear crystals, the number of paths is greater, but since two paths belong to each crystal, the simple description is appropriate. The lack of induced emission made it a "mind-boggling" experiment (Greenberger et al. 1993), but the indistinguishability of the paths along which the signal photon arrives at the detector (in fact, the biphoton arrives at the two detectors) is still held for the reason of interference. We may refer to Casado et al. (1997b), where stochastic properties of the pump beam are taken into account. Two experiments are analysed: frustrated two-photon creation by interference, and induced coherence and indistinguishability. Coincidences are not studied and a second-order interference has been obtained in the two cases. Last we mention the frustrated two-photon creation via interference (Herzog et al. 1994) restricting ourselves to the second-order interference and the monochromatic pump.

2.3.1 *Spatio-temporal Descriptions of Parametric Down-Conversion Experiments*

In the Hilbert-space representation of the light field, the electric vector is represented as a sum of two mutually conjugate operators (Casado et al. 1997a)

$$\hat{\mathbf{E}}(\mathbf{r}, t) = \hat{\mathbf{E}}^{(+)}(\mathbf{r}, t) + \hat{\mathbf{E}}^{(-)}(\mathbf{r}, t), \tag{2.161}$$

$$\hat{\mathbf{E}}^{(+)}(\mathbf{r}, t) = \mathrm{i}\sum_{\mathbf{k},\lambda}\sqrt{\frac{\hbar\omega_{\mathbf{k}}}{2L^3}}\epsilon_{\mathbf{k},\lambda}\hat{a}_{\mathbf{k},\lambda}(t)e^{\mathrm{i}\mathbf{k}\cdot\mathbf{r}}, \tag{2.162}$$

where L^3 is the normalization volume, $\hat{a}_{\mathbf{k},\lambda}(t)$ is the annihilation operator for a photon whose wave vector is $\mathbf{k}$ and whose polarization vector is $\epsilon_{\mathbf{k},\lambda}$, and $\omega_{\mathbf{k}} = c|\mathbf{k}|$. Equations (2.161) and (2.162) correspond to the Heisenberg picture, where all time dependence of the averages comes from the creation and annihilation operators $\hat{a}^{\dagger}_{\mathbf{k},\lambda}(t)$ and $\hat{a}_{\mathbf{k},\lambda}(t)$. In this picture the state of the field is represented by a time-independent statistical operator $\hat{\rho}$.

As we do not study experiments involving polarizing devices, we find it convenient to use a scalar approximation well known in classical optics. When Casado et al. (1997a) use the subscripts on the (Wigner representations of) field operators they indicate that the light beam contains frequencies within a range and that "transverse" components of wave vectors are limited by small upper values. We believe that such subscripts indicate which part of the field is considered.

The laser theory and, in general, the theory of resonators connect the quantum field with the annihilation operators not via the complex exponentials, but via more general modal functions, which are often related to the device. We suppose that such an approach can be interesting also in our study, after we find modal functions that are connected to the linear devices used and to the mirrors. Obviously, the free evolution of operators $\hat{a}_{\mathbf{k}0}$ (zeroth-order solution) is transformed into a kind of linear dynamics of the "relevant" component $\hat{E}^{(+)}_{\mathrm{ss}\,0}(\mathbf{r}, t)$ of the electric vector via the appropriate modal functions, with ss being any subscript. This process can be formalized by a quadratic Hamiltonian, which differs from the free-field Hamiltonian only by the meaning of the creation and annihilation operators. We restrict ourselves to the operator $\hat{\rho}$ that represents a vacuum state.

We suppose that one or two nonlinear crystals involved in the experiment are described in terms of interaction Hamiltonians. The action of the scattering operator on the initial field can be "guessed". The interaction lasting only for a short time and being spatially confined to the medium suggests to us an appropriate modification of the linear dynamics. We modify also the notation for the resulting field by omitting the initial subscript 0.

(i) *The process of parametric down-conversion*

We are going to study the process of parametric down-conversion of light in the Hilbert-space representation. We refer to any of our figures for a sketch of the setup used for parametric down-conversion. A nonlinear crystal is pumped by a laser beam V, producing a continuum of coloured cones around the axis defined by the pump. In experimental practice two narrow correlated beams, called "signal" $\hat{E}_{\mathrm{s}}$ and "idler" $\hat{E}_{\mathrm{i}}$, are selected by means of apertures, filters, or just the detectors. Let us take the origin of the coordinate system, $\mathbf{0} \equiv \mathbf{0}_1$, at the centre of the crystal. We treat the pump beam as an intense monochromatic wave represented, in the scalar approximation, by

$$V(\mathbf{r}, t) = V e^{\mathrm{i}(\mathbf{k}_0 \cdot \mathbf{r} - \omega_0 t)} + \text{c.c.}, \tag{2.163}$$

where V is a complex amplitude of a pump beam, ω_0 is a frequency of the pump beam, $\mathbf{k}_0$ is an appropriate wave vector, and c.c. means the complex conjugate term to the previous one. In a product with the identity operator it may be added to the electric-field operator.

Now, let us consider two narrow correlated beams, called signal and idler, with average frequencies ω_{s}, ω_{i} and wave vectors $\mathbf{k}_{\mathrm{s}}$, $\mathbf{k}_{\mathrm{i}}$, respectively, fulfilling the matching conditions

$$\omega_{\mathrm{s}} + \omega_{\mathrm{i}} = \omega_0, \quad \mathbf{k}_{\mathrm{s}} + \mathbf{k}_{\mathrm{i}} = \mathbf{k}_0. \tag{2.164}$$

The response $\hat{E}_{\mathrm{s}}^{(+)}(\mathbf{r}, t)$ and $\hat{E}_{\mathrm{i}}^{(-)}(\mathbf{r}, t)$ of a nonlinear crystal to the input fields $\hat{E}_{0\mathrm{s}}^{(+)}(\mathbf{r}, t)$ and $\hat{E}_{0\mathrm{i}}^{(-)}(\mathbf{r}, t)$ is as

$$\begin{aligned} \hat{E}_{\mathrm{s}}^{(+)}(\mathbf{r}, t) &= (\hat{\hat{1}} + g^2|V|^2\hat{\hat{J}})\hat{E}_{0\mathrm{s}}^{(+)}(\mathbf{r}, t) + e^{-\mathrm{i}\omega_0 t} gV\hat{\hat{G}}\hat{E}_{0\mathrm{i}}^{(-)}(\mathbf{r}, t), \\ \hat{E}_{\mathrm{i}}^{(-)}(\mathbf{r}, t) &= e^{\mathrm{i}\omega_0 t} gV\hat{\hat{G}}^{\dagger}\hat{E}_{0\mathrm{s}}^{(+)}(\mathbf{r}, t) + (\hat{\hat{1}} + g^2|V|^2\hat{\hat{J}})\hat{E}_{0\mathrm{i}}^{(-)}(\mathbf{r}, t), \end{aligned} \tag{2.165}$$

where g is an effective coupling constant, $\hat{\hat{1}}$ is the identity superoperator, $\hat{\hat{G}}$ and $\hat{\hat{J}}$ are antilinear and linear superoperators, respectively, which substitute expansions in annihilation (creation) operators for annihilation (creation) operators ($\hat{\hat{J}}$ yields an expansion in the annihilation operators in the first equation and $\hat{\hat{G}}$ yields an expansion in the creation operators in the same equation), and $\hat{\hat{G}}^{\dagger}\hat{E}_{0\mathrm{s}}^{(+)}(\mathbf{r}, t) \equiv [\hat{\hat{G}}\hat{E}_{0\mathrm{s}}^{(-)}(\mathbf{r}, t)]^{\dagger}$. The relation (2.165) can be modified (doubled) so that it relates output fields $\hat{E}_{\mathrm{s}_j}^{(+)}(\mathbf{r}, t)$, $\hat{E}_{\mathrm{i}_j}^{(-)}(\mathbf{r}, t)$, $j = 1, 2$, to input fields $\hat{E}_{0\mathrm{s}_j}^{(+)}(\mathbf{r}, t)$, $\hat{E}_{0\mathrm{i}_j}^{(-)}(\mathbf{r}, t)$, $j = 1, 2$. Since a pointlike crystal is considered (Casado et al. 1997b), it may be interesting to imagine Equations (2.165) at $\mathbf{r} = \mathbf{0}$ without the subscript 0 on the right-hand side. It can occur, but at the cost of other notation. The interaction does not change the field just in front of the crystal, so we can interpret the initial field as the "in" resulting field. As it is almost at the centre of the crystal, it differs negligibly from the initial field just behind the crystal, which becomes the "out" resulting field.

In order to determine the detection probabilities in the Hilbert-space representation, we adopt the correlation properties (Casado et al. 1997a). In such a work it has proved convenient to substitute slowly varying amplitudes $\hat{F}_j^{(+)}(\mathbf{r}, t)$ $[\hat{F}_j^{(-)}(\mathbf{r}, t)]$ for the amplitudes $\hat{E}_J^{(+)}(\mathbf{r}, t)$ $[\hat{E}_J^{(-)}(\mathbf{r}, t)]$, the relation between them being

$$\hat{F}_J^{(+)}(\mathbf{r}, t) = e^{i\omega_J t} \hat{E}_J^{(+)}(\mathbf{r}, t),\ J = \text{s, i}. \tag{2.166}$$

According to Casado et al. (1997a) it is essential to use the following relation, which is still an approximation:

$$\hat{F}^{(+)}(\mathbf{r}_b, t) = \hat{F}^{(+)}\left(\mathbf{r}_a, t - \frac{r_{ab}}{c}\right) \exp\left(i\omega_a \frac{r_{ab}}{c}\right), \tag{2.167}$$

where ω_a is some frequency appropriate to a light beam and $r^{ab} = \mathbf{e}_a \cdot (\mathbf{r}_b - \mathbf{r}_a)$, with $\mathbf{e}_a$ being the unit vector in the direction of propagation. Since the vectors whose dot product is taken are usually of the same direction, the magnitude of displacement vector may be evoked.

If we consider the signal beam emerging from the crystal at different times t and t', we can use the autocorrelations (Casado et al. 1997a):

$$\langle \hat{F}_J^{(-)}(\mathbf{r}, t) \hat{F}_J^{(+)}(\mathbf{r}, t') \rangle = g^2 |V|^2 \mu_J(t' - t),\ J = \text{s, i}. \tag{2.168}$$

The following autocorrelations, and their complex conjugates, vanish:

$$\langle \hat{F}_J^{(+)}(\mathbf{r}, t) \hat{F}_J^{(+)}(\mathbf{r}, t') \rangle = 0,\ J = \text{s, i}. \tag{2.169}$$

The relation holds at any point of the outgoing beam, most interestingly just behind the crystal.

With respect to the cross correlation, we prefer to characterize the signal and idler field operators just behind the crystal at different times (Casado et al. 1997a):

$$\langle \hat{F}_{\text{s out}}^{(+)}(\mathbf{0}, t) \hat{F}_{\text{i out}}^{(+)}(\mathbf{0}, t') \rangle = g V \nu(t' - t). \tag{2.170}$$

It is useful to know that

$$\langle \hat{F}_\text{s}^{(+)}(\mathbf{r}, t) \hat{F}_\text{i}^{(-)}(\mathbf{r}, t') \rangle = \langle \hat{F}_\text{s}^{(-)}(\mathbf{r}, t) \hat{F}_\text{i}^{(+)}(\mathbf{r}, t') \rangle = 0. \tag{2.171}$$

In the Hilbert-space formalism, the usual theory of detection (by photon absorption) is based on the normal ordering. The joint detection rate is given by

$$P_{ab}(\mathbf{r}_a, t; \mathbf{r}_b, t') = K' \langle 0 | \hat{E}^{(-)}(\mathbf{r}_a, t) \hat{E}^{(-)}(\mathbf{r}_b, t') \hat{E}^{(+)}(\mathbf{r}_b, t') \hat{E}^{(+)}(\mathbf{r}_a, t) | 0 \rangle \tag{2.172}$$

in the Heisenberg picture, where $K' = K_a K_b$ and K_a (K_b) is a constant related to the efficiency of the detector and the energy of a single photon. The well-known property of Gaussian random variables A, B, C, and D,

$$\langle ABCD \rangle = \langle AB \rangle \langle CD \rangle + \langle AC \rangle \langle BD \rangle + \langle AD \rangle \langle BC \rangle, \tag{2.173}$$

applies not only in the Weyl (Casado et al. 1997a) but also in the normal ordering and entails that the joint detection rate is written in three terms. The first two terms are fourth order in g, while the last term is second order in g. We may discard the first two terms (Casado et al. 1997a) and finally obtain

$$P_{ab}(\mathbf{r}_a, t; \mathbf{r}_b, t') = K'|\langle \hat{E}^{(+)}(\mathbf{r}_a, t)\hat{E}^{(+)}(\mathbf{r}_b, t')\rangle|^2. \tag{2.174}$$

We will determine the visibility $\mathcal{V}'$ of the intensity interference:

$$\mathcal{V}' = \frac{R_{ab\max} - R_{ab\min}}{R_{ab\max} + R_{ab\min}}, \tag{2.175}$$

where

$$R_{ab\max} = \int_{-\frac{w}{2}}^{\frac{w}{2}} P_{ab\max}(\tau')\,\mathrm{d}\tau',\ R_{ab\min} = \int_{-\frac{w}{2}}^{\frac{w}{2}} P_{ab\min}(\tau')\,\mathrm{d}\tau', \tag{2.176}$$

with w the coincidence window which we choose to be $w = 13 \times 10^{-9}$ s, defined in terms of the integral

$$\begin{aligned} M\left(\frac{d}{c}, \frac{h}{c}\right) &= K' \int_{-\frac{w}{2}}^{\frac{w}{2}} \left|\nu\left(\tau' + \frac{d}{c}\right)\right| \left|\nu\left(\tau' + \frac{h}{c}\right)\right| \mathrm{d}\tau' \\ &= \exp\left[-\frac{\sigma^2}{2}\left(\frac{h-d}{c}\right)^2\right] \\ &\quad \times \frac{1}{2}\left\{\mathrm{erf}\left[\frac{\sigma}{\sqrt{2}}\left(\frac{d+h}{c} + w\right)\right] \mp \mathrm{erf}\left[\mp\frac{\sigma}{\sqrt{2}}\left(w - \frac{d+h}{c}\right)\right]\right\}. \end{aligned} \tag{2.177}$$

Here

$$\mathrm{erf}(x) = \frac{2}{\sqrt{\pi}} \int_0^x e^{-t^2}\,\mathrm{d}t, \tag{2.178}$$

d, h are parameters of an experimental setup, $c = 2.998 \times 10^8\ \mathrm{ms}^{-1}$ is the speed of light, $\nu(\tau)$ is a Gaussian,

$$\nu(\tau') = |\nu(\tau')| = \sqrt{\frac{1}{K'}} \left(\frac{2}{\pi}\right)^{\frac{1}{4}} \sqrt{\sigma} e^{-\sigma^2\tau'^2}, \tag{2.179}$$

where $\sigma = 10^{12}\ \mathrm{s}^{-1}$. Let us remember that for $\sigma^{-1} \ll w$, we have $\mathrm{erf}(\pm\infty) = \pm 1$. Particularly,

$$\begin{aligned} M\left(\frac{d}{c}, \frac{d}{c}\right) &= \frac{1}{2}\left\{\mathrm{erf}\left[\frac{\sigma}{\sqrt{2}}\left(\frac{2d}{c} + w\right)\right] + \mathrm{erf}\left[\frac{\sigma}{\sqrt{2}}\left(w - \frac{2d}{c}\right)\right]\right\}, \\ M(0,0) &= \mathrm{erf}\left(\frac{\sigma}{\sqrt{2}} w\right). \end{aligned} \tag{2.180}$$

(ii) *Experiment on the interference of signal and idler photons*

Let us start with an experiment demonstrating the coherence properties of the parametric down-conversion photon pairs as proposed in Ghosh et al. (1986). It is assumed that $\omega_s = \omega_i = \frac{\omega_0}{2}$ and the signal and idler beams are directed to a screen by means of two mirrors. There is no second-order interference between the two beams. When two detectors are put on the screen one can show a fourth-order, or intensity–intensity, interference. As seen from Fig. 2.1, the tracing of the beams is not so evident, and we modify the well-known result (Casado et al. 1997a, Ghosh et al. 1986). A report on the experiment was brief (Ghosh and Mandel 1987).

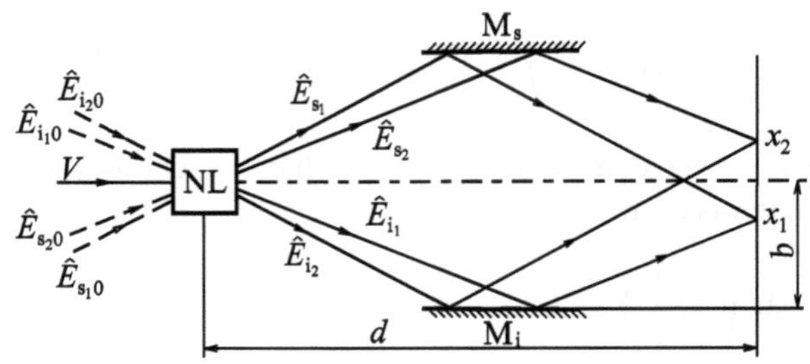

Fig. 2.1 Experimental setup on interference on a screen

In what follows, we will specify modal functions and nonlinear dynamics of field operators. We introduce the notation for the points of reflection $\mathbf{0}_{\mathrm{M}_{\mathrm{s}_j}} = (b, 0, z_{\mathrm{M}_{\mathrm{s}_j}})$, $\mathbf{0}_{\mathrm{M}_{\mathrm{i}_j}} = (-b, 0, z_{\mathrm{M}_{\mathrm{i}_j}})$, $j = 1, 2$, where

$$z_{\mathrm{M}_{\mathrm{s}_j}} = \frac{bd}{2b - x_j}, \quad z_{\mathrm{M}_{\mathrm{i}_j}} = \frac{bd}{2b + x_j}, \quad j = 1, 2, \tag{2.181}$$

with d being the distance from the centre of the crystal to the screen and b being the distance from the axis of the pumping to the mirrors.

We consider the initial electric field in the form

$$\hat{E}_0^{(+)}(\mathbf{r}, t) = V^{(+)}(\mathbf{r}, t)\hat{1} + \sum_{j=1}^{2} \left[\hat{E}_{\mathrm{s}_j 0}^{(+)}(\mathbf{r}, t) + \hat{E}_{\mathrm{i}_j 0}^{(+)}(\mathbf{r}, t) \right], \tag{2.182}$$

where $\mathbf{r} = (x, y, z)$, $\mathbf{k} = (k_x, k_y, k_z)$, and

$$\hat{E}_{\mathrm{s}_j 0}^{(+)}(\mathbf{r}, t) = \sum_{\mathbf{k} \in [\mathbf{k}]_{\mathrm{s}_j}} v_{\mathrm{s}_j \mathbf{k}}(\mathbf{r}) \hat{a}_{\mathrm{s}_j \mathbf{k} 0}(t), \tag{2.183}$$

$$\hat{E}_{\mathrm{i}_j 0}^{(+)}(\mathbf{r}, t) = \sum_{\mathbf{k} \in [\mathbf{k}]_{\mathrm{i}_j}} v_{\mathrm{i}_j \mathbf{k}}(\mathbf{r}) \hat{a}_{\mathrm{i}_j \mathbf{k} 0}(t), \tag{2.184}$$

with the orthonormal systems of functions $v_{\mathrm{s}_j\mathbf{k}}(\mathbf{r})$, $\mathbf{k} \in [\mathbf{k}]_{\mathrm{s}_j}$, $v_{\mathrm{i}_j\mathbf{k}}(\mathbf{r})$, $\mathbf{k} \in [\mathbf{k}]_{\mathrm{i}_j}$, $j = 1, 2$,

$$\|v_{\mathrm{s}_j\mathbf{k}}(\mathbf{r})\|^2 = \int |v_{\mathrm{s}_j\mathbf{k}}(\mathbf{r})|^2 \, \mathrm{d}^3\mathbf{r} = \hbar\omega_{\mathbf{k}}, \quad \mathbf{k} \in [\mathbf{k}]_{\mathrm{s}_j}, \tag{2.185}$$

$$\|v_{\mathrm{i}_j\mathbf{k}}(\mathbf{r})\|^2 = \int |v_{\mathrm{i}_j\mathbf{k}}(\mathbf{r})|^2 \, \mathrm{d}^3\mathbf{r} = \hbar\omega_{\mathbf{k}}, \quad \mathbf{k} \in [\mathbf{k}]_{\mathrm{i}_j}. \tag{2.186}$$

The $[\mathbf{k}]_{\mathrm{s}_j}$ is a set of integer multiples of the vector $\frac{2\pi}{L}\mathbf{e}_{\mathrm{s}_j}$, $\mathbf{e}_{\mathrm{s}_j}$ being a unit vector of the signal beam at the origin. Similarly for $[\mathbf{k}]_{\mathrm{i}_j}$. A formal expression for $v_{J\mathbf{k}}(\mathbf{r})$, $J = \mathrm{s}_j, \mathrm{i}_j$, $j = 1, 2$, is of the form

$$\begin{aligned} v_{J\mathbf{k}}(\mathbf{r}) &= \mathrm{i}\sqrt{\frac{\hbar\omega_{\mathbf{k}}}{\mathcal{A}L}} \exp(\mathrm{i}\mathbf{k}\cdot\mathbf{r}) \ \text{ for } \ \mathbf{r} \parallel \mathbf{k}, \ \ z < z_{\mathrm{M}_J}, \ \ \mathbf{k} \in [\mathbf{k}]_J, \\ v_{J\mathbf{k}}(\mathbf{r}) &= -\mathrm{i}\sqrt{\frac{\hbar\omega_{\mathbf{k}}}{\mathcal{A}L}} \exp[\mathrm{i}\mathbf{k}'\cdot(\mathbf{r} - \mathbf{0}'_J)] \ \text{ for } \ (\mathbf{r} - \mathbf{0}'_J) \parallel \mathbf{k}', \\ &\quad z > z_{\mathrm{M}_J}, \ \ \mathbf{k} \in [\mathbf{k}]_J, \end{aligned} \tag{2.187}$$

where $J = \mathrm{s}_j, \mathrm{i}_j$, $j = 1, 2$, $\mathcal{A}$ is the effective transverse area of the beam, $\mathbf{0}'_J = (2b, 0, 0)$ for $J = \mathrm{s}_1, \mathrm{s}_2$, $\mathbf{0}'_J = (-2b, 0, 0)$ for $J = \mathrm{i}_1, \mathrm{i}_2$, $\mathbf{k}' = (-k_x, k_y, k_z)$. In a standard fashion, we associate the signal and idler modal functions (2.187) with fields we denote as $\hat{E}_{J0}^{(+)}(\mathbf{r}, t)$, $J = \mathrm{s}_j, \mathrm{i}_j$, $j = 1, 2$.

After switching on the nonlinear interaction, part of the field is not influenced:

$$\hat{E}_{\mathrm{s}_j}^{(+)}(\mathbf{r}, t) = \hat{E}_{\mathrm{s}_j0}^{(+)}(\mathbf{r}, t), \ \ \hat{E}_{\mathrm{i}_j}^{(+)}(\mathbf{r}, t) = \hat{E}_{\mathrm{i}_j0}^{(+)}(\mathbf{r}, t) \ \text{ for } \ z < 0, \tag{2.188}$$

whereas for $z > 0$ provided that $g|V| \ll 1$, the perturbative approximation of the solution of the Heisenberg equations of motion that retains terms up to g^2 can be written as

$$\hat{E}_{\mathrm{s}_j}^{(+)}(\mathbf{r}, t) = \hat{E}_{\mathrm{s}_j0}^{(+)}(\mathbf{r}, t) + e^{-\mathrm{i}\omega_0 t} gV\hat{\hat{G}}_j \hat{E}_{\mathrm{i}_j0}^{(-)}(\mathbf{r}, t) + g^2|V|^2\hat{\hat{J}}_j \hat{E}_{\mathrm{s}_j0}^{(+)}(\mathbf{r}, t), \tag{2.189}$$

$$\begin{aligned} \hat{E}_{\mathrm{i}_j}^{(+)}(\mathbf{r}, t) &= \hat{E}_{\mathrm{i}_j0}^{(+)}(\mathbf{r}, t) + e^{-\mathrm{i}\omega_0 t} gV\hat{\hat{G}}_j \hat{E}_{\mathrm{s}_j0}^{(-)}(\mathbf{r}, t) \\ &\quad + g^2|V|^2\hat{\hat{J}}_j \hat{E}_{\mathrm{i}_j0}^{(+)}(\mathbf{r}, t), \quad j = 1, 2, \end{aligned} \tag{2.190}$$

where

$$\hat{E}_{\mathrm{s}_j0}^{(-)}(\mathbf{r}, t) = [\hat{E}_{\mathrm{s}_j0}^{(+)}(\mathbf{r}, t)]^\dagger, \ \ \hat{E}_{\mathrm{i}_j0}^{(-)}(\mathbf{r}, t) = [\hat{E}_{\mathrm{i}_j0}^{(+)}(\mathbf{r}, t)]^\dagger, \tag{2.191}$$

and $\hat{\hat{G}}_j$ and $\hat{\hat{J}}_j$ are antilinear and linear superoperators, respectively, which substitute the expansions in annihilation operators ($\hat{\hat{J}}_j$ yields an expansion in the annihilation operators) for annihilation (creation) operators ($\hat{\hat{G}}_j$ yields an expansion in creation operators). Compare Casado et al. (1997a), where G_j and J_j are appropriate linear operators acting on functions of the argument $\mathbf{r}$ for complex-valued functions. The superoperators $\hat{\hat{G}}_j$ and $\hat{\hat{J}}_j$ have the properties

$$\hat{\hat{G}}_j \hat{a}_{j0\mathbf{k}}(t) = \sum_{[\mathbf{k}']_{\mathrm{s}_j}} f(\mathbf{k}, \mathbf{k}') u \left[\frac{\Delta t}{2} (\omega_0 - \omega_{\mathbf{k}} - \omega_{\mathbf{k}'}) \right] \times \exp \left[\mathrm{i}t(\omega_0 - \omega_{\mathbf{k}} - \omega_{\mathbf{k}'}) \right] \hat{a}^{\dagger}_{j0\mathbf{k}'}(t) \text{ for } \mathbf{k} \in [\mathbf{k}]_{\mathrm{i}_j}, \tag{2.192}$$

$$\hat{\hat{J}}_j \hat{a}_{j0\mathbf{k}}(t) = \sum_{[\mathbf{k}']_{\mathrm{i}_j}} \sum_{[\mathbf{k}'']_{\mathrm{s}_j}} f(\mathbf{k}, \mathbf{k}') f^*(\mathbf{k}', \mathbf{k}'') \times u \left[\frac{\Delta t}{2} (\omega_{\mathbf{k}'} + \omega_{\mathbf{k}''} - \omega_0) \right] u \left[\frac{\Delta t}{2} (\omega_{\mathbf{k}''} - \omega_{\mathbf{k}}) \right] \exp \left[\mathrm{i}t(\omega_{\mathbf{k}''} - \omega_{\mathbf{k}}) \right] \times \hat{a}_{j0\mathbf{k}''}(t) \text{ for } \mathbf{k} \in [\mathbf{k}]_{\mathrm{s}_j}, \tag{2.193}$$

respectively, with

$$u(x) = \frac{\sin x}{x} e^{\mathrm{i}x}, \tag{2.194}$$

$$\hat{a}_{j0\mathbf{k}}(t) = \hat{a}_{j0\mathbf{k}}(0) e^{-\mathrm{i}\omega_{\mathbf{k}} t}. \tag{2.195}$$

Supposing that in the sense of classical nonlinear optics, $f(\mathbf{k}, \mathbf{k}')$ is a distribution with a support determined by the condition

$$\omega_0 - \omega_{\mathbf{k}} - \omega_{\mathbf{k}'} = 0, \tag{2.196}$$

we easily obtain that

$$\hat{\hat{G}}_j \hat{a}_{j0\mathbf{k}}(t) = \sum_{[\mathbf{k}']_{\mathrm{s}_j}} f(\mathbf{k}, \mathbf{k}') \hat{a}^{\dagger}_{j0\mathbf{k}'}(t) \text{ for } \mathbf{k} \in [\mathbf{k}]_{\mathrm{i}_j}, \tag{2.197}$$

$$\hat{\hat{J}}_j \hat{a}_{j0\mathbf{k}}(t) = \sum_{[\mathbf{k}'']_{\mathrm{s}_j}} \left[\sum_{[\mathbf{k}']_{\mathrm{i}_j}} f(\mathbf{k}, \mathbf{k}') f^*(\mathbf{k}', \mathbf{k}'') \right] \hat{a}_{j0\mathbf{k}''}(t) \text{ for } \mathbf{k} \in [\mathbf{k}]_{\mathrm{s}_j}, \tag{2.198}$$

which is a great unexpected simplification.

Further we will express the intensity correlations that have been determined in the experiment. Introducing

$$\hat{F}^{(+)}_{\mathrm{s}_j}(\mathbf{r}, t) = \exp(\mathrm{i}\omega_{\mathrm{s}} t) \hat{E}^{(+)}_{\mathrm{s}_j}(\mathbf{r}, t), \tag{2.199}$$

we express the field at a point $\mathbf{r}_j$, $j = 1, 2$, on the screen as

$$\hat{F}^{(+)}(\mathbf{r}_j, t) = \hat{F}^{(+)}_{\mathrm{s}_j\mathrm{out}} \left(\mathbf{0}, t - \frac{r_{\mathrm{s}_j}}{c} \right) \exp \left(\mathrm{i} \frac{\omega_0 r_{\mathrm{s}_j}}{2c} \right) + \hat{F}^{(+)}_{\mathrm{i}_j\mathrm{out}} \left(\mathbf{0}, t - \frac{r_{\mathrm{i}_j}}{c} \right) \exp \left(\mathrm{i} \frac{\omega_0 r_{\mathrm{i}_j}}{2c} \right), \quad j = 1, 2, \tag{2.200}$$

where

$$r_{\mathrm{s}_j} = r_{\mathrm{s}}(x_j) = \sqrt{z_{\mathrm{M}_{\mathrm{s}_j}}^2 + b^2} + \sqrt{(d - z_{\mathrm{M}_{\mathrm{s}_j}})^2 + (b - x_j)^2}\,, \quad j = 1, 2,$$

$$r_{\mathrm{i}_j} = r_{\mathrm{i}}(x_j) = \sqrt{z_{\mathrm{M}_{\mathrm{i}_j}}^2 + b^2} + \sqrt{(d - z_{\mathrm{M}_{\mathrm{i}_j}})^2 + (b + x_j)^2}\,, \tag{2.201}$$

and the subscript out indicates that the field behind the nonlinear crystal is considered. Hence, we can obtain the relations (2.209) below.

By taking into account the correlation relations

$$\begin{aligned}\langle \hat{F}^{(+)}_{\mathrm{s}_1\mathrm{out}}(\mathbf{0}, t)\hat{F}^{(+)}_{\mathrm{i}_2\mathrm{out}}(\mathbf{0}, t')\rangle &= \langle \hat{F}^{(+)}_{\mathrm{i}_1\mathrm{out}}(\mathbf{0}, t')\hat{F}^{(+)}_{\mathrm{s}_2\mathrm{out}}(\mathbf{0}, t)\rangle \\ &= gV\nu(t' - t),\end{aligned} \tag{2.202}$$

we get (cf. Casado et al. 1997a)

$$\begin{aligned}&\langle \hat{F}^{(+)}(\mathbf{r}_1, t_1)\hat{F}^{(+)}(\mathbf{r}_2, t_2)\rangle \\ &= gV\left\{\nu\left(t_2 - t_1 + \frac{r_{\mathrm{s}_1}}{c} - \frac{r_{\mathrm{i}_2}}{c}\right)\exp\left[\mathrm{i}\frac{\omega_0}{2c}(r_{\mathrm{i}_2} + r_{\mathrm{s}_1})\right]\right. \\ &\quad \left. + \nu\left(t_1 - t_2 + \frac{r_{\mathrm{s}_2}}{c} - \frac{r_{\mathrm{i}_1}}{c}\right)\exp\left[\mathrm{i}\frac{\omega_0}{2c}(r_{\mathrm{i}_1} + r_{\mathrm{s}_2})\right]\right\}.\end{aligned} \tag{2.203}$$

Assuming that the beams with different subscripts j are mutually uncorrelated, we finally get (cf. Casado et al. 1997a)

$$\begin{aligned}&P_{12}(\mathbf{r}_1, t + \tau_1; \mathbf{r}_2, t + \tau_2) \approx K' g^2 |V|^2 \\ &\times \left\{\left|\nu\left(\tau_2 - \tau_1 + \frac{r_{\mathrm{s}_1}}{c} - \frac{r_{\mathrm{i}_2}}{c}\right)\right|^2 + \left|\nu\left(\tau_1 - \tau_2 + \frac{r_{\mathrm{s}_2}}{c} - \frac{r_{\mathrm{i}_1}}{c}\right)\right|^2\right. \\ &+ 2\mathrm{Re}\left[\nu\left(\tau_2 - \tau_1 + \frac{r_{\mathrm{s}_1}}{c} - \frac{r_{\mathrm{i}_2}}{c}\right)\nu^*\left(\tau_1 - \tau_2 + \frac{r_{\mathrm{s}_2}}{c} - \frac{r_{\mathrm{i}_1}}{c}\right)\right. \\ &\left.\left.\times \exp\left[\mathrm{i}\frac{\omega_0}{2c}(r_{\mathrm{i}_2} + r_{\mathrm{s}_1} - r_{\mathrm{i}_1} - r_{\mathrm{s}_2})\right]\right]\right\},\end{aligned} \tag{2.204}$$

where K' is a constant related to the efficiency of the detectors,

$$K' = K_1 K_2, \quad K_1 = \frac{2\eta_1}{\hbar\omega_0}, \quad K_2 = \frac{2\eta_2}{\hbar\omega_0}, \tag{2.205}$$

η_J, $J = 1, 2$, is the efficiency of the detector D_J.

The visibility is expressed by the formula (2.175), where

$$R_{\text{simax}} + R_{\text{simin}} = 2g^2|V|^2$$

$$\times \left[M\left(\frac{r_{\text{s}_1} - r_{\text{i}_2}}{c}, \frac{r_{\text{s}_1} - r_{\text{i}_2}}{c}\right) + M\left(\frac{r_{\text{i}_1} - r_{\text{s}_2}}{c}, \frac{r_{\text{i}_1} - r_{\text{s}_2}}{c}\right)\right], \tag{2.206}$$

$$R_{\text{simax}} - R_{\text{simin}} = 4g^2|V|^2 M\left(\frac{r_{\text{s}_1} - r_{\text{i}_2}}{c}, \frac{r_{\text{i}_1} - r_{\text{s}_2}}{c}\right). \tag{2.207}$$

We may ask whether the visibility has its proper meaning for all x_1, x_2, whether it is associated only with the extremes of the detection rate, i.e. x_1 and x_2 for which $R_{\text{si}} = R_{\text{simax}}$ or $R_{\text{si}} = R_{\text{simin}}$. By relation (2.204) and the choice (2.179) we are interested in $C = \pm 1$, where

$$C = \cos\left[\frac{\omega_0}{2}\left(\frac{r_{\text{i}_2} + r_{\text{s}_1}}{c} - \frac{r_{\text{i}_1} + r_{\text{s}_2}}{c}\right)\right], \tag{2.208}$$

with $\omega_0 = \frac{2\pi c}{351} \times 10^9$ Hz.

The original formulae for $r_{\text{s}_j}, r_{\text{i}_j}$ (instead of the relation (2.201)) were as (Ghosh et al. 1986)

$$r_{\text{s}_j} = r_{\text{s}}(x_j) = \sqrt{(2b - x_j)^2 + d^2}\,, \quad j = 1, 2,$$

$$r_{\text{i}_j} = r_{\text{i}}(x_j) = \sqrt{(2b + x_j)^2 + d^2}\,. \tag{2.209}$$

In Fig. 2.2 a short period of the oscillations of the cosine is depicted after Ghosh et al. (1986). An equal phase is assumed on any of the lines $y_2 \equiv x_2 - x_1 =$ constant. The short oscillation period corresponds to the change in the signed distance

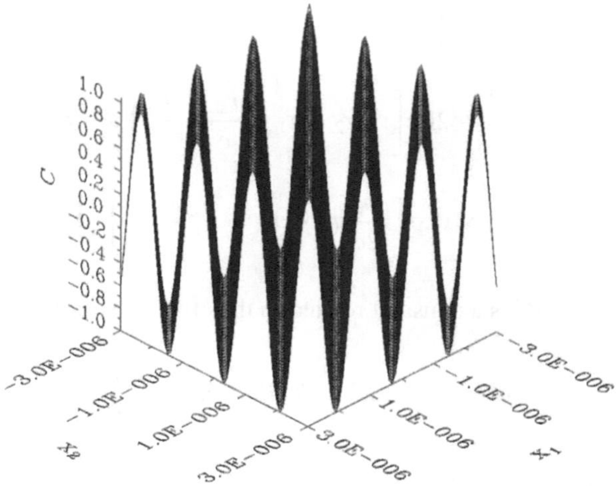

Fig. 2.2 The dependence of the cosine C of the phase on the position coordinates x_1 and x_2 for $d = 1$ and $b = 0.1$. The analysis of the setup in Fig. 2.1 after Ghosh et al. (1986). The cosine C of the phase depends only on the signed distance of the detectors

$y_2 \equiv x_2 - x_1$. As obvious from Fig. 2.3, the complement of the visibility, $1 - \mathcal{V}'$, depends only on the coordinate of the point amid the detectors D1 and D2. An equal visibility $\mathcal{V}'$ is assumed on the lines $y_1 = \frac{1}{2}(x_1 + x_2) =$ constant. For $y_1 = 0.0015$ the visibility almost vanishes.

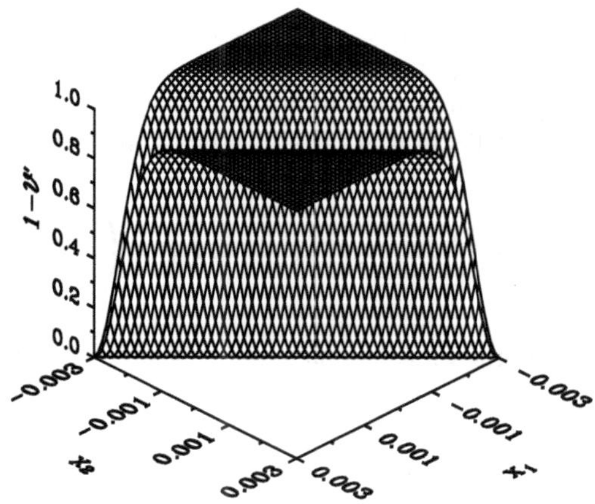

Fig. 2.3 The complement of the visibility, $1 - \mathcal{V}'$, versus the position coordinates x_1 and x_2 for $d = 1$ and $b = 0.1$. The analysis of the setup in Figure 2.1 after Ghosh et al. (1986)

(iii) *The experiment of Rarity and Tapster*

Rarity and Tapster (1990) demonstrated a violation of Bell's inequality using phase and momentum of photon pairs instead of polarization as in previous experiments. They selected two signal beams of the same colour (the frequency ω_s) and two idler beams also of the same colour (frequency $\omega_i \neq \omega_s$). They directed one of the signal beams and one of the idler beams to a mirror M_1 and another mirror M_2 (see Fig. 2.4). On the paths to the mirror M_2 they increased the phase of the signal by φ_s and that of the idler by φ_i. They coherently mixed the two signals and idlers.

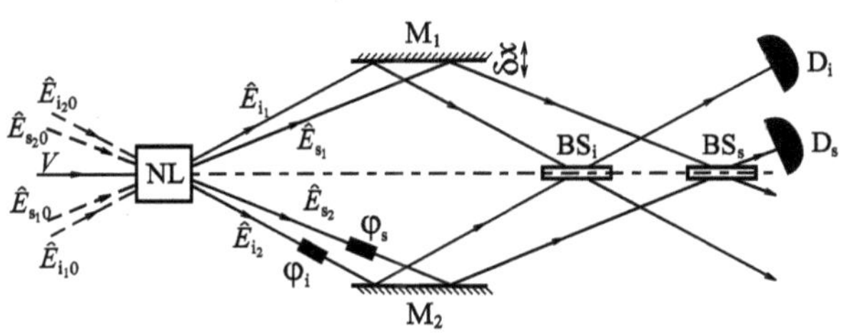

Fig. 2.4 Experimental setup of Rarity and Tapster

Now we will determine nonlinear dynamics of field operators. We assume the electric field in the form (2.161), (2.162), and (2.165), with another orthonormal system of functions $v_{s_j\mathbf{k}}(\mathbf{r})$, $\mathbf{k} \in [\mathbf{k}]_{s_j}$, $v_{i_j\mathbf{k}}(\mathbf{r})$, $\mathbf{k} \in [\mathbf{k}]_{i_j}$, $j = 1, 2$. The distinction is in the sets $[\mathbf{k}]_{s_j}$ and $[\mathbf{k}]_{i_j}$, which are appropriate to the experimental setup.

We will give a specification of $v_{J\mathbf{k}}(\mathbf{r})$, $J = \mathrm{s}_j, \mathrm{i}_j$, $j = 1, 2$. We associate points $\mathbf{0}_{\mathrm{BS_i}}$, $\mathbf{0}_{\mathrm{BS_s}}$ with the beam splitters. We introduce the notation $z_{\mathrm{BS_i}}$, $z_{\mathrm{BS_s}}$ such that

$$\mathbf{0}_{\mathrm{BS_i}} = (0, 0, z_{\mathrm{BS_i}}), \mathbf{0}_{\mathrm{BS_s}} = (0, 0, z_{\mathrm{BS_s}}). \tag{2.210}$$

We respect the phase shifters and the spots of reflection on the mirror M_2 with the notation $z_{\mathrm{PS_i}}$, $z_{\mathrm{M_{i_2}}}$, $z_{\mathrm{PS_s}}$, $z_{\mathrm{M_{s_2}}}$. We have located the phase shifter for the idler beam, the spot of reflection of the idler beam, the phase shifter for the signal beam, and the spot of reflection of the signal beam, respectively, at

$$\left(-b\frac{z_{\mathrm{PS_i}}}{z_{\mathrm{M_{i_2}}}}, 0, z_{\mathrm{PS_i}}\right), (-b, 0, z_{\mathrm{M_{i_2}}}), \left(-b\frac{z_{\mathrm{PS_s}}}{z_{\mathrm{M_{s_2}}}}, 0, z_{\mathrm{PS_s}}\right), (-b, 0, z_{\mathrm{M_{s_2}}}). \tag{2.211}$$

The origin $\mathbf{0}$ has its image $\mathbf{0}'_{\mathrm{s_2}} = \mathbf{0}'_{\mathrm{i_2}} = (-2b, 0, 0)$ in the mirror M_2.

Then we assume that the mirror M_1 is simple. We respect the spots of reflection on the mirror M_1 with the notation $z_{\mathrm{M_{i_1}}}$, $z_{\mathrm{M_{s_1}}}$. We have located the spot of reflection of the idler beam and the spot of reflection of the signal beam, respectively, at

$$(b, 0, z_{\mathrm{M_{i_1}}}), (b, 0, z_{\mathrm{M_{s_1}}}). \tag{2.212}$$

The origin $\mathbf{0}$ has its image $\mathbf{0}'_{\mathrm{s_1}} = \mathbf{0}'_{\mathrm{i_1}} = (2b, 0, 0)$ in the mirror M_1, which is assumed to be simple so far. We specify that

$$v_{J\mathbf{k}}(\mathbf{r}) = \mathrm{i}\sqrt{\frac{\hbar\omega_{\mathbf{k}}}{\mathcal{A}L}}\exp(\mathrm{i}\mathbf{k}\cdot\mathbf{r}) \;\text{ for }\; \mathbf{r} \parallel \mathbf{k}, \;\; z < z_{\mathrm{M}_J}, \;\; \mathbf{k} \in [\mathbf{k}]_J, \tag{2.213}$$

$$v_{J\mathbf{k}}(\mathbf{r}) = -\mathrm{i}\sqrt{\frac{\hbar\omega_{\mathbf{k}}}{\mathcal{A}L}}\exp\left[\mathrm{i}\mathbf{k}'\cdot(\mathbf{r}-\mathbf{0}'_J) + \frac{\omega_{J_1}\delta x}{c}\right] \;\text{ for }\; (\mathbf{r}-\mathbf{0}'_J) \parallel \mathbf{k}',$$
$$z_{\mathrm{M}_J} < z < z_{\mathrm{BS}_{J_1}}, \;\; \mathbf{k} \in [\mathbf{k}]_J, \tag{2.214}$$

where δx is a path-length difference, $J = \mathrm{s}_1, \mathrm{i}_1$, $J_1 = \mathrm{s}, \mathrm{i}$, $\mathbf{k}' = (-k_x, k_y, k_z)$,

$$v_{J\mathbf{k}}(\mathbf{r}) = \mathrm{i}\sqrt{\frac{\hbar\omega_{\mathbf{k}}}{\mathcal{A}L}}\exp(\mathrm{i}\mathbf{k}\cdot\mathbf{r}) \;\text{ for }\; \mathbf{r} \parallel \mathbf{k}, \;\; z < z_{\mathrm{PS}_{J_1}}, \;\; \mathbf{k} \in [\mathbf{k}]_J, \tag{2.215}$$

$$v_{J\mathbf{k}}(\mathbf{r}) = \mathrm{i}\sqrt{\frac{\hbar\omega_{\mathbf{k}}}{\mathcal{A}L}}\exp[\mathrm{i}(\mathbf{k}\cdot\mathbf{r} + \varphi_{J_1})] \;\text{ for }\; \mathbf{r} \parallel \mathbf{k},$$
$$z_{\mathrm{PS}_{J_1}} < z < z_{\mathrm{M}_J}, \;\; \mathbf{k} \in [\mathbf{k}]_J, \tag{2.216}$$

$$v_{J\mathbf{k}}(\mathbf{r}) = -\mathrm{i}\sqrt{\frac{\hbar\omega_{\mathbf{k}}}{\mathcal{A}L}}\exp\{\mathrm{i}[\mathbf{k}'\cdot(\mathbf{r}-\mathbf{0}'_J) + \varphi_{J_1}]\} \;\text{ for }\; (\mathbf{r}-\mathbf{0}'_J) \parallel \mathbf{k}',$$
$$z_{\mathrm{M}_J} < z < z_{\mathrm{BS}_{J_1}}, \;\; \mathbf{k} \in [\mathbf{k}]_J, \tag{2.217}$$

where $J = \mathrm{s}_2, \mathrm{i}_2$, $J_1 = \mathrm{s}, \mathrm{i}$.

Beam splitter BS_s with the transmissivities $\mathtt{t}_s$, $\mathtt{t}'_s$ and reflectivities $\mathtt{r}_s$, $\mathtt{r}'_s$: Input values are

$$v_{s_1\mathbf{k}\mathrm{in}}(\mathbf{0}_{BS_s}) = -i\sqrt{\frac{\hbar\omega_{\mathbf{k}}}{\mathcal{A}L}}\exp\left[i\mathbf{k}'\cdot(\mathbf{0}_{BS_s}-\mathbf{0}'_{s_1})+\frac{\omega_s\delta x}{c}\right],\ \mathbf{k}\in[\mathbf{k}]_{s_1}, \tag{2.218}$$

$$v_{s_2\mathbf{k}\mathrm{in}}(\mathbf{0}_{BS_s}) = -i\sqrt{\frac{\hbar\omega_{\mathbf{k}}}{\mathcal{A}L}}\exp\{i[\mathbf{k}'\cdot(\mathbf{0}_{BS_s}-\mathbf{0}'_{s_2})+\varphi_s]\},\ \mathbf{k}\in[\mathbf{k}]_{s_2}. \tag{2.219}$$

Under the assumption $\delta x = 0$ we can perform the exchange $\mathbf{k}\leftrightarrow\mathbf{k}'$ and write the output values for $\mathbf{k}\in[\mathbf{k}]_{s_1}$

$$\begin{aligned} v_{s_1\mathbf{k}\mathrm{out}}(\mathbf{0}_{BS_s}) &= -i\sqrt{\frac{\hbar\omega_{\mathbf{k}}}{\mathcal{A}L}}\exp\{i\mathbf{k}'\cdot(\mathbf{0}_{BS_s}-\mathbf{0}'_{s_1})\}\mathtt{t}_s \\ &\quad -i\sqrt{\frac{\hbar\omega_{\mathbf{k}}}{\mathcal{A}L}}\exp\{i[\mathbf{k}\cdot(\mathbf{0}_{BS_s}-\mathbf{0}'_{s_2})+\varphi_s]\}\mathtt{r}'_s, \end{aligned} \tag{2.220}$$

$$\begin{aligned} v_{s_2\mathbf{k}'\mathrm{out}}(\mathbf{0}_{BS_s}) &= -i\sqrt{\frac{\hbar\omega_{\mathbf{k}}}{\mathcal{A}L}}\exp\{i\mathbf{k}'\cdot(\mathbf{0}_{BS_s}-\mathbf{0}'_{s_1})\}\mathtt{r}_s \\ &\quad -i\sqrt{\frac{\hbar\omega_{\mathbf{k}}}{\mathcal{A}L}}\exp\{i[\mathbf{k}\cdot(\mathbf{0}_{BS_s}-\mathbf{0}'_{s_2})+\varphi_s]\}\mathtt{t}'_s. \end{aligned} \tag{2.221}$$

Performing the exchange $\mathbf{k}\leftrightarrow\mathbf{k}'$ in (2.221), we have for $(\mathbf{r}-\mathbf{0}'_{s_1})\parallel\mathbf{k}'$, $z > z_{BS_s}$, $\mathbf{k}\in[\mathbf{k}]_{s_1}$,

$$\begin{aligned} v_{s_1\mathbf{k}}(\mathbf{r}) &= -i\sqrt{\frac{\hbar\omega_{\mathbf{k}}}{\mathcal{A}L}}\exp[i\mathbf{k}'\cdot(\mathbf{r}-\mathbf{0}'_{s_1})]\mathtt{t}_s \\ &\quad - i\sqrt{\frac{\hbar\omega_{\mathbf{k}}}{\mathcal{A}L}}\exp\{i[\mathbf{k}'\cdot(\mathbf{r}-\mathbf{0}_{BS_s})+\mathbf{k}\cdot(\mathbf{0}_{BS_s}-\mathbf{0}'_{s_2})+\varphi_s]\}\mathtt{r}'_s, \end{aligned} \tag{2.222}$$

and for $(\mathbf{r}-\mathbf{0}'_{s_2})\parallel\mathbf{k}'$, $z > z_{BS_s}$, $\mathbf{k}\in[\mathbf{k}]_{s_2}$,

$$\begin{aligned} v_{s_2\mathbf{k}}(\mathbf{r}) &= -i\sqrt{\frac{\hbar\omega_{\mathbf{k}}}{\mathcal{A}L}}\exp\{i[\mathbf{k}'\cdot(\mathbf{r}-\mathbf{0}_{BS_s})+\mathbf{k}\cdot(\mathbf{0}_{BS_s}-\mathbf{0}'_{s_1})]\}\mathtt{r}_s \\ &\quad -i\sqrt{\frac{\hbar\omega_{\mathbf{k}}}{\mathcal{A}L}}\exp\{i[\mathbf{k}'\cdot(\mathbf{r}-\mathbf{0}'_{s_2})+\varphi_s]\}\mathtt{t}'_s. \end{aligned} \tag{2.223}$$

Beam splitter BS_i with the transmissivities $\mathtt{t}_i$, $\mathtt{t}'_i$ and reflectivities $\mathtt{r}_i$, $\mathtt{r}'_i$ can be described analogously: In (2.222) and (2.223) we perform the replacement s $\leftrightarrow$ i.

In a standard fashion, we associate the modal functions, e.g. (2.213), (2.214), (2.215), (2.216), (2.217), (2.222), and (2.223), with fields we denote $\hat{E}^{(+)}_{j0}(\mathbf{r},t)$, $J = s_1, s_2, i_1, i_2$. Nonlinear dynamics is described in the same way as for the experiment on the interference of signal and idler photons by relations (2.188), (2.189), (2.190),

(2.192), and (2.193). It allows the fields with $z < 0$ to stay initial and those with $z > 0$ at least to obey the same rules we have used to calculate the modal functions.

We introduce the slowly varying field operators

$$\hat{F}_J^{(+)}(\mathbf{r}, t) = \exp(\mathrm{i}\omega_{J_1} t)\hat{E}_J^{(+)}(\mathbf{r}, t), \tag{2.224}$$

where $J = \mathrm{s}_1, \mathrm{s}_2$, $J_1 = \mathrm{s}$ and $J = \mathrm{i}_1, \mathrm{i}_2$, $J_1 = \mathrm{i}$, for expressing the intensity correlations. The field operators at the signal and idler detectors placed at $\mathbf{r}_\mathrm{s}$, $\mathbf{r}_\mathrm{i}$, respectively, are

$$\hat{F}_{\mathrm{s}_2}^{(+)}(\mathbf{r}_\mathrm{s}, t) = \mathtt{t}'_\mathrm{s} \hat{F}_{\mathrm{s}_2\mathrm{out}}^{(+)}\left(\mathbf{0}, t - \frac{r_\mathrm{s}}{c} + \frac{\varphi_\mathrm{s}}{\omega_\mathrm{s}}\right) \exp\left(\mathrm{i}\frac{\omega_\mathrm{s} r_\mathrm{s}}{c} + \mathrm{i}\varphi_\mathrm{s}\right)$$
$$+ \mathtt{r}_\mathrm{s} \hat{F}_{\mathrm{s}_1\mathrm{out}}^{(+)}\left(\mathbf{0}, t - \frac{r_\mathrm{s}}{c}\right) \exp\left(\mathrm{i}\frac{\omega_\mathrm{s} r_\mathrm{s}}{c}\right), \tag{2.225}$$

$$\hat{F}_{\mathrm{i}_2}^{(+)}(\mathbf{r}_\mathrm{i}, t') = \mathtt{t}'_\mathrm{i} \hat{F}_{\mathrm{i}_2\mathrm{out}}^{(+)}\left(\mathbf{0}, t' - \frac{r_\mathrm{i}}{c} + \frac{\varphi_\mathrm{i}}{\omega_\mathrm{i}}\right) \exp\left(\mathrm{i}\frac{\omega_\mathrm{i} r_\mathrm{i}}{c} + \mathrm{i}\varphi_\mathrm{i}\right)$$
$$+ \mathtt{r}_\mathrm{i} \hat{F}_{\mathrm{i}_1\mathrm{out}}^{(+)}\left(\mathbf{0}, t' - \frac{r_\mathrm{i}}{c}\right) \exp\left(\mathrm{i}\frac{\omega_\mathrm{i} r_\mathrm{i}}{c}\right), \tag{2.226}$$

with $\mathbf{0}$ the centre of the coordinate system, r_s and r_i the path lengths of the lower signal and idler beams, respectively, to the appropriate detector. In the experiment under consideration both upper paths were modified by δx, since the upper and the lower mirrors were not at exactly the same distance from the pumping beam axis (Casado et al. 1997a) . The mirror above the pumping beam axis is not simple, but a mirror assembly which enables one to change δx (Rarity and Tapster 1990). In the following, we assume $\delta x = 0$.

We will take into account that the signal field $\hat{F}_{\mathrm{s}_1}^{(+)}(\mathbf{r}, t)$ is correlated with the idler field $\hat{F}_{\mathrm{i}_2}^{(+)}(\mathbf{r}, t)$ and also $\hat{F}_{\mathrm{s}_2}^{(+)}(\mathbf{r}, t)$ is correlated with $\hat{F}_{\mathrm{i}_1}^{(+)}(\mathbf{r}, t)$, but $\hat{F}_{\mathrm{s}_j}^{(+)}(\mathbf{r}, t)$ is uncorrelated with $\hat{F}_{\mathrm{i}_j}^{(+)}(\mathbf{r}, t)$, $j = 1, 2$, these pairs not fulfilling matching conditions. If we consider that the time intervals $\frac{r_\mathrm{s}}{c} - \frac{r_\mathrm{i}}{c} - \frac{\varphi_\mathrm{s}}{\omega_\mathrm{s}}, \frac{r_\mathrm{s}}{c} - \frac{r_\mathrm{i}}{c} + \frac{\varphi_\mathrm{i}}{\omega_\mathrm{i}}$ are small in comparison with the coherence time of signal and idler given by the function $\nu(\tau)$, we obtain

$$\langle \hat{F}_{\mathrm{s}_2}^{(+)}(\mathbf{r}_\mathrm{s}, t)\hat{F}_{\mathrm{i}_2}^{(+)}(\mathbf{r}_\mathrm{i}, t')\rangle \approx \left\{\mathtt{r}_\mathrm{i}\mathtt{t}'_\mathrm{s} \exp\left[\mathrm{i}\left(\omega_\mathrm{s}\frac{r_\mathrm{s}}{c} + \omega_\mathrm{i}\frac{r_\mathrm{i}}{c} + \varphi_\mathrm{s}\right)\right]\right.$$
$$\left. + \mathtt{r}_\mathrm{s}\mathtt{t}'_\mathrm{i} \exp\left[\mathrm{i}\left(\omega_\mathrm{s}\frac{r_\mathrm{s}}{c} + \omega_\mathrm{i}\frac{r_\mathrm{i}}{c} + \varphi_\mathrm{i}\right)\right]\right\}\nu(t' - t). \tag{2.227}$$

From this we have

$$P_{\mathrm{si}}(\mathbf{r}_\mathrm{s}, t + \tau; \mathbf{r}_\mathrm{i}, t + \tau') = K' g^2 |V|^2 |\nu(\tau' - \tau)|^2$$
$$\times\left[|\mathtt{r}_\mathrm{i}\mathtt{t}'_\mathrm{s}|^2 + |\mathtt{r}_\mathrm{s}\mathtt{t}'_\mathrm{i}|^2 + 2\mathrm{Re}\{\mathtt{r}_\mathrm{s}^*\mathtt{t}'_\mathrm{s}\mathtt{r}_\mathrm{i}\mathtt{t}_\mathrm{i}^{\prime *} \exp[\mathrm{i}(\varphi_\mathrm{s} - \varphi_\mathrm{i})]\}\right]. \tag{2.228}$$

The visibility is expressed by the formula (2.175), where

$$R_{\mathrm{simax}} + R_{\mathrm{simin}} = 2g^2|V|^2 M(0,0)\left(|\mathrm{t}'_\mathrm{s}\mathrm{r}_\mathrm{i}|^2 + |\mathrm{r}_\mathrm{s}\mathrm{t}'_\mathrm{i}|^2\right), \qquad (2.229)$$

$$R_{\mathrm{simax}} - R_{\mathrm{simin}} = 8g^2|V|^2 M(0,0)\,|\mathrm{t}'_\mathrm{s}\mathrm{r}_\mathrm{i}||\mathrm{r}_\mathrm{s}\mathrm{t}'_\mathrm{i}|. \qquad (2.230)$$

The dependence of the visibility on the beam splitters with the transmissivities $|\mathrm{t}'_\mathrm{s}| \in [0,1]$, $|\mathrm{t}'_\mathrm{i}| \in [0,1]$ and the reflectivities $|\mathrm{r}_\mathrm{s}| = \sqrt{1-|\mathrm{t}'_\mathrm{s}|^2}$, $|\mathrm{r}_\mathrm{i}| = \sqrt{1-|\mathrm{t}'_\mathrm{i}|^2}$ is plotted in Fig. 2.5.

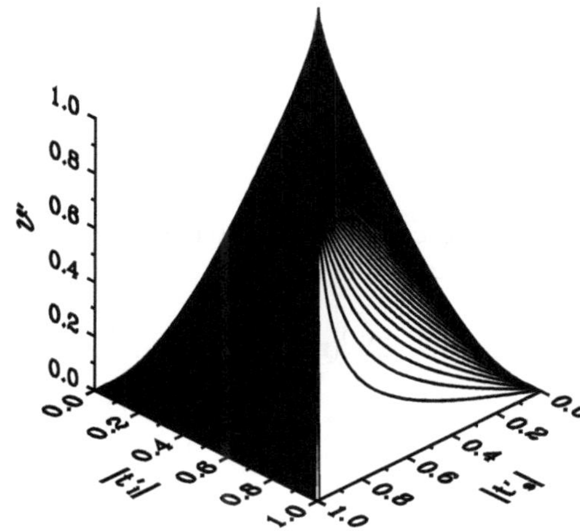

Fig. 2.5 The visibility $\mathcal{V}'$ versus moduli of the amplitude transmissivities t'_s, t'_i from the "lower side" of the beam splitters for signal and idler beams for $\delta x = 0$

Unfortunately, the quantity under consideration is not dependent on the characteristic of the nonlinear optical process. The surface plotted has a boundary condition zero. It may be equal to unity in the sense of the equality $|\mathrm{t}'_\mathrm{s}\mathrm{r}_\mathrm{i}| = |\mathrm{t}'_\mathrm{i}\mathrm{r}_\mathrm{s}|$. The maximum is attained on the line segment connecting the points $|\mathrm{t}'_\mathrm{s}| = 0,\ \ |\mathrm{t}'_\mathrm{i}| = 0$ and $|\mathrm{t}'_\mathrm{s}| = 1,\ \ |\mathrm{t}'_\mathrm{i}| = 1$. The interference manifests itself as a cosine variation of the coincidence rate with $\varphi_\mathrm{s} - \varphi_\mathrm{i}$.

(iv) *The experiment of Franson*

Franson (1989) proposed a test of "Bell's inequality for energy and time". He arranged two Mach–Zehnder interferometers and let a signal and an idler beam each pass through an interferometer (see Fig. 2.6). The experiment was originally proposed for an atom and free-space propagation.

The coincidence detection shows a fourth-order interference as a cosine dependence on $\frac{1}{c}(\omega_\mathrm{s}\Delta L_\mathrm{s} + \omega_\mathrm{i}\Delta L_\mathrm{i})$, where ΔL_s (ΔL_i) is the length difference between the long (short) route of the signal (idler) beam through the corresponding interferometer. In the past few years several groups have performed experiments of that type. In Tapster et al. (1994) Franson's experiment has been adapted to parametric down-conversion and fibres.

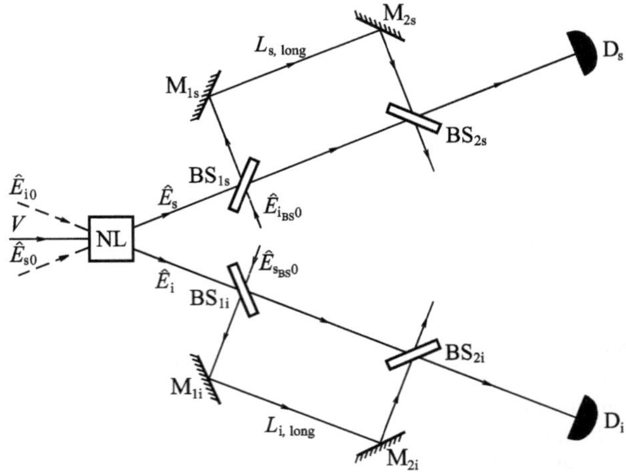

Fig. 2.6 Experimental setup of Franson's type. For simplicity $\hat{E}_{i_{BS}0} \equiv \hat{E}_{i_{BS_{1s}}0}$ and $\hat{E}_{s_{BS}0} \equiv \hat{E}_{s_{BS_{1i}}0}$

For the description of the nonlinear dynamics of field operators, we consider the initial electric-field in the form

$$\hat{E}_0^{(+)}(\mathbf{r}, t) = V^{(+)}(\mathbf{r}, t)\hat{1} + \hat{E}_{s0}^{(+)}(\mathbf{r}, t) + \hat{E}_{i_{BS_{1s}}0}^{(+)}(\mathbf{r}, t) + \hat{E}_{i0}^{(+)}(\mathbf{r}, t) + \hat{E}_{s_{BS_{1i}}0}^{(+)}(\mathbf{r}, t), \tag{2.231}$$

where

$$\hat{E}_{J0}^{(+)}(\mathbf{r}, t) = \sum_{\mathbf{k}\in[\mathbf{k}]_J} v_{J\mathbf{k}}(\mathbf{r})\hat{a}_{J\mathbf{k}0}(t), \quad J = \mathrm{s},\ \mathrm{i},\ \mathrm{i}_{BS_{1s}},\ \mathrm{s}_{BS_{1i}}. \tag{2.232}$$

The modal functions as restricted to linear segments are

$$v_{s\mathbf{k}}(\mathbf{r}) = \mathrm{i}\sqrt{\frac{\hbar\omega_{\mathbf{k}}}{\mathcal{A}L}}e^{\mathrm{i}\mathbf{k}\cdot\mathbf{r}} \text{ for } \mathbf{r} \parallel \mathbf{k},\ z < z_{BS_{1s}},\ \mathbf{k} \in [\mathbf{k}]_s, \tag{2.233}$$

$$v_{i_{BS_{1s}}\mathbf{k}}(\mathbf{r}) = \mathrm{i}\sqrt{\frac{\hbar\omega_{\mathbf{k}}}{\mathcal{A}L}}e^{\mathrm{i}\mathbf{k}\cdot\mathbf{r}} \text{ for } (\mathbf{r} - \mathbf{0}_{BS_{1s}}) \parallel \mathbf{k},\ z > z_{BS_{1s}},\ \mathbf{k} \in [\mathbf{k}]_{i_{BS_{1s}}}. \tag{2.234}$$

Here $\mathbf{0}_{BS_{1s}}$ is the centre of the beam splitter BS$_{1s}$, $z_{BS_{1s}}$ is the corresponding z-coordinate, and $[\mathbf{k}]_{i_{BS_{1s}}}$ is the set of wave vectors $\mathbf{k}$ of the beam corresponding to the unused input port of this beam splitter.

The modal functions at the output of this beam splitter are

$$v_{s\mathbf{k}}(\mathbf{r}) = \mathrm{i}\sqrt{\frac{\hbar\omega_{\mathbf{k}}}{\mathcal{A}L}}e^{\mathrm{i}\mathbf{k}\cdot\mathbf{r}}t_s + \mathrm{i}\sqrt{\frac{\hbar\omega_{\mathbf{k}}}{\mathcal{A}L}}e^{\mathrm{i}\mathbf{k}\cdot(\mathbf{r}-\mathbf{0}'_{BS_{1s}i_{BS_{1s}}})}r_s \quad \text{for } \mathbf{r} \parallel \mathbf{k},\ z_{BS_{1s}} < z < z_{BS_{2s}},\ \mathbf{k} \in [\mathbf{k}]_s, \tag{2.235}$$

$$v_{\mathrm{iBS_{1s}}\mathbf{k}}(\mathbf{r}) = \mathrm{i}\sqrt{\frac{\hbar\omega_{\mathbf{k}}}{\mathcal{A}L}}e^{\mathrm{i}\mathbf{k}\cdot(\mathbf{r}-\mathbf{0}'_{\mathrm{BS_{1s}s}})}\mathrm{r_s} + \mathrm{i}\sqrt{\frac{\hbar\omega_{\mathbf{k}}}{\mathcal{A}L}}e^{\mathrm{i}\mathbf{k}\cdot\mathbf{r}}\mathrm{t_s}$$

$$\text{for } (\mathbf{r}-\mathbf{0}_{\mathrm{BS_{1s}}}) \parallel \mathbf{k},\ z_{\mathrm{M_{1s}}} < z < z_{\mathrm{BS_{1s}}},\ \mathbf{k} \in [\mathbf{k}]_{\mathrm{iBS_{1s}}}, \tag{2.236}$$

where $\mathbf{0}'_{\mathrm{BS_{1s}iBS_{1s}}}$, $\mathbf{0}'_{\mathrm{BS_{1s}s}}$ are chosen such that

$$\mathbf{k}\cdot(\mathbf{r}-\mathbf{0}'_{\mathrm{BS_{1s}iBS_{1s}}}) = \mathbf{k}\cdot(\mathbf{r}-\mathbf{0}_{\mathrm{BS_{1s}}}) + \mathbf{k}'\cdot\mathbf{0}_{\mathrm{BS_{1s}}},\quad \mathbf{k}\in[\mathbf{k}]_{\mathrm{s}}, \tag{2.237}$$

$$\mathbf{k}\cdot(\mathbf{r}-\mathbf{0}'_{\mathrm{BS_{1s}s}}) = \mathbf{k}\cdot(\mathbf{r}-\mathbf{0}_{\mathrm{BS_{1s}}}) + \mathbf{k}'\cdot\mathbf{0}_{\mathrm{BS_{1s}}},\quad \mathbf{k}\in[\mathbf{k}]_{\mathrm{iBS_{1s}}}. \tag{2.238}$$

Here $z_{\mathrm{M_{1s}}}$ is the z-coordinate of the centre of the signal mirror and $z_{\mathrm{BS_{2s}}}$ is the z-coordinate of $\mathbf{0}_{\mathrm{BS_{2s}}}$, the centre of the beam splitter $\mathrm{BS_{2s}}$.

After the reflection from the first mirror, the modal function is

$$v_{\mathrm{iBS_{1s}}\mathbf{k}}(\mathbf{r}) = -\mathrm{i}\sqrt{\frac{\hbar\omega_{\mathbf{k}}}{\mathcal{A}L}}e^{\mathrm{i}\mathbf{k}'\cdot(\mathbf{r}-\mathbf{0}'_{\mathrm{M_{1s}BS_{1s}s}})}\mathrm{r_s}$$
$$-\mathrm{i}\sqrt{\frac{\hbar\omega_{\mathbf{k}}}{\mathcal{A}L}}e^{\mathrm{i}\mathbf{k}'\cdot(\mathbf{r}-\mathbf{0}'_{\mathrm{M_{1s}}})}\mathrm{t_s} \text{ for } (\mathbf{r}-\mathbf{0}_{\mathrm{M_{1s}}}) \parallel \mathbf{k}',$$
$$z_{\mathrm{M_{1s}}} < z < z_{\mathrm{M_{2s}}},\ \mathbf{k}\in[\mathbf{k}]_{\mathrm{iBS_{1s}}}, \tag{2.239}$$

where $z_{\mathrm{M_{2s}}}$ is the z-coordinate of the centre of the second mirror and $\mathbf{0}'_{\mathrm{M_{1s}BS_{1s}s}}$ and $\mathbf{0}'_{\mathrm{M_{1s}}}$ are chosen such that

$$\mathbf{k}'\cdot(\mathbf{r}-\mathbf{0}'_{\mathrm{M_{1s}BS_{1s}s}}) = \mathbf{k}'\cdot(\mathbf{r}-\mathbf{0}_{\mathrm{M_{1s}}}) + \mathbf{k}\cdot(\mathbf{0}_{\mathrm{M_{1s}}} - \mathbf{0}'_{\mathrm{BS_{1s}s}}),\quad \mathbf{k}\in[\mathbf{k}]_{\mathrm{iBS_{1s}}}, \tag{2.240}$$

$$\mathbf{k}'\cdot(\mathbf{r}-\mathbf{0}'_{\mathrm{M_{1s}}}) = \mathbf{k}'\cdot(\mathbf{r}-\mathbf{0}_{\mathrm{M_{1s}}}) + \mathbf{k}\cdot\mathbf{0}_{\mathrm{M_{1s}}},\quad \mathbf{k}\in[\mathbf{k}]_{\mathrm{iBS_{1s}}}. \tag{2.241}$$

After the reflection from the second mirror, the modal function is

$$v_{\mathrm{iBS_{1s}}\mathbf{k}}(\mathbf{r}) = \mathrm{i}\sqrt{\frac{\hbar\omega_{\mathbf{k}}}{\mathcal{A}L}}e^{-\mathrm{i}\mathbf{k}\cdot(\mathbf{r}-\mathbf{0}'_{\mathrm{M_{2s}M_{1s}BS_{1s}s}})}\mathrm{r_s}$$
$$+\mathrm{i}\sqrt{\frac{\hbar\omega_{\mathbf{k}}}{\mathcal{A}L}}e^{-\mathrm{i}\mathbf{k}\cdot(\mathbf{r}-\mathbf{0}'_{\mathrm{M_{2s}M_{1s}}})}\mathrm{t_s} \text{ for } (\mathbf{r}-\mathbf{0}_{\mathrm{M_{2s}}}) \parallel -\mathbf{k},$$
$$z_{\mathrm{M_{2s}}} < z < z_{\mathrm{BS_{2s}}},\ \mathbf{k}\in[\mathbf{k}]_{\mathrm{iBS_{1s}}}, \tag{2.242}$$

where $\mathbf{0}'_{\mathrm{M_{2s}M_{1s}BS_{1s}s}}$ and $\mathbf{0}'_{\mathrm{M_{2s}M_{1s}}}$ are chosen such that

$$-\mathbf{k}\cdot(\mathbf{r}-\mathbf{0}'_{\mathrm{M_{2s}M_{1s}BS_{1s}s}}) = -\mathbf{k}\cdot(\mathbf{r}-\mathbf{0}_{\mathrm{M_{2s}}})$$
$$+\mathbf{k}'\cdot(\mathbf{0}_{\mathrm{M_{2s}}} - \mathbf{0}'_{\mathrm{M_{1s}BS_{1s}s}}),\quad \mathbf{k}\in[\mathbf{k}]_{\mathrm{iBS_{1s}}}, \tag{2.243}$$

$$-\mathbf{k}\cdot(\mathbf{r}-\mathbf{0}'_{\mathrm{M_{2s}M_{1s}}}) = -\mathbf{k}\cdot(\mathbf{r}-\mathbf{0}_{\mathrm{M_{2s}}})$$
$$+\mathbf{k}'\cdot(\mathbf{0}_{\mathrm{M_{2s}}} - \mathbf{0}'_{\mathrm{M_{1s}}}),\quad \mathbf{k}\in[\mathbf{k}]_{\mathrm{iBS_{1s}}}. \tag{2.244}$$

The output modal functions for the beam to be detected are

$$
\begin{aligned}
v_{\mathrm{s}\mathbf{k}}(\mathbf{r}) &= \mathrm{i}\sqrt{\frac{\hbar\omega_{\mathbf{k}}}{\mathcal{A}L}}e^{\mathrm{i}\mathbf{k}\cdot\mathbf{r}}\mathtt{t}_{\mathrm{s}}^{2} \\
&+ \left[\mathrm{i}\sqrt{\frac{\hbar\omega_{\mathbf{k}}}{\mathcal{A}L}}e^{\mathrm{i}\mathbf{k}\cdot(\mathbf{r}-\mathbf{0}'_{\mathrm{BS}_{1\mathrm{s}}\mathrm{i}_{\mathrm{BS}_{1\mathrm{s}}}})} + \mathrm{i}\sqrt{\frac{\hbar\omega_{\mathbf{k}}}{\mathcal{A}L}}e^{\mathrm{i}\mathbf{k}\cdot(\mathbf{r}-\mathbf{0}'_{\mathrm{BS}_{2\mathrm{s}}\mathrm{M}_{2\mathrm{s}}\mathrm{M}_{1\mathrm{s}}})}\right]\mathtt{t}_{\mathrm{s}}\mathtt{r}_{\mathrm{s}} \\
&+ \mathrm{i}\sqrt{\frac{\hbar\omega_{\mathbf{k}}}{\mathcal{A}L}}e^{\mathrm{i}\mathbf{k}\cdot(\mathbf{r}-\mathbf{0}'_{\mathrm{BS}_{2\mathrm{s}}\mathrm{M}_{2\mathrm{s}}\mathrm{M}_{1\mathrm{s}}\mathrm{BS}_{1\mathrm{s}}\mathrm{s}})}\mathtt{r}_{\mathrm{s}}^{2}, \quad \text{for } \mathbf{r} \parallel \mathbf{k},\ z > z_{\mathrm{BS}_{2\mathrm{s}}},\ \mathbf{k} \in [\mathbf{k}]_{\mathrm{s}},
\end{aligned}
\tag{2.245}
$$

where $\mathbf{0}'_{\mathrm{BS}_{2\mathrm{s}}\mathrm{M}_{2\mathrm{s}}\mathrm{M}_{1\mathrm{s}}}$ and $\mathbf{0}'_{\mathrm{BS}_{2\mathrm{s}}\mathrm{M}_{2\mathrm{s}}\mathrm{M}_{1\mathrm{s}}\mathrm{BS}_{1\mathrm{s}}\mathrm{s}}$ are chosen such that

$$
\mathbf{k}\cdot(\mathbf{r}-\mathbf{0}'_{\mathrm{BS}_{2\mathrm{s}}\mathrm{M}_{2\mathrm{s}}\mathrm{M}_{1\mathrm{s}}}) = \mathbf{k}\cdot(\mathbf{r}-\mathbf{0}_{\mathrm{BS}_{2\mathrm{s}}}) - \mathbf{k}'\cdot(\mathbf{0}_{\mathrm{BS}_{2\mathrm{s}}} - \mathbf{0}'_{\mathrm{M}_{2\mathrm{s}}\mathrm{M}_{1\mathrm{s}}}),\quad \mathbf{k} \in [\mathbf{k}]_{\mathrm{s}}, \tag{2.246}
$$

$$
\begin{aligned}
\mathbf{k}\cdot(\mathbf{r}-\mathbf{0}'_{\mathrm{BS}_{2\mathrm{s}}\mathrm{M}_{2\mathrm{s}}\mathrm{M}_{1\mathrm{s}}\mathrm{BS}_{1\mathrm{s}}\mathrm{s}}) &= \mathbf{k}\cdot(\mathbf{r}-\mathbf{0}_{\mathrm{BS}_{2\mathrm{s}}}) \\
&-\mathbf{k}'\cdot(\mathbf{0}_{\mathrm{BS}_{2\mathrm{s}}} - \mathbf{0}'_{\mathrm{M}_{2\mathrm{s}}\mathrm{M}_{1\mathrm{s}}\mathrm{BS}_{1\mathrm{s}}\mathrm{s}}),\quad \mathbf{k} \in [\mathbf{k}]_{\mathrm{s}}.
\end{aligned}
\tag{2.247}
$$

The output modal functions for the second beam are

$$
\begin{aligned}
v_{\mathrm{i}_{\mathrm{BS}_{2\mathrm{s}}}\mathbf{k}}(\mathbf{r}) &= \mathrm{i}\sqrt{\frac{\hbar\omega_{\mathbf{k}}}{\mathcal{A}L}}e^{-\mathrm{i}\mathbf{k}\cdot(\mathbf{r}-\mathbf{0}'_{\mathrm{BS}_{2\mathrm{s}}\mathrm{BS}_{1\mathrm{s}}\mathrm{i}_{\mathrm{BS}_{1\mathrm{s}}}})}\mathtt{r}_{\mathrm{s}}^{2} \\
&+ \left[\mathrm{i}\sqrt{\frac{\hbar\omega_{\mathbf{k}}}{\mathcal{A}L}}e^{-\mathrm{i}\mathbf{k}\cdot(\mathbf{r}-\mathbf{0}'_{\mathrm{BS}_{2\mathrm{s}}\mathrm{s}})} + \mathrm{i}\sqrt{\frac{\hbar\omega_{\mathbf{k}}}{\mathcal{A}L}}e^{-\mathrm{i}\mathbf{k}\cdot(\mathbf{r}-\mathbf{0}'_{\mathrm{M}_{2\mathrm{s}}\mathrm{M}_{1\mathrm{s}}\mathrm{BS}_{1\mathrm{s}}})}\right]\mathtt{r}_{\mathrm{s}}\mathtt{t}_{\mathrm{s}} \\
&+ \mathrm{i}\sqrt{\frac{\hbar\omega_{\mathbf{k}}}{\mathcal{A}L}}e^{-\mathrm{i}\mathbf{k}\cdot(\mathbf{r}-\mathbf{0}'_{\mathrm{M}_{2\mathrm{s}}\mathrm{M}_{1\mathrm{s}}})}\mathtt{t}_{\mathrm{s}}^{2} \text{ for } (\mathbf{r}-\mathbf{0}_{\mathrm{BS}_{1\mathrm{s}}}) \parallel -\mathbf{k}, \\
&\qquad z > z_{\mathrm{BS}_{2\mathrm{s}}},\ \mathbf{k} \in [\mathbf{k}]_{\mathrm{i}_{\mathrm{BS}_{1\mathrm{s}}}},
\end{aligned}
\tag{2.248}
$$

where $\mathbf{0}'_{\mathrm{BS}_{2\mathrm{s}}\mathrm{BS}_{1\mathrm{s}}\mathrm{i}_{\mathrm{BS}_{1\mathrm{s}}}}$ and $\mathbf{0}'_{\mathrm{BS}_{2\mathrm{s}}\mathrm{s}}$ are chosen such that

$$
\begin{aligned}
-\mathbf{k}\cdot(\mathbf{r}-\mathbf{0}'_{\mathrm{BS}_{2\mathrm{s}}\mathrm{BS}_{1\mathrm{s}}\mathrm{i}_{\mathrm{BS}_{1\mathrm{s}}}}) &= -\mathbf{k}\cdot(\mathbf{r}-\mathbf{0}_{\mathrm{BS}_{2\mathrm{s}}}) \\
&+\mathbf{k}'\cdot(\mathbf{0}_{\mathrm{BS}_{2\mathrm{s}}} - \mathbf{0}'_{\mathrm{BS}_{1\mathrm{s}}\mathrm{i}_{\mathrm{BS}_{1\mathrm{s}}}}),\quad \mathbf{k} \in [\mathbf{k}]_{\mathrm{i}_{\mathrm{BS}_{1\mathrm{s}}}}.
\end{aligned}
\tag{2.249}
$$

$$
-\mathbf{k}\cdot(\mathbf{r}-\mathbf{0}'_{\mathrm{BS}_{2\mathrm{s}}\mathrm{s}}) = -\mathbf{k}\cdot(\mathbf{r}-\mathbf{0}_{\mathrm{BS}_{2\mathrm{s}}}) + \mathbf{k}'\cdot\mathbf{0}_{\mathrm{BS}_{2\mathrm{s}}},\quad \mathbf{k} \in [\mathbf{k}]_{\mathrm{i}_{\mathrm{BS}_{1\mathrm{s}}}}. \tag{2.250}
$$

In a standard fashion, we associate the modal functions which travel to the above detector, (2.233), (2.234), (2.235), (2.236), (2.239), (2.242), (2.245), and (2.248), with fields we denote as $\hat{E}^{(+)}_{\mathrm{s}0}(\mathbf{r},t)$, $\hat{E}^{(+)}_{\mathrm{i}_{\mathrm{BS}_{1\mathrm{s}}}0}(\mathbf{r},t)$. Exchanging s $\leftrightarrow$ i, we introduce modal functions, which travel to the lower detector. We relate them with fields we denote as $\hat{E}^{(+)}_{\mathrm{i}0}(\mathbf{r},t)$, $\hat{E}^{(+)}_{\mathrm{s}_{\mathrm{BS}_{1\mathrm{i}}}0}(\mathbf{r},t)$. Switching on the nonlinear interaction, we find the field to obey the relations (2.189) and (2.190). Counter to propagation the field stays initial and along with propagation it at least obeys the rules we have used to generate the modal functions. For simplicity, it is assumed that $\mathtt{t}_J = \mathtt{t}'_J = \mathtt{r}_J = \mathtt{r}'_J = \frac{1}{\sqrt{2}}$, $J = \mathrm{s}, \mathrm{i}$.

For the calculation of intensity correlations determined in the experiment, we introduce the slowly varying field operators

$$\hat{F}_{\mathrm{s}}^{(+)}(\mathbf{r}, t) = \exp(\mathrm{i}\omega_{\mathrm{s}} t)\hat{E}_{\mathrm{s}}^{(+)}(\mathbf{r}, t),$$
$$\hat{F}_{\mathrm{i}}^{(+)}(\mathbf{r}, t) = \exp(\mathrm{i}\omega_{\mathrm{i}} t)\hat{E}_{\mathrm{i}}^{(+)}(\mathbf{r}, t). \tag{2.251}$$

The field operators at the signal and idler detectors placed at $\mathbf{r}_{\mathrm{s}}$, $\mathbf{r}_{\mathrm{i}}$, respectively, are

$$\begin{aligned}
&\hat{F}_{\mathrm{s}}^{(+)}(\mathbf{r}_{\mathrm{s}}, t) \\
&= \frac{1}{2}\Bigg\{\Bigg[\hat{F}_{\mathrm{sout}}^{(+)}\left(\mathbf{0}, t - \frac{|\mathbf{r}_{\mathrm{s}} - \mathbf{r}_{\mathrm{BS}_{2\mathrm{s}}}|}{c} - \frac{L_{\mathrm{s,long}}}{c} - \frac{|\mathbf{r}_{\mathrm{BS}_{1\mathrm{s}}}|}{c}\right)\exp\left(\mathrm{i}\omega_{\mathrm{s}}\frac{|\mathbf{r}_{\mathrm{BS}_{1\mathrm{s}}}|}{c}\right) \\
&- \mathrm{i}\hat{F}_{\mathrm{iBS}_{1\mathrm{s}}\mathrm{in}}^{(+)}\left(\mathbf{0}_{\mathrm{BS}_{1\mathrm{s}}}, t - \frac{|\mathbf{r}_{\mathrm{s}} - \mathbf{r}_{\mathrm{BS}_{2\mathrm{s}}}|}{c} - \frac{L_{\mathrm{s,long}}}{c}\right)\Bigg]\exp\left(\mathrm{i}\omega_{\mathrm{s}}\frac{L_{\mathrm{s,long}}}{c}\right) \\
&+ \Bigg[\hat{F}_{\mathrm{sout}}^{(+)}\left(\mathbf{0}, t - \frac{|\mathbf{r}_{\mathrm{s}} - \mathbf{r}_{\mathrm{BS}_{2\mathrm{s}}}|}{c} - \frac{L_{\mathrm{s,short}}}{c} - \frac{|\mathbf{r}_{\mathrm{BS}_{1\mathrm{s}}}|}{c}\right)\exp\left(\mathrm{i}\omega_{\mathrm{s}}\frac{|\mathbf{r}_{\mathrm{BS}_{1\mathrm{s}}}|}{c}\right) \\
&+ \mathrm{i}\hat{F}_{\mathrm{iBS}_{1\mathrm{s}}\mathrm{in}}^{(+)}\left(\mathbf{0}_{\mathrm{BS}_{1\mathrm{s}}}, t - \frac{|\mathbf{r}_{\mathrm{s}} - \mathbf{r}_{\mathrm{BS}_{2\mathrm{s}}}|}{c} - \frac{L_{\mathrm{s,short}}}{c}\right)\Bigg]\exp\left(\mathrm{i}\omega_{\mathrm{s}}\frac{L_{\mathrm{s,short}}}{c}\right)\Bigg\} \\
&\times \exp\left(\mathrm{i}\omega_{\mathrm{s}}\frac{|\mathbf{r}_{\mathrm{s}} - \mathbf{r}_{\mathrm{BS}_{2\mathrm{s}}}|}{c}\right),
\end{aligned} \tag{2.252}$$

$$\hat{F}_{\mathrm{i}}^{(+)}(\mathbf{r}_{\mathrm{i}}, t) = \hat{F}_{\mathrm{s}}^{(+)}(\mathbf{r}_{\mathrm{s}}, t)\Big|_{\mathrm{s}\leftrightarrow\mathrm{i}}. \tag{2.253}$$

Let us denote $L_{\mathrm{s,short}}$ ($L_{\mathrm{i,short}}$) a length of the short arm of the interferometer for the signal (idler) beam. Supposing that $\Delta L_{\mathrm{s}} \equiv L_{\mathrm{s,long}} - L_{\mathrm{s,short}}$ ($\Delta L_{\mathrm{i}} \equiv L_{\mathrm{i,long}} - L_{\mathrm{i,short}}$) is much greater than the coherence length of the signal (idler) in order to avoid the second-order interference, we get (Casado et al. 1997a)

$$\begin{aligned}
&\langle \hat{F}_{\mathrm{s}}^{(+)}(\mathbf{r}_{\mathrm{s}}, t + \tau)\hat{F}_{\mathrm{i}}^{(+)}(\mathbf{r}_{\mathrm{i}}, t + \tau')\rangle = \frac{1}{4}gV \\
&\times \Bigg\{\nu(\tau' - \tau)\exp\Bigg[\frac{\mathrm{i}}{c}\omega_{\mathrm{s}}\left(|\mathbf{r}_{\mathrm{BS}_{1\mathrm{s}}}| + L_{\mathrm{s,long}} + |\mathbf{r}_{\mathrm{s}} - \mathbf{r}_{\mathrm{BS}_{2\mathrm{s}}}|\right) \\
&+ \frac{\mathrm{i}}{c}\omega_{\mathrm{i}}\left(|\mathbf{r}_{\mathrm{BS}_{1\mathrm{i}}}| + L_{\mathrm{i,long}} + |\mathbf{r}_{\mathrm{i}} - \mathbf{r}_{\mathrm{BS}_{2\mathrm{i}}}|\right)\Bigg] \\
&+ \nu\left(\tau' - \tau + \frac{\Delta L_{\mathrm{i}} - \Delta L_{\mathrm{s}}}{c}\right)\exp\Bigg[\frac{\mathrm{i}}{c}\omega_{\mathrm{s}}\Big(|\mathbf{r}_{\mathrm{BS}_{1\mathrm{s}}}| + L_{\mathrm{s,short}} \\
&+ |\mathbf{r}_{\mathrm{s}} - \mathbf{r}_{\mathrm{BS}_{2\mathrm{s}}}|\Big) + \frac{\mathrm{i}}{c}\omega_{\mathrm{i}}\left(|\mathbf{r}_{\mathrm{BS}_{1\mathrm{i}}}| + L_{\mathrm{i,short}} + |\mathbf{r}_{\mathrm{i}} - \mathbf{r}_{\mathrm{BS}_{2\mathrm{i}}}|\right)\Bigg]\Bigg\},
\end{aligned} \tag{2.254}$$

provided that $|\mathbf{r}_{\mathrm{BS}_{1J}}| + L_{J,\mathrm{short}} + |\mathbf{r}_J - \mathbf{r}_{\mathrm{BS}_{2J}}|$ is the same for $J = \mathrm{s}$ and $J = \mathrm{i}$. We finally obtain

$$P_{\mathrm{si}}(\mathbf{r}_\mathrm{s}, t+\tau; \mathbf{r}_\mathrm{i}, t+\tau') = \frac{1}{16} K' g^2 |V|^2 \Bigg\{ |\nu(\tau' - \tau)|^2 + \left| \nu\left(\tau' - \tau + \frac{\Delta L_\mathrm{i} - \Delta L_\mathrm{s}}{c}\right)\right|^2 - 2\mathrm{Re}\left\{ \nu(\tau'-\tau)\nu^*\left(\tau' - \tau + \frac{\Delta L_\mathrm{i} - \Delta L_\mathrm{s}}{c}\right) \times \exp\left[\frac{\mathrm{i}}{c}(\omega_\mathrm{s}\Delta L_\mathrm{s} + \omega_\mathrm{i}\Delta L_\mathrm{i})\right]\right\}\Bigg\}. \quad (2.255)$$

The visibility is given in (2.175), where

$$R_{\mathrm{simax}} + R_{\mathrm{simin}} = \frac{1}{8} g^2 |V|^2 \left[M(0,0) + M\left(\frac{\Delta L_\mathrm{i} - \Delta L_\mathrm{s}}{c}, \frac{\Delta L_\mathrm{i} - \Delta L_\mathrm{s}}{c}\right)\right], \quad (2.256)$$

$$R_{\mathrm{simax}} - R_{\mathrm{simin}} = \frac{1}{4} g^2 |V|^2 M\left(0, \frac{\Delta L_\mathrm{i} - \Delta L_\mathrm{s}}{c}\right). \quad (2.257)$$

The dependence of the visibility on the difference $(\Delta L_\mathrm{i} - \Delta L_\mathrm{s})$ is plotted in Fig. 2.7. The variation of the visibility is due to the function $M\left(\frac{d}{c}, \frac{h}{c}\right)$, but the function erf does not contribute to it. The distance between the points of inflection is $2\frac{c}{\sigma} = 6 \times 10^{-4}$ m. The interference manifests itself as a cosine variation of the coincidence rate with $\frac{1}{c}(\omega_\mathrm{s}\Delta L_\mathrm{s} + \omega_\mathrm{i}\Delta L_\mathrm{i})$.

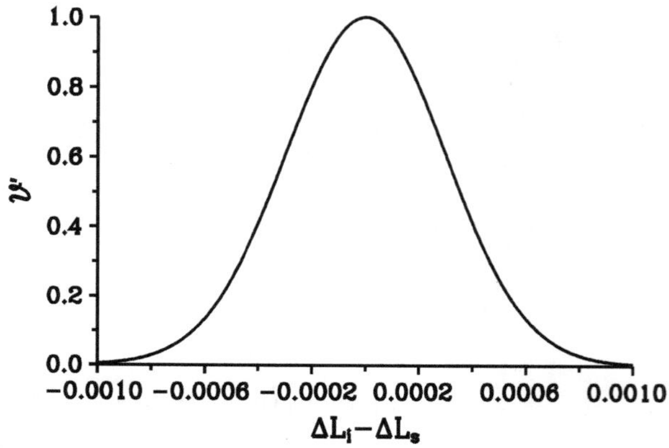

Fig. 2.7 The visibility $\mathcal{V}'$ versus the difference $(\Delta L_\mathrm{i} - \Delta L_\mathrm{s}) \in [-10^{-3}, 10^{-3}]$ measured in metres

(v) *Induced coherence and indistinguishability in two-photon interference*

Zou et al. (1991) performed an experiment in which fourth-order interference is observed in the superposition of signal photons from two coherently pumped parametric down-conversion crystals, when the paths of the idler photons are aligned. The experimental setup is outlined in Fig. 2.8, in which two nonlinear crystals NL1 and NL2 are optically pumped by two mutually coherent, classical pump waves of complex amplitudes

$$V_j(\mathbf{r}, t) = V_j \exp[\mathrm{i}(\mathbf{k}_0 \cdot \mathbf{r} - \omega_0 t)], \quad j = 1, 2. \tag{2.258}$$

We assume that $V_1 = V_2 \exp(\mathrm{i}\mathbf{k}_0 \cdot \mathbf{0}_2) = V$. On the contrary, there were similar crystals in the experiment, but we consider more general ones. The parametric down-conversion occurs at both crystals, each with the emission of a signal photon and an idler photon. We are interested in the joint detection rate of the detectors $\mathrm{D_s}$ and $\mathrm{D_i}$ when the trajectories of the two idlers i_1, i_2 are aligned and the path difference between the two signals is varied slightly. Fourth-order interference disappears when the idlers are misaligned or separated by a beam stop.

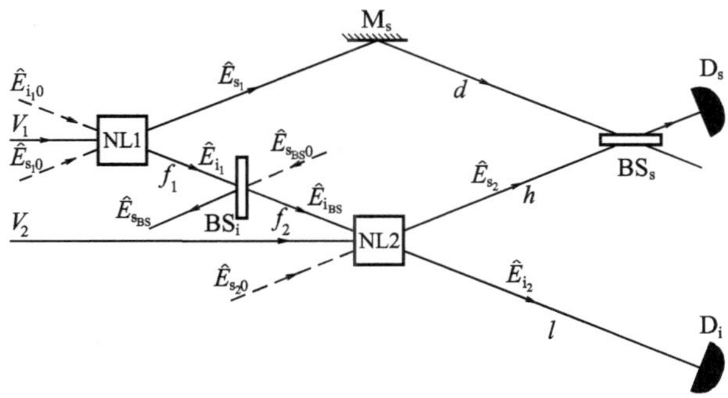

Fig. 2.8 Experimental setup on induced coherence without induced emission. For simplicity, $\hat{E}_{\mathrm{s_{BS}}0} \equiv \hat{E}_{\mathrm{s_{BS_i}}0}$

In what follows, we will specify modal functions and the nonlinear dynamics of field operators. We consider the initial electric field in the form

$$\hat{E}_0^{(+)}(\mathbf{r}, t) = V^{(+)}(\mathbf{r}, t)\hat{1} + \sum_{j=1}^{2} \hat{E}_{\mathrm{s}_j 0}^{(+)}(\mathbf{r}, t) + \hat{E}_{\mathrm{i}0}^{(+)}(\mathbf{r}, t) + \hat{E}_{\mathrm{s_{BS_i}}0}^{(+)}(\mathbf{r}, t), \tag{2.259}$$

where

$$\hat{E}_{J0}^{(+)}(\mathbf{r}, t) = \sum_{\mathbf{k} \in [\mathbf{k}]_J} v_{J\mathbf{k}}(\mathbf{r}) \hat{a}_{J\mathbf{k}0}(t), \quad J = \mathrm{s}_1, \mathrm{s}_2, \mathrm{i}, \mathrm{s_{BS_i}}. \tag{2.260}$$

The modal functions as restricted to linear segments are

$$v_{\mathrm{s}_1\mathbf{k}}(\mathbf{r}) = \mathrm{i}\sqrt{\frac{\hbar\omega_{\mathbf{k}}}{\mathcal{A}L}} e^{\mathrm{i}\mathbf{k}\cdot\mathbf{r}} \text{ for } \mathbf{r} \parallel \mathbf{k}, \; z < z_{\mathrm{M_s}}, \; \mathbf{k} \in [\mathbf{k}]_{\mathrm{s}}, \tag{2.261}$$

$$v_{\mathrm{s}_2\mathbf{k}}(\mathbf{r}) = \mathrm{i}\sqrt{\frac{\hbar\omega_{\mathbf{k}}}{\mathcal{A}L}} e^{\mathrm{i}\mathbf{k}\cdot\mathbf{r}} \text{ for } (\mathbf{r} - \mathbf{0}_2) \parallel \mathbf{k}, \; z < z_{\mathrm{BS_s}}, \; \mathbf{k} \in [\mathbf{k}]_{\mathrm{s}}, \tag{2.262}$$

$$v_{s_1\mathbf{k}}(\mathbf{r}) = -\mathrm{i}\sqrt{\frac{\hbar\omega_{\mathbf{k}}}{\mathcal{A}L}}e^{\mathrm{i}\mathbf{k}'\cdot(\mathbf{r}-\mathbf{0}'_{\mathrm{M_s}})} \text{ for } (\mathbf{r}-\mathbf{0}_{\mathrm{M_s}}) \parallel \mathbf{k}',$$
$$z_{\mathrm{M_s}} < z < z_{\mathrm{BS_s}},\ \mathbf{k} \in [\mathbf{k}]_{\mathrm{s}}. \tag{2.263}$$

Here $\mathbf{0}'_{\mathrm{M_s}}$ and $\mathbf{k}'$ are used as in the definition of modal functions related to Fig. 2.1 and $\mathbf{k}'$ has the same meaning. Here, as in the definitions related to Fig. 2.4, we specify that $\mathbf{0}'_{\mathrm{M_s}}$ has been chosen so that

$$\mathbf{k}'\cdot(\mathbf{r}-\mathbf{0}'_{\mathrm{M_s}}) = \mathbf{k}'\cdot(\mathbf{r}-\mathbf{0}_{\mathrm{M_s}}) + \mathbf{k}\cdot\mathbf{0}_{\mathrm{M_s}},\ \ \mathbf{k} \in [\mathbf{k}]_{\mathrm{s}}. \tag{2.264}$$

Similarly as in (2.222) and (2.223) for $\mathtt{t} = \mathtt{t}' = \frac{1}{\sqrt{2}}$, $\mathtt{r} = \mathtt{r}' = \frac{\mathrm{i}}{\sqrt{2}}$, we have

$$v_{s_1\mathbf{k}}(\mathbf{r}) = -\mathrm{i}\sqrt{\frac{\hbar\omega_{\mathbf{k}}}{\mathcal{A}L}}e^{\mathrm{i}\mathbf{k}'\cdot(\mathbf{r}-\mathbf{0}'_{\mathrm{M_s}})}\frac{1}{\sqrt{2}} + \mathrm{i}\sqrt{\frac{\hbar\omega_{\mathbf{k}}}{\mathcal{A}L}}e^{\mathrm{i}\mathbf{k}'\cdot(\mathbf{r}-\mathbf{0}'_{\mathrm{BS_s}})}\frac{\mathrm{i}}{\sqrt{2}}$$
$$\text{for } (\mathbf{r}-\mathbf{0}_{\mathrm{M_s}}) \parallel \mathbf{k}',\ z > z_{\mathrm{BS_s}},\ \mathbf{k} \in [\mathbf{k}]_{\mathrm{s}}, \tag{2.265}$$
$$v_{s_2\mathbf{k}}(\mathbf{r}) = -\mathrm{i}\sqrt{\frac{\hbar\omega_{\mathbf{k}}}{\mathcal{A}L}}e^{\mathrm{i}\mathbf{k}\cdot(\mathbf{r}-\mathbf{0}'_{\mathrm{BS_sM_s}})}\frac{\mathrm{i}}{\sqrt{2}} + \mathrm{i}\sqrt{\frac{\hbar\omega_{\mathbf{k}}}{\mathcal{A}L}}e^{\mathrm{i}\mathbf{k}\cdot\mathbf{r}}\frac{1}{\sqrt{2}}$$
$$\text{for } (\mathbf{r}-\mathbf{0}_2) \parallel \mathbf{k},\ z > z_{\mathrm{BS_s}},\ \mathbf{k} \in [\mathbf{k}]_{\mathrm{s}}, \tag{2.266}$$

where $\mathbf{0}'_{\mathrm{BS_s}}$ and $\mathbf{0}'_{\mathrm{BS_sM_s}}$ have been chosen so that, respectively,

$$\mathbf{k}'\cdot(\mathbf{r}-\mathbf{0}'_{\mathrm{BS_s}}) = \mathbf{k}'\cdot(\mathbf{r}-\mathbf{0}_{\mathrm{BS_s}}) + \mathbf{k}\cdot\mathbf{0}_{\mathrm{BS_s}},\ \ \mathbf{k} \in [\mathbf{k}]_{\mathrm{s}}, \tag{2.267}$$
$$\mathbf{k}\cdot(\mathbf{r}-\mathbf{0}'_{\mathrm{BS_sM_s}}) = \mathbf{k}\cdot(\mathbf{r}-\mathbf{0}_{\mathrm{BS_s}}) + \mathbf{k}'\cdot\mathbf{0}'_{\mathrm{M_s}},\ \ \mathbf{k} \in [\mathbf{k}]_{\mathrm{s}}; \tag{2.268}$$

$$v_{\mathrm{i}\mathbf{k}}(\mathbf{r}) = \mathrm{i}\sqrt{\frac{\hbar\omega_{\mathbf{k}}}{\mathcal{A}L}}e^{\mathrm{i}\mathbf{k}\cdot\mathbf{r}} \text{ for } \mathbf{r} \parallel \mathbf{k},\ z < z_{\mathrm{BS_i}},\ \mathbf{k} \in [\mathbf{k}]_{\mathrm{i}}, \tag{2.269}$$

$$v_{s_{\mathrm{BS_i}}\mathbf{k}}(\mathbf{r}) = \mathrm{i}\sqrt{\frac{\hbar\omega_{\mathbf{k}}}{\mathcal{A}L}}e^{\mathrm{i}\mathbf{k}\cdot\mathbf{r}} \text{ for } (\mathbf{r}-\mathbf{0}_{\mathrm{BS_i}}) \parallel \mathbf{k},\ z > z_{\mathrm{BS_i}},\ \mathbf{k} \in [\mathbf{k}]_{s_{\mathrm{BS_i}}}, \tag{2.270}$$

$$v_{\mathrm{i}\mathbf{k}}(\mathbf{r}) = \mathrm{i}\sqrt{\frac{\hbar\omega_{\mathbf{k}}}{\mathcal{A}L}}e^{\mathrm{i}\mathbf{k}\cdot\mathbf{r}}\mathtt{t} + \mathrm{i}\sqrt{\frac{\hbar\omega_{\mathbf{k}}}{\mathcal{A}L}}e^{\mathrm{i}\mathbf{k}\cdot(\mathbf{r}-\mathbf{0}'_{\mathrm{BS_i}s_{\mathrm{BSi}}})}\mathtt{r}'$$
$$\text{for } \mathbf{r} \parallel \mathbf{k},\ z > z_{\mathrm{BS_i}},\ \mathbf{k} \in [\mathbf{k}]_{\mathrm{i}}, \tag{2.271}$$
$$v_{s_{\mathrm{BS_i}}\mathbf{k}}(\mathbf{r}) = \mathrm{i}\sqrt{\frac{\hbar\omega_{\mathbf{k}}}{\mathcal{A}L}}e^{\mathrm{i}\mathbf{k}\cdot(\mathbf{r}-\mathbf{0}'_{\mathrm{BS_ii}})}\mathtt{r} + \mathrm{i}\sqrt{\frac{\hbar\omega_{\mathbf{k}}}{\mathcal{A}L}}e^{\mathrm{i}\mathbf{k}\cdot\mathbf{r}}\mathtt{t}'$$
$$\text{for } (\mathbf{r}-\mathbf{0}_{\mathrm{BS_i}}) \parallel \mathbf{k},\ z < z_{\mathrm{BS_i}},\ \mathbf{k} \in [\mathbf{k}]_{s_{\mathrm{BS_i}}}, \tag{2.272}$$

where $\mathbf{k}'$ is defined relative to the beam splitter $\mathrm{BS_i}$ and $\mathbf{0}'_{\mathrm{BS_ii}}$ and $\mathbf{0}'_{\mathrm{BS_i}s_{\mathrm{BSi}}}$ have been chosen so that, respectively,

$$\mathbf{k}\cdot(\mathbf{r}-\mathbf{0}'_{\mathrm{BS_i}s_{\mathrm{BSi}}}) = \mathbf{k}\cdot(\mathbf{r}-\mathbf{0}_{\mathrm{BS_i}}) + \mathbf{k}'\cdot\mathbf{0}_{\mathrm{BS_ii}} \text{ for } \mathbf{k} \in [\mathbf{k}]_{\mathrm{i}}, \tag{2.273}$$
$$\mathbf{k}\cdot(\mathbf{r}-\mathbf{0}'_{\mathrm{BS_ii}}) = \mathbf{k}\cdot(\mathbf{r}-\mathbf{0}_{\mathrm{BS_i}}) + \mathbf{k}'\cdot\mathbf{0}_{\mathrm{BS_i}} \text{ for } \mathbf{k} \in [\mathbf{k}]_{s_{\mathrm{BS_i}}}. \tag{2.274}$$

The modal functions (2.269), (2.270), (2.271), and (2.272) travel to the lower detector. We relate them with fields we denote as $\hat{E}_{\mathrm{i}0}(\mathbf{r}, t)$, $\hat{E}_{\mathrm{sBS_i}0}(\mathbf{r}, t)$. Switching on the first nonlinear interaction, we find the field to obey the relations (2.189), (2.190), and (2.191) for $j = 1$ with $\mathrm{i}_1 \to \mathrm{i}$. Counter to propagation the field remains initial and along with propagation it at least obeys the rules we have used to generate the modal functions.

Now we would like to interpret the subscript 0 not as the order of solution but as a number of the initial stage. Since the stage is followed by the first stage, we would like to modify the relations (2.189), (2.190), and (2.191) so that they confess the first-stage operators on the left-hand side, which would lead to the use of the subscript 1.

Switching on the second nonlinear interaction, we find the field to obey the relations (2.189), (2.190), and (2.191) for $j = 2$ with $\mathrm{i}_2 \to \mathrm{i}$, but the first-stage field operators to have been substituted for the operators on the right-hand side. The action of the operators $\hat{\hat{G}}_2$ and $\hat{\hat{J}}_2$ depends on the nonlinear crystal located at $\mathbf{0}_2$. It also depends on the pump beam at the same crystal. Counter to propagation the field stays first stage and along with propagation it still at least obeys the rules to generate the modal functions.

Further we will express the intensity correlations that have been determined in the experiment. Again, we introduce the operators (2.224), where $J = \mathrm{s}_1, \mathrm{s}_2, \mathrm{i}$, $J_1 = \mathrm{s}, \mathrm{s}, \mathrm{i}$. The field operators at the signal and idler detectors placed at $\mathbf{r}_\mathrm{s}$, $\mathbf{r}_\mathrm{i}$, respectively, are

$$\hat{F}_{\mathrm{s}_2}^{(+)}(\mathbf{r}_\mathrm{s}, t) = \frac{1}{\sqrt{2}}\Bigg[-\mathrm{i}\hat{F}_{\mathrm{s}_1}^{(+)}\left(\mathbf{0}_1, t - \frac{d}{c}\right)\exp\left(\mathrm{i}\omega_\mathrm{s}\frac{d}{c}\right) + \hat{F}_{\mathrm{s}_2}^{(+)}\left(\mathbf{0}_2, t - \frac{h}{c}\right)\exp\left(\mathrm{i}\omega_\mathrm{s}\frac{h}{c}\right)\Bigg], \tag{2.275}$$

$$\hat{F}_{\mathrm{i}}^{(+)}(\mathbf{r}_\mathrm{i}, t') = \hat{F}_{\mathrm{i}}^{(+)}\left(\mathbf{0}_2, t' - \frac{l}{c}\right)\exp\left(\mathrm{i}\omega_\mathrm{i}\frac{l}{c}\right). \tag{2.276}$$

We still assume different crystals and derive a slight generalization of the well-known experiment (Casado et al. 1997a). By taking into account the correlation relations

$$\langle \hat{F}_{\mathrm{s}_1}^{(+)}(\mathbf{0}_1, t)\hat{F}_{\mathrm{i}}^{(+)}(\mathbf{0}_2, t')\rangle = \mathtt{t} g V_1 \nu_1\left(t' - t - \frac{f}{c}\right)\exp\left(\mathrm{i}\omega_\mathrm{i}\frac{f}{c}\right), \tag{2.277}$$

$$\langle \hat{F}_{\mathrm{i}}^{(+)}(\mathbf{0}_2, t')\hat{F}_{\mathrm{s}_2}^{(+)}(\mathbf{0}_2, t)\rangle = g V_2 \nu_2(t' - t), \tag{2.278}$$

we get (Casado et al. 1997a)

$$\langle \hat{F}_{s_2}^{(+)}(\mathbf{r}_s, t)\hat{F}_i^{(+)}(\mathbf{r}_i, t')\rangle = \frac{gV}{\sqrt{2}}$$
$$\times \left[-\mathrm{i}\mathrm{t}\nu_1 \left(\tau' - \tau - \frac{l}{c} - \frac{f}{c} + \frac{d}{c}\right) \exp\left\{\frac{\mathrm{i}}{c}[\omega_s d + \omega_i(l+f)]\right\}\right.$$
$$\left. + \nu_2 \left(\tau' - \tau - \frac{l}{c} + \frac{h}{c}\right) \exp\left\{\frac{\mathrm{i}}{c}[\omega_s h + \omega_i l]\right\}\right]. \quad (2.279)$$

We finally obtain

$$P_{si}(\mathbf{r}_s, t+\tau; \mathbf{r}_i, t+\tau') = \frac{1}{2}K' g^2 |V|^2$$
$$\times \left\{ \left|\mathrm{t}\nu_1 \left(\tau' - \tau - \frac{l}{c} - \frac{f}{c} + \frac{d}{c}\right)\right|^2 + \left|\nu_2 \left(\tau' - \tau - \frac{l}{c} + \frac{h}{c}\right)\right|^2 \right.$$
$$+ 2\mathrm{Im}\left[\mathrm{t}\nu_1 \left(\tau' - \tau - \frac{l}{c} - \frac{f}{c} + \frac{d}{c}\right) \nu_2^* \left(\tau' - \tau - \frac{l}{c} + \frac{h}{c}\right)\right.$$
$$\left.\left. \times \exp\left\{\frac{\mathrm{i}}{c}[\omega_s(d-h) + \omega_i f]\right\}\right]\right\}. \quad (2.280)$$

We have hopefully corrected the factor, changed a sign with respect to the reflection from the mirror, and changed signs of the argument of $\nu(\tau)$ relying on the identity $\nu(\tau) = \nu(-\tau)$, where $\nu_1(\tau) = \nu_2(\tau) \equiv \nu(\tau)$.

The visibility is expressed by the formula (2.175), where for $d = l + f, l = h$

$$R_{\mathrm{simax}} + R_{\mathrm{simin}} = g^2|V|^2 M(0,0)\left(|\mathrm{t}|^2 + 1\right), \quad (2.281)$$
$$R_{\mathrm{simax}} - R_{\mathrm{simin}} = 2g^2|V|^2 M(0,0)\,|\mathrm{t}|. \quad (2.282)$$

The maximum visibility is equal to unity and, in general, it depends on the transmissivity of the beam splitter, as can be seen from Fig. 2.9. The interference manifests itself as a cosine variation of the coincidence rate with $\frac{\omega_0}{c} f$.

(vi) *Frustrated two-photon creation via interference*

Herzog et al. (1994) performed a simple experiment interpreted as showing interference of two processes. They placed three mirrors in the three beams, laser, signal, and idler, that emerge from a nonlinear crystal, NL, and put a detector into the reflected idler beams (see Fig. 2.10). In the standard quantum interpretation a pair of correlated photons can be created either by the laser beam travelling from left to right or when the reflected laser beam travels from right to left. In both cases the idler photon may arrive at the detector. As the two possibilities are indistinguishable

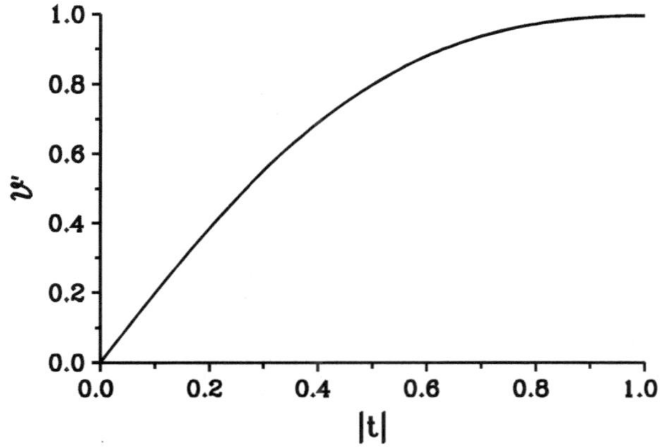

Fig. 2.9 The visibility $\mathcal{V}'$ versus the modulus of the amplitude transmissivity $|t| \in [0, 1]$ of the beam splitter BS_i. It is assumed that $d = l + f, l = h$

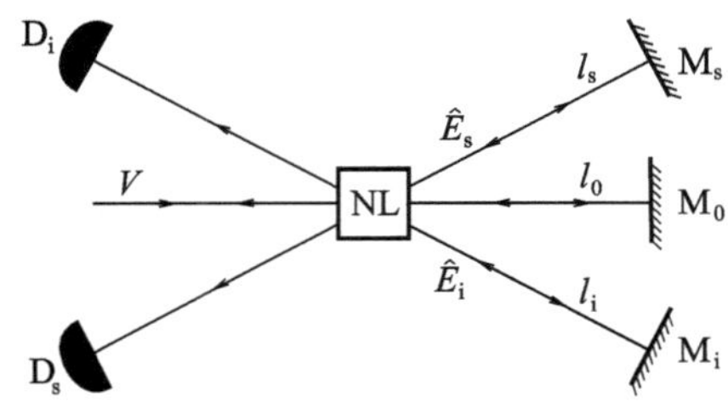

Fig. 2.10 Experimental setup on frustrated two-photon creation via interference

they interfere and the counting rate oscillates depending on the position of a chosen mirror.

Accordingly the description of the pump beam is given by

$$\begin{aligned} V^{(+)}(\mathbf{r}, t) &= V e^{i(\mathbf{k}_0\cdot\mathbf{r}-\omega_0 t)} - V e^{i[-\mathbf{k}_0\cdot(\mathbf{r}-2l_0\mathbf{e}_0)-\omega_0 t]} \\ &= V e^{i(\mathbf{k}_0\cdot\mathbf{r}-\omega_0 t)} - V e^{i\varphi_0} e^{i(-\mathbf{k}_0\cdot\mathbf{r}-\omega_0 t)} \\ &\quad \text{for } x = 0,\ y = 0,\ z < l_0, \end{aligned} \tag{2.283}$$

where $\mathbf{e}_0$ is the direction vector of the forward-propagating pump beam, $\varphi_0 = 2|\mathbf{k}_0|l_0 = 2\frac{\omega_0 l_0}{c}$.

The modal functions are

$$\begin{aligned} v_{s\mathbf{k}}(\mathbf{r}) &= i\sqrt{\frac{\hbar\omega_{\mathbf{k}}}{2\mathcal{A}L}} e^{i\mathbf{k}\cdot\mathbf{r}} - i\sqrt{\frac{\hbar\omega_{\mathbf{k}}}{2\mathcal{A}L}} e^{i\varphi_s} e^{-i\mathbf{k}\cdot\mathbf{r}} \\ &\quad \text{for } \mathbf{r} \parallel \mathbf{k},\ z < z_{M_s},\ \mathbf{k} \in [\mathbf{k}]_s, \end{aligned} \tag{2.284}$$

where $\varphi_s = 2\frac{\omega_s l_s}{c}$, $z_{M_s} = \mathbf{e}_0 \cdot \mathbf{0}_{M_s}$. The modal functions $v_{i\mathbf{k}}(\mathbf{r})$ are expressed similarly.

Associating the modal functions with quantum fields, we must consider that

$$\hat{E}_{s0}^{(+)}(\mathbf{r}, t) = \hat{E}_{sF0}^{(+)}(\mathbf{r}, t) + \hat{E}_{sB0}^{(+)}(\mathbf{r}, t), \tag{2.285}$$

where

$$\hat{E}^{(+)}_{\mathrm{sF0}}(\mathbf{r},t) = \mathrm{i}\sum_{\mathbf{k}\in[\mathbf{k}]_\mathrm{s}} \sqrt{\frac{\hbar\omega_\mathbf{k}}{2\mathcal{A}L}} e^{\mathrm{i}\mathbf{k}\cdot\mathbf{r}} \hat{a}_{\mathrm{s}\mathbf{k}0}(t),$$

$$\hat{E}^{(+)}_{\mathrm{sB0}}(\mathbf{r},t) = -\mathrm{i}\sum_{\mathbf{k}\in[\mathbf{k}]_\mathrm{s}} \sqrt{\frac{\hbar\omega_\mathbf{k}}{2\mathcal{A}L}} e^{\mathrm{i}\varphi_\mathrm{s}} e^{-\mathrm{i}\mathbf{k}\cdot\mathbf{r}} \hat{a}_{\mathrm{s}\mathbf{k}0}(t) \tag{2.286}$$

are the forward-propagating component and the backward-propagating component, respectively. The field operators $\hat{E}^{(+)}_{\mathrm{i}0}(\mathbf{r},t)$, $\hat{E}^{(+)}_{\mathrm{iF0}}(\mathbf{r},t)$, and $\hat{E}^{(+)}_{\mathrm{iB0}}(\mathbf{r},t)$ are expressed similarly. Nonlinear dynamics is described by the relations

$$\begin{aligned}\hat{E}^{(+)}_{\mathrm{sF}}(\mathbf{0}-(r=0)\mathbf{e}_\mathrm{s},t) &= \hat{E}^{(+)}_{\mathrm{sF0}}(\mathbf{0},t),\\ \hat{E}^{(+)}_{\mathrm{iF}}(\mathbf{0}-(r=0)\mathbf{e}_\mathrm{i},t) &= \hat{E}^{(+)}_{\mathrm{iF0}}(\mathbf{0},t),\end{aligned} \tag{2.287}$$

where $\mathbf{e}_j$, $j = \mathrm{s}, \mathrm{i}$, is the direction vector of the forward-propagating signal, idler beam, respectively:

$$\begin{aligned}\hat{E}^{(+)}_{\mathrm{sF}}(\mathbf{0}+(r=0)\mathbf{e}_\mathrm{s},t) &= (1+g^2|V|^2\hat{\hat{J}})\hat{E}^{(+)}_{\mathrm{sF0}}(\mathbf{0},t) + e^{-\mathrm{i}\omega_0 t} gV\hat{\hat{G}}\hat{E}^{(-)}_{\mathrm{iF0}}(\mathbf{0},t),\\ \hat{E}^{(+)}_{\mathrm{iF}}(\mathbf{0}+(r=0)\mathbf{e}_\mathrm{i},t) &= e^{-\mathrm{i}\omega_0 t} gV\hat{\hat{G}}\hat{E}^{(-)}_{\mathrm{sF0}}(\mathbf{0},t) + (1+g^2|V|^2\hat{\hat{J}})\hat{E}^{(+)}_{\mathrm{iF0}}(\mathbf{0},t),\end{aligned} \tag{2.288}$$

$$\begin{aligned}\hat{E}^{(+)}_{\mathrm{sB}}(\mathbf{0}+(r=0)\mathbf{e}_\mathrm{s},t) &= -\hat{E}^{(+)}_{\mathrm{sF}}\left(\mathbf{0}+(r=0)\mathbf{e}_\mathrm{s}, t-\frac{2l_\mathrm{s}}{c}\right)\\ &= -(1+g^2|V|^2\hat{\hat{J}})\hat{E}^{(+)}_{\mathrm{sF0}}\left(\mathbf{0}, t-\frac{2l_\mathrm{s}}{c}\right)\\ &\quad - e^{\mathrm{i}(\varphi_\mathrm{s}+\varphi_\mathrm{i})} e^{-\mathrm{i}\omega_0 t} gV\hat{\hat{G}}\hat{E}^{(-)}_{\mathrm{iF0}}\left(\mathbf{0}, t-\frac{2l_\mathrm{s}}{c}\right),\end{aligned} \tag{2.289}$$

$$\begin{aligned}\hat{E}^{(+)}_{\mathrm{iB}}(\mathbf{0}+(r=0)\mathbf{e}_\mathrm{i},t) &= -\hat{E}^{(+)}_{\mathrm{iF}}\left(\mathbf{0}+(r=0)\mathbf{e}_\mathrm{i}, t-\frac{2l_\mathrm{i}}{c}\right)\\ &= -e^{\mathrm{i}(\varphi_\mathrm{s}+\varphi_\mathrm{i})} e^{-i\omega_0 t} gV\hat{\hat{G}}\hat{E}^{(-)}_{\mathrm{sF0}}\left(\mathbf{0}, t-\frac{2l_\mathrm{i}}{c}\right)\\ &\quad - (1+g^2|V|^2\hat{\hat{J}})\hat{E}^{(+)}_{\mathrm{iF0}}\left(\mathbf{0}, t-\frac{2l_\mathrm{i}}{c}\right),\end{aligned} \tag{2.290}$$

$$\begin{aligned}\hat{E}^{(+)}_{\mathrm{sBout}}(\mathbf{0},t) &= (1+g^2|V|^2\hat{\hat{J}})\hat{E}^{(+)}_{\mathrm{sB}}(\mathbf{0}+(r=0)\mathbf{e}_\mathrm{s},t)\\ &\quad + e^{-\mathrm{i}\omega_0 t} gVe^{\mathrm{i}\varphi_0}\hat{\hat{G}}\hat{E}^{(-)}_{\mathrm{iB}}(\mathbf{0}+(r=0)\mathbf{e}_\mathrm{i},t),\\ \hat{E}^{(+)}_{\mathrm{iBout}}(\mathbf{0},t) &= e^{-\mathrm{i}\omega_0 t} gVe^{\mathrm{i}\varphi_0}\hat{\hat{G}}\hat{E}^{(-)}_{\mathrm{sB}}(\mathbf{0}+(r=0)\mathbf{e}_\mathrm{s},t)\\ &\quad + (1+g^2|V|^2\hat{\hat{J}})\hat{E}^{(+)}_{\mathrm{iB}}(\mathbf{0}+(r=0)\mathbf{e}_\mathrm{i},t).\end{aligned} \tag{2.291}$$

Further we will calculate the quantum mean intensities that have been determined in the experiment. Introducing the slowly varying field operators

$$\hat{F}_J^{(+)}(\mathbf{r},t) = e^{\mathrm{i}\omega_J t}\hat{E}_J^{(+)}(\mathbf{r},t),$$
$$\hat{F}_{J\mathrm{F}}^{(+)}(\mathbf{r},t) = e^{\mathrm{i}\omega_J t}\hat{E}_{J\mathrm{F}}^{(+)}(\mathbf{r},t),$$
$$\hat{F}_{J\mathrm{B}}^{(+)}(\mathbf{r},t) = e^{\mathrm{i}\omega_J t}\hat{E}_{J\mathrm{B}}^{(+)}(\mathbf{r},t),\ J = \mathrm{s},\mathrm{i}, \tag{2.292}$$

and $\hat{F}_J^{(-)}(\mathbf{r},t)$, $\hat{F}_{J\mathrm{F}}^{(-)}(\mathbf{r},t)$, $\hat{F}_{J\mathrm{B}}^{(-)}(\mathbf{r},t)$, we express the field operators at the signal and idler detectors placed at $\mathbf{r}_\mathrm{s}$, $\mathbf{r}_\mathrm{i}$, respectively, as

$$\hat{F}_{J\mathrm{B}}^{(+)}(\mathbf{r}_J,t) = \hat{F}_{J\mathrm{Bout}}^{(+)}\left(\mathbf{0}, t - \frac{r_J}{c}\right)\exp\left(\mathrm{i}\frac{\omega_J r_J}{c}\right),\ J = \mathrm{s},\ \mathrm{i}. \tag{2.293}$$

The quantum mean intensity or single photodetection rate in the detector D_s is

$$P_\mathrm{s}(\mathbf{r}_\mathrm{s},t) = K\left\langle 0 \left| \hat{E}^{(-)}(\mathbf{r}_\mathrm{s},t)\hat{E}^{(+)}(\mathbf{r}_\mathrm{s},t) \right| 0\right\rangle \tag{2.294}$$
$$= K\left\langle 0 \left| \hat{F}_{\mathrm{sB}}^{(-)}(\mathbf{r}_\mathrm{s},t)\hat{F}_{\mathrm{sB}}^{(+)}(\mathbf{r}_\mathrm{s},t) \right| 0\right\rangle, \tag{2.295}$$

where K is a constant related to the efficiency of the detector and the energy of a single photon. Considering the forward propagation, reflections, and the backward propagation, we obtain that

$$\left\langle \hat{F}_{\mathrm{sB}}^{(-)}(\mathbf{r}_\mathrm{s},t)\hat{F}_{\mathrm{sB}}^{(+)}(\mathbf{r}_\mathrm{s},t)\right\rangle = \left\langle \hat{F}_{\mathrm{sB}}^{(-)}\left(\mathbf{0}, t - \frac{r_\mathrm{s}}{c}\right)\hat{F}_{\mathrm{sB}}^{(+)}\left(\mathbf{0}, t - \frac{r_\mathrm{s}}{c}\right)\right\rangle$$
$$= 2g^2|V|^2\left[\mu_\mathrm{s}(0) + \mu_\mathrm{s}\left(\frac{2l_\mathrm{i}}{c} - \frac{2l_\mathrm{s}}{c}\right)\cos(\varphi_\mathrm{s} + \varphi_\mathrm{i} - \varphi_0)\right], \tag{2.296}$$

where we have relied on the identity $\mu_\mathrm{s}(\tau) = \mu_\mathrm{s}(-\tau)$. From this,

$$P_\mathrm{s}(\mathbf{r}_\mathrm{s},t) = 2Kg^2|V|^2\left[\mu_\mathrm{s}(0) + \mu_\mathrm{s}\left(\frac{2l_\mathrm{i}}{c} - \frac{2l_\mathrm{s}}{c}\right)\cos(\varphi_\mathrm{s} + \varphi_\mathrm{i} - \varphi_0)\right]. \tag{2.297}$$

The photodetection rate in the detector D_i is expressed similarly (Casado et al. 1997b).

In conclusion, we have mostly dealt with the fourth-order interference in parametric down-conversion experiments. The 1986, 1990, 1994 (adapted back to free space), and 1991 experiments were chosen according to a review article of other authors. Coincidence measurements in the various setups are essentially (or sufficiently well) described in terms of the cross correlation between the signal and the idler.

We have "promoted" the schemes of the experiments, where only paths through nonlinear and linear optical elements and the free space (with possible reflections from perfect mirrors) to detectors are drawn, to a reason of a certain neglect of the beams' divergence. We have replaced the usual assumption that the electric field is expanded in terms of an incomplete set of plane waves, which is relatively complete with respect to the expected direction of propagation, by the hypothesis that there exists an incomplete or relatively complete system of more complicated modal functions, which have still been specified only on the paths. We have performed conventional quantization by introducing annihilation operators in place of the classical complex amplitudes of the modes. We tried to choose sufficiently realistic values of the parameters for all the four experiments and to find visibilities of the intensity interference.

2.3.2 From Coupled Quantum Harmonic Oscillators Back to Interacting Fields

One of the interference experiments we have described in Section 2.3.1, the experiment of Zou et al. (1991) which has been analysed in Wang et al. (1991a,b), has attracted much attention. The arrangement of two down-converters is pumped by mutually coherent beams and the two down-converters are connected by the idler beam. The spontaneous emission from the first nonlinear crystal in the idler serves as a stimulating idler input to the second nonlinear crystal that acts as an optical amplifier. The interference of signal beams from both the crystals can be observed. A beam splitter placed between the two nonlinear crystals in the idler beam can change the strength of their connection since it attenuates the emerging field.

The parametric down-conversion in the second nonlinear crystal is stimulated by idler photons when the idler field is strong "per frequency unit". In this situation, the polychromatic theory yields results similar to those obtained by the monochromatic treatment, i.e. about multiples of the latter. When the idler field is weak per frequency unit, the second nonlinear crystal is proven to "ignore" the idler photons. The monochromatic description, even though completed by optimal scaling of its results, is far from being persuasive here. The assumption of the strong idler field is implicit in work contributing to the monochromatic theory.

In Řeháček and Peřina (1996) it has been shown that the distribution of photon-number sum in signal modes interpolates between a Bose–Einstein distribution and a convolution of two Bose–Einstein distributions. The latter distribution occurs when the idler beam is blocked. In general, the photon-number sum is distributed as if it corresponded to the number of signal degrees of freedom which varied between 1 and 2. A nonclassical distribution of photon-number sums restricted to even sums of photon numbers cannot occur, because the correlation between the photon numbers of the two signal beams is not complete. It has been found that the distribution of phase difference derived from the Q function narrows when the connection of both the down-converters via the idler mode closes up.

The monochromatic treatment associates each travelling wave with a quantum harmonic oscillator. The simple formalism of several coupled harmonic oscillators is useful for an analysis of the travelling-wave setup of interference experiment due to

Zou et al. (1991) up to suppression of the induced emission. The more complicated approach used originally for the analysis seems to be unsuitable for treating the phenomenon of induced emission. We try to formalize here a comparison between the two approaches.

When the induced emission occurs, it can be utilized. The phenomenon of induced emission makes the phase of an amplified field adopt the same phase as the incident locking field (Wang et al. 1991a, Wiseman and Mølmer 2000). The induced emission can also be used in parametric down-conversion to lock the phase of the idler and, from this, that of the signal (since the phase sum of the signal and idler is locked to the pump phase). If the field used to lock the idler of one down-converter (crystal NL2 in Fig. 2.11) is itself the idler output of another down-converter (crystal NL1 in Fig. 2.11), the two signal fields will also be locked in phase. Thus, they will have, in principle, perfect first-order coherence and so will interfere at a final beam splitter not included in Fig. 2.11. If there is no connection between the two down-converters, and hence no induced emission, the two signals will be incoherent and there will be no interference.

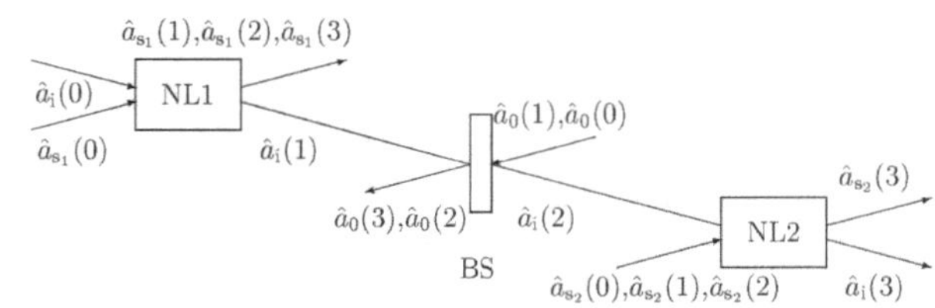

Fig. 2.11 Scheme of two parametric processes with aligned idler beams with the spatial Heisenberg picture made explicit

Zou et al. (1991) and Wang et al. (1991a) had a negligible probability of both crystals producing a down-converted photon pair and used a quantum-mechanical explanation based on indistinguishability of paths to explain the interference they observed in the experiment. To the contrary, the interference was lost when one could tell which crystal had emitted each signal photon. Using multimode analysis of the experiment, they derived that there could be no induced emission in their experiment. Nonetheless, they found that for perfect matching of idler modes, the signal fields from NL1 and NL2 show perfect interference.

The multimode approach to the analysis used by Wang et al. (1991a) yields an explanation involving a sufficient number of realistic parameters. Even though in the foregoing section the formalism yielded results similar enough to those of Wang et al. (1991a), here we try to come near their analysis. However, there exists a simple quantal description that may claim that it conforms to results of the multimode analysis. Such simple models have been published. Concerning this, we may refer to Řeháček and Peřina (1996), Wiseman and Mølmer (2000), and Peřinová et al. (2000) and provide what is a continuation of Peřinová et al. (2000).

(i) *Formalism of several modes*

We turn to the quantum analysis of the Zou–Wang–Mandel experiment. The experimental arrangement consists of two parametric down-conversion crystals with aligned idler beams, which are partially connected due to the presence of a beam splitter in between, and is illustrated in Fig. 2.11.

Restricting ourselves to the quasimonochromatic light beams (or quasimonochromatic components of these), we can describe the system by four modes, s_1 (the signal mode for crystal NL1), i (the idler modes, which are identified), s_2 (the signal mode for crystal NL2), and 0 (the escape mode for the beam splitter) (Peřinová et al. 2000). We consider the input annihilation operators $\hat{a}_{s_1}(0)$, $\hat{a}_0(1)$, $\hat{a}_{s_2}(2)$, $\hat{a}_i(0)$, the output annihilation operators $\hat{a}_{s_1}(1)$, $\hat{a}_0(2)$, $\hat{a}_{s_2}(3)$, $\hat{a}_i(3)$, and the intermediate annihilation operators $\hat{a}_i(1)$, $\hat{a}_i(2)$. Here s_1, s_2 stand for the signal mode of crystal 1 and that of crystal 2, respectively, i for the idler mode, and 0 for the "escape" mode of the beam splitter. To obtain four-mode unitary transformations between the stages 0, 1, 2, 3, we consider also the appendage input annihilation operators $\hat{a}_0(0)$, $\hat{a}_{s_2}(0)$, $\hat{a}_{s_2}(1)$ and the appendage output annihilation operators $\hat{a}_{s_1}(2)$, $\hat{a}_{s_1}(3)$, $\hat{a}_0(3)$. Of course, in the description of the dynamics below, we will have to be consistent with the identities $\hat{a}_0(0) = \hat{a}_0(1)$, $\hat{a}_{s_2}(0) = \hat{a}_{s_2}(1) = \hat{a}_{s_2}(2)$, $\hat{a}_{s_1}(1) = \hat{a}_{s_1}(2) = \hat{a}_{s_1}(3)$, $\hat{a}_0(2) = \hat{a}_0(3)$. Let us consider, in the Hilbert space of these four modes, an arbitrary operator $\hat{M}(j)$, $j = 0, 1, 2, 3$, a jth-stage operator. We will write the equation giving the transformation of $\hat{M}(j)$ from its value $\hat{M}(0)$ before the interaction to its value $\hat{M}(1)$ after the action of the first down converter, to its value $\hat{M}(2)$ after the action of the beam splitter, and to its value $\hat{M}(3)$ after the action of the second down converter.

The appropriate relations read

$$\hat{M}(j+1) = \hat{U}_{j+1}^{\dagger}(j)\hat{M}(j)\hat{U}_{j+1}(j), \quad j = 0, 1, 2. \tag{2.298}$$

Having prescribed equations of motion of operators, we have adopted a spatial modified Heisenberg picture of the dynamics. In the Heisenberg picture the input state does not change, while in the modified Heisenberg picture it changes like that of the free field. In our case of the "discrete" space (cf. $j = 0, 1, 2, 3$), the change of the state cannot be specified satisfactorily, but fortunately, we will not need it. In (2.298) $\hat{U}_{j+1}(j)$ for $j = 0, 2$ describe the down conversion in the undepleted pump approximation. The crystals are assumed to be identical or distinct and pumps are assumed to be identical so that

$$\hat{U}_{j+1}(j) = \exp\left\{\mathrm{i}\kappa_{\frac{j}{2}+1}\left[\hat{a}_{s_{\frac{j}{2}+1}}(j)\hat{a}_i(j) + \text{H.c.}\right]\right\}, \quad j = 0, 2, \tag{2.299}$$

where $\kappa_{\frac{j}{2}+1} = \frac{\chi}{c} v_{p_{\frac{j}{2}+1}} l_{\frac{j}{2}+1}$, χ is the quadratic susceptibility of the matter of which both the nonlinear crystals are made, $v_{p_{\frac{j}{2}+1}}$ are classical complex amplitudes of pumping beams, c is the speed of light, l_1 and l_2 are the lengths of the first crystal and the second crystal, respectively. In between the down converters the idler from crystal NL1 is put through a beam splitter BS and becomes the idler for crystal NL2. This process is described by

$$\hat{U}_2(1) = \exp\left\{\mathrm{i}[\bar{\omega}_0\hat{a}_0^{\dagger}(1)\hat{a}_0(1) + \bar{\omega}_i\hat{a}_i^{\dagger}(1)\hat{a}_i(1) + (\gamma^*\hat{a}_0^{\dagger}(1)\hat{a}_i(1) + \text{H.c.})]\right\}, \tag{2.300}$$

where

$$\left.\begin{matrix}\bar{\omega}_0 \\ \bar{\omega}_{\rm i}\end{matrix}\right\} = \arg\left(\frac{\mathtt{t}+\mathtt{t}'}{2}\right) \mp {\rm i}\frac{(\mathtt{t}-\mathtt{t}')}{2} f(\mathtt{t},\mathtt{t}'), \tag{2.301}$$

$$\gamma = -{\rm i}\mathtt{r} f(\mathtt{t},\mathtt{t}'), \tag{2.302}$$

with

$$f(\mathtt{t},\mathtt{t}') = \frac{\mathrm{Arccos}\left|\frac{\mathtt{t}+\mathtt{t}'}{2}\right|}{\sqrt{1-\left|\frac{\mathtt{t}+\mathtt{t}'}{2}\right|^2}} \frac{(\mathtt{t}+\mathtt{t}')^*}{|\mathtt{t}+\mathtt{t}'|}. \tag{2.303}$$

Here $\mathtt{t}$ and $\mathtt{r}$ are the transmission and reflection amplitude coefficients, respectively, for the idler mode and $\mathtt{t}'$ and $\mathtt{r}'$ are those for the "escape" mode. The modulus of the transmission amplitude coefficient $|\mathtt{t}|$ can vary between zero (where the second emission is spontaneous) and unity (where the second down conversion is stimulated in the highest degree).

Hence,

$$\begin{aligned}\hat{a}_{\rm i}(2) &= \hat{U}_2^\dagger(1)\hat{a}_{\rm i}(1)\hat{U}_2(1) = \mathtt{t}\hat{a}_{\rm i}(1) + \mathtt{r}'\hat{a}_0(1),\\ \hat{a}_0(2) &= \hat{U}_2^\dagger(1)\hat{a}_0(1)\hat{U}_2(1) = \mathtt{r}\hat{a}_{\rm i}(1) + \mathtt{t}'\hat{a}_0(1),\end{aligned} \tag{2.304}$$

and the unitarity of the transformation matrix implies that

$$|\mathtt{t}|^2 + |\mathtt{r}|^2 = 1, \quad |\mathtt{r}'|^2 + |\mathtt{t}'|^2 = 1, \quad \mathtt{t}\mathtt{r}'^* + \mathtt{r}\mathtt{t}'^* = 0. \tag{2.305}$$

It is advantageous to assume that $\mathtt{t}' = \mathtt{t}^*$, and from this $\mathtt{r} = -\mathtt{r}'^*$ (Řeháček and Peřina 1996), and that $\mathrm{Re}\,\mathtt{t} > 0$. Then

$$f(\mathtt{t},\mathtt{t}') = \frac{\mathrm{Arccos}(\mathrm{Re}\,\mathtt{t})}{\sqrt{1-(\mathrm{Re}\,\mathtt{t})^2}}, \tag{2.306}$$

$$\left.\begin{matrix}\bar{\omega}_0 \\ \bar{\omega}_{\rm i}\end{matrix}\right\} = \pm(\mathrm{Im}\,\mathtt{t}) f(\mathtt{t},\mathtt{t}'). \tag{2.307}$$

On applying the relation (2.298) at the input ($j = 0$) and at the stage 2 ($j = 2$), we obtain that

$$\begin{aligned}\hat{a}_{\rm s_1}(1) &= \hat{a}_{\rm s_1}(0)\cosh(\kappa_1) + {\rm i}\hat{a}_{\rm i}^\dagger(0)\sinh(\kappa_1),\\ \hat{a}_{\rm i}(1) &= {\rm i}\hat{a}_{\rm s_1}^\dagger(0)\sinh(\kappa_1) + \hat{a}_{\rm i}(0)\cosh(\kappa_1),\end{aligned} \tag{2.308}$$

and

$$\begin{aligned}\hat{a}_{\rm s_2}(3) &= \hat{a}_{\rm s_2}(2)\cosh(\kappa_2) + {\rm i}\hat{a}_{\rm i}^\dagger(2)\sinh(\kappa_2),\\ \hat{a}_{\rm i}(3) &= {\rm i}\hat{a}_{\rm s_2}(2)\sinh(\kappa_2) + \hat{a}_{\rm i}(2)\cosh(\kappa_2).\end{aligned} \tag{2.309}$$

Using Equations (2.304), we easily obtain the following relations:

$$\hat{a}_{s_1}(1) = \cosh(\kappa_1)\hat{a}_{s_1}(0) + \mathrm{i}\sinh(\kappa_1)\hat{a}_{\mathrm{i}}^{\dagger}(0), \tag{2.310}$$

$$\begin{aligned}\hat{a}_{s_2}(3) &= \mathtt{t}^* \sinh(\kappa_1)\sinh(\kappa_2)\hat{a}_{s_1}(0) + \mathrm{i}\mathtt{t}^* \cosh(\kappa_1)\sinh(\kappa_2)\hat{a}_{\mathrm{i}}^{\dagger}(0) \\ &\quad + \mathrm{i}\mathtt{t}^* \cosh(\kappa_2)\hat{a}_{s_2}(2) + \mathrm{i}\mathtt{r}'^* \sinh(\kappa_2)\hat{a}_0^{\dagger}(1).\end{aligned} \tag{2.311}$$

The statistical properties of the system in the Heisenberg picture can be obtained when we take into account that the initial, in fact, "permanent" statistical operator of the system is given as $\hat{\rho} \equiv \hat{\rho}(0)$ and when we average

$$\langle \hat{M}(j) \rangle = \mathrm{Tr}\{\hat{\rho}\hat{M}(j)\}. \tag{2.312}$$

Here, concretely, the statistical operator is a tensor product of separate vacuum statistical operators

$$\hat{\rho} = \prod_{j=s_1,\mathrm{i},s_2,0} |0\rangle_j \, {}_j\langle 0|. \tag{2.313}$$

We may introduce also the abbreviations $\hat{M} \equiv \hat{M}(0)$ and we consider the Schrödinger picture, where the relation (2.298) is replaced by the evolution relations

$$\hat{\rho}(j+1) = \hat{U}_{j+1}\hat{\rho}(j)\hat{U}_{j+1}^{\dagger}, \quad j = 0, 1, 2, \tag{2.314}$$

with $\hat{U}_j \equiv \hat{U}_j(0)$ given in (2.299) and (2.300). The equivalence of both the pictures can be proved and the statistical properties can be expressed in similar terms as in (2.312)

$$\langle \hat{M} \rangle(j) = \mathrm{Tr}\{\hat{\rho}(j)\hat{M}\} = \langle \hat{M}(j) \rangle. \tag{2.315}$$

Since all the initial fields are in the vacuum states, it is easy to obtain the expectation values

$$\langle \hat{a}_{s_1}^{\dagger}(1)\hat{a}_{s_1}(1) \rangle = \sinh^2(\kappa_1), \tag{2.316}$$

$$\langle \hat{a}_{s_2}^{\dagger}(3)\hat{a}_{s_2}(3) \rangle = \sinh^2(\kappa_2)[1 + |\mathtt{t}|^2 \sinh^2(\kappa_1)], \tag{2.317}$$

$$\langle \hat{a}_{s_1}^{\dagger}(1)\hat{a}_{s_2}(3) \rangle = \mathtt{t}^* \sinh(\kappa_1)\cosh(\kappa_1)\sinh(\kappa_2). \tag{2.318}$$

We will show in the Heisenberg picture that the input–output relation is connected to the SU(2,2) group. In fact,

$$\begin{pmatrix} \hat{a}_{s_1}(3) \\ \hat{a}_{\mathrm{i}}^{\dagger}(3) \\ \hat{a}_{s_2}(3) \\ \hat{a}_0^{\dagger}(3) \end{pmatrix} = \begin{pmatrix} m_{s_1s_1} & m_{s_1\mathrm{i}} & m_{s_1s_2} & m_{s_10} \\ m_{\mathrm{i}s_1} & m_{\mathrm{ii}} & m_{\mathrm{i}s_2} & m_{\mathrm{i}0} \\ m_{s_2s_1} & m_{s_2\mathrm{i}} & m_{s_2s_2} & m_{s_20} \\ m_{0s_1} & m_{0\mathrm{i}} & m_{0s_2} & m_{00} \end{pmatrix} \begin{pmatrix} \hat{a}_{s_1}(0) \\ \hat{a}_{\mathrm{i}}^{\dagger}(0) \\ \hat{a}_{s_2}(0) \\ \hat{a}_0^{\dagger}(0) \end{pmatrix}, \tag{2.319}$$

where

$$\begin{aligned} m_{s_1s_1} &= \cosh(\kappa_1), \quad m_{s_1i} = i\sinh(\kappa_1), \quad m_{s_1s_2} = m_{s_10} = 0; \\ m_{is_1} &= -it^* \sinh(\kappa_1)\cosh(\kappa_2), \quad m_{ii} = t^* \cosh(\kappa_1)\cosh(\kappa_2), \\ m_{is_2} &= -i\sinh(\kappa_2), \quad m_{i0} = r'^* \cosh(\kappa_2); \\ m_{s_2s_1} &= t^* \sinh(\kappa_1)\sinh(\kappa_2), \quad m_{s_2i} = it^* \cosh(\kappa_1)\sinh(\kappa_2), \\ m_{s_2s_2} &= \cosh(\kappa_2), \quad m_{s_20} = ir'^* \sinh(\kappa_2); \\ m_{0s_1} &= ir' \sinh(\kappa_1), \quad m_{0i} = -r' \cosh(\kappa_1), \quad m_{0s_2} = 0, \quad m_{00} = t. \end{aligned} \tag{2.320}$$

From the form of the relation (2.319) it is evident that the operator

$$\hat{N}(j) = \hat{n}_{s_1}(j) + \hat{n}_{s_2}(j) - \hat{n}_i(j) - \hat{n}_0(j), \quad j = 0, 3, \tag{2.321}$$

is independent of j. This conservation law suggests the SU(2,2) group. The coefficients of the transformation (2.319) verify the pseudoorthogonality relations

$$m_{js_1}m^*_{ks_1} + m_{js_2}m^*_{ks_2} - m_{ji}m^*_{ki} - m_{j0}m^*_{k0} = g_{jk}, \quad j, k = s_1, i, s_2, 0, \tag{2.322}$$

where

$$g_{jk} = g_{jj}\delta_{jk}, \quad g_{s_1s_1} = g_{s_2s_2} = 1, \quad g_{ii} = g_{00} = -1. \tag{2.323}$$

We observe that the antinormally ordered moments have the expression

$$\langle \hat{a}_j(3)\hat{a}^\dagger_j(3)\rangle = |m_{js_1}|^2 + |m_{js_2}|^2, \quad j = s_1, s_2, \tag{2.324}$$

and the normally ordered moments

$$\langle \hat{a}^\dagger_j(3)\hat{a}_j(3)\rangle = |m_{js_1}|^2 + |m_{js_2}|^2, \quad j = i, 0. \tag{2.325}$$

More generally,

$$\begin{aligned} \langle \hat{a}_{s_1}(3)\hat{a}^\dagger_{s_2}(3)\rangle &= m_{s_1s_1}m^*_{s_2s_1} + m_{s_1s_2}m^*_{s_2s_2}, \\ \langle \hat{a}_{s_2}(3)\hat{a}^\dagger_{s_1}(3)\rangle &= \langle \hat{a}_{s_1}(3)\hat{a}^\dagger_{s_2}(3)\rangle^*, \end{aligned} \tag{2.326}$$

$$\begin{aligned} \langle \hat{a}^\dagger_i(3)\hat{a}_0(3)\rangle &= m_{is_1}m^*_{0s_1} + m_{is_2}m^*_{0s_2}, \\ \langle \hat{a}^\dagger_0(3)\hat{a}_i(3)\rangle &= \langle \hat{a}^\dagger_i(3)\hat{a}_0(3)\rangle^*. \end{aligned} \tag{2.327}$$

Further nonvanishing moments are

$$\langle \hat{a}_j(3)\hat{a}_k(3)\rangle = m_{js_1}m^*_{ks_1} + m_{js_2}m^*_{ks_2}, \quad j = s_1, s_2, \quad k = i, 0, \tag{2.328}$$

and

$$\langle \hat{a}^\dagger_j(3)\hat{a}^\dagger_k(3)\rangle = \langle \hat{a}_j(3)\hat{a}_k(3)\rangle^*. \tag{2.329}$$

The rest second-order moments vanish:

$$\langle \hat{a}_j(3)\hat{a}_k^\dagger(3)\rangle = \langle \hat{a}_j^\dagger(3)\hat{a}_k(3)\rangle = 0, \quad j = \mathrm{s}_1, \mathrm{s}_2, \quad k = \mathrm{i}, 0, \tag{2.330}$$

$$\langle \hat{a}_j(3)\hat{a}_k(3)\rangle = \langle \hat{a}_j^\dagger(3)\hat{a}_k^\dagger(3)\rangle = 0, \quad j = \mathrm{s}_1, \mathrm{s}_2, \quad k = \mathrm{s}_1, \mathrm{s}_2, \quad \text{and} \quad j = \mathrm{i}, 0, \quad k = \mathrm{i}, 0. \tag{2.331}$$

Quantum statistics of radiation in the process under study is that of a four-mode Gaussian state, starting with the quantum characteristic function:

$$\begin{aligned} C_{\mathcal{S}}(\beta_{\mathrm{s}_1}, \beta_{\mathrm{s}_2}, \beta_{\mathrm{i}}, \beta_{0}, 3) &= \mathrm{Tr}\{\hat{\rho}(3)\hat{D}_{\mathrm{s}_1}(\beta_{\mathrm{s}_1}, 0)\hat{D}_{\mathrm{s}_2}(\beta_{\mathrm{s}_2}, 0)\hat{D}_{\mathrm{i}}(\beta_{\mathrm{i}}, 0)\hat{D}_{0}(\beta_{0}, 0)\} \\ &= \mathrm{Tr}\{\hat{\rho}\hat{D}_{\mathrm{s}_1}(\beta_{\mathrm{s}_1}, 3)\hat{D}_{\mathrm{s}_2}(\beta_{\mathrm{s}_2}, 3)\hat{D}_{\mathrm{i}}(\beta_{\mathrm{i}}, 3)\hat{D}_{0}(\beta_{0}, 3)\}, \end{aligned} \tag{2.332}$$

where the displacement operators are given by

$$\hat{D}_j(\beta_j, k) = \exp[\beta_j \hat{a}_j^\dagger(k) - \beta_j^* \hat{a}_j(k)], \quad j = \mathrm{s}_1, \mathrm{s}_2, \mathrm{i}, 0, \quad k = 0, 3. \tag{2.333}$$

By the remark above, $\hat{D}_j(\beta_j) \equiv \hat{D}_j(\beta_j, 0)$. On substituting into the relation (2.332) according to (2.319), we obtain that

$$\begin{aligned} &C_{\mathcal{S}}(\beta_{\mathrm{s}_1}, \beta_{\mathrm{s}_2}, \beta_{\mathrm{i}}, \beta_{0}, 3) \\ &= \mathrm{Tr}\{\hat{\rho}(0)\hat{D}_{\mathrm{s}_1}(\beta_{\mathrm{s}_1}(3))\hat{D}_{\mathrm{s}_2}(\beta_{\mathrm{s}_2}(3))\hat{D}_{\mathrm{i}}(\beta_{\mathrm{i}}(3))\hat{D}_{0}(\beta_{0}(3))\}, \end{aligned} \tag{2.334}$$

where

$$\begin{aligned} -\beta_{\mathrm{s}_1}^*(3) &= -\beta_{\mathrm{s}_1}^* m_{\mathrm{s}_1\mathrm{s}_1} - \beta_{\mathrm{s}_2}^* m_{\mathrm{s}_2\mathrm{s}_1} + \beta_{\mathrm{i}} m_{\mathrm{i}\mathrm{s}_1} + \beta_{0} m_{0\mathrm{s}_1}, \\ -\beta_{\mathrm{s}_2}^*(3) &= -\beta_{\mathrm{s}_1}^* m_{\mathrm{s}_1\mathrm{s}_2} - \beta_{\mathrm{s}_2}^* m_{\mathrm{s}_2\mathrm{s}_2} + \beta_{\mathrm{i}} m_{\mathrm{i}\mathrm{s}_2} + \beta_{0} m_{0\mathrm{s}_2}, \\ \beta_{\mathrm{i}}(3) &= -\beta_{\mathrm{s}_1}^* m_{\mathrm{s}_1\mathrm{i}} - \beta_{\mathrm{s}_2}^* m_{\mathrm{s}_2\mathrm{i}} + \beta_{\mathrm{i}} m_{\mathrm{i}\mathrm{i}} + \beta_{0} m_{0\mathrm{i}}, \\ \beta_{0}(3) &= -\beta_{\mathrm{s}_1}^* m_{\mathrm{s}_1 0} - \beta_{\mathrm{s}_2}^* m_{\mathrm{s}_2 0} + \beta_{\mathrm{i}} m_{\mathrm{i}0} + \beta_{0} m_{00}. \end{aligned} \tag{2.335}$$

From the known quantum characteristic function for the initial vacuum state

$$C_{\mathcal{S}}(\beta_{\mathrm{s}_1}, \beta_{\mathrm{s}_2}, \beta_{\mathrm{i}}, \beta_{0}, 0) = \exp\left\{-\frac{1}{2}\sum_{j=\mathrm{s}_1,\mathrm{s}_2,\mathrm{i},0} |\beta_j|^2\right\}, \tag{2.336}$$

we derive that

$$\begin{aligned} C_{\mathcal{S}}(\beta_{\mathrm{s}_1}, \beta_{\mathrm{s}_2}, \beta_{\mathrm{i}}, \beta_{0}, 3) &= \exp\left\{-\sum_{j=\mathrm{s}_1,\mathrm{s}_2,\mathrm{i},0} |\beta_j|^2 B_{j\mathcal{S}}\right. \\ &\left. + \left[-\beta_{\mathrm{s}_1}\beta_{\mathrm{s}_2}^* B_{\mathrm{s}_1\mathrm{s}_2}^* - \beta_{\mathrm{i}}\beta_{0}^* B_{\mathrm{i}0}^* + \sum_{j=\mathrm{s}_1,\mathrm{s}_2}\sum_{k=\mathrm{i},0} \beta_j\beta_k C_{jk}^* + \mathrm{c.c.}\right]\right\}. \end{aligned} \tag{2.337}$$

Here the coefficients $B_{j\mathcal{S}}$, B_{jk}, C_{jk} can be expressed in the form

$$\begin{aligned} B_{j\mathcal{S}} &= \langle \hat{a}_j(3)\hat{a}_j^\dagger(3)\rangle - \frac{1}{2}, \quad j = \mathrm{s}_1, \mathrm{s}_2, \\ B_{j\mathcal{S}} &= \langle \hat{a}_j^\dagger(3)\hat{a}_j(3)\rangle + \frac{1}{2}, \quad j = \mathrm{i}, 0, \\ B_{\mathrm{s}_1\mathrm{s}_2} &= \langle \hat{a}_{\mathrm{s}_1}(3)\hat{a}_{\mathrm{s}_2}^\dagger(3)\rangle, \\ B_{\mathrm{i}0} &= \langle \hat{a}_{\mathrm{i}}(3)\hat{a}_0^\dagger(3)\rangle, \\ C_{jk} &= \langle \hat{a}_j(3)\hat{a}_k(3)\rangle, \quad j = \mathrm{s}_1, \mathrm{s}_2, \quad k = \mathrm{i}, 0. \end{aligned} \tag{2.338}$$

Taking into account that $\langle \hat{a}_j(3)\rangle = 0$, $j = \mathrm{s}_1, \mathrm{s}_2, \mathrm{i}, 0$, we see that we are consistent with the more general notation

$$\begin{aligned} B_{j\mathcal{A}} &= \langle \Delta\hat{a}_j(3)\Delta\hat{a}_j^\dagger(3)\rangle, \quad j = \mathrm{s}_1, \mathrm{s}_2, \\ B_{j\mathcal{N}} &= \langle \Delta\hat{a}_j^\dagger(3)\Delta\hat{a}_j(3)\rangle, \quad j = \mathrm{i}, 0, \end{aligned} \tag{2.339}$$

where $\Delta\hat{a} = \hat{a} - \langle\hat{a}\rangle$, and with the coefficients $B_{\mathrm{s}_1\mathrm{s}_2}, B_{\mathrm{i}0}, C_{jk}, \quad j = \mathrm{s}_1, \mathrm{s}_2, \quad k = \mathrm{i}, 0,$ after similar replacement.

We confine ourselves to the study of the signal beams in what follows, which are described by the reduced statistical operator

$$\hat{\rho}_{\mathrm{signal}}(3) = \mathrm{Tr}_{\mathrm{i}}\mathrm{Tr}_0\{\hat{\rho}(3)\}, \tag{2.340}$$

where Tr_{i} and Tr_0 are partial traces over the idler and escape modes, respectively. Quantum characteristic function in the state described by the statistical operator (2.340) can easily be obtained:

$$C_{\mathcal{S}}(\beta_{\mathrm{s}_1}, \beta_{\mathrm{s}_2}) \equiv C_{\mathcal{S}}(\beta_{\mathrm{s}_1}, \beta_{\mathrm{s}_2}, 3) = C_{\mathcal{S}}(\beta_{\mathrm{s}_1}, \beta_{\mathrm{s}_2}, 0, 0, 3). \tag{2.341}$$

In Peřinová et al. (2003), the same function has been introduced as

$$\begin{aligned} C_{\mathcal{S}}(\beta_{\mathrm{s}_1}, \beta_{\mathrm{s}_2}; 1, 3) &= \mathrm{Tr}\left\{\hat{\rho}(0)\hat{D}_{\mathrm{s}_1}(\beta_{\mathrm{s}_1}, 1)\hat{D}_{\mathrm{s}_2}(\beta_{\mathrm{s}_2}, 3)\right\} \\ &= \exp\left[-\sum_{j=\mathrm{s}_1,\mathrm{s}_2} |\beta_j|^2 B_{j\mathcal{S}} + \left(-\beta_{\mathrm{s}_1}\beta_{\mathrm{s}_2}^* B_{\mathrm{s}_1\mathrm{s}_2}^* + \mathrm{c.c.}\right)\right]. \end{aligned} \tag{2.342}$$

In the following we simplify the notation s_1, s_2 for the signal modes to 1, 2, respectively. From the characteristic function

$$C_s(\beta_1, \beta_2) = \exp\left[\frac{s}{2}(|\beta_1|^2 + |\beta_2|^2)\right] C_{\mathcal{S}}(\beta_1, \beta_2), \tag{2.343}$$

where $s = 1, 0, -1$ in the subscript and also $s = \mathcal{N}, \mathcal{S}, \mathcal{A}$ denote the normal, symmetrical, and antinormal orderings of field operators, we can establish the Φ_s

quasidistribution related to the respective ordering of field operators

$$\Phi_s(\alpha_1, \alpha_2) = \frac{1}{\pi^4} \times \int C_s(\beta_1, \beta_2) \exp\left(\alpha_1\beta_1^* - \alpha_1^*\beta_1 + \alpha_2\beta_2^* - \alpha_2^*\beta_2\right) \, \mathrm{d}^2\beta_1 \, \mathrm{d}^2\beta_2. \tag{2.344}$$

After integrating, we obtain

$$\Phi_s(\alpha_1, \alpha_2) = \frac{1}{\pi^2 K_{12s}} \times \exp\left\{\frac{1}{K_{12s}}[-B_{2s}|\alpha_1|^2 - B_{1s}|\alpha_2|^2 + (B_{12}^*\alpha_1\alpha_2^* + \text{c.c.})]\right\}, \tag{2.345}$$

where

$$\begin{aligned}
B_{1\mathcal{A}} &= \cosh^2(\kappa_1),\ B_{1\mathcal{S}} = B_{1\mathcal{A}} - \frac{1}{2},\ B_{1\mathcal{N}} = B_{1\mathcal{A}} - 1,\\
B_{2\mathcal{A}} &= \cosh^2(\kappa_2) + |\mathtt{t}| \sinh^2(\kappa_2)\sinh^2(\kappa_1),\\
B_{2\mathcal{S}} &= B_{2\mathcal{A}} - \frac{1}{2},\ B_{2\mathcal{N}} = B_{2\mathcal{A}} - 1,\\
B_{12}^* &= \mathtt{t}^* \sinh(\kappa_1)\cosh(\kappa_1)\sinh(\kappa_2),
\end{aligned} \tag{2.346}$$

and

$$K_{12s} = B_{1s}B_{2s} - |B_{12}|^2. \tag{2.347}$$

Especially, for $s = -1$ it holds that

$$\Phi_{\mathcal{A}}(\alpha_1, \alpha_2) = \frac{1}{\pi^2}\langle\alpha_1, \alpha_2|\hat{\rho}_{\text{signal}}(3)|\alpha_1, \alpha_2\rangle, \tag{2.348}$$

where $|\alpha_1, \alpha_2\rangle$ is the two-mode coherent state, which yields the expansion

$$\Phi_{\mathcal{A}}(\alpha_1, \alpha_2) = \frac{1}{\pi^2}\exp(-|\alpha_1|^2 - |\alpha_2|^2) \sum_{q=-\infty}^{\infty} \sum_{m_1=\max(0,-q)}^{\infty} \times \sum_{n_2=\max(0,-q)}^{\infty} \rho(m_1+q, m_1, n_2, n_2+q) \frac{\alpha_1^{*(m_1+q)}\alpha_1^{m_1}\alpha_2^{*n_2}\alpha_2^{n_2+q}}{\sqrt{(m_1+q)!m_1!n_2!(n_2+q)!}} \tag{2.349}$$

for any $\Phi_{\mathcal{A}}$ quasidistribution that does not depend on $\alpha_1\alpha_2$, $\alpha_1^*\alpha_2^*$. Here

$$\rho(n_1, m_1, n_2, m_2) = \langle n_1, n_2|\hat{\rho}_{\text{signal}}(3)|m_1, m_2\rangle \tag{2.350}$$

are the usual matrix elements. Equating the expansion coefficients for (2.345) with $s = -1$ and those of (2.349), we arrive at the expression

$$\rho(m_1+q, m_1, n_2, n_2+q) = \sum_{p=\max(0,-q)}^{\min(m_1,n_2)} \frac{\sqrt{(m_1+q)!m_1!n_2!(n_2+q)!}}{(m_1-p)!(n_2-p)!p!(p+q)!} \times \frac{(K_{12\mathcal{A}}-B_{2\mathcal{A}})^{m_1-p}(K_{12\mathcal{A}}-B_{1\mathcal{A}})^{n_2-p} B_{12}^{*p} B_{12}^{p+q}}{K_{12\mathcal{A}}^{m_1+n_2+q+1}}, \quad (2.351)$$

while obviously

$$\rho(m_1+q_1, m_1, n_2, n_2-q_2) = 0 \;\text{ for }\; q_1 \neq -q_2. \quad (2.352)$$

(ii) *Photon-number statistics*

Numbers of photons in signal modes complete the picture of the quantum correlation between these beams. The joint photon-number distribution $p(n_1, n_2)$ can be expressed in the concise form (Řeháček and Peřina 1996). A substitution into (2.351) leads to slightly more complicated expression:

$$p(n_1, n_2) = \rho(n_1, n_1, n_2, n_2). \quad (2.353)$$

This distribution can be seen in Fig. 2.12 for $|\mathtt{t}| = 1$, $B_{1\mathcal{N}} = B_{2\mathcal{N}} = 3$, $|B_{12}| = 3$. It differs from the product of pertinent marginal distributions by larger "diagonal" probabilities. To the contrary for $|\mathtt{t}|$ small, the joint photon-number distribution is approximately the product of its marginal photon-number distributions.

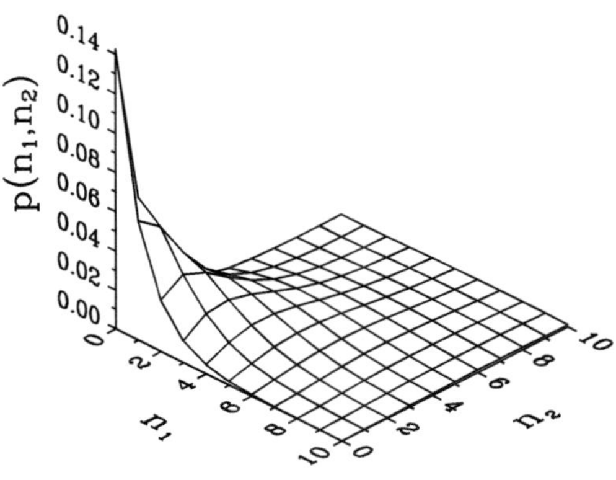

Fig. 2.12 Joint photon-number distribution for $|\mathtt{t}| = 1$; $B_{1\mathcal{N}} = B_{2\mathcal{N}} = 3$, $n_j \in [0, 10]$, $j = 1, 2$

As for the experimental arrangement under study, it depends on l_1, $\mathtt{t}$, l_2, whereas the numerical demonstration is restricted to the case when the length of the first crystal is kept fixed. Consequently, the mean photon number $B_{1\mathcal{N}}$ in the first signal mode is constant and this convenient behaviour is, for the sake of illustrations, extended also to the second one as a relationship between $|\mathtt{t}|$ and κ_2,

$$\sinh^2(\kappa_2) = \frac{B_{2\mathcal{N}}}{1 + |\mathtt{t}|^2 B_{1\mathcal{N}}}. \tag{2.354}$$

The Glauber degree of coherence (Peřina 1991) $\gamma_{12}^{(2)}$ is the complex-valued quantum correlation measure related to the normal ordering:

$$\gamma_{12}^{(2)} = \frac{B_{12}^*}{\sqrt{B_{1\mathcal{N}} B_{2\mathcal{N}}}}. \tag{2.355}$$

In the numerical illustration of the quantum correlation measures, we assume $\kappa_1 = \kappa_2 = \kappa$ and find κ_1 by the inversion of the formula

$$B_{1\mathcal{N}} = \sinh^2(\kappa_1) = \langle \hat{a}_{s_1}^\dagger(1)\hat{a}_{s_1}(1)\rangle = \overline{n}_1(1). \tag{2.356}$$

The limit case $|\mathtt{t}| = 1$ is interesting, $|\gamma_{12}^{(2)}| = R_{\mathcal{N}} = 1$, see Fig. 2.13. This maximum correlation does not correspond to a weaker correlation between the signal photon numbers. In the multimode analysis (Wang et al. 1991a) of the experiment (Zou et al. 1991), the visibility of the interference between the signal fields has been expressed, the interference manifests itself as oscillations in the counting rate I_s (see (2.412) below) when propagation times of the idler beam from NL1 to NL2, of the first signal beam from NL1 to D_s, and of the second signal beam from NL2 to D_s are incremented by $\delta\tau_0$, $\delta\tau_1$, $\delta\tau_2$, respectively. Deriving a simplified visibility $\mathcal{V}_{\text{simple}}$ for the formalism of several modes provides

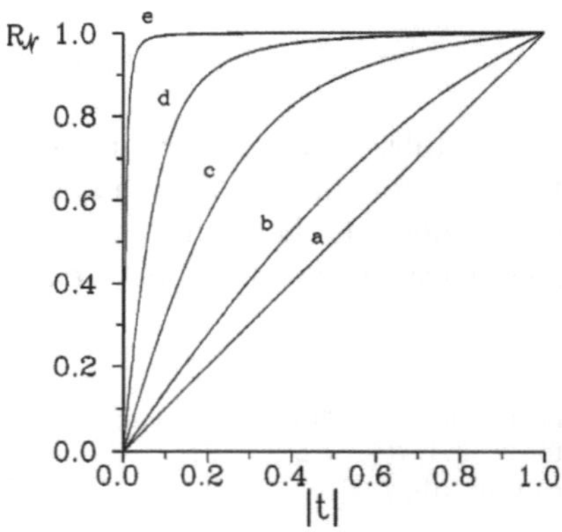

Fig. 2.13 Quantum correlation measure $R_{\mathcal{N}}$ versus the modulus of the transmission amplitude coefficient $|\mathtt{t}| \in [0, 1]$; it is assumed that $\overline{n}_1(1) = 10^{-2}, 1, 10, 100, 10^4$ (the curves a, b, c, d, e, respectively)

$$\mathcal{V}_{\text{simple}} = \frac{2R_{\mathcal{N}}\sqrt{B_{1\mathcal{N}} B_{2\mathcal{N}}}}{B_{1\mathcal{N}} + B_{2\mathcal{N}}}. \tag{2.357}$$

As $\sqrt{B_{1\mathcal{N}} B_{2\mathcal{N}}} \leq \frac{1}{2}(B_{1\mathcal{N}} + B_{2\mathcal{N}})$, the visibility cannot exceed the correlation measure $R_{\mathcal{N}}$ and the equality is attained for $B_{1\mathcal{N}} = B_{2\mathcal{N}}$. The maximum obtainable visibility between two fields in an experiment is given by the correlation measure $R_{\mathcal{N}}$, cf. Wiseman and Mølmer (2000).

On substituting (2.338) with (2.316), (2.317), and (2.318) into (2.355), we find that

$$R_{\mathcal{N}} = \frac{|\mathtt{t}|\cosh(\kappa_1)}{\sqrt{1+|\mathtt{t}|^2\sinh^2(\kappa_1)}}. \tag{2.358}$$

Noting that the idler beam, before it enters the beam splitter, has the same statistics as the output signal 1, we can rewrite (2.358) in terms of the mean photon number $\bar{n}_1(1) = \sinh^2(\kappa_1)$ as

$$R_{\mathcal{N}} = |\mathtt{t}|\sqrt{\frac{1+\overline{n}_1(1)}{1+|\mathtt{t}|^2\overline{n}_1(1)}}. \tag{2.359}$$

Wiseman and Mølmer (2000) considered the relevant limits in this form. The single-photon regime which is the regime of experiment and theory in Zou et al. (1991) and Wang et al. (1991a,b) occurs for $\overline{n}_1(1) \ll 1$. Up to the first order in the rescaled lengths κ_1, κ_2 of the crystals, we simply obtain $R_{\mathcal{N}} = |\mathtt{t}|$. The probability of a down conversion at crystal NL1 over interaction time is less than or equal to $\overline{n}_1(1)$, or it is small. The probability to have down conversions at both crystals over interaction time is less than or equal to $\left\langle \hat{a}^\dagger_{s_1}(1)\hat{a}_{s_1}(1)\hat{a}^\dagger_{s_2}(3)\hat{a}_{s_2}(3)\right\rangle = B_{1\mathcal{N}}B_{2\mathcal{N}} + |B_{12}|^2$, or it is negligible. The single-photon regime applies in the multimode analysis of Wang et al. (1991a), because each of the narrow-bandwidth signal modes $(\mathbf{k}_{s_1}, \omega_{s_1})$, $(\mathbf{k}_{s_2}, \omega_{s_2})$, with directions characterized by $\mathbf{k}_{s_j}$ and with the frequencies ω_{s_j}, $j = 1, 2$, that form broad-band signal fields, receives only a small part of the pumping photons over interaction time. The same applies to idler modes and idler fields. The signal fields s_1 and s_2 from the two down converters are allowed to come together and interfere at the detector D_s.

In the spatial interaction picture, the state of the field produced by the crystals is given by

$$|\psi(3)\rangle \equiv \hat{U}_3(0)\hat{U}_2(0)\hat{U}_1(0)|0\rangle_{s_1,i,s_2,0}, \tag{2.360}$$

where $\hat{U}_{j+1}(0)$ for $j = 1, 2, 3$ are given by relations (2.299) and (2.300), where the annihilation operators $\hat{a}_{s_{\frac{j}{2}+1}}(j) \to \hat{a}_{s_{\frac{j}{2}+1}}(0)$, $\hat{a}_i(j) \to \hat{a}_i(0)$, $\hat{a}_0(j) \to \hat{a}_0(0)$. We will drop the argument (0) at the annihilation operators in what follows. Here $|0\rangle_{s_1,i,s_2,0} \equiv |0\rangle_{s_1}|0\rangle_i|0\rangle_{s_2}|0\rangle_0$ and, in general,

$$|n_{s_1}, n_i, n_{s_2}, n_0\rangle \equiv |n_{s_1}\rangle_{s_1}|n_i\rangle_i|n_{s_2}\rangle_{s_2}|n_0\rangle_0. \tag{2.361}$$

In the Schrödinger picture, the operators do not change and in the interaction picture, which is the modified Schrödinger picture, they change like the Heisenberg picture free-field operators. An analogue of relation (2.298) for a discrete space is not used in quantum optics (the free-field propagation is absorbed in the interaction). Fortunately, we will use just the interaction-picture annihilation operators. Expanding the operators $\hat{U}_{j+1}(0)$ according to $\kappa_{\frac{j}{2}+1}$, $j = 0, 2$, we obtain that

$$|\psi(3)\rangle \simeq |0\rangle_{\mathrm{s}_1,\mathrm{i},\mathrm{s}_2,0} + \mathrm{i}\kappa_2|0, 1, 1, 0\rangle + \mathrm{i}\kappa_1(\mathtt{t}|1, 1, 0, 0\rangle + \mathtt{r}|1, 0, 0, 1\rangle), \quad (2.362)$$

when $\kappa_{\frac{j}{2}+1}$ are small. For $|\mathtt{t}| = 1$ we have a single-photon state in the idler mode and in the collection of the signal modes. In general, one can infer a conversion at crystal NL1 after a photocount in the escape mode. We introduce the probability of the detection

$$p_{1,0,0,1}(3) = |\kappa_1|^2|\mathtt{r}|^2. \quad (2.363)$$

Let us assume that one infers a conversion at crystal NL2 after no photocounts in the escape mode. We introduce the probability of a correct inference of the conversion at crystal NL2

$$p_{0,1,1,0}(3) = |\kappa_2|^2 \quad (2.364)$$

and that of such a wrong inference

$$p_{1,1,0,0}(3) = |\kappa_1|^2|\mathtt{t}|^2. \quad (2.365)$$

On a photocount in the escape mode it is certain that the conversion has happened at NL1. On no photocounts in this mode, the posterior probabilities are

$$\mathrm{Prob}(\underline{n}_{\mathrm{s}_1} = 0 \cap \underline{n}_{\mathrm{s}_2} = 1|\underline{n}_{\mathrm{i}} = 1 \cap \underline{n}_0 = 0) = \frac{|\kappa_2|^2}{|\kappa_1|^2|\mathtt{t}|^2 + |\kappa_2|^2}, \quad (2.366)$$

$$\mathrm{Prob}(\underline{n}_{\mathrm{s}_1} = 1 \cap \underline{n}_{\mathrm{s}_2} = 0|\underline{n}_{\mathrm{i}} = 1 \cap \underline{n}_0 = 0) = \frac{|\kappa_1|^2|\mathtt{t}|^2}{|\kappa_1|^2|\mathtt{t}|^2 + |\kappa_2|^2}. \quad (2.367)$$

Here the underlining means a random variable. The counting rate registered by D_{s} exhibits perfect interference when the idler fields are perfectly aligned. This may be regarded as reflecting the intrinsic impossibility of knowing whether the detected photon comes from NL1 or NL2 (Wang et al. 1991a). The multiphoton conditional states can be found in Luis and Peřina (1996a).

Let us consider the annihilation operators $\hat{a}_{\mathrm{s}_{\frac{j}{2}+1}}$, $j = 0, 2$. The action of the two operators on the state $|\psi(3)\rangle$ is asymptotically for small $\kappa_{\frac{j}{2}+1}$ expressed as

$$\hat{a}_{\mathrm{s}_1}|\psi(3)\rangle \simeq \mathrm{i}\kappa_1(\mathtt{t}|0, 1, 0, 0\rangle + \mathtt{r}|0, 0, 0, 1\rangle), \quad (2.368)$$

$$\hat{a}_{\mathrm{s}_2}|\psi(3)\rangle \simeq \mathrm{i}\kappa_2|0, 1, 0, 0\rangle. \quad (2.369)$$

Hence

$$\langle\hat{a}^\dagger_{\mathrm{s}_1}\hat{a}_{\mathrm{s}_2}\rangle \simeq \kappa_1\kappa_2\mathtt{t}^*, \quad (2.370)$$

$$\langle\hat{a}^\dagger_{\mathrm{s}_1}\hat{a}_{\mathrm{s}_1}\rangle \simeq \kappa_1^2, \quad \langle\hat{a}^\dagger_{\mathrm{s}_2}\hat{a}_{\mathrm{s}_2}\rangle \simeq \kappa_2^2. \quad (2.371)$$

It can be verified that these relations hold up to the second order in κ_1, κ_2. We can find that $\gamma_{12}^{(2)} = \mathtt{t}^*$ in the limit of small $\kappa_{\frac{j}{2}+1}$, $j = 0, 2$. Obviously, the equation holds up to the zeroth order, but it can be verified that it is valid up to the first order in κ_1, κ_2.

The opposite regime is that where $\overline{n}_1(1) \gg 1$. Here there are many photons on average in all of the down-converted beams. That is, the phase of the stage-1 or stage-2 idler mode should lock the phase of the output signal-2 mode for any nonzero transmission amplitude coefficient $\mathtt{t}$.

(iii) *Multimode formalism*

In Wang et al. (1991a) the pump beams at each crystal are represented by complex analytic signals $V_1(t)$ and $V_2(t)$ such that $|V_j(t)|^2$ is in units of photons per second ($j = 1, 2$). The multimode formalism enables one to respect that the two crystals are centred at $\mathbf{0}_1$ and $\mathbf{0}_2$. The multimode formalism views the electric fields as temporal-interaction-picture operators

$$\hat{E}_m^{(+)}(\mathbf{r}, t) = \sqrt{\frac{\delta\omega}{2\pi}} \sum_{\omega_m} \hat{a}_m(\omega_m) \exp[\mathrm{i}(\mathbf{k}_m \cdot \mathbf{r} - \omega_m t)], \quad m = \mathrm{s}_1, \mathrm{i}, \mathrm{s}_2, 0, \tag{2.372}$$

where $\delta\omega$ is the mode spacing and $\hat{a}_m(\omega_m)$ is the photon annihilation operator for narrow-bandwidth signal ($m = \mathrm{s}_j$, $j = 1, 2$), idler ($m = \mathrm{i}$), and "escape" ($m = 0$) modes $(\mathbf{k}_m, \omega_m)$ at the frequency ω_m. The Hilbert space for the multimode analysis is a tensor product of those Hilbert spaces of separate modes, whose vacuum states may be designated as $|0\rangle_m(\omega_m)$, $m = \mathrm{s}_1, \mathrm{i}, \mathrm{s}_2, 0$. We consider photon-flux amplitude operators $\hat{E}_{\mathrm{s}_j}^{(+)}(t)$, $j = 1, 2$, at the appropriate detector, $\hat{E}_{\mathrm{s}_j}^{(+)}(t) \equiv \hat{E}_{\mathrm{s}_j}(\mathbf{r}_\mathrm{s}, t) = \hat{E}_{\mathrm{s}_j}^{(+)}(\mathbf{0}_j, t - \tau_j)$, where τ_j, $j = 1, 2$, is the propagation time of s_j from NLj to D_s.

In order to compare the sophisticated multimode formalism with the simple formalism of several modes, we must present another appropriate description of the dynamics of the same down-conversion experiment. We adopt the temporal interaction picture and combine it with the spatial interaction picture. In this case, the state produced by the crystal is given by

$$|\psi(3, t)\rangle \equiv \hat{U}_3(0, t)\hat{U}_2(0, t)\hat{U}_1(0, t)|0\rangle_{\mathrm{s}_1,\mathrm{i},\mathrm{s}_2,0}. \tag{2.373}$$

Here $|0\rangle_{\mathrm{s}_1,\mathrm{i},\mathrm{s}_2,0}$ is the vacuum state of all the narrow-bandwidth signal, idler, and "escape" modes $(\mathbf{k}_{\mathrm{s}_j}, \omega_{\mathrm{s}_j})$, $j = 1, 2$, $(\mathbf{k}_\mathrm{i}, \omega_\mathrm{i})$, $(\mathbf{k}_0, \omega_0)$, respectively. $\hat{U}_{j+1}(0, t)$, $j = 0, 2$, are unitary operators:

$$\hat{U}_{j+1}(0, t) = \lim_{t_0 \to -\infty} \hat{U}_{j+1}(0, t, t_0), \tag{2.374}$$

where $\hat{U}_{j+1}(0, t, t_0)$ are unitary operators that obey the initial condition

$$\hat{U}_{j+1}(0, t, t_0)\big|_{t=t_0} = \hat{1}, \tag{2.375}$$

$\hat{U}_{j+1}(0,t,t_0)$ for $j = 0, 2$ describe the parametric down conversion and are expressed, indirectly, in terms of $\hat{E}^{(+)}_{\mathrm{s}_{\frac{j}{2}+1}}(\mathbf{r},t)$, $\hat{E}^{(+)}_{\mathrm{i}}(\mathbf{r},t)$; $\hat{U}_2(0,t)$ describes the beam splitter and is expressed as

$$\hat{U}_2(0,t) = \exp\left\{\mathrm{i}\sum_{\omega''}\left[\overline{\omega}_0\hat{a}_0^\dagger(\omega'')\hat{a}_0(\omega'') + \overline{\omega}_\mathrm{i}\hat{a}_\mathrm{i}^\dagger(\omega'')\hat{a}_\mathrm{i}(\omega'') + \left(\gamma^*\hat{a}_0^\dagger(\omega'')\hat{a}_\mathrm{i}(\omega'') + \mathrm{H.c.}\right)\right]\right\}. \tag{2.376}$$

In this point we differ from the paper by Wang et al. (1991a), who used the initial condition at $t' = t - t_1$ and they did not write down the decomposition into stages. Let $\mathcal{T}$ denote the time ordering. We will introduce the unitary operator

$$\hat{U}_3(0,t,t-t_1) = \mathcal{T}\exp\left\{\int_{t-t_1}^{t}\left[\nu_2 V_2(0)\delta\omega\sum_{\omega'}\sum_{\omega''}\phi(\omega_0-\omega'',\omega'') \times e^{-\mathrm{i}(\mathbf{k}'_{\mathrm{s}_2}+\mathbf{k}'')\cdot\mathbf{0}_2}e^{-\mathrm{i}(\omega_0-\omega'-\omega'')t'}\hat{a}^\dagger_{\mathrm{s}_2}(\omega')\hat{a}^\dagger_\mathrm{i}(\omega'') - \mathrm{H.c.}\right]\mathrm{d}t'\right\}, \tag{2.377}$$

where ν_j, $j = 1, 2$, is a constant such that $|\nu_j|^2$ gives the fraction of incident pump photons that is spontaneously down converted in the steady state, ω_0 is the frequency of the monochromatic pump beam, $\mathbf{k}'_{\mathrm{s}_2}$ ($\mathbf{k}''$) is a wave vector that is determined by the frequency ω'_{s_2} (ω'') and the direction of the second signal beam (the idler beam). To the first order in the processes the unitary operator may be expressed as

$$\hat{U}_3(0,t,t-t_1) = \hat{1} + \left\{\nu_2 V_2(0)\delta\omega\sum_{\omega'}\sum_{\omega''}\phi(\omega_0-\omega'',\omega'') \times e^{-\mathrm{i}(\mathbf{k}'_{\mathrm{s}_2}+\mathbf{k}'')\cdot\mathbf{0}_2}\frac{\sin\left[\frac{1}{2}(\omega_0-\omega'-\omega'')t_1\right]}{\frac{1}{2}(\omega_0-\omega'-\omega'')} \times \exp\left[-\mathrm{i}(\omega_0-\omega'-\omega'')\left(t-\frac{t_1}{2}\right)\right]\hat{a}^\dagger_{\mathrm{s}_2}(\omega')\hat{a}^\dagger_\mathrm{i}(\omega'') - \mathrm{H.c.}\right\}. \tag{2.378}$$

From this we obtain the vector

$$\begin{aligned}|\psi(3,t,t-t_1)\rangle &= \hat{U}_3(0,t,t-t_1)|\psi(2,t,t-t_1)\rangle \\ &= |\psi(2,t,t-t_1)\rangle + \nu_2 V_2(0)\delta\omega\sum_{\omega'}\sum_{\omega''}\phi(\omega_0-\omega'',\omega'') \\ &\quad\times e^{-\mathrm{i}(\mathbf{k}'_{\mathrm{s}_2}+\mathbf{k}'')\cdot\mathbf{0}_2}\frac{\sin\left[\frac{1}{2}(\omega_0-\omega'-\omega'')t_1\right]}{\frac{1}{2}(\omega_0-\omega'-\omega'')} \\ &\quad\times\exp\left[-\mathrm{i}(\omega_0-\omega'-\omega'')\left(t-\frac{t_1}{2}\right)\right]|\omega'\rangle_{\mathrm{s}_2}|\omega''\rangle_\mathrm{i}|0\rangle_{\mathrm{s}_1,0},\end{aligned} \tag{2.379}$$

where $|\omega'\rangle_{s_2}$ and $|\omega''\rangle_i$ are frequency eigenstates of the second signal and the idler beam, respectively, $|0\rangle_{s_1,0}$ is the vacuum state of the first signal and escape modes, and

$$
\begin{aligned}
|\psi(2,t,t-t_1)\rangle &= \hat{U}_2(0,t)\hat{U}_1(0,t,t-t_1)|0\rangle_{s_1,i,s_2,0} \\
&= \hat{U}_2(0,t)\hat{U}_1(0,t,t-t_1)\hat{U}_2^{\dagger}(0,t)|0\rangle_{s_1,i,s_2,0}.
\end{aligned} \tag{2.380}
$$

We transform the vector $|\psi(3,t,t-t_1)\rangle$ to a vector

$$
\begin{aligned}
\hat{E}_{s_2}(t)|\psi(3,t,t-t_1)\rangle &= \nu_2 V_2(0)\sqrt{\frac{\delta\omega}{2\pi}}\delta\omega\sum_{\omega'}\sum_{\omega''}\phi(\omega_0-\omega'',\omega'') \\
&\times e^{-i\mathbf{k}''\cdot\mathbf{0}_2}\frac{\sin\left[\frac{1}{2}(\omega_0-\omega'-\omega'')t_1\right]}{\frac{1}{2}(\omega_0-\omega'-\omega'')} \\
&\times \exp\left[-i(\omega_0-\omega'-\omega'')\left(\tau_2-\frac{t_1}{2}\right)\right]\exp[-i(\omega_0-\omega'')(t-\tau_2)] \\
&\times |\omega''\rangle_i|0\rangle_{s_1,s_2,0}
\end{aligned} \tag{2.381}
$$

$$
\begin{aligned}
&\simeq \nu_2 V_2(0)\sqrt{\frac{\delta\omega}{2\pi}}\sum_{\omega''}\phi(\omega_0-\omega'',\omega'') \\
&\times e^{-i\mathbf{k}''\cdot\mathbf{0}_2}\int_{-\infty}^{\infty}\frac{\sin\left[\frac{1}{2}(\omega_0-\omega'-\omega'')t_1\right]}{\frac{1}{2}(\omega_0-\omega'-\omega'')} \\
&\times \exp\left[-i(\omega_0-\omega'-\omega'')\left(\tau_2-\frac{t_1}{2}\right)\right]d\omega'\exp[-i(\omega_0-\omega'')(t-\tau_2)] \\
&\times |\omega''\rangle_i|0\rangle_{s_1,s_2,0}
\end{aligned} \tag{2.382}
$$

$$
= \nu_2 V_2(t-\tau_2)|1(\mathbf{0}_2,t-\tau_2)\rangle_i|0\rangle_{s_1,s_2,0}, \tag{2.383}
$$

where the single-photon state of the idler beam

$$
\begin{aligned}
|1(\mathbf{r},t)\rangle_i = \sqrt{2\pi\delta\omega}&\sum_{\omega''}\phi(\omega_0-\omega'',\omega'') \\
&\times e^{-i(\mathbf{k}''\cdot\mathbf{r}-\omega''t)}|\omega''\rangle_i,
\end{aligned} \tag{2.384}
$$

$\phi(\tilde{\omega}'',\omega'')$ is connected with spectral functions $\phi_j(\omega',\omega'';\omega)$ characterizing the signal and idler fields at any crystal NLj,

$$
\begin{aligned}
\phi(\tilde{\omega},\omega) &= \phi_1(\tilde{\omega},\omega) = \phi_2(\tilde{\omega},\omega), \\
\phi_j(\tilde{\omega}'',\omega'') &= \phi_j(\tilde{\omega}'',\omega'';\omega_0), \quad j=1,2.
\end{aligned} \tag{2.385}
$$

The frequency eigenstates are single-photon states:

$$
|\omega''\rangle_m = \hat{a}_m^{\dagger}(\omega'')|0\rangle_m, \tag{2.386}
$$

where

$$|0\rangle_m = \bigotimes_{\omega''} |0\rangle_m(\omega''), \quad m = \mathrm{i}, 0. \tag{2.387}$$

Unfortunately, the nonvanishing result is obtained only for

$$0 < \tau_2 < t_1. \tag{2.388}$$

To resolve this, Wang et al. (1991a) let $t_1 \to \infty$. Should t_1 mean the interaction time, it is better to change the integration limits, namely not to consider the integration interval $[t - t_1, t]$, but, for instance,

$$[t - (K_2 + 1)t_1, t - K_2 t_1], \tag{2.389}$$

for

$$K_2 t_1 < \tau_2 < (K_2 + 1)t_1 \tag{2.390}$$

to hold.

We further calculate

$$\hat{E}_{\mathrm{s}_1}(t)|\psi(3, t, t - t_1)\rangle = \hat{E}_{\mathrm{s}_1}(t)|\psi(2, t, t - t_1)\rangle. \tag{2.391}$$

We obtain the appropriate component of the vector $|\psi(2, t, t - t_1)\rangle$ by the action of the unitary operator

$$\begin{aligned} \hat{U}_2(0, t)\hat{U}_1(0, t, t - t_1)\hat{U}_2^\dagger(0, t) = \mathcal{T} \exp\Bigg\{ \int_{t-t_1}^{t} \Big[\nu_1 V_1(0)\delta\omega \\ \times \sum_{\omega'} \sum_{\omega''} \phi(\omega_0 - \omega'', \omega'') e^{-\mathrm{i}(\mathbf{k}'_{\mathrm{s}_1} + \mathbf{k}'')\cdot \mathbf{0}_1} e^{-\mathrm{i}(\omega_0 - \omega' - \omega'')t'} \\ \times \hat{a}^\dagger_{\mathrm{s}_1}(\omega')[\mathtt{t}^* \hat{a}_\mathrm{i}(\omega'') + \mathtt{r}^* \hat{a}_0(\omega'')]^\dagger - \mathrm{H.c.} \Big] \mathrm{d}t' \Bigg\}. \end{aligned} \tag{2.392}$$

The calculation proceeds similarly as in the case of NL2, but we replace $\hat{a}_\mathrm{i}^\dagger(\omega'')$ by $\mathtt{t}\hat{a}_\mathrm{i}^\dagger(\omega'') + \mathtt{r}\hat{a}_0^\dagger(\omega'')$, $|\omega''\rangle_\mathrm{i}$ by $\mathtt{t}|\omega''\rangle_\mathrm{i} + \mathtt{r}|\omega''\rangle_0$ and we change all the other subscripts that underlie to a change, so that

$$\begin{aligned} \hat{E}_{\mathrm{s}_1}(t)|\psi(3, t, t - t_1)\rangle = \nu_1 V_1(\mathbf{0}_1, t - \tau_1)\big[\mathtt{t}|1(\mathbf{0}_1, t - \tau_1)\rangle_\mathrm{i} |0\rangle_{\mathrm{s}_1,\mathrm{s}_2,0} \\ + \mathtt{r}|1(\mathbf{0}_1, t - \tau_1)\rangle_0 |0\rangle_{\mathrm{s}_1,\mathrm{i},\mathrm{s}_2}\big], \end{aligned} \tag{2.393}$$

where $|0\rangle_{\mathrm{s}_1,\mathrm{s}_2,0} \equiv |\psi_\mathrm{vac}\rangle_{\mathrm{s}_1,\mathrm{s}_2,0}$, $|0\rangle_{\mathrm{s}_1,\mathrm{i},\mathrm{s}_2} \equiv |\psi_\mathrm{vac}\rangle_{\mathrm{s}_1,\mathrm{i},\mathrm{s}_2}$ stand for vacuum states, $|1(\mathbf{r}, t)\rangle_\mathrm{i}$ is defined in (2.384), and $|1(\mathbf{r}, t)\rangle_0$ stands for single-photon state of the "escape" beam

$$|1(\mathbf{r}, t)\rangle_0 \equiv \sqrt{2\pi\delta\omega} \sum_{\omega''} \phi(\omega_0 - \omega'', \omega'') \exp[-\mathrm{i}(\mathbf{k}'' \cdot \mathbf{r} - \omega'' t)] |\omega''\rangle_0. \tag{2.394}$$

Here a nonvanishing result is obtained only for

$$0 < \tau_1 < t_1. \tag{2.395}$$

Considering a change of the integration limits as above, we see that, to the first order of the processes, no difficulties arise if we change the limits independently of NL2. We do not consider the integration interval $[t - t_1, t]$, but, for instance,

$$[t - (K_1 + 1)t_1, t - K_1 t_1], \tag{2.396}$$

for

$$K_1 t_1 < \tau_1 < (K_1 + 1)t_1 \tag{2.397}$$

to hold.

In other words, the relations (2.383) and (2.393) can be generalized to provide

$$\hat{A}_{c\tau_1}[\hat{E}^{(+)}_{s_1}(t)|\psi(3,t)\rangle] = \hat{E}_{s_1}(t)|\psi(3, t - K_1 t_1, t - (K_1 + 1)t_1)\rangle, \tag{2.398}$$

$$\hat{A}_{c(\tau_0+\tau_2)}[\hat{E}^{(+)}_{s_2}(t)|\psi(3,t)\rangle] = \hat{E}_{s_2}(t)|\psi(3, t - K_2 t_1, t - (K_2 + 1)t_1)\rangle, \tag{2.399}$$

where τ_0 is the propagation time of the idler from NL1 to NL2. $\hat{A}_f$ denotes an attenuation of the field down to the vacuum state outside an interaction length centred at the distance f from NL1 in the direction of propagation of the beam. The operator $\hat{A}_f$ compensates for the difference we have caused with the initial condition at $t' = t_0 \to -\infty$ instead of the Wang–Zou–Mandel shortening of the integration interval. The operator $\hat{A}_f$ is not unitary and is even "slightly" nonlinear. Its consideration depends on a neglect of the coherence length in comparison with the interaction length. Using such an operator we can describe, where (within which interaction length) the single-photon states are localized at the time t,

$$\hat{A}_{c\tau}[|1(\mathbf{0}_1, t - \tau)\rangle_{\mathrm{i}}|0\rangle_{\mathrm{s}_1,\mathrm{s}_2,0}] = |1(\mathbf{0}_1, t - \tau)\rangle_{\mathrm{i}}|0\rangle_{\mathrm{s}_1,\mathrm{s}_2,0}, \tag{2.400}$$

$$\hat{A}_{c\tau}[|1(\mathbf{0}_1, t - \tau)\rangle_0|0\rangle_{\mathrm{s}_1,\mathrm{i},\mathrm{s}_2}] = |1(\mathbf{0}_1, t - \tau)\rangle_0|0\rangle_{\mathrm{s}_1,\mathrm{i},\mathrm{s}_2}. \tag{2.401}$$

The angular brackets will mean averages and, when operators are involved, the brackets are supposed to average in the state $|\psi(3,t)\rangle$,

$$\langle\hat{M}\rangle = \langle\psi(3,t)|\hat{M}|\psi(3,t)\rangle, \tag{2.402}$$

with $\hat{M}$ being an operator. When the operator is situated inside an interaction length centred in the propagation distance f from NL1, it also holds that

$$\langle\hat{M}\rangle = \hat{A}_f[\langle\psi(3,t)|]\hat{M}\hat{A}_f[|\psi(3,t)\rangle]. \tag{2.403}$$

Hence, one may omit the unusual notation when no ambiguities arise.

Letting ω_s and ω_i denote the centre frequency of the signal beam and the idler beam, respectively, we have $\omega_s + \omega_i = \omega_0$. Introducing the normalized correlation function $\mu(\tau)$ of the down-converted light

$$ {}_i\langle 1(\mathbf{r}_1, t - \tau_1)|1(\mathbf{r}_2, t - \tau_2)\rangle_i = \mu(\tau_0 + \tau_2 - \tau_1)\exp[-i\omega_i(\tau_0 + \tau_2 - \tau_1)], \quad (2.404) $$

where

$$ e^{-i\omega_i\tau}\mu(\tau) = 2\pi \int_0^{\omega_0} |\phi(\tilde{\omega}, \omega)|^2 e^{-i\omega\tau}\, d\omega, \quad (2.405) $$

we obtain that the relations (2.370) and (2.371) ought to read

$$ \begin{aligned} \langle \hat{E}_{s_1}^{(-)}(t)\hat{E}_{s_2}^{(+)}(t)\rangle &\simeq \nu_1^*\nu_2 \mathtt{t}^* \langle V_1^*(t-\tau_1)V_2(t-\tau_2)\rangle \\ &\times \mu(\tau_0 + \tau_2 - \tau_1)\exp[-i\omega_i(\tau_0 + \tau_2 - \tau_1)], \quad (2.406) \\ \langle \hat{E}_{s_1}^{(-)}(t)\hat{E}_{s_1}^{(+)}(t)\rangle &\simeq |\nu_1|^2\langle |V_1(t-\tau_1)|^2\rangle, \\ \langle \hat{E}_{s_2}^{(-)}(t)\hat{E}_{s_2}^{(+)}(t)\rangle &\simeq |\nu_2|^2\langle |V_2(t-\tau_2)|^2\rangle, \quad (2.407) \end{aligned} $$

where we introduce $\hat{E}_{s_j}^{(-)}(t) \equiv \left[\hat{E}_{s_j}^{(+)}(t)\right]^\dagger$. Hence the modulus of the normalized correlation function is

$$ \begin{aligned} &\frac{|\langle \hat{E}_{s_1}^{(-)}(t)\hat{E}_{s_2}^{(+)}(t)\rangle|}{\sqrt{\langle \hat{E}_{s_1}^{(-)}(t)\hat{E}_{s_1}^{(+)}(t)\rangle\langle \hat{E}_{s_2}^{(-)}(t)\hat{E}_{s_2}^{(+)}(t)\rangle}} \\ &= \frac{\langle V_1^*(t-\tau_1)V_2(t-\tau_2)\rangle}{\sqrt{\langle |V_1(t-\tau_1)|^2\rangle\langle |V_2(t-\tau_2)|^2\rangle}} |\mu(\tau_0 + \tau_2 - \tau_1)||\mathtt{t}|. \quad (2.408) \end{aligned} $$

The maximum value is equal to $|\mathtt{t}|$, which is predicted also by equation (2.359). A linear dependence of visibility on $|\mathtt{t}|$, as seen convincingly in the original work (Zou et al. 1991, Wang et al. 1991a), is the true signature of induced coherence without induced emission.

As concerns $\hat{E}_{s_j}^{(+)}(t)|\psi(t)\rangle$, $j = 1, 2$, they are not explicitly presented in Wang et al. (1991a), but they may be derived. It emerges that the parameters of the beam splitter do not enter the relation for $\hat{E}_{s_1}^{(+)}(t)|\psi(t)\rangle$. On the contrary, $\hat{E}_{s_1}^{(+)}(t)|\psi(t)\rangle$ in Wang et al. (1991a) comprises the parameters $\mathtt{t}^*$, $\mathtt{r}'^*$. Nevertheless, the statistical properties in Peřinová et al. (2003) coincide with those in Wang et al. (1991a), because the differences under discussion resemble distinct, yet equivalent pictures. Especially, considering the photon-flux amplitude operators $\hat{E}_s^{(+)}(t)$ at the detector D_s with a quantum efficiency η_s (Wang et al. 1991a),

$$ \hat{E}_s^{(+)}(t) = \frac{1}{\sqrt{2}}\left[i\hat{E}_{s_1}^{(+)}(t) + \hat{E}_{s_2}^{(+)}(t)\right], \quad (2.409) $$

substituting into the formula for the average rate of photon counting

$$ I_s = \eta_s \langle \psi(t)|\hat{E}_s^{(-)}(t)\hat{E}_s^{(+)}(t)|\psi(t)\rangle, \quad (2.410) $$

where

$$\hat{E}_{\mathrm{s}}^{(-)}(t) = [\hat{E}_{\mathrm{s}}^{(+)}(t)]^{\dagger}, \tag{2.411}$$

and taking into account the orthogonality of single-photon states $|1(\mathbf{r}_1, t)\rangle_{\mathrm{i}}$ and $|1(\mathbf{r}_1, t)\rangle_0$ uniquely results in the relation

$$I_{\mathrm{s}} = \frac{1}{2}\eta_{\mathrm{s}}\{|\nu_1|^2\langle|V_1(t-\tau_1)|^2\rangle + |\nu_2|^2\langle|V_2(t-\tau_2)|^2\rangle$$
$$+[-\mathrm{i}\nu_1^*\nu_2\langle V_1^*(t-\tau_1)V_2(t-\tau_2)\rangle \mathrm{t}^*\mu(\tau_0+\tau_2-\tau_1)e^{-\mathrm{i}\omega_{\mathrm{i}}(\tau_0+\tau_2-\tau_1)} + \mathrm{c.c.}]\}. \tag{2.412}$$

Peřinová et al. (2000) have studied quantum statistics of radiation in signal modes of the two-mode parametric processes with aligned idler beams. They have found that the signal beams are in the correlated chaotic state. The strength of correlation depends on the degree to which the paths of the idler beams are superposed and aligned. They have compared different measures of correlation, especially the entropic or information-based measure with the modulus of the usual degree of coherence in the dependence on absolute value of the transmission amplitude coefficient of the beam splitter inserted as an attenuator of the perfect alignment. Some other measures have been introduced taking into account the symmetrical and antinormal orderings of field operators. In contrast to the normal ordering, these orderings do not indicate the maximum correlation for the perfect alignment. The situation with the photon numbers in the signal modes, whose correlation is not maximum for the perfect alignment, serves as motivation for such a more general consideration. The theory of canonical correlation has been applied to the quasidistribution of complex amplitudes related to the symmetrical ordering of field operators. They have taken into account that the quantum correlation has a significant effect on the photon-number sum, photon-number difference, and quantum phase-difference statistics. Essentially, it concerned the variances of number sum and number difference and the dispersions of quantum phase differences according to various definitions. A comparison of distributions of quantum phase difference derived from the phase-space distributions has shown that the phase-difference uncertainty increases from the normal ordering, through the symmetrical and antinormal orderings, whereas the system of canonical phase related to the antinormal ordering of exponential phase operators ranges between the symmetrical and antinormal orderings, but by no means exactly. The paper (Peřinová et al. 2000) reveals that the correlated chaotic state is the mixed partial phase-difference state. In addition to the marginal distributions, the joint number-sum and phase-difference distribution has been considered, but for the canonical quantum phase difference and the Luis–Sánchez-Soto phase difference only. The quasidistribution of number difference and phase difference has been defined with the properties that the marginal distribution of the phase difference is the canonical one. They have addressed the number sum and the quantum phase difference as simultaneously measurable observables and the number difference and the quantum phase difference as canonically conjugate observables.

Peřinová et al. (2003) have compared the simple formalism of several coupled harmonic oscillators with multimode formalism in the analysis of an interference experiment. On focusing on several modes they have been able to study phase properties of "correlated chaotic beams". Then they have assumed the single-photon regime as also previous authors did. They have indicated that, assuming several coupled harmonic oscillators, the previous authors did not try to include time delays between optical elements into the analysis. Peřinová et al. (2003) have also formally expressed, for instance, that one works with a single-photon state of some signal modes in the several-mode formalism whenever one describes the experiment with a superposition of single-photon states of modes that form the signal beam.

The utility of a simple single-mode theory has been clarified in the case where single spatial mode filters and narrow-band optical filters are used to filter the output state of parametric down-conversion Li et al. 2005).

Peřina and Křepelka (2005) have derived joint photon-number distributions in signal and idler modes and have illustrated related concepts taking into account experimental data. Peřina and Křepelka (2006) have provided the generalization of this description to stimulated parametric down-conversion. Peřina et al. (2007) have reported on a measurement of the joint signal–idler photoelectron distribution of twin beams. Parameters of the previously published model (Peřina and Křepelka 2005) have been estimated. The specific result that the joint signal–idler quasidistribution of integrated intensities can be approximated by a well-behaved function even in the case where the quasidistribution is not an ordinary function has been comprised. Peřina (2008) has shown that a nonlinear planar waveguide pumped by a beam orthogonal to its surface may serve as a versatile source of photon pairs. He considers the pump-pulse duration, pump-beam transverse width, and angular decomposition of the pump-beam frequency and their effect on characteristics of a photon pair, such as the spectral widths of signal and idler fields, the pair time duration, and the degree of entanglement between the two fields.

Chapter 3
Macroscopic Theories and Their Applications

There were several attempts at a justification of the momentum-operator approach. It is appropriate that they include quantization of the electromagnetic field at least in the one-dimensional case. A complete analysis could be provided only for the parametric processes, in which the momentum operator is effectively quadratic. It has been noted that the nonlinearity of the process may lead up to a need of a renormalization. Nevertheless, there is a modicum of papers on this theme in quantum optics.

A general approach to quantization of the electromagnetic field in a nonlinear medium enables one to compare properties of the momentum operator with those of the space–time displacement operator. We present applications of the traditional approach in quantum optics.

The spatio-temporal approach has been developed with respect to quantum solitons. An attempt has been made to take into account the frequency dispersion of a medium at least up to inclusion of the group velocity and to preserve the traditional descriptions of nonlinear processes by introducing narrow-band fields.

A mention of the quasimode description of the spectrum of squeezing will be restricted to an analysis of coupling of the cavity modes and propagating modes. The paraxial propagation of a light beam with the parabolic approximation and the asymptotic expansion of the beam has been completed with a quantized version. As an example the nonlinear process has been presented, whose description includes the renormalization.

In optical imaging with nonclassical light one wants to investigate the quantum fluctuations of light at different spatial points in the plane perpendicular to the propagation direction of the light beam. In such spatial points very small photodetectors or pixels may be located. Finally the application of one of the macroscopic approaches has led to the description of several linear optical devices and to the study of radiating atoms in a linear medium, which is a recurrent theme by the way.

3.1 Momentum-Operator Approach

Several papers devoted to macroscopic approaches to quantization of the electromagnetic field advocate the momentum operator. In general, such an operator

A. Lukš, V. Peřinová, *Quantum Aspects of Light Propagation*,
DOI 10.1007/b101766_3, © Springer Science+Business Media, LLC 2009

should be one of the space–time displacement operators. To the contrary, almost paradoxical properties of these operators have been derived. The exposition is concluded with applications of the traditional approach to the nonlinear optical couplers.

3.1.1 Temporal Modes and Their Application

Huttner et al. (1990) have developed a formalism that describes in a fully quantum-mechanical way the propagation of light in a linear and nonlinear lossless dispersive medium. At first, they assume a similar situation as Abram (1987), i.e. they consider only the one-dimensional case restricting themselves even to fields propagating in the $+z$-direction only. They take for granted that in quantum field theory there is a generator for spatial progression, i.e. that relation (2.49) holds for any operator. They remark that the change in the quantization volume pointed out by Abram (1987) is not defined when the medium is dispersive, i.e. when the refractive index depends on the frequency, but they develop Abram's idea of the use of the energy flux not being dependent on the medium (cf. Caves and Crouch 1987). In their opinion, the classical analysis of nonlinear optical processes shows that in order to obtain simple equations of propagation it is useful to introduce photon-flux amplitude, i.e., a quantity whose square is proportional to the photon flux.

At present we hesitate to accept the consequences of their approach (cf. however Ben-Aryeh et al. 1992). Specifying the state at a given point (e.g. $z = 0$) and within a time period T cannot substitute specifying the state at an initial time ($t = 0$) and within a quantization length L. Temporal modes of discrete frequencies ω_m, where $\omega_m = \frac{2m\pi}{T}$, cannot substitute the spatial modes. The equal-space commutation relations

$$[\hat{a}(z, \omega_i), \hat{a}^\dagger(z, \omega_j)] = \delta_{ij}\hat{1} \tag{3.1}$$

cannot substitute the usual equal-time commutation relations.

In comparison with Abram (1987), we see the following changes. In Huttner et al. (1990), the MKSA (SI) system of units is used. Instead of immediately reducing the unsymmetrical Maxwell stress tensor to a single component, the momentum density is first reduced. The normal ordering is used where necessary. The notation ceases to express the dependence on both z and t and states the dependence on z only. "The generalization" of the relation for the momentum-flux operator

$$\hat{g}(z, t) = c[\hat{D}^{(-)}(z, t)\hat{B}^{(+)}(z, t) + \text{H.c.}], \tag{3.2}$$

where H.c. means the Hermitian conjugate to the previous term, to the form

$$\hat{g}(z, t) = [\hat{D}^{(-)}(z, t)\hat{E}^{(+)}(z, t) + \text{H.c.}] \tag{3.3}$$

is not founded well. Its integration over T gives the momentum operator $\hat{G}(z)$,

$$\hat{G}(z) \equiv \int_{t_0}^{t_0+T} \hat{g}(z, t)\,\mathrm{d}t. \tag{3.4}$$

In the case of a linear dielectric medium in contrast with Abram (1987), the electric-field operator $\hat{E}(z,t)$ is dependent on the refractive index $n(\omega_m)$

$$\hat{E}(z,t) = \sum_m \sqrt{\frac{\hbar\omega_m}{2\epsilon_0 c T n(\omega_m)}} \left[\hat{a}(z,\omega_m)e^{-\mathrm{i}\omega_m t} + \mathrm{H.c.}\right], \tag{3.5}$$

while in Abram (1987) the operator is independent of the medium. In Abram (1987) there is not pure Heisenberg picture, so that the equivalence of the two theories (for n independent of ω) is not excluded. From relation (3.3), the momentum operator is obtained

$$\hat{G}_{\mathrm{lin}}(z) = \sum_m (\hbar k_m)\hat{a}^\dagger(z,\omega_m)\hat{a}(z,\omega_m), \tag{3.6}$$

where $k_m = \frac{n(\omega_m)\omega_m}{c}$ is the wave vector in the (linear) medium. The equal-space commutation relations are conserved. For such a medium, the equal-time commutation relations can be derived

$$\left[\hat{A}(z,t), -\hat{D}(z',t)\right] = \mathrm{i}\hbar\delta(z-z')\hat{1}. \tag{3.7}$$

Attempting at the quantization in a nonlinear medium, Huttner et al. (1990) have concentrated on the propagation of light in a multimode degenerate parametric amplifier. The postulated relation (3.3) then leads to the nonlinear part of the momentum-flux operator

$$\hat{g}_{\mathrm{nonlin}}(z,t) = \chi^{(2)}\left\{\mathcal{E}^{(+)}(z,t)\left[\hat{E}^{(-)}(z,t)\right]^2 + \mathrm{H.c.}\right\}, \tag{3.8}$$

where $\mathcal{E}^{(+)}(z,t) = |\mathcal{E}|e^{-\mathrm{i}(\omega_\mathrm{p}t - k_\mathrm{p}z)}$ is the positive-frequency part of the pump field, with the pump frequency ω_p. From relations (3.4) and (3.8), the momentum operator is obtained

$$\hat{G}_{\mathrm{nonlin}}(z) = \frac{\hbar}{4}\sum_m \lambda(\epsilon_m)\left[\hat{a}^\dagger(z,\omega_0+\epsilon_m)\hat{a}^\dagger(z,\omega_0-\epsilon_m)e^{\mathrm{i}k_\mathrm{p}z} + \mathrm{H.c.}\right], \tag{3.9}$$

where

$$\epsilon_m \equiv \omega_m - \omega_0, \quad \omega_0 = \frac{\omega_\mathrm{p}}{2} \tag{3.10}$$

and

$$\lambda(\epsilon_m) \equiv \frac{\chi^{(2)}|\mathcal{E}|}{\epsilon_0 c}\sqrt{\frac{\omega_0+\epsilon_m}{n(\omega_0+\epsilon_m)}\frac{\omega_0-\epsilon_m}{n(\omega_0-\epsilon_m)}} \tag{3.11}$$

is the coupling constant between the different modes.

It is assumed that the phase-matching condition at ω_0, $n(\omega_\text{p}) = n(\omega_0)$, is satisfied. It is found that the phase-mismatch $\Delta k(\epsilon_m)$ is proportional to ϵ_m^2. As far as $|\Delta k(\epsilon_m)| < \lambda(\epsilon_m)$, the Bogoliubov transformation for squeezing emerges and amplifying behaviour can be recognized. In the case $|\Delta k(\epsilon_m)| > \lambda(\epsilon_m)$, the evolution is not essentially different from that in a linear medium, the squeezing effect is band limited. For the equality $|\Delta k(\epsilon_m)| = \lambda(\epsilon_m)$, the amplifying is present, but the increase is only linear not exponential.

For the nonlinear medium, the equal-time commutation relations are

$$\left[\hat{A}^{(-)}(z,t), -\hat{D}^{(+)}(z',t)\right] \approx \frac{\mathrm{i}\hbar}{2}\delta(z-z')\hat{1} \tag{3.12}$$

and relation (3.7) can be recovered only approximately. In relation to the experiment, a standard two-port homodyne detection scheme is assumed, where the light is mixed at a beam splitter with a strong local oscillator $\varepsilon(z,t)$ of the frequency ω_0. For the correlation function $g_S(\tau)$ of the photocurrent difference and its Fourier transform

$$y(\eta) = \int g_S(\tau)e^{-\mathrm{i}\eta\tau}\,\mathrm{d}\tau, \tag{3.13}$$

we refer to Huttner et al. (1990). It has been shown that the values of $y(\eta)$ can be minimized uniformly enough by an adequate choice of the local oscillator phase. The result is comparable with Crouch (1988), where the usual interpretation of homodyne detection in terms of the field quadratures is used.

3.1.2 Slowly Varying Amplitude Momentum Operator

Nevertheless, there is a class of problems, for which the modal approach is very convenient. It is the cavity quantum electrodynamics. Let us mention its use in the development of the input–output formalism for nonlinear interactions in cavity (Yurke 1984, 1985, Collett and Gardiner 1984, Gardiner and Collett 1985, Carmichael 1987). The modal approach can describe many of the features of travelling-wave phenomena, but, in principal, it mixes effects related to spatial progression of the beam with the spectral manifestations of the nonlinearity. For example, for the case of the travelling-wave parametric generation (Tucker and Walls 1969), a wave vector mismatch appears as energy (frequency) nonconservation. Several authors have tried to return the quantum-mechanical propagation to direct space. One technique (Drummond and Carter 1987) involves the partition of the box of quantization into finite cells. Another technique considers the spatial progression of the temporal Fourier components of the local electric field (Yurke et al. 1987, Caves and Crouch 1987). The propagation of light in a magnetic (dielectric) medium is not considered in quantum optics.

We proceed with the field inside an effective (linear or nonlinear) medium and the direct-space formulation of the theory of quantum optics as presented by Abram and Cohen (1991). It is an alternative of the conventional reciprocal space approach to quantum optics. Their approach relies on the electromagnetic momentum operator as well as on the Hamiltonian and is restricted to the dispersionless lossless nonmagnetic dielectric medium. They have derived an operatorial wave equation that relates the temporal evolution of an electromagnetic pulse to its spatial progression. The theory is applied to squeezed light generation by the parametric down-conversion of a short laser pulse as an illustration. This approach does not use the conventional modal description of the field.

The appeal of the classical theory of optics may consist in its considering material as a continuous dielectric characterized by a set of phenomenological constants. In classical nonlinear optics the slowly varying amplitude approximation of the electromagnetic wave equation has arisen. An important simplification of quantum optics results when the microscopic description of the material is replaced by a macroscopic description, in terms of an effective linear or nonlinear polarization. In spite of the phenomenological treatment of the medium, such an effective theory still permits a quantum-mechanical description of the field (Jauch and Watson 1948, Shen 1967, Glauber and Lewenstein 1989, Glauber and Lewenstein 1991, Hillery and Mlodinow 1984, Drummond and Carter 1987).

In propagation problems, the interactions undergone by a short pulse of light are examined. Abram and Cohen (1991) simplify the geometry for the electromagnetic field so that the electric field E is polarized along the x-axis, the magnetic field B along the y-axis, while propagation occurs along the z-axis. They use the Heaviside–Lorentz units and take $\hbar = c = 1$. In this simple geometry the Maxwell equations reduce to two scalar differential equations

$$\frac{\partial E}{\partial z} = -\frac{\partial B}{\partial t}, \tag{3.14}$$

$$\frac{\partial B}{\partial z} = -\frac{\partial D}{\partial t}, \tag{3.15}$$

where the electric displacement field D is defined by

$$D = E + P, \tag{3.16}$$

with P being the polarization of the medium, which can be expressed as a converging power series in the electric field E,

$$P = \chi^{(1)}E + \chi^{(2)}E^2 + \cdots + \chi^{(n)}E^n + \cdots, \tag{3.17}$$

where $\chi^{(n)}$ is the nth-order susceptibility of the medium. The dispersion cannot be taken into account rigorously within a quantum-mechanical theory based on the effective (macroscopic) Hamiltonian formulation (Hillery and Mlodinow 1984), but it can be introduced phenomenologically (Drummond and Carter 1987). To impose the canonical structure on the field, they introduce the vector potential $\mathbf{A}$ and adopt the Coulomb gauge in which the scalar potential vanishes and $\mathbf{A}$ is transverse. In the assumed geometry, the vector potential is polarized along the x-axis and is related to the electric and magnetic fields by

$$E = -\frac{\partial A}{\partial t} \tag{3.18}$$

and

$$B = \frac{\partial A}{\partial z}. \tag{3.19}$$

The effective Lagrangian density has been chosen (Hillery and Mlodinow 1984, Drummond and Carter 1987),

$$\mathcal{L} = \frac{1}{2}(E^2 - B^2) + \frac{1}{2}\chi^{(1)}E^2 + \frac{1}{3}\chi^{(2)}E^3 + \frac{1}{4}\chi^{(3)}E^4 + \cdots, \tag{3.20}$$

which is known to be the most general density dependent only on the electric field and having the gauge invariance.

Let us note that the theory with the effective Lagrangian density (3.20) is not renormalizable (Power and Zienau 1959, Woolley 1971, Babiker and Loudon 1983, Cohen-Tannoudji et al. 1989). The canonically conjugated momentum of A with respect to the Lagrangian density (3.20) is the electric displacement

$$\Pi = \frac{\partial \mathcal{L}}{\partial \dot{A}} = -D. \tag{3.21}$$

The Lagrangian density is then transformed to some components of the energy–momentum tensor of the electromagnetic field inside a nonlinear medium $\Theta_{\mu\nu}$, namely the energy density

$$\Theta_{tt} = \Pi\frac{\partial A}{\partial t} - \mathcal{L} \tag{3.22}$$

$$= \frac{1}{2}(B^2 + A^2) + \frac{1}{2}\chi^{(1)}E^2 + \frac{2}{3}\chi^{(2)}E^3 + \frac{3}{4}\chi^{(3)}E^4 + \cdots, \tag{3.23}$$

and the momentum density

$$\Theta_{tz} = -\Pi\frac{\partial A}{\partial z} = DB. \tag{3.24}$$

In setting up the Hamiltonian functional, the electric field E is to be expressed in terms of the electric displacement, which is the canonically conjugated momentum of A according to relation (3.21). It is assumed that

$$E = \beta^{(1)}D + \beta^{(2)}D^2 + \beta^{(3)}D^3 + \cdots, \tag{3.25}$$

where the β coefficients may be expressed in terms of the susceptibilities $\chi^{(n)}$ through definition (3.16) and relation (3.17) (Hillery and Mlodinow 1984).

The Hamiltonian functional is then written as

$$H = \int_V \Theta_{tt}\,\mathrm{d}^3\mathbf{r} = \int_V \left(\frac{1}{2}B^2 + \frac{1}{2}\beta^{(1)}D^2 + \frac{1}{3}\beta^{(2)}D^3 + \frac{1}{4}\beta^{(3)}D^4 + \dots\right)\mathrm{d}^3\mathbf{r}, \tag{3.26}$$

while the momentum has the form

$$G = \int_V \Theta_{tz}\,\mathrm{d}^3\mathbf{r} = \int_V BD\,\mathrm{d}^3\mathbf{r}, \tag{3.27}$$

where the integration is over the cavity obeying periodic boundary conditions and lower and upper limits are supposed to converge to $-\infty$ and ∞, respectively.

The field can now be quantized by replacing each field variable by the corresponding operator and by replacing the Poisson bracket between the displacement D and the vector potential A by the equal-time commutator

$$[\hat{D}(\mathbf{r}, t), \hat{A}(\mathbf{r}', t)] = \mathrm{i}\delta_T(\mathbf{r} - \mathbf{r}')\hat{1}, \tag{3.28}$$

exactly by its $(-\mathrm{i})$ multiple, where the transverse $\delta_T(\mathbf{r} - \mathbf{r}')$ reduces to the ordinary δ function and where the three-dimensional position vector $\mathbf{r}$ can be replaced by the coordinate z.

The vector potential A replaced by the operator $\hat{A}$ does not appear explicitly in the Hamiltonian (3.26) and momentum (3.27) operators, but rather in terms of its spatial derivative $\hat{B}$. Taking the curl according to $\mathbf{r}'$ of both the sides of the canonical commutation relation (3.28) and using relation (3.19) in the simple geometry, we obtain that

$$[\hat{D}(z, t), \hat{B}(z', t)] = -\mathrm{i}\delta'(z - z')\hat{1}, \tag{3.29}$$

where

$$\delta'(z - z') = \frac{\mathrm{d}}{\mathrm{d}z}\delta(z - z') \tag{3.30}$$

is the derivative of the δ function.

Not caring for divergencies, Abram and Cohen (1991) consider any product of noncommuting operators that appear in an expression as fully symmetrized, i.e. including all possible permutations of the individual field operators, such as

$$BD^2 \mapsto \frac{\hat{B}\hat{D}^2 + \hat{D}\hat{B}\hat{D} + \hat{D}^2\hat{B}}{3}. \tag{3.31}$$

In contrast, more recently they carried out the renormalization, i.e. the normal ordering and an elimination of divergencies.

The description of propagative optical phenomena has been discussed within the framework of a direct-space formulation of quantum optics and the operatorial (or better commutator) equivalent of the Maxwell equations and the electromagnetic wave equation (Abram and Cohen 1991). It is emphasized that in the Hamiltonian formulation of mechanics the time variable plays a particular role.

The integrals in (3.26) and (3.27) and the equal-time commutator (3.28) correspond to the requirement that the field be specified over all space at one instant of time (e.g. at $t = 0$).

Abram and Cohen (1991) use the Kubo (1962) notation for the commutator or more exactly for a corresponding superoperator. The superoperator assigns operators to operators. Respecting this, the Heisenberg equation can be written as follows

$$\frac{\partial \hat{Q}}{\partial t} = \mathrm{i}[\hat{H}, \hat{Q}] = \mathrm{i}\hat{H}^{\times}\hat{Q}, \tag{3.32}$$

where $\hat{Q}$ is any field operator and the superscript $\times$ denotes the superoperator, namely the commutation of the operator it loads with another operator which follows. Equation (3.32) has the solution

$$\hat{Q}(t) = \exp(\mathrm{i}t\hat{H}^{\times})\hat{Q}(0) \tag{3.33}$$

$$= e^{\mathrm{i}\hat{H}t}\hat{Q}(0)e^{-\mathrm{i}\hat{H}t}. \tag{3.34}$$

The Heisenberg-like equation involving the momentum can be considered

$$\frac{\partial \hat{Q}}{\partial z} = -\mathrm{i}\hat{G}^{\times}\hat{Q}. \tag{3.35}$$

This equation has the solution

$$\hat{Q}(z) = \exp[-\mathrm{i}(z - z_0)\hat{G}^{\times}]\hat{Q}(z_0). \tag{3.36}$$

Apart from the obvious similarity of equations (3.32) and (3.35), there is also a difference. The Hamiltonian of the electromagnetic field relates the desired spatial distribution of the field at the instant $t + \mathrm{d}t$ to its spatial distribution at t, but the momentum operator $\hat{G}$ relates the translated and nontranslated fields only (at the same instant of time).

In analogy with the classical equations (3.14), (3.15) rather than with the classical equations

$$\begin{aligned} \frac{\partial(\beta^{(1)}D + \beta^{(2)}D^2 + \beta^{(3)}D^3 + \cdots)}{\partial z} &= -\frac{\partial B}{\partial t}, \\ \frac{\partial B}{\partial z} &= -\frac{\partial D}{\partial t}, \end{aligned} \tag{3.37}$$

two commutator equations can be derived

$$\hat{G}^\times \hat{E} = \hat{H}^\times \hat{B}, \tag{3.38}$$

$$\hat{G}^\times \hat{B} = \hat{H}^\times \hat{D}. \tag{3.39}$$

On assuming that the medium is homogeneous, so that the Hamiltonian and momentum operators commute with each other, that is

$$\hat{G}^\times \hat{H} = \hat{0}, \tag{3.40}$$

Equations (3.38) and (3.39) may be combined into the commutator equivalent of the electromagnetic wave equation

$$\hat{G}^\times \hat{G}^\times \hat{E} = \hat{H}^\times \hat{H}^\times \hat{D}. \tag{3.41}$$

Let us note again that it is an analogue of the wave equation

$$\frac{\partial^2 E}{\partial z^2} = \frac{\partial^2 D}{\partial t^2}, \tag{3.42}$$

not of the more complicated equation

$$\frac{\partial^2(\beta^{(1)} D + \beta^{(2)} D^2 + \beta^{(3)} D^3 + \cdots)}{\partial z^2} = \frac{\partial^2 D}{\partial t^2}. \tag{3.43}$$

In Abram and Cohen (1991) the direct-space description of propagation (i.e. without resorting to a modal decomposition of a propagating pulse) is illustrated by examining the propagation of light (of the short light pulse) through a linear medium and through a vacuum–dielectric interface.

For a linear medium the commutator wave equation (3.41) reduces to

$$(\hat{G}^\times \hat{G}^\times - \epsilon \hat{H}^\times \hat{H}^\times)\hat{E} = \hat{0}, \tag{3.44}$$

where $\epsilon = 1 + \chi^{(1)}$ is the dielectric function of the medium. It is also convenient to define $v = \frac{1}{\sqrt{\epsilon}}$, the velocity of an electromagnetic wave in a refractive medium. In the following exposition, the convention $c = 1$ will be dissolved or not used. Wave equation (3.44) enables one to rewrite equations (4.2a) and (4.2b) of Abram and Cohen (1991) in the form

$$\hat{E}(z,t) = \cosh(-\mathrm{i}vt\hat{G}^\times)\hat{E}(z,0) - \mathrm{i}v \sinh(-\mathrm{i}vt\hat{G}^\times)\hat{B}(z,0), \tag{3.45}$$

$$\hat{B}(z,t) = -\frac{\mathrm{i}}{v} \sinh(-\mathrm{i}vt\hat{G}^\times)\hat{E}(z,0) + \cosh(-\mathrm{i}vt\hat{G}^\times)\hat{B}(z,0). \tag{3.46}$$

Equations (3.45) and (3.46) indicate that the linear combination

$$\hat{W}_v^+(z,t) = \hat{E}(z,t) + v\hat{B}(z,t) \tag{3.47}$$

evolves in time as

$$\hat{W}_v^+(z,t) = \exp(\mathrm{i}vt\hat{G}^\times)\hat{W}_v^+(z,0) = \hat{W}_v^+(z-vt,0). \tag{3.48}$$

Similarly, the linear combination

$$\hat{W}_v^-(z,t) = \hat{E}(z,t) - v\hat{B}(z,t) \tag{3.49}$$

evolves as

$$\hat{W}_v^-(z,t) = \exp(-\mathrm{i}vt\hat{G}^\times)\hat{W}_v^-(z,0) = \hat{W}_v^-(z+vt,0). \tag{3.50}$$

To examine the problem of the interface, we now consider two half-spaces such that the $z \in (-\infty, 0)$ half-space is empty, while the $z \in (0, +\infty)$ half-space consists of a transparent linear dielectric. We then consider three waves, incident, $\hat{W}_c^+(z,t)$, reflected, $\hat{W}_c^-(z,t)$, and transmitted, $\hat{W}_v^+(\zeta,t)$, and the relations they obey.

Abram and Cohen derive the commutator equivalent of the slowly varying amplitude wave equation, on which the classical theory of nonlinear optics is based (Abram and Cohen 1991). Not even in classical optics, the problem of propagation of a short pulse in a nonlinear medium can be solved in the general case. In classical nonlinear optics, the assumption of a weak nonlinearity makes the slowly varying amplitude (SVA) approximation of the electromagnetic wave equation possible (Shen 1984). In Abram and Cohen (1991) further a perturbative treatment of the time evolution of the field in a nonlinear medium is examined that corresponds to propagation within the slowly varying amplitude approximation. For simplicity, a single nonlinear susceptibility $\chi^{(n)}$ is considered.

In the perturbative treatment of the nonlinear propagation it is assumed that the optical nonlinearity of the medium is absent at $t = -\infty$ and turned on adiabatically. In the absence of the nonlinearity, the electric and magnetic fields in the medium, $\hat{E}_0$ and $\hat{B}_0$, as well as the displacement field $\hat{D}_0$,

$$\hat{D}_0 = \epsilon\hat{E}_0 \tag{3.51}$$

propagate under the energy operator

$$\hat{H}_0 = \frac{1}{2}\int\left(\hat{B}_0^2 + \frac{1}{\epsilon}\hat{D}_0^2\right)\mathrm{d}^3\mathbf{r} \tag{3.52}$$

and the momentum operator

$$\hat{G}_0 = \int \hat{B}_0\hat{D}_0\,\mathrm{d}^3\mathbf{r}, \tag{3.53}$$

which are of zeroth order in the nonlinear susceptibility $\chi^{(n)}$. Following the standard perturbation theory (Itzykson and Zuber 1980), the exact field operators in the nonlinear medium, $\hat{D}$ and $\hat{B}$, can be derived from the zeroth-order fields by the unitary transformation

$$\hat{D}(z,t) = \hat{U}^{-1}(t)\hat{D}_0(z,t)\hat{U}(t) \tag{3.54}$$

and

$$\hat{B}(z,t) = \hat{U}^{-1}(t)\hat{B}_0(z,t)\hat{U}(t). \tag{3.55}$$

Here $\hat{U}(t)$ is the unitary operator, which is the solution to the differential equation

$$\frac{\partial}{\partial t}\hat{U}(t) = -\mathrm{i}\lambda\hat{\hat{H}}_1(t)\hat{U}(t), \tag{3.56}$$

with $\hat{\hat{H}}_1(t)$ the nonlinear interaction part of the Hamiltonian,

$$\hat{\hat{H}}_1(t) \equiv \frac{1}{n+1}\int \beta^{(n)}\hat{D}_0^{n+1}\,\mathrm{d}^3\mathbf{r} = -\frac{1}{n+1}\int \chi^{(n)}\hat{E}_0^{n+1}\,\mathrm{d}^3\mathbf{r} \tag{3.57}$$

or

$$\hat{\hat{H}}_1(\tau) = \exp(\mathrm{i}\tau\hat{H}_0^{\times})\hat{H}_1, \tag{3.58}$$

and obeys the initial condition

$$\hat{U}(t)\Big|_{t\to-\infty} = \hat{1}. \tag{3.59}$$

The Hamiltonian $\hat{H}_1$ is first order in the nonlinear susceptibility $\chi^{(n)}$, which is expressed also by λ, a dimensionless parameter, which has been introduced for the bookkeeping of this and higher powers of $\chi^{(n)}$. The exact Hamiltonian (3.26) can then be expressed perturbatively up to the first order in λ as

$$\hat{H} = \hat{H}_0 + \lambda\hat{H}_{1S} + O(\lambda^2), \tag{3.60}$$

where $\hat{H}_{1S}$ is the "diagonal part" of $\hat{H}_1$, S stands for stationary, which commutes with the linear Hamiltonian $\hat{H}_0$,

$$\hat{H}_{1S} \equiv \hat{H}_{1S}^{(n)} = -\frac{\chi^{(n)}}{n+1}\int \hat{S}_{n+1}\,\mathrm{d}^3\mathbf{r}, \tag{3.61}$$

with

$$\hat{S}_n = 2^{-n+1}\sum_{m=0}^{[\frac{n}{2}]}\frac{n!}{(n-2m)!(2m)!}\epsilon^{-m}\hat{B}_0^{2m}\hat{E}_0^{n-2m}, \tag{3.62}$$

which leads to the decoupling of opposite-going fields. In the context of (3.61) and (3.62), a connection with the modal approach has been mentioned in Abram and Cohen (1991). The notion of the diagonal part belongs to the perturbation theory which was treated in Sczaniecki (1983).

According to (3.33), the time evolution of the displacement operator $\hat{D}$ can be explained and cast in the form

$$\hat{D}(z,t) \approx \exp(it\lambda \hat{H}_{1S}^{\times})\hat{D}_0(z,t) + \lambda \hat{D}_1(z,t), \tag{3.63}$$

where $\lambda \equiv 1$ and

$$\hat{D}_1(z,t) = \left[\mathrm{i} \int_{-\infty}^{t} \hat{\hat{H}}_1(\tau)\,\mathrm{d}\tau \right]^{\times} \hat{D}_0(z,t) \tag{3.64}$$

is the first-order correction to the displacement field. The action of the superoperator on $\hat{D}_0$ in relation (3.63) can be compared with the multiplication of the fast-varying ("carrier") wave by a slowly varying envelope function.

On introducing the nonlinear polarization

$$\hat{P}_{\mathrm{NL}} = -\epsilon\beta^{(n)}\hat{D}_0^n = \chi^{(n)}\hat{E}_0^n, \tag{3.65}$$

it can be shown that the exact commutator wave equation (3.41) can be written up to order λ^0 as

$$(\hat{G}_0^{\times}\hat{G}_0^{\times} - \epsilon\hat{H}_0^{\times}\hat{H}_0^{\times})\hat{D}_0 = \hat{0} \tag{3.66}$$

and that to order λ^1 as

$$2\epsilon\hat{H}_0^{\times}\hat{H}_{1S}^{\times}\hat{D}_0 = -\epsilon\hat{H}_0^{\times}\hat{H}_0^{\times}\hat{D}_1 + \hat{G}_0^{\times}\hat{G}_0^{\times}\hat{D}_1 - \hat{G}_0^{\times}\hat{G}_0^{\times}\hat{P}_{\mathrm{NL}}. \tag{3.67}$$

The nonlinear polarization $\hat{P}_{\mathrm{NL}}$ consists of two parts, $\hat{P}_W$ and its complement, and $\hat{P}_W$ obeys the zeroth-order wave equation

$$(\hat{G}_0^{\times}\hat{G}_0^{\times} - \epsilon\hat{H}_0^{\times}\hat{H}_0^{\times})\hat{P}_W = \hat{0}, \tag{3.68}$$

namely

$$\hat{P}_W \equiv \hat{P}_W^{(n)} = \chi^{(n)}\hat{S}_n. \tag{3.69}$$

This partition again eliminates all terms that couple opposite-going waves in $\hat{P}_{\mathrm{NL}}$. Relying on the relation

$$\hat{0} = -\epsilon\hat{H}_0^{\times}\hat{H}_0^{\times}\hat{D}_1 + \hat{G}_0^{\times}\hat{G}_0^{\times}\hat{D}_1 - \hat{G}_0^{\times}\hat{G}_0^{\times}(\hat{P}_{\mathrm{NL}} - \hat{P}_W), \tag{3.70}$$

we can derive the commutator equation

$$2\epsilon\hat{H}_0^{\times}\hat{H}_{1S}^{\times}\hat{D}_0 = -\hat{G}_0^{\times}\hat{G}_0^{\times}\hat{P}_W, \tag{3.71}$$

which has been compared to the classical slowly varying amplitude (SVA) wave equation, which is written as Abram and Cohen (1991)

$$2ik\frac{\partial}{\partial z}\tilde{E} = \frac{\partial^2}{\partial t^2}P_W, \tag{3.72}$$

or, more often, in terms of the temporal Fourier components of $\tilde{E}$ and P_W as

$$\frac{\partial}{\partial z}\tilde{E}(\omega) = \frac{i\omega}{2\sqrt{\epsilon}}P_W(\omega), \tag{3.73}$$

where $\tilde{E}$ is the envelope function of the electric field. Also concerning $\hat{P}_W$, the connection to the modal approach has been shown in Abram and Cohen (1991).

The commutator equivalent of the slowly varying amplitude wave equation will be applied to the quantum-mechanical treatment of the propagation in a nonlinear medium. Let us consider equation (3.71) whose right-hand side does not contain $\hat{D}_0$ in contrast to the left-hand side. This problem can be remedied by defining an effective "SVA" momentum operator such that it obeys

$$\hat{G}_{\text{SVA}}^{\times}\hat{D}_0 = \frac{1}{2}\hat{G}_0^{\times}\hat{P}_W. \tag{3.74}$$

The solution $\hat{G}_{\text{SVA}}$ is the stationary part of the effective "interaction" momentum operator $\hat{G}_1$,

$$\hat{G}_1 = \frac{1}{2}\int \hat{B}_0\hat{P}_{\text{NL}}\,\mathrm{d}^3\mathbf{r}, \tag{3.75}$$

namely

$$\hat{G}_{\text{SVA}} \equiv \hat{G}_{\text{SVA}}^{(n)} = \frac{\chi^{(n)}}{n+1}\int \hat{R}_{n+1}\,\mathrm{d}^3\mathbf{r}, \tag{3.76}$$

with

$$\hat{R}_n = 2^{-n+1}\sum_{m=0}^{[\frac{n}{2}]-1}\frac{n!}{(n-2m-1)!(2m+1)!}\epsilon^{-m}\hat{B}_0^{2m+1}\hat{E}_0^{n-2m-1}. \tag{3.77}$$

Also, in the context of (3.76) and (3.77) for $\hat{G}_{\text{SVA}}$, the connection to the modal approach can be shown. With definition (3.74), the commutator wave equation (3.71) can be written as

$$(\hat{G}_{\text{SVA}}^{\times}\hat{G}_0^{\times} + \epsilon\hat{H}_{1S}^{\times}\hat{H}_0^{\times})\hat{D}_0 = \hat{0}. \tag{3.78}$$

In this form, the commutator SVA equation relates directly the slow component of the temporal evolution of a short pulse of the displacement field $\hat{D}_0$ to the long-scale modulation of its spatial progression.

In order to clarify the role of Equation (3.78), the forward (+) and backward (−) polarization waves are defined in analogy with (3.47) and (3.49),

$$\hat{V}^{\pm} = \hat{D} \pm \sqrt{\epsilon}\hat{B}, \tag{3.79}$$

which in the absence of the nonlinearity have the form

$$\hat{V}_0^{\pm} = \epsilon \hat{W}_v^{\pm}, \tag{3.80}$$

where in accordance with the perturbation theory the forward and backward electromagnetic waves are defined as

$$\hat{W}_v^{\pm} = \hat{E}_0 \pm v\hat{B}_0. \tag{3.81}$$

Relation (3.63) now becomes

$$\hat{V}^{+}(z,t) \approx \exp(\mathrm{i}t\hat{H}_{1S}^{\times})\epsilon\hat{W}_v^{+}(z,t) + \hat{V}_1^{+}(z,t), \tag{3.82}$$

$$\hat{V}^{-}(z,t) \approx \exp(\mathrm{i}t\hat{H}_{1S}^{\times})\epsilon\hat{W}_v^{-}(z,t) + \hat{V}_1^{-}(z,t), \tag{3.83}$$

where $\hat{V}_1^{\pm}$ are the first-order corrections to $\hat{V}$ given by equations analogous to (3.64). Equation (3.78) simplifies to

$$\sqrt{\epsilon}\hat{H}_{1S}^{\times}\hat{W}_v^{+} = -\hat{G}_{\mathrm{SVA}}^{\times}\hat{W}_v^{+}, \tag{3.84}$$

$$\sqrt{\epsilon}\hat{H}_{1S}^{\times}\hat{W}_v^{-} = \hat{G}_{\mathrm{SVA}}^{\times}\hat{W}_v^{-}, \tag{3.85}$$

for the forward-going and backward-going waves, respectively. These equations provide a simple rule for converting the temporal evolution of the modulation envelope to the spatial progression, i.e. relations (3.82) and (3.83) can be written as follows

$$\hat{V}^{+}(z,t) = \epsilon\exp(-\mathrm{i}vt\hat{G}_{\mathrm{SVA}}^{\times})\hat{W}_v^{+}(z-vt,0) + \hat{V}_1^{+}(z,t), \tag{3.86}$$

$$\hat{V}^{-}(z,t) = \epsilon\exp(\mathrm{i}vt\hat{G}_{\mathrm{SVA}}^{\times})\hat{W}_v^{-}(z+vt,0) + \hat{V}_1^{-}(z,t). \tag{3.87}$$

In most practical situations, the first-order terms $\hat{V}_1^{\pm}$ may be neglected and $\hat{W}_v^{\pm}$ can be introduced also on the left-hand sides of equations (3.86) and (3.87) using (3.80). Nevertheless, $\hat{V}_1^{\pm}$ play an important role in that they incorporate the coupling to the wave going in the opposite direction and do give rise to the nonlinear reflection.

As an illustration of the above quantum treatment, the travelling-wave generation of squeezed light by the parametric down-conversion of a short pulse is examined.

For the case of a classical pump, this problem was treated through a modal analysis by Tucker and Walls (1969). More recently, Yurke et al. (1987) and Caves and Crouch (1987) treated this problem by using spatial differential equations for appropriately defined creation and annihilation operators.

The full modulated wave $\hat{\overline{W}}$ can be written in terms of a carrier wave $\hat{W}$ as follows

$$\hat{\overline{W}}(z,t) = \exp(-\mathrm{i}z\hat{G}^{\times}_{\mathrm{SVA}})\hat{W}(z,t), \tag{3.88}$$

where the subscript v and the superscript $+$ have been omitted. For a medium that exhibits a second-order nonlinearity, the right-hand side of relation (3.84) can be expressed as

$$-\hat{G}^{\times}_{\mathrm{SVA}}\hat{W} = -\frac{1}{2}\frac{\chi^{(2)}}{\sqrt{\epsilon}}\hat{H}_0^{\times}\hat{W}^2. \tag{3.89}$$

It is convenient to separate the field into its positive- and negative-frequency parts:

$$\hat{W} = 2\left(\hat{E}^{(+)} + \hat{E}^{(-)}\right), \tag{3.90}$$

where the factor of 2 arises, because it is not in definition (3.47). Similarly, the modulated wave solution $\hat{\overline{W}}$ can be separated into its positive- and negative-frequency parts

$$\hat{\overline{W}} = 2\left(\hat{\overline{E}}^{(+)} + \hat{\overline{E}}^{(-)}\right). \tag{3.91}$$

In the first-order perturbative treatment, the optical frequencies retain the same sign and relation (3.88) can be modified to

$$\hat{\overline{E}}^{(+)}(z,t) = \exp(-\mathrm{i}z\hat{G}^{\times}_{\mathrm{SVA}})\hat{E}^{(+)}(z,t), \tag{3.92}$$

with a similar equation holding for the negative-frequency part.

In Abram and Cohen (1991), equations similar to familiar classical first-order differential equations have been derived (see Shen 1984). The two fields involved in parametric down-conversion are introduced: the pump field with the central pump frequency ω_P and the signal field that oscillates at approximately ω_S, $\omega_S = \frac{\omega_P}{2}$. In fact, we introduce also the notation $\hat{E}_P^{(\pm)}$, $\hat{E}_S^{(\pm)}$, $\hat{\overline{E}}_P^{(+)}$, $\hat{\overline{E}}_S^{(+)}$, and we modify relation (3.88) to

$$\begin{aligned}\hat{\overline{E}}_P^{(+)}(z,t) &= \exp(-\mathrm{i}z\hat{G}^{\times}_{\mathrm{SVA}})\hat{E}_P^{(+)}(z,t),\\ \hat{\overline{E}}_S^{(+)}(z,t) &= \exp(-\mathrm{i}z\hat{G}^{\times}_{\mathrm{SVA}})\hat{E}_S^{(+)}(z,t),\end{aligned} \tag{3.93}$$

and similar equations for the negative-frequency parts. The pump field consists of a short pulse whose duration T_P is much longer than the optical period $\frac{2\pi}{\omega_P}$. On this assumption, relation (3.89) for the signal field becomes

$$\hat{G}^{\times}_{\mathrm{SVA}}\hat{E}_S^{(+)} = -\kappa\hat{E}_P^{(+)}\hat{E}_S^{(-)}, \tag{3.94}$$

$$\hat{G}^{\times}_{\mathrm{SVA}}\hat{E}_S^{(-)} = \kappa\hat{E}_P^{(-)}\hat{E}_S^{(+)}, \tag{3.95}$$

where $\kappa = \frac{\chi^{(2)}\omega_S}{\sqrt{\epsilon}}$,

$$\hat{G}^{\times}_{\mathrm{SVA}}\hat{E}_P^{(+)} = -\kappa\hat{E}_S^{(+)}\hat{E}_S^{(+)} \tag{3.96}$$

and

$$\hat{G}^{\times}_{\mathrm{SVA}}\hat{E}_P^{(-)} = \kappa\hat{E}_S^{(-)}\hat{E}_S^{(-)}. \tag{3.97}$$

Let us observe that on the substitution $\hat{Q} = \hat{E}_S^{(+)}, \hat{E}_S^{(-)}, \hat{E}_P^{(+)}, \hat{E}_P^{(-)}$ into relation (3.35) and comparison with (3.94), (3.95), (3.96), and (3.97), we have an analogue of the classical description of the spatial progression. Within the undepleted pump assumption, the solution of (3.93) becomes

$$\begin{aligned}\hat{\overline{E}}_S^{(+)}(z,t) &= \cosh_{\mathcal{N}}\left[\kappa z\sqrt{\hat{I}_P(z,t)}\right]\hat{E}_S^{(+)}(z,t)\\ &+ \mathrm{i}\left\{\sinh\left[\kappa z\sqrt{\hat{I}_P(z,t)}\right]\frac{\hat{E}_P^{(+)}(z,t)}{\sqrt{\hat{I}_P(z,t)}}\right\}_{\mathcal{N}}\hat{E}_S^{(-)}(z,t),\end{aligned} \tag{3.98}$$

where

$$\hat{I}_P(z,t) = \hat{E}_P^{(-)}(z,t)\hat{E}_P^{(+)}(z,t) \tag{3.99}$$

is essentially the intensity operator for the pump field and the subscript $\mathcal{N}$ denotes the normal ordering of the operators $\hat{E}_P^{(\pm)}$, which means that $\hat{E}_P^{(-)}$ stands to the left from $\hat{E}_P^{(+)}$.

Deviating slightly from Abram and Cohen (1991), we formulate the interaction picture as follows:

$$|(P+S)(t)\rangle = \hat{U}(t)|P(t)\rangle, \tag{3.100}$$

where $|P(t)\rangle$ is the state of the field (pump and signal) in the remote past,

$$|P(t)\rangle = |P(t)\rangle_P \otimes |0\rangle_S. \tag{3.101}$$

We are now in a position to describe a travelling-wave experiment of parametric down-conversion. In such an experiment, a pump pulse expressed by the state $|P(t)\rangle_P$ initially traverses a nonlinear crystal that extends from $z = 0$ to L and generates a signal pulse in the course of its propagation. Using the previous approximations, we rewrite (3.100) as

$$|(P+S)(t)\rangle = \exp(\mathrm{i}\hat{G}_{\mathrm{SVA}}vt)|P(t)\rangle. \tag{3.102}$$

The dependence on z which has been introduced by the substitution $t = \frac{z}{v}$ in the previous exposition is not explicit here. The interaction picture is not needed for calculation of expectation values of the observables.

For example, a measurement of the intensity profile of the signal pulse can be expressed by the equal-time function $I_S(t)$

$$I_S(t) = \langle P(t)|\hat{\overline{E}}_S^{(-)}(L,t)\hat{\overline{E}}_S^{(+)}(L,t)|P(t)\rangle. \tag{3.103}$$

Since $\hat{\overline{E}}_S^{(+)}(L,t) = \hat{\overline{E}}_S^{(+)}(L - vt, 0)$, relation (3.103) after a simplification yields

$$I_S(t) = {}_P\left\langle P\left(t - \frac{L}{v}\right)\right|\sinh^2_{\mathcal{N}}\left[\kappa L\sqrt{\hat{I}_P\left(0, t - \frac{L}{v}\right)}\right]\left|P\left(t - \frac{L}{v}\right)\right\rangle_P \\ \times {}_S\langle 0|\hat{E}_S^{(+)}\hat{E}_S^{(-)}|0\rangle_S, \tag{3.104}$$

with $\hat{I}_P\left(0, t - \frac{L}{v}\right) = \hat{I}_P(L - vt, 0)$; relation (3.104) is simplified by omitting the terms without the antinormal ordering of signal field operators. This can be considered as legitimate, because the vacuum expectation value of the operator product ${}_S\langle 0|\hat{E}_S^{(+)}\hat{E}_S^{(-)}|0\rangle_S$ diverges. There exist results for the two-time correlation function $g_S^{(1)}(t_2, t_1)$ and for the nth-order photon-coincidence rate for the signal pulse $g_S^{(n)}$.

The numerical results have been obtained for a laser pulse that has an amplitude profile $A_P(t')$ at $z = 0$, but have been restricted to the intensity profile measured at the exit of the crystal $z = L$. Let us assume that $|P(t')\rangle$ is a coherent state with the property

$$\hat{E}_P^{(+)}(0, t')|P(t')\rangle = A_P\left(t'\right)U_P\left(t'\right)|P(t')\rangle, \tag{3.105}$$

where

$$A_P\left(t'\right) \geq 0, \ \left|U_P\left(t'\right)\right| = 1. \tag{3.106}$$

Hence,

$${}_P\left\langle P\left(t - \frac{L}{v}\right)\right|\hat{I}_P\left(0, t - \frac{L}{v}\right)\left|P\left(t - \frac{L}{v}\right)\right\rangle_P = \left[A_P\left(t - \frac{L}{v}\right)\right]^2, \tag{3.107}$$

so that (after a renormalization)

$$I_S(L,t) = \sinh^2\left[\kappa L A_P\left(t - \frac{L}{v}\right)\right]. \tag{3.108}$$

3.1.3 Space–Time Displacement Operators

Serulnik and Ben-Aryeh (1991) have discussed a general problem of the electromagnetic wave propagation through nonlinear nondispersive media. They have used the four-dimensional formalism of the field theory in order to develop an extension of the formalism introduced by Hillery and Mlodinow (1984). The complications following from the common definitions for the vector and scalar potentials are indicated. It is shown that the scalar potential can be neglected only by using alternative definitions.

First, it is shown that the conventional approach that uses the standard potentials **A** and V is not appropriate for treating the general case of nonlinear polarization when $\nabla \cdot \mathbf{P} \neq 0$, since for such cases V does not vanish. As a solution to this problem it is proposed to use vector potential ψ,

$$\mathbf{D} = -\nabla \times \psi, \tag{3.109}$$

which fulfils the relation $\nabla \cdot \mathbf{D} = 0$. This choice enables Serulnik and Ben-Aryeh (1991) to work in the new Coulomb gauge, where $\nabla \cdot \psi = 0$, so that from the condition $\nabla \cdot \mathbf{B} = 0$ it follows that the dual scalar potential ξ obeys the equation

$$\nabla^2 \xi = 0. \tag{3.110}$$

It is then consistent to assume $\xi = 0$ everywhere in a nonlinear medium and the dual scalar potential need not be taken into account.

The Lagrangian and Hamiltonian densities are derived from the Maxwell equations by using nonconventional definitions for the scalar and vector potentials. The general form of the energy–momentum tensor is derived and explicit expression for its elements is given. The relation between this tensor and the space–time description of propagation is analysed. Further the quantization is performed and the properties of space–time displacement operators are presented. The space–time is described by a Lie transform (Steinberg 1985). The displacement operators are obtained from the energy–momentum tensor developed by Serulnik and Ben-Aryeh (1991) with an alternative definition for the vector potential. It has been possible to obtain explicit expressions for all the elements of the energy–momentum tensor and to discuss their physical meaning.

In the following we will show that the relationship between the energy–momentum tensor and the space–time description of propagation is different from that derived by Serulnik and Ben-Aryeh (1991). Let us restrict ourselves to the usually treated one-dimensional case, where only the fields E_1, D_1, B_2, and Λ_2 are significant, where we use $\Lambda_2 = -\psi_2$ according to Drummond (1990, 1994). The arguments of these fields are x_3 and ct. The corresponding quantum fields obey the commutation relation

$$[\hat{\Lambda}_2(x_3, ct), \hat{B}_2(y_3, ct)] = i\hbar c\delta(x_3 - y_3)\hat{1}, \tag{3.111}$$

$$[\hat{D}_1(x_3, ct), \hat{B}_2(y_3, ct)] = -i\hbar c\delta'(x_3 - y_3)\hat{1}. \tag{3.112}$$

Considering for this case the Hamiltonian density

$$\hat{H} = \frac{1}{2}\left(\overbrace{\frac{1}{\epsilon}}^{\zeta}\hat{D}_1^2 + \hat{B}_2^2\right), \tag{3.113}$$

where the right-hand side is symmetrically ordered (cf. Abram and Cohen 1991), we obtain the equations of motion in the Heisenberg picture as follows

$$\frac{\partial \hat{\Lambda}_2}{\partial t} = -\frac{\mathrm{i}}{\hbar}\left[\hat{\Lambda}_2, \int \hat{H}\,\mathrm{d}x_3\right] = c\hat{B}_2, \tag{3.114}$$

$$\frac{\partial \hat{B}_2}{\partial t} = -\frac{\mathrm{i}}{\hbar}\left[\hat{B}_2, \int \hat{H}\,\mathrm{d}x_3\right] = -c\frac{\partial\left(\frac{1}{\epsilon}\hat{D}_1\right)}{\partial x_3}. \tag{3.115}$$

Relation (3.114) explains the role of the dual vector potential and relation (3.115) is essentially the second of the Maxwell evolution equations. We could obtain the first of them as the equation of motion for the quantum field $\hat{D}_1$.

It is a question whether the tensor element

$$\hat{T}^{03} = \frac{1}{c}(\hat{D}_1\hat{B}_2)_S \tag{3.116}$$

is a correct quantum density for generation of the displacement as Serulnik and Ben-Aryeh (1991) indicate. The presumable equations of the spatial progression are

$$\frac{\partial \hat{\Lambda}_2}{\partial x_3} = \frac{\mathrm{i}}{\hbar c}\left[\hat{\Lambda}_2, \int (\hat{D}_1\hat{B}_2)_S\,\mathrm{d}x_3\right] = -\hat{D}_1, \tag{3.117}$$

$$\frac{\partial \hat{B}_2}{\partial x_3} = \frac{\mathrm{i}}{\hbar c}\left[\hat{B}_2, \int (\hat{D}_1\hat{B}_2)_S\,\mathrm{d}x_3\right] = \frac{\partial \hat{B}_2}{\partial x_3}. \tag{3.118}$$

Relation (3.117) expresses the role of the dual vector potential and relation (3.118) is a mere tautology. The same is obtained for the quantum field $\hat{D}_1$. This failure of the application of the ordinary presentations of quantum field theory has been published by Ben-Aryeh and Serulnik (1991).

Considering, in contrast, the equal-space commutators

$$[\hat{\Lambda}_2(x_3, ct), \hat{D}_1(x_3, ct')] = \hat{0}, \tag{3.119}$$

$$[\hat{\Lambda}_2(x_3, ct), \hat{B}_2(x_3, ct')] = \mathrm{i}\hbar c\delta(ct - ct')\hat{1}, \tag{3.120}$$

$$[\hat{B}_2(x_3, ct), \hat{B}_2(x_3, ct')] = \mathrm{i}\hbar c\delta'(ct - ct')\hat{1}, \tag{3.121}$$

$$[\hat{D}_1(x_3, ct), \hat{B}_2(x_3, ct')] = \hat{0}, \tag{3.122}$$

$$[\hat{D}_1(x_3, ct), \hat{D}_1(x_3, ct')] = \mathrm{i}\hbar c\epsilon\delta'(ct - ct')\hat{1}, \tag{3.123}$$

we obtain peculiar equation of the spatial progression

$$\frac{\partial \hat{\Lambda}_2}{\partial x_3} = \frac{\mathrm{i}}{\hbar}\left[\hat{\Lambda}_2, \int (\hat{D}_1 \hat{B}_2)_S \,\mathrm{d}t\right] = -\hat{D}_1, \tag{3.124}$$

$$\frac{\partial \hat{D}_1}{\partial x_3} = \frac{\mathrm{i}}{\hbar}\left[\hat{D}_1, \int (\hat{D}_1 \hat{B}_2)_S \,\mathrm{d}t\right] = -\frac{\partial \hat{B}_2}{\partial t}. \tag{3.125}$$

Relation (3.124) expresses the role of the dual vector potential and relation (3.125) is essentially the second of the Maxwell evolution equations. We could obtain the first of them as the equation of the spatial progression for the quantum field $\hat{B}_2$. Since we must often guess the commutators never known before, the above example is a warning against excessive trust in the spatial progression technique.

For a medium with nonlinear polarization, the global nature of creation and annihilation operators is lost. By consistently following this idea, Serulnik and Ben-Aryeh (1991) have introduced the shift operators which, by their definition, are based on the energy–momentum tensor. They have followed in their treatment Peierls' solution of the problem of momentum conservation in matter (Peierls 1976, 1985) by which the atoms or the bulk matter is considered to be at rest while the electromagnetic field is propagating. They show that it is always possible to relate the external field in front of the medium to that behind it by the use of the shift operators, that is by a Lie transformation. As we can see from relations (3.114), (3.115), (3.124), and (3.125), there exists no space–time description which in addition to the so-called time displacement operators suggests the use of their space-displacement analogues.

Leonhardt (2000) has determined an energy–momentum tensor of the electromagnetic fields in quantum dielectrics. The tensor is Abraham's plus the energy–momentum of the medium characterized by a dielectric pressure and enthalpy density (Abraham 1909). While the consistency of this picture with the theory of dielectrics has been demonstrated, a direct derivation from the first principles has been announced only.

The theory of the radiation pressure on dielectric surfaces (Loudon 2002) accepts the expression for the momentum density of an electromagnetic wave in a transparent material medium due to Abraham (1909). The force of the radiation pressure is obtained similarly using the Lorentz force density operator as using the momentum density according to Abraham (1909, 1910). A colloquium has been devoted to the momentum of an electromagnetic wave in dielectric media (Pfeifer et al. 2007).

3.1.4 Generator of Spatial Progression

Theoretical methods for treating propagation in quantum optics have been developed in which the momentum operator is used in addition to the Hamiltonian. A successful quantum-mechanical analysis has been given for various physical systems which include amplification and coupling between electromagnetic modes

(Toren and Ben-Aryeh 1994). Distributed feedback lasers have been described, but the overarching generalization of both successful analyses has not been developed. The authors have drawn attention to the distributed feedback lasers (Yariv and Yeh 1984, Yariv 1989), in which contradirectional beams are amplified by an active medium and are coupled by a small periodic perturbation of a refractive index.

The energy and momentum properties of the electromagnetic field can be described, in a four-dimensional form, by the energy–momentum tensor T^{jk}, where $j,k = 0,1,2,3$ (Roman 1969),

$$T^{jk} = \begin{bmatrix} \mathcal{H} & g_x & g_y & g_z \\ S_x & \sigma_{xx} & \sigma_{xy} & \sigma_{xz} \\ S_y & \sigma_{yx} & \sigma_{yy} & \sigma_{yz} \\ S_z & \sigma_{zx} & \sigma_{zy} & \sigma_{zz} \end{bmatrix}. \tag{3.126}$$

The tensor element T^{00} represents the energy density. The density of the vectorial momentum (proportional to $\mathbf{D} \times \mathbf{B}$) is represented by the vector (g_x, g_y, g_z). Let us take further, for example, line 3. The tensor element T^{30} is the component of the Poynting vector standing for the flux of energy in the z-direction. The vector $(\sigma_{zx}, \sigma_{zy}, \sigma_{zz})$ refers to a flux of momentum in the propagation direction of z.

In the conventional approach (Roman 1969), the four-vector p^{0k} is defined as

$$p^{0k} = \int\int\int T^{0k}(x, y, z, t)\, \mathrm{d}x\, \mathrm{d}y\, \mathrm{d}z. \tag{3.127}$$

The energy p^{00} is used as the Hamiltonian for the description of time evolution, but the momentum component p^{03} rather merely translates in the z-direction. Ben-Aryeh and Serulnik (1991) have shown that for the description of the spatial progression, the four-vector p^{3k}

$$p^{3k} = \int\int\int T^{3k}(x, y, z, t) c\, \mathrm{d}t\, \mathrm{d}x\, \mathrm{d}y \tag{3.128}$$

can be used. The momentum component in the z-direction p^{33} can be used as the generator of the spatial progression, but the energy p^{30} rather merely causes translation of the field in time.

In Toren and Ben-Aryeh (1994), problems of propagation are treated by expanding the field operators in terms of mode operators associated with definite frequencies. Starting with Caves and Crouch (1987), the approach has been associated with the conservation of commutation relations for creation and annihilation operators, which are space dependent (cf. Huttner et al. 1990). Imoto (1989) has developed the basic equation of motion by using a modified procedure of canonical quantization in which time and space coordinates are interchanged in comparison with the conventional procedure. Ben-Aryeh et al. (1992) did the same by using a slightly different notation. Toren and Ben-Aryeh (1994) dissociate themselves from such an approach, but they are not very explicit about the point that the use of the integrals (3.128) is compatible with the canonical quantization in which the time coordinate plays the usual role.

Linear amplification is treated by the use of momentum for space-dependent amplification. Travelling-wave attenuators and amplifiers can be treated as continuous limits of an array of beam splitters (Jeffers et al. 1993, Ban 1994). According to Toren and Ben-Aryeh (1994), the propagating modes are coupled to a momentum reservoir. The Hamiltonian of this system is given by the relation

$$\hat{H} = \hbar\Big(\omega\hat{a}^\dagger\hat{a} - \sum_j \omega_{j\mathrm{res}}\hat{b}_j^\dagger\hat{b}_j\Big) \tag{3.129}$$

and the total momentum operator is

$$\hat{G} = \hbar\Big[\beta\hat{a}^\dagger\hat{a} - \sum_j \beta_{j\mathrm{res}}\hat{b}_j^\dagger\hat{b}_j + \sum_j(\kappa_j\hat{a}^\dagger\hat{b}_j^\dagger + \kappa_j^*\hat{a}\hat{b}_j)\Big], \tag{3.130}$$

where the subscript res stands for the reservoir and κ_j are appropriate coupling constants, $\hat{a}$ and $\hat{b}_j$ represent (in the zeroth-order) modes which are propagating in the positive direction of the z-axis.

The equations of motion obtained from the momentum operator (3.130) are

$$\frac{\mathrm{d}\hat{a}}{\mathrm{d}z} = \frac{\mathrm{i}}{\hbar}[\hat{a}, \hat{G}] = \mathrm{i}\beta\hat{a} + \mathrm{i}\sum_j \kappa_j\hat{b}_j^\dagger, \tag{3.131}$$

$$\frac{\mathrm{d}\hat{b}_j^\dagger}{\mathrm{d}z} = \frac{\mathrm{i}}{\hbar}[\hat{b}_j^\dagger, \hat{G}] = -\mathrm{i}\kappa_j^*\hat{a} + \mathrm{i}\beta_{j\mathrm{res}}\hat{b}_j^\dagger. \tag{3.132}$$

By using the spatial Wigner–Weisskopf approximation, the Heisenberg–Langevin equations can be obtained

$$\frac{\mathrm{d}\hat{a}}{\mathrm{d}z} = \left[\mathrm{i}(\beta - \Delta\beta) + \frac{1}{2}\gamma\right]\hat{a} + \hat{L}^\dagger, \tag{3.133}$$

where

$$\Delta\beta = -\mathrm{V.p.}\int_{-\infty}^{\infty} \frac{|\kappa(\beta_{\mathrm{res}})|^2\rho(\beta_{\mathrm{res}})}{\beta_{\mathrm{res}} - \beta}\,\mathrm{d}\beta_{\mathrm{res}}, \tag{3.134}$$

$$\gamma = \{2\pi|\kappa(\beta_{\mathrm{res}})|^2\,\rho(\beta_{\mathrm{res}})\}\,|_{\beta_{\mathrm{res}}=\beta}, \tag{3.135}$$

with $\rho(\beta_{\mathrm{res}})$ the density function of the reservoir wave propagation constants $\beta_{j\mathrm{res}}$, and

$$\hat{L}^\dagger = \sum_j \mathrm{i}\kappa_j\hat{b}_j^\dagger e^{\mathrm{i}\beta_{j\mathrm{res}}z}. \tag{3.136}$$

The codirectional coupling is analysed. It is assumed that two modes are propagating in the same direction and they are coupled by a periodic change in the refractive index. For a classical description, we refer to Yariv and Yeh (1984) and Yariv (1989). The Hamiltonian is given by

$$\hat{H}_0 = \hat{H} = \hbar\omega[\hat{a}_1^\dagger \hat{a}_1 + \hat{a}_2^\dagger \hat{a}_2], \tag{3.137}$$

where the classical relation $\omega_1 = \omega_2 = \omega$ has been used. The total momentum operator is

$$\hat{G} = \hbar[\beta_1 \hat{a}_1^\dagger \hat{a}_1 + \beta_2 \hat{a}_2^\dagger \hat{a}_2 + \tilde{\kappa}\hat{a}_1 \hat{a}_2^\dagger + \tilde{\kappa}^* \hat{a}_1^\dagger \hat{a}_2], \tag{3.138}$$

where β_1 and β_2 are components of the wave vectors of the two modes in the propagation direction of z and

$$\tilde{\kappa} = \kappa \exp\left(-\frac{im2\pi}{\Lambda}z\right), \tag{3.139}$$

with κ a coupling constant, m an integer, and Λ the "wavelength" of the spatial periodic change in the index of refraction (a perturbation in the dielectric constant). In this connection papers Peřinová et al. (1991) and Ben-Aryeh et al. (1992) are criticized that they do not take account of the spatial dependence (3.139).

The equations of motion obtained from the momentum operator (3.138) are

$$\frac{\mathrm{d}\hat{a}_1}{\mathrm{d}z} = \frac{\mathrm{i}}{\hbar}[\hat{a}_1, \hat{G}] = \mathrm{i}\beta_1 \hat{a}_1 + \mathrm{i}\tilde{\kappa}^* \hat{a}_2, \tag{3.140}$$

$$\frac{\mathrm{d}\hat{a}_2}{\mathrm{d}z} = \frac{\mathrm{i}}{\hbar}[\hat{a}_2, \hat{G}] = \mathrm{i}\tilde{\kappa}\hat{a}_1 + \mathrm{i}\beta_2 \hat{a}_2. \tag{3.141}$$

We define slowly varying operators of the form

$$\hat{A}_1(z) \equiv \hat{a}_1(z)e^{-\mathrm{i}\beta_1 z}, \, \hat{A}_2(z) \equiv \hat{a}_2(z)e^{-\mathrm{i}\beta_2 z}. \tag{3.142}$$

Substituting the operators (3.142) into equations (3.140), (3.141), we get

$$\frac{\mathrm{d}\hat{A}_1}{\mathrm{d}z} = i\kappa^* \hat{A}_2 e^{-\mathrm{i}\Delta\beta z}, \tag{3.143}$$

$$\frac{\mathrm{d}\hat{A}_2}{\mathrm{d}z} = \mathrm{i}\kappa \hat{A}_1 e^{\mathrm{i}\Delta\beta z}, \tag{3.144}$$

where

$$\Delta\beta \equiv \beta_1 - \beta_2 - m\frac{2\pi}{\Lambda} \tag{3.145}$$

is the mismatch. A "field" mismatch may be cancelled by a medium component. For the input–output relations we refer to Yariv and Yeh (1984), Yariv (1989), and Peřinová et al. (1991).

In Peřinová et al. (1991), $m = 0$ and $2\delta = -\Delta\beta$ has been introduced in an application to the codirectional coupler. In general, the solution to equations (3.143), (3.144) coincides with the classical solution, $\hat{A}_j \leftrightarrow A_j^*$, $j = 1, 2$, where A_1, A_2 are the amplitudes of the waves propagating in the $+z$-direction.

The counterdirectional coupling is analysed in Toren and Ben-Aryeh (1994). The total Hamiltonian is given by

$$\hat{H}_0 \equiv \hat{H} = \hbar\omega[\hat{a}_1^\dagger\hat{a}_1 - \hat{a}_2^\dagger\hat{a}_2]. \tag{3.146}$$

The momentum operator is

$$\hat{G} = \hbar[\beta_1\hat{a}_1^\dagger\hat{a}_1 - \beta_2\hat{a}_2\hat{a}_2^\dagger + \tilde{\kappa}\hat{a}_1\hat{a}_2^\dagger + \tilde{\kappa}^*\hat{a}_1^\dagger\hat{a}_2]. \tag{3.147}$$

It is reasonable that the Hamiltonian and the zeroth-order operator are related, respectively, to the flux of energy and that of the component of momentum in the z-direction. Compared to Toren and Ben-Aryeh (1994), the operators $\hat{a}_2$ and $\hat{a}_2^\dagger$ have been exchanged. They criticize our assumption $[\hat{a}_2, \hat{a}_2^\dagger] \stackrel{?}{=} -\hat{1}$, which we obtained in Peřinová et al. (1991) from the usual equal-space commutator $[\hat{a}_2, \hat{a}_2^\dagger] = \hat{1}$ by this interchange. It is tempting to have the same alternation between the opposite-going modes as can be seen in comparison of (3.524) with (3.526) (cf. Abram and Cohen 1994). The equations of motion obtained from operator (3.147) are given by

$$\frac{\mathrm{d}\hat{a}_1}{\mathrm{d}z} = \frac{\mathrm{i}}{\hbar}[\hat{a}_1, \hat{G}\big|_{\hat{a}_2\leftrightarrow\hat{a}_2^\dagger}] = \mathrm{i}\beta_1\hat{a}_1 + \mathrm{i}\tilde{\kappa}^*\hat{a}_2, \tag{3.148}$$

$$\frac{\mathrm{d}\hat{a}_2}{\mathrm{d}z} = \frac{\mathrm{i}}{\hbar}[\hat{a}_2, \hat{G}\big|_{\hat{a}_2\leftrightarrow\hat{a}_2^\dagger}] = -\mathrm{i}\tilde{\kappa}\hat{a}_1 + \mathrm{i}\beta_2\hat{a}_2. \tag{3.149}$$

Using the slowly varying operators (3.142), in contrast with Toren and Ben-Aryeh (1994), we obtain that

$$\frac{\mathrm{d}\hat{A}_1}{\mathrm{d}z} = \mathrm{i}\tilde{\kappa}^*\hat{A}_2 e^{-\mathrm{i}\Delta\beta z},$$
$$\frac{\mathrm{d}\hat{A}_2}{\mathrm{d}z} = -\mathrm{i}\tilde{\kappa}\hat{A}_1 e^{\mathrm{i}\Delta\beta z}, \tag{3.150}$$

where $\Delta\beta$ has been defined by equation (3.145).

In Peřinová et al. (1991) still $2\delta = -\Delta\beta$ holds in an application to the counterdirectional coupler. The solution to equations (3.150) coincides with the classical solution, $\hat{A}_j \leftrightarrow A_j^*$, $j = 1, 2$, where A_1, A_2 are the amplitudes of the waves propagating in the $+z$- and $-z$-directions. In Yariv and Yeh (1984), the solution to the corresponding classical equations has been obtained for the boundary conditions

$A_1(z)_{|z=0} = A_1(0)$, $A_2(z)_{|z=L} = A_2(L)$. First, however, one obtains the solution for the usual condition at $z = 0$.

In Peřinová et al. (1991), the output operators have been obtained in terms of the input ones. While we simply determine the operators $\hat{A}_1(L)$, $\hat{A}_2(0)$ from two equations for the operators $\hat{A}_1(0)$, $\hat{A}_2(0)$, $\hat{A}_1(L)$, $\hat{A}_2(L)$, we observe that in this procedure the equal-space commutator $[\hat{A}_2, \hat{A}_2^\dagger] \stackrel{?}{=} -\hat{1}$ must depend on both z and L in a complicated manner, and simplifies to $[\hat{A}_2, \hat{A}_2^\dagger] = \hat{1}$. Since the commutators correspond to the Poisson brackets, much is illustrated by the appropriate classical theory (Luis and Peřina 1996b). One must be aware of the fact that in formulating the theory, Luis and Peřina (1996b) have avoided the above considerations on the z coordinate and the generator of spatial progression and they used in the bulk of their paper the usual time dependence and the Hamiltonian function. Although still obscure in the case of commutators, the situation is clear in the classical case, when the input–output transformation is characterized by the usual Poisson brackets and the solution for the usual boundary conditions at $z = 0$ requires the noncanonical transformation $\alpha_2 \leftrightarrow \alpha_2^*$, with the complex amplitude α_2. The richness of their theory is due to nonlinearities, whereas it is shown that in the quantum case only a poor linear theory is possible.

The difficulty lies in the formulation of an appropriate dynamical operator. Tarasov (2001) has defined a map of a dynamical nonlinear operator into a dynamical superoperator. He had in mind quantum dynamics of non-Hamiltonian and dissipative systems.

A quantum-mechanical treatment of distributed feedback laser using the momentum operator in addition to the Hamiltonian is developed in Toren and Ben-Aryeh (1994). The authors start from the classical description based on two coupled equations

$$\begin{aligned} \frac{\mathrm{d}A_1}{\mathrm{d}z} &= \frac{1}{2}\gamma A_1 - \mathrm{i}\kappa A_2 e^{\mathrm{i}\Delta\beta z}, \\ \frac{\mathrm{d}A_2}{\mathrm{d}z} &= \mathrm{i}\kappa^* A_1 e^{-\mathrm{i}\Delta\beta z} - \frac{1}{2}\gamma A_2, \end{aligned} \tag{3.151}$$

where A_1, A_2 are the amplitudes of the waves propagating in the $+z$- and $-z$-directions, respectively, κ is the coupling constant, γ the amplification constant, and $\Delta\beta$ is given by (3.145), with $\beta_1 = \beta$, $\beta_2 = -\beta$. The solution of the classical equations is well known (Yariv and Yeh 1984, Yariv 1989) and it shows that on a condition the amplification becomes extremely large. The classical theory does not include the quantum noise which follows from the amplification process. Unfortunately, Toren and Ben-Aryeh (1994) did not develop the overarching generalization of the analysis of amplification and that of the counterdirectional coupling. To the best of our knowledge, such a quantum-mechanical theory is not in hand.

The treatment of parametric down-conversion and parametric up-conversion by Dechoum et al. (2000) is interesting with its use of the Wigner representation of optical fields, but it starts just from the Maxwell equations for the field operators for the lossless neutral nonlinear dielectric medium. With the use of common approximation of treating the laser pump as classical, classical equations of nonlinear optics are obtained.

3.1.5 Nonlinear Optical Couplers

Optical couplers employing evanescent waves play an important role in optics, optoelectronics, and photonics where they may be conveniently used in the switching of light beams. They also provide a means for controlling light by light. Amplitude and intensity behaviour of linear couplers has been investigated extensively (Yariv and Yeh 1984, Solimeno et al. 1986, Saleh and Teich 1991). Substantial progress in controlling light beams has been achieved after nonlinear waveguides with both linear and nonlinear coupling have been taken into account (Finlayson et al. 1988, Townsend et al. 1989, Leutheuser et al. 1990, Weinert-Raczka and Lederer 1993, Assanto et al. 1994, Hatami-Hanza and Chu 1995, Hansen 1995, Weinert-Raczka 1996).

This gave new possibilities of fast all-optical switching, including digital switching, and reduction of switching power. New ways of controlling optical beams in nonlinear couplers have been invented.

Nonlinear waveguide materials used in composing nonlinear couplers provide new possibilities in constructing switching and memory elements for all-optical devices. These elements are necessary for further development of optical processing and computing. Classically, all-optical devices are analysed from the viewpoint of their amplitude or intensity dependences. However, they can be treated fully in quantum theory. Noise of light beams in nonlinear couplers is naturally included in this quantum treatment.

Peřina, Jr., and Peřina (2000) in Section 2 indicate a consequential use of the momentum operator we have mentioned in Section 3.1.4. In nonlinear quantum systems where both directions of propagation are present this formalism confronts difficulties. The generator of spatial progression is related to commutators which do not lead to proper input (and output) commutators. The two ways of introducing photon annihilation and creation operators obeying boson commutation relations are not consistent mutually. Working with operators is not secure.

The authors work with the two algebras transparently. The commutators which are preserved in spatial progression are exploited only for the derivation of evolution equations for operators. The proper input commutators are used to the other goals.

In order to determine quantum-statistical properties of light beams, solutions of nonlinear operator equations have to be found first. One can apply a short-length approximation. Or pump modes can be assumed to be in strong coherent states and a parametric approximation can be used.

The short-length approximation is used as a tool in the treatment of nonlinear quantum systems. Calculations with operators are safe when one direction of propagation is present, because the two algebras coincide. If both directions of propagation are present it is easy to use one of the two algebras in principle (cf. Peřinová et al. 1991). Paradoxes may occur.

For instance, the operator equations of spatial evolution differ from the Heisenberg equations in interaction picture only by one or a few changes of sign. But these changes entail, in the formalism of input commutators, that operator products on the right-hand sides of the equations may have an incidental order. Elas, this problem is not present in the formalism of the generator of spatial progression. We suppose it should be better to consider boundary-value problems for equations for c-numbers in the systems where both directions of propagation underlie to description and to quantize at the end.

The parametric approximation is used as another instrument. It leads to linear evolution equations of operators. Also here a description of a quantum system is specific in which two directions of propagation are present. But linear equations comprise no products on the right-hand sides.

A number of quantum descriptions are related to two modes and can be specified as two linear equations (and their conjugates) for the operators $\hat{A}_a$, $\hat{A}_b$, $\hat{A}_a^\dagger$, $\hat{A}_b^\dagger$. The right-hand sides of these equations are often independent of time. We may let $\lambda_{1,2,3,4}$ denote the eigenvalues of the matrix of the right-hand sides. Introducing operators $\hat{A}_c$, $\hat{A}_d$ as appropriate Bogoliubov transforms of the operators $\hat{A}_a$ and $\hat{A}_b$, one can express any quantum description (a regularity is assumed) in one of the six normal forms (Williamson 1936)

$$\frac{\mathrm{d}\hat{A}_c}{\mathrm{d}t} = a\hat{A}_c^\dagger, \quad \frac{\mathrm{d}\hat{A}_d}{\mathrm{d}t} = b\hat{A}_d^\dagger \tag{3.152}$$

for $\lambda_{1,2,3,4} = \pm a, \pm b$;

$$\frac{\mathrm{d}\hat{A}_c}{\mathrm{d}t} = a\hat{A}_c^\dagger + b\hat{A}_d, \quad \frac{\mathrm{d}\hat{A}_d}{\mathrm{d}t} = -b\hat{A}_c + a\hat{A}_d^\dagger \tag{3.153}$$

for $\lambda_{1,2,3,4} = \pm a \pm \mathrm{i}b$;

$$\frac{\mathrm{d}\hat{A}_c}{\mathrm{d}t} = -\mathrm{i}\rho a\hat{A}_c, \quad \frac{\mathrm{d}\hat{A}_d}{\mathrm{d}t} = -\mathrm{i}\sigma b\hat{A}_d, \quad \rho, \sigma = \pm 1, \tag{3.154}$$

for $\lambda_{1,2,3,4} = \pm \mathrm{i}a, \pm \mathrm{i}b$;

$$\frac{\mathrm{d}\hat{A}_c}{\mathrm{d}t} = a\hat{A}_c^\dagger, \quad \frac{\mathrm{d}\hat{A}_d}{\mathrm{d}t} = -\mathrm{i}\rho b\hat{A}_d, \quad \rho = \pm 1, \tag{3.155}$$

for $\lambda_{1,2,3,4} = \pm a, \pm \mathrm{i}b$;

$$\frac{\mathrm{d}\hat{A}_c}{\mathrm{d}t} = a\hat{A}_c^\dagger + \frac{1}{2}(\hat{A}_d^\dagger - \hat{A}_d), \quad \frac{\mathrm{d}\hat{A}_d}{\mathrm{d}t} = \frac{1}{2}(\hat{A}_c^\dagger + \hat{A}_c) + a\hat{A}_d^\dagger \tag{3.156}$$

for $\lambda_{1,2,3,4} = \pm a, \pm a$;

$$\frac{\mathrm{d}\hat{A}_c}{\mathrm{d}t} = \mathrm{i}\frac{\rho}{2}(\hat{A}_c^\dagger - \hat{A}_c) - a\hat{A}_d, \quad \frac{\mathrm{d}\hat{A}_d}{\mathrm{d}t} = a\hat{A}_c + \mathrm{i}\frac{\rho}{2}(\hat{A}_d^\dagger - \hat{A}_d), \quad \rho = \pm 1 \tag{3.157}$$

for $\lambda_{1,2,3,4} = \pm ia, \pm ia$. The formulae for $\hat{A}_c$ and $\hat{A}_d$ are complicated and they are not presented here.

Peřina and Peřina, Jr. (1995a) have studied a codirectional coupler composed of one linear and one nonlinear waveguide. Second-subharmonic mode (b_1) and pump mode (b_2) nonlinearly interact in the waveguide b. The second-subharmonic mode b_1 also interacts linearly with mode a in the waveguide a. The corresponding momentum operator in interaction pictures is written in the form

$$\hat{G}_{\text{int}} = \hbar \left[-\kappa_{ab_1} \hat{A}_a \hat{A}^{\dagger}_{b_1} - \Gamma_b \hat{A}^2_{b_1} \hat{A}^{\dagger}_{b_2} \exp(\mathrm{i}\Delta k_b z) + \text{H.c.} \right], \tag{3.158}$$

where κ_{ab_1} denotes the linear coupling constant of modes a and b_1 and Γ_b is the nonlinear coupling constant between modes b_1 and b_2. The nonlinear phase mismatch Δk_b is defined as $\Delta k_b = 2k_{b_1} - k_{b_2}$ and k_{b_1} (k_{b_2}) means the wave vector of mode b_1 (b_2). In (3.158), $\hat{A}_a$, $\hat{A}_{b_1}$, and $\hat{A}_{b_2}$ stand for optical-field operators of modes a, b_1, and b_2 in interaction pictures. The conservation law

$$\hat{A}^{\dagger}_a(z)\hat{A}_a(z) + \hat{A}^{\dagger}_{b_1}(z)\hat{A}_{b_1}(z) + 2\hat{A}^{\dagger}_{b_2}(z)\hat{A}_{b_2}(z) = \text{const.} \tag{3.159}$$

is fulfilled by the solution of Heisenberg equations in the interaction picture.

Peřina (1995a,b) and Peřina and Bajer (1995) have analysed squeezing of the light in a short-length approximation. The assumption of a strong coherent field in mode b_2 with the amplitude ξ_{b_2} leads to the linearization of the operator equations of motion. The analysis leads to at least one positive eigenvalue. Amplification may occur dependent on the initial conditions. In the case where $|\Gamma_b \xi_b| < |\kappa_{ab_1}|$, oscillations also occur in the spatial development of quantities characterizing the fields. Results on squeezing of the light have been obtained (Peřina and Peřina, Jr. 1995a).

The assumption of a strong coherent field in mode b_2 and the linearization are related here to three types of behaviour, which can be distinguished also using normal forms (3.152), (3.153), and (3.156).

Peřina and Peřina, Jr. (1995b) have treated a contradirectional coupler composed again of one linear and one nonlinear waveguide. Mode a propagates against modes b_1 and b_2. The appropriate conservation law reads as

$$-\hat{A}^{\dagger}_a(z)\hat{A}_a(z) + \hat{A}^{\dagger}_{b_1}(z)\hat{A}_{b_1}(z) + 2\hat{A}^{\dagger}_{b_2}(z)\hat{A}_{b_2}(z) = \text{const}, \;\; z = 0, L; \tag{3.160}$$

i.e.

$$\begin{aligned} &\hat{A}^{\dagger}_a(0)\hat{A}_a(0) + \hat{A}^{\dagger}_{b_1}(L)\hat{A}_{b_1}(L) + 2\hat{A}^{\dagger}_{b_2}(L)\hat{A}_{b_2}(L) \\ = \;& \hat{A}^{\dagger}_a(L)\hat{A}_a(L) + \hat{A}^{\dagger}_{b_1}(0)\hat{A}_{b_1}(0) + 2\hat{A}^{\dagger}_{b_2}(0)\hat{A}_{b_2}(0). \end{aligned} \tag{3.161}$$

A phase matching ($\Delta k_b = 0$) is assumed.

A formulation of short-length approximation seems to be obvious, but it uses equal-space products of field operators. We can return to the boundary-value problem for classical equations which have the same solutions in the case of contradirectional propagation as the initial-value problem for codirectional coupler up to the second order in L provided we write $A_a(L)$ in place of $A_a(0)$. Quantization up to the second order in L can be done in the case of the contradirectional propagation by using the quantum input–output relations of the codirectional coupler where $\hat{A}_a(L)$ is written in place of $\hat{A}_a(0)$.

The assumption of a strong coherent field in mode b_2 leads to linear operator equations of motion as in the case of the codirectional coupler. Introducing $s_a = \pm 1$, $s_a = 1$ when mode a propagates along with modes b_1 and b_2 and $s_a = -1$ when it propagates counter to the latter, we can write the eigenvalues as

$$\lambda_{1,2,3,4} = \pm\left(|\Gamma_b \xi_{b_2}| \pm \sqrt{|\Gamma_b \xi_{b_2}|^2 - s_a |\kappa_{ab_1}|}\right). \quad (3.162)$$

In the case of the contradirectional coupler oscillations cannot occur. Results on squeezing of the light have been obtained (Peřina and Peřina, Jr. 1995b,c).

Let us note that one cannot assess input–output relations so easily in this case using only the eigenvalues. In the case of the codirectional coupler the input–output relations are just the solutions of the initial-value problem and their dependence on $\exp(\lambda_{1,2,3,4}L)$ is linear. In the case of the contradirectional coupler the input–output relations depend on $\exp(\lambda_{1,2,3,4}L)$ in a nonlinear way.

Peřina and Bajer (1995) have studied also a codirectional coupler with four modes. A mode of frequency ω_1 (b_1) and a mode of frequency ω_2 (b_2) nonlinearly interact in the waveguide b. The pump mode b_1 of frequency ω_1 interacts linearly with mode a_1 and the mode b_2 of frequency ω_2 is coupled linearly with mode c ($\omega_2 = 2\omega_1$ holds). The momentum operator (3.158) is modified to the form

$$\hat{G}_{\text{int}} = \hbar\left(-\kappa_{ab_1}\hat{A}_a\hat{A}^\dagger_{b_1} - \kappa_{cb_2}\hat{A}_c\hat{A}^\dagger_{b_2} + \Gamma_b\hat{A}^2_{b_1}\hat{A}^\dagger_{b_2} + \text{H.c.}\right), \quad (3.163)$$

where κ_{cb_2} is the linear coupling constant of modes c and b_2, κ_{ab_1} and Γ_b have the original meaning. The conservation law

$$\hat{A}^\dagger_a(z)\hat{A}_a(z) + \hat{A}^\dagger_{b_1}(z)\hat{A}_{b_1}(z) + 2\hat{A}^\dagger_{b_2}(z)\hat{A}_{b_2}(z) + 2\hat{A}^\dagger_c(z)\hat{A}_c(z) = \text{const.} \quad (3.164)$$

is obeyed by the solutions of Heisenberg equations in the interaction picture. Peřina and Bajer (1995) have treated squeezing of the light in the short-length approximation also in this case.

Mišta, Jr. and Peřina (1997) have investigated a codirectional coupler with five modes. Second-subharmonic modes (a_1, c_1) and pump modes (a_2, c_2) nonlinearly interact in the respective parts (a, c). The second-subharmonic mode a_1 also interacts linearly with mode c_1 via mode b in the part b. On assuming linear and nonlinear phase matching, Peřina, Jr. and Peřina (2000) describe the coupler with the following momentum operator:

$$\hat{G}_{\text{int}} = \hbar\left(\Gamma_a\hat{A}^2_{a_1}\hat{A}^\dagger_{a_2} + \Gamma_c\hat{A}^2_{c_1}\hat{A}^\dagger_{c_2} + \kappa_{a_1 b}\hat{A}^\dagger_{a_1}\hat{A}_b + \kappa_{bc_1}\hat{A}^\dagger_{c_1}\hat{A}_b + \text{H.c.}\right). \quad (3.165)$$

The solutions of Heisenberg equations in the interaction picture obey the conservation law

$$\hat{A}^{\dagger}_{a_1}(z)\hat{A}_{a_1}(z) + 2\hat{A}^{\dagger}_{a_2}(z)\hat{A}_{a_2}(z) \\ +\hat{A}^{\dagger}_{b}(z)\hat{A}_{b}(z) + \hat{A}^{\dagger}_{c_1}(z)\hat{A}_{c_1}(z) + 2\hat{A}^{\dagger}_{c_2}(z)\hat{A}_{c_2}(z) = \text{const.} \quad (3.166)$$

In a short-length approximation results on squeezing of the light have been obtained.

Peřina and Peřina, Jr. (1996) have studied a codirectional coupler composed of two nonlinear waveguides. While they have used a parametric approximation from the very beginning (Peřina, Jr. and Peřina 2000), here we present a generalization of the momentum operator (3.158)

$$\hat{G}_{\text{int}} = \hbar\big[\Gamma_a \hat{A}^2_{a_1}\hat{A}^{\dagger}_{a_2}\exp(\mathrm{i}\Delta k_a z) + \Gamma_b \hat{A}^2_{b_1}\hat{A}^{\dagger}_{b_2}\exp(\mathrm{i}\Delta k_b z) \\ + \kappa_{ab}\hat{A}_{a_1}\hat{A}^{\dagger}_{b_1} + \text{H.c.}\big], \quad (3.167)$$

where $a_1 \equiv a$, $\kappa_{ab} \equiv \kappa_{ab_1}$, and Γ_a is the nonlinear coupling constant between modes a_1 and a_2. The nonlinear phase mismatch Δk_a is defined as $\Delta k_a = 2k_{a_1} - k_{a_2}$ and k_{a_1} (k_{a_2}) means the wave vector of mode a_1 (a_2). The parametric approximation has consisted in replacements $\hat{A}_{a_2} \to \xi_{a_2}\exp(\mathrm{i}\Delta l z)$, $\hat{A}_{b_2} \to \xi_{b_2}\exp(-\mathrm{i}\Delta l z)$, where $\Delta l = \frac{1}{2}(k_{b_2} - k_{a_2})$, on assuming also some linear coupling between modes a_2 and b_2. Korolkova and Peřina (1997a) have obtained and discussed solutions on some simplifying assumptions. Karpati et al. (2000) studied all-optical switching in this system. Peřina and Peřina, Jr. (1996) and Korolkova and Peřina (1997a) have considered a contradirectional coupler as well.

Janszky et al. (1995) were first to investigate a coupler composed of two nonlinear waveguides with nondegenerate optical parametric processes. Pump (a_{P}), signal (a_{S}), and idler (a_{I}) modes in one waveguide are assumed to interact linearly with their counterparts (b_{P}, b_{S}, b_{I}) in the other waveguide. The coupler is described by the following momentum operator

$$\hat{G} = \hbar\Bigg\{\sum_{i=a_{\text{P}},a_{\text{S}},a_{\text{I}},b_{\text{P}},b_{\text{S}},b_{\text{I}}} k_i\hat{a}^{\dagger}_i\hat{a}_i + [\Gamma_a\hat{a}_{a_{\text{P}}}\hat{a}^{\dagger}_{a_{\text{S}}}\hat{a}^{\dagger}_{a_{\text{I}}} + \Gamma_b\hat{a}_{b_{\text{P}}}\hat{a}^{\dagger}_{b_{\text{S}}}\hat{a}^{\dagger}_{b_{\text{I}}} + \text{H.c.}] \\ + [\kappa_{\text{P}}\hat{a}_{a_{\text{P}}}\hat{a}^{\dagger}_{b_{\text{P}}} + \kappa_{\text{S}}\hat{a}_{a_{\text{S}}}\hat{a}^{\dagger}_{b_{\text{S}}} + \kappa_{\text{I}}\hat{a}_{a_{\text{I}}}\hat{a}^{\dagger}_{b_{\text{I}}} + \text{H.c.}]\Bigg\}, \quad (3.168)$$

where k_i denotes the wave vector of the ith mode along z-axis, Γ_a (Γ_b) is the nonlinear coupling constant of modes a_{P}, a_{S}, a_{I} (b_{P}, b_{S}, b_{I}) in waveguide a (b), and κ_{P}, κ_{S}, and κ_{I} stand for the linear coupling constants between the two pump, the two signal, and the two idler modes.

Herec (1999) has used a short-length approximation to solve the Heisenberg equations in the interaction picture and he has obtained results on squeezing of the light. Mišta, Jr. (1999) has assumed strong coherent field in pump modes, $\kappa_{\text{P}} = 0$, and phase matching, discerned three (of five) regimes in spatial evolution of the coupler, and obtained various results on the squeezing.

Optical fibres and certain organic polymers with high third-order nonlinearities may be used for the construction of couplers based on the Kerr effect. The nonlinear directional couplers are interesting as they exchange energy periodically between the guides like linear ones for low total intensities while they trap the energy in the guide into which it has been launched initially for high intensities (Jensen 1982). A coupler with two nonlinear waveguides (denoted as a and b) with Kerr nonlinearities has been considered by Chefles and Barnett (1996). It is described by the momentum operator $\hat{G}_{\rm int}$ ($k_a = k_b$ is assumed) (Korolkova and Peřina 1997b),

$$\hat{G}_{\rm int} = \hbar\Big[g\hat{A}_a^{\dagger 2}\hat{A}_a^2 + g\hat{A}_b^{\dagger 2}\hat{A}_b^2 + g_{ab}\hat{A}_a^\dagger\hat{A}_a\hat{A}_b^\dagger\hat{A}_b + \left(\kappa_{ab}\hat{A}_a\hat{A}_b^\dagger + \text{H.c.}\right)\Big]. \tag{3.169}$$

Here g means a Kerr nonlinear coupling coefficient which is the same in both waveguides. The real constant g_{ab} describes nonlinear coupling of the modes and the real constant κ_{ab} characterizes linear coupling of the modes. The operator $\hat{A}_a^\dagger\hat{A}_a + \hat{A}_b^\dagger\hat{A}_b$ is a constant of motion.

Korolkova and Peřina (1997b) have assumed that $\hat{A}_a$ and $\hat{A}_b$ are fast-oscillating operators due to the linear coupling and they have introduced the operators

$$\begin{aligned}\hat{B}_a &= \frac{1}{\sqrt{2}}[\hat{A}_a\exp(-\mathrm{i}\kappa_{ab}z) + \hat{A}_b\exp(\mathrm{i}\kappa_{ab}z)],\\ \hat{B}_b &= \frac{1}{\sqrt{2}}[\hat{A}_a\exp(-\mathrm{i}\kappa_{ab}z) - \hat{A}_b\exp(\mathrm{i}\kappa_{ab}z)].\end{aligned} \tag{3.170}$$

On an approximation a solution has been obtained. Then the operators $\hat{B}_a^\dagger\hat{B}_a$ and $\hat{B}_b^\dagger\hat{B}_b$ are conserved along z. We note that the solution is exact for $g_{ab} = 2g$.

In the interaction picture a numerical solution of the Schrödinger equation may use invariant subspaces defined as eigenspaces of the constant of motion (Chefles and Barnett 1996) and it is exact for any initial state from a finite direct sum of such subspaces. This way, Fiurášek et al. (1999a,b) have obtained interesting results on assuming also initial coherent states.

The behaviour of the coupler may exhibit a bifurcation in dependence on the parameter

$$\eta = \frac{1}{2\kappa_{ab}}\,|2g - g_{ab}|\,(|A_a|^2 + |A_b|^2) \tag{3.171}$$

in the classical regime. The threshold is at $\eta = 1$. The behaviour is more complicated in the quantum regime. An optimum energy exchange between modes a and b occurs if the difference of the phases of the complex amplitudes of modes a and b equals $\frac{\pi}{2}$ or $-\frac{\pi}{2}$.

The character of the evolution of mean photon numbers in the regions of revivals can be controlled by the z-dependent linear coupling constant $\kappa_{ab}(z)$ (Korolkova and Peřina 1997c). Switching of energy between the waveguides is achieved by a suitable profile of the coupling functions. Squeezing in a given waveguide also is preserved in such a way. In the classical regime, a nonlinear optical switching matrix has been considered (Liu et al. 2003).

Peřina, Jr. and Peřina (1997) have paid attention to the couplers which are based on Raman and Brillouin scattering. Fiurášek and Peřina (1998, 1999, 2000a,b) have continued this work. A codirectional coupler composed of two waveguides is described with the momentum operator $\hat{G}$:

$$\hat{G} = \hbar \sum_{j=a,b} \Big[\sum_{l=\mathrm{L,S,A,V}} k_{j_l} \hat{a}_{jl}^\dagger \hat{a}_{jl} + \Big(\tilde{g}_{j\mathrm{A}} \hat{a}_{j\mathrm{L}} \hat{a}_{j\mathrm{V}} \hat{a}_{j\mathrm{A}}^\dagger + \tilde{g}_{j\mathrm{S}} \hat{a}_{j\mathrm{L}} \hat{a}_{j\mathrm{V}}^\dagger \hat{a}_{j\mathrm{S}}^\dagger + \mathrm{H.c.} \Big) \Big]$$
$$+ \hbar \left[\kappa_\mathrm{S} \hat{a}_{a_\mathrm{S}} \hat{a}_{b_\mathrm{S}}^\dagger + \kappa_\mathrm{A} \hat{a}_{a_\mathrm{A}} \hat{a}_{b_\mathrm{A}}^\dagger + \mathrm{H.c.} \right], \quad (3.172)$$

where k_{j_l} are wave vectors of modes l, $l = \mathrm{L}$ (laser), S (Stokes), A (anti-Stokes), and V (phonon) in waveguides j, $j = a, b$. Here $\tilde{g}_{j\mathrm{S}}$ ($\tilde{g}_{j\mathrm{A}}$) describes the Stokes (anti-Stokes) nonlinear coupling in waveguide j, κ_S (κ_A) is the linear coupling constant between the Stokes (anti-Stokes) modes in different waveguides. Vectors characterizing phase mismatches are defined as follows: $\Delta k_{j_\mathrm{S}} = k_{j_\mathrm{L}} - k_{j_\mathrm{V}} - k_{j_\mathrm{S}}$, $\Delta k_{j_\mathrm{A}} = k_{j_\mathrm{L}} + k_{j_\mathrm{V}} - k_{j_\mathrm{A}}$, $\Delta k_\mathrm{S} = k_{a_\mathrm{S}} - k_{b_\mathrm{S}}$, and $\Delta k_\mathrm{A} = k_{a_\mathrm{A}} - k_{b_\mathrm{A}}$. A parametric approximation consists in assuming strong coherent states in pump modes a_L and b_L.

Fiurášek and Peřina (1999) have used another approximation in solving Heisenberg equations in the interaction picture. The method utilized is based on linear operator corrections to a classical solution.

Fiurášek and Peřina (2000a) have considered a Raman coupler with broad phonon spectra. They have described the phonon systems of the waveguides with multimode boson fields. Then these fields have been eliminated from the description of the coupler using the Wigner–Weisskopf approximation (Peřina 1981a,b).

Mogilevtsev et al. (1997) have treated one central waveguide (a) which interacts linearly with a greater number of mutually noninteracting waveguides (b_j) in its surroundings. It is described by the following momentum operator $\hat{G}$

$$\hat{G} = \hbar \left[k_a \hat{a}_a^\dagger \hat{a}_a + \sum_{j=1}^{N} k_{b_j} \hat{a}_{b_j}^\dagger \hat{a}_{b_j} + \sum_{j=1}^{N} g_{ab_j} (\hat{a}_{b_j} \hat{a}_a^\dagger + \hat{a}_{b_j}^\dagger \hat{a}_a) \right], \quad (3.173)$$

where k_a (k_{b_j}) is the wave vector of mode a (b_j), g_{ab_j} is the linear coupling constant between modes a and b_j, and N denotes the number of surrounding waveguides. If the central waveguide a contains a second-order nonlinear medium, the momentum operator $\hat{G}_\mathrm{n}$ is

$$\hat{G}_\mathrm{n} = \hat{G} + \frac{\hbar}{2} [\xi_{a_2} \exp(2ik_a z) \hat{a}_a^{\dagger 2} + \mathrm{H.c.}], \quad (3.174)$$

where ξ_{a_2} stands for the amplitude of the pump field in the central waveguide. The behaviour of the linear and nonlinear couplers agrees with the idea that the surrounding waveguides form a reservoir. A reservoir spectrum has been considered which has a gap.

Mogilevtsev et al. (1996) have considered a coupler composed of one waveguide with $\chi^{(2)}$ medium and the other one with $\chi^{(3)}$ medium. Only a nonlinear coupling is present. The interaction momentum operator $\hat{G}_{\rm int}$ is

$$\hat{G}_{\rm int} = \hbar\left[g_a(\hat{A}_a^{\dagger 2} + \hat{A}_a^2) + g_b\hat{A}_b^{\dagger 2}\hat{A}_b^2 + g_{ab}\hat{A}_a^{\dagger}\hat{A}_a\hat{A}_b^{\dagger}\hat{A}_b\right], \quad (3.175)$$

where g_a describes the process of second-subharmonic generation and includes the coherent pump amplitude, g_b stands for the Kerr constant, and g_{ab} means the nonlinear coupling constant between modes a and b.

Assuming the vacuum state in mode a and an incident pure state in mode b, one can distinguish two regimes essentially. The third type of behaviour could result using the superposition principle. When an initial Fock state with N_b photons in mode b is assumed, the solution for the operator $\hat{A}_b(z)$ oscillates in z for $2g_a < N_b g_{ab}$, whereas it has an exponential character for $2g_a > N_b g_{ab}$. When an initial coherent state with the amplitude ξ_b in mode b is assumed, the mean number of photons in mode a oscillates in z and the exponential terms can be neglected for $2g_a \ll |\xi_b|^2 g_{ab}$. It increases exponentially in z and the oscillating terms do not contribute significantly for $2g_a \gg |\xi_b|^2 g_{ab}$.

3.2 Dispersive Nonlinear Dielectric

The spatio-temporal quantum description has been adopted in optics in spite of its complexity due to quantum solitons. As known, the existence of the optical fields that do not change during propagation is conditioned by the frequency dispersion and the nonlinearity of the medium. A macroscopic quantization was to take into account both the properties. The nonlinearity would appear as an interaction of narrow-band fields.

3.2.1 Lagrangian of Narrow-Band Fields

Drummond (1990) has presented a technique of canonical quantization in a general dispersive nonlinear dielectric medium. Contrary to Abram and Cohen (1991), Drummond creates an arbitrary number of slightly varied copies of the vacuum electromagnetic field for the nonlinear dielectric medium, essentially the number required by the classical slowly varying amplitude approximation. But Abram and Cohen (1991) work with a single field. The paradox of the validity of both the approaches can be resolved only by a detailed microscopic theory.

Drummond (1990) generalizes the treatment of a linear homogeneous dispersive medium (Schubert and Wilhelmi 1986). Till 1990, papers by Knöll (1987), Białynicka-Birula and Białynicki-Birula (1987), and Glauber and Lewenstein (1989) could be referred to as devoted to the theory of inhomogeneous nondispersive linear dielectric. Hillery and Mlodinow (1984) were attractive with their use of the idea due to Born and Infeld (1934) for the quantization of homogeneous nonlinear nondispersive medium.

The macroscopic quantization is a route to the simplest quantum theory compatible with known dielectric properties unlike the microscopic derivation of the nonlinear quantum theory of electromagnetic propagation in a real dielectric. Drummond (1994) compares the quantum theory obtained via macroscopic quantization with the traditional quantum-field theory. He concedes that most model quantum field theories prove to be either tractable, but unphysical, or physical, but intractable. The tractable model quantum field theory ceases to be unphysical when it is tested experimentally in quantum optics. An excellent example of this is the fibre optical solitons whose quantization is given in detail in Drummond (1994). In agreement with theoretical predictions (Carter et al. 1987, Drummond and Carter 1987, Drummond et al. 1989, Shelby et al. 1990, Lai and Haus 1989, Haus and Lai 1990), experiments (Rosenbluh and Shelby 1991) led to evidence of quantum solitons. More recent experiments (Friberg et al. 1992) demonstrate that solitons can be considered to be nonlinear bound states of a quantum field. In addition to the quadrature squeezing in (Rosenbluh and Shelby 1991), quantum properties of soliton collisions were measured (Watanabe et al. 1989, Haus et al. 1989). Similar nonlinearities are encountered in photonic band-gap theory (Yablonovitch and Gmitter 1987), microcavity quantum electrodynamics (Hinds 1990), pulsed squeezing (Slusher et al. 1987), and quantum chaos (Toda et al. 1989).

In description of a nonlinear dielectric medium, tensorial notation is used which will occur also elsewhere in this book. Let **u**, **v**, **w**, ... be vectors. On using a tensorial product, a tensor of rank 2, e.g. **uv**, is formed, a tensor of rank 3, e.g. **uvw**, is constructed, etc. The scalar product denoted by the dot $\cdot$ is generalized to a contraction, i.e. a simple sum after the tensors are replaced by their components and products of corresponding components are formed. The correspondence of components is achieved by using the same notation for the last subscript of the tensor to the left as for the first subscript of the tensor to the right. Also the pieces of notation : and $\vdots$ mean contractions, i.e. a double sum and a triple sum of products of corresponding components.

It is advantageous to begin with the treatment of a classical dielectric introducing the nonlinear response function in terms of the electric displacement field **D**. Contrary to the usual description (Bloembergen 1965), which uses the dielectric permittivity tensors, the inverse expansion is necessary here. For simplicity, the dielectric of interest is regarded as having uniform linear magnetic susceptibility. The charges are assumed to occur only in the induced dipoles of polarization. The field equations are therefore

$$\nabla \times \mathbf{E}(\mathbf{x}, t) = -\frac{\partial \mathbf{B}(\mathbf{x}, t)}{\partial t},$$

$$\nabla \times \mathbf{H}(\mathbf{x}, t) = \frac{\partial \mathbf{D}(\mathbf{x}, t)}{\partial t},$$

$$\nabla \cdot \mathbf{D}(\mathbf{x}, t) = 0,$$
$$\nabla \cdot \mathbf{B}(\mathbf{x}, t) = 0, \tag{3.176}$$

where

$$\mathbf{D}(\mathbf{x}, t) = \epsilon_0 \mathbf{E}(\mathbf{x}, t) + \mathbf{P}(\mathbf{x}, t),$$
$$\mathbf{B}(\mathbf{x}, t) = \mu \mathbf{H}(\mathbf{x}, t). \tag{3.177}$$

Here

$$\begin{aligned}\mathbf{P}(\mathbf{x}, t) = &\int_0^\infty \boldsymbol{\chi}(\mathbf{x}, \tau) \cdot \mathbf{E}(\mathbf{x}, t - \tau)\, \mathrm{d}\tau \\ &+ \int_0^\infty \int_0^\infty \boldsymbol{\chi}^{(2)}(\mathbf{x}, \tau_1, \tau_2) : \mathbf{E}(\mathbf{x}, t - \tau_1)\mathbf{E}(\mathbf{x}, t - \tau_2)\, \mathrm{d}\tau_1\, \mathrm{d}\tau_2 \\ &+ \int_0^\infty \int_0^\infty \int_0^\infty \boldsymbol{\chi}^{(3)}(\mathbf{x}, \tau_1, \tau_2, \tau_3) \vdots \mathbf{E}(\mathbf{x}, t - \tau_1)\mathbf{E}(\mathbf{x}, t - \tau_2) \\ &\times \mathbf{E}(\mathbf{x}, t - \tau_3)\, \mathrm{d}\tau_1\, \mathrm{d}\tau_2\, \mathrm{d}\tau_3 \\ &+ \cdots,\end{aligned} \tag{3.178}$$

where the tensor of rank 2, in general, the $(n + 1)$th-rank susceptibility tensor read

$$\begin{aligned}\boldsymbol{\chi}(\mathbf{x}, \tau) &= \frac{1}{2\pi} \int \tilde{\boldsymbol{\chi}}(\mathbf{x}, \omega) e^{-\mathrm{i}\omega\tau}\, \mathrm{d}\omega, \\ \boldsymbol{\chi}^{(n)}(\mathbf{x}, \tau_1, \ldots, \tau_n) &= \left(\frac{1}{2\pi}\right)^n \\ &\times \int \ldots \int \tilde{\boldsymbol{\chi}}^{(n)}(\mathbf{x}, \omega^1, \ldots, \omega^n) e^{-\mathrm{i}(\omega^1\tau_1 + \cdots + \omega^n\tau_n)}\, \mathrm{d}\tau_1 \ldots \mathrm{d}\tau_n,\end{aligned} \tag{3.179}$$

respectively. After adding the vacuum electric displacement field $\epsilon_0 \mathbf{E}(\mathbf{x}, t)$ to both sides of (3.178), we express the electric vector in the form

$$\begin{aligned}\mathbf{E}(\mathbf{x}, t) = &\int_0^\infty \boldsymbol{\zeta}(\mathbf{x}, \tau) \cdot \mathbf{D}(\mathbf{x}, t - \tau)\, \mathrm{d}\tau \\ &+ \int_0^\infty \int_0^\infty \boldsymbol{\zeta}^{(2)}(\mathbf{x}, \tau_1, \tau_2) : \mathbf{D}(\mathbf{x}, t - \tau_1)\mathbf{D}(\mathbf{x}, t - \tau_2)\, \mathrm{d}\tau_1\, \mathrm{d}\tau_2 \\ &+ \int_0^\infty \int_0^\infty \int_0^\infty \boldsymbol{\zeta}^{(3)}(\mathbf{x}, \tau_1, \tau_2, \tau_3) \vdots \mathbf{D}(\mathbf{x}, t - \tau_1)\mathbf{D}(\mathbf{x}, t - \tau_2) \\ &\times \mathbf{D}(\mathbf{x}, t - \tau_3)\, \mathrm{d}\tau_1\, \mathrm{d}\tau_2\, \mathrm{d}\tau_3 \\ &+ \cdots,\end{aligned} \tag{3.180}$$

where

$$\boldsymbol{\zeta}(\mathbf{x},\tau)=\frac{1}{2\pi}\int\tilde{\boldsymbol{\xi}}(\mathbf{x},\omega)e^{-\mathrm{i}\omega\tau}\,\mathrm{d}\omega,$$
$$\boldsymbol{\zeta}^{(n)}(\mathbf{x},\tau_1,\ldots,\tau_n)=\left(\frac{1}{2\pi}\right)^n \times\int\ldots\int\tilde{\boldsymbol{\xi}}^{(n)}(\mathbf{x},\omega^1,\ldots,\omega^n)e^{-\mathrm{i}(\omega^1\tau_1+\ldots+\omega^n\tau_n)}\,\mathrm{d}\tau_1\ldots\mathrm{d}\tau_n \quad (3.181)$$

and the tensors on the right-hand sides of (3.181) can be expressed from the equations

$$\begin{aligned}
&\boldsymbol{\epsilon}(\mathbf{x},\omega)\cdot\tilde{\boldsymbol{\xi}}(\mathbf{x},\omega)=\mathbf{1},\\
&\boldsymbol{\epsilon}(\mathbf{x},\omega^1+\omega^2)\cdot\tilde{\boldsymbol{\xi}}^{(2)}(\mathbf{x},\omega^1,\omega^2)+\tilde{\boldsymbol{\chi}}^{(2)}(\mathbf{x},\omega^1,\omega^2):\tilde{\boldsymbol{\xi}}(\mathbf{x},\omega^1)\tilde{\boldsymbol{\xi}}(\mathbf{x},\omega^2)=\mathbf{0}^{(2)},\\
&\boldsymbol{\epsilon}(\mathbf{x},\omega^1+\omega^2+\omega^3)\cdot\tilde{\boldsymbol{\xi}}^{(3)}(\mathbf{x},\omega^1,\omega^2,\omega^3)\\
&+2\tilde{\boldsymbol{\chi}}^{(2)}(\mathbf{x},\omega^1,\omega^2+\omega^3):\tilde{\boldsymbol{\xi}}(\mathbf{x},\omega^1)\tilde{\boldsymbol{\xi}}^{(2)}(\mathbf{x},\omega^2,\omega^3)\\
&+\tilde{\boldsymbol{\chi}}^{(3)}(\mathbf{x},\omega^1,\omega^2,\omega^3)\vdots\tilde{\boldsymbol{\xi}}(\mathbf{x},\omega^1)\tilde{\boldsymbol{\xi}}(\mathbf{x},\omega^2)\tilde{\boldsymbol{\xi}}(\mathbf{x},\omega^3)=\mathbf{0}^{(3)},\\
&\ldots,
\end{aligned}\quad (3.182)$$

with $\boldsymbol{\epsilon}(\mathbf{x},\omega)=\epsilon_0\mathbf{1}+\tilde{\boldsymbol{\chi}}(\mathbf{x},\omega)$ the usual frequency-dependent tensor of permittivity. In particular,

$$\tilde{\boldsymbol{\xi}}(\mathbf{x},\omega)=[\boldsymbol{\epsilon}(\mathbf{x},\omega)]^{-1}. \quad (3.183)$$

Here $\mathbf{1}$ and $\mathbf{0}^{(n)}$ are the second-rank unit tensor and the $(n+1)$th-rank zero tensor, respectively. Introducing the Fourier components of the electric strength field and electric displacement field, respectively,

$$\begin{aligned}
\tilde{\mathbf{E}}(\mathbf{x},\omega)&=\int\mathbf{E}(\mathbf{x},t)e^{\mathrm{i}\omega t}\,\mathrm{d}t,\\
\tilde{\mathbf{D}}(\mathbf{x},\omega)&=\int\mathbf{D}(\mathbf{x},t)e^{\mathrm{i}\omega t}\,\mathrm{d}t,
\end{aligned}\quad (3.184)$$

and performing the Fourier transform of both sides of (3.180), we obtain that

$$\begin{aligned}
\tilde{\mathbf{E}}(\mathbf{x},\omega)=&\,\tilde{\boldsymbol{\xi}}(\mathbf{x},\omega)\cdot\tilde{\mathbf{D}}(\mathbf{x},\omega)\\
&+\int\tilde{\boldsymbol{\xi}}^{(2)}(\mathbf{x},\omega^1,\omega-\omega^1):\tilde{\mathbf{D}}(\mathbf{x},\omega^1)\tilde{\mathbf{D}}(\mathbf{x},\omega-\omega^1)\,\mathrm{d}\omega^1\\
&+\int\int\tilde{\boldsymbol{\xi}}^{(3)}(\mathbf{x},\omega^1,\omega^2,\omega-\omega^1-\omega^2)\vdots\tilde{\mathbf{D}}(\mathbf{x},\omega^1)\tilde{\mathbf{D}}(\mathbf{x},\omega^2)\\
&\times\tilde{\mathbf{D}}(\mathbf{x},\omega-\omega^1-\omega^2)\,\mathrm{d}\omega^1\,\mathrm{d}\omega^2\\
&+\cdots.
\end{aligned}\quad (3.185)$$

The inverse relation of $\tilde{\mathbf{D}}(\mathbf{x}, \omega)$ comprises $\tilde{\boldsymbol{\chi}}(\mathbf{x}, \omega) \equiv \tilde{\boldsymbol{\chi}}^{(1)}(\mathbf{x}, -\omega; \omega)$, $\tilde{\boldsymbol{\chi}}^{(2)}(\mathbf{x}, \omega^1, \omega - \omega^1) \equiv \tilde{\boldsymbol{\chi}}^{(2)}(\mathbf{x}, -\omega; \omega^1, \omega - \omega^1)$,... . A similar extension of notation is conceivable also in tensors $\tilde{\boldsymbol{\xi}}(\mathbf{x}, \omega)$, $\tilde{\boldsymbol{\xi}}^{(2)}(\mathbf{x}, \omega^1, \omega - \omega^1)$,... .

Let us note that the similar relation for $\tilde{\mathbf{P}}(\mathbf{x}, \omega)$ in Peřina (1991) comprises sums instead of the integrals. Relation (3.185) can resemble relation (2.4) in Drummond (1990) on the condition that the integrals will be replaced by the sums. Such a change does not affect only the meaning of the tensors $\tilde{\boldsymbol{\chi}}^{(n)}$ and $\tilde{\boldsymbol{\xi}}^{(n)}$ but also (and above all) the physical unit of their measurement.

We will treat the time-averaged linear dispersive energy $\langle H \rangle$ for a classical monochromatic field at nonzero frequency ω. For a permittivity $\boldsymbol{\epsilon}(\mathbf{x}, \omega)$, this can be written in terms of a complex amplitude $\mathcal{E}$ (Bloembergen 1965, Landau and Lifshitz 1960, Bleany and Bleany 1985),

$$\langle H \rangle = \int_V \left\{ \mathcal{E}^*(\mathbf{x}) \cdot \frac{\partial}{\partial \omega}[\omega \boldsymbol{\epsilon}(\mathbf{x}, \omega)] \cdot \mathcal{E}(\mathbf{x}) + \frac{1}{2\mu} \langle \mathbf{B}(\mathbf{x}, t) \cdot \mathbf{B}(\mathbf{x}, t) \rangle \right\} \mathrm{d}^3\mathbf{x}, \qquad (3.186)$$

where the angular brackets mean the time average and

$$\mathbf{E}(\mathbf{x}, t) = 2\,\mathrm{Re}\left\{ \mathcal{E}(\mathbf{x}) e^{-\mathrm{i}\omega t} \right\}. \qquad (3.187)$$

It is important to distinguish the monochromatic case from the case of quasi-monochromatic fields. In the more general case, the displacement $\mathbf{D}$ is expanded in terms of a series of complex (envelope) functions, each of which has a restricted bandwidth. The relevant nonzero central frequencies are then ω^{-N}, ..., ω^N, thus

$$\mathbf{D}(\mathbf{x}, t) = \sum_{\nu=-N}^{N} \mathbf{D}^{\nu}(\mathbf{x}, t), \qquad (3.188)$$

where

$$\mathbf{D}^{-\nu} = (\mathbf{D}^{\nu})^* \qquad (3.189)$$

and in the monochromatic case

$$\mathbf{D}^{\nu}(\mathbf{x}, t) = \mathcal{D}^{\nu}(\mathbf{x}) e^{-\mathrm{i}\omega^{\nu} t}. \qquad (3.190)$$

Here, our notation slightly differs from that in Drummond (1990). Again, the electric-field vector can be expanded as

$$\mathbf{E}(\mathbf{x}, t) = \sum_{\nu=-N}^{N} \mathbf{E}^{\nu}(\mathbf{x}, t), \qquad (3.191)$$

where

$$\mathbf{E}^{-\nu} = (\mathbf{E}^{\nu})^* \qquad (3.192)$$

and in the monochromatic case

$$\mathbf{E}^{\nu}(\mathbf{x},t) = \mathcal{E}^{\nu}(\mathbf{x})e^{-\mathrm{i}\omega^{\nu}t}. \tag{3.193}$$

In the case of quasimonochromatic fields, relations (3.190) and (3.193) should be replaced by

$$\begin{aligned}
\mathbf{D}^{\nu}(\mathbf{x},t) &= \frac{1}{2\pi}\int_{\omega^{\nu}-\delta}^{\omega^{\nu}+\delta} \tilde{\mathbf{D}}(\mathbf{x},\omega)e^{-\mathrm{i}\omega t}\,\mathrm{d}\omega,\\
\mathbf{E}^{\nu}(\mathbf{x},t) &= \frac{1}{2\pi}\int_{\omega^{\nu}-\delta}^{\omega^{\nu}+\delta} \tilde{\mathbf{E}}(\mathbf{x},\omega)e^{-\mathrm{i}\omega t}\,\mathrm{d}\omega.
\end{aligned} \tag{3.194}$$

Bloembergen (1965) presented the relation (3.186) as sufficiently accurate for such a case. If relation (3.186) is exact for monochromatic fields, it must be modified for a quasimonochromatic field as follows:

$$\begin{aligned}
\langle H(t')\rangle(t) = \int\Bigg\{&\frac{1}{2}\sum_{\nu=-1}^{1}\mathbf{E}^{-\nu}(\mathbf{x},t)\cdot\frac{\partial}{\partial\omega^{\nu}}[\omega^{\nu}\boldsymbol{\epsilon}(\mathbf{x},\omega^{\nu})]\cdot\mathbf{E}^{\nu}(\mathbf{x},t)\\
&+\frac{1}{2\mu}\langle\mathbf{B}(\mathbf{x},t')\cdot\mathbf{B}(\mathbf{x},t')\rangle(t)\Bigg\}\mathrm{d}^3\mathbf{x}.
\end{aligned} \tag{3.195}$$

By modifying the summation, we obtain the energy integral in terms of the electric displacement fields

$$\begin{aligned}
\langle H(t')\rangle(t) = \int\Bigg\{&\frac{1}{2}\sum_{\nu=-N}^{N}\mathbf{D}^{-\nu}(\mathbf{x},t)\cdot[\tilde{\boldsymbol{\xi}}(\mathbf{x},\omega^{\nu}) - \omega^{\nu}\frac{\partial}{\partial\omega^{\nu}}\tilde{\boldsymbol{\xi}}'(\mathbf{x},\omega^{\nu})]\cdot\mathbf{D}^{\nu}(\mathbf{x},t)\\
&+\frac{1}{2\mu}\langle\mathbf{B}(\mathbf{x},t')\cdot\mathbf{B}(\mathbf{x},t')\rangle(t)\Bigg\}\mathrm{d}^3\mathbf{x}.
\end{aligned} \tag{3.196}$$

To achieve a completeness, we supplement relations (3.188) and (3.191) with the expansion of the magnetic induction field

$$\mathbf{B}(\mathbf{x},t) = \sum_{\nu=-N}^{N}\mathbf{B}^{\nu}(\mathbf{x},t), \tag{3.197}$$

where

$$\mathbf{B}^{-\nu} = (\mathbf{B}^{\nu})^*, \tag{3.198}$$

$$\mathbf{B}^{\nu}(\mathbf{x},t) = \frac{1}{2\pi}\int_{\omega^{\nu}-\delta}^{\omega^{\nu}+\delta}\tilde{\mathbf{B}}(\mathbf{x},\omega)e^{-\mathrm{i}\omega t}\,\mathrm{d}\omega. \tag{3.199}$$

Next, $\tilde{\boldsymbol{\xi}}(\mathbf{x},\omega)$ can be approximated near $\omega = \omega^{\nu}$ by a quadratic Taylor polynomial,

$$\tilde{\boldsymbol{\xi}}(\mathbf{x},\omega) \approx \tilde{\boldsymbol{\xi}}_{\nu}(\mathbf{x}) + \omega\tilde{\boldsymbol{\xi}}'_{\nu}(\mathbf{x}) + \frac{1}{2}\omega^2\tilde{\boldsymbol{\xi}}''_{\nu}(\mathbf{x}) \equiv \tilde{\boldsymbol{\xi}}_{\nu}(\mathbf{x},\omega), \tag{3.200}$$

so that

$$\tilde{\xi}(\mathbf{x}, \omega) - \omega \frac{\partial}{\partial \omega} \tilde{\xi}(\mathbf{x}, \omega) \approx \tilde{\xi}_\nu(\mathbf{x}) - \frac{1}{2}\omega^2 \tilde{\xi}_\nu''(\mathbf{x}). \tag{3.201}$$

For brevity, the prime stands for the partial derivative $\frac{\partial}{\partial \omega}$. Moreover, the Taylor polynomial is not in a standard form, which comprises the brackets $(\omega - \omega^\nu)$ and $(\omega - \omega^\nu)^2$. Further explanations can be found in Drummond (1990) if necessary.

Using the notation $\dot{\mathbf{D}} \equiv \frac{\partial}{\partial t}\mathbf{D}$, we rewrite relation (3.196) in the form

$$\langle H(t') \rangle (t) = \frac{1}{2} \sum_{\nu=-N}^{N} \int \Big[\mathbf{D}^{-\nu}(\mathbf{x}, t) \cdot \tilde{\xi}_\nu(\mathbf{x}) \cdot \mathbf{D}^\nu(\mathbf{x}, t) - \frac{1}{2} \dot{\mathbf{D}}^{-\nu}(\mathbf{x}, t) \cdot \tilde{\xi}_\nu''(\mathbf{x}) \cdot \dot{\mathbf{D}}^\nu(\mathbf{x}, t) + \frac{1}{\mu} \mathbf{B}^{-\nu}(\mathbf{x}, t) \cdot \mathbf{B}^\nu(\mathbf{x}, t) \Big] \, \mathrm{d}^3\mathbf{x}. \tag{3.202}$$

Here we deviate slightly from Drummond (1990). Drummond speaks of time averages and he indicates the time average on the left-hand side and partially on the right-hand side in (3.196), but he does not remove the time dependence from the right-hand side.

A canonical theory of linear dielectric will be obtained using the causal local Lagrangian. Drummond (1990) considers a Lagrangian $L(\Lambda^{-N}, \ldots, \Lambda^N)$, which is a functional of (components of) the dual vector potential. This is defined as Λ,

$$\begin{aligned} \mathbf{D}(\mathbf{x}, t) &= \nabla \times \Lambda(\mathbf{x}, t), \\ \mathbf{B}(\mathbf{x}, t) &= \mu \dot{\Lambda}(\mathbf{x}, t). \end{aligned} \tag{3.203}$$

We introduce also

$$\tilde{\Lambda}(\mathbf{x}, \omega) = \int \Lambda(\mathbf{x}, t) e^{\mathrm{i}\omega t} \, \mathrm{d}t, \tag{3.204}$$

$$\Lambda^\nu(\mathbf{x}, t) = \frac{1}{2\pi} \int_{\omega^\nu - \delta}^{\omega^\nu + \delta} \tilde{\Lambda}(\mathbf{x}, \omega) e^{-\mathrm{i}\omega t} \, \mathrm{d}\omega. \tag{3.205}$$

Each quasimonochromatic field obeys the Maxwell equations

$$\begin{aligned} \nabla \times \mathbf{E}^\nu(\mathbf{x}, t) &= -\dot{\mathbf{B}}^\nu(\mathbf{x}, t), \\ \nabla \times \mathbf{H}^\nu(\mathbf{x}, t) &= \dot{\mathbf{D}}^\nu(\mathbf{x}, t), \\ \nabla \cdot \mathbf{D}^\nu(\mathbf{x}, t) &= 0, \\ \nabla \cdot \mathbf{B}^\nu(\mathbf{x}, t) &= 0, \end{aligned} \tag{3.206}$$

where

$$\begin{aligned} \mathbf{E}^\nu(\mathbf{x}, t) &= \tilde{\xi}_\nu(\mathbf{x}) \cdot \mathbf{D}^\nu(\mathbf{x}, t) + \mathrm{i}\tilde{\xi}_\nu'(\mathbf{x}) \cdot \dot{\mathbf{D}}^\nu(\mathbf{x}, t) - \frac{1}{2}\tilde{\xi}_\nu''(\mathbf{x}) \cdot \ddot{\mathbf{D}}^\nu(\mathbf{x}, t), \\ \mathbf{H}^\nu(\mathbf{x}, t) &= \frac{1}{\mu}\mathbf{B}^\nu(\mathbf{x}, t). \end{aligned} \tag{3.207}$$

The components of the dual vector potential fulfil linear wave equations.

On the basis of (3.202) we can infer the Hamiltonian function of the form

$$\mathcal{H} = \mathcal{H}_0 = \frac{1}{2}\int \sum_{\nu=-N}^{N} \Big\{[\nabla \times \Lambda^{-\nu}(\mathbf{x}, t)] \cdot \tilde{\xi}_\nu(\mathbf{x}) \cdot [\nabla \times \Lambda^{\nu}(\mathbf{x}, t)]$$
$$- \frac{1}{2}[\nabla \times \dot{\Lambda}^{-\nu}(\mathbf{x}, t)] \cdot \tilde{\xi}''_\nu(\mathbf{x}) \cdot [\nabla \times \dot{\Lambda}^{\nu}(\mathbf{x}, t)]$$
$$+ \mu\dot{\Lambda}^{-\nu}(\mathbf{x}, t) \cdot \dot{\Lambda}^{\nu}(\mathbf{x}, t)\Big\}\, \mathrm{d}^3\mathbf{x}. \tag{3.208}$$

In order to quantize the theory, a canonical Lagrangian must be found that corresponds to (3.208) while generating the Maxwell equations (3.206) as Hamilton's equations. It is next necessary to derive a Lagrangian whose Lagrange's variational equations correspond to obvious wave equations and whose Hamiltonian corresponds to (3.208). Since Λ^{ν} can be specified to be transverse fields, the variations can also be restricted to be transverse. The use of restricted variations can be realized using transverse functional derivatives (Power and Zienau 1959, Healey 1982).

Drummond (1994) derived the Lagrangian using the method of indeterminate coefficients in the form

$$L = L_0 = \frac{1}{2}\int \sum_{\nu=-N}^{N} \Big\{\mu\dot{\Lambda}^{-\nu}(\mathbf{x}, t)\dot{\Lambda}^{\nu}(\mathbf{x}, t)$$
$$- [\nabla \times \Lambda^{-\nu}(\mathbf{x}, t)] \cdot \tilde{\xi}_\nu(\mathbf{x}) \cdot [\nabla \times \Lambda^{\nu}(\mathbf{x}, t)]$$
$$- \mathrm{i}[\nabla \times \Lambda^{-\nu}(\mathbf{x}, t)] \cdot \tilde{\xi}'_\nu(\mathbf{x}) \cdot [\nabla \times \dot{\Lambda}^{\nu}(\mathbf{x}, t)]$$
$$- \frac{1}{2}[\nabla \times \dot{\Lambda}^{-\nu}(\mathbf{x}, t)] \cdot \tilde{\xi}''_\nu(\mathbf{x}) \cdot [\nabla \times \dot{\Lambda}^{\nu}(\mathbf{x}, t)]\Big\}\, \mathrm{d}^3\mathbf{x}. \tag{3.209}$$

The canonical momenta are

$$\Pi^{\nu}(\mathbf{x}, t) = \mathbf{B}^{-\nu}(\mathbf{x}, t) - \frac{1}{2}\nabla \times \left[\mathrm{i}\zeta'_\nu(\mathbf{x}) \cdot \mathbf{D}^{-\nu}(\mathbf{x}, t) + \zeta''_\nu\mathbf{x}) \cdot \dot{\mathbf{D}}^{-\nu}(\mathbf{x}, t)\right], \tag{3.210}$$

where we introduce for brevity the fields (3.203) again. We can rewrite also the Lagrangian of Drummond in the form

$$L = L_0 = \frac{1}{2}\int \sum_{\nu=-N}^{N} \Big\{\frac{1}{\mu}\mathbf{B}^{-\nu}(\mathbf{x}, t) \cdot \mathbf{B}^{\nu}(\mathbf{x}, t)$$
$$- \mathbf{D}^{-\nu}(\mathbf{x}, t) \cdot \tilde{\xi}_\nu(\mathbf{x}) \cdot \mathbf{D}^{\nu}(\mathbf{x}, t) - \mathrm{i}\mathbf{D}^{-\nu}(\mathbf{x}, t) \cdot \tilde{\xi}'_\nu(\mathbf{x}) \cdot \dot{\mathbf{D}}^{\nu}(\mathbf{x}, t)$$
$$- \frac{1}{2}\dot{\mathbf{D}}^{-\nu}(\mathbf{x}, t) \cdot \tilde{\xi}''_\nu(\mathbf{x}) \cdot \dot{\mathbf{D}}^{\nu}(\mathbf{x}, t)\Big\}\, \mathrm{d}^3\mathbf{x}. \tag{3.211}$$

On the contrary, the Legendre transformation, i.e. a substitution of $\dot{\Lambda}^{\nu}$ in Hamiltonian (3.208) with an expression in Π^{ν} and Λ^{ν}, was not performed in Drummond (1990). A reason is that each $\dot{\Lambda}^{\nu}$ is to be found from (3.210) considered as a partial differential equation. Also this theory simplifies a great deal if the plane wave one-dimensional propagation is considered.

The local Lagrangian method is used as the foundation of a nonlinear canonical Lagrangian and Hamiltonian. The objective is the total Lagrangian and Hamiltonian of the form

$$
\begin{aligned}
L &= L^0 - \int U^{\mathrm{N}}(\mathbf{x}, t)\,\mathrm{d}^3\mathbf{x}, \\
\mathcal{H} &= \mathcal{H}^0 + \int U^{\mathrm{N}}(\mathbf{x}, t)\,\mathrm{d}^3\mathbf{x},
\end{aligned}
\tag{3.212}
$$

where $U^{\mathrm{N}}(\mathbf{x}, t)$ is a nonlinear energy density

$$
\begin{aligned}
U^{\mathrm{N}}(\mathbf{x}, t) = & \frac{1}{3} \sum_{\nu_1=-N}^{N} \sum_{\nu_2=-N}^{N} \sum_{\nu_3=-N}^{N} \mathbf{D}^{\nu_1}(\mathbf{x}, t) \\
& \cdot \tilde{\boldsymbol{\xi}}^{(2)}(\mathbf{x}, \omega^{\nu_2}, -\omega^{\nu_1} - \omega^{\nu_2}) : \mathbf{D}^{\nu_2}(\mathbf{x}, t)\mathbf{D}^{\nu_3}(\mathbf{x}, t)\delta_{-\omega^{\nu_1}, \omega^{\nu_2}+\omega^{\nu_3}} \\
& + \frac{1}{4} \sum_{\nu_1=-N}^{N} \sum_{\nu_2=-N}^{N} \sum_{\nu_3=-N}^{N} \sum_{\nu_4=-N}^{N} \mathbf{D}^{\nu_1}(\mathbf{x}, t) \\
& \cdot \tilde{\boldsymbol{\xi}}^{(3)}(\mathbf{x}, \omega^{\nu_2}, \omega^{\nu_3}, -\omega^{\nu_1} - \omega^{\nu_2} - \omega^{\nu_3}) \vdots \mathbf{D}^{\nu_2}(\mathbf{x}, t)\mathbf{D}^{\nu_3}(\mathbf{x}, t)\mathbf{D}^{\nu_4}(\mathbf{x}, t) \\
& \times \delta_{-\omega^{\nu_1}, \omega^{\nu_2}+\omega^{\nu_3}+\omega^{\nu_4}} \\
& + \cdots .
\end{aligned}
\tag{3.213}
$$

In order to give an example, a one-dimensional case is treated and the nonlinear refractive index as the lowest nonlinearity of most universal interest. For $N = 1$,

$$
U^{\mathrm{N}}(\mathbf{x}, t) = \frac{1}{4}\tilde{\xi}^{(3)}|D^1(\mathbf{x}, t)|^4. \tag{3.214}
$$

Drummond (1990) has presented the quantization of the nonlinear medium using a treatment of modes defined relative to the new Lagrangian. The canonical momenta have the form (3.210) also in the nonlinear case. In the corresponding quantum theory, the field operators $\hat{\Lambda}^{\nu}$ and $\hat{\Pi}^{\nu}$ are introduced, which obey the transverse commutation relations of the form

$$
[\hat{\Lambda}_i^{\nu}(\mathbf{x}, t), \hat{\Pi}_j^{\mu}(\mathbf{x}', t)] = \mathrm{i}\hbar\delta_{ij}^{\perp}(\mathbf{x} - \mathbf{x}')\delta_{\mu\nu}\hat{1}. \tag{3.215}
$$

Since these operators are not Hermitian, it is also interesting to note that

$$
\hat{\Lambda}^{-\nu} = (\hat{\Lambda}^{\nu})^{\dagger}, \; \hat{\Pi}^{-\mu} = (\hat{\Pi}^{\mu})^{\dagger}. \tag{3.216}
$$

This entails that $\hat{\Lambda}_i^\nu$, $(\hat{\Pi}_j^\nu)^\dagger$ commute. Then, a set of Fourier transform fields is defined and the annihilation operators $\hat{a}_k^\nu$ and $\hat{b}_k^\nu$ are introduced. The operators $\hat{a}_k^\nu$ correspond to the normal modes while $\hat{b}_k^\nu$ generate additional necessarily vacuum modes. This feature of the theory is due to the dependence of the Hamiltonian (3.208) on both the real and imaginary parts of the components $\Lambda^\nu(\mathbf{x}, t)$ $(\dot{\Lambda}^\nu(\mathbf{x}, t))$. Conversely, the right-hand side of relation (3.202) can be completed with terms which make up the Hamiltonian dependent solely on $2\,\mathrm{Re}\{\Lambda^\nu(\mathbf{x}, t)\}$.

3.2.2 Propagation in One Dimension and Applications

Drummond (1994) discusses in detail a simplified model of a one-dimensional dielectric, where $\tilde{\boldsymbol{\xi}}(\mathbf{x}, \omega) = \tilde{\xi}(\omega)\mathbf{1}$, $\tilde{\boldsymbol{\xi}}^{(n)}(\mathbf{x}, \omega^1, \ldots, \omega^n) = \tilde{\xi}^{(n)}(\omega^1, \ldots, \omega^n)\mathbf{1}^{(n)}$, with $\mathbf{1}^{(n)}$ the $(n+1)$th-rank unit tensor for n odd and $\tilde{\boldsymbol{\xi}}^{(n)}(\mathbf{x}, \omega^1, \ldots, \omega^n) = \mathbf{0}^{(n)}$ for n even, $n \leq 3$.

Instead of the time average of energy (3.186), (3.195), Drummond (1994) has presented the total energy in the length L,

$$W(t) = \int_0^L \left[\frac{1}{2}\mu H^2(x, t) + \int_{t_0}^t E(x, \tau)\dot{D}(x, \tau)\,\mathrm{d}\tau \right] \mathrm{d}x. \tag{3.217}$$

Here $H(x, t) = \frac{1}{\mu}B(x, t)$ is the magnetic strength vector. In part of the exposition, single polarization components are considered only. The Hillery–Mlodinow theory which does not take account of dispersion (Ho and Kumar 1993) has the electric-field commutation relation with the magnetic field modified from its free-field value. Drummond (1994) points out that the solution of this commutator problem is the inclusion of the important physical property of a real dielectric. Traditionally, the description of the nonlinear medium assumes that the dispersion terms are negligible. Neglecting the unphysical modes, the dual vector potential has the expansion

$$\hat{\Lambda}^1(x, t) = \sum_k \sqrt{\frac{\hbar \frac{\partial \omega}{\partial k}}{2Lk\tilde{\xi}(\omega_k)}}\, \hat{a}_k e^{ikx}, \tag{3.218}$$

where $\hat{a}_k$, $\hat{a}_{k'}^\dagger$ have the standard commutators

$$[\hat{a}_k, \hat{a}_{k'}^\dagger] = \delta_{k,k'}\hat{1}, \tag{3.219}$$

and ω_k are solutions to the equations

$$\omega_k = k\sqrt{\frac{\tilde{\xi}(\omega_k)}{\mu}}. \tag{3.220}$$

This enables one to write

$$\hat{\mathcal{H}}_0 = \sum_k \hbar\omega_k \hat{a}_k^\dagger \hat{a}_k \tag{3.221}$$

and (reintroducing $\hat{D}^1$)

$$\hat{\mathcal{H}} = \sum_k \hbar\omega_k \hat{a}_k^\dagger \hat{a}_k + \int U^{\mathrm{N}}(\hat{D}^1)\,\mathrm{d}x. \tag{3.222}$$

When there is a nonlinear refractive index or $\tilde{\zeta}^{(3)}$ term, the free particles interact via the Hamiltonian nonlinearity. It is this coupling that leads to soliton formation. It is also possible to involve other types of nonlinearity, such as $\tilde{\zeta}^{(2)}$ terms, that lead to second harmonic and parametric interactions.

With respect to practical applications, it is necessary to define photon-density and photon-flux amplitude fields. The photon-density amplitude field reminds us of the so-called detection operator (Mandel 1964, Mandel and Wolf 1995). A polariton-density amplitude field is simply defined as

$$\hat{\Psi}(x,t) = \sqrt{\frac{1}{L}} \sum_k e^{\mathrm{i}[(k-k^1)x+\omega^1 t]} \hat{a}_k, \tag{3.223}$$

where $k^1 = k(\omega^1)$ is the centre wave number for the first envelope field. This field has an equal-time commutator of the form

$$[\hat{\Psi}(x_1,t), \hat{\Psi}^\dagger(x_2,t)] = \tilde{\delta}(x_1 - x_2)\hat{1}, \tag{3.224}$$

where $\tilde{\delta}$ is a version (L-periodic) of the usual Dirac delta function

$$\tilde{\delta}(x_1 - x_2) \equiv \frac{1}{L} \sum_{\Delta k} e^{\mathrm{i}\Delta k (x_1 - x_2)}, \tag{3.225}$$

where the range of Δk is equal to that of $k - k^1$. The total polariton number operator is

$$\hat{N} = \int \hat{\Psi}^\dagger(x,t)\hat{\Psi}(x,t)\,\mathrm{d}x. \tag{3.226}$$

A polariton-flux amplitude can also be approximately expressed as

$$\hat{\Phi}(x,t) = \sqrt{\frac{v}{L}} \sum_k e^{\mathrm{i}[(k-k^1)x+\omega^1 t]} \hat{a}_k, \tag{3.227}$$

where v is the central group velocity at the carrier frequency ω^1. This flux has an equal-time commutator of the form

$$[\hat{\Phi}(x_1,t), \hat{\Phi}^\dagger(x_2,t)] = v\tilde{\delta}(x_1 - x_2)\hat{1}. \tag{3.228}$$

Operationally, $\langle \hat{\Phi}^\dagger(x,t)\hat{\Phi}(x,t)\rangle$ is the photon-flux expectation value in units of photons/second.

A common choice is to define the dimensionless field $\hat{\phi}(x,t)$ by the scaling

$$\hat{\phi}(x,t) = \hat{\Psi}(x,t)\sqrt{\frac{vt_0}{\bar{n}}}, \tag{3.229}$$

where $\bar{n}$ is the photon-number scale and t_0 is a timescale, defined so that the expectation value $\langle \hat{\phi}^\dagger(x,t)\hat{\phi}\,(x,t)\rangle$ is appropriate for the system. This scaling transformation is accompanied by a change to a comoving coordinate frame. The first choice of an altered space variable gives

$$\xi_v = \frac{\frac{x}{v} - t}{t_0},\ \tau = \frac{vt}{x_0}. \tag{3.230}$$

Here x_0 is a spatial length scale introduced to scale the interaction times. An alternative moving frame transformation is

$$\xi = \frac{x}{x_0},\ \tau_v = \frac{t - \frac{x}{v}}{t_0}. \tag{3.231}$$

The quantization technique developed by Drummond (1990) was applied to the case of a single-mode optical fibre (Drummond 1994). On simplified assumptions, the nonlinear Hamiltonian is (cf. (3.214))

$$\hat{\mathcal{H}} = \int \hbar\omega(k)\hat{a}^\dagger(k)\hat{a}(k)\,\mathrm{d}k + \frac{1}{4}\tilde{\xi}^{(3)} \int \hat{\mathbf{D}}^4(x)\,\mathrm{d}^3\mathbf{x}. \tag{3.232}$$

Here $\omega(k)$ are the angular frequencies of modes with wave vectors k describing the linear photon or polariton excitations in the fibre including dispersion. $\hat{a}(k)$ are corresponding annihilation operators defined so that, at equal times,

$$[\hat{a}(k'), \hat{a}^\dagger(k)] = \delta(k - k')\hat{1}. \tag{3.233}$$

In terms of the waveguide, the electric displacement field $\hat{\mathbf{D}}(\mathbf{x})$ is expressed as

$$\hat{\mathbf{D}}(\mathbf{x}) = \mathrm{i}\int \sqrt{\frac{\hbar\epsilon(k)kv(k)}{4\pi}}\hat{a}(k)\mathbf{u}(k,\mathbf{r})e^{ikx}\,\mathrm{d}k + \mathrm{H.c.}, \tag{3.234}$$

where $\mathbf{x} = (x, \mathbf{r})$ and

$$\int |\mathbf{u}(k,\mathbf{r})|^2\,\mathrm{d}^2\mathbf{r} = 1. \tag{3.235}$$

Here $v(k)$ is the group velocity and $\epsilon(k)$ is the dielectric permittivity. The mode function $\mathbf{u}(k,\mathbf{r})$ is included here to show how the simplified one-dimensional quantum theory relates to vector mode theory. When the interaction Hamiltonian describing the evolution of the polariton field $\hat{\Psi}(x,t)$ in the slowly varying envelope and rotating-wave approximations is considered, the coupling constant χ_e is introduced

$$\chi_e \equiv \frac{3\hbar[\tilde{\chi}^{(3)}(\omega^1)]^2[v(k^1)]^2}{4\epsilon(k^1)c^2}\int |\mathbf{u}(\mathbf{r})|^4\,\mathrm{d}^2\mathbf{r}. \tag{3.236}$$

After taking the free evolution into account, the following Heisenberg equation of motion for the field operator propagating in the $+x$-direction can be found

$$\left[v\frac{\partial}{\partial x}+\frac{\partial}{\partial t}\right]\hat{\Psi}(x,t)=\left[\frac{\mathrm{i}\omega''}{2}\frac{\partial^2}{\partial x^2}+\mathrm{i}\chi_e\hat{\Psi}^\dagger(x,t)\hat{\Psi}(x,t)\right]\hat{\Psi}(x,t), \tag{3.237}$$

where $v=v(k^1)=\frac{\partial\omega}{\partial k}\big|_{k=k^1}$, $\omega''=\frac{\partial^2\omega}{\partial k^2}\Big|_{k=k^1}$. In a comoving reference frame, this reduces to the usual quantum nonlinear Schrödinger equation

$$\mathrm{i}\frac{\partial}{\partial t}\hat{\psi}_1(x_v,t)=\left[-\frac{\omega''}{2}\frac{\partial^2}{\partial x_v^2}-\chi_e\hat{\psi}_1^\dagger(x_v,t)\hat{\psi}_1(x_v,t)\right]\hat{\psi}_1(x_v,t), \tag{3.238}$$

where $\hat{\psi}_1(x_v,t)=\hat{\Psi}(x_v+vt,t)$. In the case of anomalous dispersion which occurs at wavelength longer than 1.5 μm, allowing solitons to form, the second derivative ω'' can be expressed as

$$\omega''=\frac{\hbar}{m}, \tag{3.239}$$

where $m=\frac{\hbar}{\omega''}$ is an effective mass of a particle. Similarly, the nonlinear term χ_e describes an interaction potential

$$V(x_v,x_v')=-\chi_e\delta(x_v-x_v'). \tag{3.240}$$

This interaction potential is attractive when χ_e is positive as it is in most Kerr media. It is known that this potential has bound states and is one of the simplest exactly soluble known quantum field theories. The repulsive and attractive cases were investigated by Yang (1967, 1968). This theory is one-dimensional and tractable and does not need renormalization, while two- and three-dimensional versions do need renormalization.

In calculations, it is preferable to substitute flux amplitude operator

$$\hat{\Psi}(x,t)=\sqrt{\frac{1}{v}}\hat{\Phi}(x,t) \tag{3.241}$$

into equation (3.237). Drummond (1994) associates an idea of the spatial progression with the flux amplitude operator. Upon modifying the time variable, he obtains an "unusual form" of the quantum nonlinear Schrödinger equation, which he reduces to a more usual form again. Since the operators there have their standard meaning, they must have equal-time commutators. In contrast, the resulting equation (Drummond 1994) appears as the quantum nonlinear Schrödinger equation with time and space interchanged. Such an interpretation means that the operators have equal-space commutators. The problem is whether these commutators are well defined.

An important physical effect in propagation is that from molecular excitations. For this reason, the nonlinear Schrödinger equation requires corrections due to refractive-index fluctuations for pulses longer than about 1 ps, especially when high enough intensities are present, and fails for pulse duration much shorter than this.

The treatment of the quantum theory can start from the classical theory developed by Gordon (1986). The Raman interaction energy of a fibre is known to be [Carter and Drummond (1991)]

$$W_{\mathrm{R}} = \sum_j \boldsymbol{\zeta}_j^{\mathrm{R}} \vdots \mathbf{D}(\bar{\mathbf{x}}^j)\mathbf{D}(\bar{\mathbf{x}}^j)\delta\mathbf{x}^j. \tag{3.242}$$

Here $\mathbf{D}(\bar{\mathbf{x}}^j)$ is the electric displacement at the jth mean atomic location $\mathbf{x}^j$, $\delta\mathbf{x}^j$ is the atomic displacement operator, and $\boldsymbol{\zeta}_j^{\mathrm{R}}$ is a Raman coupling tensor.

In order this interaction to be quantized, the existence of a corresponding set of phonon operators must be taken into account. The Raman effect can be included macroscopically through a continuum Hamiltonian term coupling photons to phonons of the form (Drummond and Hardman 1993)

$$\begin{aligned}\hat{H}_{\mathrm{R}} = \hbar \int_{-\infty}^{\infty}\int_0^{\infty} \hat{\Psi}^\dagger(z,t)\hat{\Psi}(z,t)r(z,\omega)[\hat{A}(z,\omega,t) + \hat{A}^\dagger(z,\omega,t)]\,\mathrm{d}\omega\,\mathrm{d}z \\ + \hbar \int_{-\infty}^{\infty}\int_0^{\infty} \omega\hat{A}^\dagger(z,\omega,t)\hat{A}(z,\omega,t)\,\mathrm{d}\omega\,\mathrm{d}z,\end{aligned} \tag{3.243}$$

where

$$[\hat{A}(z,\omega,t), \hat{A}^\dagger(z',\omega',t)] = \delta(z-z')\delta(\omega-\omega')\hat{1}, \tag{3.244}$$

and $r(z,\omega)$ is a macroscopic frequency-dependent coupling which can be assumed to be independent of z. Here the Raman excitations are treated as an inhomogeneously broadened continuum of modes localized at each longitudinal location z. The corresponding coupled set of nonlinear operator equations is

$$\begin{aligned}\left[v\frac{\partial}{\partial z} + \frac{\partial}{\partial t}\right]\hat{\Psi}(z,t) = \left[\frac{\mathrm{i}\omega''}{2}\frac{\partial^2}{\partial z^2} + \mathrm{i}\chi_{\mathrm{e}}\hat{\Psi}^\dagger(z,t)\hat{\Psi}(z,t)\right]\hat{\Psi}(z,t) \\ - \mathrm{i}\left[\int_0^{\infty} r(\omega)[\hat{A}(z,\omega,t) + \hat{A}^\dagger(z,\omega,t)]\,\mathrm{d}\omega\right]\hat{\Psi}(z,t)\end{aligned} \tag{3.245}$$

and

$$\frac{\partial}{\partial t}\hat{A}(z,\omega,t) = -\mathrm{i}\hat{A}(z,\omega,t) - \mathrm{i}r(\omega)\hat{\Psi}^\dagger(z,t)\hat{\Psi}(z,t). \tag{3.246}$$

The phonon operators do not have white-noise behaviour, but a coloured noise property.

In practical terms, the known exact solutions (Yang 1968) of the quantum nonlinear Schrödinger equation can be hardly utilized at typical photon numbers of 10^9. It is often more useful to employ phase-space distributions or operator distributions such as the Wigner representation (Wigner 1932) and the Glauber–Sudarshan P-representation (Glauber 1963, Sudarshan 1963). In the review (Drummond 1994), the generalized P-representation is mentioned and the positive P-representation is used. Using this method, the operator equations are transformed to complex Itô stochastic equations which involve only c-number commuting variables. In other words, an operator equation can be transformed to an equivalent pair of c-number stochastic equations for $\Psi(z,t)$ and $A(z,\omega,t)$. On the transformation (3.231), equations (3.245) and (3.246) are ready for the positive P-representation. Thus, equivalent stochastic differential equations are obtainable. Substituting the integrated phonon variables into the equations for the photon field gives the following equation for a new function $\phi(\zeta,\tau_v)$:

$$\begin{aligned}\frac{\partial}{\partial\zeta}\phi(\zeta,\tau_v) \approx {} & [\mathrm{i}f\phi^\dagger(\zeta,\tau_v)\phi(\zeta,\tau_v) \pm \frac{\mathrm{i}}{2}\frac{\partial^2}{\partial\tau_v^2}\phi(\zeta,\tau_v) \\ & + \mathrm{i}\int_{-\infty}^{\tau_v} h(\tau_v-\tau_v')\phi^\dagger(\zeta,\tau_v')\phi(\zeta,\tau_v')\,\mathrm{d}\tau_v' \\ & + \sqrt{\frac{\mathrm{i}f}{\bar{n}}}\Gamma(\zeta,\tau_v) + \mathrm{i}\Gamma_{\mathrm{R}}(\zeta,\tau_v)]\phi(\zeta,\tau_v).\end{aligned} \tag{3.247}$$

There is a corresponding Hermitian conjugate equation for $\phi^\dagger$ obtained by making the substitutions $\phi\to\phi^\dagger$, $\mathrm{i}\to-\mathrm{i}$, $\Gamma\to\Gamma^\dagger$, $\Gamma_{\mathrm{R}}\to\Gamma_{\mathrm{R}}^\dagger$, with Γ, $\Gamma^\dagger$, Γ_{R}, $\Gamma_{\mathrm{R}}^\dagger$ being independent noise terms. In relation (3.247),

$$h(\tau_v) = 2\int_0^\infty r^2\left(\frac{\nu}{t_0}\right)\sin(\nu\tau_v)\frac{\mathrm{d}\nu}{\chi},\ \bar{n} = \frac{|k''|v^2}{\chi t_0}, \tag{3.248}$$

$$\zeta = \frac{z}{z_0},\ f = \frac{\chi_{\mathrm{e}}}{\chi},\ z_0 = \frac{t_0^2}{|k''|},\ k'' = -\frac{\omega''}{v^3}. \tag{3.249}$$

The last terms appearing in Equation (3.247) are stochastic functions. Γ represents the quantum noise of a field introduced by the electronic nonlinearity and Γ_{R} is the thermal noise due to the phonon coupling. In numerical simulation, the use of an enlarged nonclassical phase space can increase computation times. For this reason, the Wigner function defined on a classical phase space is useful. The Wigner function does not have an exact stochastic equation. This is because there are third-order derivative terms in the Fokker–Planck equation for the Wigner function that have no stochastic equivalent. In sufficiently intense fields, the disagreeable terms can be neglected. The Wigner function represents symmetrically ordered operators which have a diverging vacuum noise term, because a cutoff is not considered or is taken to be infinity.

As the $\chi^{(3)}$ nonlinearity in silica optical fibres is low, Drummond and He (1997) have proposed the investigation of a quantum soliton, which occurs in parametric waveguides. Such an object consists of a superposition of a second-harmonic photon with a localized pair of subharmonic photons. The system is analogous to the quark model of the meson.

On the simplifying assumption that the medium is homogeneous and isotropic, Milonni (1995) has considered the classical expression for the field energy and lifted the restriction to the magnetic susceptibility independent on frequency (cf. (3.177)). Then he has applied the results to treat basic emission and absorption processes for atoms in dispersive dielectric host media. Using this simple approach to quantization, Milonni and Maclay (2003) have shown how radiative recoil, the Doppler effect, and spontaneous and stimulated radiation rates are set up when the radiator is embedded in a host medium having a negative index of refraction.

Matsko and Kozlov (2000) have presented an approach which absorbs the results of two previous studies (Drummond and Carter 1987, Haus and Lai 1990). The two theories have been shown to provide similar outcomes of a homodyne measurement. It has been concluded that both equal-time and equal-space commutation relations are valid for the quantum soliton description. In Matsko and Kozlov (2000) the work with physical units could be amended.

Korolkova et al. (2001) have studied a quantum soliton in a Kerr medium. They have simplified, implicitly, the classical propagation equation for the slowly varying electric-field envelope by introducing a new time measurement in dependence on a position. In changing to dimensionless variables, they make the new time a "position" and the position a "time" variable and then get a classical nonlinear Schrödinger equation.

Raymond Ooi and Scully (2007) have studied three-level extended medium, which is utilized as an amplifier. They begin with a single three-level cascade atom and with a $\chi^{(2)}$ crystal, which is described by coupled parametric amplifier equations (Boyd 2003, Yariv 1989, Shen 1984). Further they present the theory and the results of the three-level cascade scheme. They compare this model with the simple one. They calculate cross correlation of the idler $\hat{E}_1(z,t)$ and the signal $\hat{E}_2(z,t)$,

$$G_{21}^{(2)}(\tau) \equiv \langle \hat{E}_1^\dagger(z,t)\hat{E}_2^\dagger(z,t+\tau)\hat{E}_2(z,t+\tau)\hat{E}_1(z,t)\rangle. \tag{3.250}$$

The neglect of the Langevin noise seems to be admissible especially for large detuning of the pump. They calculate also reverse correlation,

$$G_{12}^{(2)}(\tau) \equiv \langle \hat{E}_2^\dagger(z,t)\hat{E}_1^\dagger(z,t+\tau)\hat{E}_1(z,t+\tau)\hat{E}_2(z,t)\rangle. \tag{3.251}$$

They complete the observation of antibunching and oscillations in the reverse correlation with an interesting physics of the three-level atomic system.

3.3 Modes of Universe and Paraxial Quantum Propagation

The laser physics and the optical engineering are typical of their calculation methods and the effort for their improvement is apparent. The laser cavity is coupled to the outer space and the mode coupling can be investigated in detail. The paraxial description of light propagation can be quantized. Detectors of radiation with a spatial resolution motivate the inclusion of the optical imaging in quantum optics.

3.3.1 Quasimode Description of Spectrum of Squeezing

Toward the end of the 1980s, it had become clear that the use of squeezed states (Walls 1983, Loudon and Knight 1987) in the interferometry can lead to the enhancement of signal-to-noise ratios. Milburn and Walls (1981) have shown that the cavity of a degenerate parametric oscillator admits only a 50% amount of squeezing (in the steady state). Yurke was first to realize that the pessimistic conclusions do not hold as the noise reduction in the transmitted field can be quite different from that in the intracavity field (Yurke 1984). As a first step one had to relate the field operators inside and outside the cavity. Whereas it was obvious that the field operators inside the cavity remain the usual quantum-mechanical annihilation operators of one or a small number of harmonic oscillators, the connection of the field operators outside the cavity with the "Langevin-noise operators" was established as late as 1980s by Collett and Gardiner (1984), Gardiner and Collett (1985), and Carmichael (1987). These authors have cleared up the relation of this subtle property of squeezed light and its generation with the concept of light propagation. Not only the interpretation but also the derivation of the Langevin-like "noise" terms was presented by Lang and Scully (1973), after they introduced and studied the "modes of the universe" (Lang 1973, Ujihara 1975, 1976, 1977). It is in order to mention a book of Scully and Zubairy (1997), where the results of Gea-Banacloche et al. (1990a) are expounded or formulated as exercises.

The modes of universe are discussed, which include the interior of the imperfect cavity of interest, and are used to define the intracavity quasimode, the incident external field mode, and the output field mode. The mutual coupling of these modes emerges naturally in this formalism. Following Lang (1973), the one-sided empty cavity is described also by the relation

$$\hat{E}_{\mathrm{cav}}^{(+)}(t) = \tilde{r}\hat{E}_{\mathrm{cav}}^{(+)}\left(t - \frac{2l}{c}\right) + \tilde{t}\hat{E}_{\mathrm{in}}^{(+)}(t), \tag{3.252}$$

where l is the cavity length, $\tilde{r}$ is a real amplitude reflection coefficient, and $\tilde{t}$ is a respective transmission coefficient, $\tilde{t} = \sqrt{1-\tilde{r}^2}$. Here $\hat{E}_{\mathrm{in}}^{(+)}(t)$ is a positive-frequency part of the input field and it fulfils the commutation relation

$$[\hat{E}_{\mathrm{in}}^{(+)}(t), \hat{E}_{\mathrm{in}}^{(-)}(s)] = K\delta(t-s)\hat{1}, \tag{3.253}$$

where $\hat{E}_{\text{in}}^{(-)}(t) = [\hat{E}_{\text{in}}^{(+)}(t)]^\dagger$ and

$$K = \frac{\hbar\Omega}{2\epsilon_0 c\mathcal{A}}, \tag{3.254}$$

with Ω being a quasimode frequency. For the full Fox–Li quasimode (Fox and Li 1961, Barnett and Radmore 1988), a single-mode annihilation operator $\hat{a}(t)$ is defined

$$\hat{a}(t) = \sqrt{\frac{2l}{Kc}}\hat{E}_{\text{cav}}^{(+)}(t)e^{\mathrm{i}\Omega t}. \tag{3.255}$$

It is convenient to use the slowly varying amplitudes $\mathcal{E}_{\text{in}}^{(+)}(t)$, $\mathcal{E}_{\text{out}}^{(+)}(t)$, and $\mathcal{E}_{\text{cav}}^{(+)}(t)$ for the input, output, and cavity fields related to the cavity frequency Ω, respectively,

$$\hat{E}_{\text{in}}^{(+)}(t) = \mathcal{E}_{\text{in}}^{(+)}(t)e^{-\mathrm{i}\Omega t}, \tag{3.256}$$
$$\hat{E}_{\text{out}}^{(+)}(t) = \mathcal{E}_{\text{out}}^{(+)}(t)e^{-\mathrm{i}\Omega t}, \tag{3.257}$$
$$\hat{E}_{\text{cav}}^{(+)}(t) = \mathcal{E}_{\text{cav}}^{(+)}(t)e^{-\mathrm{i}\Omega t}. \tag{3.258}$$

We propose to consider the definition

$$\hat{a}(t) = \sqrt{\frac{c}{K2l}}\int_{t-\frac{2l}{c}}^{t} \mathcal{E}_{\text{cav}}^{(+)}(t')\,\mathrm{d}t' \tag{3.259}$$

instead of (3.255), which is in a better agreement with the quantum field theory. To approve this change we denote the rightward and leftward travelling positive-frequency parts as $\hat{E}_{>}^{(+)}(z,t)$ and $\hat{E}_{<}^{(+)}(z,t)$ so that $\hat{E}_{\text{in}}^{(+)}(t) \equiv \hat{E}_{>}^{(+)}(-0,t)$, $\hat{E}_{\text{out}}^{(+)}(t) \equiv \hat{E}_{<}^{(+)}(-0,t)$, $\hat{E}_{\text{cav}}^{(+)}(t) \equiv \hat{E}_{>}^{(+)}(+0,t)$. The factor at the delta function in the equal-time commutator of the field is Kc and from this we can calculate the equal-space commutator ($t > 0$, $s > 0$ without loss of generality)

$$[\hat{E}_{\text{in}}^{(+)}(t), \hat{E}_{\text{in}}^{(-)}(s)] = [\hat{E}_{>}^{(+)}(-0,t), \hat{E}_{>}^{(-)}(-0,s)]$$
$$= [\hat{E}_{>}^{(+)}(-ct,0), \hat{E}_{>}^{(-)}(-cs,0)] \tag{3.260}$$
$$= Kc\delta(-ct+cs)\hat{1} = K\delta(t-s)\hat{1}. \tag{3.261}$$

Hence, definition (3.259) can be rewritten in the form,

$$\hat{a}(t) = \sqrt{\frac{1}{Kc2l}}\int_0^l \left[\mathcal{E}_{>}^{(+)}(z,t) + \mathcal{E}_{<}^{(+)}(z,t)\right]\mathrm{d}z, \tag{3.262}$$

where (cf. (3.258)) $\mathcal{E}_{>}^{(+)}(z,t)$ and $\mathcal{E}_{<}^{(+)}(z,t)$ mean the slowly varying smooth amplitude with the property

$$\mathcal{E}_{>}^{(+)}(z,t) = \hat{E}_{>}^{(+)}(z,t)e^{-\mathrm{i}k_0 z+i\Omega t},\ k_0 = \frac{\Omega}{c}, \tag{3.263}$$
$$\mathcal{E}_{<}^{(+)}(z,t) = \hat{E}_{<}^{(+)}(z,t)e^{\mathrm{i}k_0 z+\mathrm{i}\Omega t}. \tag{3.264}$$

Performing the time integration on both sides of relation (3.252), we obtain that

$$\hat{a}(t) = \tilde{r}\hat{a}\left(t - \frac{2l}{c}\right) + \tilde{t}\hat{b}(t), \tag{3.265}$$

where

$$\hat{b}(t) = \sqrt{\frac{c}{K2l}} \int_{t-\frac{2l}{c}}^{t} \mathcal{E}_{\text{in}}^{(+)}(t')\,\mathrm{d}t'. \tag{3.266}$$

Recalling the space integration, we see that the annihilation operator $\hat{a}(t)$ corresponds to the same quasimode in all times, whereas $\hat{b}(t)$ is appropriate to many distinct modes.

In the situation when it holds that

$$\tilde{r}\hat{a}\left(t - \frac{2l}{c}\right) \approx \left(1 - \frac{1}{2}\tilde{t}^2\right)\hat{a}(t) - \frac{2l}{c}\frac{\mathrm{d}}{\mathrm{d}t}\hat{a}(t), \tag{3.267}$$

in the short cavity round-trip time limit we get

$$\frac{\mathrm{d}}{\mathrm{d}t}\hat{a}(t) = -\Gamma\hat{a}(t) + \sqrt{2\Gamma}\sqrt{\frac{c}{2l}}\hat{b}(t), \tag{3.268}$$

where

$$\Gamma = \frac{c}{2l}\frac{1}{2}\tilde{t}^2. \tag{3.269}$$

Noting that

$$\sqrt{\frac{c}{2l}}\hat{b}(t) = \sqrt{\frac{1}{K}\frac{c}{2l}} \int_{t-\frac{2l}{c}}^{t} \mathcal{E}_{\text{in}}^{(+)}(t')\,\mathrm{d}t'$$
$$= \sqrt{\frac{1}{K}}\mathcal{E}_{\text{in}}^{(+)}(t), \tag{3.270}$$

where we replaced the average over the short-time interval by the value of the function (at the upper limit), we rewrite equation (3.268) as a quantum Langevin equation

$$\frac{\mathrm{d}}{\mathrm{d}t}\hat{a}(t) = -\Gamma\hat{a}(t) + \sqrt{2\Gamma}\sqrt{\frac{1}{K}}\mathcal{E}_{\text{in}}^{(+)}(t). \tag{3.271}$$

Further, Gea-Banacloche et al. (1990a) define, for arbitrary measurement times, the spectrum of squeezing of the output field via the quadrature variances. They present a microscopic effective Hamiltonian model of balanced homodyne detection.

They refer to the fundamental papers (Collett and Gardiner 1984, Gardiner and Collett 1985, Caves and Schumaker 1985, Yurke 1985), where this concept of spectral squeezing was originally treated. As an approximation, the operator is introduced

$$\hat{\hat{A}}_{\rm out}(\delta\omega) = \frac{\mathcal{N}}{\sqrt{T}} \int_0^T \mathcal{E}_{\rm out}^{(+)}(t) e^{{\rm i}\delta\omega t}\, {\rm d}t, \tag{3.272}$$

where

$$\mathcal{N} = \sqrt{\frac{1}{K}}. \tag{3.273}$$

As shown also by Yurke (1985) and Carmichael (1987), with a balanced homodyne detector one measures the combinations

$$\mathcal{E}_{{\rm out}_\theta}(t) = \frac{1}{2}\left[e^{{\rm i}\theta}\mathcal{E}_{\rm out}^{(+)}(t) + e^{-{\rm i}\theta}\mathcal{E}_{\rm out}^{(-)}(t)\right], \tag{3.274}$$

and from this the natural generalization of the single-mode quadrature concept is

$$\hat{\hat{A}}_{{\rm out}_\theta}(\delta\omega) = \frac{1}{2}\left[e^{{\rm i}\theta}\hat{\hat{A}}_{\rm out}(\delta\omega) + e^{-{\rm i}\theta}\hat{\hat{A}}_{\rm out}^{\dagger}(-\delta\omega)\right]. \tag{3.275}$$

We may wonder why a non-Hermitian operator is taken for such a generalization of the Hermitian operator. Finally, the connection between single quasimode squeezing and spectral squeezing is explored and the difference in the noise reduction inside and outside the cavity is clarified in a way that lends itself to a simple visualization.

Gea-Banacloche et al. (1990b) have first analysed measurements of small phase or frequency changes for an ordinary laser and calculated the extra cavity phase noise for a phase-locked laser. These analyses are based on the mean values and the normally ordered variances of quantum operators for which classical Langevin equations may be written down. The classical Langevin formalism is further replaced by the alternative Fokker–Planck formalism for the calculation of the spectrum of squeezing. This general Fokker–Planck formalism was applied to the two-photon correlated-spontaneous-emission laser. It has been shown that without one-photon resonance and initial atomic coherences involving the middle level, the maximum squeezing of the ultracavity mode is 50% while the detected field can be almost perfectly squeezed. Almost the exact reverse holds, however, if one-photon resonance and initial atomic coherences involving the middle level are present. In particular, the intracavity field may be perfectly squeezed while the outside field is not only unsqueezed but has, in fact, increased noise in the conjugate quadrature. Finally, the effect of finite measurement time on the quadrature variances is briefly analysed.

Dutra and Nienhuis (2000) have unified the concept of normal modes used in quantum optics and that of Fox–Li modes from semiclassical laser physics. Their one-dimensional theory solves the problem of how to describe the quantized radiation field in a leaky cavity using Fox–Li modes. In this theory, unlike conventional models, system and reservoir operators no longer commute with each other, as a consequence of natural cavity modes having been used. Aiello (2000) has derived simple relations for an electromagnetic field inside and outside an optical cavity, limiting himself to one- and two-photon states of the field. He has expressed input–output relations using a nonunitary transformation between intracavity and output operators.

Brown and Dalton (2002) have considered three-dimensional unstable optical systems. They have defined non-Hermitian modes and their adjoints in both the cavity and external regions. A number of concepts and properties resulting from the standard canonical quantization procedure have been suited to the non-Hermitian modes by the exact transformation method. The results are applied to the spontaneous decay of a two-level atom inside an unstable cavity.

3.3.2 Steady-State Propagation

In Deutsch and Garrison (1991a) it is assumed that in the case of amplifier, one is usually interested in the spatial dependence of temporally steady-state fields. It is no attempt at a reformulation of one-dimensional propagation, cf. Abram and Cohen (1991), where the temporal evolution by the Hamiltonian is supplemented by the spatial progression with the momentum operator. An alternative proposal is made that the quantum-mechanical equivalent of the classical steady-state condition is the description of the system by a stationary state of a suitable Hamiltonian. There is a formal resemblance to a nonrelativistic many-body theory for a complex scalar field (Deutsch and Garrison 1991b), which helps determine the Hamiltonian. In this theory a non-Hermitian envelope-field operator $\hat{\Psi}(z,t)$, with the property

$$[\hat{\Psi}(z,t), \hat{\Psi}^\dagger(z',t)] = \delta(z-z')\hat{1}, \tag{3.276}$$

is introduced. In the application to the optical field, the vector potential (or electric field in the lowest order) corresponding to a carrier plane wave of a given polarization $\mathbf{e}$ is expressed as follows:

$$\hat{E}^{(+)}_\omega(z,t) = \mathbf{e}\sqrt{\frac{2\pi\hbar\omega}{\mathcal{A}n^2(\omega)}}\,\hat{\Psi}(z,t)\exp[\mathrm{i}(kz-\omega t)], \tag{3.277}$$

where $\mathcal{A}$ is the beam area and $n(\omega)$ is a dispersive index of refraction.

In contrast to Deutsch and Garrison (1991a), we will make a simplification, i.e. we will not consider a carrier wave Hamiltonian. For the single wave interacting nonlinearly with matter, the total Hamiltonian can be written as

$$\hat{H} = \hat{H}_{\mathrm{env}} + \hat{H}_{\mathrm{int}}, \tag{3.278}$$

$$\hat{H}_{\mathrm{env}} = -\frac{\mathrm{i}\hbar c}{2n(\omega)} \times \int \left(\hat{\Psi}^{\dagger}(z,t)\frac{\partial \hat{\Psi}(z,t)}{\partial z} - \frac{\partial \hat{\Psi}^{\dagger}(z,t)}{\partial z}\hat{\Psi}(z,t) \right) \mathrm{d}z, \tag{3.279}$$

where $\hat{H}_{\mathrm{env}}$ is the Hamiltonian governing the free progression of the envelope and $\hat{H}_{\mathrm{int}}$ is a general interaction Hamiltonian. In fact, the generality will not be exercised and we will treat only the vacuum input and the case of degenerate parametric amplifier.

In the standard Heisenberg picture, the equation of motion for the envelope-field operator reads

$$\frac{\partial \hat{\Psi}(z,t)}{\partial t} = -\frac{\mathrm{i}}{\hbar}[\hat{\Psi}(z,t), \hat{H}], \tag{3.280}$$

or for a linear medium

$$\frac{\partial \hat{\Psi}(z,t)}{\partial t} = -\frac{c}{n(\omega)}\frac{\partial \hat{\Psi}(z,t)}{\partial z}. \tag{3.281}$$

The solution is

$$\hat{\Psi}(z,t) = \hat{\Psi}\left(z - \frac{ct}{n(\omega)}, 0\right). \tag{3.282}$$

In the standard Schrödinger picture, the state $|\Phi\rangle$ evolves by the Schrödinger equation

$$\frac{\partial|\Phi\rangle}{\partial t} = -\frac{\mathrm{i}}{\hbar}(\hat{H}_{\mathrm{env}} + \hat{H}_{\mathrm{int}})|\Phi\rangle. \tag{3.283}$$

The introducing of the carrier wave Hamiltonian has revoked the considering of the Schrödinger picture (Deutsch and Garrison 1991b), along with the envelope picture which we have confined ourselves to. Relation (3.277) is the positive-frequency component of the electric-field in the envelope picture similarly as relation (4.5b) in Caves and Schumaker (1985) is this component in the interaction picture. The envelope picture is essentially the modulation picture in Caves and Schumaker (1985).

For application under consideration there will be exact frequency matching between the carrier frequencies of the various waves which interact so that the Hamiltonian in equation (3.283) will be independent of time, thus the steady state (ss) solutions are identified with the stationary solutions to equation (3.283):

$$\left(\hat{H}_{\mathrm{env}} + \hat{H}_{\mathrm{int}}\right)|\Phi\rangle_{\mathrm{ss}} = \lambda|\Phi\rangle_{\mathrm{ss}}. \tag{3.284}$$

For the stationary solutions, the label (ss) will be omitted.

(i) In the case

$$\hat{H}_{\text{int}} = \hat{0}, \tag{3.285}$$

i.e. in the case of vacuum propagation, the stationary solutions are the translation-invariant states. To have a unitary representation of the translation, we may consider either the limits $-\infty, \infty$ in the integral on the left-hand side in (3.279) or the spatial periodicity. We prefer the latter possibility.

As an example, we can consider a coherent state corresponding to a constant one-photon wave function. We define a functional displacement operator

$$\hat{D}[\alpha] \equiv \exp\left(\int [\alpha(z)\hat{\Psi}^\dagger(z) - \alpha^*(z)\hat{\Psi}(z)]\,\mathrm{d}z\right) \tag{3.286}$$

$$= \exp(\rho\hat{a}^\dagger - \rho\hat{a}), \tag{3.287}$$

where

$$\rho = \sqrt{\int |\alpha(z)|^2\,\mathrm{d}z}, \tag{3.288}$$

$$\hat{a} \equiv \hat{a}\left[\frac{\alpha}{\rho}\right] \equiv \frac{1}{\rho}\int \alpha^*(z)\hat{\Psi}(z)\,\mathrm{d}z. \tag{3.289}$$

The coherent state is defined as the displaced vacuum

$$|\{\alpha\}\rangle \equiv \hat{D}[\alpha]|0\rangle. \tag{3.290}$$

Since

$$[\hat{a}, \hat{a}^\dagger] = \hat{1}, \tag{3.291}$$

relation (3.286) can be rewritten in the normal ordering form

$$\hat{D}[\alpha] = \exp\left(-\frac{1}{2}\int |\alpha(z)|^2\,\mathrm{d}z\right) \times \exp\left(\int \alpha(z)\hat{\Psi}^\dagger(z)\,\mathrm{d}z\right)\exp\left(-\int \alpha^*(z)\hat{\Psi}(z)\,\mathrm{d}z\right) \tag{3.292}$$

and the corresponding one-photon state is

$$\left|1, \left\{\frac{\alpha}{\rho}\right\}\right\rangle \equiv \frac{1}{\rho}\int \alpha(z)\hat{\Psi}^\dagger(z)|0\rangle\,\mathrm{d}z. \tag{3.293}$$

On substituting $|\Phi\rangle = |\{\alpha\}\rangle$ into relation (3.284) and applying then the operator $\hat{D}^\dagger[\alpha]$ from left to both sides, we derive that $\alpha(z)$ and λ should make the vacuum state the eigenstate of the operator

$$\hat{D}^\dagger[\alpha]\hat{H}_{\rm env}\hat{D}[\alpha] = -\frac{i\hbar}{2}\frac{c}{n(\omega)}\int\left\{\left[\hat{\Psi}^\dagger(z)+\alpha^*(z)\hat{1}\right]\frac{\partial\left[\hat{\Psi}(z)+\alpha(z)\hat{1}\right]}{\partial z}\right.$$
$$\left.-\frac{\partial\left[\hat{\Psi}^\dagger(z)+\alpha^*(z)\hat{1}\right]}{\partial z}\left[\hat{\Psi}(z)+\alpha(z)\hat{1}\right]\right\}dz. \tag{3.294}$$

The eigenvalue is λ again. On equating

$$\hat{D}^\dagger[\alpha]\hat{H}_{\rm env}\hat{D}[\alpha]|0\rangle = -i\hbar\frac{c}{n(\omega)}\int\hat{\Psi}^\dagger(z)\frac{d\alpha(z)}{dz}|0\rangle\,dz$$
$$-i\hbar\frac{c}{n(\omega)}\int\alpha^*(z)\frac{d\alpha(z)}{dz}\,dz|0\rangle \tag{3.295}$$

with $\lambda|0\rangle$, we see that

$$\frac{d\alpha(z)}{dz} = 0 \tag{3.296}$$

and hence

$$\lambda = 0. \tag{3.297}$$

Instead of a translation-invariant wave function, we may try one that is an eigenfunction of the translation operator. When the boson number operator commutes with the operator

$$\hat{A} = \hat{H}_{\rm env} + \hat{H}_{\rm int}, \tag{3.298}$$

on the left-hand side of (3.284), this problem can be generalized by the insertion of the number operator $\hat{N}$,

$$\hat{N} = \int\hat{\Psi}^\dagger(z)\hat{\Psi}(z)\,dz, \tag{3.299}$$

behind λ on the right-hand side. The idea takes into account that the wave function of any number of particles should be the eigenfunction of the operator $\hat{A}$

$$\exp(i\hat{A})|\Phi\rangle = \exp(i\lambda\hat{N})|\Phi\rangle \tag{3.300}$$

or

$$\exp(i(\hat{A}-\lambda\hat{N}))|\Phi\rangle = |\Phi\rangle. \tag{3.301}$$

Condition (3.301) is equivalent to

$$(\hat{A}-\lambda\hat{N})|\Phi\rangle = 0 \tag{3.302}$$

or to the relation with the insertion. On substituting again Φ into the new relation, and applying the operator $\hat{D}^\dagger[\alpha]$, we obtain the right-hand side in the form

$$\lambda \int \left[\hat{\Psi}^\dagger(z) + \alpha^*(z)\hat{1}\right]\left[\hat{\Psi}(z) + \alpha(z)\hat{1}\right] \mathrm{d}z. \tag{3.303}$$

Condition (3.296) is thus generalized,

$$-\mathrm{i}\hbar \frac{c}{n(\omega)} \frac{\mathrm{d}\alpha(z)}{\mathrm{d}z} = \lambda\alpha(z). \tag{3.304}$$

The solution to equation (3.304) reads

$$\alpha(z) = \alpha(0)\exp\left[\frac{\mathrm{i}\lambda n(\omega)}{\hbar c} z\right], \tag{3.305}$$

where λ is any real number. Expression (3.305) for the complex amplitude is sufficient for the fulfilment of the relation $\hat{A}|\Phi\rangle = \lambda\hat{N}|\Phi\rangle$.

(ii) In the case of the degenerate parametric amplifier, the interaction Hamiltonian can be written (Hillery and Mlodinow 1984) as follows:

$$\hat{H}_{\mathrm{int}} = -\frac{1}{2}\int\int\int \Big\{\chi^{(2)}(z)\mathcal{E}_{\mathrm{p}}^*(z)\exp[-\mathrm{i}(k_{\mathrm{p}}z - \omega_{\mathrm{p}}t)][\hat{E}_\omega^{(+)}(z,t)]^2 + \mathrm{H.c.}\Big\}\,\mathrm{d}x\,\mathrm{d}y\,\mathrm{d}z, \tag{3.306}$$

where ω_{p} is the pump frequency, $\omega_{\mathrm{p}} = 2\omega$, $\chi^{(2)}(z)$ is the second-order susceptibility coupling the pump to the degenerate signal and idler fields and $\mathcal{E}_{\mathrm{p}}(z)$ is the pump amplitude. Substituting for $\hat{E}_\omega^{(+)}(z,t)$ from relation (3.277) gives the interaction Hamiltonian in the envelope picture

$$\hat{H}_{\mathrm{int}} = \frac{\mathrm{i}\hbar}{2}\frac{c}{n(\omega)}\int [\kappa^*(z)\hat{\Psi}^2(z) - \kappa(z)\hat{\Psi}^{\dagger 2}(z)]\,\mathrm{d}z, \tag{3.307}$$

with

$$\kappa(z) = \frac{1}{2}g_0(z)\exp[\mathrm{i}\phi(z)], \tag{3.308}$$

$$g_0(z) = \frac{4\pi\omega}{n(\omega)c}|\chi^{(2)}(z)||\mathcal{E}_{\mathrm{p}}(z)|, \tag{3.309}$$

$$\phi(z) = -\frac{\pi}{2} + \Delta k\, z + \beta(z). \tag{3.310}$$

Here $g_0(z)$ is the standard power gain coupling constant (Yariv 1985), $\Delta k = 2k - k_{\mathrm{p}}$ is the phase mismatch at the degenerate frequency, and $\beta(z)$ is the remaining phase originating from the product $\chi^{(2)}(z)\mathcal{E}_{\mathrm{p}}^*(z)$.

To solve the time-independent Schrödinger equation, Deutsch and Garrison (1991a) assume that the eigenstate is a squeezed vacuum state corresponding to a two-photon wave function. They define a functional squeezing operator

$$\hat{S}[\xi] \equiv \exp\left\{\frac{1}{2}\int [\xi(z)\hat{\Psi}^{\dagger 2}(z) - \xi^*(z)\hat{\Psi}^2(z)]\,\mathrm{d}z\right\}, \tag{3.311}$$

with z-dependent squeezing parameter $\xi(z) = -r(z)\exp[\mathrm{i}\theta(z)]$. The squeezed vacuum is defined as

$$|0\rangle_{\{\xi\}} \equiv \hat{S}[\xi]|0\rangle. \tag{3.312}$$

Similarly as in case (i), $\xi(z)$ and λ should be solutions of the equation

$$\hat{S}^\dagger[\xi](\hat{H}_{\mathrm{env}} + \hat{H}_{\mathrm{int}})\hat{S}[\xi]|0\rangle = \lambda|0\rangle. \tag{3.313}$$

Applying the operator $\hat{\Psi}(z)$ to both the sides of (3.313) and taking into account that

$$\lambda\hat{\Psi}(z)|0\rangle = 0 = \hat{S}^\dagger[\xi](\hat{H}_{\mathrm{env}} + \hat{H}_{\mathrm{int}})\hat{S}[\xi]\hat{\Psi}(z)|0\rangle, \tag{3.314}$$

we rewrite the eigenvalue problem in the λ-independent form

$$\left[\hat{S}^\dagger[\xi](\hat{H}_{\mathrm{env}} + \hat{H}_{\mathrm{int}})\hat{S}[\xi], \hat{\Psi}(z)\right]|0\rangle_{\{\xi\}} = 0 = \hat{\hat{C}}|0\rangle, \tag{3.315}$$

where the commutator $\hat{\hat{C}}$ is

$$\begin{aligned}
\hat{\hat{C}} = \frac{\mathrm{i}\hbar c}{n(\omega)}\Bigg\{ & -\exp[\mathrm{i}\theta(z)]\Big(\exp[-\mathrm{i}\theta(z)]\cosh[r(z)]\frac{\mathrm{d}}{\mathrm{d}z}\big(\exp[\mathrm{i}\theta(z)]\sinh[r(z)]\big) \\
& - \frac{\mathrm{d}}{\mathrm{d}z}\big(\cosh[r(z)]\big)\sinh[r(z)] \\
& + \kappa^*(z)\exp[\mathrm{i}\theta(z)]\sinh^2[r(z)] - \kappa(z)\exp[-\mathrm{i}\theta(z)]\cosh^2[r(z)]\Big)\hat{\Psi}^\dagger(z) \\
& + \Big(\cosh[r(z)]\frac{\mathrm{d}}{\mathrm{d}z}\big(\cosh[r(z)]\big) \\
& - \exp[\mathrm{i}\theta(z)]\sinh[r(z)]\frac{\mathrm{d}}{\mathrm{d}z}\big(\exp[-\mathrm{i}\theta(z)]\sinh[r(z)]\big) \\
& + \kappa^*(z)\exp[\mathrm{i}\theta(z)]\sinh[r(z)]\cosh[r(z)] \\
& - \kappa(z)\exp[-\mathrm{i}\theta(z)]\sinh[r(z)]\cosh[r(z)]\Big)\hat{\Psi}(z) + \frac{\partial}{\partial z}\hat{\Psi}(z)\Bigg\}.
\end{aligned} \tag{3.316}$$

The eigenvalue condition requires that the real and imaginary parts of the coefficient at $\hat{\Psi}^\dagger$ vanish yielding the desired propagation equations

$$\frac{\mathrm{d}r}{\mathrm{d}z} = \frac{1}{2} g_0 \cos(\theta - \phi), \tag{3.317}$$

$$\frac{\mathrm{d}\theta}{\mathrm{d}z} = -g_0 \coth(2r) \sin(\theta - \phi). \tag{3.318}$$

On introducing the complex amplitude

$$\zeta(z) = -\exp[\mathrm{i}\theta(z)] \tanh[r(z)], \tag{3.319}$$

we can write equations (3.317) and (3.318) in the compact form

$$\frac{\mathrm{d}(-\zeta)}{\mathrm{d}z} = \kappa - \kappa^* \zeta^2, \tag{3.320}$$

which may be useful for guessing the boundary condition

$$r(z)\,|_{z=0} = 0, \quad \theta(z)\,|_{z=0} = \phi(0). \tag{3.321}$$

When $\beta(z) = 0$ and the phase difference $\theta(z) - \phi(0)$ is small, the squeezing parameter $r(z)$ integrates values of experimental parameter $\frac{1}{2} g_0(z')$, $z' \in [0, z]$, when moreover $g_0(\infty) > |\Delta k|$, the squeezing parameter $\theta(z)$ converges to the function of the experimental parameters $\left\{\phi(z) - \arcsin\left[\frac{\Delta k}{g_0(\infty)}\right]\right\}$.

A direct solution of (3.313) requires that the real and imaginary parts of the coefficient at $\hat{\Psi}^\dagger(z)$ vanish yielding the propagation equations (3.317) and (3.318) again. The presence of the singular operator $\hat{\Psi}(z)\hat{\Psi}^\dagger(z)$ indicates that λ has no finite value in general. Resorting to the partition of the field into finite elements of length Δz in each of which we can define local field operators, we find that $\lambda = O\left(\frac{1}{\Delta z}\right)$,

$$\lambda \simeq -\frac{1}{\Delta z} \frac{\hbar c}{n(\omega)} \int \sin(\theta - \phi) \frac{\sinh^3 r}{\cosh r} \, \mathrm{d}z. \tag{3.322}$$

3.3.3 Approximation of Slowly Varying Envelope

The macroscopic approach to the quantum propagation aims at a quantum version of the slowly varying envelope approximation. Such an envelope implies that the wave is paraxial and monochromatic. The problem of quantum propagation of paraxial fields was considered first by Graham and Haken (1968). The revived interest is indicated by Kennedy and Wright (1988). Deutsch and Garrison (1991b) begin with generalizing the results of Lax et al. (1974), which develop the classical theory of a strictly monochromatic wave in an inhomogeneous nonlinear (perhaps amplifying) medium. The generalization is made only to a quasimonochromatic wave and the quantum theory is presented in the simplest system of codirectional propagation considering only the free-field dynamics.

In the Coulomb gauge, the positive-frequency component of the vector potential satisfies the free-field wave equation

$$\nabla^2 \mathbf{A}^{(+)}(\mathbf{x}, t) - \frac{1}{c^2}\frac{\partial^2}{\partial t^2}\mathbf{A}^{(+)}(\mathbf{x}, t) = \mathbf{0}. \tag{3.323}$$

The approximation of the slowly varying envelope is introduced by expressing $\mathbf{A}^{(+)}(\mathbf{x}, t)$ as envelope modulating a carrier plane wave propagating in the z-direction with the wave number k_0 and the frequency $\omega_0 = ck_0$,

$$\mathbf{A}^{(+)}(\mathbf{x}, t) = A_0 \Psi(\mathbf{x}, t) \exp[\mathrm{i}(k_0 z - \omega_0 t)]. \tag{3.324}$$

Here, $\Psi(\mathbf{x}, t)$ is a vector-valued function, henceforth referred to as an envelope field, and A_0 is a normalization constant, which we will specify before relation (3.333). The initial positive-frequency component can be expressed as follows:

$$\mathbf{A}^{(+)}(\mathbf{x}, t = 0) = \frac{1}{(2\pi)^{\frac{3}{2}}} \int \sqrt{\frac{\hbar c}{2|\mathbf{k}|}} \sum_{\lambda=1,2} \mathbf{e}_\lambda(\mathbf{k}) \mathcal{F}_\lambda(\mathbf{k}) e^{\mathrm{i}\mathbf{k}.\mathbf{x}} \, \mathrm{d}^3\mathbf{k}. \tag{3.325}$$

Here $\mathbf{e}_\lambda(\mathbf{k})$ are the orthogonal polarization unit vectors and the reduced Planck constant is introduced in view of the possible later quantization. $\mathcal{F}_\lambda(\mathbf{k})$ are thus momentum-space wave functions.

The intuitive notion of a paraxial field is that it is composed of rays making small angles with the main propagation axis. In other words, a paraxial wave function $\{\mathcal{F}_\lambda(\mathbf{k})\}$ is concentrated in a small neighbourhood $\mathbf{k}_0 = k_0 \mathbf{e}_3$ of the wave vector of the carrier wave. We define $f_\lambda(\mathbf{q})$ by the relation

$$f_\lambda(\mathbf{q}) = \mathcal{F}_\lambda(\mathbf{q} + \mathbf{k}_0), \tag{3.326}$$

where $\mathbf{q}$ is the relative wave vector. Let us observe that $\mathbf{q} = (\mathbf{q}_\mathrm{T}, q_z)$, where $\mathbf{q}_\mathrm{T}$ is the transverse part of $\mathbf{q}$, $\mathbf{q} = \mathbf{q}_\mathrm{T} + q_z \mathbf{e}_3$.

In contrast to Deutsch and Garrison (1991b), we stress that we express the concentration in a small neighbourhood of $\mathbf{q}_0 = \mathbf{0}$, by letting the wave function $\{f_\lambda(\mathbf{q})\}$ depend on a small positive parameter θ, $f_\lambda(\mathbf{q}) \equiv f_\lambda(\mathbf{q}, \theta)$. Let us assume that

$$f_\lambda(\mathbf{q}_\mathrm{T}, q_z, \theta) = \sqrt{\mathcal{V}}\, \overline{f}_\lambda \left(\frac{\mathbf{q}_\mathrm{T}}{\theta k_0}, \frac{q_z}{\theta^2 k_0}, \theta \right), \tag{3.327}$$

where

$$\mathcal{V} = \frac{1}{\theta^4 k_0^3}, \tag{3.328}$$

and we have introduced the notational convention that an overbar indicates a dimensionless function of the scaled variables (and perhaps θ). This relates to defining a dimensionless "momentum" vector $\boldsymbol{\eta} = (\boldsymbol{\eta}_\mathrm{T}, \eta_z)$, where

$$\boldsymbol{\eta}_\mathrm{T} = \frac{\mathbf{q}_\mathrm{T}}{\theta k_0}, \tag{3.329}$$

$$\eta_z = \frac{q_z}{\theta^2 k_0}. \tag{3.330}$$

The functions of interest are those that have a convergent power-series expansion in θ,

$$\overline{f}_\lambda(\boldsymbol{\eta}, \theta) = \sum_{n=0}^{\infty} \theta^n \overline{f}_\lambda^{(n)}(\boldsymbol{\eta}). \tag{3.331}$$

In contrast to Deutsch and Garrison (1991b), we note that

$$\overline{f}_\lambda(\boldsymbol{\eta}, \theta) = \overline{f}_\lambda^{(0)}(\boldsymbol{\eta}). \tag{3.332}$$

Relations (3.329) and (3.332) lead to the wave function being θ-dependent, a difference from Deutsch and Garrison (1991b).

Substituting integral (3.325) for the envelope field defined by (3.324) at $t = 0$, with the momentum-space wave function given by equation (3.326), and choosing $A_0 = \sqrt{\frac{\hbar c}{2k_0}}$, we find that

$$\Psi(\mathbf{x}, t = 0, \theta) = \frac{1}{(2\pi)^{\frac{3}{2}}} \times \int \sqrt{\frac{k_0}{|\mathbf{q} + \mathbf{k}_0|}} \sum_{\lambda=1,2} \mathbf{e}_\lambda(\mathbf{q} + \mathbf{k}_0) f_\lambda(\mathbf{q}, \theta) e^{\mathrm{i}\mathbf{q}.\mathbf{x}} \, \mathrm{d}^3\mathbf{q}. \tag{3.333}$$

Here, the parameter θ has been introduced, which is not present in integral (3.325), where $\mathcal{F}_\lambda(\mathbf{k}) \equiv \mathcal{F}_\lambda(\mathbf{k}, \theta)$, $\mathbf{A}^{(+)}(\mathbf{x}, t) \equiv \mathbf{A}^{(+)}(\mathbf{x}, t, \theta)$.

In Deutsch and Garrison (1991b), the integro-differential form of the wave equation for $\mathbf{A}^{(+)}(\mathbf{x}, t, \theta)$ is investigated,

$$\mathrm{i}\frac{\partial}{\partial t}\mathbf{A}^{(+)}(\mathbf{x}, t, \theta) = c(-\nabla^2)^{\frac{1}{2}}\mathbf{A}^{(+)}(\mathbf{x}, t, \theta), \tag{3.334}$$

where $(-\nabla^2)^{\frac{1}{2}}$ is an integral operator defined by

$$(-\nabla^2)^{\frac{1}{2}} F(\mathbf{x}) = \frac{1}{(2\pi)^{\frac{3}{2}}} \int |\mathbf{k}| \tilde{F}(\mathbf{k}) e^{\mathrm{i}\mathbf{k}.\mathbf{x}} \, \mathrm{d}^3\mathbf{k}, \tag{3.335}$$

with $\tilde{F}(\mathbf{k})$ being the Fourier transform. Let us note that $\nabla = (\nabla_\mathrm{T}, \frac{\partial}{\partial z})$. Substituting from (3.324) into (3.334) gives

$$\mathrm{i}\frac{\partial}{\partial t}\Psi(\mathbf{x}, t, \theta) = (c\Omega - \omega_0)\Psi(\mathbf{x}, t, \theta), \tag{3.336}$$

where

$$\Omega \equiv \Omega\left(\nabla_{\mathrm{T}}, \frac{\partial}{\partial z}\right) = \left(k_0^2 - 2\mathrm{i}k_0\frac{\partial}{\partial z} - \nabla_{\mathrm{T}}^2 - \frac{\partial^2}{\partial z^2}\right)^{\frac{1}{2}}. \tag{3.337}$$

The scaled configuration-space variables $\boldsymbol{\xi} = (\boldsymbol{\xi}_{\mathrm{T}}, \zeta)$ are

$$\boldsymbol{\xi}_{\mathrm{T}} = \theta k_0 \mathbf{x}_{\mathrm{T}}, \quad \zeta = \theta^2 k_0 z, \tag{3.338}$$

and the dimensionless time variable

$$\tau = \theta^2 \omega_0 t. \tag{3.339}$$

After expressing the envelope field in the form

$$\Psi(\mathbf{x}, t, \theta) \equiv \Psi(\mathbf{x}_{\mathrm{T}}, z, t, \theta) = \frac{1}{\sqrt{\mathcal{V}}}\overline{\Psi}(\theta k_0 \mathbf{x}_{\mathrm{T}}, \theta^2 k_0 z, \theta^2 \omega_0 t, \theta), \tag{3.340}$$

we can rewrite relation (3.336) as follows:

$$\mathrm{i}\frac{\partial}{\partial \tau}\overline{\Psi}(\boldsymbol{\xi}, \tau, \theta) = \overline{\mathcal{H}}(\theta)\overline{\Psi}(\boldsymbol{\xi}, \tau, \theta), \tag{3.341}$$

where

$$\overline{\mathcal{H}}(\theta) = \frac{1}{\theta^2}[\,\overline{\Omega}(\theta) - 1\,], \tag{3.342}$$

$$\overline{\Omega}(\theta) = \frac{\Omega\left(\theta k_0 \overline{\nabla}_{\mathrm{T}}, \theta^2 k_0 \frac{\partial}{\partial \zeta}\right)}{k_0}. \tag{3.343}$$

This provides the possibility of expanding the differential operator $\overline{\mathcal{H}}(\theta)$,

$$\overline{\mathcal{H}}(\theta) = \sum_{n=0}^{\infty} \theta^n \overline{\mathcal{H}}^{(n)}, \tag{3.344}$$

where the differential operators $\overline{\mathcal{H}}^{(n)}$ are just defined by the formal expression.

The dimensionless amplitude $\overline{\Psi}(\boldsymbol{\xi}, \tau, \theta)$ has the expansion

$$\overline{\Psi}(\boldsymbol{\xi}, \tau, \theta) = \sum_{n=0}^{\infty} \theta^n \overline{\Psi}^{(n)}(\boldsymbol{\xi}, \tau). \tag{3.345}$$

It is evident that the terms satisfy the equations

$$\mathrm{i}\frac{\partial}{\partial \tau}\overline{\Psi}^{(n)}(\boldsymbol{\xi}, \tau) = \sum_{m=0}^{n} \overline{\mathcal{H}}^{(n-m)}\overline{\Psi}^{(m)}(\boldsymbol{\xi}, \tau), \quad n = 0, 1, 2, \ldots. \tag{3.346}$$

In Deutsch and Garrison (1991b), the discussion of the classical equation of motion is completed by considering the initial-value problem. We rewrite equation (3.333) as

$$\Psi(\mathbf{x},\theta)=\frac{1}{(2\pi)^{\frac{3}{2}}}\int\sum_{\lambda=1,2}\mathbf{K}_\lambda(\mathbf{q})f_\lambda(\mathbf{q},\theta)e^{\mathrm{i}\mathbf{q}\cdot\mathbf{x}}\,\mathrm{d}^3\mathbf{q}, \tag{3.347}$$

where the function $\mathbf{K}_\lambda(\mathbf{q})$ is defined by

$$\mathbf{K}_\lambda(\mathbf{q})=\sqrt{\frac{k_0}{|\mathbf{q}+\mathbf{k}_0|}}\,\tilde{\mathbf{e}}_\lambda(\mathbf{q}), \tag{3.348}$$

where

$$\tilde{\mathbf{e}}_\lambda(\mathbf{q})=\mathbf{e}_\lambda(\mathbf{q}+\mathbf{k}_0). \tag{3.349}$$

Reexpressing (3.347) in terms of the scaled variables gives

$$\overline{\Psi}(\boldsymbol{\xi},\tau=0;\theta)=\frac{1}{(2\pi)^{\frac{3}{2}}}\int\sum_{\lambda=1,2}\overline{\mathbf{K}}_\lambda(\boldsymbol{\eta},\theta)\overline{f}_\lambda(\boldsymbol{\eta})e^{\mathrm{i}\boldsymbol{\eta}\cdot\boldsymbol{\xi}}\,\mathrm{d}^3\boldsymbol{\eta}. \tag{3.350}$$

The scaled kernel function is

$$\overline{\mathbf{K}}_\lambda(\boldsymbol{\eta},\theta)=\frac{\overline{\tilde{\mathbf{e}}}_\lambda(\boldsymbol{\eta},\theta)}{\sqrt{\overline{w}(\boldsymbol{\eta},\theta)}}, \tag{3.351}$$

where

$$\overline{w}(\boldsymbol{\eta},\theta)=\sqrt{1+\theta^2(2\eta_z+\eta_T^2)+\theta^4\eta_z^2}\,. \tag{3.352}$$

Considering the expansion

$$\overline{\mathbf{K}}_\lambda(\boldsymbol{\eta},\theta)=\sum_{n=0}^{\infty}\theta^n\overline{\mathbf{K}}_\lambda^{(n)}(\boldsymbol{\eta}), \tag{3.353}$$

we obtain the initial expansion of the envelope fields

$$\overline{\Psi}^{(n)}(\boldsymbol{\xi})=\frac{1}{(2\pi)^{\frac{3}{2}}}\int\sum_{\lambda=1,2}\overline{\mathbf{K}}_\lambda^{(n)}(\boldsymbol{\eta})\overline{f}_\lambda(\boldsymbol{\eta})e^{\mathrm{i}\boldsymbol{\eta}\cdot\boldsymbol{\xi}}\,\mathrm{d}^3\boldsymbol{\eta}. \tag{3.354}$$

Deutsch and Garrison (1991b) claim that the preceding arguments can be used to identify the subspace of the photon Fock space consisting of the paraxial states of the field. They resorted to the space S, the infinitely differentiable functions that decrease, when $|\boldsymbol{\eta}| \to \infty$, faster than any power of $|\boldsymbol{\eta}|^{-1}$.

Let us recall the standard plane wave creation and annihilation operators $\hat{a}^\dagger_\lambda(\mathbf{k})$, $\hat{a}_\lambda(\mathbf{k})$, and the wave packet creation and annihilation operators

$$\hat{a}^\dagger[\mathcal{F}] = \int \sum_{\lambda=1,2} \mathcal{F}_\lambda(\mathbf{k})\hat{a}^\dagger_\lambda(\mathbf{k})\,\mathrm{d}^3\mathbf{k} \tag{3.355}$$

and its conjugate. We now define

$$\hat{c}^\dagger[f] = \int \sum_{\lambda=1,2} \hat{c}^\dagger_\lambda(\mathbf{q}) f_\lambda(\mathbf{q})\,\mathrm{d}^3\mathbf{q}, \tag{3.356}$$

where

$$\hat{c}^\dagger_\lambda(\mathbf{q}) = \hat{a}^\dagger_\lambda(\mathbf{q} + \mathbf{k}_0) \tag{3.357}$$

are the creation operators corresponding to the envelope field.

Before we generalize the definition of a state with exactly one photon present,

$$|f;1\rangle\rangle = \hat{c}^\dagger[f]|0\rangle, \tag{3.358}$$

where f is a normalized one-photon wave function, we return to the original operators. Let us proceed with the generalized commutation relations

$$\left[\hat{a}[\mathcal{F}], \hat{a}^\dagger[\mathcal{G}]\right] = (\mathcal{F}, \mathcal{G})\hat{1} = \int \sum_{\lambda=1,2} \mathcal{F}^*_\lambda(\mathbf{k})\mathcal{G}_\lambda(\mathbf{k})\,\mathrm{d}^3\mathbf{k}\,\hat{1}. \tag{3.359}$$

If $\mathcal{F}$ is normalized, then

$$\left[\hat{a}[\mathcal{F}], \hat{a}^\dagger[\mathcal{F}]\right] = \hat{1}. \tag{3.360}$$

Let us observe that

$$\hat{a}^{\dagger 2}[\mathcal{F}] = \int\int \sum_{\lambda_1}\sum_{\lambda_2} \mathcal{F}_{\lambda_1\lambda_2}(\mathbf{k}_1, \mathbf{k}_2)\hat{a}^\dagger_{\lambda_1}(\mathbf{k}_1)\hat{a}^\dagger_{\lambda_2}(\mathbf{k}_2)\,\mathrm{d}^3\mathbf{k}_1\,\mathrm{d}^3\mathbf{k}_2, \tag{3.361}$$

where

$$\mathcal{F}_{\lambda_1\lambda_2}(\mathbf{k}_1, \mathbf{k}_2) = \mathcal{F}_{\lambda_1}(\mathbf{k}_1)\mathcal{F}_{\lambda_2}(\mathbf{k}_2) \tag{3.362}$$

exemplifies a normalized symmetric function. We are led to the definition of the state with exactly two photons present

$$|\mathcal{F};2\rangle\rangle = \frac{1}{\sqrt{2}}\int\int \sum_{\lambda_1}\sum_{\lambda_2} \mathcal{F}_{\lambda_1\lambda_2}(\mathbf{k}_1, \mathbf{k}_2)\hat{a}^\dagger_{\lambda_1}(\mathbf{k}_1)\hat{a}^\dagger_{\lambda_2}(\mathbf{k}_2)|0\rangle\,\mathrm{d}^3\mathbf{k}_1\,\mathrm{d}^3\mathbf{k}_2, \tag{3.363}$$

where $\mathcal{F}$ is a normalized symmetric two-photon wave function. In general,

$$|\mathcal{F};m\rangle\rangle = \frac{1}{\sqrt{m!}}\int\cdots\int\sum_{\lambda_1}\cdots\sum_{\lambda_m}\mathcal{F}_{\lambda_1\ldots\lambda_m}(\mathbf{k}_1,\ldots,\mathbf{k}_m)$$
$$\times\,\hat{a}^\dagger_{\lambda_1}(\mathbf{k}_1)\ldots\hat{a}^\dagger_{\lambda_m}(\mathbf{k}_m)|0\rangle\,\mathrm{d}^3\mathbf{k}_1\ldots\mathrm{d}^3\mathbf{k}_m,\quad m\geq 1, \tag{3.364}$$

where $\mathcal{F}$ is this time a normalized symmetric wave function of m photons and

$$|\mathcal{F};0\rangle\rangle = \mathcal{F}|0\rangle, \tag{3.365}$$

with $\mathcal{F}$ a complex unit. This almost completes the definition of the Fock space, since any element of this space has the form

$$|\Phi\rangle = \sum_{m=0}^{\infty}|\mathcal{F}^{(m)};m\rangle\rangle, \tag{3.366}$$

where $\mathcal{F}^{(m)}$ are (unnormalized) symmetric m-photon wave functions, $m\geq 1$, and $\mathcal{F}^{(0)}$ is a complex number. The pure state $|\Phi\rangle$ is normalized if and only if the functions $\mathcal{F}^{(m)}$ are jointly normalized by

$$\sum_{m=0}^{\infty}(\mathcal{F}^{(m)},\mathcal{F}^{(m)}) = 1. \tag{3.367}$$

Similarly,

$$|f;m\rangle = \frac{1}{\sqrt{m!}}\int\cdots\int\sum_{\lambda_1}\cdots\sum_{\lambda_m}f_{\lambda_1\ldots\lambda_m}(\mathbf{q}_1,\ldots,\mathbf{q}_m)$$
$$\times\,\hat{c}^\dagger_{\lambda_1}(\mathbf{q}_1)\ldots\hat{c}^\dagger_{\lambda_m}(\mathbf{q}_m)|0\rangle\,\mathrm{d}^3\mathbf{q}_1\ldots\mathrm{d}^3\mathbf{q}_m,\quad m\geq 1, \tag{3.368}$$

where

$$f_{\lambda_1\ldots\lambda_m}(\mathbf{q}_1,\ldots,\mathbf{q}_m) = \mathcal{F}_{\lambda_1\ldots\lambda_m}(\mathbf{q}_1+\mathbf{k}_0,\ldots,\mathbf{q}_m+\mathbf{k}_0) \tag{3.369}$$

and

$$|f;0\rangle = f|0\rangle, \tag{3.370}$$

where $f=\mathcal{F}$, and any element of the Fock space has the form

$$|\Phi\rangle = \sum_{m=0}^{\infty}|f^{(m)};m\rangle, \tag{3.371}$$

where

$$f^{(m)}_{\lambda_1\ldots\lambda_m}(\mathbf{q}_1,\ldots,\mathbf{q}_m)=\mathcal{F}^{(m)}_{\lambda_1\ldots\lambda_m}(\mathbf{q}_1+\mathbf{k}_0,\ldots,\mathbf{q}_m+\mathbf{k}_0) \tag{3.372}$$

and $f^{(0)}=\mathcal{F}^{(0)}$.

For the subsequent analysis, a unitary operator $\hat{T}(\theta)$ is of interest such that

$$\hat{T}(\theta)|f';m\rangle=|f(\theta);m\rangle, \tag{3.373}$$

where (cf. (3.327))

$$f_{\lambda_1\ldots\lambda_m}(\mathbf{q}_{\mathrm{T}1},q_{z1},\ldots,\mathbf{q}_{\mathrm{T}m},q_{zm},\theta)=\mathcal{V}^{\frac{m}{2}} \times f'_{\lambda_1\ldots\lambda_m}\left(\frac{\mathbf{q}_{\mathrm{T}1}}{\theta},\frac{q_{z1}}{\theta^2},\ldots,\frac{\mathbf{q}_{\mathrm{T}m}}{\theta},\frac{q_{zm}}{\theta^2}\right). \tag{3.374}$$

In the description of the dynamics using the Schrödinger picture, the paraxial approximation means mainly the evolution of the initial state $|\Phi(t,\theta)\rangle$,

$$\frac{\partial}{\partial t}|\Phi(t,\theta)\rangle=-\frac{\mathrm{i}}{\hbar}\hat{H}|\Phi(t,\theta)\rangle, \tag{3.375}$$

where

$$|\Phi(t,\theta)\rangle|_{t=0}=\hat{T}(\theta)|\Phi'(t=0)\rangle. \tag{3.376}$$

Defining the state

$$|\Phi'(t,\theta)\rangle=\hat{T}^{\dagger}(\theta)|\Phi(t,\theta)\rangle, \tag{3.377}$$

for all times, we can rewrite (3.375) in the form

$$\frac{\partial}{\partial t}|\Phi'(t,\theta)\rangle=-\frac{\mathrm{i}}{\hbar}\hat{H}'(\theta)|\Phi'(t,\theta)\rangle, \tag{3.378}$$

where

$$\hat{H}'(\theta)=\hat{T}^{\dagger}(\theta)\hat{H}\hat{T}(\theta). \tag{3.379}$$

Using the expansion

$$\hat{H}'(\theta)=\sum_{m=0}^{\infty}\theta^m\hat{H}'^{(m)}, \tag{3.380}$$

we may expand equation (3.378) into the coupled equations for the coefficients of the series

$$|\Phi'(t,\theta)\rangle=\sum_{m=0}^{\infty}\theta^m|\Phi'^{(m)}(t)\rangle. \tag{3.381}$$

Describing the dynamics in the Heisenberg picture, we should generalize relations (3.379) and (3.380) to an arbitrary operator $\hat{M}(t)$,

$$\hat{M}'(t,\theta) = \hat{T}^{\dagger}(\theta)\hat{M}(t)\hat{T}(\theta), \tag{3.382}$$

$$\hat{M}'(t,\theta) = \sum_{m=0}^{\infty} \theta^m \hat{M}'^{(m)}(t). \tag{3.383}$$

We can then rewrite the equation of motion

$$\frac{\partial}{\partial t}\hat{M}(t) = -\frac{\mathrm{i}}{\hbar}[\hat{M}(t), \hat{H}(t)] \tag{3.384}$$

in the form

$$\frac{\partial}{\partial t}\hat{M}'(t,\theta) = -\frac{\mathrm{i}}{\hbar}[\hat{M}'(t,\theta), \hat{H}'(t,\theta)]. \tag{3.385}$$

We may expand this equation into coupled equations similarly as (3.378). Since

$$\hat{T}(1) = \hat{1}, \quad \hat{M}'(t,1) = \hat{M}(t), \tag{3.386}$$

relation (3.383) simplifies for $\theta = 1$, or any operator $\hat{M}(t)$ can be expressed as

$$\hat{M}(t) = \sum_{m=0}^{\infty} \hat{M}^{(m)}(t). \tag{3.387}$$

In both the pictures, definition (3.339) can be used whenever it is advantageous.

According to Deutsch and Garrison (1991b), we introduce the operator $\hat{\Psi}(\mathbf{x})$ by relation (3.333), with $\Psi(\mathbf{x}, t = 0, \theta) \mapsto \hat{\Psi}(\mathbf{x})$ on the left-hand side and $f_\lambda(\mathbf{q}, \theta) \mapsto \hat{c}_\lambda(\mathbf{q})$ on the right-hand side. This operator can be expanded as

$$\hat{\Psi}(\mathbf{x}) = \sum_{n=0}^{\infty} \hat{\Psi}^{(n)}(\mathbf{x}), \tag{3.388}$$

where

$$\hat{\Psi}^{(n)}(\mathbf{x}) = \frac{1}{(2\pi)^{\frac{3}{2}}} \int \sum_{\lambda=1,2} \hat{\mathbf{K}}_\lambda^{(n)}(\mathbf{q})\hat{c}_\lambda(\mathbf{q}) e^{\mathrm{i}\mathbf{q}\cdot\mathbf{x}}\, \mathrm{d}^3\mathbf{q}. \tag{3.389}$$

Using this expansion, we can compute the commutators between fields of different orders that are

$$[\hat{\Psi}_i^{(n)}(\mathbf{x}), \hat{\Psi}_j^{(m)\dagger}(\mathbf{x}')] = \frac{1}{(2\pi)^{\frac{3}{2}}} \int \sum_{\lambda=1,2} \hat{K}_{\lambda i}^{(n)}(\mathbf{q})\hat{K}_{\lambda j}^{(m)}(\mathbf{q}) e^{\mathrm{i}\mathbf{q}\cdot(\mathbf{x}-\mathbf{x}')}\, \mathrm{d}^3\mathbf{q}. \tag{3.390}$$

As expected, the nth-order commutator can be expressed as

$$[\hat{\Psi}_i(\mathbf{x}), \hat{\Psi}_j^\dagger(\mathbf{x}')]^{(n)} = \sum_{m=0}^{n} [\hat{\Psi}_i^{(n-m)}(\mathbf{x}), \hat{\Psi}_j^{(m)\dagger}(\mathbf{x}')]. \tag{3.391}$$

The equal-time commutation relations are preserved by the dynamics in each order of the approximation scheme,

$$\frac{\partial}{\partial t}[\hat{\Psi}_i(\mathbf{x}, t), \hat{\Psi}_j^\dagger(\mathbf{x}', t)]^{(n)} = \hat{0}. \tag{3.392}$$

In the zeroth order, the theory yields a quantized analogue of the classical paraxial wave equation, and formally resembles a nonrelativistic many-particle theory. This formalism is applied to show that Mandel's local-photon-number operator and Glauber's photon-counting operator reduce, in the zeroth order, to the same true number operator. In addition, it is shown that the O(θ^2)-difference between them vanishes for experiments described by stationary coherent states.

A nonperturbative quantization of a paraxial electromagnetic field has been achieved by forcing the plane waves involved in the expression for the vector-potential operator to obey paraxial wave equations at the time origin (Aiello and Woerdman 2005).

3.3.4 Optical Imaging with Nonclassical Light

In optical imaging with nonclassical light or quantum imaging it is important to know how quantum entanglement properties of light beams in the spatial domain can be exploited in order to improve the quality of processing of images and of parallel signals (Gatti 2003). In this section, we first expound some general concepts as spatially multimode squeezing and spatial entanglement, and describe some optical devices that are able to generate light beams with these properties (Kolobov 1999). Then we provide some references to interesting approaches in this field.

Kolobov (1999) enriches exposition of the usual quantum optics by new facts. The time moments are completed with space points $\boldsymbol{\rho}$. It is connected with existence of very small photodetectors or pixels. The observed quantity is the surface photocurrent density operator, an Hermitian operator, $\hat{i}(\boldsymbol{\rho}, t)$.

Let the photodetection plane be located at the point with longitudinal coordinate z normal to the z-axis. Let $\hat{E}^{(+)}(z, \boldsymbol{\rho}, t)$ mean the positive-frequency operator of the electric field of a quasiplane and a quasimonochromatic wave travelling in the $+z$-direction, where $\boldsymbol{\rho}$ is the position vector in the transverse plane of the wave. This operator can be written in terms of space- and time-dependent photon annihilation and creation operators $\hat{a}(z, \boldsymbol{\rho}, t)$ and $\hat{a}^\dagger(z, \boldsymbol{\rho}, t)$ as

$$\hat{E}^{(+)}(z, \boldsymbol{\rho}, t) = \mathrm{i}\sqrt{\frac{\hbar\omega_0}{2\epsilon_0 c}} \exp[\mathrm{i}(k_0 z - \omega_0 t)]\hat{a}(z, \boldsymbol{\rho}, t). \tag{3.393}$$

Here ω_0 is the carrier frequency of the wave and k_0 is its wave number. But $\hat{a}(z, \boldsymbol{\rho}, t)$ and $\hat{a}^\dagger(z, \boldsymbol{\rho}, t)$ are not the standard-modal annihilation and creation operators. They obey the commutation relations

$$[\hat{a}(z, \boldsymbol{\rho}, t), \hat{a}^\dagger(z, \boldsymbol{\rho}', t')] = \delta(\boldsymbol{\rho} - \boldsymbol{\rho}')\delta(t - t')\hat{1},$$
$$[\hat{a}(z, \boldsymbol{\rho}, t), \hat{a}(z, \boldsymbol{\rho}', t')] = \hat{0} \tag{3.394}$$

and are normalized so that the mean value $\langle \hat{a}^\dagger(z, \boldsymbol{\rho}, t)\hat{a}(z, \boldsymbol{\rho}, t)\rangle$ determines the mean photon-flux density in photons per cm^2 per second at point $\boldsymbol{\rho}$ and time t.

The quantum theory of photodetection provides the following expressions for the mean value of the photocurrent density operator $\langle \hat{i}(\boldsymbol{\rho}, t)\rangle$, and its space–time correlation function $\langle \frac{1}{2}\{\delta\hat{i}(\boldsymbol{\rho}, t), \delta\hat{i}(\boldsymbol{\rho}', t')\}_+\rangle$:

$$\langle \hat{i}(\boldsymbol{\rho}, t)\rangle = \eta\langle \hat{I}(\boldsymbol{\rho}, t)\rangle, \tag{3.395}$$

$$\left\langle \frac{1}{2}\{\delta\hat{i}(\boldsymbol{\rho}, t), \delta\hat{i}(\boldsymbol{\rho}', t')\}_+ \right\rangle = \langle \hat{i}(\boldsymbol{\rho}, t)\rangle\delta(\boldsymbol{\rho} - \boldsymbol{\rho}')\delta(t - t')$$
$$+ \eta^2\langle : \hat{I}(\boldsymbol{\rho}, t)\hat{I}(\boldsymbol{\rho}', t') : \rangle - \langle \hat{i}(\boldsymbol{\rho}, t)\rangle\langle \hat{i}(\boldsymbol{\rho}', t')\rangle. \tag{3.396}$$

Here $\hat{I}(\boldsymbol{\rho}, t) = \hat{a}^\dagger(z, \boldsymbol{\rho}, t)\hat{a}(z, \boldsymbol{\rho}, t)$ is the photon-flux density operator.

The second contribution to the correlation function of the photocurrent density operator is proportional to the normal- and time-ordered space–time intensity correlation function

$$G^{(2)}(\boldsymbol{\rho}, t; \boldsymbol{\rho}', t') = \langle : \hat{I}(\boldsymbol{\rho}, t)\hat{I}(\boldsymbol{\rho}', t') : \rangle. \tag{3.397}$$

This correlation function is proportional to the probability of detecting a photon at time t' and at the spatial point $\boldsymbol{\rho}'$ under the condition that the previous detection happened at time t and point $\boldsymbol{\rho}$. When the intensity of light is stationary in time and uniform in the transverse area of the light beam, this correlation function depends only on the time difference $\tau = t' - t$ and the spatial difference $\boldsymbol{\xi} = \boldsymbol{\rho}' - \boldsymbol{\rho}$ between two points, $G^{(2)}(\boldsymbol{\rho}, t; \boldsymbol{\rho}', t') = G^{(2)}(\boldsymbol{\xi}, \tau)$.

One can define the degree of second-order spatio-temporal coherence as

$$g^{(2)}(\boldsymbol{\xi}, \tau) = \frac{\langle : \hat{I}(\boldsymbol{\rho}, t)\hat{I}(\boldsymbol{\rho} + \boldsymbol{\xi}, t + \tau) : \rangle}{\langle \hat{I}(\boldsymbol{\rho}, t)\rangle^2}. \tag{3.398}$$

If the correlation function $g^{(2)}(\boldsymbol{\xi}, \tau)$ has its maximum at $\boldsymbol{\xi} = \mathbf{0}$ and at $\tau = 0$, $g^{(2)}(\mathbf{0}, 0) > g^{(2)}(\boldsymbol{\xi}, \tau)$, it is natural to speak of the bunching in space–time. Analogously, if $g^{(2)}(\mathbf{0}, 0) < g^{(2)}(\boldsymbol{\xi}, \tau)$, one may speak of the antibunching in space–time.

The antibunching in space–time is a purely quantum-mechanical phenomenon. Indeed, it follows from the Schwarz inequality that the correlation function $g^{(2)}(\boldsymbol{\xi}, \tau)$ of a classical electromagnetic field stationary in time and uniform in space must satisfy

$$g^{(2)}(\mathbf{0}, 0) \geq g^{(2)}(\boldsymbol{\xi}, \tau) \tag{3.399}$$

for arbitrary $\boldsymbol{\xi}$ and τ. Since the antibunching in space–time means an exactly opposite inequality, it cannot be explained within the framework of semiclassical theory, i.e. when the light field is treated as a c-number.

The photocurrent noise spectrum is defined as a Fourier transform of the photocurrent correlation function. As follows from relation (3.396), for a light field stationary in time and uniform in the transverse plane, the correlation function of the photocurrent density operator $\langle \frac{1}{2}\{\delta\hat{i}(\boldsymbol{\rho}, t), \delta\hat{i}(\boldsymbol{\rho}', t')\}_+\rangle$ depends only on the time difference τ and the spatial difference $\boldsymbol{\xi}$. The noise spectrum of the photocurrent density operator is the spatio-temporal Fourier transform of this correlation function,

$$\widetilde{(\delta\hat{i})^2}(\mathbf{q}, \Omega) = \int\int \left\langle \frac{1}{2}\{\delta\hat{i}(\mathbf{0}, 0), \delta\hat{i}(\boldsymbol{\rho}, t)\}_+ \right\rangle \times \exp[\mathrm{i}(\Omega t - \mathbf{q}\cdot\boldsymbol{\rho})]\, \mathrm{d}t\, \mathrm{d}^2\boldsymbol{\rho}. \tag{3.400}$$

Using the photodetection formula (3.396), we can write the noise spectrum $\widetilde{(\delta\hat{i})^2}(\mathbf{q}, \Omega)$ as follows

$$\widetilde{(\delta\hat{i})^2}(\mathbf{q}, \Omega) = \langle \hat{i}(\boldsymbol{\rho}, t)\rangle + \tilde{G}^{(2)}(\mathbf{q}, \Omega) - \langle \hat{i}(\boldsymbol{\rho}, t)\rangle^2 \delta(\Omega)\delta(\mathbf{q}). \tag{3.401}$$

Here the first contribution comes from the shot-noise term in relation (3.396), the second one is the spatio-temporal Fourier transform of the intensity correlation function,

$$\tilde{G}^{(2)}(\mathbf{q}, \Omega) = \int\int \exp[\mathrm{i}(\Omega t - \mathbf{q}\cdot\boldsymbol{\rho})] G^{(2)}(\boldsymbol{\rho}, t)\, \mathrm{d}t\, \mathrm{d}^2\boldsymbol{\rho}, \tag{3.402}$$

and the last one from the space–time-independent product of two mean photocurrent densities. One can show that in semiclassical theory the sum of the second and third contributions is always nonnegative. Therefore the semiclassical minimum value of the photocurrent density noise is given by the shot noise in space–time,

$$\widetilde{(\delta\hat{i})^2}(\mathbf{q}, \Omega) = \langle \hat{i}(\boldsymbol{\rho}, t)\rangle. \tag{3.403}$$

This formula is a generalization of the standard quantum limit for a single-mode field described by the photon annihilation and creation operators $\hat{a}(t)$ and $\hat{a}^\dagger(t)$, respectively,

$$\widetilde{(\delta\hat{i})^2}(\Omega) = \int \left\langle \frac{1}{2}\{\delta\hat{i}(0), \delta\hat{i}(t)\}_+ \right\rangle \exp(\mathrm{i}\Omega t)\, \mathrm{d}t = \langle \hat{i}(t)\rangle, \tag{3.404}$$

where $\hat{i}(t) = \hat{a}^\dagger(t)\hat{a}(t)$, $\{\cdot,\cdot\}_+$ indicates an anticommutator, from the temporal domain into the space–time one. In quantum theory the sum of the second and third terms in relation (3.401) can be negative and compensate partially or even completely for the shot-noise contribution for some frequencies Ω and spatial frequencies $\mathbf{q}$.

An opinion of many workers in quantum optics is expressed in Kolobov (1999). The difficulties associated with the quantum-mechanical description of field propagation in free space or a nonlinear medium lie in the usual procedure of field quantization. Evolution of the quantized field due, for instance, to the interaction with an atomic medium is described in terms of the Heisenberg equations for annihilation and creation operators, i.e. as purely temporal evolution. Such a description of field dynamics is not well suited to the problem of field propagation in free space or a medium. At the start of such a study, it would be more appropriate to have a quantum-mechanical analogue of the classical wave-optical propagation and diffraction theory. Such a description for transparent nonlinear media when the field interaction with atoms is described in terms of an effective Hamiltonian is much appreciated. But the simpler question of quantized field propagation in free space is considered first.

Let $\hat{E}^{(+)}(\mathbf{r}, t)$, where $\mathbf{r} = (x, y, z)$ is the spatial coordinate, be the positive-frequency operator of the electric field in a vacuum. In the continuum limit, this operator is written in the form of the modal decomposition

$$\hat{E}^{(+)}(\mathbf{r}, t) = \mathrm{i}\sqrt{\frac{\hbar}{2\epsilon_0}\frac{1}{(2\pi)^3}} \int \sqrt{\omega(k)}\hat{a}(\mathbf{k}) \times \exp[\mathrm{i}(\mathbf{k}\cdot\mathbf{r} - \omega(k)t)]\,\mathrm{d}^3\mathbf{k}. \tag{3.405}$$

Here $\hat{a}(\mathbf{k})$ and $\hat{a}^\dagger(\mathbf{k})$ are the photon annihilation and creation operators of a spatial mode with the wave vector $\mathbf{k}$; the frequency $\omega(k)$ is given by the free-space dispersion relation $\omega(k) = kc$, with $k = |\mathbf{k}|$. The operators $\hat{a}(\mathbf{k})$ and $\hat{a}^\dagger(\mathbf{k})$ obey the canonical commutation relations

$$[\hat{a}(\mathbf{k}), \hat{a}^\dagger(\mathbf{k}')] = (2\pi)^3\delta(\mathbf{k} - \mathbf{k}')\hat{1}, \quad [\hat{a}(\mathbf{k}), \hat{a}(\mathbf{k}')] = \hat{0}. \tag{3.406}$$

The factor $(2\pi)^3$ is not usual, but such particularities appear consistently in the review article. Equation (3.405) determines the Heisenberg field operator $\hat{E}^{(+)}(\mathbf{r}, t)$ in all points $\mathbf{r}$ and t of the space–time as a solution of the initial-value problem, i.e. through the modal operators $\hat{a}(\mathbf{k})$ and $\hat{a}^\dagger(\mathbf{k})$ given at time $t = 0$ as Schrödinger operators. For a complete quantum-mechanical description, we have to specify the density matrix of the field for the continuum set of modes $\mathbf{k}$. In the Heisenberg representation (3.405), this density matrix remains constant as time evolves. For a wave travelling in the $+z$-direction, we would like to have a formula that determines the field operator at any point $\boldsymbol{\rho}$ in the transverse plane at coordinate z given the field operator over the plane $z = 0$.

On comparison of relation (3.393) with relation (3.405), one sees that

$$\hat{a}(z, \boldsymbol{\rho}, t) = \frac{1}{(2\pi)^3}\int\int \sqrt{\frac{\omega(k)}{k_0}}\hat{a}(\mathbf{k}) \times \exp\{\mathrm{i}[\mathbf{q}\cdot\boldsymbol{\rho} + (k_z - k_0)z - (\omega(k) - \omega_0)t]\}\,\mathrm{d}^2\mathbf{q}\,\mathrm{d}k_z. \tag{3.407}$$

The normalization of these operators is such that the free Hamiltonian of the electromagnetic field can be written as

$$\hat{H}_0 = \frac{\hbar\omega_0}{c} \int_V \hat{a}^\dagger(z, \boldsymbol{\rho}, t)\hat{a}(z, \boldsymbol{\rho}, t)\, \mathrm{d}^3\mathbf{r}. \quad (3.408)$$

The commutation relations (3.394) are not proved in this connection, but equal-time ones are derived,

$$[\hat{a}(z, \boldsymbol{\rho}, t), \hat{a}^\dagger(z', \boldsymbol{\rho}', t)] = \tilde{\delta}(\mathbf{r} - \mathbf{r}')\hat{1}, \quad (3.409)$$

where

$$\tilde{\delta}(\mathbf{r} - \mathbf{r}') \approx \left(1 - \frac{\mathrm{i}}{k_0}\frac{\partial}{\partial z} - \frac{1}{2k_0^2}\nabla_\perp^2\right)\delta(\mathbf{r} - \mathbf{r}'), \quad (3.410)$$

with $\nabla_\perp^2$ being the transverse Laplacian with respect to $\boldsymbol{\rho}$. Here we have reproduced only the expression derived in the quasimonochromatic and paraxial approximations from the literature. In this approximation, the equation for the slowly varying operator $\hat{a}(z, \boldsymbol{\rho}, t)$ reads

$$\frac{\partial}{\partial t}\hat{a}(z, \boldsymbol{\rho}, t) = \left(-c\frac{\partial}{\partial z} + c\frac{\mathrm{i}}{2k_0}\nabla_\perp^2\right)\hat{a}(z, \boldsymbol{\rho}, t). \quad (3.411)$$

An unpublished result of Sokolov, which is related to the equation for propagation of a quantized field in a nonlinear parametric medium, is reproduced in Kolobov (1999). In part, it is based on the book of Klyshko (1988).

The positive-frequency operator of a quantized electric field in a transparent dielectric medium can be written in a form similar to that for a vacuum (Klyshko 1988),

$$\hat{E}^{(+)}(\mathbf{r}, t) = \mathrm{i}\sqrt{\frac{\hbar}{2\epsilon_0}\frac{1}{(2\pi)^3}} \int \xi(k)\sqrt{\omega(k)}\hat{a}(\mathbf{k}) \times \exp[\mathrm{i}(\mathbf{k}\cdot\mathbf{r} - \omega(k)t)]\, \mathrm{d}^3\mathbf{k}. \quad (3.412)$$

This differs from relation (3.405) in the factor $\xi(k)$, which describes the strength of the field in the medium as compared to that in a vacuum. This constant is

$$\xi^2(k) = \frac{u(k)v(k)}{c^2 \cos\rho(k)}. \quad (3.413)$$

Here $v(k) = \frac{c}{n(k)}$ is the phase velocity of light in the medium, $u(k) = \frac{\partial\omega(k)}{\partial k}$ is the group velocity, and $\rho(k)$ is the so-called generalized anisotropy angle, that is, the angle between the electric field and the induction.

The fact that the electromagnetic field is a vector field is not emphasized in Kolobov (1999), but it is mentioned in respect of the book by Klyshko (1988). Relation (3.412) yet neglects the use of and summation over the appropriate parameter ν. The dispersion relation $\omega(k)$ is not single valued, but has at least two branches. These branches should be distinguished by a parameter μ. The annihilation operators correspond to these branches. Relation (3.412) neglects the use of and summation over the parameter μ too.

Using this notation one can introduce the slowly varying operator $\hat{a}(z, \boldsymbol{\rho}, t)$ of the quantized field in the medium (cf. equation (3.393)),

$$\hat{E}^{(+)}(z, \boldsymbol{\rho}, t) = \mathrm{i}\xi\sqrt{\frac{\hbar\omega_0}{2\epsilon_0 c}}\exp[\mathrm{i}(k_1 z - \omega_0 t)]\hat{a}(z, \boldsymbol{\rho}, t). \tag{3.414}$$

Here we have denoted by k_1 the wave number of the wave in the medium. The slowly varying operator $\hat{a}(z, \boldsymbol{\rho}, t)$ is given by an equation identical to relation (3.407),

$$\begin{aligned}\hat{a}(z, \boldsymbol{\rho}, t) = \frac{1}{(2\pi)^3}\int\int\sqrt{\frac{\omega(k)}{k_0}}\hat{a}(\mathbf{k})\\ \times \exp\{\mathrm{i}[\mathbf{q}\cdot\boldsymbol{\rho} + (k_z - k_0)z - (\omega(k) - \omega_0)t]\}\,\mathrm{d}^2\mathbf{q}\,\mathrm{d}k_z,\end{aligned} \tag{3.415}$$

but here $\omega(k)$ means a dispersion relation for the medium.

One will describe the parametric interaction in the medium in terms of an effective Hamiltonian. It is assumed that a $\chi^{(2)}$ nonlinear parametric medium fills a volume V. The medium is illuminated by a monochromatic plane wave, the pump. The pump wave propagates in the $+z$-direction and has the frequency ω_{p} and wave number k_{p},

$$\hat{E}_{\mathrm{p}}^{(+)}(z, \boldsymbol{\rho}, t) = E_{\mathrm{p}}\exp[\mathrm{i}(k_p z - \omega_p t)]. \tag{3.416}$$

We choose the frequency ω_{p} of the pump wave in the form $\omega_{\mathrm{p}} = 2\omega_0$ and consider the amplitude E_{p} as a c-number, i.e. we neglect the quantum fluctuations of the pump wave.

Under usual assumptions the parametric interaction can be described by the following effective Hamiltonian

$$\hat{H}_{\mathrm{int}} = \mathrm{i}\hbar\frac{n_0 g}{c}\int_V \exp[\mathrm{i}(k_{\mathrm{p}} - 2k_1)z][\hat{a}^\dagger(z, \boldsymbol{\rho}, t)]^2\,\mathrm{d}^3\mathbf{r} + \mathrm{H.c.} \tag{3.417}$$

Here n_0 gives the density of active atoms in the parametric medium, and g is the strength constant of the parametric interaction proportional to the amplitude E_{p} of the pump wave and the susceptibility constant $\chi^{(2)}$ of the medium.

The evolution of the slowly varying amplitude operator $\hat{a}(z, \boldsymbol{\rho}, t)$ in the parametric medium is described by the following equation:

$$\frac{\partial}{\partial t}\hat{a}(z, \boldsymbol{\rho}, t) = \mathrm{i}\omega_0\hat{a}(z, \boldsymbol{\rho}, t) + \frac{\mathrm{i}}{\hbar}[\hat{H}_0 + \hat{H}_{\mathrm{int}}, \hat{a}(z, \boldsymbol{\rho}, t)]. \tag{3.418}$$

Here $\hat{H}_0$ is the free-field Hamiltonian in the medium. In terms of $\hat{a}(z, \boldsymbol{\rho}, t)$ and $\hat{a}^\dagger(z, \boldsymbol{\rho}, t)$ it is given by relation (3.408). One introduces the Fourier transform of the space–time photon annihilation operator $\hat{a}(z, \boldsymbol{\rho}, t)$,

$$\widetilde{\hat{a}}(s, \mathbf{q}, \Omega) = \int\int\int \hat{a}(z, \boldsymbol{\rho}, t)\exp(\mathrm{i}\Omega t)\exp(-isz)\exp(-\mathrm{i}\mathbf{q}\cdot\boldsymbol{\rho})\,\mathrm{d}t\,\mathrm{d}^2\boldsymbol{\rho}\,\mathrm{d}z \tag{3.419}$$

$$\equiv \int e^{-isz}\tilde{\hat{a}}(z, \mathbf{q}, \Omega)\,\mathrm{d}z.$$

We express the Fourier transform $\tilde{\hat{a}}(z, \mathbf{q}, \Omega)$ with the aid of a new operator $\hat{\epsilon}(z, \mathbf{q}, \Omega)$,

$$\tilde{\hat{a}}(z, \mathbf{q}, \Omega) = \hat{\epsilon}(z, \mathbf{q}, \Omega)\exp\{\mathrm{i}[k_z(\mathbf{q}, \Omega) - k_1]z\}, \tag{3.420}$$

where

$$k_z(\mathbf{q}, \Omega) = \sqrt{k^2(\omega_0 + \Omega) - q^2}, \tag{3.421}$$

with $q = |\mathbf{q}|$ a z-component of the wave vector with frequency $\omega_0 + \Omega$ and spatial frequency $\mathbf{q}$. For $\widetilde{\hat{\epsilon}}(s, \mathbf{q}, \Omega)$ similarly defined as $\widetilde{\hat{a}}(s, \mathbf{q}, \Omega)$ it holds that

$$\widetilde{\hat{\epsilon}}(s, \mathbf{q}, \Omega) = \widetilde{\hat{a}}(s + k_z(\mathbf{q}, \Omega) - k_1, \mathbf{q}, \Omega). \tag{3.422}$$

One introduces the mismatch function $\Delta(\mathbf{q}, \Omega)$,

$$\Delta(\mathbf{q}, \Omega) = k_z(\mathbf{q}, \Omega) + k_z(-\mathbf{q}, -\Omega) - k_\mathrm{p}. \tag{3.423}$$

One lets $u = \frac{\partial\omega(k_1)}{\partial k_1}$ mean the group velocity of the wave in the crystal. On appropriate derivations, Kolobov (1999) presents the equation of propagation for the operator $\hat{\epsilon}(z, \mathbf{q}, \Omega)$,

$$\frac{\partial}{\partial z}\hat{\epsilon}(z, \mathbf{q}, \Omega) = \sigma\hat{\epsilon}^\dagger(z, -\mathbf{q}, -\Omega)\exp[\mathrm{i}\Delta(\mathbf{q}, \Omega)z], \tag{3.424}$$

where $\sigma = \frac{2n_0 g}{u}$ is the coupling constant of the parametric interaction.

When an active nonlinear medium is placed in a resonator, a description may employ discrete transverse modes of the cavity. Lugiato and Gatti (1993), Gatti and Lugiato (1995), and Lugiato and Marzoli (1995) have adopted this approach. Let $f_l(\boldsymbol{\rho})$ mean these eigenmodes. The set of functions $f_l(\boldsymbol{\rho})$ satisfies both the condition of orthonormality

$$\int f_l^*(\boldsymbol{\rho})f_{l'}(\boldsymbol{\rho})\,\mathrm{d}^2\boldsymbol{\rho} = \delta_{ll'} \tag{3.425}$$

and completeness

$$\sum_l f_l^*(\boldsymbol{\rho})f_l(\boldsymbol{\rho}') = \delta(\boldsymbol{\rho} - \boldsymbol{\rho}'). \tag{3.426}$$

One can expand the slowly varying field operator $\hat{a}(\boldsymbol{\rho}, t)$ over the eigenmodes $f_l(\boldsymbol{\rho})$,

$$\hat{a}(\boldsymbol{\rho}, t) = \sum_l f_l(\boldsymbol{\rho})\hat{a}_l(t), \tag{3.427}$$

where $\hat{a}_l(t)$ are operator-valued expansion coefficients that have the meaning of photon annihilation operators for the lth mode. From the commutation relations (3.394) together with relation (3.425) it is easy to see that $\hat{a}_l(t)$ and $\hat{a}_l^\dagger(t)$ obey the commutation relation

$$[\hat{a}_l(t), \hat{a}_{l'}^\dagger(t')] = \delta_{ll'}\delta(t - t')\hat{1}. \tag{3.428}$$

It is noted that the derivation is valid for the field operators outside the cavity. The Fourier transforms $\widetilde{\hat{a}}_l(\Omega)$ are defined as

$$\widetilde{\hat{a}}_l(\Omega) = \int_{-\infty}^{\infty} \exp(\mathrm{i}\Omega t)\hat{a}_l(t)\,\mathrm{d}t. \tag{3.429}$$

Noise spectrum of the photocurrent density for the lth mode as an analogue of the noise spectrum $\widetilde{(\delta\hat{i})^2}(\mathbf{q}, \Omega)$ is introduced in Kolobov (1999). It is not assumed that the photocurrent density is uniform in space, but that it is stationary in time and the photocurrent density fluctuations for different eigenmodes are uncorrelated.

We can express the space–time correlation function of the photocurrent density operator (3.396) in terms of the noise spectra $\widetilde{(\delta\hat{i})^2}_l(\Omega)$ of the individual eigenmodes $f_l(\boldsymbol{\rho})$ of the cavity,

$$\left\langle \frac{1}{2}\{\delta\hat{i}(\boldsymbol{\rho}, t), \delta\hat{i}(\boldsymbol{\rho}', t')\}_+ \right\rangle = \sum_l f_l(\boldsymbol{\rho}) f_l(\boldsymbol{\rho}') \times \frac{1}{2\pi}\int_{-\infty}^{\infty} \widetilde{(\delta\hat{i})^2}_l(\Omega)\exp[-\mathrm{i}\Omega(t - t')]\,\mathrm{d}\Omega. \tag{3.430}$$

(i) Generation of multimode squeezed states of light

The generation of multimode squeezed states of light by a travelling-wave optical parametric amplifier was described by Kolobov and Sokolov (1989a,b). As a result of the parametric down-conversion, a pump photon ω_p splits into signal and idler photons, with frequencies $\omega_0 + \Omega$ and $\omega_0 - \Omega$, and wave vectors $\mathbf{k}(\mathbf{q}, \Omega)$ and $\mathbf{k}(-\mathbf{q}, -\Omega)$. Their transverse components are $\pm\mathbf{q}$ and their z-components are $k_z(\mathbf{q}, \Omega)$ and $k_z(-\mathbf{q}, -\Omega)$, respectively, by relation (3.421). The evolution of the slowly varying operator $\hat{e}(z, \mathbf{q}, \Omega)$ inside the crystal is described by equation (3.424). Solving this equation and respecting relation (3.420) between the operators $\tilde{\hat{a}}(z, \mathbf{q}, \Omega)$ and $\hat{e}(z, \mathbf{q}, \Omega)$, one arrives at the following transformation

$$\tilde{\hat{a}}(l, \mathbf{q}, \Omega) = U(\mathbf{q}, \Omega)\tilde{\hat{a}}(0, \mathbf{q}, \Omega) + V(\mathbf{q}, \Omega)\tilde{\hat{a}}^\dagger(0, -\mathbf{q}, -\Omega), \tag{3.431}$$

with coefficients $U(\mathbf{q}, \Omega)$ and $V(\mathbf{q}, \Omega)$ equal to

$$\begin{aligned} U(\mathbf{q}, \Omega) &= \exp\left\{\mathrm{i}\left[k_z(\mathbf{q}, \Omega) - k_1 - \frac{\Delta(\mathbf{q}, \Omega)}{2}\right] l\right\} \\ &\quad \times \left[\cosh(\Gamma l) + \frac{\mathrm{i}\Delta(\mathbf{q}, \Omega)}{2\Gamma} \sinh(\Gamma l)\right], \\ V(\mathbf{q}, \Omega) &= \exp\left\{\mathrm{i}\left[k_z(\mathbf{q}, \Omega) - k_1 - \frac{\Delta(\mathbf{q}, \Omega)}{2}\right] l\right\} \\ &\quad \times \frac{\sigma}{\Gamma} \sinh(\Gamma l), \end{aligned} \tag{3.432}$$

where

$$\Gamma = \sqrt{|\sigma|^2 - \frac{[\Delta(\mathbf{q}, \Omega)]^2}{4}}. \tag{3.433}$$

The functions $U(\mathbf{q}, \Omega)$ and $V(\mathbf{q}, \Omega)$ have the property

$$|U(\mathbf{q}, \Omega)|^2 - |V(\mathbf{q}, \Omega)|^2 = 1. \tag{3.434}$$

At the input to the crystal, operators $\tilde{\hat{a}}(0, \mathbf{q}, \Omega)$ and $\tilde{\hat{a}}^\dagger(0, \mathbf{q}, \Omega)$ obey the free-field commutation relation

$$[\tilde{\hat{a}}(0, \mathbf{q}, \Omega), \tilde{\hat{a}}^\dagger(0, \mathbf{q}', \Omega')] = (2\pi)^3 \delta(\mathbf{q} - \mathbf{q}')\delta(\Omega - \Omega')\hat{1}. \tag{3.435}$$

The broad-band squeezing in a three-wave interaction was discussed by Caves and Crouch (1987) and in a four-wave interaction by Levenson et al. (1985) in the case of co-propagation and by Yurke (1985) for counter-propagation. Equation (3.431) involves the spatial frequency $\mathbf{q}$. It is assumed that, along with the pump wave, a monochromatic plane wave of frequency ω_0 is incident normal to the input surface of the crystal. Upon leaving the crystal, this wave will serve as a local oscillator wave with the complex amplitude β, which first enters as a wave with complex amplitude α and $\mathbf{q} = \mathbf{0}$. Relation (3.431) is used,

$$\beta = |\beta| \exp(\mathrm{i}\varphi_\beta) = \alpha U(\mathbf{0}, 0) + \alpha^* V(\mathbf{0}, 0). \tag{3.436}$$

The type of noise modulation of the resultant field in space–time is determined by the angle

$$\theta(\mathbf{q}, \Omega) = \psi(l, \mathbf{q}, \Omega) - \varphi_\beta. \tag{3.437}$$

Phase modulation predominates for $\theta(\mathbf{q}, \Omega) = \pm\frac{\pi}{2}$ and amplitude modulation for $\theta(\mathbf{q}, \Omega) = 0, \ \pi$. The mean of the photocurrent density operator and its noise spectrum is found from relations (3.395), (3.396), and (3.400). The mean of the photocurrent density operator has the forms

$$\langle \hat{i} \rangle = \eta |\beta|^2 + \frac{\eta}{(2\pi)^3} \int\!\!\int |V(\mathbf{q}, \Omega)|^2 \, \mathrm{d}^2\mathbf{q} \, \mathrm{d}\Omega \equiv \langle \hat{i} \rangle_1 + \langle \hat{i} \rangle_\mathrm{s}, \tag{3.438}$$

where the subscript l refers to the local oscillator field and the subscript s indicates the spontaneous parametric down-conversion. One introduces the function

$$\delta(\mathbf{q}, \Omega) = |V(\mathbf{q}, \Omega)|^2. \tag{3.439}$$

The mean of the photocurrent density operator $\langle \hat{i} \rangle_{\mathrm{s}}$ can be written as

$$\langle \hat{i} \rangle_{\mathrm{s}} = \eta \frac{\delta_{\mathrm{s}}}{T_{\mathrm{c}} S_{\mathrm{c}}}, \tag{3.440}$$

where

$$\delta_{\mathrm{s}} = \frac{1}{q_{\mathrm{c}}^2 \Omega_{\mathrm{c}}} \int\int \delta(\mathbf{q}, \Omega)\, \mathrm{d}^2\mathbf{q}\, \mathrm{d}\Omega \tag{3.441}$$

is the degeneracy parameter for spontaneous parametric down-conversion (Mandel and Wolf 1995), $T_{\mathrm{c}} = \frac{2\pi}{\Omega_{\mathrm{c}}}$ is its coherence time, $S_{\mathrm{c}} = (\frac{2\pi}{q_{\mathrm{c}}})^2$ the coherence area, and Ω_{c} and q_{c} the widths of the frequency and spatial frequency spectra of spontaneous parametric down-conversion.

The noise spectrum of the photocurrent density has the form

$$\begin{aligned}
\widetilde{(\delta\hat{i})^2}(\mathbf{q}, \Omega) = {} & \langle \hat{i} \rangle + 2\eta^2 |\beta|^2 \left[\delta(\mathbf{q}, \Omega) + \mathrm{Re}\{\exp(-2\mathrm{i}\varphi_\beta) g(\mathbf{q}, \Omega)\}\right] \\
& + \frac{\eta^2}{(2\pi)^3} \int\int [\delta(\mathbf{q}', \Omega')\delta(\mathbf{q} - \mathbf{q}', \Omega - \Omega') \\
& + g^*(\mathbf{q}', \Omega') g(\mathbf{q} - \mathbf{q}', \Omega - \Omega')]\, \mathrm{d}^2\mathbf{q}'\, \mathrm{d}\Omega',
\end{aligned} \tag{3.442}$$

where

$$g(\mathbf{q}, \Omega) = U(\mathbf{q}, \Omega) V(-\mathbf{q}, -\Omega). \tag{3.443}$$

Under homodyne detection, the down-conversion waves $(\mathbf{q}, \Omega)$ and $(-\mathbf{q}, -\Omega)$ modulate the local oscillator wave in space and time. With any angle $\psi(z, \mathbf{q}, \Omega)$ for a while, slow quadrature components $\tilde{\hat{a}}_{\mu\lambda}(z, \mathbf{q}, \Omega)$, $\lambda = \mathrm{c, s}$, are introduced with the property

$$\begin{aligned}
& \tilde{\hat{a}}_{1\mathrm{c}}(z, \mathbf{q}, \Omega) + \mathrm{i}\tilde{\hat{a}}_{1\mathrm{s}}(z, \mathbf{q}, \Omega) \\
& = \exp[-\mathrm{i}\psi(z, \mathbf{q}, \Omega)]\tilde{\hat{a}}(z, \mathbf{q}, \Omega) + \exp[\mathrm{i}\psi(z, \mathbf{q}, \Omega)]\tilde{\hat{a}}^\dagger(z, -\mathbf{q}, \Omega),
\end{aligned} \tag{3.444}$$

$$\begin{aligned}
& \tilde{\hat{a}}_{2\mathrm{c}}(z, \mathbf{q}, \Omega) + \mathrm{i}\tilde{\hat{a}}_{2\mathrm{s}}(z, \mathbf{q}, \Omega) \\
& = -\mathrm{i}\left\{\exp[-\mathrm{i}\psi(z, \mathbf{q}, \Omega)]\tilde{\hat{a}}(z, \mathbf{q}, \Omega) - \exp[\mathrm{i}\psi(z, \mathbf{q}, \Omega)]\tilde{\hat{a}}^\dagger(z, -\mathbf{q}, \Omega)\right\}.
\end{aligned} \tag{3.445}$$

In other words, "complex quadrature components" are first defined. When the complex components are used, transformation (3.431) simplifies to the form

$$\begin{aligned}
& \tilde{\hat{a}}_{\mu\mathrm{c}}(l, \mathbf{q}, \Omega) + \mathrm{i}\tilde{\hat{a}}_{\mu\mathrm{s}}(l, \mathbf{q}, \Omega) \\
& = \exp[\mathrm{i}\kappa(\mathbf{q}, \Omega)] \exp[\pm r(\mathbf{q}, \Omega)][\tilde{\hat{a}}_{\mu\mathrm{c}}(0, \mathbf{q}, \Omega) + \mathrm{i}\tilde{\hat{a}}_{\mu\mathrm{s}}(0, \mathbf{q}, \Omega)],
\end{aligned} \tag{3.446}$$

where + corresponds to the component with $\mu = 1$ and − to the component with $\mu = 2$. The components $\tilde{\hat{a}}_{\mu\lambda}(0, \mathbf{q}, \Omega)$ at the input surface to the crystal are defined in the coordinate system with $\psi(0, \mathbf{q}, \Omega)$ and the components $\tilde{\hat{a}}_{\mu\lambda}(l, \mathbf{q}, \Omega)$ at the output surface of the crystal are defined with $\psi(l, \mathbf{q}, \Omega)$. These angles are

$$\psi(0, \mathbf{q}, \Omega) = \frac{1}{2} \arg[V(\mathbf{q}, \Omega)U^{-1}(\mathbf{q}, \Omega)],$$
$$\psi(l, \mathbf{q}, \Omega) = \frac{1}{2} \arg[U(\mathbf{q}, \Omega)V(-\mathbf{q}, -\Omega)]. \tag{3.447}$$

In relation (3.446), $\kappa(\mathbf{q}, \Omega)$ and $r(\mathbf{q}, \Omega)$ are two other squeezing parameters,

$$\kappa(\mathbf{q}, \Omega) = \frac{1}{2} \arg[U(\mathbf{q}, \Omega)U^{-1}(\mathbf{q}, \Omega)],$$
$$\exp[\pm r(\mathbf{q}, \Omega)] = |U(\mathbf{q}, \Omega)| \pm |V(-\mathbf{q}, -\Omega)|. \tag{3.448}$$

Leaving out the integral term in the noise spectrum of the photocurrent density (3.442), one can rewrite it in the form

$$\widetilde{(\delta\hat{i})^2}(\mathbf{q}, \Omega) = \langle\hat{i}\rangle\big\{1 - \eta + \eta\big[\cos^2[\theta(\mathbf{q}, \Omega)]\exp[2r(\mathbf{q}, \Omega)] + \sin^2[\theta(\mathbf{q}, \Omega)]\exp[-2r(\mathbf{q}, \Omega)]\big]\big\}. \tag{3.449}$$

Maximum squeezing occurs at frequencies $\mathbf{q}_\mathrm{m}$, Ω_m, which fulfil the condition $\Delta(\mathbf{q}_\mathrm{m}, \Omega_\mathrm{m}) = 0$. They are said to belong to such a phase-matching surface. To reduce shot noise to the highest extent, one chooses a complex amplitude of the local oscillator wave with $\theta(\mathbf{q}_\mathrm{m}, \Omega_\mathrm{m}) = \pm\frac{\pi}{2}$. If the phase-matching condition is not perfectly met, the phase $\theta(\mathbf{q}, \Omega)$ is affected to the first order in $\Delta(\mathbf{q}, \Omega)l_\mathrm{amp}$ and the squeezing parameter is not yet influenced to the first order.

In Kolobov (1999), the case of the frequency and angle-degenerate phase matching, $\Delta(\mathbf{0}, 0) = 0$, is considered. In the case of degenerate phase matching and $k''_\Omega < 0$ one infers that in the region of frequencies $\Omega < \Omega_\mathrm{m}$ and spatial frequencies $q < q_\mathrm{m}$ the noise of the photocurrent density operator is reduced below the shot-noise level. In space–time language this can be said as follows. The frequencies Ω_m and q_m determine the minimum time T_m and the minimum area of photodetector S_m, which are necessary for reducing fluctuations in the number of photoelectrons below the Poissonian limit.

In the case of nondegenerate phase matching in a crystal, when $\Delta(\mathbf{0}, 0) > 0$, one must pay attention to nonzero carrier frequencies $\mathbf{q}$ and Ω. In fact, $\Delta(\mathbf{q}, \Omega) \approx 0$, when (a) $\mathbf{q} \neq 0, \Omega \neq 0$, (b) $\mathbf{q} \neq 0, \Omega = 0$, and (c) $\mathbf{q} = 0, \Omega \neq 0$. Three kinds of measurement can be distinguished based on the three types of phase matching.

Yuen and Shapiro (1979) were the first to propose a degenerate mixing process as a possible source for squeezed light. A four-wave mixer can be set up either in a backward geometry, as proposed by Yuen and Shapiro (1979), or in a forward geometry, according to Kumar and Shapiro (1984). A backward four-wave mixer can produce multimode squeezed light with a much larger spatial bandwidth than a forward four-wave mixer and another scheme.

The exposition is restricted to the backward four-wave mixing. The process occurs in a transparent $\chi^{(3)}$ nonlinear medium. It is assumed that the medium has the form of a plane slab the thickness of which (distance between two surfaces parallel to the $\boldsymbol{\rho}$ plane) is equal to l. Two counterpropagating plane monochromatic pump waves E_1 and E_2 of angular frequency ω_0 and wave vectors $\mathbf{k}_1$ and $\mathbf{k}_2$, respectively, illuminate the slab at a small angle to the z-axis. A quasiplane and quasimonochromatic probe wave of carrier frequency ω_0 enters the medium from the left and propagates in the $+z$-direction. In the nonlinear interaction between the two pump waves and the probe wave, a phase conjugate wave is generated in the medium that propagates in the opposite direction to the probe wave (Fisher 1983). One describes the probe and conjugate waves by two corresponding slowly varying operators $\hat{\epsilon}_{\mathrm{p}}(z, \boldsymbol{\rho}, t)$ and $\hat{\epsilon}_{\mathrm{c}}(z, \boldsymbol{\rho}, t)$. Let $\mathbf{k}_\mu(\mathbf{q}, \Omega)$, $\mu = \mathrm{p, c}$, mean the wave vectors of the probe and conjugate waves, respectively. One introduces the Fourier transforms of these space–time operators, $\tilde{\hat{\epsilon}}_\mu(z, \mathbf{q}, \Omega)$, $\mu = \mathrm{p, c}$, as follows:

$$\tilde{\hat{\epsilon}}_\mu(z, \mathbf{q}, \Omega) = \int\int \hat{\epsilon}_\mu(z, \boldsymbol{\rho}, t) \exp[\mathrm{i}(\Omega t - \mathbf{q} \cdot \boldsymbol{\rho})] \, \mathrm{d}t \, \mathrm{d}^2\boldsymbol{\rho}. \tag{3.450}$$

These operators evolve in the nonlinear medium according to the equations

$$\frac{\partial}{\partial z} \tilde{\hat{\epsilon}}_{\mathrm{p}}(z, \mathbf{q}, \Omega) = -\mathrm{i}\kappa \tilde{\hat{\epsilon}}_{\mathrm{c}}^{\dagger}(z, -\mathbf{q}, -\Omega) \exp[-\mathrm{i}\Delta(\mathbf{q}, \Omega) z], \tag{3.451}$$

$$\frac{\partial}{\partial z} \tilde{\hat{\epsilon}}_{\mathrm{c}}(z, \mathbf{q}, \Omega) = \mathrm{i}\kappa \tilde{\hat{\epsilon}}_{\mathrm{p}}^{\dagger}(z, -\mathbf{q}, -\Omega) \exp[-\mathrm{i}\Delta(\mathbf{q}, \Omega) z]. \tag{3.452}$$

Here κ is a coupling constant proportional to the product of the two pump wave amplitudes and to the nonlinear susceptibility $\chi^{(3)}$ of the medium, $\Delta(\mathbf{q}, \Omega)$ is a phase-mismatch function given by

$$\Delta(\mathbf{q}, \Omega) = k_{\mathrm{p}z}(\mathbf{q}, \Omega) + k_{\mathrm{c}z}(-\mathbf{q}, -\Omega) - k_{1z} - k_{2z}, \tag{3.453}$$

with $k_{\mathrm{p,c}z}(\mathbf{q}, \Omega)$ being the projections of the probe and conjugate wave vectors onto the positive z-direction and $k_{1,2z}$ the corresponding projections for the pump waves.

The solution of equations (3.451) and (3.452) with the boundary conditions $\tilde{\hat{\epsilon}}_{\mathrm{p}}(z = 0, \mathbf{q}, \Omega) = \tilde{\hat{\epsilon}}_{\mathrm{p}}(0, \mathbf{q}, \Omega)$ and $\tilde{\hat{\epsilon}}_{\mathrm{c}}(z = l, \mathbf{q}, \Omega) = \tilde{\hat{\epsilon}}_{\mathrm{c}}(l, \mathbf{q}, \Omega)$ may follow the classical one (Fisher 1983). The input–output transformation, which yet differs from the solution by the inclusion of the incoupling and outcoupling beam splitters, is presented in Kolobov (1999),

$$\hat{a}_{\mathrm{out}}(\mathbf{q}, \Omega) \propto U(\mathbf{q}, \Omega) \hat{a}_{\mathrm{in}}(\mathbf{q}, \Omega) + V(\mathbf{q}, \Omega) \hat{a}_{\mathrm{in}}^{\dagger}(-\mathbf{q}, -\Omega), \tag{3.454}$$

$$\hat{b}_{\mathrm{out}}(\mathbf{q}, \Omega) \propto U(\mathbf{q}, \Omega) \hat{b}_{\mathrm{in}}(\mathbf{q}, \Omega) + V(\mathbf{q}, \Omega) \hat{b}_{\mathrm{in}}^{\dagger}(-\mathbf{q}, -\Omega). \tag{3.455}$$

Notably one obtains two independent processes of multimode squeezing. Any of them can be compared with the optical parametric amplifier.

The investigation of the difference is concentrated on comparison between the phase-mismatch functions $\Delta(\mathbf{q}, \Omega)$. (Here $\Delta \equiv \Delta_{\mathrm{OPA}}, \Delta_{\mathrm{FFWM}}, \Delta_{\mathrm{BFWM}}$.) This function depends on the spatial frequency in the optical parametric amplification and the forward four-wave mixing and does not depend on it in the backward four-wave mixing. From these properties one determines "spatial" squeezing bandwidth $q = \sqrt{\frac{k_1}{l}}$ and $q = \infty$ in the paraxial approximation.

The frequency dependence can be used to filter the probe signal. One is led to the idea of a "realistic" medium and its "quantum nature". Obviously, a nonlinearity of the equations is meant, but also a better quantization desired, which just as we saw at the beginning has not yet been employed.

Another source is based on a cavity, and it still can generate multimode squeezed states. It is a subthreshold optical parametric oscillator, concretely this scheme in a cavity with spherical mirrors (Lugiato and Marzoli 1995). Here we are provided by the literature with another example, in which situation the paraxial approximation has been used. Even though here a cavity is investigated, not a travelling wave, the cavity modes still seem to have been determined in the paraxial approximation.

For a cavity-based geometry a more natural language for the description of multimode squeezing is that of discrete eigenmodes of the resonator. In the case of the cavity with spherical mirrors, such a discrete eigenset is given by the Gauss–Laguerre modes

$$f_{pli}(r,\phi) = \tilde{f}_{pl}(r) \times \begin{cases} \cos(l\phi) & \text{for } i = 1, \\ \sin(l\phi) & \text{for } i = 2, \end{cases} \tag{3.456}$$

$$\tilde{f}_{pl}(r) = \frac{2}{\sqrt{2^{\delta_{l,0}}\pi w^2}} \sqrt{\frac{p!}{(p+l)!}} \sqrt{\frac{2r^2}{w^2}} L_p^l\left(\frac{2r^2}{w^2}\right) \exp\left(-\frac{r^2}{w^2}\right), \tag{3.457}$$

where w is the waist of the beam and $p,\ l = 0, 1, 2, \ldots$ are the radial and angular indices, respectively, $r = \sqrt{x^2 + y^2}$ is the radial and ϕ is the angular variable. The functions L_p^l are the Laguerre polynomials. The functions $f_{pli}(r,\phi)$ satisfy the conditions of orthonormality,

$$\int_0^{2\pi} \int_0^{\infty} f_{pli}(r,\phi) f_{p'l'i'}(r,\phi) r \,\mathrm{d}r \,\mathrm{d}\phi = \delta_{pp'}\delta_{ll'}\delta_{ii'}. \tag{3.458}$$

The eigenfrequencies of these modes are given by

$$\omega_{pl} = \omega_{00} + (2p + l)\zeta, \tag{3.459}$$

where ω_{00} is the lowest eigenfrequency of the resonator and the parameter ζ depends on the curvature of mirrors and the distance between them (Yariv 1989).

The source is described by the master equation,

$$\frac{\partial}{\partial t}\hat{\rho} = \frac{1}{\mathrm{i}\hbar}[\hat{H}_{\mathrm{int}}, \hat{\rho}] + \sum_{i=1}^{2} \sum_{p,l} \hat{\hat{\Lambda}}_{pli} \hat{\rho}, \tag{3.460}$$

where the term $\hat{\hat{\Lambda}}_{pli}\hat{\rho}$,

$$\hat{\hat{\Lambda}}_{pli}\hat{\rho} = \gamma\left(2\hat{a}_{pli}\hat{\rho}\hat{a}^{\dagger}_{pli} - \hat{a}^{\dagger}_{pli}\hat{a}_{pli}\hat{\rho} - \hat{\rho}\hat{a}^{\dagger}_{pli}\hat{a}_{pli}\right) \tag{3.461}$$

describes the damping of the mode *pli* due to cavity decay through the outcoupling mirror with the rate γ. The interaction Hamiltonian H_{int} is given by Lugiato and Marzoli (1995),

$$\hat{H}_{\text{int}} = \mathrm{i}\hbar\frac{\gamma}{2}A_{\mathrm{p}}\int_0^{2\pi}\int_0^{\infty}\left[\hat{A}^{\dagger 2}(r,\phi) - \hat{A}^2(r,\phi)\right] r\,\mathrm{d}r\,\mathrm{d}\phi, \tag{3.462}$$

where A_{p} is the coupling constant proportional to the nonlinear susceptibility $\chi^{(2)}$ of the medium and the amplitude of the pump wave.

Instead of solving the master equation (3.460), we can write a set of independent Langevin equations (Walls and Milburn 1994) for the annihilation and creation operators $\hat{a}_{pli}(t)$ and $\hat{a}^{\dagger}_{pli}(t)$ inside the cavity,

$$\hat{a}_{pli}(t) = -\gamma[(1+\mathrm{i}\Delta_{pl})\hat{a}_{pli}(t) - A_{\mathrm{p}}\hat{a}^{\dagger}_{pli}(t)] + \sqrt{2\gamma}\,\hat{c}_{pli}(t), \tag{3.463}$$

where

$$\Delta_{pl} = \frac{\omega_{pl} - \omega_{\mathrm{s}}}{\gamma}, \tag{3.464}$$

with ω_{pl} and ω_{s} the eigenfrequencies of the eigenmodes of the resonator and the frequencies of signal photons, respectively. Every mode is damped and the rate constant for each of the modes is the same. Such a simplification should still be explained. The method of description seems to be known in quantum optics. The operators $\hat{c}_{pli}(t)$ and $\hat{c}^{\dagger}_{pli}(t)$ correspond to the operator-valued Langevin forces and describe the vacuum fluctuations entering the cavity through the outcoupling mirrors. These operators obey the commutation relations

$$[\hat{c}_{pli}(t), \hat{c}^{\dagger}_{pli}(t')] = \delta_{pp'}\delta_{ll'}\delta_{ii'}\delta(t-t')\hat{1}. \tag{3.465}$$

In relation (3.462), a coupling constant is expressed as the product γA_{p}, from which we conclude that $0 < A_{\mathrm{p}} < 1$. This property says that a subthreshold oscillator is being investigated.

Kolobov (1999) refers to Collet and Gardiner (1984) for the input–output relations for the field operators $\hat{b}_{pli}(t)$ in the wave outgoing from the cavity, $\hat{c}_{pli}(t)$ of the vacuum fluctuations entering it, and $\hat{a}_{pli}(t)$ inside the cavity,

$$\hat{b}_{pli}(t) = \sqrt{2\gamma}\,\hat{a}_{pli}(t) - \hat{c}_{pli}(t). \tag{3.466}$$

Using the Fourier transform

$$\tilde{\hat{g}}_{pli}(\Omega) = \int_{-\infty}^{\infty}\exp(\mathrm{i}\Omega t)\hat{g}_{pli}(t)\,\mathrm{d}t,\quad g = a,\ b,\ c, \tag{3.467}$$

to equations (3.463), we arrive at the following squeezing transformation between the Fourier transforms of the incoming and outgoing operators:

$$\tilde{\hat{b}}_{pli}(\Omega) = U_{pl}(\Omega)\tilde{\hat{c}}_{pli}(\Omega) + V_{pl}(\Omega)\tilde{\hat{c}}^{\dagger}_{pli}(-\Omega), \tag{3.468}$$

with the coefficients

$$U_{pl}(\Omega) = \frac{[1 - \mathrm{i}\Delta_{pl}(-\Omega)][1 - \mathrm{i}\Delta_{pl}(\Omega)] + A_{\mathrm{p}}^2}{[1 + \mathrm{i}\Delta_{pl}(\Omega)][1 - \mathrm{i}\Delta_{pl}(-\Omega)] - A_{\mathrm{p}}^2},$$
$$V_{pl}(\Omega) = \frac{2A_{\mathrm{p}}}{[1 + \mathrm{i}\Delta_{pl}(\Omega)][1 - \mathrm{i}\Delta_{pl}(-\Omega)] - A_{\mathrm{p}}^2}, \tag{3.469}$$

with $\Delta_{pl}(\pm\Omega) = \Delta_{pl} \mp \frac{\Omega}{\gamma}$, which is a discrete equivalent of the multimode squeezing transformation (3.431).

The calculation of the photocurrent noise spectrum is not included, but the following relation is presented (Lugiato and Marzoli 1995):

$$\left\langle \frac{1}{2}\{\delta\hat{i}(\boldsymbol{\rho}, t), \delta\hat{i}(\boldsymbol{\rho}', t')\}_{+} \right\rangle = \sum_{p,l} \tilde{f}_{pl}(r)\tilde{f}_{pl}(r')\cos[l(\phi - \phi')]$$
$$\times \frac{1}{2\pi}\int_{-\infty}^{\infty} \widetilde{(\delta\hat{i})^2}_{pl}(\Omega)\exp[-\mathrm{i}\Omega(t - t')]\,\mathrm{d}\Omega, \tag{3.470}$$

which is an analogue of relation (3.430). A relation holds

$$\widetilde{(\delta\hat{i})^2}_{pl}(\Omega) = \langle \hat{i}(\boldsymbol{\rho}, t)\rangle \left\{ 1 + \frac{4A_{\mathrm{p}}}{(1 + \Delta_{pl}^2 - A_{\mathrm{p}}^2 - \tilde{\Omega}^2)^2 + 4\tilde{\Omega}^2} \right.$$
$$\left. \times [2A_{\mathrm{p}} + \mathrm{Re}\{\exp(-2\mathrm{i}\varphi_{\beta})(1 - \Delta_{pl}^2 + A_{\mathrm{p}}^2 + \tilde{\Omega}^2 - 2\mathrm{i}\Delta_{pl})\}] \right\}, \tag{3.471}$$

where φ_β is the phase of the local oscillator and $\tilde{\Omega} = \frac{\Omega}{\gamma}$ is the dimensionless frequency and $\eta = 1$ (Collett and Walls 1985, Savage and Walls 1987).

(ii) Free propagation and diffraction of multimode squeezed light

With respect to the free propagation and diffraction of multimode squeezed light it is shown that propagation in free space in general deteriorates the resolving power of low-noise measurements with squeezed light. A lens allows one to compensate for this deterioration and even further to improve the resolving power.

We will assume that the plane of photodetection lies at a distance L from the exit plane of the nonlinear crystal and is parallel to it. The slowly varying operators $\tilde{\hat{a}}(l, \mathbf{q}, \Omega)$ at the exit plane of the crystal and $\tilde{\hat{a}}(l + L, \mathbf{q}, \Omega)$ at the photodetection plane are for the free propagation related as (cf. equation (3.420)) follows:

$$\tilde{\hat{a}}(l+L, \mathbf{q}, \Omega) = \exp\{\mathrm{i}[k_z^{(0)}(\mathbf{q}, \Omega) - k_0]L\}\tilde{\hat{a}}(l, \mathbf{q}, \Omega), \tag{3.472}$$

where $k_z^{(0)}(\mathbf{q}, \Omega)$ is the z-component of the wave vector in free space. Along with the free propagation one is interested in the dependence of the field operator at the plane of photodetection on that at the input to the nonlinear crystal,

$$\tilde{\hat{a}}(l+L, \mathbf{q}, \Omega) = \tilde{U}(\mathbf{q}, \Omega)\tilde{\hat{a}}(0, \mathbf{q}, \Omega) + \tilde{V}(\mathbf{q}, \Omega)\tilde{\hat{a}}^\dagger(0, -\mathbf{q}, -\Omega), \tag{3.473}$$

with the coefficients

$$\tilde{U}(\mathbf{q}, \Omega) = \exp\{\mathrm{i}[k_z^{(0)}(\mathbf{q}, \Omega) - k_0]L\}U(\mathbf{q}, \Omega), \tag{3.474}$$

and the like for $\tilde{V}(\mathbf{q}, \Omega)$. It is a simple generalization of the above description. New quantities are provided with a tilde.

Relation

$$\begin{aligned}\tilde{\theta}(\mathbf{q}, \Omega) &= \theta(\mathbf{q}, \Omega) + \frac{L}{2}[k_z^{(0)}(\mathbf{q}, \Omega) + k_z^{(0)}(-\mathbf{q}, -\Omega) - 2k_0] \\ &\approx \theta(\mathbf{q}, \Omega) - \frac{q^2 L}{2k_0},\end{aligned} \tag{3.475}$$

where a paraxial and quasimonochromatic approximation is assumed, says that the orientation angle (i.e. phase) of the squeezing changes more rapidly in dependence on the spatial frequency than on the output from the crystal. The minimum area S_m of low-noise detection is proportional to $\frac{l_\mathrm{amp}}{k_1 + \frac{2L}{k_0}}$, where l_amp is the amplification length. The increase is related to the diffraction. The resolving power of the low-noise observation has decreased. The deterioration is reversible. Even the phase shifts produced during wave propagation inside the nonlinear crystal can be compensated for.

For a lens of focal length f, provided that the object plane has the position $-2f$ relative to the lens and the image plane has the position $2f$ relative to the lens, the optical imaging is represented by the field operators

$$\hat{a}(z+4f, \boldsymbol{\rho}, t) = \exp\left[-\mathrm{i}\frac{k_0\rho^2}{2f}\right]\hat{a}\left(z, -\boldsymbol{\rho}, t - \frac{1}{c}\left[4f + \frac{\rho^2}{2f}\right]c\right). \tag{3.476}$$

From this relation it follows that the noise spectrum $\widetilde{(\delta\hat{i})^2}(\mathbf{q}, \Omega)$ has been conserved. From the results of the analyses performed it is natural to choose $z = l$, i.e. the output plane of the crystal.

Concerns with correct quantization call attention to the proposal of imaging some plane inside the crystal onto the detection plane. This imaging is understood as a general choice $z = l + L$, where L is negative. It suffices to choose

$$L = -l_\mathrm{amp}\frac{k_0}{2k_l}, \tag{3.477}$$

for the phase $\tilde{\theta}(\mathbf{q}, \Omega)$ in the vicinity of the matching surface to become independent of spatial frequency. Thus geometrical imaging of the plane inside the crystal at the distance L given by (3.477) onto the photodetection plane broadens the range of spatial frequencies at which one has a noise reduction below the shot-noise level.

Kolobov (1999) remarks on what follows. An improvement of the frequency behaviour of the noise spectrum can be achieved by inserting into the light beam a slab of a dispersive medium with wave number $k^{(1)}(\Omega)$. The length of the slab will be

$$L^{(1)} = -l_{\mathrm{amp}} \frac{k''_{\Omega}}{2k^{(1)''}_{\Omega}}, \tag{3.478}$$

if $k^{(1)''}_{\Omega}$ has the opposite sign to k''_{Ω}.

In order to assess physical possibilities for low-noise measurements, the photoelectron number collected by a pixel with the area S_{d} during the time interval T_{d} is considered as an example. If it holds that $S_{\mathrm{d}} \geq S_{\mathrm{c}}$ and $T_{\mathrm{d}} \geq T_{\mathrm{c}}$, the result is independent of S_{d} and T_{d}. For high quantum efficiency, $\eta \approx 1$, the statistics of photoelectrons is sub-Poissonian, when squeezing is significant. Here we concede the efficiency of models simplified to several modes: The average number of photons necessary for a single low-noise measurement is given by a quantity, which we could obtain on choosing as "average" model simplified to two modes.

Whereas the coherence time T_{c} limits the number of images, which can be transmitted in a time interval T, the coherence area S_{c} limits the number of modes on an illuminated spot of area S on the input to the nonlinear crystal, even though a statistical definition of the mode is peculiar. The scheme of homodyne detection is closely related to holographic measurements.

(iii) Noiseless control of multimode squeezed light

With respect to the noiseless control of multimode squeezed light, the detection of faint phase objects as proposed by Kolobov and Kumar (1993) is described. The sub-shot-noise microscopy utilizes a Mach–Zehnder interferometer. The outgoing light from the two ports of the second beam splitter is detected by two photodetector arrays. As a natural generalization of the analysis in Caves (1981), the minimum detectable spatially varying phase change is defined.

The amplitude modulation in space is not advantageous for creation of optical images with a regular (sub-Poissonian) photon statistics. One example of nondestructive modulation in space is an opaque screen with apertures larger than the coherence area of squeezed light S_{c}. A number of references for interference mixing, which provides such nondestructive modulation in time are presented. For generalization and analogy in the case of spatial modulation, see Sokolov (1991a,b).

(iv) Spatially noiseless optical amplification of images

Noiseless amplification has been defined in phase-sensitive amplifiers (see, for example, Caves (1982)) and it should be extended to the spatial domain. Many areas of physics would benefit from the possibility of noiseless amplification of faint optical images. Astronomy and microscopy come to mind. Kolobov (1999) refers to Kolobov and Lugiato (1995) for such a proposal. One considers a ring-cavity degenerate optical parametric amplifier and monochromatic images. In general, spectral bandwidth of images should be within the bandwidth of the cavity employed. The optical parametric amplifier is combined with input and output lenses, which perform the spatial Fourier transformation and broaden a narrow region of transverse vectors **q**, which is an analogue of the band of (temporal) frequencies. The exposition is self-contained.

Let $\hat{a}(\boldsymbol{\rho}, t)$ and $\hat{a}^{\dagger}(\boldsymbol{\rho}, t)$ mean the photon annihilation and creation operators in the object plane P_1 and let $\hat{e}(\boldsymbol{\rho}, t)$ and $\hat{e}^{\dagger}(\boldsymbol{\rho}, t)$ stand for the photon annihilation and creation operators in the image plane P_4. Let $\hat{b}_{\text{in}}(\boldsymbol{\xi}, t)$ and $\hat{b}_{\text{out}}(\boldsymbol{\xi}, t)$ mean the field operators in the input and the output planes of the optical parametric oscillator, respectively. The operator $\hat{b}_{\text{in}}(\boldsymbol{\xi}, t)$ is expressed through the operator $\hat{a}(\boldsymbol{\rho}, t)$ in the object plane by the following transformation performed by the lens L_1:

$$\hat{b}_{\text{in}}(\boldsymbol{\xi}, t) = \frac{1}{\lambda f} \int \hat{a}(\boldsymbol{\rho}, t) \exp\left[-\mathrm{i}\frac{2\pi}{\lambda f}\boldsymbol{\xi} \cdot \boldsymbol{\rho}\right] \mathrm{d}^2\boldsymbol{\rho}, \tag{3.479}$$

where f is the focal length of the lens and λ is the wavelength of the light. In a paraxial approximation, the slowly varying field operator $\hat{b}(\boldsymbol{\xi}, t)$ of the cavity mode closest to resonance with input signal is described by the equation

$$\begin{aligned}\frac{\partial}{\partial t}\hat{b}(\boldsymbol{\xi}, t) - \mathrm{i}\frac{c}{2k}\nabla_{\perp}^2\hat{b}(\boldsymbol{\xi}, t) &= -(\kappa + \mathrm{i}\Delta)\hat{b}(\boldsymbol{\xi}, t) \\ &\quad + \sigma\hat{b}^{\dagger}(\boldsymbol{\xi}, t) + \sqrt{2\kappa}\,\hat{b}_{\text{in}}(\boldsymbol{\xi}, t).\end{aligned} \tag{3.480}$$

Here κ is the cavity decay constant equal to

$$\kappa = \frac{cT}{2L}, \tag{3.481}$$

where T is the intensity transmission coefficient of the cavity outcoupling mirror, L is the perimeter of the cavity, and c is the light velocity in a vacuum; the detuning parameter is defined as

$$\Delta = \omega_{\text{c}} - \omega_{\text{s}}, \tag{3.482}$$

where ω_{c} is the longitudinal cavity frequency closest to the frequency ω_{s} of the signal field. In (3.480) σ is the constant of parametric interaction proportional to the pump amplitude and k is the wave number of the travelling wave inside the cavity, $k = \frac{2\pi}{\lambda}$.

The output field operator $\hat{b}_{\text{out}}(\boldsymbol{\xi}, t)$ is the sum of two waves, one of which is reflected from and another transmitted through the outcoupling mirror of the cavity,

$$\hat{b}_{\text{out}}(\boldsymbol{\xi}, t) = \sqrt{2\kappa}\,\hat{b}(\boldsymbol{\xi}, t) - \hat{b}_{\text{in}}(\boldsymbol{\xi}, t). \tag{3.483}$$

To express the output field operators in terms of the input operators one takes the spatio-temporal Fourier transform of $\hat{b}(\boldsymbol{\xi}, t)$,

$$\tilde{\hat{b}}(\mathbf{q}, \Omega) = \int\int \hat{b}(\boldsymbol{\xi}, t) \exp[\mathrm{i}(\Omega t - \mathbf{q}\cdot\boldsymbol{\xi})]\, \mathrm{d}^2\boldsymbol{\xi}\, \mathrm{d}t. \tag{3.484}$$

The spatio-temporal Fourier transforms of $\hat{b}_{\text{in}}(\boldsymbol{\xi}, t)$ and $\hat{b}_{\text{out}}(\boldsymbol{\xi}, t)$ are similar.

The transformation of the field amplitude from the object plane P_1 to the input plane P_2 given by relation (3.479) is equivalent to the following relation between the spatio-temporal Fourier transform $\tilde{\hat{b}}_{\text{in}}(\mathbf{q}, \Omega)$ and the temporal Fourier transform $\tilde{\hat{a}}(\boldsymbol{\rho}, \Omega)$

$$\tilde{\hat{b}}_{\text{in}}(\mathbf{q}, \Omega) = \lambda f \tilde{\hat{a}}\left(-\frac{\lambda f}{2\pi}\mathbf{q}, \Omega\right), \tag{3.485}$$

where we have used

$$\tilde{\hat{a}}(\boldsymbol{\rho}, \Omega) = \int \hat{a}(\boldsymbol{\rho}, t) \exp(\mathrm{i}\Omega t)\, \mathrm{d}t. \tag{3.486}$$

Since the lens L_2 has the same focal length as L_1, we have an identical relationship between the Fourier transforms $\tilde{\hat{b}}_{\text{out}}(\mathbf{q}, \Omega)$ and $\tilde{\hat{e}}(\boldsymbol{\rho}, \Omega)$ in the output plane P_3 and in the image plane P_4,

$$\tilde{\hat{e}}(\boldsymbol{\rho}, \Omega) = \lambda f \tilde{\hat{b}}_{\text{out}}\left(-\frac{\lambda f}{2\pi}\boldsymbol{\rho}, \Omega\right), \tag{3.487}$$

$$\tilde{\hat{e}}(\boldsymbol{\rho}, \Omega) = \int \hat{e}(\boldsymbol{\rho}, t) \exp(\mathrm{i}\Omega t)\, \mathrm{d}t. \tag{3.488}$$

It can be derived that

$$\tilde{\hat{e}}(\boldsymbol{\rho}, \Omega) = u(\boldsymbol{\rho}, \Omega)\tilde{\hat{a}}(\boldsymbol{\rho}, \Omega) + v(\boldsymbol{\rho}, \Omega)\tilde{\hat{a}}^{\dagger}(-\boldsymbol{\rho}, -\Omega), \tag{3.489}$$

with $u(\boldsymbol{\rho}, \Omega)$ and $v(\boldsymbol{\rho}, \Omega)$ given by

$$\begin{aligned} u(\boldsymbol{\rho}, \Omega) &= \frac{[1 - \mathrm{i}\delta(\boldsymbol{\rho}, \Omega)][1 - \mathrm{i}\delta(\boldsymbol{\rho}, -\Omega)] + |g|^2}{[1 + \mathrm{i}\delta(\boldsymbol{\rho}, \Omega)][1 - \mathrm{i}\delta(\boldsymbol{\rho}, -\Omega)] - |g|^2}, \\ v(\boldsymbol{\rho}, \Omega) &= \frac{2g}{[1 + \mathrm{i}\delta(\boldsymbol{\rho}, \Omega)][1 - \mathrm{i}\delta(\boldsymbol{\rho}, -\Omega)] - |g|^2}. \end{aligned} \tag{3.490}$$

Here one has introduced the dimensionless coupling strength g of the parametric interaction,

$$g = \frac{\sigma}{\kappa}, \tag{3.491}$$

and the dimensionless mismatch function $\delta(\boldsymbol{\rho}, \Omega)$,

$$\delta(\boldsymbol{\rho}, \Omega) = \frac{\Delta}{\kappa} - \frac{\Omega}{\kappa} + \left(\frac{\rho}{\rho_0}\right)^2, \tag{3.492}$$

with ρ_0 defined as

$$\rho_0 = f\sqrt{\frac{\lambda T}{2\pi L}}. \tag{3.493}$$

To ensure a linear amplification regime, the coupling strength g must be $|g| < 1$.

In Kolobov (1999) it has been shown that multimode squeezed states of light come about as a natural generalization of single-mode squeezed states. They can be produced in experiments when just one spatial mode of the field is cut out by means of a high-Q optical cavity. Travelling-wave configurations are most convenient for the generation of multimode squeezed states. To observe them, one must employ a dense array of photodetectors.

Many new physical phenomena are connected with multimode squeezing. Multimode squeezed states offer a few applications including optical imaging with sub-shot-noise sensitivity, sub-shot-noise microscopy, and noiseless amplification of optical images. There are some other phenomena related to multimode squeezing, such as the similarity of homodyne detection of multimode squeezed states to the scheme of optical holography, an application of these states to optical image recognition with photon-limited images (Morris 1989), a possibility to improve a quantum limit in optical resolution with the use of nonclassical light (den Dekker and van den Bos 1997). Multimode squeezed states can be applied in the field of optical pattern formation, which studies the spatial and spatio-temporal phenomena that arise in the structure of the electromagnetic field in the plane orthogonal to the direction of propagation. For instance, the filamentation of a laser beam initiated by quantum fluctuations of light in its transverse area (Nagasako et al. 1997, Lugiato et al. 1999).

Björk et al. (2004) have shown that the use of entangled photon pairs in an imaging system can be simulated with a classically correlated source sometimes. They have considered two schemes with "bucket detection" of one of the photons. In contrast, entangled two-photon imaging may exhibit effects that cannot be mimicked by any classical source when bucket detection is not used (Strekalov et al. 1995).

Caetano and Souto Ribeiro (2004) have investigated theoretically and experimentally the transfer of the angular spectrum of the pump beam to the down-converted beams. They have demonstrated that the image of a given object placed in the pump can be formed in the twin beams by manipulating the entangled angular spectrum and performing coincidence detection.

Gatti et al. (2004) have analytically shown that it is possible to perform coherent imaging by using the classical correlation of two beams obtained by splitting thermal light. They have presented a formal analogy between two such classically correlated beams and two entangled beams produced by parametric down conversion. The classical beams can qualitatively reproduce all the imaging properties of the entangled beams. These classical beams are spatially correlated both in the near field and in the far field even though to an imperfect degree.

Bache et al. (2004) have presented a theoretical study of ghost imaging which uses balanced homodyne detection to measure signal and idler fields arising from parametric down conversion. They have used a general model describing the three-wave quantum interaction with respect to finite size and duration of the pump pulse. They have shown that the signal–idler correlations contain the full amplitude and phase information about an object located in the signal arm, both in the near-field (object image) and the far-field (object diffraction pattern) cases. One may pass from the far-field result to the near-field result by simply performing inverse Fourier transformation. The analytical results are confirmed by numerical simulations.

Bennink et al. (2004) have reported two distinct experimental demonstrations of coincidence imaging. They have shown that uncertainties of distance and mean direction of two classical fields must obey an inequality. With the use of entangled photons they formed two images whose resolution had a product that was three times better than is possible according to classical diffraction theory. For the sake of comparison, a similar experiment was performed with light in a classical mixture of states (cf. Gatti et al. 2003). While the resolution of the image was good in the far field, the uncertainty product obeyed the classical inequality in the near field.

Valencia et al. (2005) presented the first experimental demonstration of two-photon ghost imaging with a pseudothermal source. They have introduced the concepts of two-photon coherent and two-photon incoherent imaging. Similar to the case of entangled states, a two-photon Gaussian thin lens equation connects the object plane and the image plane. Specifically, the thermal source acts as a phase conjugated mirror.

Altman et al. (2005) have probed the quantum image produced by parametric down-conversion with a pump beam carrying orbital angular momentum. With one detector fixed and the other scanning, the usual single-spot coincidence pattern is predicted (Monken et al. 1998) to split into two spots, which has been demonstrated.

Mosset et al. (2005) have presented the first experimental demonstration of noiseless amplification of images that yielded spatially integrated intensity (of the photodetection process) for different lateral detector sizes. Achieving two-beam and single-beam conditions, they have compared phase-insensitive and phase-sensitive schemes with theory.

Quantum imaging is a branch of quantum optics that investigates the ultimate performance limits of optical imaging imposed by quantum mechanics. The use of quantum-optical methods enables one to solve the problems of image formation, processing and detection with sensitivity and resolution which exceeds the limits of classical imaging. The most important theoretical and experimental results in quantum imaging can be found in Kolobov (2007).

3.4 Optical Nonlinearity and Renormalization

Abram and Cohen (1994) mainly applied a travelling-wave formulation of the theory of quantum optics to the description of the self-phase modulation of a short coherent pulse of light. They seem to have been first to use a renormalization (Kubo 1962, Zinn-Justin 1989). The renormalized theory successfully describes the nonlinear chirp that the pulse undergoes in the course of its propagation and permits the calculation of the squeezing characteristics of self-phase modulation. The description of the propagation of a short coherent pulse of light inside a medium that exhibits an intensity-dependent refractive index (Kerr effect) has become relevant to optical fibre communications, all-optical switching, and optical logic gates (Agrawal 1989). Neglect of dispersion and the Raman and Brillouin scattering leads to the description of self-phase modulation. In classical theory it is derived that, in the course of its propagation, the pulse becomes chirped (i.e. different parts of the pulse acquire different central frequencies), which influences also its spectrum. Abram and Cohen (1994) have pointed out many difficulties in the investigation of the quantum noise properties of a light pulse undergoing self-phase modulation. The traditional cavity-based formalism truncates the mutual interaction among the spatial modes to a self-coupling of a single mode (or only a few modes) and cannot give a reasonable approximation to the frequency spectrum produced by self-phase modulation. In spite of the difficulties, papers based on a single-mode description of a field indicated that the slowly varying approximation can produce squeezed light (Kitagawa and Yamamoto 1986, Shirasaki et al. 1989, Shirasaki and Haus 1990, Wright 1990, Blow et al. 1991) and others treated squeezing in solitons (Drummond and Carter 1987, Shelby et al. 1990, Lai and Haus 1989), an effect that was verified experimentally (Rosenbluh and Shelby 1991). Blow et al. (1991) have shown the divergence of nonlinear phase shift, which Abram and Cohen (1994) treat through the process of renormalization.

Let us review the basic features of the quantization of the electromagnetic field in a Kerr medium and discuss the relevance of the renormalization procedure to the treatment of divergences of effective medium theories. We consider a transparent, homogeneous isotropic, and dispersionless dielectric medium that exhibits a nonlinear refractive index. We examine the situation similar to Abram and Cohen (1991). The Hamiltonian for the electromagnetic field in a Kerr medium is

$$\hat{H}(t) = \int \left\{ \frac{1}{2} \left[\hat{B}^2(z,t) + \epsilon \hat{E}^2(z,t) \right] + \frac{3}{4} \chi \hat{E}^4(z,t) \right\} \mathrm{d}z, \tag{3.494}$$

where the integration along the direction of propagation z is denoted explicitly, but integration over the transversed directions x and y will be implicit. We use the Heaviside–Lorentz units for the electromagnetic field without $\hbar = c = 1$, χ is the nonlinear (third-order) optical susceptibility. From the perspective of the substitution of (3.25), the Hamiltonian (3.494) can be written as (Hillery and Mlodinow 1984)

$$\hat{H}(t) = \int \left\{ \frac{1}{2} \left[\hat{B}^2(z,t) + \frac{1}{\epsilon} \hat{D}^2(z,t) \right] - \frac{1}{4} \frac{\chi}{\epsilon^4} \hat{D}^4(z,t) \right\} \mathrm{d}z, \tag{3.495}$$

where the displacement field

$$\hat{D}(z,t) = \epsilon \hat{E}(z,t) + \chi \hat{E}^3(z,t). \tag{3.496}$$

The canonical equal-time commutators are (cf. (3.28))

$$[\hat{A}(z_1,t), \hat{D}(z_2,t)] = -\mathrm{i}\hbar c \delta(z_1 - z_2)\hat{1}, \tag{3.497}$$

$$[\hat{B}(z_1,t), \hat{D}(z_2,t)] = -\mathrm{i}\hbar c \delta'(z_1 - z_2)\hat{1}. \tag{3.498}$$

It is convenient to adopt a slowly varying operator picture in which the zeroth-order dynamics of the field governed by the linear medium Hamiltonian are already taken into account exactly, while the optical nonlinearity can be treated within the framework of perturbation theory. In such a picture, a field operator $\hat{Q}(z,t)$ evolving inside a nonlinear medium is related to the corresponding linear medium operator $\hat{Q}_0(z,t)$ by

$$\hat{Q}(z,t) = \hat{U}^{-1}(t)\hat{Q}_0(z,t)\hat{U}(t). \tag{3.499}$$

The unitary transformation $\hat{U}(t)$ is given by

$$\hat{U}(t) = \mathcal{T} \exp\left[-\frac{\mathrm{i}}{\hbar} \int_{-\infty}^{t} \hat{H}_1(\tau)\, \mathrm{d}\tau \right], \tag{3.500}$$

where $\mathcal{T} \equiv \overleftarrow{\mathcal{T}}$ denotes the time ordering and

$$\hat{H}_1(t) = -\frac{1}{4} \frac{\chi}{\epsilon^4} \int \hat{D}_0^4(z,t)\, \mathrm{d}z \tag{3.501}$$

is the interaction Hamiltonian. The full nonlinear Hamiltonian (3.495) can be written as follows:

$$H(\hat{D}_0, \hat{B}_0) = \overbrace{H_0(\hat{D}_0, \hat{B}_0)}^{\hat{H}_0(t)} + \hat{H}_1(t), \tag{3.502}$$

where $\hat{H}_0(t)$ is the linear medium Hamiltonian. Following the traditional modal approach to relation (3.499), Kitagawa and Yamamoto (1986) developed a single-mode treatment of the self-phase modulation. Clearly, such an investigation is valid only inside an optical cavity with a sparse mode structure. In this situation, the time evolution cannot be interpreted as space progression. In developing a travelling-wave theory for self-phase modulation, Blow et al. (1991) obtained a solution similar to the single-mode solution. They encountered a nonintegrable singularity upon normal ordering, a thing what is termed in quantum field theory as "ultraviolet" divergence. To avoid an infinite nonlinear phase shift due to the Kerr interaction so simply described, Blow et al. (1991, 1992) introduced a finite response time for the nonlinear medium as regularization in the Heisenberg picture. Alternatively, Haus and Kärtner (1992) considered the group velocity dispersion for the propagation of pulses in the medium as a regularization.

At any rate, the regularization is a subsequent sophistication of the simple model as known in classical theory. The response time and the group velocity dispersion are necessary ingredients of a complete description of the propagation of the electromagnetic excitation in fibres. But they have no influence on the effects associated with the vacuum fluctuations under study. A systematic way of dealing with the vacuum fluctuations in quantum field theory is the procedure of renormalization (Itzykson and Zuber 1980, Zinn-Justin 1989). The renormalization is known also in classical field theory. In order to obtain finite results, the procedure of renormalization redefines all the quantities that enter the Hamiltonian. The renormalization point of view is that the new Hamiltonian is the only one we have access to. It contains the observable consequences of the theory and the parameters are the ones we obtain from experiments. The bare quantities are only auxiliary parameters that should be eliminated exactly from the description (Stenholm 2000). The re-defined (renormalized) quantities are able to incorporate the (infinite) effects of the vacuum fluctuations. We will provide the definitions of broad-band electromagnetic field operators and treat the propagation of light in a linear medium. The normal ordering is considered as the simplest renormalization, e.g. in the case of the effective linear Hamiltonian

$$\hat{H}_0(t) = \frac{1}{2}\int \left[\hat{B}_0^2(z,t) + \frac{1}{\epsilon}\hat{D}_0^2(z,t)\right] \mathrm{d}z. \tag{3.503}$$

To this end, the Hamiltonian $\hat{H}_0(t)$ is to be written in terms of the creation and annihilation operators. The normal ordering allows us to subtract the vacuum-field energy up to the first order from the effective Kerr Hamiltonian (3.495). However, when this Hamiltonian is used to describe propagation disregarding the richness of the quantum field theory, the normal ordering gives rise to additional divergences that can be attributed to the participation of the vacuum fields. Upon renormalization, involving also the refractive index, the divergences are removed.

As the Kerr nonlinearity involves the fourth power of the derivative $\frac{\partial}{\partial t}\hat{A}(z,t)$, it cannot be in the general case renormalized to all orders with a finite number of corrections. Inspired by nonlinear optics, the slowly varying amplitude approximation decouples counterpropagating waves and the renormalization to all orders becomes possible. All types of optical nonlinearity $\chi^{(n)}$ give rise to divergences which require the renormalization. In the treatment of parametric down conversion (Abram and

Cohen 1991), the problem of divergences and the need of normalization were not formulated.

In Abram and Cohen (1994), the broad-band electromagnetic field operators are defined and the propagation of light in a nonlinear medium is treated. In the absence of the optical nonlinearity, $\chi = 0$ and the linear-medium displacement field has the usual proportionality relationship to the electric field,

$$\hat{D}_0(z,t) = \epsilon \hat{E}_0(z,t). \tag{3.504}$$

The magnetic field and the displacement field in the linear medium obey the equal-time commutation relation

$$[\hat{B}_0(z_1,t), \hat{D}_0(z_2,t)] = -\mathrm{i}\hbar c \delta'(z_1 - z_2)\hat{1}. \tag{3.505}$$

The operators $\hat{V}_0^{\pm}$ (cf. (3.80), (3.81) of Abram and Cohen (1991)) reappear as the operators

$$\hat{\psi}_+(z,t) = \frac{1}{\sqrt{2\epsilon}} \hat{V}_0^+(z,t), \tag{3.506}$$

$$\hat{\psi}_-(z,t) = \frac{1}{\sqrt{2\epsilon}} \hat{V}_0^-(z,t). \tag{3.507}$$

For $\hat{\psi}_\pm(z,t)$, the equations of motion in the Heisenberg picture may be calculated by the use of commutator (3.505) as follows:

$$\frac{\partial}{\partial t}\hat{\psi}_\pm(z,t) = \frac{\mathrm{i}}{\hbar}\left[\hat{H}_0, \hat{\psi}_\pm(z,t)\right] = \mp v \frac{\partial}{\partial z}\hat{\psi}_\pm(z,t), \tag{3.508}$$

where $v = \frac{c}{\sqrt{\epsilon}}$ is the speed of light inside the dielectric exhibiting the refractive index ϵ. Their solutions are

$$\hat{\psi}_+(z,t) = \hat{\psi}_+(z - vt, 0), \tag{3.509}$$

$$\hat{\psi}_-(z,t) = \hat{\psi}_-(z + vt, 0). \tag{3.510}$$

The equal-time commutators of the copropagating field operators can be obtained from the definition and commutator (3.505) as follows:

$$[\hat{\psi}_+(z_1,t), \hat{\psi}_+(z_2,t)] = -\mathrm{i}\hbar v \delta'(z_1 - z_2)\hat{1}, \tag{3.511}$$

$$[\hat{\psi}_-(z_1,t), \hat{\psi}_-(z_2,t)] = \mathrm{i}\hbar v \delta'(z_1 - z_2)\hat{1}. \tag{3.512}$$

For the counter propogating fields, the corresponding operators commute with each other,

$$[\hat{\psi}_+(z_1,t), \hat{\psi}_-(z_2,t)] = \hat{0}. \tag{3.513}$$

Operators (3.506) and (3.507) permit us to express the linear medium Hamiltonian (3.503) as

$$\hat{H}_0(t) = \frac{1}{2}\int \left[\hat{\psi}_+^2(z,t) + \hat{\psi}_-^2(z,t)\right] \mathrm{d}z, \tag{3.514}$$

thus separating it into a sum of two mutually commuting partial operators, one for each direction of propagation.

In the homogeneous medium it is possible to separate the electromagnetic field operators $\hat{\psi}_\pm(z,t)$ into positive- and negative-frequency parts

$$\hat{\psi}_\pm(z,t) = \hat{\phi}_\pm(z,t) + \hat{\phi}_\pm^\dagger(z,t), \tag{3.515}$$

defined as

$$\hat{\phi}_\pm(z,t) = \frac{1}{2}\left[\hat{\psi}_\pm(z,t) \pm \frac{\mathrm{i}}{\pi}\mathrm{V.p.}\int_{-\infty}^{\infty} \frac{\hat{\psi}_\pm(z',t)}{z-z'}\,\mathrm{d}z'\right], \tag{3.516}$$

where V.p. denotes the Cauchy principal value. The operators $\hat{\phi}_\pm^\dagger(z,t)$ and $\hat{\phi}_\pm(z,t)$ can be considered as creation and annihilation operators, respectively, for a right (or left)-moving electromagnetic excitation which at time t is at point z. The equal-time commutators of $\hat{\phi}_\pm(z,t)$ are somewhat complicated,

$$\begin{aligned}
[\hat{\phi}_\pm(z_1,t), \hat{\phi}_\pm^\dagger(z_2,t)] &= \frac{\hbar v}{2}\frac{\partial}{\partial z_1}\left[\mathcal{P}\frac{1}{z_1-z_2} \mp \mathrm{i}\pi\delta(z_1-z_2)\right]\hat{1}, \\
[\hat{\phi}_+(z_1,t), \hat{\phi}_-^\dagger(z_2,t)] &= \hat{0},
\end{aligned} \tag{3.517}$$

where $\mathcal{P}$ refers to the familiar generalized function $\mathcal{P}\frac{1}{z}$. Nevertheless, an important simplification results when only unidirectional propagation is considered. On introducing the operators

$$\hat{D}_0^{(+)}(z,t) = \sqrt{\frac{\epsilon}{2}}[\hat{\phi}_+(z,t) + \hat{\phi}_-(z,t)], \quad \hat{D}_0^{(-)}(z,t) = [\hat{D}_0^{(+)}(z,t)]^\dagger, \tag{3.518}$$

$$\hat{B}_0^{(+)}(z,t) = \frac{1}{\sqrt{2}}[\hat{\phi}_+(z,t) - \hat{\phi}_-(z,t)], \quad \hat{B}_0^{(-)}(z,t) = [\hat{B}_0^{(+)}(z,t)]^\dagger \tag{3.519}$$

and considering the relations

$$\begin{aligned}
\hat{D}_0(z,t) &= \sqrt{\frac{\epsilon}{2}}[\hat{\psi}_+(z,t) + \hat{\psi}_-(z,t)], \\
\hat{B}_0(z,t) &= \frac{1}{\sqrt{2}}[\hat{\psi}_+(z,t) - \hat{\psi}_-(z,t)]
\end{aligned} \tag{3.520}$$

and relation (3.515), we verify that

$$\begin{aligned}
\hat{D}_0(z,t) &= \hat{D}_0^{(+)}(z,t) + \hat{D}_0^{(-)}(z,t), \\
\hat{B}_0(z,t) &= \hat{B}_0^{(+)}(z,t) + \hat{B}_0^{(-)}(z,t).
\end{aligned} \tag{3.521}$$

Using the new operators, the equal-time commutation relation (3.505) can suitably be modified as

$$[\hat{B}_0^{(+)}(z_1,t), \hat{D}_0^{(-)}(z_2,t)] = -\frac{\mathrm{i}}{2}\hbar c\delta'(z_1 - z_2)\hat{1}. \tag{3.522}$$

For a right-moving electromagnetic excitation, we observe that

$$\begin{aligned} \hat{\phi}_{+(+)}(z_1,t) &= \sqrt{2}\hat{B}_{0(+)}^{(+)}(z_1,t), \\ \hat{\phi}_{+(+)}(z_2,t) &= \sqrt{\frac{2}{\epsilon}}\hat{D}_{0(+)}^{(+)}(z_2,t), \end{aligned} \tag{3.523}$$

where the subscript $(+)$ refers to $k > 0$. Using (3.523), we obtain that

$$[\hat{\phi}_+(z_1,t), \hat{\phi}_+^\dagger(z_2,t)]_{(+)} = -\mathrm{i}\hbar v\delta'(z_1 - z_2)\hat{1}_{(+)}. \tag{3.524}$$

Similarly, for a left-moving electromagnetic excitation, we note that

$$\begin{aligned} \hat{\phi}_{-(-)}(z_1,t) &= -\sqrt{2}\hat{B}_{0(-)}^{(+)}(z_1,t), \\ \hat{\phi}_{-(-)}(z_2,t) &= \sqrt{\frac{2}{\epsilon}}\hat{D}_{0(-)}^{(+)}(z_2,t), \end{aligned} \tag{3.525}$$

where the subscript $(-)$ refers to $k < 0$. From this

$$[\hat{\phi}_-(z_1,t), \hat{\phi}_-^\dagger(z_2,t)]_{(-)} = \mathrm{i}\hbar v\delta'(z_1 - z_2)\hat{1}_{(-)}. \tag{3.526}$$

The electromagnetic creation and annihilation operators allow us to speak of the normal order, for instance, when we write Hamiltonian (3.514) in the form

$$\hat{H}_0(t) = \int \left[\hat{\phi}_+^\dagger(z,t)\hat{\phi}_+(z,t) + \hat{\phi}_-^\dagger(z,t)\hat{\phi}_-(z,t)\right] \mathrm{d}z. \tag{3.527}$$

We can define annihilation and creation wave-packet photon operators

$$\hat{F}_+(\bar{z},t) = \int F(z - \bar{z})\hat{\phi}_+(z,t)\,\mathrm{d}z \tag{3.528}$$

and

$$\hat{F}_+^\dagger(\bar{z},t) = \int F^*(z - \bar{z})\hat{\phi}_+^\dagger(z,t)\,\mathrm{d}z, \tag{3.529}$$

respectively. Here $F(z)$ is a complex function:

$$F(z) = \frac{1}{\sqrt{vK}}\exp(\mathrm{i}Kz)\tilde{F}(z), \tag{3.530}$$

where vK is the central (carrier) frequency and $\tilde{F}(z)$ is the wave-packet envelope function peaked at $z = 0$ and $k = 0$.

On the usual assumption of narrow bandwidth and

$$\mathrm{i}\hbar v \int F'(z)F^*(z)\,\mathrm{d}z = 1, \tag{3.531}$$

where F' denotes the spatial derivative, we obtain that the operators $\hat{F}_+$ and $\hat{F}_+^\dagger$ follow the boson commutation relation

$$\left[\hat{F}_+(\bar{z},t),\, \hat{F}_+^\dagger(\bar{z},t)\right] = \hat{1}. \tag{3.532}$$

Let us remark that the commutation relation (3.532) is relation (A5) in Milburn et al. (1984), where the formalism of the counterdirectional coupling was derived or rather this pitfall underestimated.

Now a coherent pulse can be considered whose shape is described by $\rho F(z)$ with a scaling factor ρ. A coherent state appropriate to ρF is defined as

$$|\rho F\rangle = \exp\left[\rho\left(\hat{F}_+^\dagger - \hat{F}_+\right)\right]|0\rangle. \tag{3.533}$$

It satisfies the "single-mode" eigenvalue equation

$$\hat{F}_+(\bar{z},t)|\rho F\rangle = \rho|\rho F\rangle \tag{3.534}$$

and, at the same time, it obeys the approximate quantum field eigenvalue equation

$$\hat{\phi}_+(z,t)|\rho F\rangle = \hbar\sqrt{vK}\rho\tilde{F}^*(z-\bar{z})\exp[-\mathrm{i}K(z-\bar{z})]|\rho F\rangle. \tag{3.535}$$

The approximation made in the derivation of (3.535) has kindled the interest in the Glauber factorization conditions and the theory of coherence (see also Ledinegg (1966)).

When we examine right-moving pulses, we can introduce moving-frame coordinate

$$\eta = z - vt \tag{3.536}$$

and simplify relation (3.509),

$$\hat{\phi}_+(z,t) = \hat{\phi}(\eta,0) \equiv \hat{\phi}(\eta), \tag{3.537}$$

dropping the subscript $+$, whenever we use the coordinate η explicitly. Similarly, the commutation relation (3.524) can be modified. The right-moving narrow-bandwidth wavepacket operators (3.528), (3.529) can now be written as

$$\hat{F}(\bar{\eta}) = \int F(\eta-\bar{\eta})\hat{\phi}(\eta)\,\mathrm{d}\eta, \tag{3.538}$$

where $\bar{\eta} = \bar{z} - vt$ and

$$\hat{F}^{\dagger}(\bar{\eta}) = \int F^{*}(\eta - \bar{\eta})\hat{\phi}^{\dagger}(\eta)\,\mathrm{d}\eta. \qquad (3.539)$$

In the moving-frame representation $\hat{F}(\bar{\eta}, t) = \hat{F}(\bar{\eta}, 0)$.

It is feasible to find a connection with the approaches leading to narrow-band field operators ($\hat{a}$) contained in Shirasaki and Haus (1990), Drummond (1990), and Blow et al. (1990) and used in papers by Blow et al. (1991) and Shirasaki and Haus (1990). An important feature of these operators is that their commutator is a δ function

$$[\hat{a}_{k_0}(z_1), \hat{a}_{k_0}^{\dagger}(z_2)] = \hbar v k_0 \delta(z_1 - z_2)\hat{1}. \qquad (3.540)$$

Under the same narrow-bandwidth condition, the commutator of the Abram–Cohen operators, which is a delta function derivative, can be approximated by

$$\delta'(z_1 - z_2) \approx -\mathrm{i}k_0\delta(z_1 - z_2). \qquad (3.541)$$

Abram and Cohen (1994) have analysed the approximations that enter the quantum treatment of propagation in a Kerr medium and outline the corresponding renormalization procedure. The slowly varying amplitude approximation according to Abram and Cohen (1991) is used in Abram and Cohen (1994). For a Kerr medium, the interaction Hamiltonian $\hat{H}_1(t)$ is expressed as

$$\hat{H}_1(t) = -\frac{\chi}{4\epsilon^4}\int \hat{D}_0^4(z, t)\,\mathrm{d}z. \qquad (3.542)$$

According to (3.520), the interaction Hamiltonian can be written as

$$\hat{H}_1(t) = \frac{\chi}{16\epsilon^2}\int \left[\hat{\psi}_+(z - vt) + \hat{\psi}_-(z + vt)\right]^4 \mathrm{d}z. \qquad (3.543)$$

The exact Hamiltonian (3.495) may be written up to the first order in χ as

$$\hat{H}(t) = \hat{H}_0(t) + \hat{H}_{1\mathrm{S}+}(t) + \hat{H}_{1\mathrm{S}-}(t) + O(\chi^2), \qquad (3.544)$$

where

$$\hat{H}_{1\mathrm{S}\pm}(t) = -\frac{\chi}{16\epsilon^2}\int \hat{\psi}_{\pm}^4(z \mp vt)\,\mathrm{d}z \qquad (3.545)$$

are the parts of the Hamiltonian (3.543) that commute with $\hat{H}_0(t)$.

In terms of application of (3.537), no coordinate transformation has been explained. In this case, the transformation leaves the time coordinate unchanged.

In view of approximation (3.544), the equation of motion of a right-moving field operator can be written in the interaction picture as follows:

$$\frac{\partial}{\partial t}\hat{\phi}(\eta, t) = \frac{\mathrm{i}}{\hbar}\left[\hat{H}_{1\mathrm{S}+}(t), \hat{\phi}(\eta, t)\right]. \tag{3.546}$$

This first-order approximation to the equation of motion can be solved formally using the corresponding time-evolution operator (cf. (3.499))

$$\hat{U}_{S+}(t) = \mathcal{T}\exp\left[-\frac{\mathrm{i}}{\hbar}\int_{-\infty}^{t}\hat{H}_{1\mathrm{S}+}(\tau)\,\mathrm{d}\tau\right]. \tag{3.547}$$

The classical slowly varying approximation has its quantum counterpart on a double assumption: (1) the initial state of the field is a narrow-bandwidth state and (2) the nonlinearity is weak enough so that the full nonlinear Hamiltonian (3.543) may be approximated by its first-order stationary component $\hat{H}_{1\mathrm{S}}$,

$$\hat{H}_{1\mathrm{S}} = \hat{H}_{1\mathrm{S}+} + \hat{H}_{1\mathrm{S}-}. \tag{3.548}$$

Therefore, $\hat{H}_{1\mathrm{S}+}$ will be referred to as the slowly varying amplitude Hamiltonian.

Now we turn to the renormalization. In the framework of the rotating-wave approximation, we obtain that

$$\hat{H}_{1\mathrm{S}+} = -\frac{\chi}{16\epsilon^2}\int 6\left(\hat{\phi}^\dagger\hat{\phi}^\dagger\hat{\phi}\hat{\phi}\right)_S \mathrm{d}z, \tag{3.549}$$

where

$$6\left(\hat{\phi}^\dagger\hat{\phi}^\dagger\hat{\phi}\hat{\phi}\right)_S = \hat{\phi}^\dagger\hat{\phi}^\dagger\hat{\phi}\hat{\phi} + \hat{\phi}^\dagger\hat{\phi}\hat{\phi}^\dagger\hat{\phi} + \hat{\phi}^\dagger\hat{\phi}\hat{\phi}\hat{\phi}^\dagger + \hat{\phi}\hat{\phi}^\dagger\hat{\phi}^\dagger\hat{\phi} + \hat{\phi}\hat{\phi}^\dagger\hat{\phi}\hat{\phi}^\dagger + \hat{\phi}\hat{\phi}\hat{\phi}^\dagger\hat{\phi}^\dagger. \tag{3.550}$$

Upon the normal ordering, the perturbative Hamiltonian (3.544) for the electromagnetic field in a Kerr medium can be written as

$$\begin{aligned}\hat{H}(t) = &\int \hat{\phi}^\dagger(z,t)\hat{\phi}(z,t)\,\mathrm{d}z - \frac{\kappa}{2}\int \hat{\phi}^\dagger(z,t)\hat{\phi}^\dagger(z,t)\hat{\phi}(z,t)\hat{\phi}(z,t)\,\mathrm{d}z\\ &- \hbar\kappa Z\int \hat{\phi}^\dagger(z,t)\hat{\phi}(z,t)\,\mathrm{d}z,\end{aligned} \tag{3.551}$$

where $\kappa = \frac{3\chi}{4\epsilon^2}$. Z is a function we give only asymptotically (cf. Abram and Cohen 1994)

$$Z \underset{\Lambda\to\infty}{\simeq} \frac{v}{2\pi}\Lambda^2, \tag{3.552}$$

where Λ is a high-frequency cutoff. Whereas the first two terms in equation (3.551) are familiar, the third term, which is divergent, arises in the normal ordering procedure. For Λ fixed, this last term vanishes if $\hbar \to 0$.

In the renormalization procedure, a formal series (in $\hbar$) of "counterterms" is added to the Hamiltonian in order to remove the divergences that arise upon normally ordering the results of calculation (Itzykson and Zuber 1980). The Hamiltonian itself exemplifies that it is not sufficient for removing divergences, but at the same time renormalized parameters and renormalized field operators are introduced. In particular, a renormalized Kerr Hamiltonian $\hat{H}_{\mathrm{R}}(t)$ may be defined by introducing a counterterm of order $\hbar$ as follows:

$$\hat{H}_{\mathrm{R}}(t) = \hat{H}(t) + 2\hbar\kappa Z \int \hat{\phi}^{\dagger}(z,t)\hat{\phi}(z,t)\,\mathrm{d}z. \tag{3.553}$$

The third term in equation (3.551) exchanges the sign and for $\Lambda \to \infty$ it is an infinite change in the inverse of the refractive index. The renormalized field operators

$$\hat{\phi}_{\mathrm{R}}(z,t) = \sqrt{1+\hbar\kappa Z}\,\hat{\phi}(z,t), \tag{3.554}$$
$$\hat{\phi}_{\mathrm{R}}^{\dagger}(z,t) = \sqrt{1+\hbar\kappa Z}\,\hat{\phi}^{\dagger}(z,t) \tag{3.555}$$

are further quantities which, or at least whose Hermitian parts, etc. would relate to an experiment. Such a relationship is no more required from the bare quantities. At the same time, a renormalized refractive index is defined

$$n_{\mathrm{R}} = \frac{n}{1+\hbar\kappa Z}. \tag{3.556}$$

The renormalized Kerr Hamiltonian can be written in terms of the renormalized field operators as

$$\hat{H}_{\mathrm{R}}(t) = \hat{H}_{0\mathrm{R}}(t) + \hat{H}_{1\mathrm{S}+,\mathrm{R}}(t), \tag{3.557}$$

with

$$\hat{H}_{0\mathrm{R}}(t) = \int \hat{\phi}_{\mathrm{R}}^{\dagger}(z,t)\hat{\phi}_{\mathrm{R}}(z,t)\,\mathrm{d}z, \tag{3.558}$$

$$\hat{H}_{1\mathrm{S}+,\mathrm{R}}(t) = \sum_{j=0}^{\infty} \hbar^{j}\kappa^{j+1}\hat{H}_{1\mathrm{S}+,\mathrm{R}}^{(j)}(t), \tag{3.559}$$

where

$$\hat{H}_{1\mathrm{S}+,\mathrm{R}}^{(j)}(t) = \frac{(-1)^{j+1}}{2}(j+1)Z^{j} \int \hat{\phi}_{\mathrm{R}}^{\dagger}(z,t)\hat{\phi}_{\mathrm{R}}^{\dagger}(z,t)\hat{\phi}_{\mathrm{R}}(z,t)\hat{\phi}_{\mathrm{R}}(z,t)\,\mathrm{d}z; \tag{3.560}$$

here $\kappa \hat{H}^{(0)}_{1S+,R}(t)$ is the "usual" Kerr term and $\hbar^j \kappa^{j+1} \hat{H}^{(j)}_{1S+,R}(t)$ is the jth quantum correction.

Abram and Cohen (1994) have calculated the quantum noise properties of a coherent pulse undergoing self-phase modulation in the course of its propagation by eliminating the vacuum divergences through the renormalization procedure. The one-point averages were first determined. The detection of a light pulse by a balanced homodyne detector can be expressed in a moving frame through the measured quantum operator

$$\hat{M}_\theta(\eta) = \exp(i\theta)\sqrt{K_{LO}}F_{LO}(\bar{\eta} - \eta)\hat{\phi}(\eta) + \text{H.c.}, \tag{3.561}$$

where $F_{LO}(\eta)$ is the coherent amplitude of the local oscillator (LO) pulse peaked at $\eta = \bar{\eta}$ and θ is the phase difference between the local oscillator and signal pulses. In homodyne detection, the central frequency of the local oscillator is the same as that of the incident pulse, $K_{LO} = K$. This set-up measures simultaneously the expectation values of the operators $\hat{M}_0(\eta)$ and $\hat{M}_{\frac{\pi}{2}}(\eta)$. For simplicity, we have neglected the Kerr medium here, but it is included when we write $\hat{\phi}(\eta; z) = \hat{\phi}(\eta, t = \frac{z}{v})$ instead of $\hat{\phi}(\eta)$. The instantaneous intensity of the signal pulse peaked at $\eta = 0$ is introduced as

$$I(\eta) = vK\rho^2 \tilde{F}^*(\eta)\tilde{F}(\eta). \tag{3.562}$$

It can be obtained that the operator $\hat{\phi}(\eta; z)$ has the expectation value in the coherent state

$$\begin{aligned} F^*(\eta; z) &= \langle \rho F|\hat{\phi}(\eta; z)|\rho F\rangle \\ &= vK\rho F^*(\eta)\exp[-i\Theta(\eta; z)], \end{aligned} \tag{3.563}$$

where

$$\Theta(\eta; z) = \kappa v K z I(\eta) \tag{3.564}$$

is the nonlinear phase shift produced by self-phase modulation. The nonlinear phase Θ has exactly the same value as in classical nonlinear optics. The fact that equation (3.563) obtained by invoking renormalization corresponds directly to what is observed experimentally in a propagative configuration underlines the validity of this approach.

Two-point correlation functions were examined, in order to obtain the quantum noise spectrum

$$S_0(k) = \int\int \exp[ik(\eta_1 - \eta_2)]\langle \rho F|\Delta\hat{M}_\theta(\eta_1)\Delta\hat{M}_\theta(\eta_2)|\rho F\rangle \, d\eta_2 \, d\eta_1, \tag{3.565}$$

where

$$\Delta\hat{M}_\theta(\eta) = \hat{M}_\theta(\eta) - \langle \rho F|\hat{M}_\theta(\eta)|\rho F\rangle \tag{3.566}$$

and k is the spatial frequency at which the quantum noise is measured. In experiment, such a noise is detected at frequencies several orders of magnitude below the carrier optical frequency and its spectrum is considered to be flat throughout the typical bandwidth. The low-frequency noise spectrum $S_\theta(0)$ can be decomposed into four terms

$$S_\theta(k) \approx S_\theta(0) = S_1 + S_2 + \exp(-2\mathrm{i}\theta)S_3 + \exp(2\mathrm{i}\theta)S_4, \tag{3.567}$$

with

$$S_1 = \int I_{\mathrm{LO}}(\bar{\eta} - \eta)\Theta^2(\eta; z)\,\mathrm{d}\eta, \tag{3.568}$$

$$S_2 = \int I_{\mathrm{LO}}(\bar{\eta} - \eta)[1 + \Theta^2(\eta; z)]\,\mathrm{d}\eta, \tag{3.569}$$

$$S_3 = \int I_{\mathrm{LO}}(\bar{\eta} - \eta)[\mathrm{i}\Theta(\eta; z)][1 + \mathrm{i}\Theta(\eta; z)]\exp[2\mathrm{i}\Theta(\eta; z)]\,\mathrm{d}\eta, \tag{3.570}$$

$$S_4 = \int I_{\mathrm{LO}}(\bar{\eta} - \eta)[-\mathrm{i}\Theta(\eta; z)][1 - \mathrm{i}\Theta(\eta; z)]\exp[-2\mathrm{i}\Theta(\eta; z)]\,\mathrm{d}\eta, \tag{3.571}$$

where $I_{\mathrm{LO}}(\bar{\eta} - \eta)$ is the local oscillator instantaneous intensity of the local oscillator pulse. This result is similar to that obtained by linearizing the self-phase modulation exponential operator $\exp(\mathrm{i}\gamma\hat{a}_{k0}^\dagger\hat{a}_{k0})$ around the mean field (Shirasaki and Haus 1990).

It is appropriate to give a physical interpretation of the above results and to discuss the case of squeezing that can be observed in the propagation of a coherent pulse. For narrow-bandwidth signals, the phase properties of quantum noise in equation (3.567) can be visualized by examining the field fluctuations. First, the quantum characteristic function is defined as

$$C(u, v) = \langle\alpha|\exp\left\{\mathrm{i}\int[u\hat{M}_{\frac{\pi}{2}}(\eta) + v\hat{M}_0(\eta)]\right\}|\alpha\rangle, \tag{3.572}$$

where $|\alpha\rangle$ is the continuous-wave coherent state. Using the linked cluster theorem, the characteristic function (3.572) can be written in terms of connected averages. Two lowest order connected averages are feasible and describe the moments of the Wigner distribution

$$W(q, p) = \frac{1}{4\pi^2}\int\int C(u, v)\exp[-\mathrm{i}(pu + qv)]\,\mathrm{d}u\,\mathrm{d}v. \tag{3.573}$$

According to equation (3.563), the expectation value of the Wigner distribution is

$$q_0 + \mathrm{i}p_0 = \sqrt{I}e^{\mathrm{i}\theta}, \tag{3.574}$$

while the principal squeeze variances of the quantum noise are

$$2(B_S \mp |C|) = \frac{1}{\sqrt{1+\Theta^2} \pm \Theta} \tag{3.575}$$

$$= \sqrt{1+\Theta^2} \mp \Theta, \tag{3.576}$$

where $\Theta \equiv \Theta(\eta; z)$ and we have used the characteristics of quantum noise (Peřinová et al. 1991)

$$B_S = \frac{1}{2}\sqrt{1+\Theta^2}, \quad |C| = \frac{1}{2}\Theta. \tag{3.577}$$

Moreover,

$$\arg C = 2\Theta + \arctan(\Theta) - \frac{3\pi}{2}. \tag{3.578}$$

In the Abram–Cohen theory, for the case of a coherent beam of central frequency 2×10^{15} Hz, bandwidth 100 GHz, and intensity 1 W propagating in a silica fibre, the Gaussian noise is a good approximation up to nonlinear phase shifts of the order 10^3 rad.

When the phase of the local oscillator is constant along the pulse profile, particularly when the principal quadrature is measured at the peak of the pulse, the variance will not be the same everywhere, the quadrature will not always be the principal quadrature due to the chirp. To circumvent this problem, the use of a matched local oscillator has been proposed such that its phase $\Theta(\eta)$ varies in a way that matches the signal chirp. Besides the Kerr effect, the Sagnac interferometer can be used for this purpose (Shirasaki and Haus 1990, Blow et al. 1992, Bergman and Haus 1991). Alternatively, a local oscillator pulse that is much shorter than the signal pulse can sample only the central portion of the signal in order to measure the appropriate squeezing.

Bespalov et al. (2002) have investigated the propagation of light in the $(1 + 1)$-dimensional approximation. They have paid attention to the two series expansions of the index of refraction of an isotropic optical medium in Born and Wolf (1968). On these expansions they have based two wave equations, both with and without the second space derivative term. They have presented a method to derive the nonlinear wave equations suitable for describing dynamics of extremely short pulses. Although this analysis is completely classical, it does not exclude that a quantization of the field and of the equations will be necessary in the near future.

Restriction to the transverse components and, finally, to the scalar wave equation is common. Lu et al. (2003) have studied the propagation of ultrashort pulsed beam beyond the paraxial approximation in free space. The nonparaxial corrections to an arbitrary paraxial solution are given in a series form. A comparison with rigorous nonparaxial results obtained by numerical method is carried out. Spatial and temporal distributions are considered.

In this book the "fully" relativistic quantum electrodynamics is not treated. Nevertheless, its importance for quantum optics is going to be appreciated in the nearest future. Shukla et al. (2004) have considered the nonlinear propagation of randomly distributed intense short photon pulses in a photon gas. Fragmentation of incoherent photon pulses in astrophysical contexts and in forthcoming experiments using very intense short laser pulses has been predicted.

The renormalization and the Bogoliubov renormalization group are different concepts (Shirkov and Kovalev 2001). Kovalev et al. (2000) and Tatarinova and Garcia (2007) have expounded the renormalization-group approach to the problem of light-beam self-focusing.

Tatarinova and Garcia (2007) set the problem in the framework of the classical nonlinear optics and so the renormalization is not needed. The propagation of a laser beam of intensity I in a nonlinear medium with a refractive index $n_0 + n(I)$, where n_0 is the linear refractive index, $n(I)$ is such that $n(0) = 0$, arbitrary in other respects, is studied. The case of nonlinear self-focusing accompanied by multiphoton ionization has been explicitly analysed. The procedure of analytical solution begins with an approximate transformation of the nonlinear Schrödinger equation onto eikonal equations. Irrespective of these and some other approximations, the appropriate, easy, calculation provides results which are in good agreement with numerical simulations.

3.5 Quasimode Theory

Glauber and Lewenstein (1991) have developed quantum optics of inhomogeneous media with linear susceptibilities. The topics treated have included the normal-mode expansion and the plane-wave expansion. The authors have shown that plane-wave photons can be related to the normal-mode ones within the framework of scattering theory. They have used the quantization schemes discussed to determine the fluctuation properties of various field components. They have considered excited atoms and changes in the spontaneous emission rates for both electric and magnetic dipole transitions of the atoms within or near dielectric media. They have provided a quantum description of the transition radiation emitted by a charged particle in passing from one dielectric medium to another.

Dalton et al. (1996) have carried out canonical quantization of the electromagnetic field and radiative atoms in passive, lossless, dispersionless, and linear dielectric media. The quantum Hamiltonian has been derived in a generalized multipole form. Dalton et al. (1999b) have presented a macroscopic canonical quantization of the electromagnetic field and radiating atom system in dielectric media based on expanding the vector potential in terms of quasimode functions. The quasimode functions approximate the true mode functions of a classical optics device when they are obtained on the assumption of an ideal electric permittivity function and the permittivity function describing the device does not deviate much from the ideal one. Plane waves in Glauber and Lewenstein (1991) are such quasimodes. In the coupled-mode theory (see, e.g. Section 6.2) the "ideal" waveguide modes are also such quasimodes.

Here we present part of the theory (Dalton et al. 1999b), which will be completed below. It is assumed that a classical linear optics device is described with the spatially dependent electric permittivity $\epsilon(\mathbf{R})$ (and the magnetic permeability $\mu(\mathbf{R})$). The generalized Coulomb gauge condition for the vector potential $\mathbf{A}(\mathbf{R}, t)$

$$\nabla \cdot [\epsilon(\mathbf{R})\mathbf{A}(\mathbf{R}, t)] = 0 \tag{3.579}$$

is used. In a generalization of Helmholtz's theorem (Dalton and Babiker 1997), a vector field $\mathbf{F}(\mathbf{R}, t)$ can be decomposed uniquely in the generalized transverse and longitudinal components $\mathbf{F}_{\perp}^{(\epsilon)}(\mathbf{R}, t)$, $\mathbf{F}_{\parallel}^{(\epsilon)}(\mathbf{R}, t)$ in the form

$$\mathbf{F}(\mathbf{R}, t) = \mathbf{F}_{\perp}^{(\epsilon)}(\mathbf{R}, t) + \mathbf{F}_{\parallel}^{(\epsilon)}(\mathbf{R}, t), \tag{3.580}$$

with

$$\nabla \cdot [\epsilon(\mathbf{R})\mathbf{F}_{\perp}^{(\epsilon)}(\mathbf{R}, t)] = 0, \quad \nabla \times \mathbf{F}_{\parallel}^{(\epsilon)}(\mathbf{R}, t) = 0. \tag{3.581}$$

The macroscopic Lagrangian is given by the relation

$$L'(t) = \int \mathcal{L}_{\mathrm{c}}'(\mathbf{R}, t)\, \mathrm{d}^3\mathbf{R}, \tag{3.582}$$

where the Lagrangian density $\mathcal{L}_{\mathrm{c}}'(\mathbf{R}, t)$ is given by the relation

$$\mathcal{L}_{\mathrm{c}}'(\mathbf{R}, t) = \frac{1}{2}\epsilon(\mathbf{R})\left[\frac{\partial \mathbf{A}(\mathbf{R}, t)}{\partial t}\right]^2 - \frac{1}{2\mu(\mathbf{R})}[\nabla \times \mathbf{A}(\mathbf{R}, t)]^2. \tag{3.583}$$

The conjugate momentum field $\mathbf{\Pi}(\mathbf{R}, t)$ is obtained from the Lagrangian density $\mathcal{L}_{\mathrm{c}}'(\mathbf{R}, t)$ as

$$\mathbf{\Pi}(\mathbf{R}, t) = \epsilon(\mathbf{R})\frac{\partial \mathbf{A}(\mathbf{R}, t)}{\partial t}. \tag{3.584}$$

The Hamiltonian is

$$H'(t) = \int \left\{ \frac{[\mathbf{\Pi}(\mathbf{R}, t)]^2}{2\epsilon(\mathbf{R})} + \frac{[\nabla \times \mathbf{A}(\mathbf{R}, t)]^2}{2\mu(\mathbf{R})} \right\} \mathrm{d}^3\mathbf{R}. \tag{3.585}$$

We have the electric displacement field $\mathbf{D}(\mathbf{R}, t)$,

$$\mathbf{D}(\mathbf{R}, t) = -\mathbf{\Pi}(\mathbf{R}, t). \tag{3.586}$$

The true mode functions satisfy the generalized Helmholtz equation

$$\nabla \times \frac{1}{\mu(\mathbf{R})}[\nabla \times \mathbf{A}_k(\mathbf{R})] = \omega_k^2 \epsilon(\mathbf{R})\mathbf{A}_k(\mathbf{R}), \tag{3.587}$$

where ω_k are real and positive angular frequencies and satisfy the generalized Coulomb gauge condition

$$\nabla \cdot [\epsilon(\mathbf{R})\mathbf{A}_k(\mathbf{R})] = 0. \tag{3.588}$$

The true mode functions satisfy the orthogonality and normalization conditions respecting $\epsilon(\mathbf{R})$ as a weight function,

$$\int \epsilon(\mathbf{R})\mathbf{A}_k^*(\mathbf{R}) \cdot \mathbf{A}_l(\mathbf{R})\, \mathrm{d}^3\mathbf{R} = \delta_{kl}. \tag{3.589}$$

Generalized coordinates $q_k(t)$ and generalized momenta $p_k(t)$ can be introduced by the relations

$$q_k(t) = \int \epsilon(\mathbf{R})\mathbf{A}_k^*(\mathbf{R}) \cdot \mathbf{A}(\mathbf{R}, t)\, \mathrm{d}^3\mathbf{R}, \tag{3.590}$$

$$p_k(t) = \int \mathbf{A}_k^*(\mathbf{R}) \cdot \mathbf{\Pi}(\mathbf{R}, t)\, \mathrm{d}^3\mathbf{R}. \tag{3.591}$$

These variables are complex. Expansions of the vector potential and conjugate momentum field in terms of the true modes $\mathbf{A}_k(\mathbf{R})$ are

$$\mathbf{A}(\mathbf{R}, t) = \sum_k q_k(t)\mathbf{A}_k(\mathbf{R}), \tag{3.592}$$

$$\mathbf{\Pi}(\mathbf{R}, t) = \sum_k p_k(t)\epsilon(\mathbf{R})\mathbf{A}_k(\mathbf{R}). \tag{3.593}$$

It is assumed that the exact electric permittivity and magnetic permeability functions do not deviate much from artificially chosen functions $\tilde{\epsilon}(\mathbf{R})$, $\tilde{\mu}(\mathbf{R})$. It is assumed that these functions produce quasimode functions $\mathbf{U}_\alpha(\mathbf{R})$, idealized versions of the true mode functions $\mathbf{A}_k(\mathbf{R})$. Let λ_α denote the angular frequency of the quasimode. One has equations

$$\nabla \times \frac{1}{\tilde{\mu}(\mathbf{R})}[\nabla \times \mathbf{U}_\alpha(\mathbf{R})] = \lambda_\alpha^2 \tilde{\epsilon}(\mathbf{R})\mathbf{U}_\alpha(\mathbf{R}), \tag{3.594}$$

$$\nabla \cdot [\tilde{\epsilon}(\mathbf{R})\mathbf{U}_\alpha(\mathbf{R})] = 0, \tag{3.595}$$

$$\int \tilde{\epsilon}(\mathbf{R})\mathbf{U}_\alpha^*(\mathbf{R}) \cdot \mathbf{U}_\beta(\mathbf{R})\, \mathrm{d}^3\mathbf{R} = \delta_{\alpha\beta}, \tag{3.596}$$

which are the generalized Helmholtz equation, the gauge conditions, and orthonormality conditions, respectively.

As the vector potential $\mathbf{A}(\mathbf{R}, t)$ satisfies the generalized Coulomb gauge condition (3.579), the field $\frac{\epsilon(\mathbf{R})}{\tilde{\epsilon}(\mathbf{R})}\mathbf{A}(\mathbf{R}, t)$ fulfils the generalized Coulomb gauge condition

$$\nabla \cdot [\tilde{\epsilon}(\mathbf{R})\mathbf{F}(\mathbf{R}, t)] = 0, \tag{3.597}$$

where $\mathbf{F}(\mathbf{R}, t)$, any field, can be chosen in the form $\frac{\epsilon(\mathbf{R})}{\tilde{\epsilon}(\mathbf{R})}\mathbf{A}(\mathbf{R}, t)$. Another choice is based on the gauge transformation

$$\tilde{\mathbf{A}}(\mathbf{R}, t) = \mathbf{A}(\mathbf{R}, t) - \nabla\psi(\mathbf{R}, t), \tag{3.598}$$

where $\psi(\mathbf{R}, t)$, an arbitrary function of $\mathbf{R}$ and t, must be specified (Glauber and Lewenstein 1991), and the scalar potential

$$\tilde{\Phi}(\mathbf{R}, t) = \dot{\psi}(\mathbf{R}, t) \tag{3.599}$$

need not be considered. Therefore, the expansion

$$\frac{\epsilon(\mathbf{R})}{\tilde{\epsilon}(\mathbf{R})}\mathbf{A}(\mathbf{R}, t) = \sum_{\alpha,\beta} Q_\alpha(t) K_{\alpha\beta} \mathbf{U}_\beta(\mathbf{R}) \tag{3.600}$$

exists, where the complicated form of the coefficients $\sum_\alpha Q_\alpha(t) K_{\alpha\beta}$ may and may not involve

$$Q_\alpha(t) = \int \tilde{\epsilon}(\mathbf{R}) \mathbf{U}_\alpha^*(\mathbf{R}) \cdot \tilde{\mathbf{A}}(\mathbf{R}, t)\, \mathrm{d}^3\mathbf{R} \tag{3.601}$$

and matrix elements $K_{\alpha\beta}$ of a suitable matrix $\mathbf{K}$. The vector potential is given as

$$\mathbf{A}(\mathbf{R}, t) = \sum_{\alpha,\beta} Q_\alpha(t) K_{\alpha\beta} \frac{\tilde{\epsilon}(\mathbf{R})}{\epsilon(\mathbf{R})} \mathbf{U}_\beta(\mathbf{R}). \tag{3.602}$$

Using expansion (3.602), one can write the Lagrangian (3.582), (3.583) as

$$L'(t) = \frac{1}{2} \sum_{\alpha,\beta} \dot{Q}_\alpha^*(t) (\mathbf{W}^{-1})_{\alpha\beta} \dot{Q}_\beta(t) - \frac{1}{2} \sum_{\alpha,\beta} Q_\alpha^*(t) V_{\alpha\beta} Q_\beta(t), \tag{3.603}$$

where

$$\mathbf{W} = (\mathbf{K}^T)^{-1} \mathbf{M}^{-1} (\mathbf{K}^*)^{-1}, \tag{3.604}$$

$$\mathbf{V} = \mathbf{K}^* \mathbf{H} \mathbf{K}^T, \tag{3.605}$$

and

$$M_{\alpha\beta} = \int \epsilon(\mathbf{R}) \frac{\tilde{\epsilon}(\mathbf{R})}{\epsilon(\mathbf{R})} \mathbf{U}_\alpha^*(\mathbf{R}) \cdot \frac{\tilde{\epsilon}(\mathbf{R})}{\epsilon(\mathbf{R})} \mathbf{U}_\beta(\mathbf{R})\, \mathrm{d}^3\mathbf{R}, \tag{3.606}$$

$$H_{\alpha\beta} = \int \frac{1}{\mu(\mathbf{R})} \left[\nabla \times \frac{\tilde{\epsilon}(\mathbf{R})}{\epsilon(\mathbf{R})} \mathbf{U}_\alpha^*(\mathbf{R})\right] \cdot \left[\nabla \times \frac{\tilde{\epsilon}(\mathbf{R})}{\epsilon(\mathbf{R})} \mathbf{U}_\beta(\mathbf{R})\right] \mathrm{d}^3\mathbf{R}. \tag{3.607}$$

Making a specific choice for $\mathbf{K}$ (Dalton et al. 1999b)

$$\mathbf{K} = (\mathbf{M}^*)^{-1}, \tag{3.608}$$

one has

$$\mathbf{W} = \mathbf{M}, \tag{3.609}$$

$$\mathbf{V} = \mathbf{M}^{-1}\mathbf{H}\mathbf{M}^{-1}. \tag{3.610}$$

The generalized momentum coordinates $P_\alpha(t)$ for the electromagnetic field are

$$P_\alpha(t) = \sum_\beta (\mathbf{M}^{-1})_{\alpha\beta}\dot{Q}_\beta(t). \tag{3.611}$$

The Hamiltonian is given by the relation

$$H'(t) = \frac{1}{2}\sum_{\alpha,\beta} P_\alpha^*(t)W_{\alpha\beta}P_\beta(t) + \frac{1}{2}\sum_{\alpha,\beta} Q_\alpha^*(t)V_{\alpha\beta}Q_\beta(t). \tag{3.612}$$

As the same Lagrangian is used and definition (3.584) does not depend on the gauge condition (3.588), the conjugate momentum field $\mathbf{\Pi}(\mathbf{R}, t)$ is still expressed by equation (3.584). As from (3.602) it follows that

$$\mathbf{A}(\mathbf{R}, t) = \sum_{\alpha,\beta} Q_\alpha(t)(\mathbf{M}^{-1})_{\beta\alpha}\frac{\tilde{\epsilon}(\mathbf{R})}{\epsilon(\mathbf{R})}\mathbf{U}_\beta(\mathbf{R}), \tag{3.613}$$

respecting (3.584) and exchanging $\alpha \leftrightarrow \beta$, we obtain that

$$\mathbf{\Pi}(\mathbf{R}, t) = \sum_\alpha P_\alpha(t)\tilde{\epsilon}(\mathbf{R})\mathbf{U}_\alpha(\mathbf{R}). \tag{3.614}$$

The classical generalized coordinates $Q_\alpha(t)$ and generalized momenta $P_\alpha(t)$ are replaced by quantum operators according to the prescriptions

$$Q_\alpha(t) \to \hat{Q}_\alpha(t),\ Q_\alpha^*(t) \to \hat{Q}_\alpha^\dagger(t), \tag{3.615}$$

$$P_\alpha(t) \to \hat{P}_\alpha(t),\ P_\alpha^*(t) \to \hat{P}_\alpha^\dagger(t). \tag{3.616}$$

The nonzero equal-time commutators are

$$[\hat{Q}_\alpha(t), \hat{P}_\beta^\dagger(t)] = \mathrm{i}\hbar\delta_{\alpha\beta}\hat{1} = [\hat{Q}_\alpha^\dagger(t), \hat{P}_\beta(t)]. \tag{3.617}$$

The vector potential $\mathbf{A}(\mathbf{R}, t)$ and conjugate momentum field $\mathbf{\Pi}(\mathbf{R}, t)$ now become field operators $\hat{\mathbf{A}}(\mathbf{R}, t)$, $\hat{\mathbf{\Pi}}(\mathbf{R}, t)$. Let us imagine the replacements (3.615) ((3.616)) in the expression (3.602) ((3.614)) when the quantum operator $\hat{\mathbf{A}}(\mathbf{R}, t)$ ($\hat{\mathbf{\Pi}}(\mathbf{R}, t)$) is introduced. Similarly, the Hamiltonian given through equation (3.612) now becomes a quantum Hamiltonian $\hat{H}'(t)$.

It has the form $\hat{H}'(t) = \hat{H}_Q(t) + \hat{V}_{Q-Q}(t)$, where

$$\hat{H}_Q(t) = \frac{1}{2}\sum_{\alpha}\left[\hat{P}_\alpha^\dagger(t)W_{\alpha\alpha}\hat{P}_\alpha(t) + \hat{Q}_\alpha^\dagger(t)V_{\alpha\alpha}\hat{Q}_\alpha(t)\right], \tag{3.618}$$

$$\hat{V}_{Q-Q}(t) = \frac{1}{2}\sum_{\substack{\alpha,\beta \\ \alpha\neq\beta}}\left[\hat{P}_\alpha^\dagger(t)W_{\alpha\beta}\hat{P}_\beta(t) + \hat{Q}_\alpha^\dagger(t)V_{\alpha\beta}\hat{Q}_\beta(t)\right]. \tag{3.619}$$

Dalton et al. (1999b) have defined an effective quasimode angular frequency μ_α as follows:

$$\mu_\alpha = \sqrt{W_{\alpha\alpha}V_{\alpha\alpha}}. \tag{3.620}$$

It may differ from λ_α in (3.594). As usual with quantum harmonic oscillators, annihilation and creation operators for each of the quasimodes are introduced as usual linear combinations

$$\hat{A}_\alpha(t) = \sqrt{\frac{\eta_\alpha}{2\hbar}}\hat{Q}_\alpha(t) + \mathrm{i}\sqrt{\frac{1}{2\hbar\eta_\alpha}}\hat{P}_\alpha(t), \tag{3.621}$$

$$\hat{A}_\alpha^\dagger(t) = \sqrt{\frac{\eta_\alpha}{2\hbar}}\hat{Q}_\alpha^\dagger(t) - \mathrm{i}\sqrt{\frac{1}{2\hbar\eta_\alpha}}\hat{P}_\alpha^\dagger(t), \tag{3.622}$$

where

$$\eta_\alpha = \sqrt{\frac{V_{\alpha\alpha}}{W_{\alpha\alpha}}}. \tag{3.623}$$

The nonzero equal-time commutators are standard, $[\hat{A}_\alpha(t), \hat{A}_\beta^\dagger(t)] = \delta_{\alpha\beta}\hat{1}$. Definitions (3.621) and (3.622) can be completed with the equations for $\hat{A}_{-\alpha}(t)$, $\hat{A}_{-\alpha}^\dagger(t)$, where $-\alpha$ denotes that the operators are associated with the quasimode function $\hat{\mathbf{U}}_\alpha^*(\mathbf{R})$.

The relationship between the annihilation, creation operators $\hat{a}_k$, $\hat{a}_k^\dagger$ for the true modes (see Dalton et al. (1996)) and the quantities just introduced can be obtained from the expansions of two sets of functions $\epsilon(\mathbf{R})\mathbf{A}_k$ and $\tilde{\epsilon}(\mathbf{R})\mathbf{U}_\alpha(\mathbf{R})$ in terms of the other. It is shown to involve a Bogoliubov transformation (Dalton et al. (1999c).

From Equations (3.621) and (3.622) one expresses the generalized coordinates and momenta $\hat{Q}_\alpha(t)$, $\hat{P}_\alpha(t)$ as follows:

$$\hat{Q}_\alpha(t) = \sqrt{\frac{\hbar}{2\eta_\alpha}}\left[\hat{A}_\alpha(t) + \hat{A}_{-\alpha}^\dagger(t)\right], \tag{3.624}$$

$$\hat{P}_\alpha(t) = \frac{1}{\mathrm{i}}\sqrt{\frac{\hbar\eta_\alpha}{2}}\left[\hat{A}_\alpha(t) - \hat{A}^\dagger_{-\alpha}(t)\right]. \tag{3.625}$$

On substituting (3.624) and (3.625) into the Hamiltonians (3.618) and (3.619), the Hamiltonians become

$$\hat{H}_Q(t) = \hbar\sum_\alpha\left[\hat{A}^\dagger_\alpha(t)\hat{A}_\alpha(t) + \frac{1}{2}\hat{1}\right]\mu_\alpha, \tag{3.626}$$

$$\hat{V}_{Q-Q}(t) = \hat{V}^{\mathrm{RWA}}_{Q-Q}(t) + \hat{V}^{\mathrm{non-RWA}}_{Q-Q}(t), \tag{3.627}$$

where $\hat{V}^{\mathrm{RWA}}_{Q-Q}(t)$ means the rotating-wave contribution,

$$\hat{V}^{\mathrm{RWA}}_{Q-Q}(t) = \frac{\hbar}{2}\sum_{\substack{\alpha,\beta\\ \alpha\neq\beta}}\left(\sqrt{\eta_\alpha\eta_\beta}M_{\alpha\beta} + \frac{V_{\alpha\beta}}{\sqrt{\eta_\alpha\eta_\beta}}\right)\hat{A}^\dagger_\alpha(t)\hat{A}_\beta(t), \tag{3.628}$$

and $\hat{V}^{\mathrm{non-RWA}}_{Q-Q}(t)$ stands for the nonrotating wave correction term,

$$\hat{V}^{\mathrm{non-RWA}}_{Q-Q}(t) = \frac{\hbar}{4}\sum_{\substack{\alpha,\beta\\ \alpha\neq\beta}}\left(-\sqrt{\eta_\alpha\eta_\beta}M_{\alpha,-\beta} + \frac{V_{\alpha,-\beta}}{\sqrt{\eta_\alpha\eta_\beta}}\right)\hat{A}^\dagger_\alpha(t)\hat{A}^\dagger_\beta(t) + \mathrm{H.c.} \tag{3.629}$$

Similarly, the field operators $\hat{\mathbf{A}}(\mathbf{R},t)$ and $\hat{\mathbf{\Pi}}(\mathbf{R},t)$ should be expressed in terms of annihilation and creation operators. One finds that

$$\hat{\mathbf{A}}(\mathbf{R},t) = \sum_{\alpha,\beta}\sqrt{\frac{\hbar}{2\eta_\alpha}\frac{\tilde{\epsilon}(\mathbf{R})}{\epsilon(\mathbf{R})}}\left[K_{\alpha\beta}\hat{A}_\alpha(t)\mathbf{U}_\beta(\mathbf{R}) + K^*_{\alpha\beta}\hat{A}^\dagger_\alpha(t)\mathbf{U}^*_\beta(\mathbf{R})\right], \tag{3.630}$$

$$\hat{\mathbf{\Pi}}(\mathbf{R},t) = \sum_\alpha\frac{1}{\mathrm{i}}\sqrt{\frac{\hbar\eta_\alpha}{2}}\tilde{\epsilon}(\mathbf{R})\left[\hat{A}_\alpha(t)\mathbf{U}_\alpha(\mathbf{R}) - \hat{A}^\dagger_\alpha(t)\mathbf{U}^*_\alpha(\mathbf{R})\right]. \tag{3.631}$$

The quantum Hamiltonian in the rotating wave approximation is $\hat{H}'_{\mathrm{RWA}}(t) = \hat{H}_Q(t) + \hat{V}^{\mathrm{RWA}}_{Q-Q}(t)$. Dalton et al. (1999b) have considered also an approximate form of this Hamiltonian. A simplification is related to the approximation

$$\mu_\alpha \approx \lambda_\alpha \tag{3.632}$$

or

$$\mu_\alpha \approx \lambda_\alpha + v_{\alpha\alpha}, \tag{3.633}$$

where we have added a term. Further,

$$\hat{V}^{\mathrm{RWA}}_{Q-Q}(t) \approx \hbar\sum_{\substack{\alpha,\beta\\ \alpha\neq\beta}} v_{\alpha\beta}\hat{A}^\dagger_\alpha(t)\hat{A}_\beta(t), \tag{3.634}$$

where

$$v_{\alpha\beta} = \frac{1}{2}\left(\sqrt{\lambda_\alpha \lambda_\beta}(M_1)_{\alpha\beta} + \frac{(H_1)_{\alpha\beta}}{\sqrt{\lambda_\alpha \lambda_\beta}}\right), \tag{3.635}$$

$$(M_1)_{\alpha\beta} = \int \tilde{\epsilon}(\mathbf{R})\left[\frac{\tilde{\epsilon}(\mathbf{R})}{\epsilon(\mathbf{R})} - 1\right]^2 \mathbf{U}_\alpha^*(\mathbf{R}) \cdot \mathbf{U}_\beta(\mathbf{R})\,\mathrm{d}^3\mathbf{R}, \tag{3.636}$$

$$(H_1)_{\alpha\beta} = \int \frac{1}{\mu_0}\left\{\nabla \times \left[\frac{\tilde{\epsilon}(\mathbf{R})}{\epsilon(\mathbf{R})} - 1\right]\mathbf{U}_\alpha^*(\mathbf{R})\right\} \cdot \left\{\nabla \times \left[\frac{\tilde{\epsilon}(\mathbf{R})}{\epsilon(\mathbf{R})} - 1\right]\mathbf{U}_\beta(\mathbf{R})\right\}\,\mathrm{d}^3\mathbf{R}. \tag{3.637}$$

3.5.1 Relation to Quantum Scattering Theory

The phenomenon of scattering occurs in various situations in optics. In the classical particle mechanics, the scattering theory is a natural continuation and generalization of the analysis of collisions. It must be and has been reconstructed for wave functions in quantum mechanics. So the formulation in the Schrödinger picture is appropriate. The methods developed apply also to optical and acoustic scattering in classical physics (Reed and Simon 1979).

In quantum field theory, the scattering theory resembles its simplified form for quantum mechanics, but with wave functions replaced by field operators. Accordingly, it is formulated in the Heisenberg picture. In spite of simplifications this graduation is present in quantum optics. First we will review basics of the Schrödinger picture approach to the scattering theory and then outline the Heisenberg picture approach to this theory (Dalton et al. 1999a).

The single-channel scattering theory may be adequate for many applications in quantum optics. In this theory the Hamiltonian $\hat{H}(t, t')$ is written as a sum of an unperturbed Hamiltonian $\hat{H}_0(t, t')$ and an interaction term $\hat{V}(t, t')$, $t' = 0, t$. We should denote more exactly $\hat{H}(t, t)$ in the Heisenberg picture and $\hat{H}(t, 0)$ in the Schrödinger picture. The first variable denotes the explicit time dependence of the Hamiltonian. For simplicity it is assumed that $\hat{H}_0(t, t')$, $\hat{V}(t, t')$, and $\hat{H}(t, t')$ are t-independent and the notation $\hat{H}_0(t')$, $\hat{V}(t')$, and $\hat{H}(t')$, $t' = 0, t$, is used.

The state vector $|\psi(t)\rangle$ evolves as

$$|\psi(t)\rangle = \hat{U}(t)|\psi(0)\rangle, \tag{3.638}$$

where the evolution operator $\hat{U}(t)$ given by

$$\hat{U}(t) = \exp\left[-\frac{\mathrm{i}}{\hbar}\hat{H}(0)t\right] \tag{3.639}$$

is unitary.

When a scattering experiment is described, the state vector $|\psi(t)\rangle$ should approach freely evolving state vectors as $t \to \pm\infty$, which are based on the so-called input states $|\psi_{\rm in}\rangle$ and output states $|\psi_{\rm out}\rangle$. So with the unitary free evolution operator $\hat{U}_0(t)$ given by

$$\hat{U}_0(t) = \exp\left[-\frac{\rm i}{\hbar}\hat{H}_0(0)t\right], \tag{3.640}$$

we have the so-called asymptotic conditions (Taylor 1972, Newton 1966)

$$\left.\begin{array}{l}\hat{U}_0(t)|\psi_{\rm in}\rangle \\ \hat{U}_0(t)|\psi_{\rm out}\rangle\end{array}\right\} - |\psi(t)\rangle \to 0 \text{ as } t \to \left\{\begin{array}{l}-\infty, \\ +\infty.\end{array}\right. \tag{3.641}$$

The conditions (Taylor 1972, Newton 1966) that are sufficient for the asymptotic conditions to hold are that ($\|\cdots\|$ are the norms of the state vectors)

$$\int_{-\infty}^{0} \|\hat{V}\hat{U}_0(\tau)|\psi_{\rm in}\rangle\|\, d\tau < \infty \text{ for a dense set of } |\psi_{\rm in}\rangle, \tag{3.642}$$

$$\int_{0}^{\infty} \|\hat{V}\hat{U}_0(\tau)|\psi_{\rm out}\rangle\|\, d\tau < \infty \text{ for a dense set of } |\psi_{\rm out}\rangle. \tag{3.643}$$

If the asymptotic conditions hold, then the Møller wave operators $\hat{\Omega}_\pm$ exist which map $|\psi_{\rm in}\rangle$ and $|\psi_{\rm out}\rangle$ onto the state vector at $t = 0$:

$$\begin{aligned}|\psi(0)\rangle &= \hat{\Omega}_+|\psi_{\rm in}\rangle \\ &= \hat{\Omega}_-|\psi_{\rm out}\rangle\end{aligned} \tag{3.644}$$

and are defined through the relation

$$\begin{aligned}\hat{\Omega}_+ &= \lim_{t\to-\infty}[\hat{U}^\dagger(t)\hat{U}_0(t)] \\ &= \lim_{t\to-\infty}\left\{\exp\left[\frac{\rm i}{\hbar}\hat{H}(0)t\right]\exp\left[-\frac{\rm i}{\hbar}\hat{H}_0(0)t\right]\right\}\end{aligned} \tag{3.645}$$

and

$$\begin{aligned}\hat{\Omega}_- &= \lim_{t\to+\infty}[\hat{U}^\dagger(t)\hat{U}_0(t)] \\ &= \lim_{t\to+\infty}\left\{\exp\left[\frac{\rm i}{\hbar}\hat{H}(0)t\right]\exp\left[-\frac{\rm i}{\hbar}\hat{H}_0(0)t\right]\right\}.\end{aligned} \tag{3.646}$$

The Møller wave operators are isometric. They satisfy ($\hat{\Omega} = \hat{\Omega}_+$ or $\hat{\Omega}_-$) the relation

$$\hat{\Omega}^\dagger\hat{\Omega} = \hat{1}, \tag{3.647}$$

but may not satisfy $\hat{\Omega}\hat{\Omega}^\dagger = \hat{1}$ (see below). So they may not be unitary.

The scattering operator $\hat{S}$ maps the input vector $|\psi_{\text{in}}\rangle$ onto the output vector $|\psi_{\text{out}}\rangle$,

$$|\psi_{\text{out}}\rangle = \hat{S}|\psi_{\text{in}}\rangle, \tag{3.648}$$

and from equations (3.644) and (3.647) it is obvious that it involves two Møller operators

$$\hat{S} = \hat{\Omega}_{-}^{\dagger}\hat{\Omega}_{+}. \tag{3.649}$$

It could be easily verified that the scattering operator $\hat{S}$ is unitary,

$$\hat{S}^{\dagger}\hat{S} = \hat{S}\hat{S}^{\dagger} = \hat{1}. \tag{3.650}$$

Mapping (3.648) can be inverted,

$$|\psi_{\text{in}}\rangle = \hat{S}^{\dagger}|\psi_{\text{out}}\rangle. \tag{3.651}$$

The Møller wave operators satisfy the important intertwining relation

$$\hat{H}(0)\hat{\Omega}_{\pm} = \hat{\Omega}_{\pm}\hat{H}_0(0). \tag{3.652}$$

From this relation and its Hermitian conjugate it follows that $\hat{S}$ and $\hat{H}_0(0)$ commute,

$$\hat{S}\hat{H}_0(0) = \hat{H}_0(0)\hat{S}. \tag{3.653}$$

In other words, the unperturbed Hamiltonian is invariant under the unitary transformation $\hat{S}$ or the unperturbed energy is conserved in a scattering process.

The Møller wave operators are not unitary if there exist bound energy eigenstates for the Hamiltonian $\hat{H}$. It can be derived that

$$\hat{\Omega}\hat{\Omega}^{\dagger} = \hat{1} - \hat{\Lambda}, \tag{3.654}$$

$\hat{\Lambda}$ is called unitary deficiency and is a sum of all the projectors onto the bound states of $\hat{H}(0)$.

Some of the previous results simplify in the interaction picture. In this picture state vector is defined through the equation

$$|\psi_{\text{I}}(t)\rangle = \hat{U}_0^{\dagger}(t)|\psi(t)\rangle. \tag{3.655}$$

In physical systems the limits $|\psi_{\text{I}}(\mp\infty)\rangle$ may exist. On this assumption the simplification occurs. Namely equations (3.641) and (3.648) become

$$|\psi_{\text{I}}(\mp\infty)\rangle = \begin{cases} |\psi_{\text{in}}\rangle, \\ |\psi_{\text{out}}\rangle \end{cases} \tag{3.656}$$

and

$$|\psi_{\rm I}(+\infty)\rangle = \hat{S}|\psi_{\rm I}(-\infty)\rangle. \tag{3.657}$$

Schrödinger picture operators are transformed to interaction-picture ones via the equation

$$\hat{A}_{\rm I}(t) = \hat{U}_0^\dagger(t)\hat{A}\hat{U}_0(t), \tag{3.658}$$

where $\hat{A}$ is any Schrödinger operator. A possible time dependence of the operator $\hat{A}$ has not been designated. If this operator is time independent it is found that

$$\mathrm{i}\hbar\frac{\mathrm{d}\hat{A}_{\rm I}(t)}{\mathrm{d}t} = [\hat{A}_{\rm I}(t), \hat{H}_{0\rm I}(t)], \tag{3.659}$$

where $\hat{H}_{0\rm I}(t) = \hat{U}_0^\dagger(t)\hat{H}_0(0)\hat{U}_0(t)$. Especially, the Møller wave operators are associated with the interaction-picture operators $\hat{\Omega}^{\rm I}_\pm(t)$,

$$\hat{\Omega}^{\rm I}_\pm(t) = \hat{U}_0^\dagger(t)\hat{\Omega}_\pm\hat{U}_0(t) = \exp\left[\frac{\mathrm{i}}{\hbar}\hat{H}_0(0)t\right]\exp\left[-\frac{\mathrm{i}}{\hbar}\hat{H}(0)t\right]\hat{\Omega}_\pm, \tag{3.660}$$

where we have used the intertwining relation. Taking the limits as $t \to \pm\infty$ one sees that (Taylor 1972, Newton 1966)

$$\begin{aligned}\hat{\Omega}^{\rm I}_+(+\infty) &= \hat{S}, \quad \hat{\Omega}^{\rm I}_+(-\infty) = \hat{1},\\ \hat{\Omega}^{\rm I}_-(+\infty) &= \hat{1}, \quad \hat{\Omega}^{\rm I}_-(-\infty) = \hat{S}^\dagger.\end{aligned} \tag{3.661}$$

In the case of $\hat{\Omega}^{\rm I}_+(t)$ equation (3.659) becomes

$$\mathrm{i}\hbar\frac{\mathrm{d}\hat{\Omega}^{\rm I}_+(t)}{\mathrm{d}t} = \hat{V}_{\rm I}(t)\hat{\Omega}^{\rm I}_+(t), \tag{3.662}$$

where we have used both the intertwining relation and the equation $\hat{V}_{\rm I}(t) = \hat{H}_{\rm I}(t) - \hat{H}_{0\rm I}(t)$. The formal solution of the problem consisting in this equation with the boundary conditions (3.661) provides one with a Dyson expression (Taylor 1972, Newton 1966) for the scattering operator,

$$\hat{S} = \mathcal{T}\exp\left[-\frac{\mathrm{i}}{\hbar}\int_{-\infty}^{\infty}\hat{V}_{\rm I}(t_1)\,\mathrm{d}t_1\right], \tag{3.663}$$

where $\mathcal{T}$ means the time ordering.

In the Schrödinger picture, scattering processes are often spoken of in terms of transitions between initial and final states that are eigenstates of $\hat{H}_0(0)$. So an initial state $|i\rangle$ and a final state $|f\rangle$ have the properties

$$\hat{H}_0(0)|i\rangle = \hbar\omega_i|i\rangle, \quad \hat{H}_0(0)|f\rangle = \hbar\omega_f|f\rangle, \tag{3.664}$$

where ω_i and ω_f are frequencies. As conservation of the unperturbed energy holds, the matrix element $\langle f|\hat{S}|i\rangle$ is zero unless $\omega_f = \omega_i$. This fact is expressed using the transition operator $\hat{T}(z)$ (Taylor 1972, Newton 1966), where z is a complex energy variable. So

$$\langle f|\hat{S}|i\rangle = \langle f|i\rangle - \frac{2\pi \mathrm{i}}{\hbar}\delta(\omega_f - \omega_i)\langle f|\hat{T}(\hbar\omega_i + \mathrm{i}0)|i\rangle. \tag{3.665}$$

The $\hat{T}(z)$ operator is defined through the relation

$$\hat{T}(z) = \hat{V}(0) + \hat{V}(0)\hat{G}(z)\hat{V}(0), \tag{3.666}$$

where

$$\hat{G}(z) = [z\hat{1} - \hat{H}(0)]^{-1} \tag{3.667}$$

is the resolvent operator. It obeys the Lippmann–Schwinger integral equation

$$\hat{T}(z) = \hat{V}(0) + \hat{V}(0)\hat{G}_0(z)\hat{T}(z), \tag{3.668}$$

where

$$\hat{G}_0(z) = [z\hat{1} - \hat{H}_0(0)]^{-1} \tag{3.669}$$

is the resolvent operator associated with $\hat{H}_0(0)$.

The Møller wave operators can also be related to the $\hat{T}(z)$ operator. So

$$\begin{aligned}\hat{\Omega}_+ &= \hat{1} + \int_{-\infty}^{\infty} \hat{G}_0(\hbar\omega + \mathrm{i}0)\hat{T}(\hbar\omega + \mathrm{i}0)\delta\left(\omega\hat{1} - \frac{\hat{H}_0(0)}{\hbar}\right) \mathrm{d}\omega,\\ \hat{\Omega}_- &= \hat{1} + \int_{-\infty}^{\infty} \hat{G}_0(\hbar\omega - \mathrm{i}0)\hat{T}(\hbar\omega - \mathrm{i}0)\delta\left(\omega\hat{1} - \frac{\hat{H}_0(0)}{\hbar}\right) \mathrm{d}\omega.\end{aligned} \tag{3.670}$$

Glauber's theory of photodetection assumes multitime quantum correlation functions of the form (Glauber 1965)

$$G(t_1, \ldots, t_n) = \langle\psi(0)|\hat{b}_1(t_1)\hat{b}_2(t_2)\ldots\hat{b}_n(t_n)|\psi(0)\rangle. \tag{3.671}$$

We will define the input and output operators through the relations (Dalton et al. 1999a)

$$\begin{aligned}\hat{b}_k^{\mathrm{in}}(t) &= \hat{\Omega}_+^{\dagger}\hat{b}_k(t)\hat{\Omega}_+,\\ \hat{b}_k^{\mathrm{out}}(t) &= \hat{\Omega}_-^{\dagger}\hat{b}_k(t)\hat{\Omega}_-.\end{aligned} \tag{3.672}$$

Therefore, we read relations (8) and (10) in Dalton et al. (1999a) as follows:

$$\begin{aligned}G(t_1, \ldots, t_n) &= \langle\psi_{\mathrm{in}}|\hat{b}_1^{\mathrm{in}}(t_1)\hat{b}_2^{\mathrm{in}}(t_2)\ldots\hat{b}_n^{\mathrm{in}}(t_n)|\psi_{\mathrm{in}}\rangle,\\ G(t_1, \ldots, t_n) &= \langle\psi_{\mathrm{out}}|\hat{b}_1^{\mathrm{out}}(t_1)\hat{b}_2^{\mathrm{out}}(t_2)\ldots\hat{b}_n^{\mathrm{out}}(t_n)|\psi_{\mathrm{out}}\rangle.\end{aligned} \tag{3.673}$$

Further from the relations

$$(\hat{1}-\hat{\Lambda})\hat{b}_k(t)(\hat{1}-\hat{\Lambda}) = \hat{\Omega}_+\hat{b}_k^{\rm in}(t)\hat{\Omega}_+^\dagger,$$
$$\hat{b}_k^{\rm out}(t) = \hat{\Omega}_-^\dagger\hat{b}_k(t)\hat{\Omega}_- = \hat{\Omega}_-^\dagger(\hat{1}-\hat{\Lambda})\hat{b}_k(t)(\hat{1}-\hat{\Lambda})\hat{\Omega}_-, \quad (3.674)$$

it holds on substitution that

$$\hat{b}_k^{\rm out}(t) = \hat{\Omega}_-^\dagger\hat{\Omega}_+\hat{b}_k^{\rm in}(t)\hat{\Omega}_+^\dagger\hat{\Omega}_-. \quad (3.675)$$

Using definition (3.649), we may write relation (3.675) in the form

$$\hat{b}_k^{\rm out}(t) = \hat{S}\hat{b}_k^{\rm in}(t)\hat{S}^\dagger. \quad (3.676)$$

We may summarize that

$$\hat{b}_k^{\rm in}(t) - \hat{U}_0^\dagger(t)\hat{b}_k(0)\hat{U}_0(t) \to \hat{0} \text{ as } t \to -\infty,$$
$$\hat{b}_k^{\rm out}(t) - \hat{U}_0^\dagger(t)\hat{b}_k(0)\hat{U}_0(t) \to \hat{0} \text{ as } t \to +\infty. \quad (3.677)$$

Let us recall that operators in the interaction picture are introduced as

$$\hat{b}_k^{\rm I}(t) = \hat{U}_0^\dagger(t)\hat{b}_k(0)\hat{U}_0(t). \quad (3.678)$$

The input and output operators are related to the interaction-picture operators for long times,

$$\hat{b}_k^{\rm in}(t) - \hat{b}_k^{\rm I}(t) \to \hat{0} \text{ as } t \to -\infty, \quad \hat{b}_k^{\rm in}(t) - \hat{S}^\dagger\hat{b}_k^{\rm I}(t)\hat{S} \to \hat{0} \text{ as } t \to +\infty,$$
$$\hat{b}_k^{\rm out}(t) - \hat{S}\hat{b}_k^{\rm I}(t)\hat{S}^\dagger \to \hat{0} \text{ as } t \to -\infty, \quad \hat{b}_k^{\rm out}(t) - \hat{b}_k^{\rm I}(t) \to \hat{0} \text{ as } t \to +\infty. \quad (3.679)$$

Dalton et al. (1999b) have outlined quasimode theory of the lossless beam splitter and Dalton et al. (1999d) have continued and extended it in relation to scattering theory. They have found references to the true mode and quasimode theories of the beam splitter (see there).

They assume that the device consists of two trihedral pieces of glass of refractive index n ($n > \sqrt{2}$) separated by a thin air gap of width d. The coordinate axes are chosen such that refractive index equals n for $|z| > \frac{d}{2}$ and it equals unity for $|z| \leq \frac{d}{2}$.

To obtain quasimodes they choose $\tilde{\epsilon}(\mathbf{R}) = n^2\epsilon_0$, $\tilde{\mu}(\mathbf{R}) = \mu_0$ everywhere. In this case the quasimode functions are plane waves. For the sake of quantization, they assume that the field is contained in a box of volume $V = L^3$ and the quasimode functions

$$\mathbf{U}_\alpha(\mathbf{R}) = \frac{1}{\sqrt{n^2\epsilon_0 V}}\mathbf{e}_\alpha \exp(i\mathbf{k}_\alpha\cdot\mathbf{R}), \quad (3.680)$$

where $\mathbf{e}_\alpha$ are polarization vectors and $\mathbf{k}_\alpha$ are wave vectors ($\mathbf{e}_\alpha \cdot \mathbf{k}_\alpha = 0$). The angular frequencies are

$$\lambda_\alpha = \frac{c}{n}\|\mathbf{k}_\alpha\|. \tag{3.681}$$

For the treatment to be simple, the authors may consider only two directions of propagation: one along $\frac{1}{\sqrt{2}}(\mathbf{e}_y - \mathbf{e}_z)$ and the other along $\frac{1}{\sqrt{2}}(\mathbf{e}_y + \mathbf{e}_z)$. In general, the beam splitter is described by the Hamiltonian

$$\hat{H}'(t) = \hat{H}_Q(t) + \hat{V}_{Q-Q}(t), \tag{3.682}$$

where the unperturbed Hamiltonian $\hat{H}_Q(t)$ and the coupling Hamiltonian $\hat{V}_{Q-Q}(t)$ in the rotating-wave approximation are given by the relations

$$\hat{H}_Q(t) = \hbar \sum_\alpha \mu_\alpha \hat{A}_\alpha^\dagger(t)\hat{A}_\alpha(t), \tag{3.683}$$

$$\hat{V}_{Q-Q}(t) = \hbar \sum_{\substack{\alpha,\beta \\ \alpha\neq\beta}} \upsilon_{\alpha\beta} \hat{A}_\alpha^\dagger(t)\hat{A}_\beta(t). \tag{3.684}$$

The wave vectors $\mathbf{k}_\alpha$ for the quasimodes are

$$k_{\alpha j} = \nu_{\alpha j}\frac{2\pi}{L}, \quad j = x, y, z, \tag{3.685}$$

where $\nu_{\alpha x}$, $\nu_{\alpha y}$, $\nu_{\alpha z}$ are integers. The interesting directions are represented by $\nu_{\alpha x} = 0$, $\nu_{\alpha y} > 0$, $\nu_{\alpha z} = \mp\nu_{\alpha y}$. On calculating the matrix elements $(M_1)_{\alpha\beta}$, $(H_1)_{\alpha\beta}$ one obtains that $\upsilon_{\alpha\beta}$ is zero unless

$$\nu_{\alpha x} = \nu_{\beta x} \equiv \nu_x, \, \nu_{\alpha y} = \nu_{\beta y} \equiv \nu_y \tag{3.686}$$

and the polarization type ($\|$, $\perp$) is the same. On the simplifying assumption it follows that $\nu_{\alpha z} = \mp\nu_y$, $\nu_{\beta z} = \mp\nu_y$, and $\lambda_\alpha = \frac{c}{n}\nu_y\frac{2\pi}{L}\sqrt{2} = \lambda_\beta$. So only the quasimodes of the same angular frequency are coupled.

The complete expression for the coupling constant (see Dalton et al. (1999d)) simplifies

$$\upsilon_{\alpha\beta} = \frac{1}{2}\frac{(n^2-1)}{2L}\sin\left(\frac{2\pi}{L}\nu_y d\right) \times \frac{c}{n}\sqrt{2}\begin{cases} (n^2-1) & (\text{poln}(\alpha) = \text{poln}(\beta) = \|) \\ 1 & (\text{poln}(\alpha) = \text{poln}(\beta) = \perp) \end{cases} \tag{3.687}$$

for the appropriate values of α and β.

Sums over quasimodes α with the same ν_x, ν_y reduce to sums over ν_y and the sign of ν_z. The sums over ν_y can be converted to integral over k_y using the prescription

$$\sum_{\nu_y} \to \frac{L}{2\pi} \int_0^\infty \mathrm{d}k_y. \tag{3.688}$$

Let us note that

$$\frac{L}{2\pi} v_{\alpha\beta} \to \frac{1}{2} \frac{(n^2-1)}{4\pi} \sin\left(k_y d\right) \frac{c}{n} \sqrt{2} \begin{cases} (n^2-1) & (\text{poln} = \|) \\ 1 & (\text{poln} = \perp) \end{cases}. \tag{3.689}$$

The application of quantum scattering theory to the beam splitter is justified in the usual situation where integrated one-photon and two-photon detection rates are finite for incident light field states of interest (Dalton et al. 1999d). With modification made above we have not yet rederived the results of these authors. Also we are afraid that the appropriate operators do not converge in the rotating-wave approximation when only two directions of the incident light are considered.

3.5.2 *Mode Functions for Fabry–Perot Cavity*

Dalton and Knight (1999a,b) have given a justification of the standard model of cavity quantum electrodynamics in terms of a quasimode theory of macroscopic canonical quantization. The quasimodes are treated for the representative case of the three-dimensional Fabry–Perot cavity. The form of the travelling and trapped mode functions for this cavity is derived in Dalton and Knight (1999a) and the mode–mode coupling constants are calculated in Dalton and Knight (1999b). The weak dependence of the coupling constants on the mode frequency difference demonstrates that the conditions for Markovian damping of the cavity quasimode are satisfied. We will speak of the atom–field interaction in the following subsection.

A standard model used in cavity quantum electrodynamics and laser physics may be pictured as follows. An optical cavity is produced by a perfect mirror and a semi-transparent mirror. Radiative atoms are located in the optical cavity. The atoms are coupled directly to a cavity quasimode, whose mode function is nonzero inside the cavity and zero outside, with an atom–cavity coupling constant g. The cavity quasimode decays via Markovian damping with a rate constant Γ_c to certain external quasimodes, the mode functions of which are nonzero outside the cavity and zero inside, and which have the same axial wave vector as the cavity quasimode. Also the atom can decay directly via the Markovian damping with a rate constant Γ_0 to certain external quasimodes, with nonaxial wave vectors.

The standard model may be specified as a typical cavity model, the three-dimensional planar Fabry–Perot cavity. The cavity region I lies between a perfect mirror in the $z = +l$ plane and a thin layer of dielectric material with dielectric constant $\kappa = n^2$ of thickness d, located between the $z = 0$ and $z = -d$ planes (region II). The external region III lies between the dielectric layer plane at $z = -d$ and a second perfect mirror in the $z = -(L+d)$ plane. The external region length L is much greater than the cavity length l, and both are larger than the dielectric layer thickness d.

It is assumed that the three regions constitute a rectangular cuboid with boundaries also at $x = \pm\frac{L'}{2}$, $y = \pm\frac{L'}{2}$. The mode functions and the necessary partial derivatives of these functions must have the period L' in x and y, i.e. be invariant to transition from the plane $x = -\frac{L'}{2}$ to the plane $x = +\frac{L'}{2}$ and from the plane $y = -\frac{L'}{2}$ to the plane $y = +\frac{L'}{2}$.

The electric permittivity function $\epsilon(\mathbf{R})$ for the true cavity is given as

$$\epsilon(\mathbf{R}) = \begin{cases} \epsilon_0 & \text{for } -(L+d) \le z < -d, \\ \kappa\epsilon_0 & \text{for } \quad -d \le z < 0, \\ \epsilon_0 & \text{for } \quad 0 \le z \le l. \end{cases} \tag{3.690}$$

An artificial cavity is described by a modified thickness $\tilde{d}$ and a modified refractive index $\tilde{n}$. For the quasi-cavity the electric permittivity function $\tilde{\epsilon}(\mathbf{R})$ is given as

$$\tilde{\epsilon}(\mathbf{R}) = \begin{cases} \epsilon_0 & \text{for } -(L+\tilde{d}) \le z < -\tilde{d}, \\ \tilde{\kappa}\epsilon_0 & \text{for } \quad -\tilde{d} \le z < 0, \\ \epsilon_0 & \text{for } \quad 0 \le z \le l, \end{cases} \tag{3.691}$$

where the dielectric constant $\tilde{\kappa}$ is related to the refractive index $\tilde{n}$ through $\tilde{\kappa} = \tilde{n}^2$. One makes the thin, strong dielectric approximation (Dalton and Knight 1999a). In both cases $\mu = \mu_0$ everywhere, as there are no magnetic media involved. Obviously, the general form of the true mode functions and the quasimode functions is the same. We may interpret the notation of the true mode functions as general as far as it is useful.

Each wave vector $\mathbf{k}$ is written in terms of its axial component $k_z\mathbf{e}_z$ and its transverse component $\mathbf{k}_\tau$,

$$\mathbf{k}_\tau = k_x\mathbf{e}_x + k_y\mathbf{e}_y. \tag{3.692}$$

It can be derived that the mode functions have the form

$$\mathbf{A}_k(\mathbf{R}) = \exp[\mathrm{i}\mathbf{k}_\tau \cdot (x\mathbf{e}_x + y\mathbf{e}_y)]\mathbf{Z}_k(z). \tag{3.693}$$

Here it has been assumed that in (3.692)

$$\begin{aligned} k_x &= \nu_x \frac{2\pi}{L'}, \quad \nu_x = 0, \pm 1, \pm 2, \ldots, \\ k_y &= \nu_y \frac{2\pi}{L'}, \quad \nu_y = 0, \pm 1, \pm 2, \ldots. \end{aligned} \tag{3.694}$$

In (3.693)

$$\mathbf{Z}_k(\mathbf{z}) = \begin{cases} \alpha_\mathrm{i}\mathbf{e}_\mathrm{i}\exp(\mathrm{i}k_{\mathrm{i}z}z) + \alpha_\mathrm{r}\mathbf{e}_\mathrm{r}\exp(\mathrm{i}k_{\mathrm{r}z}z) & \text{region III}, \\ \beta_\mathrm{t}\mathbf{e}_\mathrm{t}\exp(\mathrm{i}k_{\mathrm{t}z}z) + \beta_\mathrm{s}\mathbf{e}_\mathrm{s}\exp(\mathrm{i}k_{\mathrm{s}z}z) & \text{region II}, \\ \gamma_\mathrm{u}\mathbf{e}_\mathrm{u}\exp(\mathrm{i}k_{\mathrm{u}z}z) + \gamma_\mathrm{v}\mathbf{e}_\mathrm{v}\exp(\mathrm{i}k_{\mathrm{v}z}z) & \text{region I}, \end{cases} \tag{3.695}$$

where α_i, α_r, β_t, β_s, γ_u, and γ_v are coefficients.

The quantities, which are contained in relations (3.693) and (3.695), are defined with respect to the $\|$ and $\perp$ polarizations. Dalton and Knight (1999a,b) investigated not only the travelling modes but also the trapped modes. We refer to the original work for the latter.

In the $\|$ polarization case we have the following parameters. In region III ($\mathbf{n} = \mathbf{e}_z$) the wave vector of the forward-propagating wave is

$$\mathbf{k}_\mathrm{i} = k_\omega[\boldsymbol{\tau}\sin(\theta_1) + \mathbf{n}\cos(\theta_1)], \tag{3.696}$$

where

$$k_\omega = \frac{\omega_k}{c}, \tag{3.697}$$

$$\boldsymbol{\tau} = \mathbf{e}_x\cos\phi + \mathbf{e}_y\sin\phi \tag{3.698}$$

and that of the backward-propagating wave is

$$\mathbf{k}_\mathrm{r} = k_\omega[\boldsymbol{\tau}\sin(\theta_1) - \mathbf{n}\cos(\theta_1)]. \tag{3.699}$$

The polarization vectors are

$$\mathbf{e}_\mathrm{i} = \boldsymbol{\tau}\cos(\theta_1) - \mathbf{n}\sin(\theta_1), \tag{3.700}$$

$$\mathbf{e}_\mathrm{r} = -\boldsymbol{\tau}\cos(\theta_1) - \mathbf{n}\sin(\theta_1). \tag{3.701}$$

The coefficients are

$$\begin{pmatrix} \alpha_\mathrm{i} \\ \alpha_\mathrm{r} \end{pmatrix} = \begin{pmatrix} a_\mathrm{i} \\ a_\mathrm{r} \end{pmatrix} \alpha_0, \tag{3.702}$$

where

$$\begin{pmatrix} a_\mathrm{i} \\ a_\mathrm{r} \end{pmatrix} = \frac{1}{2\mathrm{i}} \begin{pmatrix} \exp[\mathrm{i}k_\omega(L+d)\cos(\theta_1)] \\ \exp[-\mathrm{i}k_\omega(L+d)\cos(\theta_1)] \end{pmatrix} \tag{3.703}$$

and α_0 will be characterized below.

In region II the wave vectors are

$$\mathbf{k}_\mathrm{t} = nk_\omega[\boldsymbol{\tau}\sin(\theta_2) + \mathbf{n}\cos(\theta_2)], \tag{3.704}$$

$$\mathbf{k}_\mathrm{s} = nk_\omega[\boldsymbol{\tau}\sin(\theta_2) - \mathbf{n}\cos(\theta_2)], \tag{3.705}$$

where Snell's law holds,

$$n\sin(\theta_2) = \sin(\theta_1). \tag{3.706}$$

The polarization vectors are

$$\mathbf{e}_{\rm t} = \boldsymbol{\tau}\cos(\theta_2) - \mathbf{n}\sin(\theta_2), \tag{3.707}$$
$$\mathbf{e}_{\rm s} = -\boldsymbol{\tau}\cos(\theta_2) - \mathbf{n}\sin(\theta_2). \tag{3.708}$$

The coefficients are

$$\begin{pmatrix} \beta_{\rm t} \\ \beta_{\rm s} \end{pmatrix} = \begin{pmatrix} b_{\rm t} \\ b_{\rm s} \end{pmatrix} \beta_0 \exp(\mathrm{i}\xi_0), \tag{3.709}$$

where

$$\begin{pmatrix} b_{\rm t} \\ b_{\rm s} \end{pmatrix} = \frac{1}{2}\begin{pmatrix} \exp(\mathrm{i}\phi_0) \\ -\exp(-\mathrm{i}\phi_0) \end{pmatrix} \tag{3.710}$$

and $\beta_0 \exp(\mathrm{i}\xi_0)$ ($\beta_0 \geq 0$) will be characterized below.

In region I the wave vectors are

$$\mathbf{k}_{\rm u} = \mathbf{k}_{\rm i}, \quad \mathbf{k}_{\rm v} = \mathbf{k}_{\rm r}. \tag{3.711}$$

The polarization vectors are

$$\mathbf{e}_{\rm u} = \mathbf{e}_{\rm i}, \mathbf{e}_{\rm v} = \mathbf{e}_{\rm r}. \tag{3.712}$$

The coefficients are

$$\begin{pmatrix} \gamma_{\rm u} \\ \gamma_{\rm v} \end{pmatrix} = \begin{pmatrix} g_{\rm u} \\ g_{\rm v} \end{pmatrix} \gamma_0, \tag{3.713}$$

where

$$\begin{pmatrix} g_{\rm u} \\ g_{\rm v} \end{pmatrix} = \frac{1}{2\mathrm{i}}\begin{pmatrix} \exp[-\mathrm{i}k_\omega l\cos(\theta_1)] \\ \exp[\mathrm{i}k_\omega l\cos(\theta_1)] \end{pmatrix} \tag{3.714}$$

and γ_0 will be characterized in what follows.

We have concentrated on the quasimodes. We have calculated the dependence of α_0 and β_0 on γ_0 and that of γ_0 and β_0 on α_0. We have assumed that the γ_0 dependence is useful for $\gamma_0 \gg \alpha_0$ and the α_0 dependence is useful for $\alpha_0 \gg \gamma_0$.

We obtain that

$$\begin{aligned}\frac{\alpha_0}{\gamma_0} \approx{}& \cos[k_\omega(L+d+l)\cos(\theta_1)] \\ &- \Lambda\cos(\theta_1)\cos[k_\omega(L+d)\cos(\theta_1)]\sin[k_\omega l\cos(\theta_1)],\end{aligned} \tag{3.715}$$

where $\Lambda = n^2 k_\omega d \gg 1$,

$$\frac{\beta_0 \exp(\mathrm{i}\xi_0)}{\gamma_0} \approx -\frac{\cos(\theta_1)}{\cos(\theta_2)} \sin[k_\omega l \cos(\theta_1)]. \tag{3.716}$$

We should also assume that $\sin[k_\omega l \cos(\theta_1)] \approx 0$ or a resonant field.

We obtain that

$$\frac{\gamma_0}{\alpha_0} \approx \cos[k_\omega(l + L + d)\cos(\theta_1)] \\ - \Lambda \cos(\theta_1) \cos[k_\omega l \cos(\theta_1)] \sin[k_\omega(L + d)\cos(\theta_1)], \tag{3.717}$$

$$\frac{\beta_0 \exp(\mathrm{i}\xi_0)}{\alpha_0} \approx \frac{\cos(\theta_1)}{\cos(\theta_2)} \sin[k_\omega(L + d)\cos(\theta_1)]. \tag{3.718}$$

We should also assume that $\cos[k_\omega l \cos(\theta_1)] \approx 0$ or an external field far from resonance.

In the $\perp$ polarization case we have the following parameters. In region III the wave vectors are given in relations (3.696) and (3.699). The polarization vectors are

$$\mathbf{e}_\mathrm{i} = \mathbf{e}_\mathrm{r} = \boldsymbol{\sigma}, \tag{3.719}$$

where

$$\boldsymbol{\sigma} = \boldsymbol{\tau} \times \mathbf{n} \tag{3.720}$$

$$= \mathbf{e}_x \sin\phi - \mathbf{e}_y \cos\phi. \tag{3.721}$$

The coefficients are given in relation (3.702), where

$$\begin{pmatrix} a_\mathrm{i} \\ a_\mathrm{r} \end{pmatrix} = \frac{1}{2\mathrm{i}} \begin{pmatrix} \exp[\mathrm{i}k_\omega(L + d)\cos(\theta_1)] \\ -\exp[-\mathrm{i}k_\omega(L + d)\cos(\theta_1)] \end{pmatrix} \tag{3.722}$$

and α_0 will be characterized below.

In region II the wave vectors are given in relations (3.704) and (3.705). The polarization vectors are

$$\mathbf{e}_\mathrm{t} = \mathbf{e}_\mathrm{s} = \boldsymbol{\sigma}. \tag{3.723}$$

The coefficients are given in relation (3.709), where

$$\begin{pmatrix} b_\mathrm{t} \\ b_\mathrm{s} \end{pmatrix} = \frac{1}{2} \begin{pmatrix} \exp(\mathrm{i}\phi_0) \\ \exp(-\mathrm{i}\phi_0) \end{pmatrix} \tag{3.724}$$

and $\beta_0 \exp(\mathrm{i}\xi_0)$ ($\beta_0 \geq 0$) will be characterized below.

In region I the wave vectors are given in relation (3.711). The polarization vectors are

$$\mathbf{e}_\mathrm{u} = \mathbf{e}_\mathrm{v} = \mathbf{e}_\mathrm{i}. \tag{3.725}$$

The coefficients are introduced in relation (3.713), where

$$\begin{pmatrix} g_u \\ g_v \end{pmatrix} = \frac{1}{2i} \begin{pmatrix} \exp[-ik_\omega l \cos(\theta_1)] \\ -\exp[ik_\omega l \cos(\theta_1)] \end{pmatrix} \tag{3.726}$$

and γ_0 will be characterized in what follows.

We obtain that

$$\frac{\alpha_0}{\gamma_0} \approx \cos[k_\omega(L+d+l)\cos(\theta_1)] - \Lambda \frac{[\cos(\theta_2)]^2}{\cos(\theta_1)} \cos[k_\omega(L+d)\cos(\theta_1)] \sin[k_\omega l \cos(\theta_1)], \tag{3.727}$$

$$\frac{\beta_0 \exp(i\xi_0)}{\gamma_0} \approx -\sin[k_\omega l \cos(\theta_1)]. \tag{3.728}$$

We should also assume that $\sin[k_\omega l \cos(\theta_1)] \approx 0$ or a resonant field.

We obtain that

$$\frac{\gamma_0}{\alpha_0} \approx \cos[k_\omega(l+L+d)\cos(\theta_1)] - \Lambda \frac{[\cos(\theta_2)]^2}{\cos(\theta_1)} \cos[k_\omega l \cos(\theta_1)] \sin[k_\omega(L+d)\cos(\theta_1)], \tag{3.729}$$

$$\frac{\beta_0 \exp(i\xi_0)}{\alpha_0} \approx \sin[k_\omega(L+d)\cos(\theta_1)]. \tag{3.730}$$

We should also assume that $\cos[k_\omega l \cos(\theta_1)] \approx 0$ or an external field far from resonance.

These values have been obtained using wave optics, in which, for instance, the coefficients in region III are connected to those in region I by the relation

$$\begin{pmatrix} \alpha_i \\ \alpha_r \end{pmatrix} = \mathsf{T} \begin{pmatrix} \gamma_u \\ \gamma_v \end{pmatrix}, \tag{3.731}$$

with

$$\mathsf{T} \approx \begin{pmatrix} 1 - \frac{1}{2} i\Lambda \cos(\theta_1) & \frac{1}{2} i\Lambda \cos(\theta_1) \\ -\frac{1}{2} i\Lambda \cos(\theta_1) & 1 + \frac{1}{2} i\Lambda \cos(\theta_1) \end{pmatrix} \tag{3.732}$$

for $\parallel$ polarization and

$$\mathsf{T} \approx \begin{pmatrix} 1 - \frac{1}{2} i\Lambda \frac{[\cos(\theta_2)]^2}{\cos(\theta_1)} & -\frac{1}{2} i\Lambda \frac{[\cos(\theta_2)]^2}{\cos(\theta_1)} \\ \frac{1}{2} i\Lambda \frac{[\cos(\theta_2)]^2}{\cos(\theta_1)} & 1 + \frac{1}{2} i\Lambda \frac{[\cos(\theta_2)]^2}{\cos(\theta_1)} \end{pmatrix} \tag{3.733}$$

for $\perp$ polarization. The field must fulfil the boundary conditions

$$\left.\begin{array}{l} a_r\alpha_i - a_i\alpha_r = 0, \\ g_v\gamma_u - g_u\gamma_v = 0 \end{array}\right\} \text{ for both polarizations.} \tag{3.734}$$

The eigenfrequencies ω_k are given as solutions of the transcendental equation

$$\begin{pmatrix} -a_\mathrm{r} & a_\mathrm{i} \end{pmatrix} \mathsf{T} \begin{pmatrix} g_\mathrm{u} \\ g_\mathrm{v} \end{pmatrix} = 0, \tag{3.735}$$

where relation (3.697) and the relations

$$\cos(\theta_2) = \sqrt{1 - \frac{|\mathbf{k}_\tau|^2}{n^2 k_\omega^2}}, \quad \cos(\theta_1) = \sqrt{1 - \frac{|\mathbf{k}_\tau|^2}{k_\omega^2}} \tag{3.736}$$

may be mentioned, and $|\mathbf{k}_\tau|^2$ is a parameter.

To achieve an approximate normalization of the near-resonant mode, we put

$$|\gamma_0| = \sqrt{\frac{2}{(L')^2 l \epsilon_0}} \tag{3.737}$$

independent of the polarization. When the external field is off a resonance, we put

$$|\alpha_0| = \sqrt{\frac{2}{(L')^2 L \epsilon_0}} \tag{3.738}$$

independent of the polarization.

The coupling constants between different quasimodes are calculated according to relations (3.636) and (3.637). The notation $\mathbf{U}_\alpha(\mathbf{R})$ should be used for the quasimode functions which $\mathbf{A}_\alpha(\mathbf{R})$ still denotes.

Dalton and Knight (1999b) have found that

$$M_{\alpha\beta} = H_{\alpha\beta} = 0 \text{ if } \nu_{\alpha x} \neq \nu_{\beta x} \text{ or } \nu_{\alpha y} \neq \nu_{\beta y}. \tag{3.739}$$

They have also found that $M_{\alpha\beta}$ and $H_{\alpha\beta}$ are zero for quasimodes of different polarizations. They have shown that the coupling problem for modes in a three-dimensional Fabry–Perot cavity is equivalent to a similar problem in a one-dimensional Fabry–Perot cavity. The selection rules allow coupling between axial cavity quasimodes and axial external quasimodes.

Coupling constants between a single axial cavity quasimode and axial external quasimodes depend on the external quasimode frequency slowly. One may conclude that the conditions for irreversible Markovian damping of the cavity quasimode are satisfied.

Analyses of cavities have been mentioned in the beginning of Section 3.3.1. Barnett and Radmore (1988) have shown that even the mode-strength function which characterizes the true modes may be approximated using quasimodes and a phenomenological coupling. They have concluded that the approximation is good if the cavity is of sufficiently high quality and if the precise spatial dependence of the field does not weigh.

In the work of Garraway (1997a,b), atom-true field mode couplings are used as a basis for the pseudomode model. In certain situations quasimodes can be associated with pseudomodes (Dalton et al. 2003).

3.5.3 Atom–Field Interaction Within Cavity

First we complete what is needed for the description of a system of radiative atoms interacting with the electromagnetic field in the presence of a neutral dielectric medium on the assumptions made before. We add that the radiative atoms are stationary and electrically neutral.

The radiative atoms possess charge density $\rho_{\mathrm{L}}(\mathbf{R}, t)$, current density $\mathbf{j}_{\mathrm{L}}(\mathbf{R}, t)$, polarization density $\mathbf{P}_{\mathrm{L}}(\mathbf{R}, t)$, and magnetization density $\mathbf{M}_{\mathrm{L}}(\mathbf{R}, t)$ given in terms of the positions $\mathbf{r}_{\xi\alpha}(t)$ and velocities $\dot{\mathbf{r}}_{\xi\alpha}(t)$ of the charged particles forming the radiative atoms,

$$\begin{aligned}\rho_{\mathrm{L}}(\mathbf{R}, t) &= \sum_{\xi,\alpha} q_{\xi\alpha}\delta\Big(\mathbf{R} - \mathbf{r}_{\xi\alpha}(t)\Big),\\ \mathbf{j}_{\mathrm{L}}(\mathbf{R}, t) &= \sum_{\xi,\alpha} q_{\xi\alpha}\delta\Big(\mathbf{R} - \mathbf{r}_{\xi\alpha}(t)\Big)\dot{\mathbf{r}}_{\xi\alpha}(t).\end{aligned} \tag{3.740}$$

Here $\xi = 1, 2, \ldots$ lists different radiative atoms and $\alpha = 1, 2, \ldots$ lists different particles within atom ξ. $q_{\xi\alpha}$, $M_{\xi\alpha}$ are the charge and mass for the $\xi\alpha$ particle, respectively. One defines

$$\begin{aligned}\mathbf{P}_{\mathrm{L}}(\mathbf{R}, t) = \sum_{\xi,\alpha} q_{\xi\alpha} \int_0^1 &[\mathbf{r}_{\xi\alpha}(t) - \mathbf{R}_{\xi}(t)]\\ &\times \delta\Big(\mathbf{R} - \mathbf{R}_{\xi}(t) - u[\mathbf{r}_{\xi\alpha}(t) - \mathbf{R}_{\xi}(t)]\Big)\,\mathrm{d}u,\end{aligned} \tag{3.741}$$

$$\begin{aligned}\mathbf{M}_{\mathrm{L}}(\mathbf{R}, t) = \sum_{\xi,\alpha} q_{\xi\alpha} \int_0^1 &u[\mathbf{r}_{\xi\alpha}(t) - \mathbf{R}_{\xi}(t)] \times [\dot{\mathbf{r}}_{\xi}(t) - \dot{\mathbf{R}}_{\xi}(t)]\\ &\times \delta\Big(\mathbf{R} - \mathbf{R}_{\xi}(t) - u[\mathbf{r}_{\xi\alpha}(t) - \mathbf{R}_{\xi}(t)]\Big)\,\mathrm{d}u\\ &+ \sum_{\xi,\alpha} q_{\xi\alpha} \int_0^1 [\mathbf{r}_{\xi\alpha}(t) - \mathbf{R}_{\xi}(t)] \times \dot{\mathbf{R}}_{\xi}(t)\\ &\times \delta\Big(\mathbf{R} - \mathbf{R}_{\xi}(t) - u[\mathbf{r}_{\xi\alpha}(t) - \mathbf{R}_{\xi}(t)]\Big)\,\mathrm{d}u,\end{aligned} \tag{3.742}$$

where $\mathbf{R}_{\xi}(t)$ is the position of the centre of mass of the atom ξ, whose mass is M_{ξ}. In the generalized Coulomb gauge condition, the scalar potential $\phi(\mathbf{R}, t)$ satisfies a generalized Poisson equation

$$\nabla \cdot [\epsilon(\mathbf{R})\nabla\phi(\mathbf{R}, t)] = -\rho_{\mathrm{L}}(\mathbf{R}, t). \tag{3.743}$$

The macroscopic Lagrangian is given by the relation

$$L'(t) = \frac{1}{2}\sum_{\xi,\alpha} M_{\xi\alpha}\dot{\mathbf{r}}^2_{\xi\alpha}(t) - V_{\text{coul}}(t) + \int \mathcal{L}'_{\text{c}}(\mathbf{R}, t)\, \mathrm{d}^3\mathbf{R}, \tag{3.744}$$

where the Lagrangian density $\mathcal{L}'_{\text{c}}(\mathbf{R}, t)$ is given by the relation

$$\begin{aligned}\mathcal{L}'_{\text{c}}(\mathbf{R}, t) = &\frac{1}{2}\epsilon(\mathbf{R})\left[\frac{\partial \mathbf{A}(\mathbf{R}, t)}{\partial t}\right]^2 - \frac{1}{2\mu(\mathbf{R})}[\nabla \times \mathbf{A}(\mathbf{R}, t)]^2 \\ &- \mathbf{P}'_{\text{L}}(\mathbf{R}, t)\cdot \frac{\partial \mathbf{A}(\mathbf{R}, t)}{\partial t} + \mathbf{M}_{\text{L}}(\mathbf{R}, t)\cdot \nabla \times \mathbf{A}(\mathbf{R}, t).\end{aligned} \tag{3.745}$$

In the Lagrangian $V_{\text{coul}}(t)$ is the Coulomb energy given by the relation

$$V_{\text{coul}}(t) = \int \frac{[\epsilon(\mathbf{R})\nabla\phi(\mathbf{R}, t)]^2}{2\epsilon(\mathbf{R})}\, \mathrm{d}^3\mathbf{R} \tag{3.746}$$

and $\mathbf{P}'_{\text{L}}(\mathbf{R}, t)$ is the reduced polarization density:

$$\mathbf{P}'_{\text{L}}(\mathbf{R}, t) = \mathbf{P}_{\text{L}}(\mathbf{R}, t) - \epsilon(\mathbf{R})\nabla\phi(\mathbf{R}, t). \tag{3.747}$$

The conjugate momentum field $\mathbf{\Pi}(\mathbf{R}, t)$ is obtained from the Lagrangian density $\mathcal{L}'_{\text{c}}(\mathbf{R}, t)$ as

$$\mathbf{\Pi}(\mathbf{R}, t) = \epsilon(\mathbf{R})\frac{\partial \mathbf{A}(\mathbf{R}, t)}{\partial t} - \mathbf{P}'_{\text{L}}(\mathbf{R}, t) \tag{3.748}$$

and the particle momenta are obtained from $L'(t)$ as

$$\begin{aligned}\mathbf{p}_{\xi\alpha}(t) = &M_{\xi\alpha}\dot{\mathbf{r}}_{\xi\alpha}(t) \\ &+ q_{\xi\alpha}\int_0^1 u\mathbf{B}\Big(\mathbf{R}_\xi(t) + u[\mathbf{r}_{\xi\alpha}(t) - \mathbf{R}_\xi(t)], t\Big)\, \mathrm{d}u \times [\mathbf{r}_{\xi\alpha}(t) - \mathbf{R}_\xi(t)].\end{aligned} \tag{3.749}$$

The multipolar Hamiltonian is

$$\begin{aligned}H'(t) = &\sum_{\xi,\alpha}\frac{\mathbf{p}^2_{\xi\alpha}(t)}{2M_{\xi\alpha}} + V_{\text{coul}}(t) + \int \frac{\mathbf{P}'^{\,2}_{\text{L}}(\mathbf{R}, t)}{2\epsilon(\mathbf{R})}\, \mathrm{d}^3\mathbf{R} \\ &+ \int \left\{\frac{\mathbf{\Pi}^2(\mathbf{R}, t)}{2\epsilon(\mathbf{R})} + \frac{[\nabla \times \mathbf{A}(\mathbf{R}, t)]^2}{2\mu(\mathbf{R})}\right\} \mathrm{d}^3\mathbf{R} \\ &+ \int \frac{\mathbf{\Pi}(\mathbf{R},\mathrm{t})\cdot \mathbf{P}'_{\text{L}}(\mathbf{R}, t)}{\epsilon(\mathbf{R})}\, \mathrm{d}^3\mathbf{R} - \int [\nabla \times \mathbf{A}(\mathbf{R}, t)]\cdot \mathbf{M}'_{\text{L}}(\mathbf{R}, t)\, \mathrm{d}^3\mathbf{R} \\ &+ \sum_{\xi,\alpha}\frac{q^2_{\xi\alpha}}{2M_{\xi\alpha}}\left\{\int_0^1 u\nabla \times \mathbf{A}\Big(\mathbf{R}_\xi(t) + u\left[\mathbf{r}_{\xi\alpha}(t) - \mathbf{R}_\xi(t)\right], t\Big)\, \mathrm{d}u \right. \\ &\left. \times \left[\mathbf{r}_{\xi\alpha}(t) - \mathbf{R}_\xi(t)\right]\right\}^2,\end{aligned} \tag{3.750}$$

where the reduced magnetization density $\mathbf{M}'_{\mathrm{L}}(\mathbf{R},t)$ is given as

$$\mathbf{M}'_{\mathrm{L}}(\mathbf{R},t) = \sum_{\xi,\alpha} q_{\xi\alpha} \int_0^1 u[\mathbf{r}_{\xi\alpha}(t) - \mathbf{R}_\xi(t)] \times \frac{\mathbf{p}_{\xi\alpha}(t)}{M_{\xi\alpha}} \times \delta\Big(\mathbf{R} - \mathbf{R}_\xi(t) - u[\mathbf{r}_{\xi\alpha}(t) - \mathbf{R}_\xi(t)]\Big)\,\mathrm{d}u. \tag{3.751}$$

We have property (3.586).

The reduced polarization density is given in terms of true mode functions as

$$\mathbf{P}'_{\mathrm{L}}(\mathbf{R},t) = \sum_k \epsilon(\mathbf{R})\mathbf{A}_k(\mathbf{R}) \int \mathbf{P}_{\mathrm{L}}(\mathbf{R}',t)\cdot\mathbf{A}_k^*(\mathbf{R}')\,\mathrm{d}^3\mathbf{R}'. \tag{3.752}$$

Using expansion (3.602), one can write Lagrangian (3.744), with the Lagrangian density (3.745) as follows:

$$\begin{aligned}
L'(t) = {} & \frac{1}{2}\sum_{\alpha,\xi} M_{\alpha\xi}\dot{\mathbf{r}}^2_{\alpha\xi}(t) - V_{\mathrm{coul}}(t) \\
& + \sum_{\alpha,\xi} q_{\alpha\xi}\dot{\mathbf{r}}_{\alpha\xi}(t)\cdot \int_0^1 u\mathbf{B}\Big(\mathbf{R}_\xi(t) + u\left[\mathbf{r}_{\xi\alpha}(t) - \mathbf{R}_\xi(t)\right], t\Big)\,\mathrm{d}u \times \left[\mathbf{r}_{\xi\alpha}(t) - \mathbf{R}_\xi(t)\right] \\
& + \frac{1}{2}\sum_{\alpha,\beta} \dot{Q}^*_\alpha(t)(\mathbf{W}^{-1})_{\alpha\beta}\dot{Q}_\beta(t) - \frac{1}{2}\sum_{\alpha,\beta} Q^*_\alpha(t)V_{\alpha\beta}Q_\beta(t) \\
& - \sum_\alpha \dot{Q}^*_\alpha(t)N_\alpha(t),
\end{aligned} \tag{3.753}$$

where

$$\begin{aligned}
\mathbf{N}(t) &= \mathbf{K}^*\mathbf{L}(t), \\
L_\alpha(t) &= \int \frac{\tilde{\epsilon}(\mathbf{R})}{\epsilon(\mathbf{R})}\mathbf{U}^*_\alpha(\mathbf{R})\cdot\mathbf{P}_{\mathrm{L}}(\mathbf{R},t)\,\mathrm{d}^3\mathbf{R}.
\end{aligned} \tag{3.754}$$

Making the choice (3.608) for $\mathbf{K}$ (Dalton et al. 1999b), one has

$$\mathbf{N}(t) = \mathbf{M}^{-1}\mathbf{L}(t). \tag{3.755}$$

The generalized momentum coordinates $P_\alpha(t)$ for the electromagnetic field are

$$P_\alpha(t) = \sum_\beta (\mathbf{M}^{-1})_{\alpha\beta}\dot{Q}_\beta(t) - N_\alpha(t). \tag{3.756}$$

The Hamiltonian is given by the relation

$$
\begin{aligned}
H'(t) = {} & \sum_{\alpha,\xi} \frac{\mathbf{p}^2_{\alpha\xi}(t)}{2M_{\alpha\xi}} + V_{\text{coul}}(t) + \frac{1}{2}\sum_{\alpha,\xi} N^*_\alpha(t) M_{\alpha\xi} N_\xi(t) \\
& + \frac{1}{2}\sum_{\alpha,\beta} P^*_\alpha(t) M_{\alpha\beta} P_\beta(t) + \frac{1}{2}\sum_{\alpha,\beta} Q^*_\alpha(t) V_{\alpha\beta} Q_\beta(t) \\
& + \sum_{\alpha,\beta} P^*_\alpha(t) M_{\alpha\beta} N_\beta(t) + \frac{1}{2}\sum_{\alpha,\beta} Q^*_\alpha(t) (\mathbf{M}^{-1})_{\alpha\beta} R_\beta(t) \\
& + \frac{1}{2}\sum_{\alpha,\beta} Q^*_\alpha(t) X_{\alpha\beta}(t) Q_\beta(t),
\end{aligned}
\tag{3.757}
$$

where

$$
\mathbf{X}(t) = \mathbf{M}^{-1}\mathbf{D}(t)\mathbf{M}^{-1},
\tag{3.758}
$$

with

$$
R_\beta(t) = -\int \mathbf{M}'_{\text{L}}(\mathbf{R},t)\cdot\left[\nabla\times\frac{\tilde{\epsilon}(\mathbf{R})}{\epsilon(\mathbf{R})}\mathbf{U}^*_\beta(\mathbf{R})\right] \mathrm{d}^3\mathbf{R},
\tag{3.759}
$$

$$
\begin{aligned}
D_{\alpha\beta}(t) = {} & \sum_{\mu,\xi} \frac{q^2_{\mu\xi}}{M_{\mu\xi}} \int\int\int_0^1 u \int_0^1 u'\delta\Big(\mathbf{R} - \mathbf{R}_\xi(t) - u\left[\mathbf{r}_{\mu\xi}(t) - \mathbf{R}_\xi(t)\right]\Big) \\
& \times \delta\Big(\mathbf{R}' - \mathbf{R}'_\xi(t) - u'\left[\mathbf{r}_{\mu\xi}(t) - \mathbf{R}_\xi(t)\right]\Big)\,\mathrm{d}u'\mathrm{d}u \\
& \times \Bigg\{\left[\mathbf{r}_{\mu\xi}(t) - \mathbf{R}_\xi(t)\right]^2 \left[\nabla\times\frac{\tilde{\epsilon}(\mathbf{R})}{\epsilon(\mathbf{R})}\mathbf{U}^*_\alpha(\mathbf{R})\right]\cdot\left[\nabla\times\frac{\tilde{\epsilon}(\mathbf{R})}{\epsilon(\mathbf{R})}\mathbf{U}_\beta(\mathbf{R})\right]' \\
& - \left[\mathbf{r}_{\mu\xi}(t) - \mathbf{R}_\xi(t)\right]\cdot\left[\nabla\times\frac{\tilde{\epsilon}(\mathbf{R})}{\epsilon(\mathbf{R})}\mathbf{U}^*_\alpha(\mathbf{R})\right]\left[\mathbf{r}_{\mu\xi}(t) - \mathbf{R}_\xi(t)\right] \\
& \cdot\left[\nabla\times\frac{\tilde{\epsilon}(\mathbf{R})}{\epsilon(\mathbf{R})}\mathbf{U}_\beta(\mathbf{R})\right]'\Bigg\}\,\mathrm{d}^3\mathbf{R}\,\mathrm{d}^3\mathbf{R}',
\end{aligned}
\tag{3.760}
$$

where $\left[\nabla\times\frac{\tilde{\epsilon}(\mathbf{R})}{\epsilon(\mathbf{R})}\mathbf{U}_\beta(\mathbf{R})\right]'$ means that the vector $\mathbf{R}$ and the derivatives with respect to its components are replaced by the vector $\mathbf{R}'$ and the derivatives with respect to the components of $\mathbf{R}'$, respectively. The terms in the Hamiltonian are the particle kinetic energy, Coulomb energy, polarization energy, radiative energy (two terms), electric interaction energy, magnetic interaction energy, and diamagnetic energy.

The reduced polarization density can be expanded in terms of the quasimode functions $\tilde{\epsilon}(\mathbf{R})\mathbf{U}_\alpha(\mathbf{R})$ as

$$
\mathbf{P}'_{\text{L}}(\mathbf{R},t) = \sum_\alpha (\mathbf{M}^{-1}\mathbf{L}(t))_\alpha \tilde{\epsilon}(\mathbf{R})\mathbf{U}_\alpha(\mathbf{R}).
\tag{3.761}
$$

The quantization for the radiative atom charged particles is the familiar prescriptions involving Hermitian operators

$$\mathbf{r}_{\alpha\xi}(t) \rightarrow \hat{\mathbf{r}}_{\alpha\xi}(t),\ \mathbf{p}_{\alpha\xi}(t) \rightarrow \hat{\mathbf{p}}_{\alpha\xi}(t), \tag{3.762}$$

with the usual commutation rules applying.

The full quantum multipolar Hamiltonian is

$$\begin{aligned}
\hat{H}'(t) &= \sum_{\xi\alpha} \frac{\hat{\mathbf{p}}^2_{\xi\alpha}(t)}{2M_{\xi\alpha}} + \hat{V}_{\text{coul}}(t) + \int \frac{\hat{\mathbf{P}}'^{\,2}_{\text{L}}(\mathbf{R}, t)}{2\epsilon(\mathbf{R})}\, \mathrm{d}^3\mathbf{R} \\
&+ \hbar \sum_{\alpha} \mu_\alpha \left[\hat{A}^\dagger_\alpha(t)\hat{A}_\alpha(t) + \frac{1}{2}\hat{1} \right] \\
&+ \frac{\hbar}{2} \sum_{\substack{\alpha,\beta \\ \alpha\neq\beta}} \left(\sqrt{\eta_\alpha\eta_\beta} M_{\alpha\beta} + \frac{V_{\alpha\beta}}{\sqrt{\eta_\alpha\eta_\beta}} \right) \hat{A}^\dagger_\alpha(t)\hat{A}_\beta(t) \\
&+ \frac{\hbar}{4} \sum_{\substack{\alpha,\beta \\ \alpha\neq\beta}} \left(-\sqrt{\eta_\alpha\eta_\beta} M_{\alpha,-\beta} + \frac{V_{\alpha,-\beta}}{\sqrt{\eta_\alpha\eta_\beta}} \right) \hat{A}^\dagger_\alpha(t)\hat{A}^\dagger_\beta(t) + \text{H.c.} \\
&+ \int \frac{\hat{\Pi}(\mathbf{R}, t) \cdot \hat{\mathbf{P}}'^{\,2}_{\text{L}}(\mathbf{R}, t)}{\epsilon(\mathbf{R})}\, \mathrm{d}^3\mathbf{R} - \int \left[\nabla \times \hat{\mathbf{A}}(\mathbf{R}, t) \right] \cdot \hat{\mathbf{M}}'_{\text{L}}(\mathbf{R}, t)\, \mathrm{d}^3\mathbf{R} \\
&+ \sum_{\xi,\alpha} \frac{q^2_{\xi\alpha}}{2M_{\xi\alpha}} \left\{ \int_0^1 u\nabla \times \hat{\mathbf{A}}\Big(\mathbf{R}_\xi(t)\hat{1} + u\left[\hat{\mathbf{r}}_{\xi\alpha}(t) - \mathbf{R}_\xi(t)\hat{1} \right], t \Big)\, \mathrm{d}u \right. \\
&\left. \times \left[\hat{\mathbf{r}}_{\xi\alpha}(t) - \mathbf{R}_\xi(t)\hat{1} \right] \right\}^2,
\end{aligned} \tag{3.763}$$

where $\hat{\mathbf{P}}'_{\text{L}}(\mathbf{R}, t)$, $\hat{\mathbf{M}}'_{\text{L}}(\mathbf{R}, t)$ are given by equations (3.761) and (3.751) in the operator form.

The polarization energy term and the Coulomb energy term can be combined to be equal to the sum of intra-atomic Coulomb and polarization energy terms plus intra-atomic contact energy terms. One has (Dalton and Babiker 1997)

$$\hat{V}_{\text{coul}}(t) + \int \frac{\hat{\mathbf{P}}'^{\,2}_{\text{L}}(\mathbf{R}, t)}{2\epsilon(\mathbf{R})}\, \mathrm{d}^3\mathbf{R} = \hat{V}^{\text{IA}}_{\text{coul}}(t) + \hat{V}^{\text{IA}}_{\text{pol}}(t) + \hat{V}_{\text{cont}}(t). \tag{3.764}$$

One defines $\rho_{\text{L}\xi}(\mathbf{R}, t)$ by relation (3.740) with the sum over ξ omitted and $\mathbf{P}_{\text{L}\xi}(\mathbf{R}, t)$ by (3.741) with the same modification. One modifies (3.743) and (3.747) as

$$\nabla \cdot \left[\epsilon(\mathbf{R}) \nabla\phi_\xi(\mathbf{R}, t) \right] = -\rho_{\text{L}\xi}(\mathbf{R}, t) \tag{3.765}$$

and

$$\mathbf{P}'_{\text{L}\xi}(\mathbf{R}, t) = \mathbf{P}_{\text{L}\xi}(\mathbf{R}, t) - \left[\epsilon(\mathbf{R}) \nabla\phi_\xi(\mathbf{R}, t) \right], \tag{3.766}$$

respectively. In (3.764)

$$\hat{V}^{\mathrm{IA}}_{\mathrm{coul}}(t) = \sum_{\xi} \int \frac{\left[\epsilon(\mathbf{R})\nabla\hat{\phi}_{\xi}(\mathbf{R},t)\right]\cdot\left[\epsilon(\mathbf{R})\nabla\hat{\phi}_{\xi}(\mathbf{R},t)\right]}{2\epsilon(\mathbf{R})}\,\mathrm{d}^3\mathbf{R}, \tag{3.767}$$

$$\hat{V}^{\mathrm{IA}}_{\mathrm{pol}}(t) = \sum_{\xi} \int \frac{\hat{\mathbf{P}}'_{\mathrm{L}\xi}(\mathbf{R},t)\cdot\hat{\mathbf{P}}'_{\mathrm{L}\xi}(\mathbf{R},t)}{2\epsilon(\mathbf{R})}\,\mathrm{d}^3\mathbf{R}, \tag{3.768}$$

$$\hat{V}_{\mathrm{cont}}(t) = \sum_{\substack{\xi,\eta \\ \xi\neq\eta}} \int \frac{\hat{\mathbf{P}}_{\mathrm{L}\xi}(\mathbf{R},t)\cdot\hat{\mathbf{P}}_{\mathrm{L}\eta}(\mathbf{R},t)}{2\epsilon(\mathbf{R})}\,\mathrm{d}^3\mathbf{R}. \tag{3.769}$$

To obtain the electric dipole approximation one neglects the magnetic and diamagnetic interaction energy terms. The polarization density is given in its dipolar approximation

$$\hat{\mathbf{P}}_{\mathrm{L}}(\mathbf{R},t) = \sum_{\xi} \hat{\boldsymbol{\mu}}_{\xi}(t)\delta\Big(\mathbf{R}-\mathbf{R}_{\xi}(t)\Big), \tag{3.770}$$

where $\hat{\boldsymbol{\mu}}_{\xi}(t)$ is the dipolar moment for the ξ atom. The reduced polarization density becomes

$$\hat{\mathbf{P}}'_{\mathrm{L}}(\mathbf{R},t) = \sum_{\xi,\alpha,\beta} (\mathbf{M}^{-1})_{\alpha\beta} \frac{\tilde{\epsilon}\Big(\mathbf{R}_{\xi}(t)\Big)}{\epsilon\Big(\mathbf{R}_{\xi}(t)\Big)} \hat{\boldsymbol{\mu}}_{\xi}(t)\cdot\mathbf{U}^{*}_{\beta}\Big(\mathbf{R}_{\xi}(t)\Big)\tilde{\epsilon}(\mathbf{R})\mathbf{U}_{\alpha}(\mathbf{R}). \tag{3.771}$$

The atom–electromagnetic field interaction energy then assumes the forms

$$\begin{aligned}\hat{V}^{\mathrm{E1}}_{\mathrm{A-F}}(t) &= \sum_{\xi,\alpha} \frac{1}{\mathrm{i}}\sqrt{\frac{\hbar\eta_{\alpha}}{2}} \frac{\tilde{\epsilon}\Big(\mathbf{R}_{\xi}(t)\Big)}{\epsilon\Big(\mathbf{R}_{\xi}(t)\Big)} \\ &\quad\times\left[\hat{A}_{\alpha}(t)\hat{\boldsymbol{\mu}}_{\xi}(t)\cdot\mathbf{U}_{\alpha}\Big(\mathbf{R}_{\xi}(t)\Big) - \hat{A}^{\dagger}_{\alpha}(t)\hat{\boldsymbol{\mu}}_{\xi}(t)\cdot\mathbf{U}^{*}_{\alpha}\Big(\mathbf{R}_{\xi}(t)\Big)\right], \qquad (3.772)\\ &= \sum_{\xi} \frac{\hat{\boldsymbol{\mu}}_{\xi}(t)\cdot\hat{\boldsymbol{\Pi}}\Big(\mathbf{R}_{\xi}(t),t\Big)}{\epsilon\Big(\mathbf{R}_{\xi}(t)\Big)}. \qquad (3.773)\end{aligned}$$

The quantum Hamiltonian in the electric dipole approximation and rotating wave approximation is

$$\hat{H}'_{\mathrm{E1,RWA}}(t) = \hat{H}_{\mathrm{A}}(t) + \hat{H}_{\mathrm{Q}}(t) + \hat{V}^{\mathrm{E1}}_{\mathrm{A-F}}(t) + \hat{V}^{\mathrm{RWA}}_{\mathrm{Q-Q}}(t), \tag{3.774}$$

with

$$\hat{H}_{\mathrm{A}}(t) = \sum_{\xi,\alpha} \frac{\hat{\mathbf{p}}^2_{\xi\alpha}(t)}{2M_{\xi\alpha}} + \hat{V}^{\mathrm{IA}}_{\mathrm{coul}}(t) + \hat{V}^{\mathrm{IA}}_{\mathrm{pol}}(t). \tag{3.775}$$

The coupling constants describing energy exchange processes between a radiative atom placed in the cavity and nonaxial external quasimodes vary slowly with the external quasimode frequency. It follows that Markovian spontaneous emission damping occurs for the radiative atoms. On the contrary, their coupling with the (axial) cavity quasimodes consists in reversible photon exchanges as characterized through one-photon Rabi frequencies.

In the analysis, the standard model in cavity quantum electrodynamics has been considered. In the model the basic processes are described by a cavity damping rate, a radiative atom spontaneous decay rate, and an atom–cavity mode coupling constant. This model has been justified in terms of the quasimode theory of macroscopic canonical quantization (Dalton and Knight 1999a,b).

3.5.4 Several Sets of Quasimodes

The quasimode theory of macroscopic quantization (Dalton et al. 1999b,c) has been generalized (Brown and Dalton 2001a,b). The generalization allows for the case where two or more quasipermittivities are introduced, along with their associated sets of quasimode functions. This suggests problems such as reflection and refraction at a dielectric boundary, the linear coupler, and the coupling of two optical cavities. The theory comprises the above relations (3.579), (3.744), (3.745), (3.746), (3.747), (3.748), (3.586), and (3.750).

In some situations, such as a single laser cavity or a beam splitter, it suffices to consider just a single quasipermittivity function in order to obtain suitable quasimode functions (Dalton et al. 1999b,c). A full quasimode treatment of the beam splitter has been given in Dalton et al. (1999d). In other situations, the linear coupler (Lai et al. 1991) being an example, it is appropriate to construct quasimode functions via the introduction of two distinct quasipermittivities, each with its own set of associated mode functions.

We assume N sets of quasimode functions $\mathbf{U}_\alpha^{(l)}(\mathbf{R})$ $(l = 1, \ldots, N)$, which are defined as the solutions of N separate Helmholtz equations involving the quasipermittivities and quasipermeabilities $\tilde{\varepsilon}^{(l)}(\mathbf{R})$, $\tilde{\mu}^{(l)}(\mathbf{R})$, respectively. With $\lambda_\alpha^{(l)}$ the angular frequency of the (l, α) quasimode, relations (3.594), (3.595), and (3.596) are generalized,

$$\nabla \times \frac{1}{\tilde{\mu}^{(l)}(\mathbf{R})}[\nabla \times \mathbf{U}_\alpha^{(l)}(\mathbf{R})] = (\lambda_\alpha^{(l)})^2 \tilde{\varepsilon}^{(l)}(\mathbf{R})\mathbf{U}_\alpha^{(l)}(\mathbf{R}), \tag{3.776}$$

$$\nabla \cdot \tilde{\varepsilon}^{(l)}(\mathbf{R})\mathbf{U}_\alpha^{(l)}(\mathbf{R}) = 0, \tag{3.777}$$

$$\int \tilde{\varepsilon}^{(l)}(\mathbf{R})\mathbf{U}_\alpha^{(l)*}(\mathbf{R}) \cdot \mathbf{U}_\beta^{(l)}(\mathbf{R})\, \mathrm{d}^3\mathbf{R} = \delta_{\alpha\beta}. \tag{3.778}$$

Expansion of the vector potential $\mathbf{A}(\mathbf{R})$ directly in terms of the quasimode functions $\mathbf{U}_\alpha^{(l)}(\mathbf{R})$ is not possible. Instead we can write

$$\frac{\epsilon(\mathbf{R})}{\tilde{\epsilon}(\mathbf{R})}\mathbf{A}(\mathbf{R}, t) = \sum_{l,m} \sum_{\alpha,\beta} Q_\alpha^{(l)}(t) K_{\alpha\beta}^{(l,m)} \mathbf{U}_\beta^{(m)}(\mathbf{R}), \tag{3.779}$$

which involves a double sum over all quasimodes. The square matrix **K** has become a block matrix **K**, composed of $\mathbf{K}^{(l,m)}$.

The quasimode functions $\mathbf{U}_\alpha^{(l)}(\mathbf{R})$ for $N > 1$ are not all linearly independent. The set of quasimodes arising from N different quasipermittivities is overcomplete. It is solved by following an analogy with the theory of linear combinations of atomic orbitals (Coulson 1952). In that theory only the lower energy atomic orbitals are included. Here only quasimodes with the frequencies are retained that are important for the quantum optics system.

When applying the theory to situations where the true or quasi permittivities and permeabilities contain discontinuities, space integrals are cured with excluding infinitesimal volumes containing these discontinuities from their domain.

Relation (3.753) is generalized,

$$\begin{aligned} L'(t) &= \frac{1}{2}\sum_{\alpha,\xi} M_{\alpha\xi}\dot{\mathbf{r}}_{\alpha\xi}^2(t) - V_{\mathrm{coul}}(t) \\ &+ \sum_{\alpha,\xi} q_{\alpha\xi}\dot{\mathbf{r}}_{\alpha\xi}(t)\cdot\int_0^1 u\mathbf{B}\Big(\mathbf{R}_\xi(t)+u[\mathbf{r}_{\xi\alpha}(t)-\mathbf{R}_\xi(t)],t\Big)\mathrm{d}u \\ &\times [\mathbf{r}_{\xi\alpha}(t)-\mathbf{R}_\xi(t)] \\ &+\frac{1}{2}\sum_{l,m}\sum_{\alpha,\beta}\dot{Q}_\alpha^{(l)*}(t)(\mathbf{W}^{-1})_{\alpha\beta}^{(l,m)}\dot{Q}_\beta^{(m)}(t) - \frac{1}{2}\sum_{\alpha,\beta} Q_\alpha^{(l)*}(t)V_{\alpha\beta}^{(l,m)}Q_\beta^{(m)}(t) \\ &-\sum_\alpha \dot{Q}_\alpha^{(l)*}(t)N_\alpha^{(l)}(t). \end{aligned} \tag{3.780}$$

Some of the matrices become block matrices of the form (given for an arbitrary case **Y**)

$$\mathbf{Y} = \begin{pmatrix} \mathbf{Y}^{(1,1)} & \mathbf{Y}^{(1,2)} & \ldots & \mathbf{Y}^{(1,N)} \\ \mathbf{Y}^{(2,1)} & \mathbf{Y}^{(2,2)} & \ldots & \mathbf{Y}^{(2,N)} \\ \vdots & \vdots & \ddots & \vdots \\ \mathbf{Y}^{(N,1)} & \mathbf{Y}^{(N,2)} & \ldots & \mathbf{Y}^{(N,N)} \end{pmatrix} \tag{3.781}$$

and column block matrices of the form (given for an arbitrary case **C**)

$$\mathbf{C} = \begin{pmatrix} \mathbf{C}^{(1,N)} \\ \mathbf{C}^{(2,N)} \\ \vdots \\ \mathbf{C}^{(N,N)} \end{pmatrix}. \tag{3.782}$$

A matrix **Y** becomes the block matrix **Y** and a matrix **C** becomes the column block matrix **C**, but the notation is not changed. Relations (3.606), (3.607), and (3.754) are generalized

$$M_{\alpha\beta}^{(l,m)} = \int \epsilon(\mathbf{R})\frac{\tilde{\epsilon}^{(l)}(\mathbf{R})}{\epsilon(\mathbf{R})}\mathbf{U}_\alpha^{(l)*}(\mathbf{R})\cdot\frac{\tilde{\epsilon}^{(m)}(\mathbf{R})}{\epsilon(\mathbf{R})}\mathbf{U}_\beta^{(m)}(\mathbf{R})\,\mathrm{d}^3\mathbf{R}, \tag{3.783}$$

$$H_{\alpha\beta}^{(l,m)} = \int \frac{1}{\mu(\mathbf{R})} \left[\nabla \times \frac{\tilde{\epsilon}^{(l)}(\mathbf{R})}{\epsilon(\mathbf{R})} \mathbf{U}_\alpha^{(l)*}(\mathbf{R})\right] \cdot \left[\nabla \times \frac{\tilde{\epsilon}^{(m)}(\mathbf{R})}{\epsilon(\mathbf{R})} \mathbf{U}_\beta^{(m)}(\mathbf{R})\right] \mathrm{d}^3\mathbf{R}, \tag{3.784}$$

$$L_\alpha^{(l)}(t) = \int \frac{\tilde{\epsilon}^{(l)}(\mathbf{R})}{\epsilon(\mathbf{R})} \mathbf{U}_\alpha^{(l)*}(\mathbf{R}) \cdot \mathbf{P}_\mathrm{L}(\mathbf{R}, t)\, \mathrm{d}^3\mathbf{R}. \tag{3.785}$$

Relation (3.756) is generalized,

$$P_\alpha^{(l)}(t) = \sum_j \sum_\beta (\mathbf{M}^{-1})_{\alpha\beta}^{(l,m)} \dot{Q}_\beta^{(m)}(t) - N_\alpha^{(l)}(t). \tag{3.786}$$

Relation (3.757) is generalized:

$$\begin{aligned}
H'(t) = {} & \sum_{\alpha,\xi} \frac{\mathbf{p}_{\alpha\xi}^2(t)}{2M_{\alpha\xi}} + V_\mathrm{coul}(t) + \frac{1}{2} \sum_{l,m} \sum_{\alpha,\xi} N_\alpha^{(l)*}(t) M_{\alpha\xi}^{(l,m)} N_\beta^{(m)}(t) \\
& + \frac{1}{2} \sum_{l,m} \sum_{\alpha,\beta} P_\alpha^{(l)*}(t) M_{\alpha\beta}^{(l,m)} P_\beta^{(m)}(t) + \frac{1}{2} \sum_{l,m} \sum_{\alpha,\beta} Q_\alpha^{l*}(t) V_{\alpha\beta}^{(l,m)} Q_\beta^{(m)}(t) \\
& + \sum_{l,m} \sum_{\alpha,\beta} P_\alpha^{(l)*}(t) M_{\alpha\beta}^{(l,m)} N_\beta^{(m)}(t) \\
& + \frac{1}{2} \sum_{l,m} \sum_{\alpha,\beta} Q_\alpha^{(l)*}(t) (\mathbf{M}^{-1})_{\alpha\beta}^{(l,m)*} R_\beta^{(m)}(t) \\
& + \frac{1}{2} \sum_{l,m} \sum_{\alpha,\beta} Q_\alpha^{(l)*}(t) X_{\alpha\beta}^{(l,m)}(t) Q_\beta^{(m)}(t).
\end{aligned} \tag{3.787}$$

The block matrix $\mathbf{X}(t)$ is given by (3.758), where the notation has the actual meaning. Relations (3.760) and (3.759) are generalized,

$$\begin{aligned}
D_{\alpha\beta}^{(l,m)}(t) = {} & \sum_{\mu,\xi} \frac{q_{\mu\xi}^2}{M_{\mu,\xi}} \int \int \int_0^1 u \int_0^1 u' \delta\Big(\mathbf{R} - \mathbf{R}_\xi(t) - u[\mathbf{r}_{\mu\xi}(t) - \mathbf{R}_\xi(t)]\Big) \\
& \times \delta\Big(\mathbf{R}' - \mathbf{R}'_\xi(t) - u'[\mathbf{r}_{\mu\xi}(t) - \mathbf{R}_\xi(t)]\Big)\, \mathrm{d}u'\, \mathrm{d}u \\
& \times \Bigg\{ [\mathbf{r}_{\mu\xi}(t) - \mathbf{R}_\xi(t)]^2 \left[\nabla \times \frac{\tilde{\epsilon}^{(l)}(\mathbf{R})}{\epsilon(\mathbf{R})} \mathbf{U}_\alpha^{(l)*}(\mathbf{R})\right] \cdot \left[\nabla \times \frac{\tilde{\epsilon}^{(m)}(\mathbf{R})}{\epsilon(\mathbf{R})} \mathbf{U}_\beta^{(m)}(\mathbf{R})\right]' \\
& - [\mathbf{r}_{\mu\xi}(t) - \mathbf{R}_\xi(t)] \cdot \left[\nabla \times \frac{\tilde{\epsilon}^{(l)}(\mathbf{R})}{\epsilon(\mathbf{R})} \mathbf{U}_\alpha^{(l)*}(\mathbf{R})\right] [\mathbf{r}_{\mu\xi}(t) - \mathbf{R}_\xi(t)] \\
& \cdot \left[\nabla \times \frac{\tilde{\epsilon}^{(m)}(\mathbf{R})}{\epsilon(\mathbf{R})} \mathbf{U}_\beta^{(m)}(\mathbf{R})\right]' \Bigg\} \mathrm{d}^3\mathbf{R}\, \mathrm{d}^3\mathbf{R}',
\end{aligned} \tag{3.788}$$

$$R_\alpha^{(l)}(t) = - \int \mathbf{M}'_\mathrm{L}(\mathbf{R}, t) \cdot \left[\nabla \times \frac{\tilde{\epsilon}^{(l)}(\mathbf{R})}{\epsilon(\mathbf{R})} \mathbf{U}_\alpha^{(l)*}(\mathbf{R})\right] \mathrm{d}^3\mathbf{R}. \tag{3.789}$$

Relation (3.761) is generalized:

$$\mathbf{P}'_{\mathrm{L}}(\mathbf{R},t)=\sum_l\sum_\alpha(\mathbf{M}^{-1}\mathbf{L}(t))_\alpha^{(l)}\tilde{\epsilon}^{(l)}(\mathbf{R})\mathbf{U}_\alpha^{(l)}(\mathbf{R}). \tag{3.790}$$

Replacements (3.615), (3.616) are adapted,

$$Q_\alpha^{(l)}(t)\to\hat{Q}_\alpha^{(l)}(t),\ Q_\alpha^{(l)*}(t)\to\hat{Q}_\alpha^{(l)\dagger}(t), \tag{3.791}$$
$$P_\alpha^{(l)}(t)\to\hat{P}_\alpha^{(l)}(t),\ P_\alpha^{(l)*}(t)\to\hat{P}_\alpha^{(l)\dagger}(t). \tag{3.792}$$

The nonzero equal-time commutators are (cf. (3.617))

$$[\hat{Q}_\alpha^{(l)}(t),\hat{P}_\beta^{(m)\dagger}(t)]=\mathrm{i}\hbar\delta_{lm}\delta_{\alpha\beta}\hat{1}=[\hat{Q}_\alpha^{(l)\dagger}(t),\hat{P}_\beta^{(m)}(t)]. \tag{3.793}$$

The electromagnetic field is real, which somewhat complicates the quantization (Brown and Dalton 2001a).

Relations (3.618) and (3.619) are generalized,

$$\hat{H}_Q(t)=\frac{1}{2}\sum_l\sum_\alpha\left[\hat{P}_\alpha^{(l)\dagger}(t)W_{\alpha\alpha}^{(l,l)}\hat{P}_\alpha^{(l)}(t)+\hat{Q}_\alpha^{(l)\dagger}(t)V_{\alpha\alpha}^{(l,l)}\hat{Q}_\alpha^{(l)}(t)\right], \tag{3.794}$$

$$\hat{V}_{Q-Q}(t)=\frac{1}{2}\times\sum_{\substack{l,m\\(l,\alpha)\neq(m,\beta)}}\sum_{\alpha,\beta}\left[\hat{P}_\alpha^{(l)\dagger}(t)W_{\alpha\beta}^{(l,m)}\hat{P}_\beta^{(m)}(t)+\hat{Q}_\alpha^{(l)\dagger}(t)V_{\alpha\beta}^{(l,m)}\hat{Q}_\beta^{(m)}(t)\right]. \tag{3.795}$$

Relations (3.621), (3.622), and (3.623) are generalized,

$$\hat{A}_\alpha^{(l)}(t)=\sqrt{\frac{\eta_\alpha^{(l)}}{2\hbar}}\hat{Q}_\alpha^{(l)}(t)+\mathrm{i}\sqrt{\frac{1}{2\hbar\eta_\alpha^{(l)}}}\hat{P}_\alpha^{(l)}(t), \tag{3.796}$$

$$\hat{A}_\alpha^{(l)\dagger}(t)=\sqrt{\frac{\eta_\alpha^{(l)}}{2\hbar}}\hat{Q}_\alpha^{(l)\dagger}(t)-\mathrm{i}\sqrt{\frac{1}{2\hbar\eta_\alpha^{(l)}}}\hat{P}_\alpha^{(l)\dagger}(t), \tag{3.797}$$

where

$$\eta_\alpha^{(l)}=\sqrt{\frac{V_{\alpha\alpha}^{(l,l)}}{W_{\alpha\alpha}^{(l,l)}}}. \tag{3.798}$$

It follows simply from (3.793) that these annihilation and creation operators obey the following equal-time nonzero commutation relations:

$$[\hat{A}_\alpha^{(l)}(t),\hat{A}_\beta^{(m)\dagger}(t)]=\delta_{lm}\delta_{\alpha\beta}\hat{1}. \tag{3.799}$$

Relations (3.624) and (3.625) are generalized,

$$\hat{Q}_\alpha^{(l)}(t) = \sqrt{\frac{\hbar}{2\eta_\alpha^{(l)}}} \left[\hat{A}_\alpha^{(l)}(t) + \hat{A}_{-\alpha}^{(l)\dagger}(t)\right], \tag{3.800}$$

$$\hat{P}_\alpha^{(l)}(t) = -\mathrm{i}\sqrt{\frac{\hbar\eta_\alpha^{(l)}}{2}} \left[\hat{A}_\alpha^{(l)}(t) - \hat{A}_{-\alpha}^{(l)\dagger}(t)\right]. \tag{3.801}$$

Relation (3.626) is generalized,

$$\hat{H}_Q(t) = \hbar \sum_l \sum_\alpha \left[\hat{A}_\alpha^{(l)\dagger}(t)\hat{A}_\alpha^{(l)}(t) + \frac{1}{2}\hat{1}\right] \mu_\alpha^{(l)}, \tag{3.802}$$

Relation (3.620) is generalized,

$$\mu_\alpha^{(l)} = \sqrt{W_{\alpha\alpha}^{(l,l)} V_{\alpha\alpha}^{(l,l)}}. \tag{3.803}$$

Let us recall that

$$\hat{V}_{Q-Q}(t) = \hat{V}_{Q-Q}^{\mathrm{RWA}}(t) + \hat{V}_{Q-Q}^{\mathrm{non-RWA}}(t), \tag{3.804}$$

where $\hat{V}_{Q-Q}^{\mathrm{RWA}}(t)$ is generalized,

$$\hat{V}_{Q-Q}^{\mathrm{RWA}}(t) = \frac{\hbar}{2} \times \sum_{\substack{l,m \\ (l,\alpha)\neq(m,\beta)}} \sum_{\alpha,\beta} \left(\sqrt{\eta_\alpha^{(l)}\eta_\beta^{(m)}} M_{\alpha\beta}^{(l,m)} + \frac{V_{\alpha\beta}^{(l,m)}}{\sqrt{\eta_\alpha^{(l)}\eta_\beta^{(m)}}}\right) \hat{A}_\alpha^{(l)\dagger}(t)\hat{A}_\beta^{(m)}(t), \tag{3.805}$$

and $\hat{V}_{Q-Q}^{\mathrm{non-RWA}}(t)$ is also generalized,

$$\hat{V}_{Q-Q}^{\mathrm{non-RWA}}(t) = \frac{\hbar}{4} \sum_{\substack{l,m \\ (l,\alpha)\neq(m,\beta)}} \sum_{\alpha,\beta} \times \left[\left(-\sqrt{\eta_\alpha^{(l)}\eta_\beta^{(m)}} M_{\alpha,-\beta}^{(l,m)} + \frac{V_{\alpha,-\beta}^{(l,m)}}{\sqrt{\eta_\alpha^{(l)}\eta_\beta^{(m)}}}\right) \hat{A}_\alpha^{(l)\dagger}(t)\hat{A}_\beta^{(m)\dagger}(t) + \mathrm{H.c.}\right]. \tag{3.806}$$

Relations (3.630) and (3.631) are generalized,

$$\hat{\mathbf{A}}(\mathbf{R}, t) = \sum_{l,m} \sum_{\alpha,\beta} \sqrt{\frac{\hbar}{2\eta_\alpha^{(l)}}} \frac{\tilde{\epsilon}^{(m)}(\mathbf{R})}{\epsilon(\mathbf{R})} \times \left[K_{\alpha\beta}^{(l,m)} \hat{A}_\alpha^{(l)}(t)\mathbf{U}_\beta^{(m)}(\mathbf{R}) + K_{\alpha\beta}^{(l,m)*} \hat{A}_\alpha^{(l)\dagger}(t)\mathbf{U}_\beta^{(m)*}(\mathbf{R})\right], \tag{3.807}$$

$$\hat{\mathbf{\Pi}}(\mathbf{R},t) = -\mathrm{i}\sum_{\alpha}\sqrt{\frac{\hbar\eta_\alpha^{(l)}}{2}}\tilde{\epsilon}^{(l)}(\mathbf{R}) \times \left[\hat{A}_\alpha^{(l)}(t)\mathbf{U}_\alpha^{(l)}(\mathbf{R}) - \hat{A}_\alpha^{(l)\dagger}(t)\mathbf{U}_\alpha^{(l)*}(\mathbf{R})\right]. \tag{3.808}$$

The theory comprises the above relations (3.764), (3.767), (3.768), and (3.769). Relation (3.772), which comprises the annihilation and creation operators, is generalized,

$$\hat{V}_{\mathrm{A-F}}^{\mathrm{E1}}(t) = -\mathrm{i}\sum_{\xi}\sum_{l,\alpha}\sqrt{\frac{\hbar\eta_\alpha^{(l)}}{2}}\left[\frac{\tilde{\epsilon}^{(l)}(\mathbf{R})}{\epsilon(\mathbf{R})}\right]_{\mathbf{R}_\xi(t)} \times \left[\hat{A}_\alpha^{(l)}(t)\hat{\boldsymbol{\mu}}_\xi(t)\cdot\mathbf{U}_\alpha^{(l)}\left(\mathbf{R}_\xi(t)\right) - \hat{A}_\alpha^{(l)\dagger}(t)\hat{\boldsymbol{\mu}}_\xi(t)\cdot\mathbf{U}_\alpha^{(l)*}\left(\mathbf{R}_\xi(t)\right)\right]. \tag{3.809}$$

Let us recall that in the rotating wave and electric dipole approximations we can write the quantum Hamiltonian as

$$\hat{H}'_{\mathrm{E1,RWA}}(t) = \hat{H}_{\mathrm{A}}(t) + \hat{H}_{\mathrm{Q}}(t) + \hat{V}_{\mathrm{A-F}}^{\mathrm{E1}}(t) + \hat{V}_{\mathrm{Q-Q}}^{\mathrm{RWA}}(t). \tag{3.810}$$

The coupling constants $M_{\alpha\beta}^{(l,m)}$, $V_{\alpha\beta}^{(l,m)}$ can be calculated from the matrices **M** and **H** using relations (3.609) and (3.610) (Brown and Dalton 2001a). For the usual case where $\tilde{\mu}^{(l)}(\mathbf{R})$=$\mu(\mathbf{R})$ and the overlap between the set l and the set m of mode functions is small, to good accuracy we have

$$M_{\alpha\beta}^{(l,m)} \approx \begin{cases} 1, & i = j, \alpha = \beta, \\ (M_1)_{\alpha\beta}^{(l,m)} + (M_2)_{\alpha\beta}^{(l,m)}, & \text{otherwise,} \end{cases} \tag{3.811}$$

$$V_{\alpha\beta}^{(l,m)} \approx \begin{cases} \lambda_\alpha^{(l)2}, & i = j, \alpha = \beta, \\ (H_1)_{\alpha\beta}^{(l,m)} + (H_2)_{\alpha\beta}^{(l,m)}, & \text{otherwise.} \end{cases} \tag{3.812}$$

Relations (3.636), (3.637) are generalized and also modified,

$$(M_1)_{\alpha\beta}^{(l,m)} = \int \tilde{\epsilon}(\mathbf{R})\left[\frac{\tilde{\epsilon}^{(l)}(\mathbf{R})}{\epsilon(\mathbf{R})} - 1\right]\left[\frac{\tilde{\epsilon}^{(m)}(\mathbf{R})}{\epsilon(\mathbf{R})} - 1\right]\mathbf{U}_\alpha^{(l)*}(\mathbf{R}) \cdot \mathbf{U}_\beta^{(m)}(\mathbf{R})\,\mathrm{d}^3\mathbf{R}, \tag{3.813}$$

$$(M_2)_{\alpha\beta}^{(l,m)} = \int \tilde{\epsilon}(\mathbf{R})\left[\frac{\tilde{\epsilon}^{(l)}(\mathbf{R})}{\epsilon(\mathbf{R})} + \frac{\tilde{\epsilon}^{(m)}(\mathbf{R})}{\epsilon(\mathbf{R})} - 1\right]\mathbf{U}_\alpha^{(l)*}(\mathbf{R}) \cdot \mathbf{U}_\beta^{(m)}(\mathbf{R})\,\mathrm{d}^3\mathbf{R}, \tag{3.814}$$

$$(H_1)_{\alpha\beta}^{(l,m)} = \int \frac{1}{\mu(\mathbf{R})} \left\{ \nabla \times \left[\frac{\tilde{\epsilon}^{(l)}(\mathbf{R})}{\epsilon(\mathbf{R})} - 1 \right] \mathbf{U}_\alpha^{(l)*}(\mathbf{R}) \right\}$$
$$\cdot \left\{ \nabla \times \left[\frac{\tilde{\epsilon}^{(m)}(\mathbf{R})}{\epsilon(\mathbf{R})} - 1 \right] \mathbf{U}_\beta^{(m)}(\mathbf{R}) \right\} \, \mathrm{d}^3\mathbf{R}, \tag{3.815}$$

$$(H_2)_{\alpha\beta}^{(l,m)} = -\frac{1}{2} \int \tilde{\epsilon}(\mathbf{R}) \left[\lambda_\alpha^{(m)2} \frac{\tilde{\epsilon}^{(l)}(\mathbf{R})}{\epsilon(\mathbf{R})} + \lambda_\beta^{(m)2} \frac{\tilde{\epsilon}^{(m)}(\mathbf{R})}{\epsilon(\mathbf{R})} \right]$$
$$\times \mathbf{U}_\alpha^{(l)*}(\mathbf{R}) \cdot \mathbf{U}_\beta^{(m)}(\mathbf{R}) \, \mathrm{d}^3\mathbf{R}. \tag{3.816}$$

Relation (3.632) is generalized,

$$\mu_\alpha^{(l)} \approx \lambda_\alpha^{(l)} \tag{3.817}$$

and relation (3.633) is also generalized,

$$\dot{\mu}_\alpha^{(l)} \approx \lambda_\alpha^{(l)} + v_{\alpha\alpha}^{(l,l)}. \tag{3.818}$$

Relation (3.634) is generalized,

$$\hat{V}_{Q-Q}^{\mathrm{RWA}}(t) \approx \hbar \sum_{l,m} \sum_{\substack{\alpha,\beta \\ (l,\alpha)\neq(m,\beta)}} v_{\alpha\beta}^{(l,m)} \hat{A}_\alpha^{(l)\dagger}(t) \hat{A}_\beta^{(m)}(t). \tag{3.819}$$

Relation (3.635) is generalized,

$$v_{\alpha\beta}^{(l,m)} = \frac{1}{2} \left\{ \sqrt{\lambda_\alpha^{(l)} \lambda_\beta^{(m)}} \left[(M_1)_{\alpha\beta}^{(l,m)} + (M_2)_{\alpha\beta}^{(l,m)} \right] \right.$$
$$\left. + \frac{(H_1)_{\alpha\beta}^{(l,m)} - (H_2)_{\alpha\beta}^{(l,m)}}{\sqrt{\lambda_\alpha^{(l)} \lambda_\beta^{(m)}}} \right\}. \tag{3.820}$$

The foregoing theory has been applied to reflection and refraction at a dielectric interface (Brown and Dalton 2001b). The true mode approach has continued the previous literature, e.g. Allen and Stenholm (1992) or Carniglia and Mandel (1971). The analysis has been very thorough including the quantum scattering theory in the Heisenberg picture. The behaviour of the intensity for a localized one-photon wave packet has been examined, which has exhibited agreement with the classical laws of reflection and refraction. Such an accord is described also by the quantum theory based on a microscopic model of the dielectric media (Hynne and Bullough 1990).

Here we will expound the quasimode approach in part. We shall assume that space has been divided into two regions. Region 1, which is formed by the points with $z \geq 0$, is assumed to be filled with linear, homogeneous dielectric material of refractive index n_1. Region 2, which consists of the points with $z < 0$, is assumed to contain material obeying the same restrictions, but with refractive index n_2. The permittivity function for the system, $\epsilon(z)$, is then

$$\epsilon(z) = \begin{cases} n_1^2\epsilon_0, & z \geq 0, \\ n_2^2\epsilon_0, & z < 0. \end{cases} \tag{3.821}$$

We could try to use two quasipermittivity functions, one being $n_1^2\epsilon_0$ in all space and the other being $n_2^2\epsilon_0$ in all space. With the two values, two sets of plane waves are associated. The union of these sets does not enjoy the mutual orthogonality of all functions. Instead we choose two sets of quasimodes, each set effectively being restricted to just one of the regions. At a closer look, the functions are not confined in one region, but are evanescent in the other region. An effective mutual orthogonality is present. A completeness of the union of these sets is also available. The spatially confined nature of these types of mode functions is also used when applying a quantum scattering theory approach to energy transfer from one region to another.

For the reflection and refraction problems we choose the two quasipermittivity functions (Brown and Dalton 2001b),

$$\tilde{\epsilon}^{(1)}(z) = \begin{cases} n_1^2\epsilon_0, & z \geq 0, \\ (\tilde{n}_2)^2\epsilon_0, & z < 0, \end{cases} \tag{3.822}$$

$$\tilde{\epsilon}^{(2)}(z) = \begin{cases} (\tilde{n}_1)^2\epsilon_0, & z \geq 0, \\ n_2^2\epsilon_0, & z < 0, \end{cases} \tag{3.823}$$

where the quasirefractive indices $\tilde{n}_1$ and $\tilde{n}_2$ are positive constants which fulfil $\tilde{n}_1, \tilde{n}_2 \ll 1$. The vanishingly small refractive index in one region means that all incident waves in the other region except some with angles of incidence smaller than the critical angle produce only an evanescent wave in the region with negligible refractive index.

From the generalized Helmholtz equation (3.776), we can determine the form of the quasimode functions $\mathbf{U}_\alpha^{(1)}(\mathbf{R})$ and $\mathbf{U}_\alpha^{(2)}(\mathbf{R})$, which are associated with the quasipermittivities $\tilde{\epsilon}^{(1)}(z)$ and $\tilde{\epsilon}^{(2)}(z)$, respectively. We will treat the case of out-of-plane polarization. The quasimode functions are to an excellent approximation given by the formulae

$$\begin{aligned} \mathbf{U}_\alpha^{(1)}(\mathbf{R}) \approx{}& \mathcal{N}_\alpha^{(1)} \\ &\times \left\{ \left[\sqrt{\tilde{r}_1^*} \exp(\mathrm{i}\mathbf{k}_{\alpha\mathrm{i}}^{(1)} \cdot \mathbf{R}) + \sqrt{\tilde{r}_1} \exp(\mathrm{i}\mathbf{k}_{\alpha\mathrm{r}}^{(1)} \cdot \mathbf{R}) \right] \Theta(z) \right. \\ &\left. + \tilde{t}_1 \sqrt{\tilde{r}_1^*} \exp(\mathrm{i}\mathbf{k}_{\alpha\mathrm{t}}^{(1)} \cdot \mathbf{R})[1 - \Theta(z)] \right\} \boldsymbol{\sigma}, \end{aligned} \tag{3.824}$$

$$\begin{aligned} \mathbf{U}_\alpha^{(2)}(\mathbf{R}) \approx{}& \mathcal{N}_\alpha^{(2)} \\ &\times \left\{ \left[\sqrt{\tilde{r}_2^*} \exp(\mathrm{i}\mathbf{k}_{\alpha\mathrm{i}}^{(2)} \cdot \mathbf{R}) + \sqrt{\tilde{r}_2} \exp(\mathrm{i}\mathbf{k}_{\alpha\mathrm{r}}^{(2)} \cdot \mathbf{R}) \right] [1 - \Theta(z)] \right. \\ &\left. + \tilde{t}_2 \sqrt{\tilde{r}_2^*} \exp(\mathrm{i}\mathbf{k}_{\alpha\mathrm{t}}^{(2)} \cdot \mathbf{R}) \Theta(z) \right\} \boldsymbol{\sigma}, \end{aligned} \tag{3.825}$$

where

$$\sqrt{\tilde{r}_l} = \frac{1 + \mathrm{i}\tan(\tilde{\theta}_{\alpha\mathrm{i}}^{(l)})\tanh(\tilde{\theta}_{\alpha\mathrm{t}}^{(l)\prime\prime})}{\left|1 - \mathrm{i}\tan(\tilde{\theta}_{\alpha\mathrm{i}}^{(l)})\tanh(\tilde{\theta}_{\alpha\mathrm{t}}^{(l)\prime\prime})\right|},$$

$$\tilde{t}_l = \frac{2}{1 - \mathrm{i}\tan(\tilde{\theta}_{\alpha\mathrm{i}}^{(l)})\tanh(\tilde{\theta}_{\alpha\mathrm{t}}^{(l)\prime\prime})}, \tag{3.826}$$

with

$$\tan(\tilde{\theta}_{\alpha\mathrm{i}}^{(l)}) = \frac{|(\mathbf{k}_{\alpha\mathrm{i}}^{(l)})_\tau|}{(\mathbf{k}_{\alpha\mathrm{i}}^{(l)})_z},$$

$$\tanh(\tilde{\theta}_{\alpha\mathrm{t}}^{(l)\prime\prime}) = \pm\frac{\sqrt{|(\mathbf{k}_{\alpha\mathrm{i}}^{(l)})_\tau|^2 - |\mathbf{k}_{\alpha\mathrm{t}}^{(l)}|^2}}{|(\mathbf{k}_{\alpha\mathrm{i}}^{(l)})_\tau|}, \; + \text{ for } l = 1, \; - \text{ for } l = 2,$$

$$|\mathbf{k}_{\alpha\mathrm{t}}^{(1)}| = \frac{\tilde{n}_2}{n_1}|\mathbf{k}_{\alpha\mathrm{i}}^{(1)}|, \; |\mathbf{k}_{\alpha\mathrm{t}}^{(2)}| = \frac{\tilde{n}_1}{n_2}|\mathbf{k}_{\alpha\mathrm{i}}^{(2)}|. \tag{3.827}$$

Here $\Theta(z)$ is the step function and $\mathcal{N}_\alpha^{(l)}$ are normalization constants appropriate to the case where the evanescent wave has been neglected. We note that $\tilde{r}_l$ are complex units and $\tilde{t}_l\sqrt{\tilde{r}_l^*} = |\tilde{t}_l|$, which may justify an alternative phase factor.

The usual formulae are obtained by multiplying the right-hand sides by $\sqrt{\tilde{r}_l}$, which also simplifies them. Approximate expressions are obtained also by considering $\tilde{r}_l = -1$ and $\tilde{t}_l = 0$. On the modification the two sets of the functions are continuous contrary to the notation. It is obvious that the subscript α should be replaced by a variable $\mathbf{k}_\mathrm{i}^{(l)}$. The corresponding z-component should be negative for $l = 1$ and it should be positive for $l = 2$.

In the case of the continuous sets, the scalar product of the functions simplifies to a Dirac delta function when we choose

$$\mathcal{N}_\alpha^{(l)} = \left(\frac{1}{2\pi}\right)^{\frac{3}{2}} \frac{1}{n_l\sqrt{\epsilon_0}}. \tag{3.828}$$

As this value does not depend on $\mathbf{k}_\mathrm{i}^{(l)}$, the subscript α has been preserved. We have concluded that the discrete sets are not always obtained so easily as desired. The continuous sets of the mode functions in the case where the evanescent waves have been neglected are discretized easily. The use of quantization box of volume L^3 is immediate. We assume that the propagation vectors $\mathbf{k}_{\alpha\mathrm{i}}^{(l)}$ and $\mathbf{k}_{\alpha\mathrm{r}}^{(l)}$ obey

$$k_{\alpha\mathrm{i},X}^{(l)} = k_{\alpha\mathrm{r},X}^{(l)} = N_{\alpha,X}^{(l)}\frac{2\pi}{L}, \tag{3.829}$$

where $X = x, y$ and $N_{\alpha,X}^{(l)}$ are integers. We should complete the case $X = z$.

For $X = z$ the quantization box should not suggest the periodic boundary condition. We assume that the mode function vanish for $z = 0$ and $z = L$. Then of course

$$k_{\alpha\mathrm{i},z}^{(l)} = -k_{\alpha\mathrm{r},z}^{(l)} = N_{\alpha,z}^{(l)} \frac{\pi}{L}. \tag{3.830}$$

The integer $N_{\alpha,z}^{(l)}$ should be negative for $l = 1$ and it should be positive for $l = 2$. The modes with in-plane polarization are not considered for simplicity (Brown and Dalton 2001b).

Chapter 4
Microscopic Theories

A divergence from the macroscopic theories emerges, when the polarization of the medium is described by separate equations. In the framework of this approach the electric permittivity of the medium can be derived. The description of the fields can be quantized. It seems that the separation of the equations for the medium polarization is not a sufficient ground for the theory to be considered microscopic, but we adopt this nomenclature. It is important that the motion of the medium polarization may be damped and losses may be included. A quantum noise is considered for the field commutators not to depend on the time.

Many papers have been devoted to the Green-function approach to the quantization of the electromagnetic field in a medium. As this theory rather begins with a quantum noise, it differs formally from the method of continua of harmonic oscillators. The equivalence between the Green-function approach and the method of continua for media suitable for both approaches has been demonstrated, however. The Green-function approach has been elaborated on for various media, only the inclusion of a nonlinearity of the medium was under development in the course of writing this book.

The magnetic properties are usually neglected, but they must be included in the phenomenological quantum description of negative-index materials. Even though the Casimir effect is not regularly connected with the propagation, an expression of the noise which is quantal in essence fits in the framework of the electromagnetic-field quantization.

4.1 Method of Continua of Harmonic Oscillators

Many scientists would call the following exposition a macroscopic theory for a lossy medium, whereas we refer to a microscopic theory. A standard microscopic approach is expected from the quantum theory of solids. Still, in the quantum theory of solids, continua of harmonic oscillators have been considered, on which we shall concentrate ourselves in what follows. In the framework of this model, one can see the presence and correlation of fluctuations of the electric-field strength in the vacuum state of the field and the ground state of the matter.

A. Lukš, V. Peřinová, *Quantum Aspects of Light Propagation*,
DOI 10.1007/b101766_4, © Springer Science+Business Media, LLC 2009

4.1.1 Dispersive Lossy Homogeneous Linear Dielectric

Huttner and Barnett (1992a) have started from the observation that the macroscopic approach to the theory of the electromagnetic field in a medium is a quantization scheme that does accept dispersion, but not losses. So it does not deal with a fundamental property of the susceptibility, the Kramers–Kronig relations. Losses in quantum mechanics are treated by coupling to a reservoir, and thus a quantization scheme to describe the losses must introduce the medium explicitly.

A rigorous treatment is contained in the book of Klyshko (1988). Huttner and Barnett (1992a) use the model of Hopfield (1958) and Fano (1956), having first treated the quantization of light in a purely dispersive dielectric (Huttner et al. 1991) using a simple version of this model (Kittel 1987). Their analysis is restricted to a one-dimensional model and to transverse electromagnetic fields. After introducing the Lagrangian densities, the effect of choice of the type of coupling between light and matter on the definition of the conjugate variables for the components of the vector potential is discussed. The matter is not quite identical with the reservoir, but couplings rather form a chain, the radiation is coupled to the matter (it is a field again) and the matter is coupled to the reservoir (it is a field of the dimension increased by unity). Diagonalization by the Fano technique is performed (Fano 1961, Barnett and Radmore 1988), cf., (Rosenau da Costa et al. 2000).

Huttner and Barnett (1992a) work, as usual, with fields in the reciprocal space. Only at the very beginning the radiation and the matter in the direct space, and the reservoir in the Cartesian product of the direct and a reciprocal space are considered. The total transition to a direct space for the reservoir is not usual, but is conceivable.

The description of the matter and reservoir is first diagonalized. This diagonalization gives rise to the (dressed) matter-field $\hat{B}(k, \omega)$, whose operator exhibits the same dependence on the wave vector and the frequency as the operator of the reservoir field. It is proven that also the coupling constant dependent on at least the frequency of the reservoir "elementary" mode fulfils the assumptions for further diagonalization. This diagonalization gives origin to the field $\hat{C}(k, \omega)$ for polaritons. The operator of this field shares the dependence on the wave vector and the frequency with the operator of the reservoir field. In contrast with the vacuum theory (the theory of the electromagnetic field in a vacuum) a macroscopic field emerges this way whose operator depends also on the frequency.

The vector potential depends on the spatial coordinate and the time as usual and it has the form of the integral of the vector potential for a unit density of polaritons with the wave vector k and the frequency ω multiplied by the polariton operator $\hat{C}(k, \omega)$. The appropriate relation contains the complex relative permittivity of the medium $\epsilon(\omega)$ as a linear transform of the coupling constant $g(\omega)$ between the light and the dressed matter-field $\hat{B}(k, \omega)$. The complex relative permittivity $\epsilon(\omega)$ fulfils the Kramers–Kronig relations.

Taking into account the frequency decompositions of the fields $\hat{E}(x, t)$ and $\hat{B}(x, t)$ (see, Huttner and Barnett (1992b)), one can introduce, in an "almost" conventional manner, the positive and negative propagating components. These differ almost negligibly due to the imaginary part of the refractive index (see (3.506), (3.507), or (3.79)). That is to say that the fields $\hat{c}_+(x,\omega)$ and $\hat{c}_-(x,\omega)$ are introduced. Respecting the frequency decomposition of the field $\hat{D}(x,t)$, the fields $\hat{c}_\pm(x,\omega)$ are used and the spatial Langevin force $\hat{f}(x,\omega)$ is introduced. Using these definitions, two Maxwell equations are transformed onto two spatial Langevin equations. The equal-space commutation relations between the operators at the frequencies ω and ω' can also be derived. From this, simple equal-space commutation relations between the operators in the "application" times s and s' follow,

$$\hat{c}_{\pm\mathrm{dir}}(x,s) = \frac{1}{\sqrt{2\pi}} \int_0^\infty \hat{c}_\pm(x,\omega) \exp(-\mathrm{i}\omega s)\,\mathrm{d}\omega, \tag{4.1}$$

and that may be why Huttner and Barnett (1992a) name the papers devoted to the phenomenological approach to quantization (Levenson et al. 1985, Potasek and Yurke 1987, Caves and Crouch 1987, Lai and Haus 1989, Huttner et al. 1990).

Let us note that Huttner and Barnett (1992b) in the introduction mention also the popular approach (Huttner et al. 1990), in which spatial progression equations are derived and quantization of the field is performed imposing the equal-space commutation relations. In contrast with the macroscopic theories this technique is not derived from a Lagrangian and has not been justified in terms of a canonical scheme.

In Huttner and Barnett (1992b) the derivation of such equal-space commutation relations is provided in the case of a linear dielectric. The canonical scheme and losses cannot be easily unified, but this has been solved in Huttner and Barnett (1992b). The one-dimensional model has been expanded to three dimensions.

The Hamiltonian is first derived, then diagonalized, and the expansions of the field operators are transformed. The propagation of light in the dielectric is analysed, the field is expressed in terms of space-dependent amplitudes, and their spatial equations of evolution are obtained.

Huttner and Barnett (1992b) have started the canonical quantization from a Lagrangian density

$$\mathcal{L} = \mathcal{L}_{\mathrm{em}} + \mathcal{L}_{\mathrm{mat}} + \mathcal{L}_{\mathrm{res}} + \mathcal{L}_{\mathrm{int}}, \tag{4.2}$$

where

$$\mathcal{L}_{\mathrm{em}} = \frac{\epsilon_0}{2} \overbrace{(\dot{\mathbf{A}} + \nabla U)}^{-\mathbf{E}}{}^2 - \frac{1}{2\mu_0} \overbrace{(\nabla \times \mathbf{A})}^{\mathbf{B}}{}^2 \tag{4.3}$$

is the electromagnetic part which is expressed in terms of the vector potential $\mathbf{A}$ and the scalar potential U,

$$\mathcal{L}_{\mathrm{mat}} = \frac{\rho}{2}(\dot{\mathbf{X}}^2 - \omega_0^2 \mathbf{X}^2) \tag{4.4}$$

is the polarization part, modelled by a harmonic-oscillator field $\mathbf{X}$ of frequency ω_0 (the polarization field),

$$\mathcal{L}_{\mathrm{res}} = \frac{\rho}{2} \int_0^{\infty} (\dot{\mathbf{Y}}_\omega^2 - \omega^2 \mathbf{Y}_\omega^2)\, \mathrm{d}\omega \tag{4.5}$$

is the reservoir part, comprising a field $\mathbf{Y}_\omega$ of the continua of harmonic oscillators of frequencies ω, used to model the losses (reservoirs), and

$$\mathcal{L}_{\mathrm{int}} = -\alpha(\mathbf{A} \cdot \dot{\mathbf{X}} + U \nabla \cdot \mathbf{X}) - \mathbf{X} \cdot \int_0^{\infty} v(\omega) \dot{\mathbf{Y}}_\omega \, \mathrm{d}\omega \tag{4.6}$$

is the interaction part with coupling constants α and $v(\omega)$. The interaction between the light and the polarization field has the coupling constant α and the interaction between the polarization field and other oscillator fields used to model the losses has the coupling constant $v(\omega)$. In general, α could be a tensor.

The displacement field is defined by

$$\mathbf{D}(\mathbf{r}, t) = \epsilon_0 \mathbf{E}(\mathbf{r}, t) - \alpha \mathbf{X}(\mathbf{r}, t). \tag{4.7}$$

As $\dot{U}$ does not appear in the Lagrangian, U is not a proper dynamical variable, but it can be written in terms of the proper dynamical variable $\mathbf{X}$. The former has an integral expression and that is why we go to the reciprocal space. For example the electric field is written as

$$\mathbf{E}(\mathbf{r}, \mathbf{t}) = \frac{1}{(2\pi)^{\frac{3}{2}}} \int \underline{\mathbf{E}}(\mathbf{k}, t) e^{\mathrm{i}\mathbf{k}\cdot\mathbf{r}} \, \mathrm{d}^3\mathbf{k}. \tag{4.8}$$

We shall underline the newly introduced quantities in order to differentiate between the quantities in real and reciprocal spaces. Let us recall that $\underline{\mathbf{E}}^*(\mathbf{k}, t) = \underline{\mathbf{E}}(-\mathbf{k}, t)$. It comprises both the annihilation and the creation operators, see below.

The total Lagrangian can be written in the form

$$L = \int' (\underline{\mathcal{L}}_{\mathrm{em}} + \underline{\mathcal{L}}_{\mathrm{mat}} + \underline{\mathcal{L}}_{\mathrm{res}} + \underline{\mathcal{L}}_{\mathrm{int}}) \, \mathrm{d}^3\mathbf{k}, \tag{4.9}$$

where the prime means that the integration is restricted to half the reciprocal space and the Lagrangian densities become

$$\begin{aligned}
\underline{\mathcal{L}}_{\mathrm{em}} &= \epsilon_0 (|\underline{\mathbf{E}}|^2 - c^2 |\underline{\mathbf{B}}|^2), \\
\underline{\mathcal{L}}_{\mathrm{mat}} &= \rho (|\dot{\underline{\mathbf{X}}}|^2 - \omega_0^2 |\underline{\mathbf{X}}|^2), \\
\underline{\mathcal{L}}_{\mathrm{res}} &= \rho \int_0^{\infty} (|\dot{\underline{\mathbf{Y}}}_\omega|^2 - \omega^2 |\underline{\mathbf{Y}}_\omega|^2) \, \mathrm{d}\omega, \\
\underline{\mathcal{L}}_{\mathrm{int}} &= -\alpha [\underline{\mathbf{A}}^* \cdot \dot{\underline{\mathbf{X}}} + \underline{\mathbf{A}} \cdot \dot{\underline{\mathbf{X}}}^* + \mathrm{i}\mathbf{k} \cdot (\underline{U}^* \underline{\mathbf{X}} - \underline{U} \underline{\mathbf{X}}^*)] \\
&\quad - \int_0^{\infty} v(\omega) (\underline{\mathbf{X}}^* \cdot \dot{\underline{\mathbf{Y}}}_\omega + \underline{\mathbf{X}} \cdot \dot{\underline{\mathbf{Y}}}_\omega^*) \, \mathrm{d}\omega.
\end{aligned} \tag{4.10}$$

As usual in quantum optics, we choose the Coulomb gauge, $\mathbf{k} \cdot \underline{\mathbf{A}}(\mathbf{k}, t) = 0$, so that the vector potential $\mathbf{A}$ is a purely transverse field. The scalar potential in the reciprocal space

$$\underline{U}(\mathbf{k}, t) = \mathrm{i}\frac{\alpha}{\epsilon_0}\left(\frac{\boldsymbol{\kappa} \cdot \underline{\mathbf{X}}(\mathbf{k}, t)}{k}\right), \tag{4.11}$$

where $\boldsymbol{\kappa}$ is a unit vector in the direction of $\mathbf{k}$. The polarization field $\mathbf{X}$ and other oscillator fields $\mathbf{Y}_\omega$ (the matter fields) are decomposed into transverse and longitudinal parts. For example $\underline{\mathbf{X}}$ can be written as

$$\underline{\mathbf{X}}(\mathbf{k}, t) = \underline{\mathbf{X}}^{\perp}(\mathbf{k}, t) + \underline{X}^{\parallel}(\mathbf{k}, t)\boldsymbol{\kappa} \tag{4.12}$$

and $\underline{\mathbf{Y}}_\omega$ can be expressed similarly. The total Lagrangian can then be written as the sum of two independent parts. The transverse part contains only transverse fields and is

$$L^{\perp} = \int' (\underline{\mathcal{L}}_{\mathrm{em}}^{\perp} + \underline{\mathcal{L}}_{\mathrm{mat}}^{\perp} + \underline{\mathcal{L}}_{\mathrm{res}}^{\perp} + \underline{\mathcal{L}}_{\mathrm{int}}^{\perp})\, \mathrm{d}^3\mathbf{k}, \tag{4.13}$$

where

$$\begin{aligned}
\underline{\mathcal{L}}_{\mathrm{em}}^{\perp} &= \epsilon_0(|\dot{\underline{\mathbf{A}}}|^2 - c^2k^2|\underline{\mathbf{A}}|^2), \\
\underline{\mathcal{L}}_{\mathrm{mat}}^{\perp} &= \rho(|\dot{\underline{\mathbf{X}}}^{\perp}|^2 - \omega_0^2|\underline{\mathbf{X}}^{\perp}|^2), \\
\underline{\mathcal{L}}_{\mathrm{res}}^{\perp} &= \rho\int_0^\infty (|\dot{\underline{\mathbf{Y}}}_\omega^{\perp}|^2 - \omega^2|\underline{\mathbf{Y}}_\omega^{\perp}|^2)\, \mathrm{d}\omega, \\
\underline{\mathcal{L}}_{\mathrm{int}}^{\perp} &= -\left[\alpha\underline{\mathbf{A}} \cdot \dot{\underline{\mathbf{X}}}^{\perp *} + \int_0^\infty v(\omega)\underline{\mathbf{X}}^{\perp *} \cdot \dot{\underline{\mathbf{Y}}}_\omega^{\perp}\, \mathrm{d}\omega\right] + \mathrm{c.\,c.}
\end{aligned} \tag{4.14}$$

The longitudinal part, containing only longitudinal fields, is also given in Huttner and Barnett (1992b). It can be derived that $\mathbf{D}$ is a purely transverse field. For convenience, one can restrict oneself to transverse components of other fields and omit the superscript $^{\perp}$.

Unit polarization vectors $\mathbf{e}_\lambda(\mathbf{k})$, $\lambda = 1, 2$, are introduced, which are orthogonal to $\mathbf{k}$ and to one another, and the transverse fields are decomposed along them to get

$$\underline{\mathbf{A}}(\mathbf{k}, t) = \sum_{\lambda=1,2} \underline{A}^{\lambda}(\mathbf{k}, t)\mathbf{e}_\lambda(\mathbf{k}) \tag{4.15}$$

and similar expressions for the other fields. $\underline{\mathcal{L}}$ can now be used to obtain the components of the conjugate variables for the fields

$$-\epsilon_0\underline{E}^{\lambda} \equiv \frac{\partial\underline{\mathcal{L}}}{\partial\dot{\underline{A}}^{\lambda *}} = \epsilon_0\dot{\underline{A}}^{\lambda}, \tag{4.16}$$

$$\underline{P}^{\lambda} \equiv \frac{\partial \mathcal{L}}{\partial \dot{\underline{X}}^{\lambda *}} = \rho \dot{\underline{X}}^{\lambda} - \alpha \underline{A}^{\lambda}, \tag{4.17}$$

$$\underline{Q}_{\omega}^{\lambda} \equiv \frac{\delta \mathcal{L}}{\delta \dot{\underline{Y}}_{\omega}^{\lambda *}} = \rho \dot{\underline{Y}}_{\omega}^{\lambda} - v(\omega) \underline{X}^{\lambda}. \tag{4.18}$$

The famous ambiguity is worth mentioning: The conjugate of $\underline{\mathbf{A}}$ can be $-\epsilon_0 \underline{\mathbf{E}}$ (with the coupling $\frac{\alpha}{\rho} \underline{\mathbf{A}} \cdot \underline{\mathbf{P}}$), as well as $-\underline{\mathbf{D}}$ (with the coupling $\underline{\mathbf{E}} \cdot \underline{\mathbf{X}}$). Thus any gauge determines a type of coupling.

The Hamiltonian for the transverse fields is

$$H = \int' (\underline{\mathcal{H}}_{\text{em}} + \underline{\mathcal{H}}_{\text{mat}} + \underline{\mathcal{H}}_{\text{int}}) \, \text{d}^3\mathbf{k}, \tag{4.19}$$

where

$$\underline{\mathcal{H}}_{\text{em}} = \epsilon_0 \left(|\underline{\mathbf{E}}|^2 + c^2 \tilde{k}^2 |\underline{\mathbf{A}}|^2 \right) \tag{4.20}$$

is the electromagnetic energy density, $\tilde{k}$ being defined by $\tilde{k} = \sqrt{k^2 + k_c^2}$ with $k_c \equiv \frac{\omega_c}{c} = \sqrt{\frac{\alpha^2}{\rho c^2 \epsilon_0}}$,

$$\underline{\mathcal{H}}_{\text{mat}} = \frac{|\underline{\mathbf{P}}|^2}{\rho} + \rho \tilde{\omega}_0^2 |\underline{\mathbf{X}}|^2 + \int_0^{\infty} \left[\frac{|\underline{\mathbf{Q}}_{\omega}|^2}{\rho} + \rho \omega^2 |\underline{\mathbf{Y}}_{\omega}|^2 + \frac{v(\omega)}{\rho} (\underline{\mathbf{X}}^* \cdot \underline{\mathbf{Q}}_{\omega} + \text{c. c.}) \right] \text{d}\omega \tag{4.21}$$

is the energy density of the matter fields, including the interaction between the polarization and the reservoirs and $\tilde{\omega}_0^2 \equiv \omega_0^2 + \int_0^{\infty} \frac{[v(\omega)]^2}{\rho^2} \, \text{d}\omega$ is the renormalized frequency of the polarization field,

$$\underline{\mathcal{H}}_{\text{int}} = \frac{\alpha}{\rho} (\underline{\mathbf{A}}^* \cdot \underline{\mathbf{P}} + \text{c. c.}) \tag{4.22}$$

is the interaction energy between the electromagnetic field and the polarization. Part of the interaction energy with the matter, namely $\frac{\alpha^2}{\rho} |\underline{\mathbf{A}}|^2$, has already been classified into (4.20).

Fields are quantized in a standard fashion (Cohen-Tannoudji et al. 1989) by postulating equal-time commutation relations between the variables and their conjugates

$$[\hat{\underline{A}}^{\lambda}(\mathbf{k}, t), \hat{\underline{E}}^{\lambda' *}(\mathbf{k}', t)] = -\frac{\text{i}\hbar}{\epsilon_0} \delta_{\lambda\lambda'} \delta(\mathbf{k} - \mathbf{k}') \hat{1}, \tag{4.23}$$

$$[\hat{\underline{X}}^{\lambda}(\mathbf{k}, t), \hat{\underline{P}}^{\lambda' *}(\mathbf{k}', t)] = \text{i}\hbar \delta_{\lambda\lambda'} \delta(\mathbf{k} - \mathbf{k}') \hat{1}, \tag{4.24}$$

$$[\hat{\underline{Y}}_{\omega}^{\lambda}(\mathbf{k}, t), \hat{\underline{Q}}_{\omega'}^{\lambda' *}(\mathbf{k}', t)] = \text{i}\hbar \delta_{\lambda\lambda'} \delta(\mathbf{k} - \mathbf{k}') \delta(\omega - \omega') \hat{1}, \tag{4.25}$$

where all quantized operators are denoted by a caret. As usual, the annihilation operators are introduced.

$$\hat{a}(\lambda, \mathbf{k}, t) = \sqrt{\frac{\epsilon_0}{2\hbar\tilde{k}c}} \left[\tilde{k}c\underline{\hat{A}}^{\lambda}(\mathbf{k}, t) - \mathrm{i}\underline{\hat{E}}^{\lambda}(\mathbf{k}, t)\right], \tag{4.26}$$

$$\hat{b}(\lambda, \mathbf{k}, t) = \sqrt{\frac{\rho}{2\hbar\tilde{\omega}_0}} \left[\tilde{\omega}_0\underline{\hat{X}}^{\lambda}(\mathbf{k}, t) + \frac{\mathrm{i}}{\rho}\underline{\hat{P}}^{\lambda}(\mathbf{k}, t)\right], \tag{4.27}$$

$$\hat{b}_{\omega}(\lambda, \mathbf{k}, t) = \sqrt{\frac{\rho}{2\hbar\tilde{\omega}}} \left[-\mathrm{i}\omega\underline{\hat{Y}}^{\lambda}_{\omega}(\mathbf{k}, t) + \frac{1}{\rho}\underline{\hat{Q}}^{\lambda}_{\omega}(\mathbf{k}, t)\right] \tag{4.28}$$

From the equal-time commutation relations for the fields (4.23), (4.24), (4.25), the equal-time commutation relations for the creation and annihilation operators

$$\begin{aligned}
[\hat{a}(\lambda, \mathbf{k}, t), \hat{a}^{\dagger}(\lambda', \mathbf{k}', t)] &= \delta_{\lambda\lambda'}\delta(\mathbf{k} - \mathbf{k}')\hat{1}, \\
[\hat{b}(\lambda, \mathbf{k}, t), \hat{b}^{\dagger}(\lambda', \mathbf{k}', t)] &= \delta_{\lambda\lambda'}\delta(\mathbf{k} - \mathbf{k}')\hat{1}, \\
[\hat{b}_{\omega}(\lambda, \mathbf{k}, t), \hat{b}^{\dagger}_{\omega'}(\lambda', \mathbf{k}', t)] &= \delta_{\lambda\lambda'}\delta(\omega - \omega')\delta(\mathbf{k} - \mathbf{k}')\hat{1}
\end{aligned} \tag{4.29}$$

are obtained. The normally ordered Hamiltonian for the transverse fields is

$$\hat{H} = \hat{H}_{\text{em}} + \hat{H}_{\text{mat}} + \hat{H}_{\text{int}}, \tag{4.30}$$

where

$$\hat{H}_{\text{em}} = \int \sum_{\lambda=1,2} \hbar\tilde{k}c\hat{a}^{\dagger}(\lambda, \mathbf{k}, t)\hat{a}(\lambda, \mathbf{k}, t)\,\mathrm{d}^3\mathbf{k}, \tag{4.31}$$

$$\begin{aligned}
\hat{H}_{\text{mat}} = \int \sum_{\lambda=1,2} \Bigg\{ &\hbar\tilde{\omega}_0\hat{b}^{\dagger}(\lambda, \mathbf{k}, t)\hat{b}(\lambda, \mathbf{k}, t) + \int_0^{\infty} \hbar\omega\hat{b}^{\dagger}_{\omega}(\lambda, \mathbf{k}, t)\hat{b}_{\omega}(\lambda, \mathbf{k}, t)\,\mathrm{d}\omega \\
&+ \frac{\hbar}{2}\int_0^{\infty} V(\omega)\big[\hat{b}(\lambda, \mathbf{k}, t)\hat{b}^{\dagger}_{\omega}(\lambda, \mathbf{k}, t) \\
&+ \hat{b}^{\dagger}(\lambda, -\mathbf{k}, t)\hat{b}^{\dagger}_{\omega}(\lambda, \mathbf{k}, t) + \text{H. c.}\big]\,\mathrm{d}\omega\Bigg\}\,\mathrm{d}^3\mathbf{k},
\end{aligned} \tag{4.32}$$

$$\begin{aligned}
\hat{H}_{\text{int}} = \mathrm{i}\frac{\hbar}{2}\int \sum_{\lambda=1,2} &\Lambda(\mathbf{k})\big[\hat{a}(\lambda, \mathbf{k}, t)\hat{b}^{\dagger}(\lambda, \mathbf{k}, t) \\
&+ \hat{a}^{\dagger}(\lambda, -\mathbf{k}, t)\hat{b}^{\dagger}(\lambda, \mathbf{k}, t) + \text{H. c.}\big]\,\mathrm{d}^3\mathbf{k},
\end{aligned} \tag{4.33}$$

where $V(\omega) = \frac{v(\omega)}{\rho}\sqrt{\frac{\omega}{\tilde{\omega}_0}}$, $\Lambda(\mathbf{k}) \equiv \sqrt{\frac{\tilde{\omega}_0 c k_c^2}{\tilde{k}}}$, and the $\mathbf{k}$ integration has been restored to the full reciprocal space. It is worth mentioning that the Maxwell–Lorentz equations can be derived from the Hamiltonian.

It is important that the matter can be formally decoupled from the reservoir by the Fano technique and a dressed matter field obtained. Following Fano (1961), the polarization and reservoir parts of the Hamiltonian can be diagonalized. The dressed matter field creation and annihilation operators $\hat{B}^\dagger(\lambda, \mathbf{k}, \omega, t)$ and $\hat{B}(\lambda, \mathbf{k}, \omega, t)$ are introduced, respectively, which satisfy the usual equal-time commutation relations,

$$[\hat{B}(\lambda, \mathbf{k}, \omega, t), \hat{B}^\dagger(\lambda', \mathbf{k}', \omega', t)] = \delta_{\lambda\lambda'}\delta(\mathbf{k}-\mathbf{k}')\delta(\omega-\omega')\hat{1}, \tag{4.34}$$

$$\begin{aligned}\hat{B}(\lambda, \mathbf{k}, \omega, t) &= \alpha_0(\omega)\hat{b}(\lambda, \mathbf{k}, t) + \beta_0(\omega)\hat{b}^\dagger(\lambda, -\mathbf{k}, t) \\ &\quad + \int_0^\infty \Big[\alpha_1(\omega, \omega')\hat{b}_{\omega'}(\lambda, \mathbf{k}, t) \\ &\quad + \beta_1(\omega, \omega')\hat{b}^\dagger_{\omega'}(\lambda, -\mathbf{k}, t)\Big]\, \mathrm{d}\omega'.\end{aligned} \tag{4.35}$$

The coefficients $\alpha_0(\omega)$, $\beta_0(\omega)$, $\alpha_1(\omega, \omega')$, $\beta_1(\omega, \omega')$ are defined as follows. It is interesting that the diagonalization is performed once for the polarization and reservoir parts of the Hamiltonian and once for the total Hamiltonian. The Hamiltonian expressed in the modal annihilation operators is considered. From relation (4.35) it can be seen that the diagonalization is performed independently for every pair of the counterpropagating modes of the polarization field ("the modes" here are only formally similar to those of the electromagnetic field) and that it is performed using a Bogoliubov transformation. A useful definition of an "eigenoperator" is presented Barnett and Radmore (1988),

$$[\hat{B}(\lambda, \mathbf{k}, \omega, t), \hat{H}_{\mathrm{mat}}] = \hbar\omega\hat{B}(\lambda, \mathbf{k}, \omega, t). \tag{4.36}$$

The coefficients of the Bogoliubov transformation are calculated, that is to say the formulae

$$\alpha_0(\omega) = \left(\frac{\omega + \tilde{\omega}_0}{2}\right)\frac{V(\omega)}{\omega^2 - \tilde{\omega}_0^2 z(\omega)}, \tag{4.37}$$

where $z(\omega)$ is defined by

$$z(\omega) = 1 - \frac{1}{2\tilde{\omega}_0}\left[\lim_{\varepsilon\to+0}\int_{-\infty}^{\infty}\frac{\mathcal{V}(\omega')}{\omega' - \omega + \mathrm{i}\varepsilon}\,\mathrm{d}\omega'\right], \tag{4.38}$$

$$\beta_0(\omega) = \left(\frac{\omega - \tilde{\omega}_0}{2}\right)\frac{V(\omega)}{\omega^2 - \tilde{\omega}_0^2 z(\omega)}, \tag{4.39}$$

$$\alpha_1(\omega, \omega') = \delta(\omega - \omega') + \left(\frac{\tilde{\omega}_0}{2}\right)\left(\frac{V^*(\omega')}{\omega - \omega' - \mathrm{i}0}\right)\frac{V(\omega)}{\omega^2 - \tilde{\omega}_0^2 z(\omega)}, \tag{4.40}$$

and

$$\beta_1(\omega, \omega') = \left(\frac{\tilde{\omega}_0}{2}\right)\left(\frac{V(\omega')}{\omega + \omega'}\right)\frac{V(\omega)}{\omega^2 - \tilde{\omega}_0^2 z(\omega)} \tag{4.41}$$

are derived. In the study no constant $\mathcal{V}(\omega) \equiv |V(\omega)|^2$ occurs.

As usual with the substitutions, we are also interested in the inverse transformation. It is given by the relations

$$\hat{b}(\lambda, \mathbf{k}, t) = \int_0^\infty \left[\alpha_0^*(\omega)\hat{B}(\lambda, \mathbf{k}, \omega, t) - \beta_0(\omega)\hat{B}^\dagger(\lambda, -\mathbf{k}, \mathbf{t}, \omega)\right]\mathrm{d}\omega \tag{4.42}$$

and

$$\hat{b}_\omega(\lambda, \mathbf{k}, t) = \int_0^\infty \left[\alpha_1^*(\omega', \omega)\hat{B}(\lambda, \mathbf{k}, \omega', t) - \beta_1(\omega', \omega)\hat{B}^\dagger(\lambda, -\mathbf{k}, \omega', t)\right] \mathrm{d}\omega'. \tag{4.43}$$

The conditions

$$I \equiv \int_0^\infty \left[|\alpha_0(\omega)|^2 - |\beta_0(\omega)|^2\right] \mathrm{d}\omega = 1, \tag{4.44}$$

$$I(\omega, \omega') \equiv \int_0^\infty \left[\alpha_1^*(\nu, \omega)\alpha_1(\nu, \omega') - \beta_1(\nu, \omega)\beta_1^*(\nu, \omega')\right] \mathrm{d}\nu = \delta(\omega - \omega') \tag{4.45}$$

for the coefficients of the Bogoliubov transformation seem to be familiar. It has been shown that the diagonalization cannot be performed on the common assumption of white noise (the Markov-type coupling). It is commented on free charges and a conducting medium being beyond the scope of Huttner and Barnett (1992a). We need not grieve for the assumption of the white noise. Without it, we are farther from the original Lorentzian formulation, nothing more.

The diagonalization of the total Hamiltonian is formally very similar to the diagonalization of its matter part. A dimensionless coupling constant $\zeta(\omega) = \mathrm{i}\sqrt{\tilde{\omega}_0}[\alpha_0(\omega) + \beta_0(\omega)]$ is defined and the annihilation operators are introduced (by a Fano type of technique),

$$\begin{aligned}\hat{C}(\lambda, \mathbf{k}, \omega, t) &= \tilde{\alpha}_0(k, \omega)\hat{a}(\lambda, \mathbf{k}, t) + \tilde{\beta}_0(k, \omega)\hat{a}^\dagger(\lambda, -\mathbf{k}, t) \\ &\quad + \int_0^\infty \Big[\tilde{\alpha}_1(k, \omega, \omega')\hat{B}(\lambda, \mathbf{k}, \omega', t) \\ &\quad + \tilde{\beta}_1(k, \omega, \omega')\hat{B}^\dagger(\lambda, -\mathbf{k}, \omega', t)\Big] \mathrm{d}\omega',\end{aligned} \tag{4.46}$$

where the coefficients $\tilde{\alpha}_0(k, \omega)$, $\tilde{\beta}_0(k, \omega)$, $\tilde{\alpha}_1(k, \omega, \omega')$, and $\tilde{\beta}_1(k, \omega, \omega')$ are rather complicated and are derived in the form

$$\tilde{\alpha}_0(k, \omega) = \sqrt{\frac{\omega_\mathrm{c}^2}{\tilde{k}c}} \left(\frac{\omega + \tilde{k}c}{2}\right) \frac{\zeta(\omega)}{\omega^2 - \tilde{k}^2c^2\tilde{z}(k, \omega)}, \tag{4.47}$$

where

$$\tilde{z}(k, \omega) = 1 - \frac{\omega_\mathrm{c}^2}{2(\tilde{k}c)^2} \left[\lim_{\varepsilon \to +0} \int_{-\infty}^\infty \frac{|\zeta(\omega')|^2}{\omega' - \omega + \mathrm{i}\varepsilon}\right] \mathrm{d}\omega', \tag{4.48}$$

or alternatively

$$\tilde{\alpha}_0(k, \omega) = \sqrt{\frac{\omega_\mathrm{c}^2}{\tilde{k}c}} \left(\frac{\omega + \tilde{k}c}{2}\right) \frac{\zeta(\omega)}{\epsilon^*(\omega)\omega^2 - k^2c^2}, \tag{4.49}$$

where the complex relative permittivity $\epsilon(\omega)$ is introduced

$$\epsilon(\omega) = 1 + \frac{1}{\omega^2}\left[k^2c^2 - (k^2c^2 + \omega_\mathrm{c}^2)\tilde{z}^*(k, \omega)\right], \text{ independent of } k, \tag{4.50}$$

$$\tilde{\beta}_0(k,\omega) = \sqrt{\frac{\omega_c^2}{\tilde{k}c}}\left(\frac{\omega - \tilde{k}c}{2}\right)\frac{\zeta(\omega)}{\epsilon^*(\omega)\omega^2 - k^2c^2}, \tag{4.51}$$

$$\tilde{\alpha}_1(k,\omega,\omega') = \delta(\omega-\omega') + \frac{\omega_c^2}{2}\left(\frac{\zeta^*(\omega')}{\omega-\omega'-\mathrm{i}0}\right)\frac{\zeta(\omega)}{\epsilon^*(\omega)\omega^2 - k^2c^2}, \tag{4.52}$$

and

$$\tilde{\beta}_1(k,\omega,\omega') = \frac{\omega_c^2}{2}\left(\frac{\zeta^*(\omega')}{\omega-\omega'-\mathrm{i}0}\right)\frac{\zeta(\omega)}{\epsilon^*(\omega)\omega^2 - k^2c^2}. \tag{4.53}$$

The operators $\hat{C}(\lambda,\mathbf{k},\omega,t)$ and $\hat{C}^\dagger(\lambda,\mathbf{k},\omega,t)$ also satisfy the usual commutation relations,

$$[\hat{C}(\lambda,\mathbf{k},\omega,t), \hat{C}^\dagger(\lambda',\mathbf{k}',\omega',t)] = \delta_{\lambda\lambda'}\delta(\mathbf{k}-\mathbf{k}')\delta(\omega-\omega')\hat{1} \tag{4.54}$$

and being operators for eigenmodes,

$$[\hat{C}(\lambda,\mathbf{k},\omega,t), \hat{H}] = \hbar\omega\hat{C}(\lambda,\mathbf{k},\omega,t), \tag{4.55}$$

they have a harmonic time dependence

$$\hat{C}(\lambda,\mathbf{k},\omega,t) = \hat{C}(\lambda,\mathbf{k},\omega,0)e^{-\mathrm{i}\omega t}. \tag{4.56}$$

The vector potential is now given by

$$\hat{\mathbf{A}}(\mathbf{r},t) = \frac{1}{(2\pi)^{\frac{3}{2}}}\int\sqrt{\frac{\hbar\omega_c^2}{2\epsilon_0}}\sum_{\lambda=1,2}\mathbf{e}_\lambda(\mathbf{k}) \times \int_0^\infty\left[\frac{\zeta^*(\omega)}{\omega^2\epsilon(\omega)-k^2c^2}\hat{C}(\lambda,\mathbf{k},\omega,0)e^{-\mathrm{i}(\omega t-\mathbf{k}\cdot\mathbf{r})} + \mathrm{H.\,c.}\right]\mathrm{d}\omega\,\mathrm{d}^3\mathbf{k}. \tag{4.57}$$

Relation (4.8) being modified for the operators expresses the operators $\hat{\mathbf{A}}(\mathbf{r},t)$, $\hat{\mathbf{E}}(\mathbf{r},t)$,... in terms of the operators $\underline{\hat{\mathbf{A}}}(\mathbf{k},t)$, $\underline{\hat{\mathbf{X}}}(\mathbf{k},t)$, $\underline{\hat{\mathbf{Y}}}_\omega(\mathbf{k},t)$, $\underline{\hat{\mathbf{E}}}(\mathbf{k},t)$, $\underline{\hat{\mathbf{P}}}(\mathbf{k},t)$, $\underline{\hat{\mathbf{Q}}}_\omega(\mathbf{k},t)$.

Let us note that on the substitution into (4.8) for $\underline{\hat{\mathbf{A}}}(\mathbf{k},t)$, $\underline{\hat{\mathbf{E}}}(\mathbf{k},t)$,... by the relations

$$\begin{aligned}\underline{\hat{\mathbf{A}}}(\mathbf{k},t) &= \sqrt{\frac{\hbar}{2\tilde{k}c\epsilon_0}}[\hat{\mathbf{a}}(\mathbf{k},t) + \hat{\mathbf{a}}^\dagger(-\mathbf{k},t)],\\ \underline{\hat{\mathbf{E}}}(\mathbf{k},t) &= \mathrm{i}\sqrt{\frac{\hbar\tilde{k}c}{2\epsilon_0}}[\hat{\mathbf{a}}(\mathbf{k},t) + \hat{\mathbf{a}}^\dagger(-\mathbf{k},t)],\\ \dots\quad &\dots\qquad\quad \dots,\end{aligned} \tag{4.58}$$

the annihilation and creation operators $\hat{\mathbf{a}}(\mathbf{k}, t)$, $\hat{\mathbf{b}}(\mathbf{k}, t)$, $\hat{\mathbf{b}}_\omega(\mathbf{k}, t)$ and $\hat{\mathbf{a}}^\dagger(\mathbf{k}, t)$, $\hat{\mathbf{b}}^\dagger(\mathbf{k}, t)$, $\hat{\mathbf{b}}^\dagger_\omega(\mathbf{k}, t)$, respectively, are introduced. On the substitution into the intermediate result for the operators $\hat{\mathbf{b}}(\mathbf{k}, t)$ and $\hat{\mathbf{b}}_\omega(\mathbf{k}, t)$ by relations (4.42) and (4.43), the operators $\hat{\mathbf{B}}(\mathbf{k}, \omega, t)$ are introduced. On the substitution into the intermediate result for the operators $\hat{\mathbf{a}}(\mathbf{k}, t)$ by the relation (the slightly modified relation (4.2) from Huttner and Barnett (1992b))

$$\hat{\mathbf{a}}(\mathbf{k}, t) = \int_0^\infty \left[\tilde{\alpha}_0^*(k, \omega)\hat{\mathbf{C}}(\mathbf{k}, \omega, t) - \tilde{\beta}_0(k, \omega)\hat{\mathbf{C}}^\dagger(-\mathbf{k}, \omega, t)\right] \mathrm{d}\omega \tag{4.59}$$

and for the operators $\hat{\mathbf{B}}(\mathbf{k}, \omega, t)$ by the relation

$$\hat{\mathbf{B}}(\mathbf{k}, \omega, t) = \int_0^\infty \left[\tilde{\alpha}_1^*(k, \omega, \omega')\hat{\mathbf{C}}(\mathbf{k}, \omega, t) - \tilde{\beta}_1(k, \omega', \omega)\hat{\mathbf{C}}^\dagger(-\mathbf{k}, \omega, t)\right] \mathrm{d}\omega', \tag{4.60}$$

the operators $\hat{\mathbf{C}}(\mathbf{k}, \omega, t)$ are introduced, which have the time dependence (4.56).

On the substitution into the intermediate result for $\hat{\mathbf{C}}(\mathbf{k}, \omega, 0)$ by relation (4.46), the operators $\hat{\mathbf{a}}(\mathbf{k}, 0)$ and $\hat{\mathbf{B}}(\mathbf{k}, \omega, 0)$ are introduced and on the substitution into the intermediate result for the operators $\hat{\mathbf{B}}(\mathbf{k}, \omega, 0)$ by relation (4.35), the operators $\hat{\mathbf{b}}(\mathbf{k}, 0)$ and $\hat{\mathbf{b}}_\omega(\mathbf{k}, 0)$ are introduced. On the substitution into the intermediate result for these operators by the formulae (4.26), (4.27), and (4.28), the operators $\underline{\hat{\mathbf{A}}}(\mathbf{k}, 0)$, $\underline{\hat{\mathbf{X}}}(\mathbf{k}, 0)$, $\underline{\hat{\mathbf{Y}}}_\omega(\mathbf{k}, 0)$, $\underline{\hat{\mathbf{E}}}(\mathbf{k}, 0)$, $\underline{\hat{\mathbf{P}}}(\mathbf{k}, 0)$, $\underline{\hat{\mathbf{Q}}}_\omega(\mathbf{k}, 0)$ are introduced. On the substitution into the intermediate result for these operators by the relations

$$\begin{aligned}
\underline{\hat{\mathbf{A}}}(\mathbf{k}, t) &= \frac{1}{(2\pi)^{\frac{3}{2}}} \int \hat{\mathbf{A}}(\mathbf{r}, t) e^{-\mathrm{i}\mathbf{k}\cdot\mathbf{r}}\, \mathrm{d}^3\mathbf{r}, \\
\underline{\hat{\mathbf{E}}}(\mathbf{k}, t) &= \frac{1}{(2\pi)^{\frac{3}{2}}} \int \hat{\mathbf{E}}(\mathbf{r}, t) e^{-\mathrm{i}\mathbf{k}\cdot\mathbf{r}}\, \mathrm{d}^3\mathbf{r}, \\
\ldots \quad & \ldots \qquad \ldots,
\end{aligned} \tag{4.61}$$

the operators $\hat{\mathbf{A}}(\mathbf{r}, 0)$, $\hat{\mathbf{E}}(\mathbf{r}, 0)$,... are introduced. These substitutions solve the Cauchy or initial problem.

Huttner and Barnett (1992b) restrict themselves first to a one-dimensional case when describing the propagation in the dielectric. The vector potential is considered in the simpler form

$$\hat{A}(x, t) = \frac{1}{\sqrt{4\pi}} \int_0^\infty \mathcal{A}(\omega) \left[\hat{c}_+(x, \omega, 0) e^{-\mathrm{i}\omega t} + \hat{c}_-(x, \omega, 0) e^{-\mathrm{i}\omega t} + \mathrm{H.\,c.}\right] \mathrm{d}\omega, \tag{4.62}$$

where

$$\mathcal{A}(\omega) = \sqrt{\frac{\hbar\eta(\omega)}{\epsilon_0 S c \omega |n(\omega)|^2}}, \tag{4.63}$$

$\mathcal{S}$ is a cross-sectional area, $n(\omega)$ is the complex refractive index defined as the square root of the relative permittivity $\epsilon(\omega)$ with a positive real part $\eta(\omega)$, and the operators $\hat{c}_{\pm}(x, \omega, t)$ are introduced

$$\hat{c}_{\pm}(x, \omega, t) = \sqrt{\frac{\text{Im}\{K(\omega)\}}{\pi}} e^{\text{i}\phi(\omega)} \int_{-\infty}^{\infty} \frac{\hat{C}(k, \omega, t) e^{\text{i}kx}}{K(\omega) \mp k} \, \text{d}k, \tag{4.64}$$

where the complex wave number $K(\omega)$ and the phase factor $e^{\text{i}\phi(\omega)}$ are expressed as

$$K(\omega) = \frac{n(\omega)\omega}{c}, \tag{4.65}$$

$$e^{\text{i}\phi(\omega)} = \frac{\zeta^*(\omega)}{|\zeta(\omega)|} \frac{|n(\omega)|}{n(\omega)}, \tag{4.66}$$

and $\text{Im}\{K(\omega)\} > 0$.

Since the magnetic field can be expressed similarly as the vector potential, the spatial quantum Langevin equations of progression can be obtained as

$$\frac{\partial \hat{c}_{\pm}(x, \omega, t)}{\partial x} = \pm \text{i} K(\omega) \hat{c}_{\pm}(x, \omega, t) \pm \sqrt{2\text{Im}\{K(\omega)\}} \, \hat{f}(x, \omega, t), \tag{4.67}$$

where the Langevin-noise operator is

$$\hat{f}(x, \omega, t) = -\frac{\text{i}}{\sqrt{2\pi}} e^{\text{i}\phi(\omega)} \int_{-\infty}^{\infty} \hat{C}(k, \omega, t) e^{\text{i}kx} \, \text{d}k \tag{4.68}$$

and it also enters a rather similar expression for the electric displacement operator. Equation (4.67) have been obtained from the Maxwell equations for the monochromatic fields.

Huttner and Barnett (1992b) remind of the simple commutation relations

$$[\hat{f}(x, \omega, t), \hat{f}^{\dagger}(x', \omega', t)] = \delta(x - x')\delta(\omega - \omega')\hat{1}, \tag{4.69}$$

further of the equal-space commutation relations

$$\begin{aligned} [\hat{c}_{\pm}(x, \omega, t), \hat{c}_{\pm}^{\dagger}(x, \omega', t)] &= \delta(\omega - \omega')\hat{1}, \\ [\hat{c}_{\pm}(x, \omega, t), \hat{c}_{\mp}^{\dagger}(x, \omega', t)] &= \hat{0} \end{aligned} \tag{4.70}$$

and, finally, that $\hat{c}_{+}(x, \omega, t)$ commutes with all the Langevin operators $\hat{f}(x', \omega', t)$ and $\hat{f}^{\dagger}(x', \omega', t)$ for all $x' > x$, while $\hat{c}_{-}(x, \omega, t)$ commutes with all the Langevin operators $\hat{f}(x', \omega', t)$ and $\hat{f}^{\dagger}(x', \omega', t)$ for all $x' < x$.

Jeffers and Barnett (1994) modelled the propagation of squeezed light through an absorbing dispersive dielectric medium. Hradil (1996) considered "lossless" dispersive dielectrics, i.e. dielectrics with a thin absorption line. He formulated a canonical quantization of the electromagnetic field in a closed Fabry–Pérot resonator with a dispersive slab.

Wubs and Suttorp (2001) have solved the initial-value problem for the damped-polariton model formulated by Huttner and Barnett (1992a,b) and have found that for long times all field operators can be expressed in terms of the initial reservoir operators. They have investigated the transient dynamics of the spontaneous-emission rate of a guest atom in an absorbing medium.

Hillery and Drummond (2001) have studied the scattering of the quantized electromagnetic field from a linear dispersive dielectric in the limit of "thin" absorption lines. The field is represented by means of the dual vector potential. Input–output relations are unitary and no additional quantum-noise terms are required. Equations specialized to the case of a dielectric layer with a uniform density of oscillators are usual expressions.

Janowicz et al. (2003) analyse radiative heat transfer between two dielectric bodies. Quantization of the electromagnetic field in inhomogeneous, dispersive, and lossy dielectrics is performed with the help of a procedure which is still attributed to Huttner and Barnett (1992b). Expectation value of the Poynting vector operator is computed. To this end, two techniques suitable for nonequilibrium processes are utilized: the Heisenberg equation of motion and the diagrammatic Keldysh procedure. It is remarked that in nonlinear models the Keldysh formalism provides a framework for the perturbation expansion. The calculations fit into the development of the theory of thermal scanning microscopy.

4.1.2 Correlation of Ground-State Fluctuations

The quantization of the radiation imbedded in a dielectric with a space-dependent refractive index has been expounded in the book by Vogel and Welsch (1994). A canonical quantization scheme for radiation fields in linear dielectrics with a space-dependent refractive index has been developed by Knöll et al. (1987) and later by Glauber and Lewenstein (1991). For application, see, for example, Knöll et al. (1986, 1990, 1991), Knöll and Welsch (1992) and a related work (Knöll and Leonhardt 1992).

Gruner and Welsch (1995) have contributed to the stream of papers aiming at a description of quantum properties of the dispersive and lossy dielectrics including the vacuum fluctuations, i.e. fluctuations of radiation field in the ground state of the coupled light–matter system. They study it in terms of a symmetrized correlation function. They try to expound and supplement the paper by Huttner and Barnett (1992b) from the point of view of the quantization of the phenomenological Maxwell theory.

First, the quantization of radiation in a dispersive and lossy dielectric is performed. This begins from the classical Maxwell equations (3.176) with (3.177) and a constitutive relation comprising an integral term,

$$\mathbf{D}(\mathbf{r}, t) = \epsilon_0 \left[\mathbf{E}(\mathbf{r}, t) + \int_0^\infty \chi(\tau) \mathbf{E}(\mathbf{r}, t-\tau)\, \mathrm{d}\tau \right], \tag{4.71}$$

is transformed into the Fourier space to yield

$$\underline{\mathbf{D}}(\mathbf{r},\omega) = \epsilon_0\epsilon(\omega)\underline{\mathbf{E}}(\mathbf{r},\omega) \tag{4.72}$$

and the Helmholtz equation is presented.

The Huttner–Barnett quantization scheme is introduced with a diagonalized Hamiltonian

$$\hat{H} = \sum_{\lambda=1,2} \int \int_0^\infty \hbar\omega \hat{C}^\dagger(\lambda,\mathbf{k},\omega)\hat{C}(\lambda,\mathbf{k},\omega)\,\mathrm{d}\omega\,\mathrm{d}^3\mathbf{k}, \tag{4.73}$$

which comprises a sum over λ which is absent from relation (3.14) of Huttner and Barnett (1992b). The effect of the medium is entirely determined by the complex permittivity $\epsilon(\omega)$. It still has no tensorial character. Let us remember relations (4.3), (4.5), (4.6), and (4.7) of Huttner and Barnett (1992b). In these relations, $\hat{C}(\mathbf{k},\omega)$ should be replaced by $\hat{C}(\lambda,\mathbf{k},\omega)$ and the identity $\sqrt{\frac{\omega_c^2}{2}}\zeta^*(\omega) = \omega\sqrt{\mathrm{Im}\{\epsilon(\omega)\}}$ utilized.

The frequency-dependent field operators are introduced in the three-dimensional case. Not only the equal-time commutation relations, but even the most general ones are presented. The vector-field operators $\hat{\mathbf{a}}(\mathbf{r},\omega)$ and $\hat{\mathbf{f}}(\mathbf{r},\omega)$ have been introduced, the vector $\hat{\mathbf{a}}(\mathbf{r},\omega)$ being a generalization of the component $\hat{c}(x,\omega)$ from Huttner and Barnett (1992b). The definition of the transverse δ function is presented as

$$\delta_{ij}^{\perp}(\mathbf{r}) = \frac{1}{(2\pi)^3}\int_{-\infty}^{\infty}\left(\delta_{ij} - \frac{k_ik_j}{k^2}\right)e^{\mathrm{i}\mathbf{k}\cdot\mathbf{r}}\,\mathrm{d}^3\mathbf{k}, \tag{4.74}$$

which can be interpreted also as a transverse projection of the columns of 3×3 identity matrix multiplied by the δ function. The transverse projection of the columns of an identity matrix multiplied by other functions has proved to be useful, for example, the commutators $[\hat{a}_i(\mathbf{r},\omega), \hat{a}_j(\mathbf{r}',\omega')]$ can be expressed in terms of such projections. Here, the operator Δ_{ij},

$$\Delta_{ij}F(\mathbf{r}) = \int_{-\infty}^{\infty}\delta_{ij}^{\perp}(\mathbf{r}-\mathbf{r}')F(\mathbf{r}')\,\mathrm{d}^3\mathbf{r}', \tag{4.75}$$

is applied (the putting of an identity matrix to be multiplied by a function and subsequent transverse projection of the columns), but very complicated expressions are obtained.

Continuing the use of the operators $\hat{\mathbf{a}}(\mathbf{r},\omega)$ and $\hat{\mathbf{f}}(\mathbf{r},\omega)$, an analogue of relation (5.21) of Huttner and Barnett (1992b) (cf. (4.62) here) has been written. Then, analogues of their frequency decompositions of the vector-potential operators, electric-field strength operators, etc. have been presented. The operator constitutive equation (in the Fourier space)

$$\hat{\underline{\mathbf{D}}}(\mathbf{r},\omega) = \epsilon_0\epsilon(\omega)\hat{\underline{\mathbf{E}}}(\mathbf{r},\omega) - \epsilon_0\mathcal{F}(\omega)\hat{\mathbf{f}}(\mathbf{r},\omega), \tag{4.76}$$

where

$$\epsilon_0 \mathcal{F}(\omega) = \sqrt{\frac{\hbar}{\pi} \epsilon_0 \mathrm{Im}\{\epsilon(\omega)\}} \tag{4.77}$$

differs from the classical equation (4.72) by an additional term. On substituting into the phenomenological Maxwell equations, the partial differential equation for the operator $\hat{\mathbf{a}}(\mathbf{r}, \omega)$ is obtained which is a Helmholtz equation with a right-hand side. In the three-dimensional case, there exists no decomposition into first-order equations. The canonical commutation relations are

$$[\hat{A}_i(\mathbf{r}, t), \hat{E}_j(\mathbf{r}', t)] = -\frac{\mathrm{i}\hbar}{\epsilon_0} \delta_{ij}^{\perp}(\Delta\mathbf{r}) \hat{1}, \tag{4.78}$$

with the abbreviation

$$\Delta\mathbf{r} = \mathbf{r} - \mathbf{r}'. \tag{4.79}$$

A test of the consistency of the theory in the limit $\epsilon(\omega) \to 1$ has been accomplished.

Let us recall the usual annihilation operators $\hat{a}(\lambda, \mathbf{k})$, which satisfy the commutation relations

$$[\hat{a}(\lambda, \mathbf{k}), \hat{a}^\dagger(\lambda', \mathbf{k}')] = \delta_{\lambda\lambda'} \delta(\mathbf{k} - \mathbf{k}') \hat{1} \tag{4.80}$$

and enter the expansion for $\hat{\mathbf{A}}(\mathbf{r}, 0)$.

The operators $\hat{\mathbf{a}}(\mathbf{r}, \omega)$ and $\hat{\mathbf{f}}(\mathbf{r}, \omega)$ derived from $\hat{C}(\lambda, \mathbf{k}, \omega)$ are not independent operators. Cf., Huttner and Barnett (1992b) who in the one-dimensional case introduce forward and backward-propagating fields and show that such a definition ensures the causal (one-sided) independence of the respective operators of the operator $\hat{\mathbf{f}}(\mathbf{r}, \omega)$. In the three-dimensional case, there exists no generalization of relation (4.64) and no equation for such quantities.

The theory is applied to the determination of the correlation of the ground-state fluctuations of the electric-field strength. The symmetric correlation function of the electric-field strength

$$\begin{aligned} K_{mn}(\Delta\mathbf{r}, \tau) = \frac{1}{2} \langle 0 | \big[& \hat{E}_m(\mathbf{r}, t + \tau) \hat{E}_n(\mathbf{r} + \Delta\mathbf{r}, t) \\ & + \hat{E}_n(\mathbf{r} + \Delta\mathbf{r}, t) \hat{E}_m(\mathbf{r}, t + \tau) \big] | 0 \rangle \end{aligned} \tag{4.81}$$

is considered. We remark that

$$\begin{aligned} K_{mn}(\Delta\mathbf{r}, \tau) = & \frac{\hbar}{4\pi^2 c^2 \epsilon_0} n_{\mathrm{R}}(\omega) \\ & \times \int_0^\infty \cos(\omega\tau) \Delta_{ij} \frac{\exp\left[-\frac{\omega}{c} n_{\mathrm{I}}(\omega)\right]}{n_{\mathrm{R}}(\omega) \Delta r} \sin\left[\frac{\omega}{c} n_{\mathrm{R}}(\omega) \Delta r\right] \mathrm{d}\omega, \end{aligned} \tag{4.82}$$

where $\Delta r = |\Delta \mathbf{r}|$ and

$$n_{\mathrm{I}}(\omega) = \mathrm{Im}\{\sqrt{\epsilon(\omega)}\} = \mathrm{Im}\{n(\omega)\}, \tag{4.83}$$

$$n_{\mathrm{R}}(\omega) = \mathrm{Re}\{\sqrt{\epsilon(\omega)}\} = \mathrm{Re}\{n(\omega)\}. \tag{4.84}$$

Restricting attention to optical frequencies within an interval of the width $2\Delta\omega$,

$$\omega_0 - \Delta\omega < \omega < \omega_0 + \Delta\omega, \tag{4.85}$$

$$\frac{\Delta\omega}{\omega_0} \ll 1, \tag{4.86}$$

where ω_0 is an appropriately chosen centre frequency and assuming that dispersion and absorption are small on lengths of order of β^{-1},

$$\beta = \frac{\omega}{c} n_{\mathrm{R}}(\omega), \tag{4.87}$$

and times of order of ω^{-1}, Gruner and Welsch (1995) let

$$(\beta \Delta r)^{-1} \ll 1, \tag{4.88}$$

$$(\omega\tau)^{-1} \ll 1. \tag{4.89}$$

Further, they assume a transparent medium, such as a fibre, for which it may be justified to put approximately

$$n_{\mathrm{R}}(\omega) \approx n_{\mathrm{R}0} + n_{\mathrm{R}1}\frac{\omega}{\omega_0}, \tag{4.90}$$

$$n_{\mathrm{I}}(\omega) \approx n_{\mathrm{I}}(\omega_0) \equiv n_{\mathrm{I}0}. \tag{4.91}$$

The influence of absorption, phase, and group velocities and group velocity dispersion on the dynamics of the field fluctuations within a frequency interval (4.85) has been studied. The absorption causes a spatial decay of the correlation of the field fluctuations. The light cone of strong correlation, which in empty space is determined by the speed of light in vacuum, is now given by the group velocity in the medium provided that the spatial distance is not too large. With increasing distance, also the dispersion of the group velocity needs a consideration.

4.2 Green-Function Approach

On allowing for a frequency-dependent complex permittivity that is consistent with the Kramers–Kronig relations and introducing a random operator noise source associated with the absorption of radiation, the classical Maxwell equations can be considered as quantum operator equations. Their solution based on a Green-function expansion of the vector-potential operator seems to be a natural generalization of the mode expansion applicable to source-free radiation in nearly lossless dielectrics.

4.2.1 Dispersive Lossy Linear Inhomogeneous Dielectric

Gruner and Welsch (1996a) have expounded a quantization scheme which starts with phenomenological Maxwell equations instead of Lagrangian densities and is consistent with the Kramers–Kronig relations and the familiar (equal-time) canonical commutation relations for the vector potential and electric field. This is realized for homogeneous and inhomogeneous, especially, multilayered dielectrics.

In the phenomenological classical Maxwell theory, the equations comprise $\epsilon(\omega)$, the frequency-dependent complex relative permittivity introduced phenomenologically. This function has the analytical continuation in the upper complex half-plane, $\epsilon(\Omega)$, which satisfies the relation

$$\epsilon(-\Omega^*) = \epsilon^*(\Omega). \tag{4.92}$$

The real and imaginary parts of the relative permittivity satisfy the well-known Kramers–Kronig relations

$$\mathrm{Re}\{\epsilon(\omega)\} - 1 = \frac{1}{\pi}\mathrm{V.p.}\int_{-\infty}^{\infty} \frac{\mathrm{Im}\{\epsilon(\omega')\}}{\omega' - \omega}\,\mathrm{d}\omega', \tag{4.93}$$

$$\mathrm{Im}\{\epsilon(\omega)\} = -\frac{1}{\pi}\mathrm{V.p.}\int_{-\infty}^{\infty} \frac{\mathrm{Re}\{\epsilon(\omega')\} - 1}{\omega' - \omega}\,\mathrm{d}\omega', \tag{4.94}$$

where V.p. is the principal value of the integral.

The quantization scheme is based on the Helmholtz equation with the source term

$$\Delta\hat{\hat{\mathbf{A}}}(\mathbf{r}, \omega) + \mathbf{K}^2(\omega)\hat{\hat{\mathbf{A}}}(\mathbf{r}, \omega) = \hat{\hat{\mathbf{j}}}_{\mathrm{n}}(\mathbf{r}, \omega), \tag{4.95}$$

where $\hat{\hat{\mathbf{A}}}(\mathbf{r}, \omega)$ is the "Fourier transform" of the (known) operator vector-potential $\hat{\mathbf{A}}(\mathbf{r}, t)$ and $\hat{\hat{\mathbf{j}}}_{\mathrm{n}}(\mathbf{r}, \omega)$ is the "Fourier transform" of the operator-noise current. In fact, from the exposition it can be seen that the vector-potential operator is introduced by the relation

$$\hat{\mathbf{A}}(\mathbf{r}, 0) = \int_0^{\infty} \hat{\hat{\mathbf{A}}}(\mathbf{r}, \omega, \mathbf{0})\,\mathrm{d}\omega + \mathrm{H.\,c.}, \tag{4.96}$$

where quantum mechanically also the frequency-dependent operators can be time dependent, $t = 0$.

When $\mathrm{Im}\{\epsilon(\omega)\} > 0$, a hypothetical addition of a nontrivial solution of the homogeneous Helmholtz equation would violate the boundary condition at infinity.

Hence, the operator $\hat{\hat{\mathbf{A}}}(\mathbf{r}, \omega) \equiv \hat{\hat{\mathbf{A}}}(\mathbf{r}, \omega, \mathbf{t})$ is uniquely determined by a linear transformation of the source operator $\hat{\hat{\mathbf{j}}}_{\mathrm{n}}(\mathbf{r}, \omega) \equiv \hat{\hat{\mathbf{j}}}_{\mathrm{n}}(\mathbf{r}, \omega, \mathbf{t})$. This operator can be chosen in the form (cf., Gruner and Welsch (1995))

$$\hat{\hat{\mathbf{j}}}_{\mathrm{n}}(\mathbf{r}, \omega) = \mathcal{F}(\omega)\frac{\omega}{c^2}\hat{\mathbf{f}}(\mathbf{r}, \omega), \tag{4.97}$$

with $\mathcal{F}(\omega)$ given in (4.77). The Hamiltonian $\hat{H}$ is diagonal in the operators $\hat{\mathbf{f}}(\mathbf{r}, \omega)$,

$$\hat{H} = \int \int_0^\infty \hbar\omega \hat{\mathbf{f}}^\dagger(\mathbf{r}, \omega) \cdot \hat{\mathbf{f}}(\mathbf{r}, \omega)\, \mathrm{d}\omega\, \mathrm{d}^3\mathbf{r}, \tag{4.98}$$

and these operators have the usual properties

$$[\hat{f}_i(\mathbf{r}, \omega), \hat{f}_j^\dagger(\mathbf{r}', \omega')] = \delta_{ij}^\perp(\mathbf{r} - \mathbf{r}')\delta(\omega - \omega')\hat{1}, \tag{4.99}$$

$$[\hat{f}_i(\mathbf{r}, \omega), \hat{f}_j(\mathbf{r}', \omega')] = [\hat{f}_i^\dagger(\mathbf{r}, \omega), \hat{f}_j^\dagger(\mathbf{r}', \omega')] = \hat{0}. \tag{4.100}$$

From the foregoing considerations it follows that (when all appropriate conditions are fulfilled) the operator of the vector potential can be defined by the relation

$$\hat{\mathbf{A}}(\mathbf{r}, 0) = \int_0^\infty \int G(\mathbf{r}, \mathbf{r}', \omega)\hat{\hat{\mathbf{j}}}_{\mathrm{n}}(\mathbf{r}', \omega, \mathbf{0})\, \mathrm{d}^3\mathbf{r}'\, \mathrm{d}\omega + \mathrm{H.\,c.}, \tag{4.101}$$

where the Green function $G(\mathbf{r}, \mathbf{r}', \omega)$ satisfies the equation

$$\Delta G(\mathbf{r}, \mathbf{r}', \omega) + K^2(\omega)G(\mathbf{r}, \mathbf{r}', \omega) = \delta(\mathbf{r} - \mathbf{r}') \tag{4.102}$$

and the boundary condition that it vanishes at infinity. Another required property is

$$\hat{\mathbf{E}}(\mathbf{r}, 0) = -\dot{\hat{\mathbf{A}}}(\mathbf{r}, \mathbf{0}) \tag{4.103}$$

and the canonical field commutation relations

$$[\hat{A}_i(\mathbf{r}, 0), \hat{E}_j(\mathbf{r}', 0)] = -\frac{\mathrm{i}\hbar}{\epsilon_0}\delta_{ij}^\perp(\mathbf{r} - \mathbf{r}')\hat{1}. \tag{4.104}$$

Relation (4.104) must be verified by straightforward calculation. For the sake of clarity, Gruner and Welsch (1996a) illustrate this procedure in linearly polarized radiation propagating in the x direction. Relation (4.104) are replaced by the relation

$$[\hat{A}(x, 0), \hat{E}(x', 0)] = -\frac{\mathrm{i}\hbar}{\mathcal{A}\epsilon_0}\delta(x - x')\hat{1}, \tag{4.105}$$

where $\mathcal{A}$ is the normalization area perpendicular to the x direction. It is shown that when losses in the dielectric may be disregarded, $\mathrm{Im}\{\epsilon(\omega)\} \to 0$, the concept of quantization through the mode expansion can be recognized. The operators $\hat{f}(x,\omega)$ are replaced by the operators $\hat{a}_{\pm}(x,\omega)$ (it would be possible to introduce the operators $\hat{a}(x, k = \pm\mathrm{Re}\{n(\omega)\}\frac{\omega}{c})$), which satisfy the commutation relations

$$[\hat{a}_{\pm}(x,\omega), \hat{a}^{\dagger}_{\pm}(x',\omega')] = \exp\left[-\mathrm{Im}\{n(\omega)\}\frac{\omega}{c}|x-x'|\right]\delta(\omega-\omega')\hat{1}, \tag{4.106}$$

$$\begin{aligned}[\hat{a}_{\pm}(x,\omega), \hat{a}^{\dagger}_{\mp}(x',\omega')] = {} & 2\mathrm{Im}\{n(\omega)\}\frac{\omega}{c}\exp\left[\mp \mathrm{i}\beta(\omega)\frac{\omega}{c}(x+x')\right] \\ & \times \exp\left[-\mathrm{Im}\{n(\omega)\}\frac{\omega}{c}|x-x'|\right]\frac{\sin\left[\mathrm{Re}\{n(\omega)\}\frac{\omega}{c}|x-x'|\right]}{\mathrm{Re}\{n(\omega)\}\frac{\omega}{c}} \\ & \times \theta[\pm(x-x')]\delta(\omega-\omega')\hat{1},\end{aligned} \tag{4.107}$$

where $\theta(x)$ is the Heaviside function. These operators become independent of x in the limit $\mathrm{Im}\{n(\omega)\}\frac{\omega}{c}|x-x'| \to 0$. As the commutation relation (4.105) is in an obvious contradiction with a macroscopic approach, it is important that Gruner and Welsch (1996a) have derived the relation

$$[\hat{A}_{\Delta\omega}(x,0), \hat{E}_{\Delta\omega}(x',0)] = -\frac{\mathrm{i}\hbar}{\mathcal{A}\epsilon_{\mathrm{R}}(\omega_{\mathrm{c}})\epsilon_0}\delta(x-x')\hat{1}, \tag{4.108}$$

where ω_{c} is the centre frequency for suitably defined operators, $\hat{A}_{\Delta\omega}(x,0)$, $\hat{E}_{\Delta\omega}(x,0)$. The theory further reveals that the weak absorption gives rise to space-dependent mode operators that spatially progress according to quantum Langevin equations in the direct space. As could be expected, the operators $\hat{a}_{\pm}(x,\omega)$, as the forward- and backward-propagating fields, are governed by quantum Langevin equations, but it holds that the operator-valued Langevin noise is space dependent,

$$\hat{F}_{\pm}(x,\omega) = \pm\frac{1}{\mathrm{i}}\sqrt{2\mathrm{Im}\{n(\omega)\}\frac{\omega}{c}}\exp\left[\mp \mathrm{i}\mathrm{Re}\{n(\omega)\}\frac{\omega}{c}x\right]\hat{f}(x,\omega). \tag{4.109}$$

In other words, the operators $\hat{a}_{\pm}(x,\omega)$ progress in space.

As an example of inhomogeneous structure, two bulk dielectrics with a common interface are considered. The problem of determining a classical Green function reappears. The verification of the commutation relation (4.105) is performed by straightforward calculation, which is more complicated. A general proof of this relation is not present, causality reasons are only pointed out. There exists a straightforward generalization of the quantization method based on a mode expansion (Khosravi and Loudon 1991, 1992, Agarwal 1975). The behaviour of short light pulses propagating in a dispersive absorbing linear dielectric with a special attention to squeezed pulses has been studied (Schmidt et al. 1996).

4.2.2 Dispersive Lossy Nonlinear Inhomogeneous Dielectric

Emphasizing the important differences from the linear model, the Lagrangian and Hamiltonian for the nonlinear dielectric are introduced by Schmidt et al. (1998). The Lagrangian density (4.2) has been denoted by $\mathcal{L}_{\rm l}(\mathbf{r})$ and this relation with $\mathcal{L}$ replaced by $\mathcal{L}_{\rm l}(\mathbf{r})$ has been utilized in the Lagrangian density in the relation

$$\mathcal{L}(\mathbf{r}) = \mathcal{L}_{\rm l}(\mathbf{r}) + \mathcal{L}_{\rm nl}(\mathbf{r}), \tag{4.110}$$

where moreover

$$\mathcal{L}_{\rm nl}(\mathbf{r}) = f[\mathbf{X}(\mathbf{r})]. \tag{4.111}$$

While in the linear case it is sufficient to quantize only the transverse fields, in the nonlinear case such a procedure would result in a loss of generality. The result of the substitution from relations (4.31), (4.32), (4.33) into relation (4.30) which we have denoted as $\hat{H}$, we denote here as $\hat{H}_{\rm l}^{\perp}$. The total Hamiltonian can be written as

$$\hat{H} = \hat{H}_{\rm l} + \hat{H}_{\rm nl}, \tag{4.112}$$

where the nonlinear interaction term $\hat{H}_{\rm nl}$ is given by

$$\hat{H}_{\rm nl} = -\int f[\mathbf{X}(\mathbf{r})]\, {\rm d}^3\mathbf{r} \tag{4.113}$$

and the Hamiltonian $\hat{H}_{\rm l}$ that governs the linear dynamics can be written as

$$\hat{H}_{\rm l} = \hat{H}_{\rm l}^{\parallel} + \hat{H}_{\rm l}^{\perp}, \tag{4.114}$$

where

$$\begin{aligned}\hat{H}_{\rm l}^{\parallel} &= \int \left(\hbar\omega_0 \hat{b}^\dagger(\parallel, \mathbf{k}, t)\hat{b}(\parallel, \mathbf{k}, t) + \int_0^\infty \hbar\omega \hat{b}_\omega^\dagger(\parallel, \mathbf{k}, t)\hat{b}_\omega(\parallel, \mathbf{k}, t)\, {\rm d}\omega \right) {\rm d}^3\mathbf{k} \\ &\quad + \frac{\hbar}{2}\int\int_0^\infty V(\omega)[\hat{b}^\dagger(\parallel, -\mathbf{k}, t) + \hat{b}(\parallel, \mathbf{k}, t)][\hat{b}_\omega^\dagger(\parallel, \mathbf{k}, t)\hat{b}_\omega(\parallel, -\mathbf{k}, t)]\, {\rm d}\omega\, {\rm d}^3\mathbf{k}\end{aligned} \tag{4.115}$$

and the components $\hat{b}(\parallel, \mathbf{k}, t)$, $\hat{b}_\omega(\parallel, -\mathbf{k}, t)$ must be appropriately defined (see, Schmidt et al. (1998)) for $\hat{b}_{\parallel}(\mathbf{k})$, $\hat{b}_{\parallel}(\mathbf{k}, \omega)$). In general, $\hat{H}_{\rm nl}$ couples the transverse and longitudinal fields, cf., relation (4.12).

Schmidt et al. (1998) have derived evolution equations for the field operators and shown that additional noise sources appear in the nonlinear terms. Linear relationships between quantum (operator-valued) fields are introduced following Huttner and Barnett (1992b) as well as Gruner and Welsch (1995). The relations hold for all times and both in linear and in nonlinear cases.

Schmidt et al. (1998) do not attempt at diagonalization of the nonquadratical Hamiltonian $\hat{H}$, relation (4.112), the notation of which is still the same as of the Hamiltonian in (4.30). They avoid the difficulty with the generalization of the definitions (4.46) and (4.35) to the nonlinear functions of the operators $\hat{a}(\mathbf{k})$, $\hat{B}(\mathbf{k}, \omega)$, $\hat{b}(\mathbf{k})$, $\hat{b}_\omega(\mathbf{k})$.

We now approach the following representations of the matter fields. The longitudinal matter field $\hat{\mathbf{X}}^{\parallel}(\mathbf{r})$ can be expressed in terms of the field $\hat{\mathbf{f}}^{\parallel}(\mathbf{r}, \omega)$ as

$$\hat{\mathbf{X}}^{\parallel}(\mathbf{r}) = \sqrt{\frac{\hbar}{2\rho\tilde{\omega}_0}} \int_0^\infty [\alpha_0^*(\omega) - \beta_0^*(\omega)]\hat{\mathbf{f}}^{\parallel}(\mathbf{r}, \omega)\, d\omega + \text{H. c.}, \tag{4.116}$$

the transverse matter field $\hat{\mathbf{X}}^{\perp}(\mathbf{r})$ can be expressed in terms of the field $\hat{\mathbf{f}}(\mathbf{r}, \omega)$ as

$$\hat{\mathbf{X}}^{\perp}(\mathbf{r}) = \int_0^\infty \hat{\underline{\mathbf{X}}}^{\perp}(\mathbf{r}, \omega)\, d\omega + \text{H. c.}, \tag{4.117}$$

where

$$\hat{\underline{\mathbf{X}}}^{\perp}(\mathbf{r}, \omega) = \frac{\epsilon_0}{\alpha} \left[-i\omega[\epsilon(\omega) - 1]\hat{\underline{\mathbf{A}}}(\mathbf{r}, \omega) + \sqrt{\frac{\hbar}{\pi\epsilon_0} \text{Im}\{\epsilon(\omega)\}}\, \hat{\mathbf{f}}(\mathbf{r}, \omega) \right], \tag{4.118}$$

with $\hat{\underline{\mathbf{A}}}(\mathbf{r}, \omega)$ being connected with the field $\hat{\mathbf{f}}(\mathbf{r}, \omega)$ as the solution of equation (4.95) and the explicit relation (4.97), and the vector-potential field

$$\hat{\mathbf{A}}(\mathbf{r}) = \int_0^\infty \hat{\underline{\mathbf{A}}}(\mathbf{r}, \omega)\, d\omega + \text{H. c.} \tag{4.119}$$

If the validity of the expressions (4.119), (4.116), and (4.117) is related to the time evolution of the kind of (4.56), we may be afraid that this correctness will not endure the change to the nonlinear case. This change is reflected in the equations of motion for the basic fields and the vector-potential field in the Heisenberg picture,

$$i\hbar\frac{\partial}{\partial t}\hat{\mathbf{f}}^{\parallel}(\mathbf{r}, \omega) = [\hat{\mathbf{f}}^{\parallel}(\mathbf{r}, \omega), \hat{H}] = \hbar\omega\hat{\mathbf{f}}^{\parallel}(\mathbf{r}, \omega) + [\hat{\mathbf{f}}^{\parallel}(\mathbf{r}, \omega), \hat{H}_{\text{nl}}], \tag{4.120}$$

$$i\hbar\frac{\partial}{\partial t}\hat{\mathbf{f}}(\mathbf{r}, \omega) = [\hat{\mathbf{f}}(\mathbf{r}, \omega), \hat{H}] = \hbar\omega\hat{\mathbf{f}}(\mathbf{r}, \omega) + [\hat{\mathbf{f}}(\mathbf{r}, \omega), \hat{H}_{\text{nl}}], \tag{4.121}$$

$$i\hbar\frac{\partial}{\partial t}\hat{\underline{\mathbf{A}}}(\mathbf{r}, \omega) = [\hat{\underline{\mathbf{A}}}(\mathbf{r}, \omega), \hat{\text{H}}] = \hbar\omega\hat{\underline{\mathbf{A}}}(\mathbf{r}, \omega) + [\hat{\underline{\mathbf{A}}}(\mathbf{r}, \omega), \hat{\text{H}}_{\text{nl}}]. \tag{4.122}$$

To the number of the relations, which nevertheless hold in linear and nonlinear cases, the nonhomogeneous Helmholtz equation belongs

$$\Delta\hat{\underline{\mathbf{A}}}(\mathbf{r}, \omega) + \text{K}^2(\omega)\hat{\underline{\mathbf{A}}}(\mathbf{r}, \omega) = \frac{\omega}{\text{c}^2}\sqrt{\frac{\hbar}{\pi\epsilon_0}\text{Im}\{\epsilon(\omega)\}}\, \hat{\mathbf{f}}(\mathbf{r}, \omega). \tag{4.123}$$

Here it is noticed that $\hat{\mathbf{f}}(\mathbf{r},\omega) \equiv \hat{\mathbf{f}}(\mathbf{r},\omega,t)$, $\hat{\hat{\mathbf{A}}}(\mathbf{r},\omega) \equiv \hat{\hat{\mathbf{A}}}(\mathbf{r},\omega,\mathrm{t})$. Respecting the notation $K^2(\omega) = c^{-2}\omega^2\epsilon(\omega)$ (cf., (4.65)), we can see that

$$K^2(\omega)\hat{\hat{\mathbf{A}}}(\mathbf{r},\omega,\mathrm{t}) = \mathrm{K}^2(\omega\hat{\hat{1}})\hat{\hat{\mathbf{A}}}(\mathbf{r},\omega,\mathrm{t}), \tag{4.124}$$

where $\hat{\hat{1}}$ is the identity superoperator and relation (4.122) implies that

$$\omega\hat{\hat{1}} = \hat{\hat{1}}\frac{\partial}{\partial t} + \frac{1}{\hbar}\hat{H}_{\mathrm{nl}}^{\times}, \tag{4.125}$$

where we, for the sake of clarity, write $\frac{\partial}{\partial t}$ to the right from the notation $\hat{\hat{1}}$ and the action of $\hat{H}_{\mathrm{nl}}^{\times}$ on an operator $\hat{O}$ is defined by

$$\hat{H}_{\mathrm{nl}}^{\times}\hat{O} \equiv [\hat{H}_{\mathrm{nl}}, \hat{O}]. \tag{4.126}$$

Relation (4.125) can be written in the form

$$\Delta\hat{\hat{\mathbf{A}}}(\mathbf{r},\omega,\mathrm{t}) + \mathrm{K}^2\left(\hat{\hat{1}}\frac{\partial}{\partial \mathrm{t}} + \frac{1}{\hbar}\hat{\mathrm{H}}_{\mathrm{nl}}^{\times}\right)\hat{\hat{\mathbf{A}}}(\mathbf{r},\omega,\mathrm{t}) = \frac{\omega}{\mathrm{c}^2}\sqrt{\frac{\hbar}{\pi\epsilon_0}\mathrm{Im}\{\epsilon(\omega)\}}\,\hat{\mathbf{f}}(\mathbf{r},\omega,\mathrm{t}), \tag{4.127}$$

where the elimination of the field $\hat{\mathbf{X}}(\mathbf{r})$ using relations (4.116) and (4.118) indicates new noise sources. All of the fields $\hat{\mathbf{f}}^{\parallel}(\mathbf{r},\omega)$, $\hat{\mathbf{f}}(\mathbf{r},\omega)$, $\hat{\hat{\mathbf{A}}}(\mathbf{r},\omega)$ obey the nonlinear dynamics. By integration of (4.127) over ω, an equation adequate to the linear and nonlinear cases is obtained. The wealth of operator-valued fields serves the expression of the dispersion and absorption in the nonlinear medium. The basic equations are applied to the one-dimensional case and propagation equations for the slowly varying field amplitudes of pulse-like radiation are derived. The scheme is related to the familiar model of classical susceptibilities and applied to the problem of propagation of quantized radiation in a dispersive and lossy Kerr medium. In the linear theory it is possible to separate the two transverse polarization directions from each other and from the longitudinal direction. As has already been stated, this is not possible for nonlinear media. In practice, in a single-mode optical fibre, only one transverse polarization direction will be excited. Then the total Hamiltonian (4.112) can be reduced to a one-dimensional single-polarization form.

Let us consider the propagation in the x direction of plane waves polarized in the y direction. The one dimensionality of the problem permits one to decompose the field $\hat{\hat{A}}(x,\omega)$ into components $\hat{\hat{A}}_+(x,\omega)$ and $\hat{\hat{A}}_-(x,\omega)$, respectively, propagating in the positive and negative x-directions,

$$\hat{\hat{A}}(x,\omega) = \hat{\hat{A}}_+(x,\omega) + \hat{\hat{A}}_-(x,\omega), \tag{4.128}$$

where $\hat{\hat{A}}_\pm(x,\omega)$ are the solutions of spatial equations of progression

$$\frac{\partial}{\partial x}\hat{\hat{\mathrm{A}}}_\pm(x,\omega) = \pm\mathrm{i}K(\omega)\hat{\hat{A}}_\pm(x,\omega) \mp \mathrm{i}\mathcal{N}\sqrt{\frac{\mathrm{Im}\{\epsilon(\omega)\}}{\epsilon(\omega)}}\,\hat{f}(x,\omega), \tag{4.129}$$

with the normalization factor $\mathcal{N} = \sqrt{\frac{\hbar}{4\pi\epsilon_0 \mathcal{A} c^2}}$. We remark that $\hat{\hat{A}}_\pm(x,\omega) \equiv \hat{\hat{A}}_\pm(x,\omega,t)$. Similarly as from relations (4.123) and (4.127), one can arrive from relation (4.129) at relation

$$\frac{\partial}{\partial x}\hat{\hat{A}}_\pm(x,\omega,t) = \pm \mathrm{i}K\left(\hat{\hat{1}}\frac{\partial}{\partial t} + \frac{1}{\hbar}\hat{H}_{\mathrm{nl}}^{\times}\right)\hat{\hat{A}}_\pm(x,\omega,t)$$
$$\mp \mathrm{i}\mathcal{N}\sqrt{\frac{\mathrm{Im}\{\epsilon(\omega)\}}{\epsilon(\omega)}}\,\hat{f}(x,\omega,t). \tag{4.130}$$

In analogy to (4.119), the operators $\hat{A}_\pm^{(+)}(x)$ can be introduced,

$$\hat{A}_\pm^{(+)}(x) = \int_0^\infty \hat{\hat{A}}_\pm(x,\omega)\,\mathrm{d}\omega. \tag{4.131}$$

Integrating (4.130) over ω, an equation appropriate to the linear and nonlinear cases is obtained.

Adequately to the derived equations which we consider to be mere approximations in the nonlinear case, Schmidt et al. (1998) study the narrow-bandwidth field components and narrow-bandwidth pulses. The theory has been applied to narrow-bandwidth pulses propagating in a dielectric with a Kerr-like nonlinearity.

4.2.3 Elaboration of Linear Theory

Dung et al. (1998) have developed three-dimensional quantization presented in part in Gruner and Welsch (1996a) concerning dispersive and absorbing inhomogeneous dielectric medium. The approach directly starts with the Maxwell equations in the frequency domain for the macroscopic electromagnetic field. It is shown that the classical Maxwell equations together with the constitutive relations except relation (4.71) can be transferred to quantum theory. On considering the charge and current densities, one concentrates oneself on the noise-charge and noise-current densities. The operator-valued noise-charge density $\hat{\hat{\rho}}$ and the operator-valued noise-current density $\hat{\hat{\mathbf{j}}}$ are introduced, which are related to the operator-valued noise polarization $\hat{\hat{\mathbf{P}}}$,

$$\hat{\hat{\rho}}(\mathbf{r},\omega) = -\nabla\cdot\hat{\hat{\mathbf{P}}}(\mathbf{r},\omega), \tag{4.132}$$
$$\hat{\hat{\mathbf{j}}}(\mathbf{r},\omega) = -\mathrm{i}\omega\hat{\hat{\mathbf{P}}}(\mathbf{r},\omega). \tag{4.133}$$

It follows from relations (4.132) and (4.133) that $\hat{\hat{\rho}}$ and $\hat{\hat{\mathbf{j}}}$ fulfil the equation of continuity:

$$\nabla\cdot\hat{\hat{\mathbf{j}}}(\mathbf{r},\omega) = \mathrm{i}\omega\hat{\hat{\rho}}(\mathbf{r},\omega). \tag{4.134}$$

The source term $\hat{\hat{\mathbf{j}}}$ is related to a bosonic vector field $\hat{\mathbf{f}}$ by the relation like (4.97). The commutation relation (4.100) remains valid and relation (4.99) must be modified to the form

$$[\hat{f}_i(\mathbf{r}, \omega), \hat{f}_j^{\dagger}(\mathbf{r}', \omega')] = \delta_{ij}\delta(\mathbf{r} - \mathbf{r}')\delta(\omega - \omega')\hat{1}. \tag{4.135}$$

It is pointed out that the current density $\hat{\hat{\mathbf{j}}}$ is not transverse, because the whole electromagnetic field is considered. Hence, the vector field $\hat{\mathbf{f}}$ assumed here is not transverse as well and the spatial δ function in relation (4.135) is an ordinary δ function instead of a transverse δ function.

Relation (4.96) is an integral representation of the vector-potential operator. Dung et al. (1998) start from the partial differential equation

$$\nabla \times \nabla \times \hat{\hat{\mathbf{E}}}(\mathbf{r}, \omega) - \frac{\omega^2}{c^2}\epsilon(\mathbf{r}, \omega)\hat{\hat{\mathbf{E}}}(\mathbf{r}, \omega) = \mathrm{i}\omega\mu_0\hat{\hat{\mathbf{j}}}(\mathbf{r}, \omega), \tag{4.136}$$

whose solution can be represented as (here and in part of what follows we use a different notation)

$$\hat{\hat{\mathbf{E}}}(\mathbf{r}, \omega) = \mathrm{i}\omega\mu_0 \int \mathbf{G}(\mathbf{r}, \mathbf{s}, \omega) \cdot \hat{\hat{\mathbf{j}}}(\mathbf{s}, \omega)\, \mathrm{d}^3\mathbf{s}, \tag{4.137}$$

where $\mathbf{G}(\mathbf{r}, \mathbf{s}, \omega)$ is the tensor-valued Green function of the classical problem. It satisfies the equation

$$\left[\nabla_{\mathbf{r}}\nabla_{\mathbf{r}} - \mathbf{1}\left(\Delta_{\mathbf{r}} + \frac{\omega^2}{c^2}\epsilon(\mathbf{r}, \omega)\right)\right] \cdot \mathbf{G}(\mathbf{r}, \mathbf{s}, \omega) = \delta(\mathbf{r} - \mathbf{s})\mathbf{1} \tag{4.138}$$

together with appropriate boundary conditions. Dung et al. (1998) have derived commutation relations

$$[\hat{E}_i(\mathbf{r}), \hat{B}_k(\mathbf{r}')] = \frac{\hbar}{\pi\epsilon_0}\varepsilon_{kmj}\frac{\partial}{\partial x'_m}\int_{-\infty}^{\infty}\frac{\omega}{c^2}G_{ij}(\mathbf{r}, \mathbf{r}', \omega)\,\mathrm{d}\omega, \tag{4.139}$$

where ε_{kmj} is the Levi-Cività tensor and

$$G_{ij}(\mathbf{r}, \mathbf{r}', \omega) = \mathbf{e}_i \cdot \mathbf{G}(\mathbf{r}, \mathbf{r}', \omega) \cdot \mathbf{e}_j, \tag{4.140}$$

and

$$[\hat{E}_i(\mathbf{r}), \hat{E}_k(\mathbf{r}')] = \hat{0} = [\hat{B}_i(\mathbf{r}), \hat{B}_k(\mathbf{r}')]. \tag{4.141}$$

In the sense of the Helmholtz theorem there exists a unique decomposition of the electric field $\hat{\hat{\mathbf{E}}}$ into a transverse part $\hat{\hat{\mathbf{E}}}^{\perp}$ and a longitudinal part, $\hat{\hat{\mathbf{E}}}^{\parallel}$, i.e. the Coulomb gauge can be introduced, where $\hat{\hat{\mathbf{E}}}^{\perp} = \mathrm{i}\omega\hat{\hat{\mathbf{A}}}$ and $\hat{\hat{\mathbf{E}}}^{\parallel} = -\nabla\hat{\hat{\varphi}}$. In the Coulomb gauge, the vector and scalar potentials $\hat{\hat{\mathbf{A}}}$ and $\hat{\hat{\varphi}}$, respectively, are related to the electric field as

$$\hat{\hat{A}}_i(\mathbf{r}, \omega) = \frac{1}{\mathrm{i}\omega} \int \delta^{\perp}_{ij}(\mathbf{r} - \mathbf{s}) \hat{\hat{E}}_j(\mathbf{s}, \omega)\, \mathrm{d}^3\mathbf{s}, \tag{4.142}$$

$$\frac{\partial}{\partial x_i} \hat{\hat{\varphi}}(\mathbf{r}, \omega) = - \int \delta^{\parallel}_{ij}(\mathbf{r} - \mathbf{s}) \hat{\hat{E}}_j(\mathbf{s}, \omega)\, \mathrm{d}^3\mathbf{s}, \tag{4.143}$$

where $\delta^{\perp}_{ij}$ and $\delta^{\parallel}_{ij}$ are the components of the transverse and longitudinal tensor-valued δ functions

$$\boldsymbol{\delta}^{\perp}(\mathbf{r}) = \delta(\mathbf{r})\mathbf{1} + \nabla\nabla(4\pi|\mathbf{r}|)^{-1}, \tag{4.144}$$

$$\boldsymbol{\delta}^{\parallel}(\mathbf{r}) = -\nabla\nabla(4\pi|\mathbf{r}|)^{-1}. \tag{4.145}$$

It is recalled that $\hat{\mathbf{A}}(\mathbf{r})$ and $\epsilon_0\dot{\hat{\mathbf{A}}}(\mathbf{r})$ are canonically conjugated field variables. On the contrary, the complexity of the commutation relation (4.139) suggests that the "canonical" commutators are not so simple as we would expect by the definition. The commutation relation between the vector potential and the scalar potential is as complicated, when one and only one of these quantities is differentiated with respect to the time or comprises such a derivative. The simple commutation relations are

$$[\hat{A}_i(\mathbf{r}), \hat{A}_j(\mathbf{r}')] = \hat{0} \;=\; [\dot{\hat{A}}_i(\mathbf{r}), \dot{\hat{A}}_j(\mathbf{r}')], \tag{4.146}$$

$$[\hat{\varphi}(\mathbf{r}), \hat{\varphi}(\mathbf{r}')] = \hat{0} \;=\; [\hat{\varphi}(\mathbf{r}), \dot{\hat{A}}_i(\mathbf{r}')]. \tag{4.147}$$

Then, the theory is applied to the bulk dielectric such that the dielectric function can be assumed to be independent of space, $\epsilon(\mathbf{r}, \omega) = \epsilon(\omega)$ for all $\mathbf{r}$. In this case, the solution of equation (4.138) that satisfies the boundary condition at infinity is (cf., Tomaš 1995)

$$\mathbf{G}(\mathbf{r}, \mathbf{r}', \omega) = \left[\nabla_{\mathbf{r}}\nabla_{\mathbf{r}} + K^2(\omega)\mathbf{1}\right] K^{-2}(\omega) g(|\mathbf{r} - \mathbf{r}'|, \omega), \tag{4.148}$$

where

$$g(r, \omega) = \frac{\exp[\mathrm{i}K(\omega)r]}{4\pi r}. \tag{4.149}$$

Relation (4.139) can be simplified as

$$[\hat{E}_i(\mathbf{r}), \hat{B}_k(\mathbf{r}')] = -\frac{\mathrm{i}\hbar}{\epsilon_0} \varepsilon_{ikm} \frac{\partial}{\partial x_m} \delta(\mathbf{r} - \mathbf{r}')\hat{1}, \tag{4.150}$$

and the "canonical" commutator corresponds to the definition

$$[\hat{A}_i(\mathbf{r}), \dot{\hat{A}}_j(\mathbf{r}')] = \frac{\mathrm{i}\hbar}{\epsilon_0} \delta^{\perp}_{ij}(\mathbf{r} - \mathbf{r}')\hat{1}. \tag{4.151}$$

Moreover,

$$[\hat{\varphi}(\mathbf{r}), \hat{A}_j(\mathbf{r}')] = \hat{0}. \tag{4.152}$$

The commutation relations presented are equal-time Heisenberg picture ones and therefore it is emphasized that they are conserved. To make contact with the earlier work, Dung et al. (1998) define the vectors

$$\mathbf{f}^{\perp}(\mathbf{r}, \omega) = \int \boldsymbol{\delta}^{\perp}(\mathbf{r} - \mathbf{s}) \cdot \mathbf{f}(\mathbf{s}, \omega)\, \mathrm{d}^3\mathbf{s}, \tag{4.153}$$

$$\mathbf{f}^{\parallel}(\mathbf{r}, \omega) = \int \boldsymbol{\delta}^{\parallel}(\mathbf{r} - \mathbf{s}) \cdot \mathbf{f}(\mathbf{s}, \omega)\, \mathrm{d}^3\mathbf{s}. \tag{4.154}$$

The commutation relations (4.135) and (4.100) imply that

$$[\hat{f}_i^{\perp(\parallel)}(\mathbf{r}, \omega), (\hat{f}_j^{\perp(\parallel)}(\mathbf{r}', \omega'))^{\dagger}] = \delta_{ij}^{\perp(\parallel)}(\mathbf{r} - \mathbf{r}')\delta(\omega - \omega')\hat{1}, \tag{4.155}$$

$$[\hat{f}_i^{\perp(\parallel)}(\mathbf{r}, \omega), \hat{f}_j^{\perp(\parallel)}(\mathbf{r}', \omega')] = [\hat{f}_i^{\perp}(\mathbf{r}, \omega), (\hat{f}_j^{\parallel}(\mathbf{r}', \omega'))^{\dagger}] = \hat{0}. \tag{4.156}$$

The representation of transverse vector potential simplifies to

$$\hat{\tilde{\mathbf{A}}}(\mathbf{r}, \omega, 0) = \mu_0 \int \mathbf{g}(|\mathbf{r} - \mathbf{r}'|, \omega)\hat{\tilde{\mathbf{j}}}^{\perp}(\mathbf{r}', \omega, 0)\, \mathrm{d}^3\mathbf{r}'. \tag{4.157}$$

It can be derived that the scalar potential operator

$$\hat{\tilde{\varphi}}(\mathbf{r}, \omega, 0) = \frac{1}{4\pi\epsilon_0\epsilon(\omega)} \int \frac{\hat{\tilde{\rho}}(\mathbf{s}, \omega, 0)}{|\mathbf{r} - \mathbf{s}|}\, \mathrm{d}^3\mathbf{s}, \tag{4.158}$$

where $\hat{\tilde{\rho}}(\mathbf{r}, \omega, 0) = (\mathrm{i}\omega)^{-1}\nabla \cdot \hat{\tilde{\mathbf{j}}}^{\parallel}(\mathbf{r}, \omega, 0)$.

Another application is the quantization of the electromagnetic field in an inhomogeneous medium that consists of two bulk dielectrics with a common interface. The determination of the tensor-valued Green function for three-dimensional configuration of dielectric bodies is a very involved problem, in general. Dung et al. (1998) return to the simple configuration which was mentioned in Gruner and Welsch (1996a). It is referred to Tomaš (1995) for the classical treatment of multilayer structures. It is shown that for the configuration under study, the commutation relations (4.150), (4.151), and (4.152) hold.

The necessity of a new calculation of the quantum electrodynamical commutation relations for a new three-dimensional configuration (cf., Dung et al. 1998) is not absolute. Scheel et al. (1998) have proven that the fundamental equal-time commutation relations of quantum electrodynamics are preserved for an arbitrarily space-dependent Kramers–Kronig dielectric function.

Let us recall that the complex-valued dielectric function $\epsilon(\mathbf{r}, \omega)$ depends on frequency and space,

$$\epsilon(\mathbf{r}, \omega) \to 1, \text{ if } \omega \to \infty. \tag{4.159}$$

It is assumed that the real part (responsible for dispersion) and the imaginary part (responsible for absorption) are related to each other according to the Kramers–Kronig relations, because of causality. This also implies that $\epsilon(\mathbf{r}, \omega)$ is a holomorfic function in the upper complex half-plane of frequency

$$\frac{\partial}{\partial \omega^*}\epsilon(\mathbf{r}, \omega) = 0, \quad \mathrm{Im}\,\omega > 0. \tag{4.160}$$

Scheel et al. (1998) study relation (4.139). By comparison of the right-hand sides of this relation and relation (4.150), they arrive at the identity to be proven

$$-\int_{-\infty}^{\infty} \frac{\omega}{c^2}\mathbf{G}(\mathbf{r}, \mathbf{r}', \omega)\,\mathrm{d}\omega \times \overleftarrow{\nabla}_{\mathbf{r}'} = -\mathrm{i}\pi \mathbf{1}\delta(\mathbf{r} - \mathbf{r}') \times \overleftarrow{\nabla}_{\mathbf{r}'}. \tag{4.161}$$

Here the left arrow means that the operators $\frac{\partial}{\partial x'_m}$ will first be written as $\frac{\overleftarrow{\partial}}{\partial x'_m}$ in the expansion of the Hamilton operator, ∇, with this upper limit.

Based on the partial differential equation (4.138) for the tensor-valued Green function, an integral equation will be presented in what follows. The partial differential equation and the boundary condition at infinity determine the Green function uniquely. By comparison of relation (4.137) with a constitutive relation, we could derive that $\mathrm{i}\mu_0\omega G_{ij}(\mathbf{r}, \mathbf{s}, \omega)$ are holomorphic functions of ω in the upper complex half-plane, i.e.

$$\frac{\partial}{\partial \omega^*}\left[\omega G_{kj}(\mathbf{r}, \mathbf{s}, \omega)\right] = 0, \quad \mathrm{Im}\,\omega > 0, \tag{4.162}$$

with

$$\omega G_{kj}(\mathbf{r}, \mathbf{s}, \omega) \to 0 \text{ if } |\omega| \to \infty. \tag{4.163}$$

Second derivation of the Cauchy–Riemann equation (4.162) consists in the application of $\frac{\partial}{\partial \omega^*}$ to relation (4.138). The left-hand side of relation (4.162) is then the unique solution of the homogeneous problem.

Knöll and Leonhardt (1992) calculate the time dependent, let us say a direct-space Green function. This could be useful in the combination with a time-dependent (direct-space) noise $\mathbf{f}_{\mathrm{dir}}(\mathbf{r}, s)$. Scheel et al. (1998) have derived the relation

$$\mathrm{i}\mu_0\omega G_{ij}(\mathbf{r}, \mathbf{s}, \omega) = \tilde{D}_{ij}(\mathbf{r}, \mathbf{s}, \omega) = \int_0^{\infty} e^{\mathrm{i}\omega\tau} D_{ij}(\mathbf{r}, \mathbf{s}, \tau)\,\mathrm{d}\tau, \tag{4.164}$$

where $D_{ij}(\mathbf{r}, \mathbf{s}, \tau)$ are components of the tensor-valued response function that causally relates the electric field $\mathbf{E}(\mathbf{r}, t)$ to an external current $\mathbf{j}_{\mathrm{ext}}(\mathbf{s}, t - \tau)$, so that

$$\begin{aligned} D_{ij}(\mathbf{r}, \mathbf{s}, \tau) &= \frac{1}{2\pi}\int_{-\infty}^{\infty} e^{-\mathrm{i}\omega\tau}\tilde{D}_{ij}(\mathbf{r}, \mathbf{s}, \omega)\,\mathrm{d}\omega \\ &= -\mu_0\frac{\partial}{\partial \tau}G_{ij\,\mathrm{dir}}(\mathbf{r}, \mathbf{s}, \tau), \end{aligned} \tag{4.165}$$

where

$$G_{ij\text{dir}}(\mathbf{r}, \mathbf{s}, \tau) \equiv \frac{1}{2\pi} \int_{-\infty}^{\infty} e^{-\mathrm{i}\omega\tau} G_{ij}(\mathbf{r}, \mathbf{s}, \omega)\, \mathrm{d}\omega \tag{4.166}$$

is the direct-space Green function.

From the theory of partial differential equations it is known (see, e.g. Garabedian (1964)) that there exists an equivalent formulation of the problem in terms of an integral equation. On introducing $\epsilon_0(\omega) \equiv \frac{\int \epsilon(\mathbf{r},\omega)\, \mathrm{d}^3\mathbf{r}}{\int \mathrm{d}^3\mathbf{r}}$, an appropriately space-averaged reference relative permittivity, the integral equation for the tensor-valued Green function can be written as

$$\mathbf{G}(\mathbf{r}, \mathbf{s}, \omega) = \mathbf{G}^{(0)}(\mathbf{r}, \mathbf{s}, \omega) + \int \mathbf{K}(\mathbf{r}, \mathbf{v}, \omega) \cdot \mathbf{G}(\mathbf{v}, \mathbf{s}, \omega)\, \mathrm{d}^3\mathbf{v}, \tag{4.167}$$

where

$$\mathbf{G}^{(0)}(\mathbf{r}, \mathbf{s}, \omega) = [\mathbf{1} - \nabla_{\mathbf{r}}\nabla_{\mathbf{s}} K^{-2}(\mathbf{s}, \omega)]g(|\mathbf{r} - \mathbf{s}|, \omega), \tag{4.168}$$

$$\begin{aligned}\mathbf{K}(\mathbf{r}, \mathbf{v}, \omega) &= [\nabla_{\mathbf{r}} g(|\mathbf{r} - \mathbf{v}|, \omega)][\nabla_{\mathbf{v}} \ln\left(K^2(\mathbf{v}, \omega)\right)] \\ &\quad + [K^2(\mathbf{v}, \omega) - K_0^2(\omega)]g(|\mathbf{r} - \mathbf{v}|, \omega)]\mathbf{1}.\end{aligned} \tag{4.169}$$

Here $g(r, 0) \equiv g_0(r, 0)$ is given by (4.149), where $K(\omega) \equiv K_0(\omega)$,

$$K^2(\mathbf{r}, \omega) = \frac{\omega^2}{c^2}\epsilon(\mathbf{r}, \omega), \tag{4.170}$$

$$K_0^2(\omega) = \frac{\omega^2}{c^2}\epsilon_0(\omega). \tag{4.171}$$

It can be seen that the components of the kernel $K_{ik}(\mathbf{r}, \mathbf{v}, \omega)$ are holomorphic functions of ω in the upper complex half-plane, with

$$K_{ik}(\mathbf{r}, \mathbf{v}, \omega) \to 0 \text{ if } |\omega| \to \infty. \tag{4.172}$$

To prove the fundamental commutation relation (4.150), we first decompose the tensor-valued Green function into two parts,

$$\mathbf{G}(\mathbf{r}, \mathbf{s}, \omega) = \mathbf{G}_1(\mathbf{r}, \mathbf{s}, \omega) + \mathbf{G}_2(\mathbf{r}, \mathbf{s}, \omega), \tag{4.173}$$

where $\mathbf{G}_1(\mathbf{r}, \mathbf{s}, \omega)$ satisfies the integral equation

$$\mathbf{G}_1 = \mathbf{G}_1^{(0)} + \int \mathbf{K} \cdot \mathbf{G}_1\, \mathrm{d}^3\mathbf{v}, \tag{4.174}$$

with

$$\mathbf{G}_1^{(0)}(\mathbf{r}, \mathbf{s}, \omega) = g(|\mathbf{r} - \mathbf{s}|, \omega)\mathbf{1}, \tag{4.175}$$

$$\mathbf{G}_2(\mathbf{r}, \mathbf{s}, \omega) = \Gamma(\mathbf{r}, \mathbf{s}, \omega)\overleftarrow{\nabla}_{\mathbf{s}}. \tag{4.176}$$

In relation (4.176) Γ is the solution of the integral equation

$$\Gamma = \Gamma^{(0)} + \int \mathbf{K} \cdot \Gamma \, \mathrm{d}^3\mathbf{v}, \tag{4.177}$$

with

$$\Gamma^{(0)}(\mathbf{r}, \mathbf{s}, \omega) = -\nabla_{\mathbf{r}}[K^{-2}(\mathbf{s}, \omega) g(|\mathbf{r} - \mathbf{s}|, \omega)]. \tag{4.178}$$

Scheel et al. (1998) derive that $\mathrm{i}\mu_0\omega\mathbf{G}_1$ and $\mu_0\omega^2\Gamma$ are the Fourier transforms of the response functions to the noise-current density and the noise-charge density, respectively.

Combining relations (4.173) and (4.176) and recalling that $\overleftarrow{\nabla}_{\mathbf{r}'} \times \overleftarrow{\nabla}_{\mathbf{r}'} = \mathbf{0}$, we see that the left-hand side of relation (4.161) can be rewritten as

$$-\int_{-\infty}^{\infty} \frac{\omega}{c^2} \mathbf{G}(\mathbf{r}, \mathbf{r}', \omega) \, \mathrm{d}\omega \times \overleftarrow{\nabla}_{\mathbf{r}'} = -\int_{-\infty}^{\infty} \frac{\omega}{c^2} \mathbf{G}_1(\mathbf{r}, \mathbf{r}', \omega) \, \mathrm{d}\omega \times \overleftarrow{\nabla}_{\mathbf{r}'}. \tag{4.179}$$

Thus, only the noise-current response function $\mathrm{i}\mu_0\omega\mathbf{G}_1$ contributes to commutator (4.139). Multiplying the integral equation (4.174) by the function $\frac{\omega}{c^2}$ and integrating over ω, we obtain as the derivation of relation (4.105) from the holomorphic properties of the tensors $\mathbf{K}$ and $\omega\mathbf{G}_1$ that

$$\int_{-\infty}^{\infty} \frac{\omega}{c^2} \mathbf{G}_1(\mathbf{r}, \mathbf{r}', \omega) \, \mathrm{d}\omega = \mathrm{i}\pi \mathbf{1} \delta(\mathbf{r} - \mathbf{r}'). \tag{4.180}$$

The outer product of this equation and the operator $(-\overleftarrow{\nabla}_{\mathbf{r}'})$ can be taken and together with relation (4.179) implies relation (4.161).

In addition, it is shown that the scheme also applies to media with both absorption and amplification (in a bounded region of space). An extension of the quantization scheme to linear media with bounded regions of amplification is given and the problem of anisotropic media is briefly addressed, for which the permittivity is a symmetric complex tensor-valued function of ω,

$$\epsilon_{ij}(\mathbf{r}, \omega) = \epsilon_{ji}(\mathbf{r}, \omega). \tag{4.181}$$

Extensions of previous work on the propagation in absorbing dielectrics took linear amplification into account (Jeffers et al. 1996, Matloob et al. 1997, Artoni and Loudon 1998).

Knöll et al. (1999) investigated quantum-state transformation by dispersive and absorbing four-port devices. Under the usual assumptions on the dielectric permittivity, quantization of the Hamiltonian formalism of the electromagnetic field using a method close to the microscopic approach was performed by Tip (1998). A proper definition of band gaps in the periodic case and a new continuity equation for energy flow were obtained, and an S-matrix formalism for scattering from absorbing objects was worked out. In this way the generation of Čerenkov and transition radiation have been investigated. A path-integral formulation of quantum electrodynamics in a dispersive and absorbing dielectric medium has been presented by Bechler (1999) and has been applied on the microscopic level to the quantum theory of electromagnetic fields in dielectric media. Results concerning quantum electrodynamics in dispersing and absorbing dielectric media have been reviewed by Knöll et al. (2001). Tip et al. (2001) have proven the equivalence of two methods for quantization of the electromagnetic field in general dispersing and absorbing linear dielectrics: the Langevin-noise-current method and the auxiliary field method.

Petersson and Smith (2003) have illustrated the role of evanescent waves in power calculations for counterpropagating beams. In classical optics the field of a beam can be represented in terms of its plane-wave spectrum (Smith 1997, Clemmow 1996). Counterpropagating and "counter-evanescent" plane waves are defined relative to a selected plane. When a line current is placed over a dielectric slab, it is appropriate to insert a plane between the line current and the slab. The time-average power passing through a plane is a sum of powers contributed by the propagating plane waves and by "counter-evanescent" pairs of plane waves with the same transverse components of their wave vectors.

Suttorb and Wubs (2004) have provided a microscopic justification of the phenomenological quantization scheme for the electromagnetic field in inhomogeneous dielectric due to Gruner and Welsch (1995) (cf., references in subsection 4.2.3). Matloob (2004a) has paid attention to a damped harmonic oscillator. He has used a macroscopic Langevin equation for it. A canonical quantization scheme for the Langevin equation has been provided. A macroscopic electromagnetic field has been quantized in a homogeneous linear isotropic dielectric by the association of a damped quantum-mechanical harmonic oscillator with each mode of the radiation field. Matloob (2004b) has introduced a particular damped harmonic oscillator. He has used an appropriate form of the macroscopic Langevin equation. The canonical quantization scheme has been followed. A macroscopic electromagnetic field has been quantized in a linear isotropic permeable dielectric medium by associating a damped quantum-mechanical oscillator with each mode of the radiation field.

In Matloob (2005) a homogeneous medium is assumed that is isotropic in its rest frame. One works with positive frequency parts of the fields. The Minkowski relations are presented which are generalized constitutive relations for uniformly moving media. The electric induction vector depends also on the magnetic strength vector and the magnetic-induction vector depends also on the electric strength vector. Using the Minkowski relations the Maxwell–Minkowski equations are derived. The field vectors **E** and **B** are expressed in terms of the vector potential in the Weyl gauge. A time-independent wave equation with the noise polarization and noise magnetization for the vector potential is derived. It is shown that the constitutive relations may be convenient, with the electric induction vector independent of the magnetic strength vector and the magnetic-induction vector independent of the electric strength vector, on considering the anisotropy of the material in the laboratory frame. The Green tensor is studied in reciprocal and spatial coordinate space. The fields are quantized by expressing the noise-current density in terms of two infinite sets of appropriately chosen bosonic field operators. The vacuum field fluctuation is expressed.

4.2.4 Optical Field at Dielectric Devices

Matloob et al. (1995) provided expressions for the electromagnetic-field operators for three geometries: an infinite homogeneous dielectric, a semi-infinite dielectric, and a dielectric slab. A microscopic derivation has shown that a canonical quantum theory of light at the dielectric–vacuum interface is possible Barnett, Matloob, and Loudon (1995).

A simple quantum theory of the beam splitter, which can be applied to a Fabry–Pérot resonator, was introduced by Barnett et al. (1996) and developed by Barnett et al. (1998).

Artoni and Loudon (1997) applied the Huttner–Barnett scheme for quantization of the electromagnetic field in dispersive and absorbing dielectrics for the calculations of the effects of perpendicular propagation in a dielectric slab and to the properties of the incident light pulse. Their approach has provided a deeper understanding of antibunching (Artoni and Loudon 1999). Brun and Barnett (1998) considered an experimental set-up using a two-photon interferometer, where insertion of a dielectric into one or both arms of the interferometer is essential.

Suggestive is a comparative study of fermion and boson beam splitters (Loudon 1998). Fermions can be studied in analogy with bosons (Cahill and Glauber 1999).

Di Stefano et al. (1999) extended the field quantization to these material systems whose interaction with light is described, near a medium boundary, by a nonlocal susceptibility. Di Stefano et al. (2000) have developed a quantization scheme for the electromagnetic field in dispersive and lossy dielectrics with planar interface, including propagation in all the spatial directions, and considering both the transverse electric and the transverse magnetic polarized fields. Di Stefano et al. (2001a) have presented a one-dimensional scheme for the electromagnetic field in arbitrary planar dispersing and absorbing dielectrics, taking into account their finite extent. They have derived that the complete form of the electric-field operator includes a part that corresponds to the free fields incident from the vacuum towards the medium and a particular solution which can be expressed by using the classical Green-function integral representation of the electromagnetic field. By expressing the classical Green function in terms of the classical light modes, they have obtained a generalization of the method of modal expansion (e.g. Knöll et al. (1987)) to absorbing media. Di Stefano et al. (2001b) have based an electromagnetic-field quantization scheme on a microscopic linear two-band model. They have derived for the first time a noise-current operator for general anisotropic and/or spatially nonlocal media, which can be described only in terms of an appropriate frequency-dependent susceptibility.

The Green-tensor formalism is well suited to studying the behaviour of the quantized electromagnetic field in the presence of dispersing and absorbing bodies. Especially, it has been applied successfully to the study of input–output relations (Gruner and Welsch 1996b). As a continuation of this work, Khanbekyan et al. (2003) studied the quantized field in the presence of a dispersing and absorbing multilayered planar structure (shortly, multilayer plate). Three-dimensional input–output relations have been derived for frequency components of the electric-field operator in the transverse reciprocal space. Input–output relations for frequency components of this operator in the direct space have been given as well. The conditions have been stated, under which the input–output relations can be expressed in terms of bosonic operators. These relations have been discussed for the case of the plate being surrounded by vacuum. The theory applies to effectively free fields and those created by active atomic sources inside and/or outside the plate.

Khanbekyan et al. (2003) consider $n-1$ layers with thicknesses d_j, $j = 1, \ldots, n-1$, the region on the left of the plate ($j = 0$), and the region on the right of the plate ($j = n$). The permittivity is

$$\epsilon(z, \boldsymbol{\rho}, \omega) = \sum_{j=0}^{n} \lambda_j(z)\epsilon_j(\omega), \text{ independent of } \boldsymbol{\rho}, \tag{4.182}$$

where $\boldsymbol{\rho} = (x, y)$ and

$$\lambda_j(z) = \begin{cases} 1, \text{ if } z \in j\text{th region}, \\ 0, \text{ otherwise.} \end{cases} \tag{4.183}$$

For simplicity, we shift the fields $\mathbf{G}(\mathbf{r}, \mathbf{r}', \omega)$, $\hat{\mathbf{E}}(\mathbf{r}, \omega)$, and $\hat{\mathbf{f}}(\mathbf{r}, \omega)$ and all the regions along the z-axis such that the jth region has the left boundary plane going through the origin ($j > 0$) or the right boundary plane going through the origin ($j = 0$). The result of the shift of respective fields will be denoted by $\mathbf{G}^{(jj')}(\mathbf{r}, \mathbf{r}', \omega)$, $\hat{\mathbf{E}}^{(j)}(\mathbf{r}, \omega)$, and $\hat{\mathbf{f}}^{(j)}(\mathbf{r}, \omega)$. If $\Theta(z)$ is the unit-step function, then

$$[\Theta(z - z')]^{(jj')} = \begin{cases} \Theta(j - j') & \text{for } j \neq j', \\ \Theta(z - z') & \text{for } j = j', \end{cases} \tag{4.184}$$

$$[\Theta(z' - z)]^{(jj')} = \begin{cases} \Theta(j' - j) & \text{for } j \neq j', \\ \Theta(z' - z) & \text{for } j = j'. \end{cases} \tag{4.185}$$

Since the Green tensor depends only on the difference $\boldsymbol{\rho}-\boldsymbol{\rho}'$, it can be represented as a two-dimensional Fourier integral

$$\mathbf{G}^{(jj')}(\mathbf{r}, \mathbf{r}', \omega) = \frac{1}{(2\pi)^2} \int e^{\mathrm{i}\mathbf{k}\cdot(\boldsymbol{\rho}-\boldsymbol{\rho}')} \mathbf{G}^{(jj')}(z, z', \mathbf{k}, \omega)\, \mathrm{d}^2\mathbf{k}, \tag{4.186}$$

where $\mathbf{k} = (k_x, k_y)$ is the wave vector parallel to the interfaces and

$$\mathbf{G}^{(jj')}(z, z', \mathbf{k}, \omega) = \int e^{-\mathrm{i}\mathbf{k}\cdot\boldsymbol{\sigma}}\, \mathbf{G}^{(jj')}(z, z', \boldsymbol{\sigma} \equiv \boldsymbol{\rho} - \boldsymbol{\rho}', \omega)\, \mathrm{d}^2\boldsymbol{\sigma}. \tag{4.187}$$

The electric field operator $\hat{\mathbf{E}}^{(j)}(\mathbf{r}, \omega)$ may be written as a twofold Fourier transform:

$$\hat{\mathbf{E}}^{(j)}(\mathbf{r}, \omega) = \frac{1}{(2\pi)^2} \int e^{\mathrm{i}\mathbf{k}\cdot\boldsymbol{\rho}}\, \hat{\mathbf{E}}^{(j)}(z, \mathbf{k}, \omega)\, \mathrm{d}^2\mathbf{k}, \tag{4.188}$$

where

$$\hat{\mathbf{E}}^{(j)}(z, \mathbf{k}, \omega) = \int e^{-\mathrm{i}\mathbf{k}\cdot\boldsymbol{\rho}}\, \hat{\mathbf{E}}^{(j)}(z, \boldsymbol{\rho}, \omega)\, \mathrm{d}^2\boldsymbol{\rho}, \tag{4.189}$$

and the bosonic field operator may be written in the integral form as

$$\hat{\mathbf{f}}^{(j)}(\mathbf{r}, \omega) = \frac{1}{(2\pi)^2} \int e^{\mathrm{i}\mathbf{k}\cdot\boldsymbol{\rho}}\, \hat{\mathbf{f}}^{(j)}(z, \mathbf{k}, \omega)\, \mathrm{d}^2\mathbf{k}, \tag{4.190}$$

where

$$\hat{\mathbf{f}}^{(j)}(z, \mathbf{k}, \omega) = \int e^{-\mathrm{i}\mathbf{k}\cdot\boldsymbol{\rho}}\, \hat{\mathbf{f}}^{(j)}(z, \boldsymbol{\rho}, \omega)\, \mathrm{d}^2\boldsymbol{\rho}. \tag{4.191}$$

The Green tensor $\mathbf{G}^{(jj')}(z, z', \mathbf{k}, \omega)$ by the paper (Tomaš 1995) may be written as

$$\mathbf{G}^{(jj')}(z, z', \mathbf{k}, \omega) = -\mathbf{e}_z \frac{\delta_{jj'}}{k_j^2} \mathbf{e}_z \delta(z - z') + \mathbf{g}^{(jj')}(z, z', \mathbf{k}, \omega), \tag{4.192}$$

where

$$\begin{aligned}\mathbf{g}^{(jj')}(z, z', \mathbf{k}, \omega) = \frac{\mathrm{i}}{2} \sum_{q=\mathrm{p,s}} \sigma_q \big\{ &\boldsymbol{\mathcal{E}}_q^{j>}(z, \mathbf{k}, \omega) \Xi_q^{jj'} \boldsymbol{\mathcal{E}}_q^{j'<}(z', -\mathbf{k}, \omega) [\Theta(z - z')]^{(jj')} \\ &+ \boldsymbol{\mathcal{E}}_q^{j<}(z, \mathbf{k}, \omega) \Xi_q^{j'j} \boldsymbol{\mathcal{E}}_q^{j'>}(z', -\mathbf{k}, \omega) [\Theta(z' - z)]^{(jj')} \big\},\end{aligned} \tag{4.193}$$

($\sigma_\mathrm{p} = 1, \sigma_\mathrm{s} = -1$). In equation (4.193)

$$\boldsymbol{\mathcal{E}}_q^{j>}(z, \mathbf{k}, \omega) = \mathbf{e}_{q+}^{(j)}(\mathbf{k}) e^{\mathrm{i}\beta_j (z - d_j)} + r_{j/n}^q \mathbf{e}_{q-}^{(j)}(\mathbf{k}) e^{-\mathrm{i}\beta_j (z - d_j)}, \tag{4.194}$$

$$\boldsymbol{\mathcal{E}}_q^{j<}(z, \mathbf{k}, \omega) = \mathbf{e}_{q-}^{(j)}(\mathbf{k}) e^{-\mathrm{i}\beta_j z} + r_{j/0}^q \mathbf{e}_{q+}^{(j)}(\mathbf{k}) e^{\mathrm{i}\beta_j z} \tag{4.195}$$

and

$$\Xi_q^{jj'} = \frac{1}{\beta_n t_{0/n}^q} \frac{t_{0/j}^q e^{\mathrm{i}\beta_j d_j}}{D_{qj}} \frac{t_{0/j'}^q e^{\mathrm{i}\beta_{j'} d_{j'}}}{D_{qj'}}, \tag{4.196}$$

with

$$D_{qj} = 1 - r_{j/0}^q r_{j/n}^q e^{2\mathrm{i}\beta_j d_j}, \tag{4.197}$$

$(d_0 = d_n = 0)$ and

$$\beta_j = \sqrt{k_j^2 - k^2} = \beta_j' + \mathrm{i}\beta_j'' \ \ (\beta_j',\ \beta_j'' \geq 0) \tag{4.198}$$

$(k = |\mathbf{k}|)$, where

$$k_j = \sqrt{\varepsilon_j(\omega)}\frac{\omega}{c} = k_j' + \mathrm{i}k_j'' \ \ (k_j', k_j'' \geq 0), \tag{4.199}$$

and $t^q_{j/j'} = \frac{\beta_j}{\beta_{j'}} t^q_{j'/j}$ and $r^q_{j/j'}$, respectively, the transmission and reflection coefficients between the regions j' and j. The unit vectors $\mathbf{e}^{(j)}_{q\pm}(\mathbf{k})$ in equations (4.194) and (4.195) are the polarization unit vectors for transverse electric $(q = \mathrm{s})$ and transverse magnetic $(q = \mathrm{p})$ waves

$$\mathbf{e}^{(j)}_{\mathrm{s}\pm}(\mathbf{k}) = \frac{\mathbf{k}}{k} \times \mathbf{e}_z, \tag{4.200}$$

$$\mathbf{e}^{(j)}_{\mathrm{p}\pm}(\mathbf{k}) = \frac{1}{k_j}\left(\mp\beta_j \frac{\mathbf{k}}{k} + k\mathbf{e}_z\right). \tag{4.201}$$

If both the incoming fields incident on the two boundary planes of the plate are known, as well as the fields generated inside the plate, one can calculate the fields outgoing from the two boundary planes by means of input–output relations. These relations are valid also for evanescent-field components.

To obtain generally valid input–output relations, we restrict our attention to the electric-field operator. This operator in front of the structure (superscript 0) and behind the structure (superscript n) is decomposed in terms of input and output amplitude operators

$$\hat{\mathbf{E}}^{(0)}(z, \mathbf{k}, \omega) = \sum_{q=\mathrm{p,s}} \left[\mathbf{e}^{(0)}_{q+}(\mathbf{k})\hat{E}^{(0)}_{q\mathrm{in}}(z, \mathbf{k}, \omega) + \mathbf{e}^{(0)}_{q-}(\mathbf{k})\hat{E}^{(0)}_{q\mathrm{out}}(z, \mathbf{k}, \omega)\right], \tag{4.202}$$

$$\hat{\mathbf{E}}^{(n)}(z, \mathbf{k}, \omega) = \sum_{q=\mathrm{p,s}} \left[\mathbf{e}^{(n)}_{q-}(\mathbf{k})\hat{E}^{(n)}_{q\mathrm{in}}(z, \mathbf{k}, \omega) + \mathbf{e}^{(n)}_{q+}(\mathbf{k})\hat{E}^{(n)}_{q\mathrm{out}}(z, \mathbf{k}, \omega)\right]. \tag{4.203}$$

Here, the operators

$$\hat{E}^{(0)}_{q\mathrm{in}}(z, \mathbf{k}, \omega) = -\frac{\mu_0\omega}{2\beta_0} e^{\mathrm{i}\beta_0 z} \int_{-\infty}^{z} e^{-\mathrm{i}\beta_0 z'}\, \hat{\mathbf{j}}^{(0)}(z', \mathbf{k}, \omega) \cdot \mathbf{e}^{(0)}_{q+}(\mathbf{k})\, \mathrm{d}z' \tag{4.204}$$

and

$$\hat{E}^{(n)}_{q\mathrm{in}}(z, \mathbf{k}, \omega) = -\frac{\mu_0\omega}{2\beta_n} e^{-\mathrm{i}\beta_n z} \int_{z}^{\infty} e^{\mathrm{i}\beta_n z'}\, \hat{\mathbf{j}}^{(n)}(z', \mathbf{k}, \omega) \cdot \mathbf{e}^{(n)}_{q-}(\mathbf{k})\, \mathrm{d}z' \tag{4.205}$$

are input amplitude operators, where the Green function is of a very simple form. The operators

$$\hat{E}^{(0)}_{q\mathrm{out}}(z, \mathbf{k}, \omega) = e^{-\mathrm{i}\beta_0 z}\hat{E}^{(0)}_{q\mathrm{out}}(\mathbf{k}, \omega) + \frac{\mu_0\omega}{2\beta_0} e^{-\mathrm{i}\beta_0 z} \int_z^0 e^{\mathrm{i}\beta_0 z'}\, \hat{\mathbf{j}}^{(0)}(z', \mathbf{k}, \omega) \cdot \mathbf{e}^{(0)}_{q-}(\mathbf{k})\, \mathrm{d}z' \tag{4.206}$$

and

$$\hat{E}^{(n)}_{q\mathrm{out}}(z, \mathbf{k}, \omega) = e^{\mathrm{i}\beta_n z}\,\hat{E}^{(n)}_{q\mathrm{out}}(\mathbf{k}, \omega) + \frac{\mu_0\omega}{2\beta_n} e^{\mathrm{i}\beta_n z} \int_0^z e^{-\mathrm{i}\beta_n z'}\, \hat{\mathbf{j}}^{(n)}(z', \mathbf{k}, \omega) \cdot \mathbf{e}^{(n)}_{q+}(\mathbf{k})\, \mathrm{d}z' \tag{4.207}$$

are output amplitude operators, where the Green function has the complicated form and is "hidden" in the input–output relations

$$\begin{pmatrix} \hat{E}^{(0)}_{q\mathrm{out}}(\mathbf{k}, \omega) \\ \hat{E}^{(n)}_{q\mathrm{out}}(\mathbf{k}, \omega) \end{pmatrix} = \begin{pmatrix} r^q_{0/n}(\mathbf{k}, \omega) & t^q_{n/0}(\mathbf{k}, \omega) \\ t^q_{0/n}(\mathbf{k}, \omega) & r^q_{n/0}(\mathbf{k}, \omega) \end{pmatrix} \begin{pmatrix} \hat{E}^{(0)}_{q\mathrm{in}}(\mathbf{k}, \omega) \\ \hat{E}^{(n)}_{q\mathrm{in}}(\mathbf{k}, \omega) \end{pmatrix} + \sum_{j=1}^{n-1} \begin{pmatrix} \phi^{(j)}_{q0+}(\mathbf{k}, \omega) & \phi^{(j)}_{q0-}(\mathbf{k}, \omega) \\ \phi^{(j)}_{qn+}(\mathbf{k}, \omega) & \phi^{(j)}_{qn-}(\mathbf{k}, \omega) \end{pmatrix} \begin{pmatrix} \hat{E}^{(j)}_{q+}(\mathbf{k}, \omega) \\ \hat{E}^{(j)}_{q-}(\mathbf{k}, \omega) \end{pmatrix}. \tag{4.208}$$

They relate the output amplitude operators at the boundary planes of the plate to the input amplitude operators at these planes,

$$\hat{E}^{(0)}_{q\mathrm{in},out}(\mathbf{k}, \omega) = \hat{E}^{(0)}_{q\mathrm{in},out}(z, \mathbf{k}, \omega)\Big|_{z=0^-}, \tag{4.209}$$

$$\hat{E}^{(n)}_{q\mathrm{in},out}(\mathbf{k}, \omega) = \hat{E}^{(n)}_{q\mathrm{in},out}(z, \mathbf{k}, \omega)\Big|_{z=0^+}, \tag{4.210}$$

and to the amplitude operators associated with the layers

$$\hat{E}^{(j)}_{q\pm}(\mathbf{k}, \omega) = -\frac{\mu_0\omega}{2\beta_j} \int_0^{d_j} e^{\mp\mathrm{i}\beta_j z'}\, \hat{\mathbf{j}}^{(j)}(z', \mathbf{k}, \omega) \cdot \mathbf{e}^{(j)}_{q\pm}(\mathbf{k})\, \mathrm{d}z', \tag{4.211}$$

$j = 1, 2, \ldots, n-1$, which may have been better denoted say by $\hat{F}^{(j)}_{q\pm}(\mathbf{k}, \omega)$, since

$$\hat{E}^{(j)}_{q+}(\mathbf{k}, \omega) \neq \hat{E}^{(j)}_{q+}(z, \mathbf{k}, \omega)\Big|_{z=d_j^-}. \tag{4.212}$$

$$\hat{E}^{(j)}_{q-}(\mathbf{k}, \omega) \neq \hat{E}^{(j)}_{q-}(z, \mathbf{k}, \omega)\Big|_{z=0^+}. \tag{4.213}$$

In equation (4.208), the coefficients at the operators (4.211) read

$$\phi^{(j)}_{q0+} = \frac{t^q_{j/0} e^{2i\beta_j d_j}}{D_{qj}} r^q_{j/n}, \quad \phi^{(j)}_{q0-} = \frac{t^q_{j/0}}{D_{qj}}, \tag{4.214}$$

$$\phi^{(j)}_{qn+} = \frac{t^q_{j/n} e^{i\beta_j d_j}}{D_{qj}}, \quad \phi^{(j)}_{qn-} = \frac{t^q_{j/0} e^{i\beta_j d_j}}{D_{qj}} r^q_{j/n}. \tag{4.215}$$

It is worth noting that any two planes $z = z^{(0)} \leq 0^-$ and $z = z^{(n)} \geq 0^+$ for $j = 0$ and $j = n$, respectively, also can be used in principle for a formulation of the input–output relations. The theory applies to effectively free fields and those created by active atomic sources inside and/or outside the plate.

4.2.5 Modification of Spontaneous Emission by Dielectric Media

Scheel et al. (1999a) have found quantum local-field corrections appropriate to the spontaneous emission by an excited atom. Dung et al. (2000) have developed a formalism for studying spontaneous decay of an excited two-level atom in the presence of arbitrary dispersing and absorbing dielectric bodies. They have shown how the minimal-coupling Hamiltonian simplifies to a Hamiltonian in the dipole approximation. The formalism is based on a source-quantity representation of the electromagnetic field in terms of the tensor-valued Green function of the classical problem and appropriately chosen bosonic quantum fields. All relevant information about the bodies such as form and dispersion and absorption properties is contained in the tensor-valued Green function. This function has been available for various configurations such as planarly, spherically, and cylindrically multilayered media (Chew 1995).

The theory has been applied to the spontaneous decay of a two-level atom placed at the centre of a three-layer spherical microcavity, the wall being modelled by a Lorentz dielectric. The tensor-valued Green function of the configuration has been known (Li et al. 1994). The calculations have been performed on the assumption of a dielectric with a single resonance. For simplicity, it has been assumed that the atom is positioned at the centre of the cavity. Weak and strong couplings are studied and in the study of the strong couplings both the normal-dispersion range and the anomalous-dispersion range associated with the band gap are considered. Whereas in the range of normal dispersion, the cavity input–output coupling dominates the strength of the atom–field interaction, the significant effect within the band gap is the photon absorption by the wall material.

Dung et al. (2001) have studied nonclassical decay of an excited atom near a dispersing and absorbing microsphere of given complex permittivity that satisfies the Kramers–Krönig relations laying emphasis on a Drude–Lorentz permittivity. Among others, they have found a condition on which the decay becomes purely non-radiative. Dung et al. (2002a,b) have given a rigorous quantum-mechanical derivation of the rate of intermolecular energy transfer in the presence of dispersing and absorbing media with spatially varying permittivity. They have applied the theory to bulk material, multislab planar structures, where they also have made comparison with experiments, and to microspheres. They have shown that the minimal-coupling scheme and the multipolar-coupling scheme yield exactly the same form of the rate formula.

Tip (2004) has used his auxiliary field method to obtain various equivalent Hamiltonians for charged particles interacting with absorptive dielectrics. In two steps, the representations cease to manifest the generalized Coulomb gauge used, but it remains in one term, concentrated in a wave operator. It has also been shown for excited atoms in a photon crystal with transition frequency in a band gap that their states do not decay radiatively.

For a transparent dielectric, theoretical studies can take a traditional approach. Inoue and Hori (2001) have developed a formalism of quantization of electromagnetic fields including evanescent waves based on the detector-mode functional defined in terms of those for the widely used triplet modes. They have evaluated atomic and molecular radiation near a dielectric boundary surface. Matloob and Pooseh (2000) have discussed a fully quantum-mechanical theory of the scattering of coherent light by a dissipative dispersive slab. Matloob and Falinejad (2001) have calculated the Casimir force between two dielectric slabs by using the notion of the radiation pressure associated with the quantum electromagnetic vacuum. Specifically, they have used the fact that only the field correlation functions are needed for the evaluation of vacuum radiation pressure on an interface. Matloob (2001) has postulated an electromagnetic-field Lagrangian density at each point of space–time to be of an unfamiliar form comprising the noise-current density. He has expressed the displacement $D(r, t)$ merely in terms of the electric-field $E(r, t')$, $t' \leq t$, without adding a noise polarization term.

In the framework of a semiclasical approach, Paspalakis and Kis (2002) have studied the propagation dynamics of N laser pulses interacting with an $(N+1)$-level quantum system (one upper state and N lower states). Assuming the system to be in a superposition state of all of the lower levels initially they have determined the conditions of complete opacity or transparency of the medium. The coupling of pulses is most interesting in the limit of parametric generation.

A simplified approach to the quantization is sufficient for the theory of the radiation pressure on dielectric surfaces (Loudon 2002). Loudon (2003) continues (Loudon 2002) with two changes or extensions. Instead of a planar pulse he considers Laguerre–Gaussian light beams. He considers the transfer of angular momentum to a dielectric. He may issue from the book (Allen et al. 2003) and arrive at the paper (Padgett et al. 2003).

New forces are produced by a pulse of Laguerre–Gaussian light in comparison with a plane-wave pulse, which produces only a longitudinal force. These are radial and azimuthal forces. A simplification is achieved by assuming that the modal function has zero radial index. The pulse is assumed to contain a single photon. In case an interface is considered the propagation direction of the pulse is assumed to be perpendicular to the surface. If the pulse is propagated into a dielectric, then it is assumed that the dielectric is – weakly – attenuating to ensure that the model need not complicate by including any exit or reflection, or finiteness of the dielectric in the direction of propagation.

Loudon (2003) gives Laguerre–Gaussian light in terms of the Lorentz-gauge vector potential. He does not speak of the associated Laguerre polynomials, but it is obvious that the degree of a polynomial is zero, when he restricts himself to the radial index $p = 0$.

The theory is quantized. The single-photon pulse is represented by the state vector $|1\rangle$. The spectrum of the photon wave packet is a narrow-band Gaussian function. The author introduces the normal-order Poynting operator $:\hat{\mathbf{S}}(\mathbf{r}, t):$, with z-component denoted by $:\hat{S}_z(\mathbf{r}, t):$. If the dispersion is ignored, the author can write the expectation value

$$\langle 1|:\hat{S}_z(\mathbf{r}, t):|1\rangle = \frac{\hbar\omega_0 c}{L}\sqrt{\frac{2}{\pi}} \exp\left[-\frac{2c^2}{L^2}\left(t - \frac{\eta z}{c}\right)^2\right]|u|^2, \tag{4.216}$$

where L is a conventional length of the pulse, ω_0 is a central frequency of the wave packet, $\eta \equiv \eta(\omega_0)$ is a refractive index, and $u \equiv u_{k_0,l}(\mathbf{r})$ is the modal function, $k_0 = \frac{\eta(\omega_0)\omega_0}{c}$ is an angular wave number, and l is the orbital angular-momentum quantum number. The time integral of equation (4.216) is

$$\int_{-\infty}^{\infty} \langle 1|:\hat{S}_z(\mathbf{r}, t):|1\rangle \, \mathrm{d}t = \hbar\omega_0 |u|^2. \tag{4.217}$$

Further, the author constructs the normally ordered angular-momentum density operator. He introduces the Lorentz force-density operator $:\hat{\mathbf{f}}(\mathbf{r}, t):$ and lets $:\hat{f}_z(\mathbf{r}, t):$ denote the longitudinal component of this operator. He determines that

$$\begin{aligned}\langle 1|:\hat{f}_z(\mathbf{r}, t):|1\rangle = &-\frac{2\hbar\omega_0 c(\eta^2 - 1)}{L^3}\sqrt{\frac{2}{\pi}}\left(t - \frac{\eta z}{c}\right) \\ &\times \exp\left[-\frac{2c^2}{L^2}\left(t - \frac{\eta z}{c}\right)^2\right]|u|^2.\end{aligned} \tag{4.218}$$

Similarly, he lets $:\hat{f}_\rho(\mathbf{r}, t):$ denote the radial component of the operator $:\hat{\mathbf{f}}(\mathbf{r}, t):$. As usual, $\rho = \sqrt{x^2 + y^2}$. He determines that

$$\langle 1|:\hat{f}_\rho(\mathbf{r}, t):|1\rangle = \frac{2\hbar\omega_0(\eta^2 - 1)}{\sqrt{2\pi}\eta L}\left(\frac{|l|}{\rho} - \frac{2\rho}{w_0^2}\right) \exp\left[-\frac{2c^2}{L^2}\left(t - \frac{\eta z}{c}\right)^2\right]|u|^2. \tag{4.219}$$

The radial force compresses the dielectric towards the cylinder of radius $\rho_0 = w_0\sqrt{\frac{|l|}{2}}$, where w_0 is the beam waist.

Similarly, he lets $:\hat{f}_\phi(\mathbf{r},t):$ denote the azimuthal component of the operator $:\hat{\mathbf{f}}(\mathbf{r},t):$. As usual, $\phi = \arg(x + \mathrm{i}y)$. He determines that

$$\langle 1|:\hat{f}_\phi(\mathbf{r},t):|1\rangle = -\frac{4\hbar c^2(\eta^2-1)}{\eta L^3}\sqrt{\frac{2}{\pi}}\left(t-\frac{\eta z}{c}\right) \times \exp\left[-\frac{2c^2}{L^2}\left(t-\frac{\eta z}{c}\right)^2\right]\left(\frac{l}{\rho}-\frac{\sigma|l|}{\rho}+\frac{2\sigma\rho}{w_0^2}\right)|u|^2, \quad (4.220)$$

where σ is the spin angular-momentum quantum number of the beam.

The author extends the theory to the case, where space is divided into two regions with a dielectric of real refractive index $\eta_0(\omega)$ at $z < 0$ and a dielectric of a complex refractive index

$$n(\omega) = \eta(\omega) + i\kappa(\omega) \quad (4.221)$$

at $z > 0$. The modification of the result (4.217) at $z > 0$ is

$$\int_{-\infty}^{\infty} \langle 1|:\hat{S}_z(\mathbf{r},t):|1\rangle\,\mathrm{d}t = \hbar\omega_0 \exp\left(-\frac{2\omega_0\kappa z}{c}\right)\frac{4\eta_0\eta}{(\eta_0+\eta)^2+\kappa^2}|u|^2, \quad (4.222)$$

where η_0, η, and κ are evaluated at frequency ω_0.

The author gives particular attention to the transfer of longitudinal and angular momentum to the dielectric from light incident from free space. He lets the total force on dielectric in $\{z > 0\}$ and at time t be represented by the force operator

$$\hat{\mathbf{F}}(t) = \int_{z>0} \hat{\mathbf{f}}(\mathbf{r},t)\,\mathrm{d}\mathbf{r} \quad (4.223)$$

and lets $\hat{F}_z(t)$ denote the longitudinal component of this operator. The time-integrated force, or the total linear momentum transfer to the dielectric, is

$$\int_{-\infty}^{\infty} \langle 1|:\hat{F}_z(t):|1\rangle\,\mathrm{d}t = \frac{2\hbar\omega_0}{c}\left\{\underbrace{\frac{\eta^2+\kappa^2-1}{(\eta+1)^2+\kappa^2}}_{\text{surface}} + \underbrace{\frac{2}{(\eta+1)^2+\kappa^2}}_{\text{bulk}}\right\} = \frac{2\hbar\omega_0}{c}\underbrace{\frac{\eta^2+1+\kappa^2}{(\eta+1)^2+\kappa^2}}_{\text{total}}. \quad (4.224)$$

In pursuit of the torque the author first introduces the operator that represents the density of the z-component of the torque on the dielectric $:\hat{g}_z(\mathbf{r},t):$. Next he lets the total torque on the dielectric in $\{z > 0\}$ and at time t be represented by the torque operator

$$\hat{G}_z(t) = \int_{z>0} \hat{g}_z(\mathbf{r},t)\,\mathrm{d}\mathbf{r}. \quad (4.225)$$

The time-integrated torque, or total angular-momentum transfer on dielectric, is

$$\int_{-\infty}^{\infty} \langle 1|{:}\hat{G}_z(t){:}|1\rangle \,\mathrm{d}t = \frac{4\hbar(l+\sigma)\eta}{(\eta+1)^2+\kappa^2}\left\{\underbrace{\frac{\eta^2+\kappa^2-1}{\eta^2+\kappa^2}}_{\text{surface}}+\underbrace{\frac{1}{\eta^2+\kappa^2}}_{\text{bulk}}\right\}$$
$$= \underbrace{\frac{4\hbar(l+\sigma)\eta}{\eta^2+1+\kappa^2}}_{\text{total}}. \tag{4.226}$$

It is shown that it is meaningful to divide the total transfer of linear momentum into surface reflected, surface transmitted, and bulk transmitted contributions provided a photon has passed through. From this the shift of a slab of mass M may be calculated, due to a normally incident single-photon wave packet. Similarly (but for $\kappa = 0$ only) the angular rotation of the dielectric slab, whose moment of inertia around the z-axis is denoted I may be calculated, due to such a wave packet.

A scheme for transferring quantum states from the propagating light fields to a macroscopic, collective vibrational degree of freedom of a massive mirror has been proposed (Zhang et al. 2003). The proposal may realize an Einstein–Podolsky–Rosen state in position and momentum for a pair of massive mirrors at distinct locations by exploiting a nondegenerate optical parametric amplifier. Loudon et al. (2005) have paid attention to the photon drag effect. The momentum transfer from light to a dielectric material has been calculated by evaluation of the relevant Lorentz force. The photon drag effect is named after the generation of currents or electric fields in semiconductors.

Leonhardt and Piwnicki (2001) have analysed the propagation of slow light in moving media in the case where the light is monochromatic in the laboratory frame. The extremely low group velocity is caused by the electromagnetically induced transparency of an atomic transition. Lombardi (2002) has re-examined the physical significance of different velocities which can be introduced for a wave train. Oughstun and Cartwright (2005) have compared the group velocity with the instantaneous centroid velocity of the pulse Poynting vector for an ultrashort Gaussian pulse. Very long pulses that are well tuned to a region of anomalous dispersion do not have superluminal peak velocity of a real physical significance. Yanik and Fan (2005) have formulated basic principles that underlie stopping and storing light coherently in many different physical systems. Following a brief discussion of one of known atomic stopping-light schemes, an all-optical scheme has been analysed in detail.

Tip (2007) has studied the properties of atoms close to an absorptive dielectric using his quantized form of the phenomenological Maxwell equations. The author has treated the coupling of atoms with longitudinal modes in detail. The atomic interaction potential changes from the Coulomb one to a static potential, i.e. one that obeys a Poisson equation with zero-frequency limit of the permittivity. The longitudinal interactions of atoms with absorptive dielectrics are responsible for nonradiative decay of the atoms. It has been found that the Hamiltonian used by

Dung et al. (2002b) is unitarily equivalent to a special case of the Hamiltonian used by Tip (2007).

4.2.6 Left-Handed Materials

The interest in "left-handed" materials is reflected both in the theory and in the experiment. Veselago (1967, 1968) was the first to address the question of propagation of the electromagnetic waves in the medium with both the permittivity $\epsilon_0\epsilon$ and the permeability $\mu_0\mu$ negative. Veselago has shown that, although such materials are not available in the nature, their existence is not excluded by the Maxwell equations Markoš 2005).

First we summarize the basic electromagnetic properties of left-handed materials. The propagation of the electromagnetic wave, $\mathbf{E} = \mathbf{E}_0 e^{i(\mathbf{k}\cdot\mathbf{r}-\omega t)}$, is described by the wave equation

$$\left[\mathbf{k}^2 - \frac{\omega^2}{c^2}\epsilon\mu\right]\mathbf{E}(\mathbf{r}, t) = 0, \tag{4.227}$$

where $\mathbf{k}$ is the wave vector and ω is the frequency,

$$\mathbf{k}^2 = \frac{\omega^2}{c^2}\epsilon\mu. \tag{4.228}$$

Propagation is possible if $\mathbf{k}^2 > 0$. This is always true in dielectrics, where $\mu = 1$ and ϵ is real, while no propagation is possible in metals, the permittivity of which is negative. Materials with both ϵ and μ negative allow the propagation of electromagnetic waves.

The Maxwell equations simplify to the form

$$\mathbf{k} \times \mathbf{E} = \omega\mu_0\mu\mathbf{H}, \mathbf{k} \times \mathbf{H} = -\omega\epsilon_0\epsilon\mathbf{E}. \tag{4.229}$$

Using them in the identities

$$\mathbf{k} \cdot (\mathbf{E} \times \mathbf{H}) = \mathbf{H} \cdot (\mathbf{k} \times \mathbf{E}) = \mathbf{E} \cdot (\mathbf{H} \times \mathbf{k}), \tag{4.230}$$

we can derive that

$$\mathbf{k} \cdot (\mathbf{E} \times \mathbf{H}) = \omega\mu_0\mu\mathbf{H}^2 = \omega\epsilon_0\epsilon\mathbf{E}^2, \tag{4.231}$$

or

$$\mathbf{k} \cdot \mathbf{S} < 0, \tag{4.232}$$

where

$$\mathbf{S} = \mathbf{E} \times \mathbf{H} \tag{4.233}$$

is the Poynting vector. It means that the wave vector **k** and the Poynting vector **S** have opposite directions.

The three vectors **E**, **H**, and **k** are a left-handed set of vectors, contrary to conventional materials, where they are a right-handed set. This inspired Veselago to name the materials with both ϵ and μ negative left-handed materials.

Veselago also has pointed out that the left-handed material must be dispersive, ϵ or μ must depend on the frequency of a monochromatic field, since the time-averaged energy density of the electromagnetic field,

$$\langle U\rangle = \frac{1}{2}\left\{\epsilon_0 \frac{\partial\left[\omega_0\epsilon(\omega_0)\right]}{\partial\omega_0}\langle\mathbf{E}^2\rangle + \mu_0\frac{\partial\left[\omega_0\mu(\omega_0)\right]}{\partial\omega_0}\langle\mathbf{H}^2\rangle\right\}, \tag{4.234}$$

would be negative. Here the electromagnetic field is assumed to be quasimonochromatic, with a centre or carrier frequency $\omega_0 > 0$. The time averaging is done over the period of the carrier, $\frac{2\pi}{\omega_0}$. The original Gaussian units of measurement can be respected by the replacements $\epsilon_0 \to \frac{1}{4\pi}$, $\mu_0 \to \frac{1}{4\pi}$. From the Kramers–Kronig relations it follows that ϵ and μ must be complex. Strictly speaking, we should speak of the medium with both $\operatorname{Re}\epsilon$ and $\operatorname{Re}\mu$ negative. Many results have been derived on the assumption of a transparent medium, i.e. that $\operatorname{Im}\epsilon$ and $\operatorname{Im}\mu$ may be neglected.

The energy density (4.234) is positive, since (Landau et al. 1984)

$$\frac{\partial(\omega\epsilon)}{\partial\omega} > 0, \quad \frac{\partial(\omega\mu)}{\partial\omega} > 0. \tag{4.235}$$

The permittivity and permeability determine the index of refraction, $n \equiv n(\omega)$, and the relative impedance, $Z \equiv Z(\omega)$, by the relations

$$n = \pm\sqrt{\epsilon\mu}, \quad Z = \sqrt{\frac{\mu}{\epsilon}}. \tag{4.236}$$

It is assumed that the complex square root has a positive real part, or nonnegative imaginary part if the real part vanishes. The plus sign in (4.236) is used when the square root has a positive imaginary part and the minus sign in (4.236) is used when the square root has a negative imaginary part. Even though the assumption that $\operatorname{Im}\epsilon = \operatorname{Im}\mu = 0$ is frequently used and it invalidates the rule of the sign, we may assume $\operatorname{Im}\epsilon > 0$ or $\operatorname{Im}\mu > 0$ small and apply the rule. Due to continuity, the minus sign is used in (4.236) for $\epsilon < 0$ and $\mu < 0$.

We introduce a phase velocity vector

$$\mathbf{v}_{\mathrm{p}} \equiv \frac{\omega}{\mathbf{s}\cdot\mathbf{k}}\mathbf{s}, \tag{4.237}$$

where **s** is the direction of the Poynting vector **S**. As the wave vector

$$\mathbf{k} = \frac{\omega}{c}n\mathbf{s}, \tag{4.238}$$

we have let $\mathbf{v}_\mathrm{p}$ denote $\frac{c}{n}\mathbf{s}$. On writing the wave vector $\mathbf{k}$ in the forms

$$\mathbf{k} = \pm\frac{\omega}{c}\sqrt{\epsilon\mu}\,\mathbf{s}, \tag{4.239}$$

we obtain that

$$\frac{\partial\mathbf{k}}{\partial\omega} = \frac{1}{2c}\left[Z\frac{\partial(\omega\epsilon)}{\partial\omega} + Z^{-1}\frac{\partial(\omega\mu)}{\partial\omega}\right]\mathbf{s}, \quad \mathbf{s}\cdot\frac{\partial\mathbf{k}}{\partial\omega} > 0, \tag{4.240}$$

where Z still means the relative impedance, not the coordinate. Now we introduce a group-velocity vector

$$\mathbf{v}_\mathrm{g} = \frac{1}{\mathbf{s}\cdot\frac{\partial\mathbf{k}}{\partial\omega}}\mathbf{s}, \quad \mathbf{s}\cdot\mathbf{v}_\mathrm{g} > 0. \tag{4.241}$$

It has the same direction as the Poynting vector not only in the right-handed media, but also for the left-handed materials.

For $n < 0$ or, more generally, $\mathrm{Re}\,n < 0$, the negative refraction occurs. Let us illustrate that the derivation of the Snell law is valid also for a negative refractive index. We consider a planar interface in the plane $z = 0$ between the half-spaces $z < 0$ and $z > 0$. The half-space $z < 0$ is free and the half-space $z > 0$ is filled with a left-handed material. We introduce $n_1 = 1$ and $n_2 = n$. We assume an incident monochromatic wave with $\mathbf{E}_1^+ = \mathbf{E}_{10}^+ e^{\mathrm{i}(\mathbf{k}_1^+\cdot\mathbf{r}-\omega t)}$, a reflected wave with $\mathbf{E}_1^- = \mathbf{E}_{10}^- e^{\mathrm{i}(\mathbf{k}_1^-\cdot\mathbf{r}-\omega t)}$, and a transmitted one with $\mathbf{E}_2^+ = \mathbf{E}_{20}^+ e^{\mathrm{i}(\mathbf{k}_2^+\cdot\mathbf{r}-\omega t)}$. Here $\mathbf{k}_1^+$, $\mathbf{k}_1^-$, and $\mathbf{k}_2^+$ are the respective wave vectors. The respective Poynting vectors may be denoted by $\mathbf{S}_1^+$, $\mathbf{S}_1^-$, and $\mathbf{S}_2^+$. Let $\mathbf{s}_1^+$, $\mathbf{s}_1^-$, and $\mathbf{s}_2^+$ mean their directions. Quite reasonably, we suppose that $s_{1z}^+ > 0$, $s_{1z}^- < 0$, and $s_{2z}^+ > 0$ both for the right-handed media and the left-handed materials.

We consider $k_{1y}^+ = 0$. Then $k_{1y}^- = 0$ and $k_{2y}^+ = 0$ by the isotropy. Still it holds that $k_{1x}^- = k_{1x}^+, k_{2x}^+ = k_{1x}^+$. Let us note that $k_{1x}^+ = \frac{\omega}{c}n_1 s_{1x}^+ = \frac{\omega}{c}s_{1x}^+, k_{2x}^+ = \frac{\omega}{c}n_2 s_{2x}^+ = \frac{\omega}{c}n s_{2x}^+$. From this, $n s_{2x}^+ = s_{1x}^+$. Therefore, the negative refraction occurs when the refractive index n is negative.

A planar slab of a material with $\epsilon = -1$ and $\mu = -1$ can be compared with a lens. Its imaging is described by the equation $a + b = l$, where $a > 0$ is the distance from the object to the front plane, $b > 0$ is the distance from the rear plane to the image, and l is the thickness of the slab. The image is real and direct, though not amplified.

It is worth noting that the left-handed material can enhance incident evanescent waves (Pendry 2000). As we will show below, the slab of a material with $\epsilon = \mu = -1$ does not reflect light. These remarkable properties have led to the term a perfect lens for the planar slab.

Artificial structures were first proposed, which have negative permittivity and permeability in the microwave region of frequencies. A periodic array of very thin metallic wires has the negative permittivity. We let a mean the spatial period of the lattice and r mean the radius of wires. Pendry et al. (1996) have expressed the response of this medium to the external electric field parallel to the wires (of a two-dimensional lattice) using the effective permittivity (a three-dimensional lattice has been considered first)

$$\epsilon_{\text{eff}} \equiv \epsilon_{\text{eff}}(\omega) = 1 - \frac{\omega_{\text{p}}^2}{\omega(\omega + \mathrm{i}\gamma_{\text{e}})}, \tag{4.242}$$

where

$$\omega_{\text{p}} = \frac{\sqrt{2\pi}c}{a\sqrt{\ln(\frac{a}{r})}} \tag{4.243}$$

is the plasma angular frequency and γ_{e} is the absorption parameter.

Similarly, it was predicted in Pendry et al. (1999) that a periodic array of split-ring resonators behaves as a medium with negative magnetic permeability. The response of the regular lattice to the external magnetic field perpendicular to the plane of the ring is given by the effective permeability

$$\mu_{\text{eff}} \equiv \mu_{\text{eff}}(\omega) = 1 - \frac{F\omega^2}{\omega^2 - \omega_0^2 + \mathrm{i}\Gamma\omega}. \tag{4.244}$$

Here ω_0 is the resonant frequency and Γ is the absorption parameter. The parameter F is the filling factor for the split ring. Combination of both structures gives rise to the material with both negative permittivity and permeability—the left-handed material.

The reports on first experiments, e.g. (Shelby et al. 2001) were followed by some criticism. For instance, it was argued that the negative refraction is ruled out by the causality principle (Valanju et al. 2002). Absorption was suggested as an alternative explanation of the experiment with negative refraction (Sanz et al. 2003).

Numerical simulations of the transmission of the electromagnetic waves through the left-handed medium offered an independent possibility to verify the theoretical predictions (Ziolkowski and Heyman 2001, Markoš and Soukoulis 2002a,b). Formulae for the transmission amplitude, $\mathtt{t}$, and the reflection amplitude, $\mathtt{r}$, of the electromagnetic wave through a homogeneous slab read

$$\frac{1}{\mathtt{t}} = \cos(nkl) - \frac{\mathrm{i}}{2}(Z + Z^{-1})\sin(nkl), \tag{4.245}$$

$$\frac{\mathtt{r}}{\mathtt{t}} = -\frac{\mathrm{i}}{2}(Z - Z^{-1})\sin(nkl), \tag{4.246}$$

where $k = \frac{\omega}{c}$ conventionally. Especially, $\mathtt{t} = \exp(\mathrm{i}nkl)$, $\mathtt{r} = 0$ for $\epsilon = \mu = -1$. On the assumption that the wavelength of the electromagnetic wave is much larger than the spatial period of the left-handed structure and on using relations (4.245) and (4.246), the index of refraction and the impedance have been derived from the numerical data (Smith et al. 2002). The origin of absorption has been traced up (Markoš et al. 2002).

For applications it would be very interesting to pass from the microwave to the optical frequencies, or from the GHz to THz region. Pendry (2003) has provided an approval of the contemporary reports on experimental proofs of some properties of the materials with negative refractive index. In 2007, a progress has been reported on Dolling et al. (2007).

Photonic crystals have been analysed in theory and numerical calculations (Notomi 2000, Foteinopoulou et al. 2003, Foteinopoulou and Soukoulis 2003) and used in experiments (Cubukcu et al. 2003). Optical frequencies have been assumed, but it has not been asked whether the effective permittivity and the effective permeability may be defined.

Foteinopoulou et al. (2003) present in fact all the interesting results of Veselago (1968) in the lossless case. We cite only the time-averaged energy flux $\langle \mathbf{S} \rangle = \langle U \rangle \mathbf{v}_g$ and the time-averaged momentum density $\langle \mathbf{p} \rangle = \frac{\langle U \rangle}{\omega}\mathbf{k}$.

They concentrated themselves on clarification of some of the controversial issues. Their numerical calculations have described a two-dimensional photonic crystal. For the photonic crystal system a frequency range exists for which the effective refractive index is negative. So, a wave hitting the photonic crystal interface for that frequency will undergo negative refraction similar to a wave hitting the interface of a homogeneous medium with negative index n. It is worth noting that the effective medium is two-dimensionally isotropic. Otherwise, an effect similar to the negative refraction may occur in a right-handed medium.

In their simulations, a finite extent line source was placed outside a photonic crystal at an angle of 30°. The source starts emitting at $t = 0$ an almost monochromatic TE wave. The source is adjusted to generate a Gaussian beam.

Foteinopoulou et al. (2003) have used finite-difference time-domain simulations to study the time evolution of an electromagnetic wave as it reaches the interface. The wave is trapped temporarily at the interface, reorganizes, and, after a long time, the wave exibits the negative refraction.

Shen (2004) has defined a frequency-independent effective rest mass of a photon. Letting m_{eff} denote this mass, we may find it from the relation

$$\frac{m_{\text{eff}}^2 c^4}{\hbar^2} = - \operatorname*{res}_{\omega=0} \omega n^2(\omega). \tag{4.247}$$

If the left-handed medium can be modelled as two-time derivative Lorentz material (Ziolkowski 2001), then $m_{\text{eff}}^2 \geq 0$ for the electromagnetic parameters characteristic of (Ruppin 2000).

Ruppin (2002) has obtained a modification of formula (4.234) for the time-averaged energy density, which respects the absorption in the medium, but is restricted to a relative permittivity $\epsilon(\omega)$ and a relative permeability $\mu(\omega)$ of the following forms:

$$\epsilon \equiv \epsilon(\omega) = 1 + \frac{\omega_{\text{p}}^2}{\omega_{\text{r}}^2 - \omega^2 - \mathrm{i}\Gamma_e \omega}, \quad \Gamma_e > 0,$$

$$\mu \equiv \mu(\omega) = 1 - \frac{F\omega_0^2}{\omega^2 - \omega_0^2 + \mathrm{i}\Gamma_h\omega}, \quad \Gamma_h > 0. \tag{4.248}$$

At the same time, it is a generalization of the result obtained by Loudon (1970) in the case of no magnetic dispersion. The time-averaged energy density is

$$\langle U\rangle = \frac{1}{2}\left[\epsilon_0\left(\epsilon' + \frac{2\omega\epsilon''}{\Gamma_e}\right)\langle|\mathbf{E}|^2\rangle + \mu_0\left(\mu' + \frac{2\omega\mu''}{\Gamma_h}\right)\langle|\mathbf{H}|^2\rangle\right], \tag{4.249}$$

where $\epsilon' = \mathrm{Re}\,\epsilon$, $\epsilon'' = \mathrm{Im}\,\epsilon$, $\mu' = \mathrm{Re}\,\mu$, $\mu'' = \mathrm{Im}\,\mu$, conventionally.

Veselago (2002) has presented a miniature review of the progress in negative-index materials. The subject of that paper which comprises such a review is the formulation of Fermat's principle. The formulation stating that the optical length is stationary is correct.

For simplicity, we may consider differentials only for a Euclidean space. The differential of the optical length is the differential of the Euclidean length multiplied by the refractive index. If a variation of a path is restricted to the positive-index medium, the optical length of the path taken by a ray of light in travelling between two points is a local minimum. If a variation is restricted to the negative-index medium, the optical length of the sought path is a local maximum.

Pendry (2003) has provided an approval of the contemporary reports on experimental proofs of some properties of the materials with negative refractive index.

Naqvi and Abbas (2003) have noted that principle of duality has a somewhat different form in negative-index materials. Engheta (1998) has expressed a continuous transition from the original solution ($\alpha = 0$) to the dual solution ($\alpha = 1$) in the case of the positive-index medium

$$\begin{aligned} \mathbf{E}_{\mathrm{fd}} &= \mathbf{E}\cos\left(\alpha\frac{\pi}{2}\right) + Z_0 Z\mathbf{H}\sin\left(\alpha\frac{\pi}{2}\right), \\ Z_0 Z\mathbf{H}_{\mathrm{fd}} &= -\mathbf{E}\sin\left(\alpha\frac{\pi}{2}\right) + Z_0 Z\mathbf{H}\cos\left(\alpha\frac{\pi}{2}\right). \end{aligned} \tag{4.250}$$

For the negative refractive index, a form of Maxwell's equations suggests a different transformation

$$\begin{aligned} \mathbf{E}_{\mathrm{fd}} &= \mathbf{E}\cos\left(\alpha\frac{\pi}{2}\right) - Z_0 Z\mathbf{H}\sin\left(\alpha\frac{\pi}{2}\right), \\ Z_0 Z\mathbf{H}_{\mathrm{fd}} &= \mathbf{E}\sin\left(\alpha\frac{\pi}{2}\right) + Z_0 Z\mathbf{H}\cos\left(\alpha\frac{\pi}{2}\right). \end{aligned} \tag{4.251}$$

The two transformations are identified when, in the version for the negative-index material, the replacement $\alpha \leftrightarrow -\alpha$ is made.

Marqués et al. (2002) have measured transmission in a hollow metallic waveguide. They have found that it behaves similarly as the periodic array of thin metallic wires with its negative effective permittivity both when unloaded and when loaded. After a similarity to the left-handed medium had been achieved, the waveguide transmitted waves at about 6 GHz. A standard analysis of a metallic waveguide which still utilizes the effective magnetic permeability does not admit a radiation mode and contradicts the experiment. Another calculation has associated a mode with the split-ring-resonator medium anisotropy (Kondrat'ev and Smirnov 2003). The transmission may have been radiationless. Marqués et al. (2002) worked with the wavelength of 5 cm, while they had the longest waveguide with $l = 36$ mm.

Quite remarkably, the hollow waveguide behaves as a one-dimensional plasma with effective permittivity $\epsilon_{\text{eff}} \equiv \epsilon_{\text{eff}}(\omega) = \epsilon_0 \left(1 - \frac{\omega_c^2}{\omega^2}\right)$. The assertion that the transmission of electromagnetic waves occurs due to the split-ring-resonator medium anisotropy has been dismissed by experiment (Marqués et al. 2003).

Zharov et al. (2003) have noticed that so far properties of left-handed materials in the nonlinear regime of wave propagation have not been studied. They assume that a metallic structure is embedded into a nonlinear dielectric with a permittivity which depends on the strength of the electric field in a general way, $\epsilon_{\text{D}} \equiv \epsilon_{\text{D}}(|\mathbf{E}|^2)$. As an application they consider a linear dependence which corresponds to the Kerr nonlinearity. The effective nonlinear dielectric permittivity $\epsilon_{\text{eff}}(|\mathbf{E}|^2)$ is found to be a sum of the earlier result (4.242) and a third-order nonlinear term, $\epsilon_{\text{D}}(|\mathbf{E}|^2)$.

The effective magnetic permeability of the composite structure (for $F \ll 1$)

$$\mu_{\text{eff}}(\mathbf{H}) = 1 + \frac{F\omega^2}{\omega_{0\text{NL}}^2(\mathbf{H}) - \omega^2 + \text{i}\Gamma\omega} \tag{4.252}$$

differs from the earlier result (4.244) by the dependence of the eigenfrequencies of oscillations on the magnetic field. It holds that

$$\omega_{0\text{NL}}(\mathbf{H}) = \omega_0 X, \tag{4.253}$$

where X is one of the stable roots of the equation

$$|H|^2 = \alpha A^2 E_c^2 \frac{(1 - X^2)[(X^2 - \Omega^2)^2 + \Omega^2\gamma^2]}{X^6}, \tag{4.254}$$

where H is an appropriate component of the field $\mathbf{H}$, $\alpha = \pm 1$ stands for a focusing or defocusing nonlinearity, respectively, E_c is a characteristic electric field, $A^2 = 16\frac{\epsilon_{\text{D}0}^3\omega_0^2 h^2}{c^2}$, $\epsilon_{\text{D}0} = \epsilon_{\text{D}}(0)$, and $\gamma = \frac{\Gamma}{\omega_0}$. h is the width of the ring.

Huang and Gao (2003) have been motivated by the paper (Chui and Hu 2002). They have investigated the effective refractive index spectra of a granular composite, in which metallic magnetic inclusions are embedded into the host medium. They calculate the effective permittivity and the effective permeability as well based on the Clausius–Mossotti relation (Grimes and Grimes 1991). Numerical results show that, by controlling the volume fraction of dispersive spherical particles in nondispersive host medium, a composite medium which is left handed in a certain frequency region can be prepared. They investigate a three-phase composite. Especially, by embedding dielectric and magnetic granules into the host medium and controlling the volume fractions of the two sorts of the granules, the left-handed composite medium can be realized.

A wave packet description allows an easy grasp of negative refraction (Huang and Schaich 2004). The interesting properties of the left-handed materials have been illustrated on the realistic assumption of losses and incidence of a Gaussian beam (Cui et al. 2004). It has been demonstrated that a negative-index material allows an ultrashort pulse to propagate with minimal dispersion (D'Aguanno et al. 2005). The negative-index material has been described with a lossy Drude model

$$\epsilon(\tilde{\omega}) = 1 - \frac{1}{\tilde{\omega}(\tilde{\omega} + i\tilde{\gamma}_e)}, \quad \mu(\tilde{\omega}) = 1 - \left(\frac{\omega_{pm}}{\omega_{pe}}\right)^2 \frac{1}{\tilde{\omega}(\tilde{\omega} + i\tilde{\gamma}_m)}, \tag{4.255}$$

where $\tilde{\omega} = \frac{\omega}{\omega_{pe}}$ is the normalized frequency, ω_{pe} and ω_{pm} are the respective electric and magnetic plasma frequencies, and $\tilde{\gamma}_e = \frac{\gamma_e}{\omega_{pe}}$ and $\tilde{\gamma}_m = \frac{\gamma_m}{\omega_{pe}}$ are the respective electric and magnetic loss terms normalized with respect to the electric plasma frequency. Attention has been paid to the group-velocity dispersion parameter $\beta_2 = \frac{d^2k}{d\omega^2}$ [Agrawal (1995)]. The frequencies for which $\beta_2 = 0$ are plotted as zero group-velocity dispersion points in figures. It has been noted that these points are related to $\omega < \omega_{pe}$ and that no zero group-velocity dispersion point is present when $\frac{\omega_{pm}}{\omega_{pe}} = 1$.

For a macroscopic quantization, Milonni (1995) selects any frequency region, where absorption is negligible, i.e. away from absorption resonances. He assumes a uniform (or homogeneous) dielectric medium. The formulation simplifies and, although restricted in their range of validity, the results are applicable to a wide range of interesting and practical situations. He derives the electromagnetic energy density in a form which resembles the time-averaged linear dispersive energy (3.186). But the magnetic term of the energy density is expressed using the frequency-dependent magnetic permeability, i.e. at the same level of generality as the electric term.

For some frequency ω a mode function $\mathbf{F}_\omega(\mathbf{r})$ can be considered which satisfies the transversality condition and the Helmholtz equation

$$\nabla \cdot \mathbf{F}_\omega(\mathbf{r}) = 0, \tag{4.256}$$

$$\nabla^2 \mathbf{F}_\omega(\mathbf{r}) + \frac{\omega^2}{c^2}\epsilon(\omega)\mu(\omega)\mathbf{F}_\omega(\mathbf{r}) = \mathbf{0} \tag{4.257}$$

and appropriate boundary conditions. It is assumed that the mode function is normalized,

$$\int |\mathbf{F}_\omega(\mathbf{r})|^2 \, d^3\mathbf{r} = 1. \tag{4.258}$$

The monochromatic components at frequency ω of the fields are

$$\mathbf{E}(\mathbf{r}, t) = \mathrm{Re}\{C_\omega \alpha(t) \mathbf{F}_\omega(\mathbf{r})\}, \tag{4.259}$$

$$\mathbf{B}(\mathbf{r}, t) = \mathrm{Re}\left\{-i\frac{c}{\omega} C_\omega \alpha(t) \nabla \times \mathbf{F}_\omega(\mathbf{r})\right\}, \tag{4.260}$$

$$\mathbf{D}(\mathbf{r},t) = \mathrm{Re}[\epsilon(\omega)C_\omega\alpha(t)\mathbf{F}_\omega(\mathbf{r})], \tag{4.261}$$

$$\mathbf{H}(\mathbf{r},t) = \mathrm{Re}\left\{-\frac{\mathrm{i}}{\mu(\omega)}\frac{c}{\omega}C_\omega\alpha(t)\nabla\times\mathbf{F}_\omega(\mathbf{r})\right\}, \tag{4.262}$$

where $\alpha(t) = \alpha(0)\exp(-\mathrm{i}\omega t)$. The fields are expressed with these relations as in the classical optics, with an amplitude $\alpha(t)$, but also with a normalization constant C_ω for the electric strength field. The energy is determined as the integrated energy density in the form

$$\langle H\rangle = \frac{n(\omega)}{8\pi\,\mu(\omega)}|C_\omega|^2|\alpha(t)|^2\frac{\mathrm{d}}{\mathrm{d}\omega}[n(\omega)\omega]. \tag{4.263}$$

The choice

$$C_\omega = \sqrt{\frac{4\pi\,\mu(\omega)}{n(\omega)\frac{\mathrm{d}[n(\omega)\omega]}{\mathrm{d}\omega}}} \tag{4.264}$$

of the normalization constant gives the energy in the form

$$\langle H\rangle = \frac{1}{2}|\alpha(t)|^2, \tag{4.265}$$

which corresponds to a harmonic oscillator. For the quantization of this oscillator we do the replacement $\alpha(t)\to\sqrt{2\hbar\omega}\,\hat{a}(t)$ where $\hat{a}(t)$ is an annihilation operator, in (4.259)–(4.262).

Milonni (1995) does mention (Drummond 1990), but he does not expound, or comment on, the first and the second time derivative of the electric displacement which we obtain in (3.206) on substituting the first of equations (3.207) for $\mathbf{E}^\nu$.

The approach of Drummond (1990) enables one to respect the inhomogeneity of the medium as noted in (Milonni 1995). Let us remark that this approach leads also to twice as many annihilation and creation operators as expected. Ignoring that also the number of such narrow-band fields may be greater than one, this ratio reminds us of the microscopic model assuming an oscillator medium with a single resonance.

As different systems of units have been utilized we will compare some formulae of Milonni (1995) with those of Drummond (1990). Restricting himself to a single carrier frequency ($\nu = 1$), Drummond (1990) has presented the following expansions:

$$\hat{\mathbf{D}}^1(\mathbf{x},t) = \mathrm{i}\sum_{\mathbf{k},\lambda}\sqrt{\frac{\hbar\frac{\partial\omega_k^1}{\partial k}}{2Vk\tilde{\xi}_1(\omega_k^1)}}\left(\mathbf{k}\times\mathbf{e}_{\mathbf{k}\lambda}^1\hat{a}_{\mathbf{k}\lambda}^1e^{\mathrm{i}\mathbf{k}\cdot\mathbf{x}-\mathrm{i}\omega_k^1t} - \mathrm{H.\,c.}\right), \tag{4.266}$$

$$\hat{\mathbf{B}}^1(\mathbf{x},t) = -\mathrm{i}\sum_{\mathbf{k},\lambda}\mu\omega_k^1\sqrt{\frac{\hbar\frac{\partial\omega_k^1}{\partial k}}{2Vk\tilde{\xi}_1(\omega_k^1)}}\left(\mathbf{e}_{\mathbf{k}\lambda}^1\hat{a}_{\mathbf{k}\lambda}^1e^{\mathrm{i}\mathbf{k}\cdot\mathbf{x}-\mathrm{i}\omega_k^1t} - \mathrm{H.\,c.}\right), \tag{4.267}$$

where V is the quantization volume and (cf., (3.220))

$$\omega_k^1 = k\sqrt{\frac{\tilde{\zeta}_1(\omega_k^1)}{\mu}}. \tag{4.268}$$

Regarding wide bandwidths or a simplification merely, Drummond (1990) has completed the following expansion:

$$\hat{\mathbf{A}}(\mathbf{x}, t) = \sum_{\mathbf{k},\lambda} \sqrt{\frac{\hbar \frac{\partial \omega_k}{\partial k}}{2Vk\tilde{\zeta}(\omega_k)}} \left(\hat{a}_{\mathbf{k}\lambda}\mathbf{e}_{\mathbf{k}\lambda}e^{\mathrm{i}\mathbf{k}\cdot\mathbf{x}-\mathrm{i}\omega_k t} + \mathrm{H.\,c.}\right), \tag{4.269}$$

where

$$\frac{\partial \omega_k}{\partial k} = \frac{\omega_k}{k}\left[1 - \frac{\omega_k \tilde{\zeta}'(\omega_k)}{2\tilde{\zeta}(\omega_k)}\right]^{-1}. \tag{4.270}$$

Using relation (4.266), abandoning the Taylor expansion as in relation (4.269) and dividing by ϵ, we obtain a normalization constant in the form

$$\frac{1}{2}C_\omega\sqrt{2\hbar\omega} = \sqrt{\frac{\hbar n(\omega)\omega}{2\epsilon \frac{\mathrm{d}[n(\omega)\omega]}{\mathrm{d}\omega}}}. \tag{4.271}$$

Since $\frac{\mu_r}{n} = \frac{n}{\epsilon_r}$ and for different units in the relations under consideration, the replacement $2\pi \leftrightarrow \frac{1}{2\epsilon_0}$ is performed, we have also

$$\frac{1}{2}C_\omega\sqrt{2\hbar\omega} = \sqrt{\frac{\mu(\omega)2\pi\hbar\omega}{n(\omega)\frac{\mathrm{d}[n(\omega)\omega]}{\mathrm{d}\omega}}} \tag{4.272}$$

in conformity with Milonni (1995).

The normalization constant is obvious from the expression

$$\hat{\mathbf{E}}(\mathbf{r}, t) = \frac{1}{2}C_\omega\sqrt{2\hbar\omega}\left[\hat{a}_\omega\mathbf{F}_\omega(\mathbf{r}) + \hat{a}_\omega^\dagger\mathbf{F}_\omega^*(\mathbf{r})\right]. \tag{4.273}$$

For a multimode field, the electric-field operator (4.273) is replaced by

$$\hat{\mathbf{E}}(\mathbf{r}, t) = \sum_\beta \sqrt{\frac{2\pi\hbar\omega_\beta\mu(\omega_\beta)}{n(\omega_\beta)\frac{\mathrm{d}[n(\omega_\beta)\omega_\beta]}{\mathrm{d}\omega_\beta}}}\left[\hat{a}_\beta(t)\mathbf{F}_\beta(\mathbf{r}) + \hat{a}_\beta^\dagger(t)\mathbf{F}_\beta^*(\mathbf{r})\right], \tag{4.274}$$

where $\mathbf{F}_\beta(\mathbf{r})$ is a mode function for mode β obeying the transversality condition and the Helmholtz equation

$$\nabla^2\mathbf{F}_\beta(\mathbf{r}) + \frac{\omega_\beta^2}{c^2}\epsilon(\omega_\beta)\mu(\omega_\beta)\mathbf{F}_\beta(\mathbf{r}) = \mathbf{0}. \tag{4.275}$$

For an effectively unbounded medium plane-wave modes can be used,

$$\mathbf{F}_\beta(\mathbf{r}) \to \frac{\mathrm{i}}{\sqrt{V}}\mathbf{e}_{\mathbf{k}\lambda} \exp(\mathrm{i}\mathbf{k}\cdot\mathbf{r}), \tag{4.276}$$

with V a quantization volume and $\mathbf{e}_{\mathbf{k}\lambda}$ ($\lambda = 1, 2$) a unit polarization vector orthogonal to $\mathbf{k}$.

On introducing the notation

$$\gamma_k = \frac{\mathrm{d}[n(\omega_k)\omega_k]}{\mathrm{d}\omega_k} = \frac{c}{v_\mathrm{g}(\omega_k)} \tag{4.277}$$

and assuming $\mathbf{e}_{\mathbf{k}\lambda}$ to be real for simplicity, we can write the operators for **E**, **B**, **D**, and **H** fields in the forms

$$\hat{\mathbf{E}}(\mathbf{r}, t) = \mathrm{i}\sum_{\mathbf{k},\lambda}\sqrt{\frac{2\pi\hbar\omega_k\mu(\omega_k)}{n(\omega_k)\gamma_k V}} \times [\hat{a}_{\mathbf{k}\lambda}(t)\exp(\mathrm{i}\mathbf{k}\cdot\mathbf{r}) - \hat{a}^\dagger_{\mathbf{k}\lambda}(t)\exp(-\mathrm{i}\mathbf{k}\cdot\mathbf{r})]\mathbf{e}_{\mathbf{k}\lambda}, \tag{4.278}$$

$$\hat{\mathbf{B}}(\mathbf{r}, t) = \mathrm{i}\sum_{\mathbf{k},\lambda}\sqrt{\frac{2\pi\hbar\mu(\omega_k)c^2}{\omega_k n(\omega_k)\gamma_k V}} \times [\hat{a}_{\mathbf{k}\lambda}(t)\exp(\mathrm{i}\mathbf{k}\cdot\mathbf{r}) - \hat{a}^\dagger_{\mathbf{k}\lambda}(t)\exp(-\mathrm{i}\mathbf{k}\cdot\mathbf{r})]\mathbf{k}\times\mathbf{e}_{\mathbf{k}\lambda}, \tag{4.279}$$

$$\hat{\mathbf{D}}(\mathbf{r}, t) = \mathrm{i}\sum_{\mathbf{k},\lambda}\sqrt{\frac{2\pi\hbar\omega_k n(\omega_k)\epsilon(\omega_k)}{\gamma_k V}} \times [\hat{a}_{\mathbf{k}\lambda}(t)\exp(\mathrm{i}\mathbf{k}\cdot\mathbf{r}) - \hat{a}^\dagger_{\mathbf{k}\lambda}(t)\exp(-\mathrm{i}\mathbf{k}\cdot\mathbf{r})]\mathbf{e}_{\mathbf{k}\lambda}, \tag{4.280}$$

$$\hat{\mathbf{H}}(\mathbf{r}, t) = \mathrm{i}\sum_{\mathbf{k},\lambda}\sqrt{\frac{2\pi\hbar c^2}{\omega_k n(\omega_k)\mu(\omega_k)\gamma_k V}} \times [\hat{a}_{\mathbf{k}\lambda}(t)\exp(\mathrm{i}\mathbf{k}\cdot\mathbf{r}) - \hat{a}^\dagger_{\mathbf{k}\lambda}(t)\exp(-\mathrm{i}\mathbf{k}\cdot\mathbf{r})]\mathbf{k}\times\mathbf{e}_{\mathbf{k}\lambda}. \tag{4.281}$$

A comparison with the paper (Huttner et al. 1991) can be made. It is not so easy as the previous excercise, because sums are to be compared with integral expressions. In the sum (4.280) the factor

$$\sqrt{\frac{2\pi\hbar\omega_k n(\omega_k)\epsilon(\omega_k)v_\mathrm{g}(\omega_k)}{cV}} = \sqrt{\frac{2\pi\hbar\omega_k\epsilon(\omega_k)v_\mathrm{g}(\omega_k)}{v_\mathrm{p}(\omega_k)V}} \tag{4.282}$$

is used. In the integral expression, $n_\pm^2$ is involved instead of $\epsilon(\omega_k)$ (for $\mu(\omega_k) = 1$). Again the replacement $2\pi \leftrightarrow \frac{\epsilon_0}{2}$ must be done. The macroscopic quantization (Milonni 1995) leads only to optical polaritons, while the microscopic quantization, although it is frequently named also a "macroscopic one", introduces also acoustical polaritons.

Milonni and Maclay (2003) have applied the theory of Milonni (1995) to the effects of a negative-index medium on an excited guest atom, the Doppler effect, radiative recoil, and spontaneous and stimulated radiation rates, and also to the spectral density of thermal radiation.

A quantization scheme for the electromagnetic field interacting with atomic systems in the presence of dispersing and absorbing magnetodielectric media is contained in Dung et al. (2003). The magnetodielectric media include left-handed material.

The spontaneous decay of an excited two-level atom is influenced by the environment. The atom embedded in a homogeneous, purely electric medium has the decay rate

$$\Gamma = n\Gamma_0, \tag{4.283}$$

where $n = \sqrt{\epsilon}$ is the refractive index and Γ_0 is the decay rate in free space. It follows also from Fermi's golden rule. Now the magnetodielectric medium has the decay rate

$$\Gamma = \mu n\Gamma_0, \tag{4.284}$$

where the refractive index $n = \sqrt{\epsilon\mu}$. This should be verified for the material with both μ and n negative. Moreover, the expression (4.284) should be generalized to include the magnetodielectric absorption.

It is assumed that the magnetodielectric medium is causal and linear. It is characterized by a relative permittivity $\epsilon(\mathbf{r}, \omega)$ and a relative permeability $\mu(\mathbf{r}, \omega)$. For instance,

$$\epsilon(\mathbf{r}, \omega) = 1 + \frac{\omega_{\mathrm{Pe}}^2(\mathbf{r})}{\omega_{\mathrm{Te}}^2(\mathbf{r}) - \omega^2 - \mathrm{i}\omega\gamma_{\mathrm{e}}(\mathbf{r})}, \tag{4.285}$$

$$\mu(\mathbf{r}, \omega) = 1 + \frac{\omega_{\mathrm{Pm}}^2(\mathbf{r})}{\omega_{\mathrm{Tm}}^2(\mathbf{r}) - \omega^2 - \mathrm{i}\omega\gamma_{\mathrm{m}}(\mathbf{r})}, \tag{4.286}$$

where $\omega_{\mathrm{Pe}}(\mathbf{r})$, $\omega_{\mathrm{Pm}}(\mathbf{r})$ are the coupling strengths, $\omega_{\mathrm{Te}}(\mathbf{r})$, $\omega_{\mathrm{Tm}}(\mathbf{r})$ are the transverse resonance frequencies, and $\gamma_{\mathrm{e}}(\mathbf{r})$, $\gamma_{\mathrm{m}}(\mathbf{r})$ are the absorption parameters. For notational convenience, the spatial argument has been omitted.

The quantization has been performed by generalizing the theory expounded also in this book, Section 4.1.2. Let $\hat{\underline{E}}(\mathbf{r}, \omega), \ldots$ be the operators of the electric strength, etc. in frequency space. Especially, let $\hat{\underline{P}}(\mathbf{r}, \omega)$ and $\hat{\underline{M}}(\mathbf{r}, \omega)$, respectively, be the operators of the polarization and the magnetization in frequency space.

The operator-valued Maxwell equations are very similar to the classical ones. We present the electric constitutive relation

$$\hat{\underline{\mathbf{P}}}(\mathbf{r}, \omega) = \epsilon_0[\epsilon(\mathbf{r}, \omega) - 1]\hat{\underline{\mathbf{E}}}(\mathbf{r}, \omega) + \hat{\underline{\mathbf{P}}}_{\mathrm{N}}(\mathbf{r}, \omega), \tag{4.287}$$

with $\hat{\mathbf{P}}_{\mathrm{N}}(\mathbf{r}, \omega)$ being the noise polarization associated with the electric losses due to material absorption, and the magnetic constitutive relation

$$\hat{\mathbf{M}}(\mathbf{r}, \omega) = \kappa_0[1 - \kappa(\mathbf{r}, \omega)]\hat{\mathbf{B}}(\mathbf{r}, \omega) + \hat{\mathbf{M}}_{\mathrm{N}}(\mathbf{r}, \omega), \tag{4.288}$$

where $\kappa_0 = \mu_0^{-1}$, $\kappa(\mathbf{r}, \omega) = \mu^{-1}(\mathbf{r}, \omega)$, and $\hat{\mathbf{M}}_{\mathrm{N}}(\mathbf{r}, \omega)$ is the noise magnetization associated with magnetic losses. From the viewpoint of the theory the noise terms guarantee that the field commutators do not depend on $\mathrm{Re}\{\epsilon(\mathbf{r}, \omega)\}$ and $\mathrm{Re}\{\kappa(\mathbf{r}, \omega)\}$.

We present also the wave equation for $\hat{\mathbf{E}}(\mathbf{r}, \omega)$, the right-hand side of which includes the noise polarization $\hat{\mathbf{P}}_{\mathrm{N}}(\mathbf{r}, \omega)$ and the noise magnetization $\hat{\mathbf{M}}_{\mathrm{N}}(\mathbf{r}, \omega)$,

$$\nabla \times \kappa(\mathbf{r}, \omega)\nabla \times \hat{\mathbf{E}}(\mathbf{r}, \omega) - \frac{\omega^2}{c^2}\epsilon(\mathbf{r}, \omega)\hat{\mathbf{E}}(\mathbf{r}, \omega) = \mathrm{i}\omega\mu_0\hat{\mathbf{j}}_{\mathrm{N}}(\mathbf{r}, \omega), \tag{4.289}$$

where

$$\hat{\mathbf{j}}_{\mathrm{N}}(\mathbf{r}, \omega) = -\mathrm{i}\omega\hat{\mathbf{P}}_{\mathrm{N}}(\mathbf{r}, \omega) + \nabla \times \hat{\mathbf{M}}_{\mathrm{N}}(\mathbf{r}, \omega) \tag{4.290}$$

is the noise current.

Dung et al. (2003) consider the solution

$$\hat{\mathbf{E}}(\mathbf{r}, \omega) = \mathrm{i}\omega\mu_0 \int \mathbf{G}(\mathbf{r}, \mathbf{r}', \omega)\hat{\mathbf{j}}_{\mathrm{N}}(\mathbf{r}', \omega)\,\mathrm{d}^3\mathbf{r}', \tag{4.291}$$

where $\mathbf{G}(\mathbf{r}, \mathbf{r}', \omega)$ is the classical Green tensor obeying the equation

$$\nabla \times \kappa(\mathbf{r}, \omega)\nabla \times \mathbf{G}(\mathbf{r}, \mathbf{r}', \omega) - \frac{\omega^2}{c^2}\epsilon(\mathbf{r}, \omega)\mathbf{G}(\mathbf{r}, \mathbf{r}', \omega) = \delta(\mathbf{r} - \mathbf{r}')\mathbf{1}. \tag{4.292}$$

Cf., (4.136), (4.137), and (4.138) above. In Section 4.2.7 will be referred to calculations, in which the following property of the Green tensor is used, $\mathbf{G}(\mathbf{r}, \mathbf{r}', -\omega^*) = \mathbf{G}^*(\mathbf{r}, \mathbf{r}', \omega)$.

All of the commutation relations follow from the choice

$$\hat{\mathbf{P}}_{\mathrm{N}}(\mathbf{r}, \omega) = \mathrm{i}\sqrt{\frac{\hbar\epsilon_0}{\pi}\,\mathrm{Im}\{\epsilon(\mathbf{r}, \omega)\}}\,\hat{\mathbf{f}}_{\mathrm{e}}(\mathbf{r}, \omega) \tag{4.293}$$

and

$$\hat{\mathbf{M}}_{\mathrm{N}}(\mathbf{r}, \omega) = \sqrt{-\frac{\hbar\kappa_0}{\pi}\,\mathrm{Im}\{\kappa(\mathbf{r}, \omega)\}}\,\hat{\mathbf{f}}_{\mathrm{m}}(\mathbf{r}, \omega) \tag{4.294}$$

and from the commutation relations for the fundamental bosonic vector fields ($\lambda, \lambda' = \mathrm{e}, \mathrm{m}$)

$$[\hat{f}_{\lambda i}(\mathbf{r}, \omega), \hat{f}^{\dagger}_{\lambda' j}(\mathbf{r}', \omega')] = \delta_{\lambda\lambda'}\delta_{ij}\delta(\mathbf{r} - \mathbf{r}')\delta(\omega - \omega')\hat{1}, \tag{4.295}$$

$$[\hat{f}_{\lambda i}(\mathbf{r}, \omega), \hat{f}_{\lambda' j}(\mathbf{r}', \omega')] = \hat{0}. \tag{4.296}$$

So far the time dependence has not been considered, but the complete notation should be as in (4.157) and (4.158) above. We may note that the vector potential

$$\hat{\hat{\mathbf{A}}}(\mathbf{r}, \omega, 0) = \frac{1}{\mathrm{i}\omega}\hat{\hat{\mathbf{E}}}^{\perp}(\mathbf{r}, \omega, 0), \tag{4.297}$$

where $\hat{\hat{\mathbf{E}}}^{\perp}(\mathbf{r}, \omega, 0)$ means the transverse part of $\hat{\hat{\mathbf{E}}}(\mathbf{r}, \omega, 0)$, cf. (4.142). We consider also the scalar potential with the property as quantized in (4.143)

$$-\nabla\hat{\hat{\varphi}}(\mathbf{r}, \omega, 0) = \hat{\hat{\mathbf{E}}}^{\parallel}(\mathbf{r}, \omega, 0), \tag{4.298}$$

where $\hat{\hat{\mathbf{E}}}^{\parallel}(\mathbf{r}, \omega, 0)$ means the longitudinal part of $\hat{\hat{\mathbf{E}}}(\mathbf{r}, \omega, 0)$, cf. (4.143).

Having defined $\hat{\hat{\mathbf{A}}}(\mathbf{r}, \omega, 0)$, we may introduce also

$$\hat{\hat{\mathbf{B}}}(\mathbf{r}, \omega, 0) = \nabla \times \hat{\hat{\mathbf{A}}}(\mathbf{r}, \omega, 0). \tag{4.299}$$

On dividing by $\mathrm{i}\omega\mu_0$, equation (4.289) yields

$$\nabla \times \hat{\hat{\mathbf{H}}}(\mathbf{r}, \omega, 0) + \mathrm{i}\omega\hat{\hat{\mathbf{D}}}(\mathbf{r}, \omega, 0) = \hat{0}, \tag{4.300}$$

where we have used also the two constitutive relations. The definition of the operator of the magnetic induction may be rewritten as an operator-valued Maxwell equation

$$\nabla \times \hat{\hat{\mathbf{E}}}(\mathbf{r}, \omega, 0) = \mathrm{i}\omega\hat{\hat{\mathbf{B}}}(\mathbf{r}, \omega, 0). \tag{4.301}$$

On the scalar multiplication of equation (4.289) with the ∇ operator from the left, we obtain that

$$-\frac{\omega^2}{c^2}\nabla \cdot \left[\epsilon(\mathbf{r}, \omega)\hat{\hat{\mathbf{E}}}(\mathbf{r}, \omega)\right] = \mathrm{i}\omega\mu_0(-\mathrm{i}\omega)\nabla \cdot \hat{\hat{\mathbf{P}}}_{\mathrm{N}}(\mathbf{r}, \omega), \tag{4.302}$$

or

$$\nabla \cdot \left[\epsilon_0\epsilon(\mathbf{r}, \omega)\hat{\hat{\mathbf{E}}}(\mathbf{r}, \omega) + \hat{\hat{\mathbf{P}}}_{\mathrm{N}}(\mathbf{r}, \omega)\right] = \hat{0}, \tag{4.303}$$

or

$$\nabla \cdot \hat{\hat{\mathbf{D}}}(\mathbf{r}, \omega) = \hat{0}, \tag{4.304}$$

where

$$\hat{\hat{\mathbf{D}}}(\mathbf{r}, \omega) = \epsilon_0\epsilon(\mathbf{r}, \omega)\hat{\hat{\mathbf{E}}}(\mathbf{r}, \omega) + \hat{\hat{\mathbf{P}}}_{\mathrm{N}}(\mathbf{r}, \omega). \tag{4.305}$$

By definition,

$$\nabla \cdot \hat{\hat{\mathbf{B}}}(\mathbf{r}, \omega) = \hat{0}. \tag{4.306}$$

It can be shown that the equal-time commutation relations

$$[\hat{E}_i(\mathbf{r}, t), \hat{E}_j(\mathbf{r}', t)] = \hat{0} = [\hat{B}_i(\mathbf{r}, t), \hat{B}_j(\mathbf{r}', t)], \tag{4.307}$$

$$[\epsilon_0\hat{E}_i(\mathbf{r}, t), \hat{B}_j(\mathbf{r}', t)] = -\mathrm{i}\hbar\varepsilon_{ijk}\partial_k\delta(\mathbf{r} - \mathbf{r}')\hat{1} \tag{4.308}$$

are preserved, where ∂_k means the partial derivative with respect to the kth component of the vector $\mathbf{r}$.

The interaction of charged particles with the medium-assisted electromagnetic field is studied using Hilbert space on which the components of the bosonic fields $\hat{f}_{\lambda i}(\mathbf{r}, \omega)$ and operators $\hat{\mathbf{r}}_\alpha$ and $\hat{\mathbf{p}}_\alpha$ act. Here $\hat{\mathbf{r}}_\alpha$ and $\hat{\mathbf{p}}_\alpha$ are, respectively, the position and the canonical momentum operator of the αth particle of mass m_α and charge q_α. The charge density

$$\hat{\rho}_\mathrm{A}(\mathbf{r}, t) = \sum_\alpha q_\alpha \delta\left(\mathbf{r}\hat{1} - \hat{\mathbf{r}}_\alpha(t)\right) \tag{4.309}$$

and the scalar potential of the particles

$$\hat{\varphi}_\mathrm{A}(\mathbf{r}, t) = \int \frac{\hat{\rho}_\mathrm{A}(\mathbf{r}', t)}{4\pi\epsilon_0|\mathbf{r} - \mathbf{r}'|}\,\mathrm{d}^3\mathbf{r}' \tag{4.310}$$

are introduced.

In the minimal-coupling scheme and for nonrelativistic particles, the total Hamiltonian reads

$$\begin{aligned}\hat{H}(t) = \hbar \sum_{\lambda=\mathrm{e,m}} \int\int_0^\infty \omega \hat{\mathbf{f}}_\lambda^\dagger(\mathbf{r}, \omega, t)\cdot\hat{\mathbf{f}}_\lambda(\mathbf{r}, \omega, t)\,\mathrm{d}\omega\,\mathrm{d}^3\mathbf{r}\\ + \sum_\alpha \frac{1}{2m_\alpha}\left[\hat{\mathbf{p}}_\alpha(t) - q_\alpha\hat{\mathbf{A}}(\hat{\mathbf{r}}_\alpha(t), t)\right]^2\\ + \frac{1}{2}\int \hat{\rho}_\mathrm{A}(\mathbf{r}, t)\hat{\varphi}_\mathrm{A}(\mathbf{r}, t)\,\mathrm{d}^3\mathbf{r} + \int \hat{\rho}_\mathrm{A}(\mathbf{r}, t)\hat{\varphi}(\mathbf{r}, t)\,\mathrm{d}^3\mathbf{r}.\end{aligned} \tag{4.311}$$

The third and last terms can be written in the forms

$$\frac{1}{2}\int \hat{\rho}_\mathrm{A}(\mathbf{r}, t)\hat{\varphi}_\mathrm{A}(\mathbf{r}, t)\,\mathrm{d}^3\mathbf{r} = \frac{1}{2}\sum_\alpha \sum_{\substack{\alpha' \\ \alpha\neq\alpha'}} \frac{q_\alpha q_{\alpha'}}{4\pi\epsilon_0|\hat{\mathbf{r}}_\alpha(t) - \hat{\mathbf{r}}_{\alpha'}(t)|}, \tag{4.312}$$

$$\int \hat{\rho}_\mathrm{A}(\mathbf{r}, t)\hat{\varphi}(\mathbf{r}, t)\,\mathrm{d}^3\mathbf{r} = \sum_\alpha q_\alpha \hat{\varphi}\left(\hat{\mathbf{r}}_\alpha(t), t\right). \tag{4.313}$$

In this new situation the old operators should be denoted as $\hat{\mathbf{E}}_0(\mathbf{r}, t)$, $\hat{\mathbf{B}}_0(\mathbf{r}, t)$, $\hat{\mathbf{D}}_0(\mathbf{r}, t)$, $\hat{\mathbf{H}}_0(\mathbf{r}, t)$, except the bosonic vector fields $\hat{\mathbf{f}}_\lambda(\mathbf{r}, t)$. Then we introduce the field operators in the presence of charge particles

$$\hat{\mathbf{E}}(\mathbf{r}, t) = \hat{\mathbf{E}}_0(\mathbf{r}, t) - \nabla\hat{\varphi}_\mathrm{A}(\mathbf{r}, t), \quad \hat{\mathbf{B}}(\mathbf{r}, t) = \hat{\mathbf{B}}_0(\mathbf{r}, t), \tag{4.314}$$

$$\hat{\mathbf{D}}(\mathbf{r}, t) = \hat{\mathbf{D}}_0(\mathbf{r}, t) - \epsilon_0\nabla\hat{\varphi}_\mathrm{A}(\mathbf{r}, t), \quad \hat{\mathbf{H}}(\mathbf{r}, t) = \hat{\mathbf{H}}_0(\mathbf{r}, t), \tag{4.315}$$

the new operators. The exposition in the literature may have been more explicit, because the old operators have not been renamed and the new operators are $\hat{E}(\mathbf{r}, t)$, etc.

These operators obey the time-independent and time-dependent Maxwell equations, where

$$\hat{\mathbf{j}}_{\mathrm{A}}(\mathbf{r}, t) = \frac{1}{2}\sum_{\alpha} q_{\alpha}\left[\left(\hat{\mathbf{r}}_{\alpha}(t)\right)^{\cdot}\delta\left(\mathbf{r}\hat{1} - \hat{\mathbf{r}}_{\alpha}(t)\right) + \delta\left(\mathbf{r}\hat{1} - \hat{\mathbf{r}}_{\alpha}(t)\right)\left(\hat{\mathbf{r}}_{\alpha}(t)\right)^{\cdot}\right] \quad (4.316)$$

is the operator of the current density of the particles.

In the literature it has been shown that the Hamiltonian (4.311) generates the time-dependent Maxwell equations and the Newton equations of motion for the charged particles.

The authors treat the spontaneous decay of an excited two-level atom. They solve the problem of the time development of the state $|\psi(t)\rangle$, $|\psi(0)\rangle = |\{0\}\rangle|u\rangle$, whose alternative are the quantum states $|\mathbf{1}_{\lambda}(\mathbf{r}, \omega)\rangle|l\rangle$. Here $|l\rangle$ is the lower state whose energy is set equal to zero and $|u\rangle$ is the upper state of energy $\hbar\omega_{\mathrm{A}}$. ω_{A} is a transition frequency, $|\mathbf{1}_{\lambda}(\mathbf{r}, \omega)\rangle \equiv \hat{\mathbf{f}}_{\lambda}^{\dagger}(\mathbf{r}, \omega)|\{0\}\rangle$. In a formal expression of the state $|\psi(t)\rangle$, $C_u(t)$, $\mathbf{C}_{el}(\mathbf{r}, \omega, t)$ and $\mathbf{C}_{ml}(\mathbf{r}, \omega, t)$ are the respective coefficients. These coefficients satisfy linear differential equations and the initial conditions $C_u(0) = 1$ and $\mathbf{C}_{\lambda l}(\mathbf{r}, \omega, 0) = 0$. In the differential equations, both the tensor-valued Green function and the vector $\mathbf{d}_{\mathrm{A}}$, the transition dipole moment, occur.

As expected, the solution reminds us of the Weisskopf–Wigner theory. In the integro-differential equation the vector $\mathbf{d}_{\mathrm{A}}$ and the tensor-valued Green function still occur in a relatively simple fashion. The decay rate Γ can be expressed in terms of them and the shifted transition frequency $\tilde{\omega}_{\mathrm{A}}$ is utilized.

The case of nonabsorbing bulk material is treated first. On this assumption

$$\Gamma = \mathrm{Re}[\mu(\tilde{\omega}_{\mathrm{A}})n(\tilde{\omega}_{\mathrm{A}})]\Gamma_0, \quad (4.317)$$

where

$$\Gamma_0 = \frac{\tilde{\omega}_{\mathrm{A}}^3 d_{\mathrm{A}}^2}{3\hbar\pi\epsilon_0 c^3} \quad (4.318)$$

is the free-space decay rate, but taken at the shifted transition frequency.

When $\epsilon(\tilde{\omega}_{\mathrm{A}})$ and $\mu(\tilde{\omega}_{\mathrm{A}})$ have opposite signs, then the refractive index is purely imaginary and $\Gamma = 0$. In contrast, for nonabsorbing left-handed material, the rate Γ is given by relation (4.317) without the notation Re.

The case of atom in a spherical cavity is treated second. The cavity may be conceived as a way of removing the singularity of the tensor-valued Green function. Forms of $\frac{\Gamma}{\Gamma_0}$ valid for an atom at an arbitrary position inside a spherical free-space cavity surrounded by an arbitrary spherical multilayer material environment may be applied (Dung et al. 2003, Li et al. 1994, Tai 1994).

In fact, those formulae are specialized for the atom situated at the centre of the cavity and the otherwise homogeneous material environment. Results of Scheel et al. (1999a, 2000) have been amended. The ratio of the decay rates has been calculated in the literature as a function of the (shifted) atomic transition frequency $\tilde{\omega}_A$. Large cavities are considered first. Then smaller ones are characterized. The number of clear-cut cavity resonances decreases as the radius of the cavity decreases. A cavity is considered whose radius is much smaller than the transition wavelength. Every time a comparison is made between (a) dielectric matter, (b) magnetic matter, and (c) magnetodielectric matter. The decay rate in the cases (a) and (b) inside a band gap is low and, in the case (a), it also increases much due to each cavity resonance. In the case (c) this rate is low, if the band gap may belong either to the permittivity or to the permeability. Also the role of the resonances is similar to the case (a) or (b). But in the overlap of the electric and magnetic band gaps, the decay rate increases.

Using their results, the authors may address also the local-field corrections. The description of an atom should not be derived using the macroscopic field, even if the description, such as relation (4.317), does not involve the field. The theory of the authors directly applies to the real-cavity model. But simplifying assumptions are too weak to work for the magnetodielectric matter. The complexity relies on the dependence on $z = R\frac{\tilde{\omega}_A}{c}$. The expansion in powers of z must begin with a term proportional to R^{-3}, continue with R^{-1}, and an already constant term and the $O(R)$ term. If the material absorption may be disregarded, or when ϵ and μ may be taken for real numbers, the terms proportional to R^{-3} and R^{-1} may be left out. Hence

$$\Gamma \approx \left[\frac{3\epsilon(\tilde{\omega}_A)}{1+2\epsilon(\tilde{\omega}_A)}\right]^2 \mathrm{Re}[\mu(\tilde{\omega}_A)n(\tilde{\omega}_A)]\Gamma_0. \tag{4.319}$$

For strong dielectric absorption and R small we have a different approximation

$$\Gamma \approx \frac{9\,\mathrm{Im}\{\epsilon(\tilde{\omega}_A)\}}{|1+2\epsilon(\tilde{\omega}_A)|^2}\left(\frac{c}{\tilde{\omega}_A R}\right)^3 \Gamma_0. \tag{4.320}$$

The decay may be regarded as being purely radiationless (Dung et al. 2003).

Felbacq and Bouchitté (2005) have found a unified theoretical approach to the left-handed materials. They have used a renormalization group analysis, which takes into account the coupling between each resonator. They have checked the theoretical results numerically. They can explain the result by Pokrovsky and Efros (2002) that, by embedding wires in a medium with negative μ, one does not get a left-handed medium.

Ozbay et al. (2007) have provided 17 references to reports on metamaterials appropriate to a wide range of operating frequencies such as radio, microwave, millimetre wave, far-infrared, mid-infrared, near-infrared frequencies, and even visible wavelengths. In that article results obtained from experiments at the microwave frequencies have been reported.

Roppo et al. (2007) have studied the pulsed second-harmonic generation in positive- and negative-index media. In the positive-index media and in the presence of phase mismatch two forward-propagating components of the second-harmonic are generated (Bloembergen and Pershan 1962). In the pulsed generation, the second harmonic signal comprises a pulse which walks off (and is recognized in much work) and a second pulse which is "captured" and propagates under the pump pulse.

4.2.7 Application to Casimir Effect

In this subsection we will pay attention to papers which apply the quantization of the electromagnetic field in dispersive and absorbing media to the Casimir effect. Casimir's work had its origin in a problem of colloidal chemistry, namely, the stability of hydrophobic suspensions of particles in dilute aqueous electrolytes (Sparnaay 1989, Milonni 1994). Such suspensions are said to be stable if the particles do not coagulate. The particles are charged. Each particle is surrounded by ions of opposite charge. We expect that a repulsive force between the particles separated by a distance d increases more rapidly than an attractive force as $d \to L_{\mathrm{D}} + 0$, where L_{D} is a Debye length. We realize also that the repulsive force between these particles decreases more rapidly than the attractive force as $d \to \infty$. The attractive force should be obtained by integrating the pairwise forces between atoms, assuming an interatomic force given by the London–van der Waals interaction (London 1930).

Now we mention the original idea of Casimir. In 1948 he gave expression for the attractive force per unit area

$$F_{\mathrm{C}}(d) = \frac{\hbar c \pi^2}{240 d^4}, \tag{4.321}$$

where d is a distance between two uncharged, perfectly conducting parallel plates. Milonni (1994) has reviewed a standard calculation of the Casimir force. It is the calculation of the difference between the zero-point field energies for finite and infinite plate separations. This difference is interpreted as the potential energy of the system. In calculating this energy, the Euler–Maclaurin summation formula is used. The initial difference between a divergent sum over modes of the confined field and a divergent integral is modified to a difference between a convergent sum and a convergent integral, featuring a formal dependence on a function $f(k)$. Here k is the wavenumber of a mode. Finally it emerges that the result does not depend on the function $f(k)$, if $f(k)$ satisfies some conditions. The attractive force between the plates is then obtained as the derivative of the potential energy with respect to the distance d on changing the sign.

The same result has been obtained by considering the radiation pressure (Milonni 1988). In calculating it, the Euler–Maclaurin summation formula has been applied. Let us note that only a d-independent factor is determined here. The same expression can be obtained also by a modification of the standard calculation, so the derivative with respect to the distance and with the changed sign can be taken independent of finding the constant. As the function $f(k)$ depends on $k_{\rm m} \approx \frac{1}{a_0}$, where a_0 is the Bohr radius, we see that a different function $f(x)$ depends on $k_{\rm m}d \approx \frac{d}{a_0}$, a number of the Bohr radii spanning the distance between the plates.

These calculations are presented in Chapters 2 and 3 in Milonni (1994). Much later, in Chapter 7, he mentions forces between dielectric slabs. In the early 1950s, predictions of microscopic theories did not agree with experimental results. Milonni (1994) mentions the Lifshitz macroscopic theory (Lifshitz 1956). He does not expound this theory in fact. He indicates that some results follow the Casimir approach. The force between two semi-infinite dielectric slabs separated by a different dielectric medium or vacuum is derived. The case of vacuum between slabs has been treated by Lifshitz. The general case of a dielectric medium between slabs has been treated by Schwinger et al. (1978).

The physical basis of Lifshitz's calculations is not so difficult to understand. He was first to use a random field, $\mathbf{K}(\mathbf{r}, t)$, corresponding to some real or complex noise polarization. At zero temperature, this field satisfies the fluctuation–dissipation relation (the Gaussian system)

$$\langle K_i(\mathbf{r}, t) K_j(\mathbf{r}', t)\rangle = 2\hbar\, \mathrm{Im}\{\epsilon(\omega)\}\delta_{ij}\delta(\mathbf{r} - \mathbf{r}')\,. \tag{4.322}$$

Let us recall that the (Kronecker and Dirac) delta functions in the commutator between noise polarizations are multiplied by $\frac{\hbar}{\pi}\epsilon_0\, \mathrm{Im}\{\epsilon(\omega)\}$. From relation (7.69), p. 233 in Milonni (1994), it can be seen that $\mathbf{K}(\mathbf{r}, t) = 4\pi\mathbf{P}(\mathbf{r}, t)$ in the Gaussian system of units. Therefore, it would be appropriate to convert $\mathbf{K}(\mathbf{r}, t)$ into the International System of units similarly as $\mathbf{D}(\mathbf{r}, t)$, using the factor $\sqrt{\frac{\epsilon_0}{4\pi}}$. Then the fluctuation–dissipation relation becomes

$$\langle K_i(\mathbf{r}, t) K_j(\mathbf{r}', t)\rangle = \frac{\hbar\epsilon_0}{2\pi}\, \mathrm{Im}\{\epsilon(\omega)\}\delta_{ij}\delta(\mathbf{r} - \mathbf{r}') \qquad \text{(SI)}. \tag{4.323}$$

We recognize the right-hand side as a half of the appropriate commutator ($\mathbf{K}(\mathbf{r}, t) = \mathbf{P}(\mathbf{r}, t)$ in the SI). Exactly, $\mathbf{K}(\mathbf{r}, t) \equiv \mathbf{K}(\mathbf{r}, \omega, t)$ and we miss the functional factor $\delta(\omega - \omega')$ in the Lifshitz theory. Lifshitz (1956) then calculates the force in terms of the Maxwellian stress tensor, which we present here in the form

$$\begin{aligned}\mathbf{T}(\mathbf{r}, t) = {} & \epsilon_0\mathbf{E}(\mathbf{r}, t)\mathbf{E}(\mathbf{r}, t) + \frac{1}{\mu_0}\mathbf{B}(\mathbf{r}, t)\mathbf{B}(\mathbf{r}, t) \\ & - \frac{1}{2}\left[\epsilon_0\mathbf{E}^2(\mathbf{r}, t) + \frac{1}{\mu_0}\mathbf{B}^2(\mathbf{r}, t)\right]\mathbf{1}. \quad \text{(SI)}.\end{aligned} \tag{4.324}$$

In Kupiszewska and Mostowski (1990) and Kupiszewska (1992), the Casimir effect is studied on a restriction to the one-dimensional version. This means that only wave vectors normal to the surface are taken into account in the calculations. It is also assumed that the temperature is zero, $T = 0$.

In Kupiszewska and Mostowski (1990), the Casimir effect for the case of two non absorbing dielectric slabs has been studied in detail. The electromagnetic field has been quantized in the presence of a dielectric medium. The use of the Maxwellian stress tensor has been considered. It has been noted that the value of the appropriate component of the stress tensor is equal to the energy density for the one-dimensional calculation. An infinite expression has been regularized by means of an exponential cutoff function. The Casimir force has been calculated in the limit of semi-infinite slabs. A result has been provided for any slab thickness, but for a small reflection coefficient.

In Kupiszewska (1992), the absorbing dielectric slabs have been considered. The medium has been modelled as a continuous field of quantum harmonic oscillators interacting with a heat bath. The atoms and the electromagnetic field have been described with equations of motion and the long-time solution has been found. As a generalization of the previous result, two contributions to the Casimir effect have been distinguished.

Weigert (1996) has considered several modes between the perfectly conducting metallic plates to be in a squeezed vacuum state. At a given time instant the smallest possible expectation value of the energy in a neighbourhood of one of the mirrors is obtained through a calculation. It has been proposed to generate squeezed modes inside such a cavity and to measure an increase of the Casimir force. Weigert (1996) admits that the state of the system is not stationary, but he does not consider just the Lorentz force, only the stress tensor. He calls it energy stress tensor, but he calculates only with a stress tensor.

The theoretical work on the Casimir force outweighs experimental work. On regarding dielectric measurement, an accuracy of better than 10% has been reported (Lamoreaux 1997). An introductory guide to the literature on the Casimir force has been published (Lamoreaux 1999).

Electromagnetic field quantization in an absorbing medium has been readdressed, and the Casimir effect both for two lossy dispersive dielectric slabs and between two conducting plates was analysed by Matloob (1999a,b) and by Matloob et al. (1999). Matloob and Falinejad (2001) have investigated the Casimir effect between two dispersive absorbing slabs in three dimensions. The dielectric function of the slabs has been assumed to be an arbitrary complex function of frequency satisfying the Kramers–Kronig relations. The Maxwellian stress tensor has been used to evaluate the vacuum radiation pressure of the electromagnetic field on each slab in terms of vacuum expectation values. These averages have been expressed using the fluctuation–dissipation theorem and Kubo's formula (Landau and Lifshitz 1980). A simple relation to the imaginary part of a tensor-valued Green function has been recognized. So, the infinities of the stress tensor and the regular expression which diverges to them have been obvious. No explicit electromagnetic field quantization has been made. In a certain step of calculations the infinities cancel. Attention has been paid to various limits of the general expression and to the Lorentz model of the dielectric function. The effect of finite temperature on the Casimir force between two dielectric slabs has also been considered.

Kurokawa and Wakayama (2002) have introduced a Casimir energy for a compact Riemann surface of genus at least 2 and have related it to the Selberg zeta function. The scope of the paper has been the application of the Selberg trace formula to such a Riemann surface similar to methods of mathematical physics and quantum chaos.

da Silva et al. (2002) have generalized the so-called thermofield dynamics via an analytic continuation of the Bogoliubov transformations. It has been achieved that a field in arbitrary confined regions of space and time is described. In the case of an electromagnetic field, the energy-momentum tensor has been subjected to the generalized Bogoliubov transformation. The Casimir effect has been calculated for zero and nonzero temperature. The generalized Bogoliubov transformation has been applied also to the description of the field fulfilling the Dirichlet boundary conditions (at a conducting plate) and the Neumann boundary conditions at a permeable plate (the Casimir–Boyer model).

Tomaš (2002) has considered the Casimir effect in a dispersive and absorbing multilayered system using the Minkowski stress tensor method. He has calculated the Casimir force in a lossless dispersive layer of an otherwise absorbing multilayer by employing the quantized field operators as emerge from the scheme expounded in this chapter. He has presented the expression obtained and has compared it with the result of Zhou and Spruch (1995) who had applied the surface mode summation method to purely dispersive media. As an illustration he has calculated the Casimir force on a dielectric slab in a planar cavity with realistic mirrors. The difference between Casimir energies in two distinct layers has been established and the difference between Casimir forces in two such layers has been presented provided that their refractive indices are equal.

Boyer (2003) has presented a model, where physical ideas are transparent and the calculations allow easy numerical evaluation. The model has no direct connection with experiment. One-dimensional analogues of three-dimensional concepts and their properties are studied. A simplified thermodynamics is evoked. He assumes a one-dimensional box of length L at zero temperature $T = 0$. But the one-dimensionality assumption reaches so far that all the virtual photons, if considered, should have the same, or just the opposite direction of the wave vector. He introduces the Casimir energy $\Delta U_{\mathrm{zp}}(x, L)$ for the case, where a partition is present in the box at a position x. He considers boundary conditions, let us say for intervals $(0, x)$ and (x, L), namely, the Dirichlet or Neumann boundary conditions. The Dirichlet condition corresponds to a perfectly conducting boundary condition describing a perfectly conducting material in three spatial dimensions and the Neumann condition is simplified from an infinitely permeable boundary condition describing an infinitely permeable medium for electromagnetic waves. The boundary conditions at $x' = 0$ and $x' = L$ are enforced by the walls. In fact, x cannot be used for the coordinate, which is denoted by x' instead. The boundary condition at $x' = x$ is enforced by the partition.

Boyer (2003) lets α be 0 or 1, where $\alpha = 0$ for like boundary conditions for partition and walls, and $\alpha = 1$ for unlike boundary conditions. He finds that

$$\Delta U_{\mathrm{zp}}(x, L) = -\pi \hbar c \left(\frac{1}{24} - \frac{\alpha}{16} \right) \left(\frac{1}{x} + \frac{1}{L - x} - \frac{4}{L} \right). \tag{4.325}$$

For $\alpha = 0$, we may state that, off the centre of the box, forces act which repel the partition from the nearest wall. He speaks of an attractive force between the partition and the walls. For $\alpha = 1$, we may say that, from the centre of the box, attractive forces act. He mentions a repelling force between the partitions and the walls. The force between a conducting plate and a permeable plate was given, e.g. in Boyer (1974). The zero-point-energy limit is contrasted by the high-temperature energy-equipartition limit. This corresponds to the Rayleigh–Jeans spectrum of radiation. Then

$$\Delta U_{\mathrm{RJ}}(x, L, T) = 0. \tag{4.326}$$

He discusses also the Casimir forces at finite temperature.

Emig (2003) has developed a novel approach for calculating the Casimir forces between periodically deformed objects. Theories for realistic geometries have been developed in response to high-precision measurements (Mohideen and Roy 1998, Chan et al. 2001a,b, Bressi et al. 2002). The theories do not comprise rigorous, non-perturbative methods for calculating the force. The simplest and commonly used approximation is the proximity force theorem. For corrugated metal plates, it fails at a small corrugation length. A different approximation is the pairwise summation of renormalized retarded van der Walls forces. However, Lifshitz's theory for dielectric bodies demonstrates that, in general, the interaction cannot be obtained from a pairwise summation. The results do not agree with those from the zeta-function method (Barton 2001) in a situation, where this method can be used. Emig (2003) has considered the force between a rectangular corrugated plate and a flat one. This geometry cannot be treated by perturbation theory due to the rectangular edges. The force has been found by the non-perturbative method.

It was respected that in the most precise experiments the Casimir force between rough metallic plates was measured (Genet et al. 2003). It has been only one of the real conditions which differ from the ideal situation and assumptions of the theory. Others are imperfect reflection, nonzero temperature, and a geometry different from the parallel plates. The temperature effect has been neglected, because it is significant at large distances, while roughness corrections are more necessary at the smallest distances typical of the experiments. The proximity force approximation has been tested on the case where the force is measured between a plane and a sphere. The approximation leads to correct results when the radius of the sphere is much larger than the distance of closest approach. In the case of metallic plates, the proximity force approximation is only valid for the roughness spectrum containing small enough wave numbers. While mean number of waves spanning the interplate distance (multiplied by 2π) may be informative of the accuracy of the approximation, Genet et al. (2003) have proposed a specific roughness sensitivity and have considered its expectation value.

Many problems are formulated when the perfectly conducting plates of Casimir are replaced by other perfectly conducting surfaces. It can be utilized that the Casimir problem is not modified or generalized to dielectric media at the same time. For example, the rectangular cavity has been considered by Lukosz (1971) and Maclay (2000). The generalization to a system of conducting shells has also been realized, cf., (Plunien et al. 1986). The rectangular cavity of sides (a_1, a_2, a_3) depends on the three parameters, $\Lambda \equiv (a_1, a_2, a_3)$. The system of conducting shells depends on another system of parameters Λ.

Mazzitelli et al. (2003) have computed the Casimir interaction energy between two concentric cylinders. To this end they have used approximate semiclassical methods and the exact mode-by-mode summation method. They characterize a method according to Schaden and Spruch (1998, 2000). In this method the zero-point radiation is described with trajectories of a particle, and so as a real radiation. They mention the well-known decomposition of the spectral density into a smooth term and the oscillating contribution. Periodic orbit theory relates oscillations in the quantum level density of a given Hamiltonian to the periodic orbits in the corresponding classical system. They derive an energy approximation using the periodic orbit theory,

$$E^{\text{sem}} = -\ell \frac{\hbar c}{4\pi a^2} \sqrt{\frac{b}{a}} \sum_{w \geq 0} \sum_{v \geq \tilde{v}(w)} f_{vw} \frac{1}{v^4} N\left(\frac{b}{a}, v, w\right), \tag{4.327}$$

where ℓ is the "quantization" length, a, b, $a < b$, are radii of the cylinders, $\tilde{v}(w)$ is the least positive integer v such that $\cos\left(\pi \frac{w}{v}\right) > \frac{a}{b}$, f_{vw} is 1 for $v = 2$, $w = 0$ and is 2 otherwise,

$$N(\alpha, v, w) \equiv \frac{\sqrt{\left[\alpha - \cos\left(\pi \frac{w}{v}\right)\right]\left[\alpha \cos\left(\pi \frac{w}{v}\right) - 1\right]}}{\left[1 + \alpha^2 - 2\alpha \cos\left(\pi \frac{w}{v}\right)\right]^2}. \tag{4.328}$$

They further write

$$E^{\text{sem}} = E^{\text{sem}}_{w=0} + E^{\text{sem}}_{w\geq 1}, \tag{4.329}$$

where $E^{\text{sem}}_{w=0}$ ($E^{\text{sem}}_{w\geq 1}$) are obtained from relation (4.327), where condition $w \geq 0$ is replaced by the condition $w = 0$ ($w \geq 1$). They show that, for $\frac{b}{a} \sim 1$, $E^{\text{sem}} \sim E^{\text{sem}}_{w=0}$, where

$$E^{\text{sem}}_{w=0} = -\ell \frac{\hbar c \pi^3}{360} \frac{\sqrt{ab}}{(b-a)^3}. \tag{4.330}$$

Mazzitelli et al. (2003) mention the proximity theorem (Derjaguin and Abrikosova 1957, Derjaguin 1960). For its application they assume two parallel plates of area A, not of different areas. This difference does not suggest decision whether the larger or the smaller area should be chosen. Still the relation

$$E_{\text{P}} = -\frac{\hbar c \pi^2}{720} \frac{A}{(b-a)^3} \tag{4.331}$$

is applied to the plates "wound" into a cylinder. Here $A \stackrel{?}{=} 2\pi \ell a, 2\pi \ell b$. Obviously, the theory of periodic trajectories suggests the choice $A = 2\pi \ell \sqrt{ab}$. But only the numerical calculation shows that the approximation is relatively good for $1 < \frac{b}{a} < 4$. Further they compute the exact Casimir energy for the coaxial cylinders using the mode-by-mode summation method (Nesterenko and Pirozhenko 1997). The final result has the form

$$E_{\text{ex}} = E_{12} - 0.01356 \left(\frac{1}{a^2} + \frac{1}{b^2} \right) \ell \hbar c, \tag{4.332}$$

where

$$\begin{aligned} E_{12} = &-\frac{\ell \hbar c}{2\pi^2 a^2} \\ &\times \int_0^\infty \sum_n \int_{k_z}^\infty \sqrt{k_z^2 - y^2} \frac{\mathrm{d}}{\mathrm{d}y} \ln\left[F_{n12}(\mathrm{i}y, 1, \alpha) \right] \mathrm{d}y\, \mathrm{d}k_z, \end{aligned} \tag{4.333}$$

with

$$F_{n12}(\mathrm{i}y, 1, \alpha) = \left[1 - \frac{I_n(y) K_n(\alpha y)}{I_n(\alpha y) K_n(y)} \right] \left[1 - \frac{I_n'(y) K_n'(\alpha y)}{I_n'(\alpha y) K_n'(y)} \right], \tag{4.334}$$

$\alpha = \frac{b}{a}$, $I_n(z)$ and $K_n(z)$ are the modified Bessel functions and the MacDonald functions, respectively. Again on the condition $\alpha \sim 1$ the relations simplify

$$E_{\text{ex}} \sim E_{12} \sim -\ell \frac{\hbar c \pi^3}{360 a^2} \frac{1}{(\alpha - 1)^3} + O\left[\frac{1}{(\alpha - 1)^2} \right]. \tag{4.335}$$

The semiclassical approximation is valid. In contrast, the semiclassical energy for an isolated cylinder vanishes and the exact energy for a cylinder of radius a is

$$E_{\text{C}} = -0.01356 \frac{\ell \hbar c}{a^2}. \tag{4.336}$$

They present also numerical results.

Ahmedov and Duru (2003) have calculated the Casimir energies with respect to the previous work such as Mazzitelli et al. (2003) and Høye et al. (2001). Let us consider the region between two close coaxial cylinders. We assume that the cylinders have the radii $r_0 < r_1$. Then the Casimir energy per unit height is

$$E_{\text{cyl}} = -\frac{\pi^3 R}{720\Delta^3}\left(1 + \frac{15}{2\pi^2}\frac{\Delta^2}{R^2}\right), \tag{4.337}$$

where $\Delta \equiv r_1 - r_0$, $R = \sqrt{r_0 r_1}$.

Let us imagine a flat space which is periodic in the z coordinate unlike the Euclidean space. Let us consider the region between two cylinders or tori in this space provided that the axis of these cylinders is the z axis. Then the Casimir energy is

$$E_{\text{tor}} = -\frac{\pi^3 RL}{720\Delta^3}\left(1 + \frac{15}{2\pi^2}\frac{\Delta^2}{R^2}\right), \tag{4.338}$$

where L is the length of the flat space measured parallel to the z axis.

Let us analyse the ring with a rectangular cross-section. The Casimir energy is

$$E_{\text{box}} \approx -\frac{\pi^3 RL}{720\Delta^3} + \frac{R\zeta(3)}{16\Delta^2}, \tag{4.339}$$

where L is the height of the cylinders and $\zeta(z)$ is the Riemann ζ-function.

Let us consider two close concentric spheres. We assume that the spheres have the radii $r_0 < r_1$. Then the Casimir energy is

$$E_{\text{sph}} = -\frac{\pi^3 R^2}{360\Delta^3}\left(1 + \frac{5\Delta^2}{4\pi^2 R^2}\right). \tag{4.340}$$

Let us evaluate two close coaxial cones. We assume that the cones have the apex angles $\theta_0 < \theta_1 \leq \frac{\pi}{2}$. Then the Casimir energy per unit volume is

$$E \approx -\frac{\Theta\pi^3}{720 r^4 \Delta^3}, \tag{4.341}$$

where $\Delta \equiv \theta_1 - \theta_0$, $\Theta \equiv \sqrt{\sin\theta_0 \sin\theta_1}$, and r is the distance from the common vertex. Dividing the right-hand side by $2\pi\Theta\Delta$ to "correct" the energy density (Ahmedov and Duru 2003), we obtain that

$$E = -\frac{\pi^2}{1440 r^4 \Delta^4} + \mathrm{O}(\Delta^{-3}). \tag{4.342}$$

This is similar to the solution of the wedge problem (Deutsch and Candelas 1979),

$$E = -\frac{1}{1440 r^4 \Delta^2}\left(\frac{\pi^2}{\Delta^2} - \frac{\Delta^2}{\pi^2}\right), \tag{4.343}$$

where Δ is the angle between the half-planes of the boundary.

Geyer et al. (2003) have begun with the state of the research of the Casimir effect. They have also mentioned some difficulty with calculations of the temperature effect on the Casimir force between real metals of finite conductivity. They distinguish five different approaches. According to the fifth approach, the description of the thermal Casimir force can be obtained by the Leontovich surface impedance boundary condition.

Three domains of frequencies are distinguished: The region of the normal skin effect for low frequencies, the region of the anomalous skin effect, or relaxation domain for higher frequencies, and the region of the infrared optics for yet higher frequencies. In the region of the anomalous skin effect and in the relaxation domain metal cannot be described by any dielectric permittivity depending only on the frequency. The space dispersion is also essential.

Otherwise, the theoretical basis is as follows. Boundary conditions are introduced

$$\mathbf{E}_{\mathrm{t}} = Z(\omega)\mathbf{B}_{\mathrm{t}} \times \mathbf{n}, \tag{4.344}$$

where $Z(\omega)$ is the surface impedance of the conductor, $\mathbf{E}_{\mathrm{t}}$ and $\mathbf{B}_{\mathrm{t}}$ are the tangential components of the (Fourier transformed) electric and magnetic fields, and $\mathbf{n}$ is the unit normal vector to the surface (pointing inside the metal). For an ideal metal we have $Z \equiv 0$ and for real nonmagnetic metals $|Z| \ll 1$ holds (Landau et al. 1984).

The surface impedance is determined over the whole frequency axis, even though it is different in each of the three domains. It is respected that the main contribution to the Casimir free energy and force is given by the frequency region centred around the so-called characteristic frequency $\omega_{\mathrm{c}} = \frac{c}{2a}$, where a is the space separation between two metal plates. Relation (43) in Geyer et al. (2003) is an analogue of relation (8.62) in the book (Milonni 1994). An approximate expression (45), which is not reproduced here, has been derived for the case of a sphere above a plate made of a real metal.

Geyer et al. (2003) remind of the fact that at the temperature $T = 0$ only the anomalous skin effect and infrared optics occur. For Au the transition frequency $\Omega = 6.36 \times 10^{13}$ rad/s is obtained. They determine the characteristic frequency ω_{c} at each separation distance, and then they fix the proper impedance function. In a figure, which is not reproduced here, graphs of both impedance functions are plotted. A "transition" impedance function does not exist evidently. The correction factors $\frac{E(a)}{E^{(0)}(a)}$ to the Casimir energy agree quite well in the region of the infrared optics and in the transition region. In the region of the anomalous skin effect, the results due to the right and wrong choice differ significantly. They further present numerical results for $T = 70$ K and $T = 300$ K. At these temperatures, the normal skin effect occurs already, but only at larger separations. The relative thermal correction (Geyer et al. 2003) is calculated for $0 < a \leq 5\ \mu$m and only by the use of the impedance of infrared optics and of anomalous skin effect.

Raabe et al. (2003) have underscored that one-dimensional quantization schemes are not rigorous enough when the Casimir force between absorbing multilayer dielectrics is calculated. At the beginning they warn that the "mode summation" method, which was employed by H. Casimir himself, cannot be generalized to the case of absorbing bodies, because in such bodies there are no modes. Then they characterize three procedures:

(1) The electromagnetic field and the material bodies are treated macroscopically. Explicit field quantization is not performed, but the field correlation functions are written down in conformity with statistical thermodynamics.

(2) The electromagnetic field and the material bodies are quantized at a microscopic level. The bodies are described by appropriate model systems. Simplifying assumptions are made.
(3) The electromagnetic field and the material bodies are described macroscopically as in the first procedure. But the medium-assisted electromagnetic field is quantized by using an infinite set of appropriately chosen bosonic basic fields.

Raabe et al. (2003) have reserved the first method for Lifshitz (1955, 1956). In this context they have mentioned (Schwinger et al. 1978) and have characterized the paper (Matloob and Falinejad 2001). The mentions about the second method comprise the note that the calculations were carried out only for one-dimensional systems. Let us refer only to Kupiszewska and Mostowski (1990), Kupiszewska (1992). The third method is used in Raabe et al. (2003), but the authors also refer to Tomaš (2002). Then they add two further methods. One is the surface-mode approach in the nonretarded limit (van Kampen et al. 1968) and including retardation (see references in the cited paper). The other is the scattering approach (Jaekel and Reynaud 1991).

Raabe et al. (2003, 2004) have reproduced the essential traits of their quantization scheme. Then they describe the multilayer structure. They consider $n-1$ layers of thicknesses $d_l > 0$, $l = 1, \ldots, n-1$. These layers have the boundaries z_l, $l = 1, \ldots, n$, which have the properties $z_{l+1} = z_l + d_l$, $l = 1, \ldots, n-1$. They introduce $z_0 = -\infty$, $z_{n+1} = +\infty$, and so, inclusive of the substrate and the superstrate, there are $n+1$ layers. The permittivity is

$$\epsilon(\mathbf{r}, \omega) = \epsilon_l(\omega) \text{ for } z_l < z < z_{l+1}, l = 0, \ldots, n. \tag{4.345}$$

For the tensor-valued Green function, we are referred to the paper (Tomaš 1995). From the expression of the Green tensor it follows that it conserves its form on the intervals $(z_l, z_{l+1}) \times (z_{l'}, z_{l'+1})$, $l' = 0, \ldots, n$. If both spatial arguments are in the same layer, $l = l'$, we let $\mathbf{G}_l(\mathbf{r}, \mathbf{r}', \omega)$ denote the form $\mathbf{G}(\mathbf{r}, \mathbf{r}', \omega)$. We introduce the scattering part

$$\mathbf{G}_l^{\mathrm{scat}}(\mathbf{r}, \mathbf{r}', \omega) = \mathbf{G}_l(\mathbf{r}, \mathbf{r}', \omega) - \mathbf{G}_l^{\mathrm{bulk}}(\mathbf{r}, \mathbf{r}', \omega) \text{ for } \mathbf{r}' \neq \mathbf{r}, \tag{4.346}$$

where $\mathbf{G}_l^{\mathrm{bulk}}(\mathbf{r}, \mathbf{r}', \omega)$, the bulk part, is the solution for the case that the medium of the lth layer fills up the whole space. The values of the scattering part of the Green tensor for $\mathbf{r}' = \mathbf{r}$ are obtained in the coincidence limit of the position vectors. It is assumed that a "cavity", which separates walls, is the jth layer, $1 < j < n-1$. The walls are composed of $j-1$ layers $l = 1, \ldots, j-1$ and of $n-1-j$ layers $l = j+1, \ldots, n-1$. If some of the walls are semi-infinite, the numbering may differ a little.

Before the Casimir force is calculated from the stress tensor, a more general tensor

$$\begin{aligned} \mathbf{T}(\mathbf{r}, \mathbf{r}', t) = {} & \mathbf{T}_{\mathrm{e}}(\mathbf{r}, \mathbf{r}', t) + \mathbf{T}_{\mathrm{m}}(\mathbf{r}, \mathbf{r}', t) \\ & - \frac{1}{2}\mathbf{1}\,\mathrm{Tr}\{\mathbf{T}_{\mathrm{e}}(\mathbf{r}, \mathbf{r}', t) + \mathbf{T}_{\mathrm{m}}(\mathbf{r}, \mathbf{r}', t)\} \end{aligned} \tag{4.347}$$

is defined, where **1** is the second-rank unit tensor and

$$\mathbf{T}_{\rm e}(\mathbf{r},\mathbf{r}',t) = \langle \hat{\mathbf{D}}(\mathbf{r},t)\hat{\mathbf{E}}(\mathbf{r}',t)\rangle, \tag{4.348}$$

$$\mathbf{T}_{\rm m}(\mathbf{r},\mathbf{r}',t) = \langle \hat{\mathbf{B}}(\mathbf{r},t)\hat{\mathbf{H}}(\mathbf{r}',t)\rangle (\mathbf{r} \neq \mathbf{r}'). \tag{4.349}$$

The expectation values are calculated in thermal equilibrium, a stationary state of the field.

(i) Basic equation.

For finite temperatures T, they employ the statistical operator

$$\hat{\rho} = Z^{-1} \exp\left(-\frac{\hat{H}}{k_{\rm B}T}\right), \tag{4.350}$$

where

$$Z = \mathrm{Tr}\left\{\exp\left(-\frac{\hat{H}}{k_{\rm B}T}\right)\right\}, \tag{4.351}$$

and $k_{\rm B}$ is the Boltzmann constant. In relations (4.348) and (4.349), $\hat{\rho}$ may be written explicitly. On substituting relation (4.350) into the modified relation (4.348), we get

$$\mathbf{T}_{\rm e}(\mathbf{r},\mathbf{r}') = \frac{\hbar}{\pi}\int_0^\infty \coth\left(\frac{\hbar\omega}{2k_{\rm B}T}\right)\frac{\omega^2}{c^2}\,\mathrm{Im}\{\epsilon(\mathbf{r},\omega)\mathbf{G}(\mathbf{r},\mathbf{r}',\omega)\}\,\mathrm{d}\omega \tag{4.352}$$

and on substituting relation (4.350) into the modified relation (4.349), we obtain

$$\mathbf{T}_{\rm m}(\mathbf{r},\mathbf{r}') = -\frac{\hbar}{\pi}\int_0^\infty \coth\left(\frac{\hbar\omega}{2k_{\rm B}T}\right)\nabla\times\mathrm{Im}\{\mathbf{G}(\mathbf{r},\mathbf{r}',\omega)\}\times\overleftarrow{\nabla}'\,\mathrm{d}\omega. \tag{4.353}$$

On the left-hand side of relations (4.352) and (4.353), the time argument t has been dropped, since the right-hand sides do not depend on t.

Although the generalization to a "cavity" containing dielectric medium has been known, Raabe et al. (2003, 2004) have restricted themselves to the free space between two stacks. The Casimir force (per unit area) is given by the zz-component of the stress tensor (4.324).

We may modify relations (4.347), (4.352), and (4.353) by the way of writing the superscripts scat. Then we introduce the stress tensor

$$\mathbf{T}^{\rm scat}(\mathbf{r}) = \lim_{\mathbf{r}'\to\mathbf{r}} \mathbf{T}^{\rm scat}(\mathbf{r},\mathbf{r}'). \tag{4.354}$$

With respect to a layer, it is suitable to introduce the tensor

$$\mathbf{T}_j^{\rm scat}(\mathbf{r}) = \lim_{\mathbf{r}'\to\mathbf{r}} \mathbf{T}_j^{\rm scat}(\mathbf{r},\mathbf{r}'). \tag{4.355}$$

Now a number of concepts and pieces of notation are introduced. Propagation constants

$$\beta_l = \beta_l(q,\omega) = \sqrt{\frac{\omega^2 \epsilon_l(\omega)}{c^2} - q^2} \tag{4.356}$$

and reflection coefficients for σ-polarized waves at the top (+) and bottom (−) of the jth layer are defined that are

$$r^{\sigma}_{n+} = 0, \quad \sigma = \mathrm{s},\ \mathrm{p} \tag{4.357}$$

and are calculated from the recurrence relations

$$r^{\mathrm{s}}_{l+} = \frac{\left(\frac{\beta_l}{\beta_{l+1}} - 1\right) + \left(\frac{\beta_l}{\beta_{l+1}} + 1\right) \exp\left(2\mathrm{i}\beta_{l+1}d_{l+1}\right) r^{\mathrm{s}}_{(l+1)+}}{\left(\frac{\beta_l}{\beta_{l+1}} + 1\right) + \left(\frac{\beta_l}{\beta_{l+1}} - 1\right) \exp\left(2\mathrm{i}\beta_{l+1}d_{l+1}\right) r^{\mathrm{s}}_{(l+1)+}}, \tag{4.358}$$

and

$$r^{\mathrm{p}}_{l+} = \frac{\left(\frac{\beta_l}{\beta_{l+1}} - \frac{\epsilon_l}{\epsilon_{l+1}}\right) + \left(\frac{\beta_l}{\beta_{l+1}} + \frac{\epsilon_l}{\epsilon_{l+1}}\right) \exp\left(2\mathrm{i}\beta_{l+1}d_{l+1}\right) r^{\mathrm{p}}_{(l+1)+}}{\left(\frac{\beta_l}{\beta_{l+1}} + \frac{\epsilon_l}{\epsilon_{l+1}}\right) + \left(\frac{\beta_l}{\beta_{l+1}} - \frac{\epsilon_l}{\epsilon_{l+1}}\right) \exp\left(2\mathrm{i}\beta_{l+1}d_{l+1}\right) r^{\mathrm{p}}_{(l+1)+}}. \tag{4.359}$$

The coefficients r^{σ}_{l-} are

$$r^{\sigma}_{0-} = 0 \tag{4.360}$$

and the recurrencies for the others are analogous, which are formally obtained from relations (4.358) and (4.359) on replacements

$$l \mapsto l,\, l+1 \mapsto l-1 \tag{4.361}$$

and on the change of the subscript + of the reflection coefficient to −. Also denominators of the fractions for multiple reflections

$$D_{\sigma l} = D_{\sigma l}(q,\omega) = 1 - r^{\sigma}_{l+} r^{\sigma}_{l-} \exp\left(2\mathrm{i}\beta_l d_l\right) \tag{4.362}$$

are introduced. Finally,

$$T^{\mathrm{scat}}_{zz,j} = -\frac{\hbar}{2\pi^2} \int_0^{\infty} \coth\left(\frac{\hbar\omega}{2k_{\mathrm{B}}T}\right) \times \mathrm{Re}\left\{ \int_0^{\infty} q\beta_j \exp\left(2\mathrm{i}\beta_j d_j\right) \sum_{\sigma} D^{-1}_{\sigma j} r^{\sigma}_{j-} r^{\sigma}_{j+}\, \mathrm{d}q\, \mathrm{d}\omega \right\}. \tag{4.363}$$

As $T^{\mathrm{scat}}_{zz,j}$ does not depend on the space point in the jth layer, the argument $\mathbf{r}$ has been dropped.

(ii) Imaginary frequencies.

We introduce

$$\xi_m = 2m\pi \frac{k_B T}{\hbar}, \quad m \text{ integer}. \tag{4.364}$$

Since the permittivity is positive on the positive imaginary frequency axis, we introduce

$$\kappa_j = \sqrt{\frac{\xi^2 \epsilon_j(\mathrm{i}\xi)}{c^2} + q^2}. \tag{4.365}$$

Exploiting the analytical properties of the ω integrand in relation (4.363), we arrive at an expression of the integral with respect to ω by a residue series. Finally

$$T_{zz,j}^{\mathrm{scat}} = \frac{k_B T}{\pi} \sum_{m=0}^{\infty} \left(1 - \frac{1}{2}\delta_{m0}\right) \times \left[\int_0^{\infty} q\kappa_j \exp\left(-2\kappa_j d_j\right) \sum_{\sigma} D_{\sigma j}^{-1} r_{j-}^{\sigma} r_{j+}^{\sigma} \, \mathrm{d}q\right]_{\omega=\mathrm{i}\xi_m}, \tag{4.366}$$

which may be regarded as a generalization of the famous Lifshitz formula Lifshitz (1955), (1956). For $m = 0$, the term with $\omega = 0$ is peculiar and it should be replaced by the limit $\omega \to 0+$. To obtain $T_{zz,j}^{\mathrm{scat}}$ in the zero-temperature limit, we may simply repeat the derivation from relation (4.363). Of course, replacement $\sum_{m=0}^{\infty} \to \frac{\hbar}{2\pi k_B T} \, \mathrm{d}\xi$ can be realized in relation (4.366).

(iii) One-dimensional systems.

A comparison with the three-dimensional case is made only for $T = 0$. Contrary to the three-dimensional description, the sum with respect to σ is omitted, since in the one-dimensional system normal incidence occurs and the description can be restricted to a single polarization. Further one of the integrals is replaced by a multiplication with a constant,

$$\frac{1}{4\pi^2} \int \mathrm{d}^2\mathbf{q} \mapsto \frac{1}{\mathcal{A}} \sum_{\mathbf{q}}, \tag{4.367}$$

where $\mathcal{A}$ is the normalization area.

Also analytical expressions for specific distance laws in the zero-temperature limit are derived. For example, it is shown that the Casimir force between two single-slab walls behaves asymptotically as d^{-6} instead of d^{-4} in the large-distance asymptotic regime. Results for single-slab walls for periodic multilayer wall structure are illustrated in figures.

Chen et al. (2003) study the difference of the thermal Casimir forces at different temperatures between real metals. If the temperatures are fixed, the difference of the Casimir forces increases with a decrease of the separation distance. The configurations of two parallel plates and a sphere above a plate are considered.

In the case of two parallel plates, they utilize a perturbation result from the paper Bordag et al. (2000) to express the thermal Casimir force denoted by $F_{\rm pp}(a, T)$, where a is the separation distance and T is a temperature. They concentrate on the difference

$$\Delta F_{\rm pp} \equiv \Delta F_{\rm pp}(a, T_1, T_2) = F_{\rm pp}(a, T_2) - F_{\rm pp}(a, T_1), \tag{4.368}$$

where T_1 and T_2 are temperatures. For example, for Au, $T_1 = 300$ K and $T_2 = 350$ K, $|\Delta F_{\rm pp}|$ decreases with an increase of a. For an ideal metal $|\Delta F_{\rm pp}|$ does not depend on a.

In the case of a sphere above a plate, they use a perturbation result after Klimchitskaya and Mostepanenko (2001). The thermal Casimir force is denoted by $F_{\rm ps}(a, T)$. They study the difference

$$\Delta F_{\rm ps} \equiv \Delta F_{\rm ps}(a, T_1, T_2) = F_{\rm ps}(a, T_2) - F_{\rm ps}(a, T_1). \tag{4.369}$$

For example, for Au, $T_1 = 300$ K and $T_2 = 350$ K, $|\Delta F_{\rm ps}|$ decrease with an increase of a. For an ideal metal $|\Delta F_{\rm ps}|$ is a linear function of a. Then Chen et al. (2003) compare the chosen approach with that, e.g. after the paper (Brevik et al. 2002) (further references see Chen et al. 2003). They fix $a = 0.5$ μm and $T_1 = 300$ K, while $T_1 \leq T_2 \leq 350$ K. The chosen approach exhibits an increase $|\Delta F_{\rm ps}|$ with the temperature T_2. The difference is negative both for a real and for an ideal metal. The alternative approach provides a negative difference for an ideal metal, but a positive difference (more than six times larger at $T_2 = 350$) for a real metal.

Iannuzzi and Capasso (2003) have published a comment on the paper (Kenneth et al. 2002). They believed that, at distances relevant to Casimir force measurements and to nanomachinery, the Casimir force between two slabs in vacuum was always attractive. They have referred also to (Bruno 2002), a paper devoted to an attractive Casimir magnetic force.

Kenneth et al. (2003) have replied to the comment (Iannuzzi and Cappasso 2003). They have declared the consensus that exploring the possible existence or design of materials with nontrivial magnetic properties for obtaining a repulsive Casimir force is important.

Action of the Casimir force on magnetodielectric bodies embedded in media has been analysed in (Raabe and Welsch 2005). The consistency of expressions derived in the framework of the macroscopic theory with microscopic harmonic-oscillator models is shown. It could be startling that here Raabe and Welsch (2005) declare the macroscopic quantum electrodynamics themselves. We have chosen in this book that their theory is named microscopic just as the theory due to Hopfield (1958) and

Huttner and Barnett (1992a,b). It is consensual, even though with a reservation, which can be found below relation (68), which is not reproduced here, namely, that the level of representation is rather a mesoscopic one (Raabe and Welsch 2005). It may be controversial that they do not use micro- or mesoscopic for the model with the two auxiliary fields $\mathbf{f}_e(\mathbf{r}, \omega)$, $\mathbf{f}_m(\mathbf{r}, \omega)$.

The exposition begins with the classical Maxwell equations with charges and currents, but without the constitutive relations. It is worthwhile to mention that the Lorentz force density

$$\mathbf{f}(\mathbf{r}) = \rho(\mathbf{r})\mathbf{E}(\mathbf{r}) + \mathbf{j}(\mathbf{r}) \times \mathbf{B}(\mathbf{r}) \tag{4.370}$$

is written as

$$\mathbf{f}(\mathbf{r}) = \nabla \cdot \mathbf{T}(\mathbf{r}) - \epsilon_0 \frac{\partial}{\partial t}[\mathbf{E}(\mathbf{r}) \times \mathbf{B}(\mathbf{r})], \tag{4.371}$$

where $\mathbf{T}(\mathbf{r})$ is the stress tensor,

$$\mathbf{T}(\mathbf{r}) = \epsilon_0 \mathbf{E}(\mathbf{r})\mathbf{E}(\mathbf{r}) + \frac{1}{\mu_0}\mathbf{B}(\mathbf{r})\mathbf{B}(\mathbf{r}) - \frac{1}{2}\left[\epsilon_0 \mathbf{E}^2(\mathbf{r}) + \frac{1}{\mu_0}\mathbf{B}^2(\mathbf{r})\right]\mathbf{1}. \tag{4.372}$$

The integral of the Lorentz force density $\mathbf{f}(\mathbf{r})$ over some space region (volume) V gives the total electromagnetic force $\mathbf{F}$ acting on the matter inside V

$$\mathbf{F} = \int_V \mathbf{f}(\mathbf{r})\, \mathrm{d}^3\mathbf{r}. \tag{4.373}$$

On integrating both sides of equation (4.371) over V we obtain that

$$\mathbf{F} = \int_{\partial V} \mathbf{T}(\mathbf{r}) \cdot \mathbf{da}(\mathbf{r}) - \epsilon_0 \frac{\mathrm{d}}{\mathrm{d}t} \int_V \mathbf{E}(\mathbf{r}) \times \mathbf{B}(\mathbf{r})\, \mathrm{d}^3\mathbf{r}, \tag{4.374}$$

where $\mathbf{da}(\mathbf{r})$ is an infinitesimal surface element. If the volume integral on the right-hand side of this equation does not depend on time, then the total force reduces to the surface integral

$$\mathbf{F} = \int_{\partial V} \mathrm{d}\mathbf{F}(\mathbf{r}), \tag{4.375}$$

where

$$\mathrm{d}\mathbf{F}(\mathbf{r}) = \mathbf{da}(\mathbf{r}) \cdot \mathbf{T}(\mathbf{r}) = \mathbf{T}(\mathbf{r}) \cdot \mathbf{da}(\mathbf{r}). \tag{4.376}$$

The tensor $\mathbf{T}(\mathbf{r})$ may be decreased by a constant term, i.e. a constant, space independent tensor (it suffices to consider the position on the surface ∂V). Raabe and Welsch (2005) have commented on the role of Minkowski's stress tensor, which is considered in much work devoted to the related topic, be it under this name or only as a "stress tensor".

Relation (4.371) can be generalized to characterize the density of a generalized Lorentz force

$$\begin{aligned}\mathbf{f}(\mathbf{r},\mathbf{r}')+\mathbf{f}(\mathbf{r}',\mathbf{r}) &= \nabla_{\frac{\mathbf{r}+\mathbf{r}'}{2}}\cdot\{\mathbf{T}(\mathbf{r},\mathbf{r}')+\mathbf{T}(\mathbf{r}',\mathbf{r})\}\\ &\quad -\epsilon_0\frac{\partial}{\partial t}\left[\mathbf{E}(\mathbf{r})\times\mathbf{B}(\mathbf{r}')+\mathbf{E}(\mathbf{r}')\times\mathbf{B}(\mathbf{r})\right], \end{aligned}\tag{4.377}$$

where $\mathbf{f}(\mathbf{r},\mathbf{r}')$ means the density of a generalized Lorentz force

$$\mathbf{f}(\mathbf{r},\mathbf{r}')=\rho(\mathbf{r})\mathbf{E}(\mathbf{r}')+\mathbf{j}(\mathbf{r})\times\mathbf{B}(\mathbf{r}'), \tag{4.378}$$

$$\nabla_{\frac{\mathbf{r}+\mathbf{r}'}{2}}=\nabla+\nabla'\equiv\nabla_{\mathbf{r}}+\nabla_{\mathbf{r}'}, \tag{4.379}$$

and $\mathbf{T}(\mathbf{r},\mathbf{r}')$ is a generalized stress tensor

$$\begin{aligned}\mathbf{T}(\mathbf{r},\mathbf{r}') &= \epsilon_0\left[\mathbf{E}(\mathbf{r})\mathbf{E}(\mathbf{r}')-\frac{1}{2}\mathbf{1}\mathbf{E}(\mathbf{r})\cdot\mathbf{E}(\mathbf{r}')\right]\\ &\quad +\frac{1}{\mu_0}\left[\mathbf{B}(\mathbf{r})\mathbf{B}(\mathbf{r}')-\frac{1}{2}\mathbf{1}\mathbf{B}(\mathbf{r})\cdot\mathbf{B}(\mathbf{r}')\right]. \end{aligned}\tag{4.380}$$

Raabe and Welsch (2005) expound the quantum theory of the electromagnetic field as described also in this book in Section 4.2.6. They have presented also commutation relations between the charge density operator, the current density operator, and the electromagnetic-field operators. They have calculated correlation functions of some operators in thermal states of the field. Calculation of the expectation value of the Lorentz force, which is not reproduced here, follows.

In our opinion, the operator of the Lorentz force has not been presented in such an explicit form as its expectation value. The latter is denoted by the same notation as the corresponding classical stress tensor. As the expectation value of the stress tensor operator is infinite before a quantum correction, the notation is generalized to the form $\mathbf{T}(\mathbf{r},\mathbf{r}')$ (just the same notation as in the classical theory),

$$\begin{aligned}\mathbf{T}(\mathbf{r},\mathbf{r}') &= \epsilon_0\langle\hat{\mathbf{E}}(\mathbf{r})\hat{\mathbf{E}}(\mathbf{r}')\rangle+\frac{1}{\mu_0}\langle\hat{\mathbf{B}}(\mathbf{r})\hat{\mathbf{B}}(\mathbf{r}')\rangle\\ &\quad -\frac{1}{2}\mathbf{1}\left[\epsilon_0\langle\hat{\mathbf{E}}(\mathbf{r})\cdot\hat{\mathbf{E}}(\mathbf{r}')\rangle+\frac{1}{\mu_0}\langle\hat{\mathbf{B}}(\mathbf{r})\cdot\hat{\mathbf{B}}(\mathbf{r}')\rangle\right]. \end{aligned}\tag{4.381}$$

In other words, we can quantize relation (4.380) to introduce a generalized stress tensor $\hat{\mathbf{T}}(\mathbf{r},\mathbf{r}')$ and to write relation (4.381) in the form

$$\mathbf{T}(\mathbf{r},\mathbf{r}')=\langle\hat{\mathbf{T}}(\mathbf{r},\mathbf{r}')\rangle. \tag{4.382}$$

To make contact with microscopic approaches, Raabe and Welsch (2005) consider a harmonic-oscillator medium and derive that the (steady-state) Lorentz force acting on such a medium in some space region V is

$$\mathbf{F} = \lim_{t\to\infty} \int_V \langle \hat{\rho}(\mathbf{r}, t)\hat{\mathbf{E}}(\mathbf{r}, t) + \hat{\mathbf{j}}(\mathbf{r}, t) \times \hat{\mathbf{B}}(\mathbf{r}, t)\rangle \, \mathrm{d}^3\mathbf{r}, \tag{4.383}$$

where again the charge density operator $\hat{\rho}(\mathbf{r}, t)$ and the current density operator $\hat{\mathbf{j}}(\mathbf{r}, t)$ are appropriately expressed. They apply the theory to a planar magnetodielectric structure. Its definition is specific in that homogeneity of the dielectric in an interspace ("cavity") $0 < z < d_j$ is required, where the subscript j has been used in conformity with Raabe et al. (2003, 2004).

Raabe and Welsch (2005) give the relevant stress tensor element $T_{zz}(\mathbf{r})$ in the interspace $0 < z < d_j$. Their relations (75) and (76) for this element, which are not reproduced here, are rather complicated. They include also a criticism of basing the calculations on Minkowski's stress tensor (Tomaš (2002), which leads to a relatively simple relation (81), which is not repeated here too. It can be believed that the warning is helpful, since one prefers simpler formulae to more complicated ones, provided that the simpler ones are not wrong.

To calculate the Casimir force on a plate in a nonempty cavity, Raabe and Welsch (2005) choose five regions, finite (1, 2, 3) or semi-infinite (0, 4). Let us assume that the two walls and the plate are almost perfectly reflecting. The generalization of Casimir's well-known formula is

$$F = \frac{\hbar c \pi^2}{240}\sqrt{\frac{\mu}{\epsilon}}\left(\frac{2}{3} + \frac{1}{3\epsilon\mu}\right)\left(\frac{1}{d_3^4} - \frac{1}{d_1^4}\right), \tag{4.384}$$

where d_k are thicknesses of regions $k = 1, 3$. Let us also assume that $\mu = 1$. Then the counterpart of the previous formula based on the Minkowski stress tensor is

$$F^{(\mathrm{M})} = \frac{\hbar c \pi^2}{240}\frac{1}{\sqrt{\epsilon}}\left(\frac{1}{d_3^4} - \frac{1}{d_1^4}\right), \tag{4.385}$$

which is just one of the formulae underlying the critique.

Pitaevskii (2006) defends the validity of the paper (Dzyaloshinskii et al. 1960), which has been disqualified or underestimated by the criticism in Raabe and Welsch (2005). It is important for the theory of the van der Waals–Casimir forces inside a dielectric fluid. The tensor of the van der Waals–Casimir forces was obtained by summation of an appropriate set of Feynman diagrams for the free energy and its variation with respect to the density (Dzyaloshinskii and Pitaevskii 1959). On the condition of mechanical equilibrium this tensor differs from a Minkowski-like one by a constant tensor. Dzyaloshinskii et al. (1960) obtained the same force between solid bodies, separated by a dielectric fluid, as Barash and Ginzburg (1975) and Schwinger et al. (1978). Pitaevskii (2006) discusses the reason why, in his opinion, the approach of Raabe and Welsch (2005) is incorrect. Raabe and Welsch (2006) maintain their position that the Casimir force should be calculated on the basis of the Lorentz force.

The paper (Leonhardt and Philbin 2007a) is interesting for its use of the notion of a transformation medium. This notion seems to belong to classical optics essentially and to be formed after the general relativity theory. A concise quantum theory of light in spatial transformation media has been developed in (Leonhardt and Philbin 2007b). Leonhardt and Philbin (2007a) calculate the Casimir force for a dispersive medium in their set-up inspired by Casimir's original idea. They consider two perfect conductors with a metamaterial sandwiched in between. The repulsive Casimir force of a left-handed material may balance the weight of one of the mirrors, letting it levitate on zero-point fluctuations. The simple formula for the Casimir force has been compared with the result of the more sofisticated Lifshitz theory.

Chen et al. (2006) have measured the Casimir force between a gold-coated sphere and two Si samples of higher and lower resistivity. The lowering of resistivity corresponded to enhancement of carrier density by several orders of magnitude. Each measurement was compared with theoretical results using the Lifshitz theory with different dielectric permittivities (Bordag et al. 2001, Chen et al. 2005, 2006, Lamoreaux 2005) and found to be consistent with this theory.

Lenac and Tomaš (2007) have considered the Casimir effect between metallic plates assuming them to be dispersive and lossless and separated by a medium with the (Gaussian) unit permittivity. They have taken two very different permittivities for media outside the plates, i.e. $\epsilon_2 = 1$ or $\epsilon_2 = \infty$ (perfect conductor). They have analysed the contributions of system eigenmodes with great attention to surface plasmon polariton modes. When the separation between the metallic plates is small, the surface plasmon polariton modes influence the Casimir effect dominantly except the case of thin layers that are supported by a highly reflective medium.

Messina and Passante (2007a) have calculated the Casimir–Polder force density on an uncharged, perfectly conducting plate placed in front of a neutral atom. To this aim first-order perturbation theory and the quantum operator associated to the classical electromagnetic stress tensor have been used. The result of Casimir and Polder (1948) has been rederived by integration of the force density. This integration is not an argument against the well-known nonadditivity of the Casimir–Polder forces (Milonni 1994 and references therein), and it has been discussed appropriately.

Munday and Capasso (2007) have performed precision measurements of the Casimir–Lifshitz force between two metal surfaces (gold) separated by a fluid (ethanol). For this situation, the measured force is attractive and is approximately 80% smaller than the force predicted for ideal metals in vacuum. The results were found to be consistent with Lifshitz's theory.

There exists a geometry well suited to the aim of an accurate theory–experiment comparison, namely, that with parallel and periodic corrugations of the metallic surfaces. The Casimir force is a superposition of the usual normal component and a lateral one in this situation. In general, vacuum-induced torque is present (Rodrigues et al. 2006a).

Rodrigues et al. (2007a) have studied the lateral Casimir force arising between two corrugated metallic plates. They assume that corrugations are imprinted on both plates with the same period and along the same direction, but with a spatial mismatch. They have used the scattering theory in a perturbative expansion in powers of the corrugation amplitudes. The result is valid provided that these amplitudes are smaller than L (mean separation distance), λ_C (corrugation wavelength), and λ_P (plasma wavelength). Limiting cases such as the proximity-force approximation limit and the perfect reflection limit are recovered when the length scales L, λ_C, and λ_P obey some specific orderings.

In the development of ever smaller atomic magnetic traps carbon nanotubes have been considered to become the elementary building blocks. It is well known that an atom held in a magnetic trap near an absorbing dielectric surface will undergo thermally induced spin–flip transitions. Some of these transitions lead to trapping losses. Fermani et al. (2007) have calculated atomic spin-flip lifetimes and have estimated tunneling lifetime corresponding to the sum of the Casimir–Polder potential and the magnetic trapping potential. Their analysis indicates that the Casimir–Polder force is the dominant loss agent.

Fulling (2007) have presented results on the Casimir force in one-dimensional piston models. These models are applications of quantum graphs (Roth 1985, Kuchment 2004). They have characterized the quantum star graphs mainly. A finite quantum graph consists of B one-dimensional undirected bonds or edges of length L_j ($j = 1, \ldots, B$) and some vertices. Either end of each bond ends at one of these vertices, and the valence of a vertex is defined as the number of bonds meeting there. At the univalent vertices either a Dirichlet or a Neumann boundary condition is imposed. For instance, the space may consist of B one-dimensional rays of large length L attached to a central vertex. In each ray a piston is located a distance a from the vertex. At the central vertex the field has the Kirchhoff (generalized Neumann) behaviour. In fact, the pistons are treated as univalent vertices. If at each piston the field obeys the Neumann boundary condition, then the force is ($\hbar = 1 = c$)

$$F = \frac{(B-3)\pi}{48a^2}. \tag{4.386}$$

When $B = 1$ or 2, the result is related to an ordinary Neumann interval of length a or $2a$, respectively. When $B > 3$, the force is repulsive. The pistons will tend to move outward. Fulling et al. (2007) have discussed a periodic-orbit approach to calculations of the Casimir forces. They have also numerically examined the rate of convergence of the periodic-orbit expansion.

Rodriguez et al. (2007a,b) have developed a numerical method to compute the Casimir forces in arbitrary geometries, for arbitrary dielectric and metallic materials. They have based their approach on the familiar result due to Lifshitz and Pitaevskii (1980), Dzyaloshinskii et al. (1961), and Pitaevskii (2006). The Casimir force is obtained in terms of the stress tensor integrated over space and imaginary frequency. The vacuum expectation value of the stress tensor is calculated in terms of the Green function, which is automatically regularized on application of the finite-difference method to solve for the Green function. The geometries that have been considered have the property that the bodies have not a contact and they are in the free space.

Then it is innocent to evoke the Minkowski stress tensor, since a contour or surface around the body of interest lies in the free space. But also the Maxwellian stress tensor is named.

Messina and Passante (2007b) have paid attention to fluctuations of the Casimir–Polder force between a neutral atom and a perfectly conducting wall. They have made use of the method of time-averaged operators introduced by Barton and widely used by him in his papers on fluctuations of the Casimir forces for macroscopic bodies (Barton 1991a,b). They have also calculated the Casimir–Polder force fluctuations for an atom between two conducting walls. This situation has been investigated already by Barton (1987). The force operator has been derived from an effective interaction Hamiltonian (Passante et al. 1998). To this end the effective interaction energy operator is differentiated with respect to the distance from the atom to the wall.

Intravaia et al. (2007) remind that the Casimir effect, at short distances, is dominated by the coupling between the surface plasmons that are present on two metallic mirrors (Van Kampen et al. 1968). The Casimir energy is calculated in terms of quasielectrostatic (or nonretarded) field modes. When the mirror separation increases, retardation must be taken into account. Intravaia et al. (2007) use the method of Schram (1973). They choose the dielectric function

$$\epsilon(\omega) = 1 - \frac{\omega_{\mathrm{p}}^2}{\omega^2} \tag{4.387}$$

to describe the metal. They calculate dispersion relations for the relevant modes numerically. They distinguish bulk modes, propagating cavity modes, and evanescent modes. For the TE polarization all modes are propagating, but for the TM polarization two modes are evanescent in at least some range of wave vectors. These modes are referred to as "plasmonic". A plasmonic contribution to the Casimir energy is denoted by E_{p}. The short-distance asymptotics is $-\frac{A\hbar\omega_{\mathrm{p}}}{L^2}$, where A is the area of the mirrors. The large-distance asymptotics is $+\frac{A\hbar\sqrt{\omega_{\mathrm{p}}c}}{L^{5/2}}$. This is balanced in the total Casimir energy by the contribution of photonic modes (cavity and bulk modes), which yields the negative, binding, energy again.

Passante and Spagnolo (2007) have evaluated the Casimir–Polder potential between two atoms in the presence of an infinite perfectly conducting plate and at nonzero temperature. They assume the wall located at $z = 0$ and let $\mathbf{r}_A$ and $\mathbf{r}_B$ denote the positions of atoms A and B, respectively. First they outline the method used by reproducing the Casimir–Polder potential energy between two atoms in a thermal field

$$W_{AB}(R) = \frac{\hbar c}{\pi} \int_0^\infty k^3 \alpha_A(k)\alpha_B(k) \coth\left(\frac{\hbar c k}{2k_{\mathrm{B}}T}\right) \times \mathbf{V}(k, \mathbf{R}) : \boldsymbol{\tau}(k, \mathbf{R})\,\mathrm{d}k, \tag{4.388}$$

(cf. Wennerström et al. 1999), where $R = |\mathbf{R}|$ is the distance between the two atoms, $\mathbf{R} = \mathbf{r}_B - \mathbf{r}_A$, k is a wavenumber, $\alpha_A(k)$ ($\alpha_B(k)$) is the dynamical polarizability of the atom A (B) (Power and Thirunamachandran 1993),

$$\mathbf{V}(k,\mathbf{R}) = \frac{1}{R^3}\left\{\left(\mathbf{1} - 3\frac{\mathbf{R}}{R}\frac{\mathbf{R}}{R}\right)[\cos(kR) + kR\sin(kR)] - \left(\mathbf{1} - \frac{\mathbf{R}}{R}\frac{\mathbf{R}}{R}\right)k^2R^2\cos(kR)\right\}, \tag{4.389}$$

$$\boldsymbol{\tau}(k,\mathbf{R}) = \left(\mathbf{1} - \frac{\mathbf{R}}{R}\frac{\mathbf{R}}{R}\right)\frac{\sin(kR)}{kR} + \left(\mathbf{1} - 3\frac{\mathbf{R}}{R}\frac{\mathbf{R}}{R}\right)\left[\frac{\cos(kR)}{k^2R^2} - \frac{\sin(kR)}{k^3R^3}\right]. \tag{4.390}$$

Then they derive and discuss their results for the retarded atom–atom Casimir–Polder interaction when both a thermal field and a boundary condition are present. The interaction energy is

$$W_{AB}(R,\bar{R}) = W_{AB}(R) + W_{AB}(\bar{R}) - \frac{\hbar c}{\pi}\int_0^\infty k^3\alpha_A(k)\alpha_B(k)\coth\left(\frac{\hbar ck}{2k_\mathrm{B}T}\right) \times \boldsymbol{\sigma} : \left[\boldsymbol{\tau}(k,\mathbf{R})\cdot\mathbf{V}(k,\bar{\mathbf{R}}) + \mathbf{V}(k,\mathbf{R})\cdot\boldsymbol{\tau}(k,\bar{\mathbf{R}})\right]\,\mathrm{d}k, \tag{4.391}$$

where $\bar{R} = |\bar{\mathbf{R}}|$ is the distance between one atom and the image of the other atom reflected on the plate, $\bar{\mathbf{R}} = \mathbf{r}_B - \boldsymbol{\sigma}\cdot\mathbf{r}_A$, and $\boldsymbol{\sigma}$ is the reflection tensor on the conducting plate, supposed orthogonal to the z axis.

The analysis of most Casimir force experiments using a sphere-plate geometry has relied on the proximity-force approximation (PFA), which expresses the Casimir force between a sphere and a flat plate in terms of the Casimir energy between two parallel plates. Krause et al. (2007) have conducted an experimental assessment of the range of applicability of the proximity-force approximation. They have measured the Casimir force and force gradient between a gold-coated plate and five gold-coated spheres with different radii using a microelectromechanical torsion oscillator. Specifically, according to the proximity-force approximation, the Casimir force between a sphere of radius R and a flat plate separated by a distance $z \ll R$ can be written as

$$F(z) \approx F_\mathrm{PFA}(z) \equiv 2\pi R\mathcal{E}^\mathrm{pp}(z), \tag{4.392}$$

where $\mathcal{E}^\mathrm{pp}(z)$ is the Casimir energy per unit area between two parallel plates separated by the distance z.

If the bodies are smooth and perfectly conducting, the exact Casimir force may be expanded in powers of $\frac{z}{R}$ (Scardicchio and Jaffe 2006),

$$F_\mathrm{Casimir}(z,R) = 2\pi R\mathcal{E}^\mathrm{pp}(z)\left[1 + \beta\frac{z}{R} + O\left(\frac{z^2}{R^2}\right)\right], \tag{4.393}$$

where β is a dimensionless parameter and the Landau notation $O(f(x))$ means that $\frac{O(f(x))}{f(x)}$ is bounded for $x \to 0$. An effective pressure $P^{\mathrm{eff}}(z, R)$ may be expanded similarly, but a new dimensionless parameter is denoted by β'. The roughness and conductivity effects can be included and the modified notation, $\beta(z)$ and $\beta'(z)$, respects a general dependence on z. For separations $z < 300$ nm, Krause et al. (2007) have found that $|\beta'(z)| < 0.4$ at the 95% confidence level.

Rodrigues et al. (2006b) have presented a novel theoretical approach to the lateral Casimir force beyond the regime of validity of the proximity-force approximation. They have related the results of the new approach to the measured values (Chen et al. 2002a,b). Unfortunately, the novel approach also has its region of validity and the illustration chosen does not fit into it. Besides, the complete proximity-force approximation has led to a happy coincidence of the theoretical value of 0.33 pN with the average of measured values of 0.32 pN, while the expectation value belongs to the interval 0.32 ± 0.077 pN (at 95% confidence). This situation is reflected in the comment (Chen et al. 2007) and the reply (Rodrigues et al. 2007b).

Emig (2007) has explored the lateral Casimir force between two parallel periodically patterned metal surfaces. It is assumed that the surfaces are set into relative oscillatory motion so that their normal distance is a periodic function of time. This scenario resembles the so-called ratchet systems (Reimann (2002). It is demonstrated that the system allows for directed lateral motion of the surfaces. These results show that Casimir interactions offer contactless translational actuation schemes for nanomechanical systems.

Emig et al. (2007) have developed a systematic method for computing the Casimir energy between arbitrary compact dielectric objects. Casimir interactions are completely characterized by the scattering matrices of the individual bodies. As an example they compute the Casimir energy between two identical dielectric spheres at any separation.

The thermal part of the Casimir force was subject to discussions (Milton 2004). On the assumption of real metals the Lifshitz formula is used. For LT large compared with $\frac{\hbar c}{k_{\mathrm{B}}}$, where L is the distance between the parallel plates and T is the temperature of the system, the assumption of ideal metals leads to a result which is twice the Lifshitz one.

Svetovoy (2007) has analysed the repulsive thermal Casimir force between two metals and a metal and a high-permittivity dielectric. The repulsion discussed in such work has the meaning of a negative thermal correction to the force at zero temperature, but the total force is always attractive. The force is calculated using the Lifshitz formula written via real frequencies (Landau and Lifshitz 1963). Two contributions of the fluctuating fields, propagating waves and evanescent waves, are distinguished. For both material configurations the repulsive s-polarized evanescent-wave contribution dominates for LT small. Here L is the distance between parallel plates and T is the temperature of the system. In this case, the force between ideal metals is attractive and small (Mehra 1967, Brown and Maclay 1969). The ideal metal is rather the limit case of a superconductor than of a normal metal (Antezza et al. 2006).

Rizzuto (2007) considers a neutral two-level atom uniformly accelerated in a direction parallel to an infinite mirror and calculates the atom-wall Casimir–Polder interaction between the accelerated atom and the mirror. The mirror is modelled as Dirichlet boundary conditions on a massless scalar field.

The author evaluates the vacuum fluctuation (vf) and radiation reaction (rr) contributions to the atom-wall Casimir–Polder interaction energy. First she expresses only the contributions to the radiative shifts of the atomic levels.

Let us assume the atom to be at rest. The Casimir–Polder interaction energy between the atom at rest and the wall is obtained by considering only the z_0-dependent terms, $E_{\rm CP}^{\rm (vf)}$ and $E_{\rm CP}^{\rm (rr)}$, in the vacuum fluctuation and in the radiation reaction contributions, respectively,

$$E_{\rm CP}^{\rm (vf)} = \frac{\mu^2}{64\pi^2 c^2 z_0}\left[2f\left(\omega_0, \frac{2z_0}{c}\right) - \pi \cos\left(\frac{2\omega_0 z_0}{c}\right)\right], \tag{4.394}$$

$$E_{\rm CP}^{\rm (rr)} = \frac{\mu^2}{64\pi c^2 z_0}\cos\left(\frac{2\omega_0 z_0}{c}\right), \tag{4.395}$$

where ω_0 corresponds to the energy difference of the levels of the atomic system, $\hbar\omega_0$, z_0 is the distance of the atom from the mirror, and

$$f(\alpha, \beta) = \int_0^\infty \frac{\sin(\alpha x)}{x+\beta}\,\mathrm{d}x. \tag{4.396}$$

In the case of a uniformly accelerated atom, with the acceleration in a direction parallel to the reflecting plate, a generalization of relations (4.394) and (4.395) yields

$$\begin{aligned} E_{\rm CP}^{\rm (vf)} = \frac{\mu^2}{64\pi^2 c^2 z_0 N}\Bigg\{&2f\left(\omega_0, \frac{2c}{a}\sinh^{-1}\left(\frac{z_0 a}{c^2}\right)\right) \\ &- \pi\cos\left(\frac{2\omega_0 c}{a}\sinh^{-1}\left(\frac{z_0 a}{c^2}\right)\right) \\ &- \frac{a}{\omega_0 c}\left[1 - \cos\left(\frac{2\omega_0 c}{a}\sinh^{-1}\left(\frac{z_0 a}{c^2}\right)\right)\right]\Bigg\}, \end{aligned} \tag{4.397}$$

$$E_{\rm CP}^{\rm (rr)} = \frac{\mu^2}{64\pi c^2 z_0 N}\cos\left(\frac{2\omega_0 c}{a}\sinh^{-1}\left(\frac{z_0 a}{c^2}\right)\right), \tag{4.398}$$

where $N = \sqrt{1 + \left(\frac{z_0 a}{c^2}\right)^2}$ and a is the proper acceleration of the atom.

Chapter 5
Microscopic Models as Related to Macroscopic Descriptions

The models expounded in Chapter 4 are often labelled as macroscopic, since apparently they do not allow one to consider the Clausius–Mossotti or the Lorentz–Lorenz relation. Even though we shall not, even here, expound this interesting subject, which, however, is notorious in the classical theory, we shall mention the quantum models, whose microscopic character cannot be doubted. The role of macroscopic averages, which is analyzed in the classical theory as well, is discussed from the quantal viewpoint.

5.1 Quantum Optics in Oscillator Media

A quantum-optical experimental setup may consist of active and passive devices, active devices to generate light of certain properties (e.g. nonclassical light) and passive ones to modify and apply it. It is an interesting and nontrivial problem to study how quantum statistical properties of light are influenced by passive optical devices like mirrors, resonators, beam splitters, or filters.

Knöll and Leonhardt (1992) have continued also the paper (Knöll et al. 1987), where the medium is nondispersive and lossless, but they now intend to consider dispersion and losses.

On introducing the Hamiltonian for the complete system, the Heisenberg equations of motion for field operators and medium (not source) quantities are derived. The complete system under consideration consists of the subsystems: optical field, medium atoms, and sources. The field is described by the electric-field-strength operator $\hat{E}(x,t)$ and the electromagnetic vector-potential operator $\hat{A}(x,t)$ in the Coulomb gauge. The medium is modelled by the damped harmonic oscillators $\{\hat{q}_\mu(t),\ \hat{p}_\mu(t)\}$, namely, the oscillators coupled to reservoirs composed of bath oscillators $\{\hat{q}_{\mu B}(t),\ \hat{p}_{\mu B}(t)\}$, the quanta of whose energy may be, for example, phonons. The medium oscillators are localized at x_μ, they have the same mass m and the elasticity (force) constant k. The bath oscillators are characterized by masses m_B and the elasticity constants k_B and the coupling is expressed by the coupling constants σ_B. The atomic sources are described by a current operator $\hat{j}(x,t)$, but its dynamics

A. Lukš, V. Peřinová, *Quantum Aspects of Light Propagation*,
DOI 10.1007/b101766_5,

need not be specified, e is the electron charge. For simplicity, a one-dimensional model is considered only.

The Hamiltonian of the complete system is

$$\hat{H}(t) = \hat{H}_{\mathrm{R}}(t) + \hat{H}_{\mathrm{M}}(t) + \hat{H}_{\mathrm{RS}}(t) + \hat{H}_{\mathrm{S}}(t), \tag{5.1}$$

where $\hat{H}_{\mathrm{R}}(t)$ is the Hamiltonian of the optical field, $\hat{H}_{\mathrm{M}}(t)$ is that of the medium atoms, and $\hat{H}_{\mathrm{RS}}(t)$ describes the interaction between the optical field and sources,

$$\hat{H}_{\mathrm{R}}(t) = \int \frac{\epsilon_0}{2} \left[\hat{E}^2(x,t) + c^2 \left(\frac{\partial \hat{A}(x,t)}{\partial x} \right)^2 \right] \mathrm{d}x, \tag{5.2}$$

$$\begin{aligned} \hat{H}_{\mathrm{M}}(t) = \sum_{\mu} \Bigg\{ & \frac{[\hat{p}_{\mu}(t) - e\hat{A}(x_{\mu},t)]^2}{2m} + \frac{k}{2}\hat{q}_{\mu}^2(t) \\ & + \sum_{B} \left[\frac{\hat{p}_{\mu B}^2(t)}{2m_B} + \frac{k_B}{2}\hat{q}_{\mu B}^2(t) + \frac{\sigma_B}{2}\left[\hat{q}_{\mu}(t) - \hat{q}_{\mu B}(t)\right]^2 \right] \Bigg\}, \end{aligned} \tag{5.3}$$

$$\hat{H}_{\mathrm{RS}}(t) = -\int \hat{\jmath}(x,t)\hat{A}(x,t)\,\mathrm{d}x, \tag{5.4}$$

and $\hat{H}_{\mathrm{S}}(t)$ is the Hamiltonian of the atomic sources which is left unspecified. The usual commutation rules are

$$\begin{aligned} [\hat{A}(x,t), -\epsilon_0 \hat{E}(x',t)] &= \mathrm{i}\hbar\delta(x-x')\hat{1}, \\ [\hat{q}_{\mu}(t), \hat{p}_{\mu'}(t)] &= \mathrm{i}\hbar\delta_{\mu\mu'}\hat{1}, \\ [\hat{q}_{\mu B}(t), \hat{p}_{\mu' B'}(t)] &= \mathrm{i}\hbar\delta_{\mu\mu'}\delta_{BB'}\hat{1}. \end{aligned} \tag{5.5}$$

The Heisenberg equations of motion for field operators, medium operators, and bath operators have been obtained. As a result of a Wigner–Weisskopf approximation for the interaction of medium oscillators with the bath operators, quantum Langevin equations have been obtained. By eliminating the medium quantities from equations for field operators and further by a usual procedure, a generalized wave equation for the vector potential is obtained. Using a Green function, this wave equation is solved.

The decomposition of the time-ordered quantum correlation functions into time-ordered correlation functions of the source operators and the free-field operators has been derived. The notation of positive time-ordering $\mathcal{T} \equiv \mathcal{T}_+$ (see, (2.110)) and the ordering symbol $\mathcal{O} \equiv \mathcal{O}_+$ are used. The property (2.120) and the consequence for the time-ordered quantum correlations have been recalled.

The time-dependent Green function for a dielectric layer as the simplest optical device is calculated. The field behind the layer is discussed and represented by the negative-frequency part of the field, its expectation value and the normally ordered quadrature variance determined for the sake of squeezing analysis.

5.2 Problem of Macroscopic Averages

Assuming that the medium is constituted by atoms, which do not form a continuum, we can aproximate a continuum using the macroscopic averaging. This idea is illustrated by a simple model. It is indicated that many material constants would require appropriately complicated derivations. Inclusion of losses enhances requirements on the microscopic model as well, even though we will restrict our discourse on a one-dimensional standing wave in a cavity. The macroscopic averaging can be generalized even to this case.

5.2.1 Conservative Oscillator Medium

Dutra and Furuya (1997) have investigated a simple microscopic model for the interaction between an atom and radiation in a linear lossless medium. It is a guest two-level atom inside a single-mode cavity with a host medium composed of other two-level atoms that are approximated by harmonic oscillators. There is an intention to show, in general, that the ordinary quantum electrodynamics suffices, at least in principle, and there is no need to quantize the phenomenological classical Maxwell equations. If a macroscopic description is possible, it should appear as an approximation to the fundamental microscopic theory under certain conditions. Such a "macroscopic" approximation is obtained and conditions of its validity are derived.

All of the medium harmonic oscillators are represented by a collective harmonic oscillator, then two modes of a polariton field are defined. The macroscopic average is regarded as filtering out higher spatial frequencies. The field that influences the guest atom is modified and the characteristics of the effect of such a microscopic field are calculated. A condition is pointed out under which the contribution of the atoms to the quantum noise appears only through a frequency-dependent dielectric constant. An effective description is obtained by leaving out the polariton mode which is approximately equal to the collective mode of the medium.

Dutra and Furuya (1997) have introduced a microscopic model of a material medium that they have adopted, N two-level atoms having the same resonance frequency ω_0 in a single-mode cavity of the resonance frequency ω. They consider a guest two-level atom of the resonance frequency ω_a and strongly coupled to the field so that it will not be approximated by a harmonic oscillator. The operator of the displacement field in the cavity is given by the relation

$$\hat{D}(x,t) = \sqrt{\frac{\hbar\omega\epsilon_0}{L}} \sin\left(\frac{\omega}{c}x\right) [\hat{a}(t) + \hat{a}^\dagger(t)], \tag{5.6}$$

where L is the length of the cavity. It is assumed that ω and L for the single-mode cavity are in the relation

$$\frac{\omega}{c} = \frac{\pi}{L}. \tag{5.7}$$

The operator of the polarization of the medium is given by

$$\hat{P}(x,t) = \sum_{j=1}^{N} d_j \delta(x - x_j)[\hat{b}_j(t) + \hat{b}_j^\dagger(t)], \tag{5.8}$$

where

$$d_j = \sqrt{\frac{\hbar}{2\omega_0 m_0}} q_j \tag{5.9}$$

are electric dipole moments in eigenstates $|1, \mathrm{s}\rangle_j$ of the operators $[\hat{b}_{j\mathrm{s}}(t) + \hat{b}_{j\mathrm{s}}^\dagger(t)]$, large s, weakly converging to $[\hat{b}_j(t) + \hat{b}_j^\dagger(t)]$ for $\mathrm{s} \to \infty$ (s, the greatest number of quanta, has been introduced on the normalization ground). In (5.9), m_0 is an effective mass, q_j are effective charges, products $d_j[\hat{b}_j(t) + \hat{b}_j^\dagger(t)]$ are the electric-dipole-moment operators of the atoms of the medium that are located at x_j. The operators $\hat{a}(t)$, $\hat{a}^\dagger(t)$, $\hat{b}_j(t)$, and $\hat{b}_j^\dagger(t)$ satisfy the bosonic commutation relations, in particular,

$$[\hat{b}_j(t), \hat{b}_{j'}^\dagger(t)] = \delta_{jj'} \hat{1}, \tag{5.10}$$

$$[\hat{b}_j(t), \hat{b}_{j'}(t)] = \hat{0}, \tag{5.11}$$

and $\hat{a}(t)$, $\hat{a}^\dagger(t)$ commute with $\hat{b}_j(t)$, $\hat{b}_j^\dagger(t)$.

The Hamiltonian is given by the relation

$$\hat{H}(t) = \hbar\omega \hat{a}^\dagger(t)\hat{a}(t) + \hbar\omega_0 \sum_{j=1}^{N} \hat{b}_j^\dagger(t)\hat{b}_j(t) - \frac{1}{\epsilon_0}\int \hat{D}(x,t)\hat{P}(x,t)\,\mathrm{d}x$$
$$+ \frac{\hbar\omega_\mathrm{a}}{2}\hat{\sigma}_z(t) + \hbar\Omega[\hat{a}(t) + \hat{a}^\dagger(t)][\hat{\sigma}(t) + \hat{\sigma}^\dagger(t)], \tag{5.12}$$

where $\hat{\sigma}_z(t)$ and $\hat{\sigma}(t)$ are the pseudo-spin operators and

$$\Omega = -d\sqrt{\frac{\omega m_\mathrm{a}}{\epsilon_0 L\hbar}} \sin\left(\frac{\omega}{c}x_\mathrm{a}\right) \tag{5.13}$$

is the Rabi frequency of the guest atom located at x_a whose electric dipole moment (in the eigenstate $|1\rangle$ of the operator $[\hat{\sigma}(t) + \hat{\sigma}^\dagger(t)]$) is d and the effective mass m_a. We notice that

$$-\frac{1}{\epsilon_0}\int \hat{D}(x,t)\hat{P}(x,t)\,\mathrm{d}x = \hbar\sum_{j=1}^{N} g_j(\omega)[\hat{a}(t) + \hat{a}^\dagger(t)][\hat{b}_j(t) + \hat{b}_j^\dagger(t)], \tag{5.14}$$

where

$$g_j(\omega) = -d_j\sqrt{\frac{\omega m_0}{\epsilon_0 L\hbar}}\sin\left(\frac{\omega}{c}x_j\right). \tag{5.15}$$

In simplifying the Hamiltonian, Dutra and Furuya (1997) denote

$$G(\omega) = \sqrt{\sum_{j'=1}^{N}[g_{j'}(\omega)]^2}, \tag{5.16}$$

the coupling constant between the field and the collective harmonic oscillator described by the annihilation operator

$$\hat{B}(t) = \frac{1}{G(\omega)}\sum_{j=1}^{N} g_j(\omega)\hat{b}_j(t). \tag{5.17}$$

In the Hamiltonian (5.12), the self-energy terms (Cohen-Tannoudji et al. 1989) have been neglected. This results in the condition $4[G(\omega)]^2 \leq \omega\omega_0$. Further, the original problem is reduced to the case of a single atom coupled to two polariton modes. The frequencies of these modes are denoted by Ω_1 and Ω_2 so that $\Omega_1 \simeq \omega\sqrt{1 - 4\frac{[G'(\omega)]^2}{\omega_0^2}}$, ω and $\Omega_2 \to \omega_0$ when $\omega \to 0, \infty$, respectively, with

$$G'(\omega) = \sqrt{\frac{\omega_0}{\omega}}G(\omega). \tag{5.18}$$

In other words, $\Omega_1 < \Omega_2$ for $\omega < \omega_0$, $\Omega_1 > \Omega_2$ for $\omega > \omega_0$. The dressed operators are denoted by $\hat{c}_k(t)$ and $\hat{c}_k^\dagger(t)$, $k = 1, 2$, and the operators $\hat{a}(t)$ and $\hat{B}(t)$ are expressed in their terms (Chizhov et al. 1991).

The problem of the extra quantum noise introduced by the atoms of the medium is discussed. In the case when the atoms of the medium are only weakly coupled to the field, i.e. $G'(\omega)$, $\omega \ll \omega_0$, it holds that (Glauber and Lewenstein 1991)

$$\langle\{\Delta[\hat{a}(t) + \hat{a}^\dagger(t)]\}^2\rangle \approx \sqrt{\epsilon_r}, \tag{5.19}$$

where $\Delta[\hat{a}(t) + \hat{a}^\dagger(t)] = \hat{a}(t) + \hat{a}^\dagger(t) - \langle\hat{a}(t) + \hat{a}^\dagger(t)\rangle\hat{1}$, and ϵ_r is the relative permittivity

$$\epsilon_r \approx 1 + 4\frac{[G'(\omega)]^2}{\omega_0^2}. \tag{5.20}$$

From the relation (5.19), the variance of the operator of the electric displacement field $\hat{D}(x, t)$ (cf., (Glauber and Lewenstein 1991)) can be calculated. The adoption of the continuous distribution of atoms in the medium instead of the realistic discrete one implies a greater variance (Rosewarne 1991).

Let us address the problem of macroscopic averages. The macroscopic theories of quantum electrodynamics in nonlinear media, have often, by the "definition" avoided discussing the macroscopic averaging procedure. The quantum-mechanical averaging advocated by Schram (1960) removes the quantum fluctuations from the macroscopic theory. The problem of the macroscopic theory what should be the macroscopic averaging procedure resisted, for many years, the solution. It was Lorentz who in the beginning of the twentieth century first tried such a derivation, cf., chapters (de Groot 1969, van Kranendonk and Sipe 1977). Robinson (1971, 1973) has proposed a different kind of macroscopic average. He regards a macroscopic description as a description where the spatial Fourier components of the field variables above some limiting spatial frequency k_0 are irrelevant. Dutra and Furuya (1997) take the Fourier components with the spatial frequencies above $\frac{\omega}{c}$ for irrelevant in a macroscopic description. They arrive at the following expression for the macroscopic polarization

$$\hat{\hat{P}}(x,t) = -2G(\omega)\sqrt{\frac{\hbar\epsilon_0}{\omega L m_0}}\sin\left(\frac{\omega}{c}x\right)[\hat{B}(t)+\hat{B}^\dagger(t)]. \tag{5.21}$$

The macroscopic "average" does not change the operator of the electric displacement field $\hat{D}(x,t)$ and the macroscopic electric-field-strength operator is given by the relation

$$\hat{\hat{E}}(x,t) = \frac{1}{\epsilon_0}[\hat{D}(x,t) - \hat{\hat{P}}(x,t)]. \tag{5.22}$$

The calculation of the variance of a quantity typical of the operator $\hat{\hat{E}}(x,t)$ is larger than $\epsilon_r^{-\frac{3}{2}}$, which agrees with Rosewarne's result (Rosewarne 1991). Thus, it has been shown that the contribution from the atoms to the quantum noise of the field does not restrict itself to inclusion of the dielectric constant. We are going to report the suitable macroscopic theory of electrodynamics in a material medium which does not suffer from the problems which are discussed here.

It is shown that under certain conditions a macroscopic description incorporating the frequency dependence of the relative permittivity provides a good approximation. In this domain, Milonni's result has been recovered (Milonni 1995). The guest atom is not affected by the polariton mode if the frequency of the atom is far from Ω_2. An analysis of the probability of this mode inducing transitions shows that such a probability is negligible when

$$|\Omega_2^2 - \omega^2|\frac{\Omega}{\Omega_2}\sqrt{\frac{\Omega_2\omega_0}{(\Omega_2^2-\omega^2)^2 + 4\omega_0\omega G^2}} \ll |\Omega_2 - \omega_a|. \tag{5.23}$$

In the regime described by the relation (5.23), the leaving out of the polariton mode described by $\hat{c}_2$ and $\hat{c}_2^\dagger$ and the macroscopic averaging leads to the macroscopic Hamiltonian

$$\hat{H}_{\mathrm{mac}}(t) = \hbar\Omega_1\hat{c}_1^\dagger(t)\hat{c}_1(t) + \frac{\hbar\omega_{\mathrm{a}}}{2}\hat{\sigma}_z(t) - \frac{d}{\epsilon_0}\hat{D}_{\mathrm{mac}}(x_{\mathrm{a}}, t)[\hat{\sigma}(t) + \hat{\sigma}^\dagger(t)], \tag{5.24}$$

where

$$\hat{D}_{\mathrm{mac}}(x, t) = \sqrt{\frac{\hbar\Omega_1\epsilon_0\epsilon_{\mathrm{r}}\sqrt{\epsilon_{\mathrm{r}}}}{L\gamma}} \sin\left(\sqrt{\epsilon_{\mathrm{r}}}\frac{\Omega_1}{c}x\right)[\hat{c}_1(t) + \hat{c}_1^\dagger(t)], \tag{5.25}$$

with

$$\gamma = \frac{\mathrm{d}}{\mathrm{d}\Omega_1}(\Omega_1\sqrt{\epsilon_{\mathrm{r}}}), \tag{5.26}$$

is the macroscopic displacement field. By the relation (5.26), γ is the ratio between the speed of light in the vacuum and the group velocity in the medium. The macroscopic polarization $\hat{P}_{\mathrm{mac}}(x, t)$ is given by the relation (5.21) simplified by leaving out its $\hat{c}_2(t)$, $\hat{c}_2^\dagger(t)$ polariton component. Then, from the relation

$$\hat{E}_{\mathrm{mac}}(x, t) = \frac{1}{\epsilon_0}\left[\hat{D}_{\mathrm{mac}}(x, t) - \hat{P}_{\mathrm{mac}}(x, t)\right], \tag{5.27}$$

the macroscopic electric field is obtained. It is stated that the results of Dutra and Furuya (1997) for the macroscopic fields coincide with those derived by Milonni (1995) for the case of one and more modes.

de Lange and Raab (2006) recall series decompositions of **D** and **H** comprising macroscopic densities of multipole moments

$$D_i = \epsilon_0 E_i + P_i - \frac{1}{2}\nabla_j Q_{ij} + \frac{1}{6}\nabla_k\nabla_j Q_{ijk} + \ldots, \tag{5.28}$$

$$H_i = \mu_0^{-1} B_i - M_i + \frac{1}{2}\nabla_j M_{ij} + \ldots, \tag{5.29}$$

where P_i $(Q_{ij}, Q_{ijk}, \ldots)$ is an electric dipole (quadrupole, octupole, ...) density and M_i $(M_{ij}, \ldots)$ is a magnetic dipole (quadrupole, ...) density, and, for monochromatic fields and on concentration on nonmagnetic media,

$$P_i = \alpha_{ij}E_j + \frac{1}{2}a_{ijk}\nabla_k E_j + \ldots + \frac{1}{\omega}G'_{ij}\dot{B}_j + \ldots, \tag{5.30}$$

$$Q_{ij} = a_{kij}E_k + \ldots, \quad Q_{ijk} \approx 0, \tag{5.31}$$

$$M_i = -\frac{1}{\omega}G'_{ji}\dot{E}_j + \ldots + \chi_{ij}B_j + \ldots, \quad M_{ij} \approx 0, \tag{5.32}$$

where $\alpha_{ij}, a_{ijk}, \ldots, G'_{ij}, \ldots$ are material constants. For harmonic fields, these decompositions can be written as

$$D_i = \epsilon_0 E_i + \alpha_{ij} E_j + \frac{1}{2} \mathrm{i} k_k a_{ijk} E_j - \frac{1}{2} \mathrm{i} k_j a_{kij} E_k + \ldots \\ - \mathrm{i} G'_{ij} B_j + \ldots, \tag{5.33}$$

$$H_i = -\mathrm{i} G'_{ji} E_j + \ldots + \mu_0^{-1} B_i - \chi_{ij} B_j + \ldots. \tag{5.34}$$

The authors show a lack of translational invariance. On using the Maxwell equations, these series can be recast into the form

$$D_i = \epsilon_0 E_i + \alpha_{ij} E_j + \ldots - \mathrm{i} \left(G'_{ij} - \frac{1}{2} \omega \varepsilon_{jkl} a_{kli} \right) B_j + \ldots, \tag{5.35}$$

$$H_i = -\mathrm{i} \left(G'_{ji} - \frac{1}{2} \omega \varepsilon_{ikl} a_{klj} \right) E_j + \ldots + \mu_0^{-1} B_i + \ldots, \tag{5.36}$$

which contains the origin-independent material constants.

For monochromatic fields and to describe magnetic media, it is necessary that relations (5.30), (5.31), and (5.32) are extended by other terms,

$$P_i = \frac{1}{\omega} \alpha'_{ij} \dot{E}_j + \frac{1}{2\omega} a'_{ijk} \nabla_k \dot{E}_j + \ldots + G_{ij} B_j + \ldots, \tag{5.37}$$

$$Q_{ij} = -\frac{1}{\omega} a'_{kij} \dot{E}_k + \ldots, \tag{5.38}$$

$$M_i = G_{ji} E_j + \ldots + \frac{1}{\omega} \chi'_{ij} \dot{B}_j + \ldots. \tag{5.39}$$

The original terms are included in the ellipses. Here α'_{ij}, α'_{ijk}, …, G_{ij}, χ'_{ij}, … are other material constants. For harmonic fields, the above decompositions can be extended as

$$D_i = -\mathrm{i} \alpha'_{ij} E_j + \frac{1}{2} k_k a'_{ijk} E_j + \frac{1}{2} k_j a'_{kij} E_k + \ldots + G_{ij} B_j + \ldots, \tag{5.40}$$

$$H_i = -G_{ji} E_j + \ldots + \mathrm{i} \chi'_{ij} B_j + \ldots. \tag{5.41}$$

The translational invariance is assessed also here. Using the Maxwell equations, these series can be recast into the form

$$D_i = -\mathrm{i} \alpha'_{ij} E_j + \frac{1}{3} k_k (a'_{ijk} + a'_{jki} + a'_{kij}) E_i + \ldots \\ + \left(G_{ij} - \frac{1}{3} G_{ll} \delta_{ij} - \frac{1}{6} \omega \varepsilon_{jkl} a'_{kli} \right) B_j + \ldots, \tag{5.42}$$

$$H_i = \left(-G_{ji} + \frac{1}{3}G_{ll}\delta_{ij} + \frac{1}{6}\omega\varepsilon_{ikl}a'_{klj}\right)E_j + \dots$$
$$+ (\mathrm{i}\chi'_{ij} + ?)B_j + \dots, \quad (5.43)$$

where ? means nonmagnetic polarizability densities of electric multipole order 8 and magnetic multipole order 4. This form comprises the origin-independent material constants.

5.2.2 Kramers–Kronig Dielectric

Dutra and Furuya (1998a) have pointed out that the Huttner–Barnett model at the stage after the diagonalization of the Hamiltonian for the polarization field and reservoirs can operate with a larger class of dielectric functions than that admitted by the original microscopic model. At this stage, the relative permittivity is expressed in dependence on the dimensionless coupling function $\zeta(\omega)$,

$$\epsilon(\omega) = 1 + \frac{\omega_c^2}{2\omega}\lim_{\varepsilon\to+0}\int_{-\infty}^{\infty}\frac{|\zeta(\omega')|^2}{(\omega' - \omega - \mathrm{i}\varepsilon)\omega'}\,\mathrm{d}\omega', \quad (5.44)$$

where the notation from Section 4.1.1 has been used. For example, the permittivity obtained in the Lorentz oscillator model (Klingshirn 1995) can be recovered by adopting

$$\zeta(\omega) = \frac{\pm\mathrm{i}2\omega\sqrt{\omega}}{\omega^2 - \tilde{\omega}_0^2 - \mathrm{i}2\kappa\omega}\frac{\sqrt{\kappa}}{\sqrt{\pi}}, \quad (5.45)$$

where $\tilde{\omega}_0$ means the frequency of the damped oscillations and κ is the frequency-independent absorption rate. The relative permittivity for the original Huttner–Barnett microscopic model is of the form

$$\epsilon(\omega) = 1 - \frac{\omega_c^2}{\omega^2 - \tilde{\omega}_0^2 + \frac{\tilde{\omega}_0}{2}F(\omega)}, \quad (5.46)$$

where

$$F(\omega) \equiv \lim_{\varepsilon\to+0}\int_{-\infty}^{\infty}\frac{|V(\omega')|^2}{\omega' - \omega - \mathrm{i}\varepsilon}\,\mathrm{d}\omega'. \quad (5.47)$$

It is indicated that, in the Lorentz oscillator model, equation (5.46) yields the solution

$$F(\omega) = \mathrm{i}\frac{4\omega\kappa}{\tilde{\omega}_0}, \quad (5.48)$$

but the integral equation (5.47) with this left-hand side ($\equiv$ replaced by $=$) is not solvable to yield a coupling function $V(\omega)$. This is the main difficulty, because from the relation

$$\zeta(\omega) = \mathrm{i}\sqrt{\tilde{\omega}_0}\frac{\omega V(\omega)}{\omega^2 - \tilde{\omega}_0^2 + \frac{\tilde{\omega}_0}{2}F^*(\omega)}, \tag{5.49}$$

we obtain that

$$|V(\omega)|^2 = \frac{4\omega\kappa}{\pi\tilde{\omega}_0} \tag{5.50}$$

or we could determine

$$v(\omega) = \rho\sqrt{\frac{\tilde{\omega}_0}{\omega}}V(\omega) = \pm\rho\frac{\sqrt{\kappa}}{\sqrt{\pi}}. \tag{5.51}$$

This presents a restriction of the Huttner–Barnett microscopic model, which is nevertheless mentioned by Huttner and Barnett (1992a,b).

5.2.3 Dissipative Oscillator Medium

Let us recall that in the microscopic model, the electromagnetic-field operators are given by integrals both over k and ω. Huttner and Barnett (1992a) say themselves that they lose the relationship between the frequency and the wave vector **k**. This observation is relative to the macroscopic theories, where the Dirac delta function suitable for the expression of such a relationship is never replaced by another (generalized) function. The quantities such as (4.49), (4.51), (4.52), and (4.53) are formulated in dependence on the relative permittivity $\epsilon(\omega)$. Dutra and Furuya (1998b) suggest a simplification of the expression

$$\epsilon(\omega) = 1 + \frac{\omega_\mathrm{c}^2}{2\omega}\lim_{\varepsilon\to+0}\int_{-\infty}^{\infty}\frac{\xi(\omega')}{\omega' - \omega - \mathrm{i}\varepsilon}\,\mathrm{d}\omega', \tag{5.52}$$

where (Dutra and Furuya 1998b)

$$\xi(\omega) = \tilde{\omega}_0\omega\frac{|V(\omega)|^2}{|\omega^2 - \tilde{\omega}_0^2 + \frac{\tilde{\omega}_0}{2}F(\omega)|^2}, \tag{5.53}$$

with $F(\omega)$ defined by the relation (5.47). They try to calculate the relative permittivity for the Huttner–Barnett microscopic model by means of classical electrodynamics. The Huttner–Barnett approach is applied to the particular case, where the coupling strength is a slowly varying function of frequency.

In continuation of Dutra and Furuya (1997), a modified version of a simple model takes account of absorption. The inclusion of losses necessarily introduces a continuum of modes in the model. Consequences are minimized by the adoption of the standard elimination of the reservoir.

The interaction between the radiation field and the medium is described by a dipole-coupling Hamiltonian, where the canonically conjugated field is the displacement field instead of a minimal-coupling Hamiltonian, where the canonically conjugated field is the electric field. For simplicity, a Lorentzian shape for $|V(\omega)|^2$ is assumed, given by the relation

$$V(\omega) = \frac{\mathrm{i}\Delta}{\omega - \omega_0 + \mathrm{i}\Delta}\sqrt{\frac{\kappa}{\pi}}, \tag{5.54}$$

where $\omega_0 \gg \Delta \gg \kappa$. The Hamiltonian incorporating absorption is assumed to be

$$\begin{aligned}\hat{H}(t) &= \hbar\omega_{\mathrm{c}}\hat{a}^{\dagger}(t)\hat{a}(t) + \hbar\omega_0 \sum_{j=1}^{N} \hat{b}_j^{\dagger}(t)\hat{b}_j(t) - \frac{1}{\epsilon}\int \hat{D}(x,t)\hat{P}(x,t)\,\mathrm{d}x \\ &+ \hbar\int_0^{\infty}\sum_{j=1}^{N} \Omega\hat{W}_j^{\dagger}(\Omega,t)\hat{W}_j(\Omega,t)\,\mathrm{d}\Omega \\ &+ \hbar\int_0^{\infty}\sum_{j=1}^{N}\left[V(\Omega)\hat{b}_j^{\dagger}(t)\hat{W}_j(\Omega,t) + V^*(\Omega)\hat{W}_j^{\dagger}(\Omega,t)\hat{b}_j(t)\right]\mathrm{d}\Omega,\end{aligned} \tag{5.55}$$

where ω_{c} means the same as ω in the relations (5.6), (5.7), etc., $\hat{W}_j^{\dagger}(\Omega,t)$, $\hat{W}_j(\Omega,t)$ are reservoir creation and annihilation operators that commute with every other operator except for the commutation relation

$$[\hat{W}_j(\Omega,t), \hat{W}_{j'}^{\dagger}(\Omega',t)] = \delta_{jj'}\delta(\Omega-\Omega')\hat{1}. \tag{5.56}$$

Substituting the relations (5.6) and (5.8) into the Hamiltonian (5.55) and introducing appropriate collective operators, the total Hamiltonian becomes a sum of two uncoupled Hamiltonians

$$\hat{H}(t) = \hat{H}_1(t) + \hat{H}_2(t). \tag{5.57}$$

The second Hamiltonian $\hat{H}_2(t)$ describes $(N-1)$ damped collective excitations of the medium to which the field does not couple. The field and the single damped collective excitations of the medium, to which the field couples, are described by the Hamiltonian $\hat{H}_1(t)$ alone. This Hamiltonian is given by the relation

$$\hat{H}_1(t) = \hat{H}_{\mathrm{em}}(t) + \hat{H}_{\mathrm{mat}}(t) + \hat{H}_{\mathrm{int}}(t), \tag{5.58}$$

where

$$\hat{H}_{\mathrm{em}}(t) = \hbar\omega_{\mathrm{c}}\hat{a}^{\dagger}(t)\hat{a}(t) \tag{5.59}$$

is the Hamiltonian of the field,

$$\hat{H}_{\text{mat}}(t) = \hbar\omega_0 \hat{B}^\dagger(t)\hat{B}(t) + \hbar \int_0^\infty \Omega \hat{Y}^\dagger(\Omega, t)\hat{Y}(\Omega, t)\,\mathrm{d}\Omega$$
$$+ \hbar \int_0^\infty \left[V(\Omega)\hat{B}^\dagger(t)\hat{Y}(\Omega, t) + V^*(\Omega)\hat{Y}^\dagger(\Omega, t)\hat{B}(t)\right] \mathrm{d}\Omega \qquad (5.60)$$

is the Hamiltonian of medium, and

$$\hat{H}_{\text{int}}(t) = \hbar G(\omega_c)\left[\hat{a}^\dagger(t) + \hat{a}(t)\right]\left[\hat{B}(t) + \hat{B}^\dagger(t)\right] \qquad (5.61)$$

is their interaction Hamiltonian. The collective annihilation operators $\hat{B}(t)$ and $\hat{Y}(\Omega, t)$ are given by the relations

$$\hat{B}(t) = \sum_{j=1}^{N} \phi_j \hat{b}_j(t) \qquad (5.62)$$

and

$$\hat{Y}(\Omega, t) = \sum_{k=1}^{N} \phi_k \hat{W}_k(\Omega, t), \qquad (5.63)$$

where

$$\phi_j = \frac{g_j(\omega_c)}{G(\omega_c)} \qquad (5.64)$$

and $g_j(\omega_c)$ and $G(\omega_c)$ are given by equations (5.15) and (5.16), respectively, with ω replaced by ω_c. Dutra and Furuya (1998b) have a (conventional) strictly microscopic model, where the medium is not continuous, but discrete. They admit the practicality of the macroscopic average of the physical quantities. Following Robinson (1971, 1973), they understand the macroscopic averaging as filtering out of higher spatial frequencies.

A classical Hamiltonian is considered which is identical to the relation (5.58) except that $\hat{a}(t)$, $\hat{B}(t)$, and $\hat{Y}(\Omega, t)$ will be replaced by $\frac{a(t)}{\sqrt{\hbar}}$, $\frac{B(t)}{\sqrt{\hbar}}$, and $\frac{Y(\Omega,t)}{\sqrt{\hbar}}$. The standard elimination of the reservoir variables (which corresponds to the standard treatment in the quantum theory, but with the bonus that the reservoir not being initially excited leads to a great simplification) is performed and the real variables

$$\mathcal{D}(t) = a(t) + a^*(t), \qquad (5.65)$$
$$\mathcal{B}(t) = -\mathrm{i}[a(t) - a^*(t)], \qquad (5.66)$$
$$\mathcal{P}(t) = -\mathrm{i}[B(t) - B^*(t)], \qquad (5.67)$$
$$\mathcal{X}(t) = B(t) + B^*(t) \qquad (5.68)$$

are introduced. The variables $\mathcal{D}(t)$, $\mathcal{B}(t)$, and $\mathcal{X}(t)$ are related to the electric displacement field $D(x,t)$, the magnetic field $M(x,t)$, and the polarization field $P(x,t)$ by

$$D(x,t) = \sqrt{\frac{\epsilon_0 \omega_c}{L}} \sin\left(\frac{\omega_c}{c}x\right) \mathcal{D}(t), \tag{5.69}$$

$$M(x,t) = -c\sqrt{\frac{\epsilon_0 \omega_c}{L}} \cos\left(\frac{\omega_c}{c}x\right) \mathcal{B}(t), \tag{5.70}$$

$$P(x,t) = -2G'(\omega_c)\sqrt{\frac{\epsilon_0}{\omega_0 L}} \sin\left(\frac{\omega_c}{c}x\right) \mathcal{X}(t). \tag{5.71}$$

From the classical Hamiltonian, the classical equations of motion for $\mathcal{X}(t)$ and $\mathcal{P}(t)$ are obtained,

$$\frac{\mathrm{d}}{\mathrm{d}t}\mathcal{X}(t) = -\kappa \mathcal{X}(t) + \omega_0 \mathcal{P}(t), \tag{5.72}$$

$$\frac{\mathrm{d}}{\mathrm{d}t}\mathcal{P}(t) = -\omega_0 \mathcal{X}(t) - \kappa \mathcal{P}(t) + 2G\mathcal{D}(t). \tag{5.73}$$

These equations along with equations of motion for $\mathcal{D}(t)$ and $\mathcal{B}(t)$, which are not given here, admit a solution oscillating at the frequency ω, with the property that $\frac{\mathrm{d}\mathcal{X}(t)}{\mathrm{d}t} = -\mathrm{i}\omega\mathcal{X}(t)$, $\frac{\mathrm{d}\mathcal{P}(t)}{\mathrm{d}t} = -\mathrm{i}\omega\mathcal{P}(t)$, $\frac{\mathrm{d}\mathcal{D}(t)}{\mathrm{d}t} = -\mathrm{i}\omega\mathcal{D}(t)$. All the solutions have the property

$$\mathcal{X}(t) = -\frac{2\omega_0 G(\omega_c)}{\omega_0^2 + \kappa^2 - \omega^2 - \mathrm{i}2\kappa\omega}\mathcal{D}(t). \tag{5.74}$$

From the relation

$$D(x,t) = \epsilon(\omega)[D(x,t) - P(x,t)], \tag{5.75}$$

we obtain that

$$\epsilon(\omega) = 1 + \frac{4[G'(\omega_c)]^2}{\omega_0'^2 - \omega^2 - \mathrm{i}2\kappa\omega}, \tag{5.76}$$

where

$$\omega_0'^2 = \omega_0^2 + \kappa^2 - 4[G'(\omega_c)]^2 \tag{5.77}$$

is the modified resonance frequency of the medium.

The further topic in (Dutra and Furuya 1998b) is essentially the relation (4.76) due to Gruner and Welsch (1995). In particular, it is shown that also in the case of the simple microscopic model of the medium used by Dutra and Furuya (1998b), it is possible to diagonalize first $\hat{H}_{\mathrm{mat}}(t)$ and then the total Hamiltonian (5.58). The diagonal form of the Hamiltonian $\hat{H}_{\mathrm{mat}}(t)$ is achieved in terms of the continuous operators $\hat{B}(\nu, t)$,

$$\hat{B}(\nu, t) = \alpha(\nu)\hat{B}(t) + \int_0^\infty \beta(\nu, \Omega)\hat{Y}(\Omega, t)\,\mathrm{d}\Omega, \tag{5.78}$$

where $\alpha(\nu)$ and $\beta(\nu, \Omega)$ are some coefficients. The diagonal form of the total Hamiltonian (5.58) is achieved in terms of the continuous operators $\hat{A}(\omega, t)$,

$$\begin{aligned}\hat{A}(\omega, t) &= \alpha_1(\omega)\hat{a}(t) + \alpha_2(\omega)\hat{a}^\dagger(t) \\ &\quad + \int_0^\infty \left[\beta_1(\omega, \nu)\hat{B}(\nu, t) + \beta_2(\omega, \nu)\hat{B}^\dagger(\nu, t)\right] \mathrm{d}\nu,\end{aligned} \tag{5.79}$$

where $\alpha_1(\omega)$, $\alpha_2(\omega)$, $\beta_1(\omega, \nu)$, $\beta_2(\omega, \nu)$ are some coefficients. Suitable operators, namely, those of the electric displacement field and of the macroscopic electric strength field are defined, such that

$$\begin{aligned}\hat{\hat{E}}(x, t) &= \frac{1}{\epsilon_0\epsilon(\omega)}\hat{D}(x, t) + 2\sqrt{\frac{\hbar}{\omega_0\epsilon_0 L}}\sin\left(\frac{\omega_c}{c}x\right) \\ &\quad \times G'(\omega_c)\left[\int_0^\infty \alpha^*(\omega)\hat{A}(\omega, t)\,\mathrm{d}\omega + \mathrm{H.\,c.}\right].\end{aligned} \tag{5.80}$$

The difference from the relation (4.76) arises, because Dutra and Furuya (1998b) have only a single mode of the field, use a dipole-coupling Hamiltonian instead of a minimal-coupling Hamiltonian, and have defined their field operators in terms of different quadratures of the annihilation and creation operators.

Using the microscopic approach, Hillery and Mlodinow (1997) devoted themselves to the standard optical interactions and derived an effective Hamiltonian describing counterpropagating modes in a nonlinear medium. On considering multipolar coupled atoms interacting with an electromagnetic field, a quantum theory of dispersion has been obtained whose dispersion relations are equivalent to the standard Sellmeir equations for the description of a dispersive transparent medium (Drummond and Hillery 1999).

Independently, the theory of light propagation in a Bose–Einstein condensate and a zero-temperature noninteracting Fermi–Dirac gas has been developed (Javanainen et al. 1999). Ruostekoski (2000) has theoretically studied the optical properties of a Fermi–Dirac gas in the presence of a superfluid state. He also considered diffraction of atoms by means of light-stimulated transitions of photons between two intersecting laser beams. Optical properties could possibly signal the presence of the superfluid state and determine the value of the Bardeen–Cooper–Schrieffer parameter in dilute atomic Fermi–Dirac gases. Crenshaw and Bowden (2000a,b) have derived effects of the Lorentz local fields on spontaneous emission in dielectric media. Bloch–Langevin operator equations have been obtained for two-level atoms embedded in a host dielectric medium using the macroscopic and microscopic quantizations and the macroscopic formulation has been criticized (Crenshaw and Bowden 2002).

Crenshaw (2003) has presented a real-space derivation of the macroscopic quantum Hamiltonian from the microscopic quantum electrodynamic model of a dielectric. Crenshaw (2004) has transformed the macroscopic real-space Hamiltonian to momentum space. The microscopic model has been reduced to the macroscopic Hamiltonian (Ginzburg 1940) by way of the reciprocal-space model of the field in a dielectric (Hopfield 1958).

Cerboneschi et al. (2002) have shown that the electromagnetically induced transparency is related to very small group velocities for the probe pulse also in an open system. The modifications of the atomic momentum produced by laser interactions have been taken into account.

Tanaka et al. (2003) have observed a negative delay (positive advance) of the peak of an optical pulse in case the pulse is tuned to the anomalous dispersion region. The negative velocity of the peak is not the velocity of energy flow.

5.3 Single-Photon Models

Guo (2007) has discussed propagation of one incident photon through a medium as the multiple-scattering process from the medium. The medium is assumed to be an ensemble of identical two-level atoms. Interaction with a two-level test atom outside the medium is considered as well. It is assumed that all the atoms are in the ground state when $t = 0$. Initially, there is one photon in a mode. The system is treated in the Schrödinger picture. An integral form of the evolution is presented.

As in Guo (2005), the kernel of the integral transformation is expanded. It is assumed that counter-rotating terms are superfluous and will be ignored. Photon propagation through the atomic ensemble results in states S_1, $S_{1\to 2}$, $S_{1\to 2\to 3}$,..., which come from the first-order scattering of the incident photon, from the second- and third-order scatterings of the incident photon, respectively. The photon is scattered into the states (modes) $|1_\alpha\rangle$, $|1_\beta\rangle$, $|1_\gamma\rangle$,... in the time order. The author refers to classical works (Mandel and Wolf 1995) and (Loudon 2000) for an ambiguity of the photon phase. But a possibility for the photon to become entangled with the atom has been declared.

The author expresses the time-dependent probability amplitude A for the test atom to transit from its ground state to an excited state. On the assumption of isotropy and uniformity of the medium, the conservation of polarization and wavenumber of the photon is derived. On these conditions A may be written in a form reminding of the expression for the electric field $E(\mathbf{r}, t)$ in Guo (2002). The factor $4\pi\alpha_{\text{atom}}k^2 n_0$ reappears as $4\pi n k_0^2 P_{\text{atom}}$, where $k_0 = \frac{\omega_0}{c}$, $\hbar\omega_0$ is the energy difference between the energy states of a two-level medium atom, n is the density of the atoms, and P_{atom} is the resonant component of the polarizability of the atoms in the ensemble.

Berman (2007) has considered the problem of a source atom radiating into a medium of dielectric atoms using a microscopic model. The model system consists of a source atom embedded in a dielectric. The source atom is modelled as a two-level atom with a ground, lower, state $|1\rangle$ and an excited, upper, state $|2\rangle$. These states are separated in frequency by ω_0. The dielectric is contained within a sphere of radius R_0, $\omega_0 \frac{R_0}{c} \gg 1$. This medium consists of a uniform distribution of atoms. A bath density may be introduced and N is let to denote this density. Bath atom j is modelled as a four-level atom in the inverted tripod configuration with a ground state $|g\rangle^{(j)}$ and three excited states $|m\rangle^{(j)}$, $m = -1, 0, 1$. The frequency separation between the ground state and any of the three excited states is ω. It is assumed that $\frac{|\Delta|}{\omega} \ll 1$, where $\Delta = \omega_0 - \omega$, $\Delta \neq 0$. The positive frequency component of the electric-field operator at the space point $\mathbf{R}$, $|\mathbf{R}| = R$, is

$$\hat{\mathbf{E}}_+(\mathbf{R}) = \mathrm{i} \sum_{\mathbf{k},\lambda} \sqrt{\frac{\hbar\omega_k}{2\epsilon_0 V}} \mathbf{e}_{\mathbf{k}}^{(\lambda)} \hat{a}_{\mathbf{k}\lambda} e^{\mathrm{i}\mathbf{k}\cdot\mathbf{R}}, \tag{5.81}$$

where $\hat{a}_{\mathbf{k}\lambda}$ is an annihilation operator for a photon having the propagation vector $\mathbf{k}$ and the polarization $\mathbf{e}_{\mathbf{k}}^{(\lambda)}$, $\omega_k = kc$, V is the quantization volume, and

$$\begin{aligned} \mathbf{e}_{\mathbf{k}}^{(1)} &\equiv \mathbf{e}^{(1)}(\mathbf{k}) = \cos(\theta_k)\cos(\phi_k)\mathbf{e}_x + \cos(\theta_k)\sin(\phi_k)\mathbf{e}_y - \sin(\theta_k)\mathbf{e}_z, \\ \mathbf{e}_{\mathbf{k}}^{(2)} &\equiv \mathbf{e}^{(2)}(\mathbf{k}) = -\sin(\phi_k)\mathbf{e}_x + \cos(\phi_k)\mathbf{e}_y \end{aligned} \tag{5.82}$$

are unit polarization vectors, with

$$\begin{aligned} \theta_k &\equiv \theta(\mathbf{k}) = \cos^{-1}\left(\mathbf{e}_z \cdot \frac{\mathbf{k}}{|\mathbf{k}|}\right), \\ \phi_k &\equiv \phi(\mathbf{k}) = \arg\left[(\mathbf{e}_x + \mathrm{i}\mathbf{e}_y)\cdot\mathbf{k}\right]. \end{aligned} \tag{5.83}$$

In relation (5.81) the time is not written, since the Schrödinger approach has been adopted. The source atom is excited by a pulse of a classical driving field, and the average field amplitude and the intensity radiated by the source atom are determined to the first order in the dielectric density N. The expectation value of the field is denoted as $\langle\hat{\mathbf{E}}_+(\mathbf{R}, t)\rangle$, where the time is given along with the position again. This expectation value is

$$\langle\hat{\mathbf{E}}_+(\mathbf{R}, t)\rangle = \mathrm{i} \sum_{\mathbf{k},\lambda} \sqrt{\frac{\hbar\omega_k}{2\epsilon_0 V}} \mathbf{e}_{\mathbf{k}}^{(\lambda)} e^{\mathrm{i}(\mathbf{k}\cdot\mathbf{R}-\omega_k t)} b_{\mathbf{k}\lambda}(t) b_{1,0}^*(t), \tag{5.84}$$

where $b_{1,0}(t)$ is the amplitude to find all atoms in their ground states and to find no photons in the field and $b_{\mathbf{k}\lambda}(t)$ is the amplitude to find all atoms in their ground states and a photon having the wave vector $\mathbf{k}$ and the polarization λ in the field. It is assumed that the exciting field is weak enough to approximate

$$b_{1,0}(t) \approx 1. \tag{5.85}$$

Berman (2007) has found the average field

$$\langle \hat{\mathbf{E}}_+(\mathbf{R},t)\rangle \approx -\frac{\mu\omega_0^2(\sin\theta)\mathbf{e}^{(1)}e^{\mathrm{i}(k_0R-\omega_0t)}}{4\pi\epsilon_0c^2R} \times\left\{1+2\mathrm{i}\frac{1}{\omega_0}\frac{\partial}{\partial t}+\mathrm{i}\delta nk_0R_0-\delta n\frac{R_0}{c}\frac{\partial}{\partial t}\right\}b_{2,0}^{(0)}\left(t-\frac{R}{c}\right), \tag{5.86}$$

where $\theta\equiv\theta(\mathbf{R})$, $\mathbf{e}^{(1)}\equiv\mathbf{e}^{(1)}(\mathbf{R})$, $k_0=\frac{\omega_0}{c}$, $\delta n=-\frac{1}{4\pi\epsilon_0}\left(\frac{2\pi N\mu'^2}{\hbar\Delta}\right)$, μ (μ') is a characteristic of the source atom (a bath atom), and $b_{2,0}^{(0)}(t)$ is the excited state amplitude in the absence of the medium (Berman 2004).

Macroscopic theories are expected to give

$$\langle \hat{\mathbf{E}}_+(\mathbf{R},t)\rangle_{\mathrm{d}} = \frac{\mu(\sin\theta)\mathbf{e}^{(1)}e^{\mathrm{i}k_0(R+\delta nR_0)}}{4\pi\epsilon_0c^2R} \times\frac{\partial^2}{\partial t^2}\left[b_{2,0}^{(0)}\left(t-\frac{R}{c}-\frac{\delta nR_0}{c}\right)e^{-\mathrm{i}\omega_0t}\right], \tag{5.87}$$

where the subscript d stands for dielectric, $\delta n=n-1$, and n is the index of refraction in the medium. If $|k_0\delta nR_0|\ll 1$ and $|\left(\frac{\delta nR_0}{c}\right)\frac{\partial}{\partial t}b_{2,0}^{(0)}\left(t-\frac{R}{c}\right)|\ll 1$, this expression can be expanded to the first order in δn as

$$\langle \hat{\mathbf{E}}_+(\mathbf{R},t)\rangle_{\mathrm{d}} \approx -\frac{\mu\omega_0^2(\sin\theta)\mathbf{e}^{(1)}e^{\mathrm{i}(k_0R-\omega_0t)}}{4\pi\epsilon_0c^2R} \times\left\{1+2\mathrm{i}\frac{1}{\omega_0}\frac{\partial}{\partial t}+\mathrm{i}\delta nk_0R_0-3\delta n\frac{R_0}{c}\frac{\partial}{\partial t}\right\}b_{2,0}^{(0)}\left(t-\frac{R}{c}\right), \tag{5.88}$$

which does not differ seriously from relation (5.86). Using relation (5.85), Berman (2007) has derived that the average field intensity is equal to

$$\langle \hat{\mathbf{E}}_-(\mathbf{R},t)\cdot\hat{\mathbf{E}}_+(\mathbf{R},t)\rangle = |\langle\hat{\mathbf{E}}_+(\mathbf{R},t)\rangle|^2, \tag{5.89}$$

where $\hat{\mathbf{E}}_-(\mathbf{R},t)=\hat{\mathbf{E}}_+^\dagger(\mathbf{R},t)$ and $\langle\hat{\mathbf{E}}_+(\mathbf{R},t)\rangle$ is given by relation (5.86). He has given a microscopic derivation of the fact that the field intensity propagates with a reduced velocity in the medium. He has tried to elucidate the nature of the retardation mechanism and has shown that the modifications of the emitted field are closely correlated with nearly forward scattering in the medium.

Chapter 6
Periodic and Disordered Media

The periodic and disordered media have first been studied in the condensed-matter physics. In this physics field, electrons are paid attention to. The concepts of periodic and disordered media depend on whether electrons or photons are investigated. Nevertheless there is an analogy between the wave function of a single electron and the classical electromagnetic field. Therefore it was possible to achieve many results for periodic and disordered media in photonics using this similarity.

The periodic media are conceptually simpler. Here we shall mainly speak of the macroscopic approach to quantization of the electromagnetic field in a periodic medium, but we shall also mention the papers, which have utilized specific approaches. Finite periodic media can be coupling media of free-space modes. In applications the corrugated waveguides are important. Photonic crystals are infinite or finite periodic media. The literature on one-dimensional periodic media abounds, since also fabrication of such media is simpler.

We shall mention the quantization of the electromagnetic field in a disordered medium mainly in connection with various physical studies. The macroscopic approach to the quantization of the electromagnetic field suffices usually, but a description of the disordered or random medium is not easy.

We shall cite the papers, whose authors have restricted themselves to the quantum input–output relations and the detection theory. This is also related to application of results of further fields of the quantum physics. Many papers have reported the random lasers. Even though we intend to review rather the theory, we see that we can only provide a partial review. In the essence of the matter, it is that much theoretical work is not published as the optical physics.

6.1 Quantization in Periodic Media

Media whose dielectric constant is periodic are fabricated as microstructured materials with promising photonic band-gap structures. A number of methods for studying these media originate from the solid-state theory, where they are used in investigation of ordinary, electronic crystals. This is a possible explanation why periodic media are called photonic crystals.

A. Lukš, V. Peřinová, *Quantum Aspects of Light Propagation*,
DOI 10.1007/b101766_6,

The idea of photonic crystals was tested by the early experiments of Yablonovitch and Gmitter (1989). Since then the periodic media have been treated not only in the books devoted to optics (Born and Wolf 1999, Yeh 1988, Pedrotti and Pedrotti 1993) but also in monographs on the photonic crystals, e.g. (Joannapoulos 1995).

For illustration, we will treat a one-dimensional photonic crystal as Mishra and Satpathy (2003), whose contribution consists in an interesting method of solution, which is not reproduced here. We combine the assumption of a periodic medium with that of a rectangular waveguide.

Waveguides are useful optical devices. An optical circuit can be made using them and various optical couplers and switches. Classical theory of optical waveguides and couplers has been elaborated in 1970s (Yariv and Yeh 1984), nonclassical light has been proposed as a source for improving performance, and the quantum theory has been gaining importance. Recently, quantum entanglement has been pointed out as another resource. Quantum descriptions may be very simple, but essentially, they ought to be based on a perfect knowledge of quantization. By way of paradox, quantization is based on classical normal modes. Therefore, it is appropriate to concentrate ourselves on normal modes of rectangular mirror waveguide. It will be assumed that the waveguide is filled with homogeneous refractive medium. As this has the only effect of changing the speed of light, it will be assumed that a nonhomogeneity in a finite segment of the waveguide is present to model a coupler.

6.1.1 Classical Description of Electromagnetic Field

Vast literature has been devoted to the solution of the Maxwell equations, and their value for the wave and quantum optics cannot be denied. Depending on the system of physical units used, the Maxwell equations have several forms. Let us mention only two of them, appropriate to the SI units and the Gaussian units. The time-dependent vector fields, which enter these equations, are $\mathbf{E}(x, y, z, t)$, the electric strength vector field, and $\mathbf{B}(x, y, z, t)$, the magnetic induction vector field. In fact, other two fields are used, but they can also be eliminated through the so-called constitutive relations. Saying this we make some simplifying assumptions, but we are tacit about them. The so-called monochromaticity assumption

$$\begin{aligned}\mathbf{E}(x, y, z, t) &= \mathbf{E}(x, y, z; \omega) \exp(\mathrm{i}\omega t),\\ \mathbf{B}(x, y, z, t) &= \mathbf{B}(x, y, z; \omega) \exp(\mathrm{i}\omega t)\end{aligned} \tag{6.1}$$

allows one to treat the time-independent Maxwell equations.

As announced, we restrict ourselves to a rectangular mirror waveguide. We assume that it has an infinite length, a width $2a_x$, and the height $2a_y$. The coordinate system is chosen so that the z-axis is the axis of the waveguide and the x-, y-axes are parallel with sides of the waveguide.

We deal with nonvanishing solutions of the time-independent Maxwell equations

$$\nabla \times \frac{1}{\mu(x, y, z; \omega)} \mathbf{B}(x, y, z; \omega) - \mathrm{i}\omega\epsilon(x, y, z; \omega)\mathbf{E}(x, y, z; \omega) = \mathbf{0}, \tag{6.2}$$

$$\nabla \times \mathbf{E}(x, y, z; \omega) + i\omega \mathbf{B}(x, y, z; \omega) = \mathbf{0}, \tag{6.3}$$
$$\nabla \cdot [\epsilon(x, y, z; \omega)\mathbf{E}(x, y, z; \omega)] = 0, \tag{6.4}$$
$$\nabla \cdot \mathbf{B}(x, y, z; \omega) = 0, \tag{6.5}$$

where $\mathbf{E}(x, y, z; \omega)$ and $\mathbf{B}(x, y, z; \omega)$ are vector-valued functions in a domain

$$G = \{(x, y, z) : -a_x < x < a_x, -a_y < y < a_y, -\infty < z < \infty\} \tag{6.6}$$

and $\omega > 0$ is a parameter. The desired solutions are to obey the boundary conditions

$$\mathbf{n}(x, y, z) \times \mathbf{E}(x, y, z; \omega) = \mathbf{0}, \tag{6.7}$$
$$\mathbf{n}(x, y, z) \cdot \mathbf{B}(x, y, z; \omega) = 0, \tag{6.8}$$

where $\mathbf{n}(x, y, z)$ is any unit outer-pointing normal vector at the point $(x, y, z) \in \partial G$, with

$$\partial G = \{(x, y, z) : \left(-a_x \leq x \leq a_x \wedge |y| = a_y \wedge -\infty < z < \infty\right) \\ \vee \left(|x| = a_x \wedge -a_y \leq y \leq a_y \wedge -\infty < z < \infty\right)\}. \tag{6.9}$$

The boundary conditions (6.7) and (6.8) are a formal expression of the fact that the walls of the waveguide are perfect mirrors.

Here $\mu(x, y, z; \omega) = \mu_0$, $\epsilon(x, y, z; \omega)$ is a function defined up to a finite number of z-values such that

$$\begin{aligned} \epsilon(x, y, z; \omega) &= \epsilon_0\epsilon_{r0}, \quad \text{for } z < 0, \quad z > L, \\ &= \epsilon_0\bar{\epsilon}_r(z), \quad \text{for } 0 < z < L, \end{aligned} \tag{6.10}$$

with $\mu_0 > 0$, $\mu_0 = 4\pi \times 10^{-7}$ Hm^{-1} the free-space magnetic permeability, $\epsilon_0 > 0$ the free-space electric permittivity, $\epsilon_{r0} > 0$, $\bar{\epsilon}_r(z)$ are relative electric permittivities of the medium. The medium electric permittivity $\bar{\epsilon}(z) = \epsilon_0\bar{\epsilon}_r(z)$ has a period Λ, $\Lambda \mid L$, or $\bar{\epsilon}(z) = \bar{\epsilon}(z + \Lambda)$, and is a positive function, $\int_0^L \epsilon(z)\, dz = L\epsilon_0\epsilon_{r0}$. It is a formal expression of the idea that a plate with a periodically modulated permittivity is contained in the waveguide.

6.1.2 Modal Functions

(i) Homogeneous refractive medium

We assume that the waveguide is filled with a homogeneous nonmagnetic refractive medium, which is also nondispersive and lossless for simplicity. This assumption holds on infinite intervals $(-\infty, 0)$ and (L, ∞) in the z-coordinate. On the finite interval $(0, L)$, the medium is not homogeneous, but it is periodic. On average, its electric permittivity equals to that of the homogeneous medium assumed.

For illustration, we will consider examples where solutions have finite norms, in Section 6.1.4. Let us assume that the relation $\epsilon(x, y, z; \omega) = \epsilon_0 \epsilon_{r0}$ holds everywhere in G. We will express the solution in the form

$$\begin{aligned} \mathbf{E}(x, y, z; \omega) &= \mathbf{E}(x, y) \exp(-ik_z z), \\ \mathbf{B}(x, y, z; \omega) &= \mathbf{B}(x, y) \exp(-ik_z z), \end{aligned} \tag{6.11}$$

with $k_z \neq 0$. In analogy with the electromagnetic-field theory, from equations (6.2), (6.3), (6.4), and (6.5) we derive the time-independent wave equation

$$\left(\Delta + \frac{\omega^2}{v^2} \right) \mathbf{C} = 0, \tag{6.12}$$

where

$$v = \frac{1}{\sqrt{\epsilon_0 \epsilon_{r0} \mu_0}}, \tag{6.13}$$

and $\mathbf{C} \equiv \mathbf{C}(x, y, z; \omega)$ stands for $\mathbf{E}$ and $\mathbf{B}$ substitutionally. Respecting (6.11), we may rewrite (6.12) in the form

$$\left(\frac{\partial^2}{\partial x^2} + \frac{\partial^2}{\partial y^2} \right) \mathbf{C} + \left(\frac{\omega^2}{v^2} - k_z^2 \right) \mathbf{C} = 0. \tag{6.14}$$

Introducing the notation $E_r(x, y)$, $B_r(x, y)$, $r = x, y, z$, for the components of the vectors $\mathbf{E}(x, y)$, $\mathbf{B}(x, y)$, respectively, we have (Greiner 1998, p. 366)

$$\begin{aligned} E_x(x, y) &= \alpha \cos\left(\frac{m\pi}{2a_x}(x + a_x) \right) \sin\left(\frac{n\pi}{2a_y}(y + a_y) \right), \\ E_y(x, y) &= \beta \sin\left(\frac{m\pi}{2a_x}(x + a_x) \right) \cos\left(\frac{n\pi}{2a_y}(y + a_y) \right), \\ E_z(x, y) &= i\gamma \sin\left(\frac{m\pi}{2a_x}(x + a_x) \right) \sin\left(\frac{n\pi}{2a_y}(y + a_y) \right), \end{aligned} \tag{6.15}$$

$$\begin{aligned} B_x(x, y) &= i\alpha' \sin\left(\frac{m\pi}{2a_x}(x + a_x) \right) \cos\left(\frac{n\pi}{2a_y}(y + a_y) \right), \\ B_y(x, y) &= i\beta' \cos\left(\frac{m\pi}{2a_x}(x + a_x) \right) \sin\left(\frac{n\pi}{2a_y}(y + a_y) \right), \\ B_z(x, y) &= \gamma' \cos\left(\frac{m\pi}{2a_x}(x + a_x) \right) \cos\left(\frac{n\pi}{2a_y}(y + a_y) \right), \end{aligned} \tag{6.16}$$

where $m, n = 0, 1, \ldots, \infty$. Equation (6.14) yields the relation among m, n, k_z, and ω,

$$\left(\frac{m\pi}{2a_x}\right)^2 + \left(\frac{n\pi}{2a_y}\right)^2 + k_z^2 = \frac{\omega^2}{v^2}. \tag{6.17}$$

No solution of this kind exists if $\omega < \omega_g$, where

$$\omega_g = \omega_{mn} = v\sqrt{\left(\frac{m\pi}{2a_x}\right)^2 + \left(\frac{n\pi}{2a_y}\right)^2}. \tag{6.18}$$

We can specify two linearly independent solutions for $m, n = 0, 1, \ldots, \infty$, $m + n \geq 1$,

$$\begin{aligned} \alpha_{\mathrm{TE}} &= \mathrm{i}\omega\frac{k_y}{k_x^2 + k_y^2}\gamma'_{\mathrm{TE}}, \quad \alpha'_{\mathrm{TE}} = \frac{k_x k_z}{k_x^2 + k_y^2}\gamma'_{\mathrm{TE}}, \\ \beta_{\mathrm{TE}} &= -\mathrm{i}\omega\frac{k_x}{k_x^2 + k_y^2}\gamma'_{\mathrm{TE}}, \quad \beta'_{\mathrm{TE}} = \frac{k_y k_z}{k_x^2 + k_y^2}\gamma'_{\mathrm{TE}}, \\ \gamma_{\mathrm{TE}} &= 0, \quad \gamma'_{\mathrm{TE}} = \gamma'_{\mathrm{TE}}, \end{aligned} \tag{6.19}$$

$$\begin{aligned} \alpha_{\mathrm{TM}} &= \frac{k_x k_z}{k_x^2 + k_y^2}\gamma_{\mathrm{TM}}, \quad \alpha'_{\mathrm{TM}} = \frac{1}{v^2}\frac{\mathrm{i}\omega k_y}{k_x^2 + k_y^2}\gamma_{\mathrm{TM}}, \\ \beta_{\mathrm{TM}} &= \frac{k_y k_z}{k_x^2 + k_y^2}\gamma_{\mathrm{TM}}, \quad \beta'_{\mathrm{TM}} = \frac{1}{v^2}\frac{-\mathrm{i}\omega k_x}{k_x^2 + k_y^2}\gamma_{\mathrm{TM}}, \\ \gamma_{\mathrm{TM}} &= \gamma_{\mathrm{TM}}, \quad \gamma'_{\mathrm{TM}} = 0. \end{aligned} \tag{6.20}$$

Here TE means transverse electric and TM transverse magnetic. For $m = 0$, a TE solution exists, but no TM solutions exist; for $n = 0$, the same occurs and otherwise, both solutions exist. In (6.19) and (6.20), k_x, k_y are abbreviations, $k_x \equiv \frac{m\pi}{2a_x}$, $k_y \equiv \frac{n\pi}{2a_y}$. Let us remark that γ'_{TE} and γ_{TM} are complex parameters.

(ii) Plate with a periodically modulated permittivity

We generalize the first of relations (6.11) to the form

$$\mathbf{E}(x, y, z; \omega) = \mathbf{E}(x, y)u(z), \tag{6.21}$$

where $u(z)$ is an unknown function, but $\mathbf{E}(x, y)$ is given by relations (6.15). We distinguish the case of a TE solution from the case of a TM solution. In the latter case, this form cannot be required, since in the case of a TM solution, $\mathbf{E}(x, y)$ depends on k_z. In fact the components k_x, k_y are meaningful in the inhomogeneous medium with the z-dependence of the dielectric constant, not the component k_z.

In the case of a TE solution, the relation

$$\nabla \cdot \mathbf{E}(\mathbf{x}; \omega) = 0 \tag{6.22}$$

holds, where $\mathbf{x} \equiv (x, y, z)$. The Maxwell equations are equivalent to a Helmholtz equation. It is derived that the function $u(z)$ obeys the ordinary differential equation

$$\frac{\mathrm{d}^2}{\mathrm{d}z^2}u(z) + \tilde{k}_z^2(z)u(z) = 0, \quad -\infty < z < \infty, \tag{6.23}$$

where $\tilde{k}_z^2(z) = \tilde{k}^2(z) - (k_x^2 + k_y^2)$, with $\tilde{k}^2(z) \equiv \frac{\omega^2}{c^2}\epsilon_\mathrm{r}(z)$. A solution will be "sewn" of a solution of the equation

$$\frac{\mathrm{d}^2}{\mathrm{d}z^2}u(z) + k_z^2 u(z) = 0, \; -\infty < z < \infty \tag{6.24}$$

and a solution of the equation

$$\frac{\mathrm{d}^2}{\mathrm{d}z^2}u(z) + \bar{k}_z^2(z)u(z) = 0, \; -\infty < z < \infty, \tag{6.25}$$

where $\bar{k}_z^2(z) = \bar{k}^2(z) - (k_x^2 + k_y^2)$, with $\bar{k}^2(z) \equiv \frac{\omega^2}{c^2}\bar{\epsilon}_\mathrm{r}(z)$.

The general solution of equation (6.25) has the form

$$u(z) = c_\mathrm{B} f_\mathrm{B}(z)e^{\mathrm{i}k_\mathrm{B}z} + c_\mathrm{F} f_\mathrm{F}(z)e^{-\mathrm{i}k_\mathrm{B}z}, \tag{6.26}$$

where c_B and c_F are arbitrary complex numbers, $e^{\mathrm{i}k_\mathrm{B}\Lambda}$ is a Bloch factor, with either $0 < k_\mathrm{B} < \frac{\pi}{\Lambda}$ or $\mathrm{Im}\, k_\mathrm{B} > 0$. The functions $f_\mathrm{B}(z)$ and $f_\mathrm{F}(z)$ fulfil the conditions

$$f_\mathrm{B}(0) = f_\mathrm{F}(0) = 1 \tag{6.27}$$

and have the period Λ. On differentiating relation (6.26) with respect to z, it can be seen that also $\frac{\mathrm{d}}{\mathrm{d}z}u(z)$ has the same form in principle. And so the Bloch functions

$$u(z) = f_\mathrm{B}(z)e^{\mathrm{i}k_\mathrm{B}z}, \quad f_\mathrm{F}(z)e^{-\mathrm{i}k_\mathrm{B}z} \tag{6.28}$$

are to be determined as the solutions of Equation (6.25) having along with the respective derivatives some initial data $u(0)$, $\left.\frac{\mathrm{d}u}{\mathrm{d}z}\right|_0$ and obeying the conditions

$$u(\Lambda) = \lambda u(0), \quad \left.\frac{\mathrm{d}u}{\mathrm{d}z}\right|_{z=\Lambda} = \lambda \left.\frac{\mathrm{d}u}{\mathrm{d}z}\right|_{z=0}, \tag{6.29}$$

where λ is a complex number. From the fact that equation (6.25) does not contain $\frac{\mathrm{d}}{\mathrm{d}z}u(z)$, it can be derived that the transition matrix from the initial data $u(0)$, $\left.\frac{\mathrm{d}u(z)}{\mathrm{d}z}\right|_{z=0}$ to the respective values at $z = \Lambda$ is unimodular. And so $\lambda = \exp(\pm\mathrm{i}k_\mathrm{B}\Lambda)$. Since the transition matrix is real, k_B is either real or pure imaginary.

For illustration we may consider a Kronig–Penney dielectric (Mishra and Satpathy 2003). In this case, it holds

$$\bar{k}_z(z) = \begin{cases} k_{1z}, \; 0 < z < \frac{\Lambda}{2}, \\ k_{2z}, \; \frac{\Lambda}{2} < z < \Lambda. \end{cases} \tag{6.30}$$

The equation for k_{B} is obtained in the form

$$\begin{aligned} \cos(k_{\mathrm{B}}\Lambda) = \cos\left(k_{1z}\frac{\Lambda}{2}\right)\cos\left(k_{2z}\frac{\Lambda}{2}\right) \\ -\frac{1}{2}\left(\frac{k_{1z}}{k_{2z}} + \frac{k_{2z}}{k_{1z}}\right)\sin\left(k_{1z}\frac{\Lambda}{2}\right)\sin\left(k_{2z}\frac{\Lambda}{2}\right). \end{aligned} \tag{6.31}$$

We assume that the general solution of equation (6.23) has the form

$$u(z) = \begin{cases} c_{\mathrm{B}}^{(-)} e^{\mathrm{i}k_z z} + c_{\mathrm{F}}^{(-)} e^{-\mathrm{i}k_z z}, \quad \text{for } z < 0, \\ \\ c_{\mathrm{B}}^{(0)} f_{\mathrm{B}}(z) e^{\mathrm{i}k_{\mathrm{B}} z} + c_{\mathrm{F}}^{(0)} f_{\mathrm{F}}(z) e^{-\mathrm{i}k_{\mathrm{B}} z}, \quad \text{for } 0 < z < L, \\ \\ c_{\mathrm{B}}^{(+)} e^{\mathrm{i}k_z z} + c_{\mathrm{F}}^{(+)} e^{-\mathrm{i}k_z z}, \quad \text{for } z > L. \end{cases} \tag{6.32}$$

Here $f_{\mathrm{B}}(z)$, $f_{\mathrm{F}}(z)$ are periodic functions such that $f_{\mathrm{B}}(0) = f_{\mathrm{F}}(0) = 1$. The solution along with its first derivative is continuous at $z = 0, L$, which can be expressed as relations between $c_{\mathrm{B}}^{(-)}$, $c_{\mathrm{F}}^{(-)}$, $c_{\mathrm{B}}^{(0)}$, $c_{\mathrm{F}}^{(0)}$, $c_{\mathrm{B}}^{(+)}$, $c_{\mathrm{F}}^{(+)}$. These coefficients are arbitrary otherwise.

The continuity of the (original) function at $z = 0, L$ can be written as

$$c_{\mathrm{B}}^{(-)} + c_{\mathrm{F}}^{(-)} = c_{\mathrm{B}}^{(0)} + c_{\mathrm{F}}^{(0)}, \; c_{\mathrm{B}}^{(0)} e^{ik_{\mathrm{B}}L} + c_{\mathrm{F}}^{(0)} e^{-ik_{\mathrm{B}}L} = c_{\mathrm{B}}^{(+)} e^{ik_z L} + c_{\mathrm{F}}^{(+)} e^{-ik_z L}. \tag{6.33}$$

The continuity of the derivative of the function at $z = 0, L$ can be expressed as the relations

$$\begin{aligned} k_z c_{\mathrm{B}}^{(-)} - k_z c_{\mathrm{F}}^{(-)} &= \tilde{f}_{\mathrm{B}} c_{\mathrm{B}}^{(0)} + \tilde{f}_{\mathrm{F}} c_{\mathrm{F}}^{(0)}, \\ \tilde{f}_{\mathrm{B}} c_{\mathrm{B}}^{(0)} e^{ik_{\mathrm{B}}L} + \tilde{f}_{\mathrm{F}} c_{\mathrm{F}}^{(0)} e^{-ik_{\mathrm{B}}L} &= k_z c_{\mathrm{B}}^{(+)} e^{ik_z L} - k_z c_{\mathrm{F}}^{(+)} e^{-ik_z L}, \end{aligned} \tag{6.34}$$

where

$$\tilde{f}_{\mathrm{B}} = -\mathrm{i}\frac{\mathrm{d}}{\mathrm{d}z}\left[f_{\mathrm{B}}(z)e^{\mathrm{i}k_{\mathrm{B}}z}\right]\Big|_0, \quad \tilde{f}_{\mathrm{F}} = -\mathrm{i}\frac{\mathrm{d}}{\mathrm{d}z}\left[f_{\mathrm{F}}(z)e^{-\mathrm{i}k_{\mathrm{B}}z}\right]\Big|_0. \tag{6.35}$$

One of parametrizations of the solution of equation (6.23) is a dependence on $c_{\mathrm{B}}^{(-)}$, $c_{\mathrm{F}}^{(-)}$, which is

$$\begin{pmatrix} c_{\mathrm{B}}^{(0)} \\ c_{\mathrm{F}}^{(0)} \end{pmatrix} = \frac{1}{\tilde{f}_{\mathrm{F}} - \tilde{f}_{\mathrm{B}}} \times \begin{pmatrix} \tilde{f}_{\mathrm{F}} - k_z & \tilde{f}_{\mathrm{F}} + k_z \\ -\tilde{f}_{\mathrm{B}} + k_z & -\tilde{f}_{\mathrm{B}} - k_z \end{pmatrix} \begin{pmatrix} c_{\mathrm{B}}^{(-)} \\ c_{\mathrm{F}}^{(-)} \end{pmatrix}, \tag{6.36}$$

$$\begin{pmatrix} c_{\mathrm{B}}^{(+)} \\ c_{\mathrm{F}}^{(+)} \end{pmatrix} = \begin{pmatrix} s_{\mathrm{BB}} & s_{\mathrm{BF}} \\ s_{\mathrm{FB}} & s_{\mathrm{FF}} \end{pmatrix} \begin{pmatrix} c_{\mathrm{B}}^{(-)} \\ c_{\mathrm{F}}^{(-)} \end{pmatrix}, \tag{6.37}$$

where

$$s_{\mathrm{BB}} = \frac{2}{D} e^{-\mathrm{i}k_z L} \cos(k_{\mathrm{B}} L) k_z \left(-\tilde{f}_{\mathrm{F}} + \tilde{f}_{\mathrm{B}}\right) + \mathrm{i}\frac{2}{D} e^{-\mathrm{i}k_z L} \sin(k_{\mathrm{B}} L) \left(k_z^2 - \tilde{f}_{\mathrm{B}} \tilde{f}_{\mathrm{F}}\right), \tag{6.38}$$

$$s_{\mathrm{FF}} = \frac{2}{D} e^{\mathrm{i}k_z L} \cos(k_{\mathrm{B}} L) k_z \left(-\tilde{f}_{\mathrm{F}} + \tilde{f}_{\mathrm{B}}\right) - \mathrm{i}\frac{2}{D} e^{\mathrm{i}k_z L} \sin(k_{\mathrm{B}} L) \left(k_z^2 - \tilde{f}_{\mathrm{B}} \tilde{f}_{\mathrm{F}}\right), \tag{6.39}$$

$$s_{\mathrm{BF}} = -\mathrm{i}\frac{2}{D} e^{-\mathrm{i}k_z L} \sin(k_{\mathrm{B}} L) \left(k_z + \tilde{f}_{\mathrm{B}}\right) \left(k_z + \tilde{f}_{\mathrm{F}}\right), \tag{6.40}$$

$$s_{\mathrm{FB}} = \mathrm{i}\frac{2}{D} e^{\mathrm{i}k_z L} \sin(k_{\mathrm{B}} L) \left(k_z - \tilde{f}_{\mathrm{B}}\right) \left(k_z - \tilde{f}_{\mathrm{F}}\right), \tag{6.41}$$

with

$$D = 2k_z \left(-\tilde{f}_{\mathrm{F}} + \tilde{f}_{\mathrm{B}}\right). \tag{6.42}$$

Another of parametrizations of the solution (6.32) is a dependence on $c_{\mathrm{B}}^{(+)}$, $c_{\mathrm{F}}^{(-)}$. Of remaining four coefficients, we first determine only $c_{\mathrm{B}}^{(-)}$, $c_{\mathrm{F}}^{(+)}$. Their form is interesting as an input–output relation

$$\begin{pmatrix} c_{\mathrm{B}}^{(-)} \\ c_{\mathrm{F}}^{(+)} \end{pmatrix} = \begin{pmatrix} t' & r \\ r' & t \end{pmatrix} \begin{pmatrix} c_{\mathrm{B}}^{(+)} \\ c_{\mathrm{F}}^{(-)} \end{pmatrix}, \tag{6.43}$$

where

$$r = -\frac{s_{\mathrm{BF}}}{s_{\mathrm{BB}}}, \quad r' = \frac{s_{\mathrm{FB}}}{s_{\mathrm{BB}}}, \quad t' = \frac{1}{s_{\mathrm{BB}}}, \tag{6.44}$$

$$t = \frac{s_{\mathrm{BB}} s_{\mathrm{FF}} - s_{\mathrm{FB}} s_{\mathrm{BF}}}{s_{\mathrm{BB}}} = \frac{1}{s_{\mathrm{BB}}}, \tag{6.45}$$

since the determinant of the "scattering matrix" is

$$\begin{vmatrix} s_{\mathrm{BB}} & s_{\mathrm{BF}} \\ s_{\mathrm{FB}} & s_{\mathrm{FF}} \end{vmatrix} = 1. \tag{6.46}$$

Finally, we may also express $c_{\mathrm{B}}^{(0)}$ and $c_{\mathrm{F}}^{(0)}$,

$$\begin{pmatrix} c_{\mathrm{B}}^{(0)} \\ c_{\mathrm{F}}^{(0)} \end{pmatrix} = \frac{t}{\tilde{f}_{\mathrm{F}} - \tilde{f}_{\mathrm{B}}} \begin{pmatrix} \tilde{f}_{\mathrm{F}} - k_z & e^{-i(k_{\mathrm{B}}+k_z)L}\left(\tilde{f}_{\mathrm{F}} + k_z\right) \\ -\tilde{f}_{\mathrm{B}} + k_z & e^{i(k_{\mathrm{B}}-k_z)L}\left(-\tilde{f}_{\mathrm{B}} - k_z\right) \end{pmatrix} \begin{pmatrix} c_{\mathrm{B}}^{(+)} \\ c_{\mathrm{F}}^{(-)} \end{pmatrix}. \tag{6.47}$$

In the case of a TM solution, we proceed quite generally. We introduce unknown functions $\alpha(z)$, $\beta(z)$, $\gamma(z)$ such that

$$\begin{aligned} E_x(x, y, z; \omega) &= \alpha(z) \cos\left[k_x(x + a_x)\right] \sin\left[k_y(y + a_y)\right], \\ E_y(x, y, z; \omega) &= \beta(z) \sin\left[k_x(x + a_x)\right] \cos\left[k_y(y + a_y)\right], \\ E_z(x, y, z; \omega) &= \mathrm{i}\gamma(z) \sin\left[k_x(x + a_x)\right] \sin\left[k_y(y + a_y)\right]. \end{aligned} \tag{6.48}$$

The functions $\alpha(z)$ and $\beta(z)$ are linearly dependent,

$$-k_y\alpha(z) + k_x\beta(z) = 0, \quad -\infty < z < \infty. \tag{6.49}$$

Equation (6.23) is replaced by

$$\begin{aligned} \frac{\mathrm{d}}{\mathrm{d}z}\left[k_x\alpha(z) + k_y\beta(z)\right] &= -\mathrm{i}\gamma(z)\tilde{k}_z^2(z), \\ \frac{\mathrm{d}}{\mathrm{d}z}\left[\tilde{k}^2(z)\gamma(z)\right] &= -\mathrm{i}\tilde{k}^2(z)\left[k_x\alpha(z) + k_y\beta(z)\right], \end{aligned} \tag{6.50}$$

where $\tilde{k}(z) = k$, $\tilde{k}_z(z) = k_z$ for $z < 0$, $z > L$ and $\tilde{k}(z) = \bar{k}(z)$, $\tilde{k}_z(z) = \bar{k}_z(z)$ for $0 < z < L$, where $\bar{k}(z)$, $\bar{k}_z(z)$ have the period Λ. The equations

$$\begin{aligned} \frac{\mathrm{d}}{\mathrm{d}z}\left[k_x\alpha(z) + k_y\beta(z)\right] &= -\mathrm{i}\gamma(z)k_z^2, \\ \frac{\mathrm{d}}{\mathrm{d}z}\left[k^2\gamma(z)\right] &= -\mathrm{i}k^2\left[k_x\alpha(z) + k_y\beta(z)\right] \end{aligned} \tag{6.51}$$

have a general solution

$$\begin{aligned} k_x\alpha(z) + k_y\beta(z) &= c_{\mathrm{B}}k_z e^{\mathrm{i}k_z z} + c_{\mathrm{F}}k_z e^{-\mathrm{i}k_z z}, \\ k^2\gamma(z) &= -c_{\mathrm{B}}k^2 e^{\mathrm{i}k_z z} + c_{\mathrm{F}}k^2 e^{-\mathrm{i}k_z z}, \end{aligned} \tag{6.52}$$

where c_{B} and c_{F} are arbitrary complex numbers. Let us note properties of the equations

$$\frac{\mathrm{d}}{\mathrm{d}z}\left[k_x\alpha(z)+k_y\beta(z)\right]=-\mathrm{i}\gamma(z)\bar{k}_z^2(z),$$
$$\frac{\mathrm{d}}{\mathrm{d}z}\left[\bar{k}^2(z)\gamma(z)\right]=-\mathrm{i}\bar{k}^2(z)\left[k_x\alpha(z)+k_y\beta(z)\right]. \tag{6.53}$$

The general solution of equations (6.53) has the form

$$k_x\alpha(z)+k_y\beta(z)=c_\mathrm{B}f_{\mathrm{B}\perp}(z)e^{\mathrm{i}k_\mathrm{B}z}+c_\mathrm{F}f_{\mathrm{F}\perp}(z)e^{-\mathrm{i}k_\mathrm{B}z},$$
$$\bar{k}^2(z)\gamma(z)=c_\mathrm{B}f_{\mathrm{B}3}(z)e^{\mathrm{i}k_\mathrm{B}z}+c_\mathrm{F}f_{\mathrm{F}3}(z)e^{-\mathrm{i}k_\mathrm{B}z}, \tag{6.54}$$

where c_B and c_F are arbitrary complex numbers. The functions $f_{\mathrm{B}\perp}(z)$, $f_{\mathrm{F}\perp}(z)$ fulfil the conditions

$$f_{\mathrm{B}\perp}(0)=f_{\mathrm{F}\perp}(0)=k_z. \tag{6.55}$$

It can be expected, e.g., that the inclusion of magnetic properties, with the neglect of frequency dependence, should lead to a small complication and a similar treatment of the TE solution.

For $\alpha(z)$, $\beta(z)$, $\gamma(z)$, the Bloch factor $e^{\mathrm{i}k_\mathrm{B}\Lambda}$ is of importance. For the Kronig–Penney dielectric, the Bloch factor can be determined from the equation

$$\cos(k_\mathrm{B}\Lambda)=\cos\left(k_{1z}\frac{\Lambda}{2}\right)\cos\left(k_{2z}\frac{\Lambda}{2}\right)$$
$$-\frac{1}{2}\left(\frac{k_1^2k_{2z}}{k_2^2k_{1z}}+\frac{k_2^2k_{1z}}{k_1^2k_{2z}}\right)\sin\left(k_{1z}\frac{\Lambda}{2}\right)\sin\left(k_{2z}\frac{\Lambda}{2}\right). \tag{6.56}$$

It is well known (Mishra and Salpathy 2003) that the relations (6.31) and (6.56) can be written as a single one when appropriate notation is adopted. In fact, we observe the change

$$k_{1z}\to\frac{k_1^2}{k_{1z}},\quad k_{2z}\to\frac{k_2^2}{k_{2z}}. \tag{6.57}$$

We assume a general solution, which is sewn from a solution of equations (6.51) for $z<0$, from a solution of equations (6.53) for $0<z<L$, and from another solution of equations (6.51) for $z>L$. Then we obtain modified formulae (6.44) and (6.45) adopting the changes

$$k_z\to\frac{k^2}{k_z},\quad \tilde{f}_\mathrm{B}\to-\frac{1}{k_z}f_{\mathrm{B}3}(0),\quad \tilde{f}_\mathrm{F}\to-\frac{1}{k_z}f_{\mathrm{F}3}(0). \tag{6.58}$$

Quantization of the electromagnetic field propagating in an ideal uniform waveguide is (for a fixed transverse mode) assumed in the article (Marinescu 1992), and it is shown that the appropriate scalar field obeys a Klein–Gordon type equation. Even a Dirac-type equation for the waveguides is considered in that article.

6.1.3 Method of Coupled Modes

To any $\omega > \min\{\omega_{10}, \omega_{01}\}$, there exists only a finite number of modal functions (6.11), with (6.15) and (6.16). We will let $J(\omega)$, $J(\omega) \subset J$, denote the corresponding index set. We can partition this set into disjoint sets,

$$J(\omega) = J_+(\omega) \cup J_0(\omega) \cup J_-(\omega), \tag{6.59}$$

where $J_+(\omega)$ is the set of indices that have $k_z > 0$, $J_0(\omega)$ is the set of indices that have $k_z = 0$, and $J_-(\omega)$ is the set of indices that have $k_z < 0$.

To any modal function $\mathbf{E}_{j_{\rm in}}(\mathbf{x})$ with $j_{\rm in} \in J_+(\omega)$, one (on the basis of physical knowledge of propagation) looks for a solution of the problem formulated in Section 6.1.2 as follows. If $z < 0$ then

$$\mathbf{E}(\mathbf{x}) = \sum_{j \in \{j_{\rm in}\} \cup J_-(\omega)} A_j(0)\mathbf{E}_j(\mathbf{x}), \tag{6.60}$$

$$\mathbf{B}(\mathbf{x}) = \sum_{j \in \{j_{\rm in}\} \cup J_-(\omega)} B_j(0)\mathbf{B}_j(\mathbf{x}), \tag{6.61}$$

where $A_j(0) = B_j(0) = 1$ for $j = j_{\rm in}$. If $z > L$ then

$$\mathbf{E}(\mathbf{x}) = \sum_{j \in J_+(\omega)} A_j(L)\mathbf{E}_j(\mathbf{x}), \tag{6.62}$$

$$\mathbf{B}(\mathbf{x}) = \sum_{j \in J_+(\omega)} B_j(L)\mathbf{B}_j(\mathbf{x}). \tag{6.63}$$

The coupled-mode method determines a form of the solution even for $z \in [0, L]$ and it finds complex numbers $A_j(0)$, $j \in J_-(\omega)$, and $A_j(L)$, $j \in J_+(\omega)$. This method is approximate. If $0 \leq z \leq L$ then

$$\mathbf{E}(\mathbf{x}) = \sum_{j \in J_+(\omega) \cup J_-(\omega)} A_j(z)\mathbf{E}_j(\mathbf{x}), \tag{6.64}$$

$$\mathbf{B}(\mathbf{x}) = \sum_{j \in J_+(\omega) \cup J_-(\omega)} B_j(z)\mathbf{B}_j(\mathbf{x}). \tag{6.65}$$

When the dependence of $A_j(z)$, $B_j(z)$ on the z-coordinate is weak (their variation is slow), we may consider

$$\mathbf{E}_{\rm F,B}(\mathbf{x}) = \sum_{j \in J_\pm(\omega)} A_j(z)\mathbf{E}_j(\mathbf{x}), \tag{6.66}$$

$$\mathbf{B}_{\rm F,B}(\mathbf{x}) = \sum_{j \in J_\pm(\omega)} B_j(z)\mathbf{B}_j(\mathbf{x}). \tag{6.67}$$

Here

$$A_j(z) = \left[\int \left|\mathbf{E}_j(\mathbf{x}_\perp, z)\right|^2 \mathrm{d}^2\mathbf{x}_\perp\right]^{-1} \int \mathbf{E}_j^*(\mathbf{x}_\perp, z) \cdot \mathbf{E}_{\mathrm{F,B}}(\mathbf{x})\mathrm{d}^2\mathbf{x}_\perp. \tag{6.68}$$

Similarly $B_j(z)$ is determined.

We assume TE waves. A possible use of the method for TM waves means another, rough, approximation. We concentrate on the electric field and, therefore, we solve the Helmholtz equation

$$\nabla^2\mathbf{E}(\mathbf{x}) + \epsilon_\mathrm{r}(z)\frac{\omega^2}{c^2}\mathbf{E}(\mathbf{x}) = \mathbf{0}. \tag{6.69}$$

We will expound a means of describing the transformation

$$\mathbf{E}(\mathbf{x}) \mapsto [\epsilon_\mathrm{r}(z) - \epsilon_{\mathrm{r}0}]\mathbf{E}(\mathbf{x}), \quad 0 \le z \le L. \tag{6.70}$$

In fact we assume that

$$[\epsilon_\mathrm{r}(z) - \epsilon_{\mathrm{r}0}]\mathbf{E}(\mathbf{x}) = \mathbf{E}'(\mathbf{x}) = \sum_{j\in J_+(\omega)\cup J_-(\omega)} A_j'(z)\mathbf{E}_j(\mathbf{x}), \tag{6.71}$$

$$\{[\epsilon_\mathrm{r}(z) - \epsilon_{\mathrm{r}0}]\,\mathbf{E}(\mathbf{x})\}_{\mathrm{F,B}} = \mathbf{E}'_{\mathrm{F,B}}(\mathbf{x}) = \sum_{j\in J_\pm(\omega)} A_j'(z)\mathbf{E}_j(\mathbf{x}), \tag{6.72}$$

where

$$A_j'(z) = \left[\int \left|\mathbf{E}_j(\mathbf{x}_\perp, z)\right|^2 \mathrm{d}^2\mathbf{x}_\perp\right]^{-1} \int \mathbf{E}_j^*(\mathbf{x}_\perp, z) \cdot \mathbf{E}'_{\mathrm{F,B}}(\mathbf{x})\mathrm{d}^2\mathbf{x}_\perp. \tag{6.73}$$

Particularly,

$$[\epsilon_\mathrm{r}(z) - \epsilon_{\mathrm{r}0}]\mathbf{E}_j(\mathbf{x}) = \sum_{j\in J_+(\omega)\cup J_-(\omega)} K_{jj'}(z)\mathbf{E}_{j'}(\mathbf{x}), \tag{6.74}$$

$$\left\{\mathbf{E}_j'(\mathbf{x})\right\}_{\mathrm{F,B}} = \sum_{j'\in J_\pm(\omega)} K_{jj'}(z)\mathbf{E}_{j'}(\mathbf{x}), \tag{6.75}$$

where

$$K_{jj'}(z) = \left[\int \left|\mathbf{E}_{j'}(\mathbf{x}_\perp, z)\right|^2 \mathrm{d}^2\mathbf{x}_\perp\right]^{-1} \int \mathbf{E}_{j'}^*(\mathbf{x}_\perp, z) \cdot \left\{\mathbf{E}_j'(\mathbf{x})\right\}_{\mathrm{F,B}} \mathrm{d}^2\mathbf{x}_\perp, \tag{6.76}$$

i.e.,

$$A_j'(z) = \sum_{j'\in J_+(\omega)\cup J_-(\omega)} K_{j'j}(z)A_{j'}(z). \tag{6.77}$$

Further, the coupled-mode theory comprises an approximate expression (Yariv and Yeh 1984, Section 6.4, p. 197)

$$\nabla^2 \mathbf{E}(\mathbf{x}) = \sum_{j \in J_+(\omega) \cup J_-(\omega)} \mathbf{h}_j(\mathbf{x}), \tag{6.78}$$

where

$$\begin{aligned} \mathbf{h}_j(\mathbf{x}) &= \nabla^2 A_j(z) \mathbf{E}_j(\mathbf{x}) + 2\nabla A_j(z) \cdot \nabla \mathbf{E}_j(\mathbf{x}) + A_j(z) \nabla^2 \mathbf{E}_j(\mathbf{x}) \\ &\approx -2\mathrm{i} k_{jz} \frac{\mathrm{d}}{\mathrm{d}z} A_j(z) \mathbf{E}_j(\mathbf{x}) - \epsilon_{\mathrm{r}0} \frac{\omega^2}{c^2} A_j(z) \mathbf{E}_j(\mathbf{x}), \end{aligned} \tag{6.79}$$

or

$$\nabla^2 \mathbf{E}(\mathbf{x}) \approx -2\mathrm{i} \sum_{j \in J_+(\omega) \cup J_-(\omega)} k_{jz} \frac{\mathrm{d}}{\mathrm{d}z} A_j(z) \mathbf{E}_j(\mathbf{x}) - \epsilon_{\mathrm{r}0} \frac{\omega^2}{c^2} \mathbf{E}(\mathbf{x}). \tag{6.80}$$

On substituting it into (6.69) and using (6.71), (6.77), we obtain a system of differential equations

$$-2\mathrm{i} k_{jz} \frac{\mathrm{d}}{\mathrm{d}z} A_j(z) = -\frac{\omega^2}{c^2} \sum_{j' \in J_+(\omega) \cup J_-(\omega)} K_{j'j}(z) A_{j'}(z), \tag{6.81}$$

to be solved on the boundary conditions

$$A_j(0) = \begin{cases} 1 \text{ for } j = j_{\mathrm{in}}, \\ 0 \text{ for } j \in J_+(\omega) \setminus \{j_{\mathrm{in}}\}, \end{cases} \tag{6.82}$$

$$A_j(L) = 0 \text{ for } j \in J_-(\omega). \tag{6.83}$$

We will reduce the equations to the form

$$\frac{k_{jz}}{|k_{jz}|} \frac{\mathrm{d}}{\mathrm{d}z} A_j(z) = -\frac{\mathrm{i}}{2} \frac{\omega^2}{c^2} \sum_{j' \in J_+(\omega) \cup J_-(\omega)} \frac{K_{j'j}(z)}{|k_{jz}|} A_{j'}(z). \tag{6.84}$$

On physical grounds, it is expected that

$$\sum_{j' \in J_+(\omega) \cup J_-(\omega)} \frac{k_{jz}}{|k_{jz}|} |A_j(z)|^2 \tag{6.85}$$

is independent of z. We have

$$\frac{k_{jz}}{|k_{jz}|}\frac{\mathrm{d}}{\mathrm{d}z}A_j(z)A_j^*(z) + \text{c.c.}$$

$$= -\frac{\mathrm{i}}{2}\frac{\omega^2}{c^2}\sum_{j'\in J_+(\omega)\cup J_-(\omega)}\frac{K_{j'j}(z)}{|k_{jz}|}A_{j'}(z)A_j^*(z) + \text{c.c.} \equiv 0, \tag{6.86}$$

where c.c. means complex conjugation, or the expected property is

$$\frac{K_{j'j}(z)}{|k_{jz}|} = \frac{K_{jj'}^*(z)}{|k_{j'z}|}. \tag{6.87}$$

6.1.4 Normalized Modes of the Electromagnetic Field

The normalization of modal functions of the electromagnetic field, which is made for the sake of quantization, can be based, in optics, on a simple connection of the vector potential with the electric-field strength vector. This connection follows from the use of the so-called Coulomb gauge. First, we assume that only the electromagnetic field is present in the cavity (in the so-called nonrelativistic approximation). In quantum optics, the simple instance of a perfectly closed cavity or resonator is often considered, which can be described mathematically in terms of the subset G of the usual Euclidean space R^3 with the boundary ∂G.

In optics, the quantization is a definition of the vector potential operator $\hat{\mathbf{A}}(\mathbf{x}, t)$ by the relation

$$\hat{\mathbf{A}}(\mathbf{x}, t) = \sum_{j\in J}\left[\mathbf{A}_j^{(\mathrm{phot})}(\mathbf{x}, t)\hat{a}_j + \mathbf{A}_j^{(\mathrm{phot})*}(\mathbf{x}, t)\hat{a}_j^\dagger\right], \tag{6.88}$$

where J is an index set, and the photon annihilation and creation operators $\hat{a}_j$ and $\hat{a}_j^\dagger$ in the jth mode fulfil the commutation relations

$$[\hat{a}_j, \hat{a}_{j'}^\dagger] = \delta_{jj'}\hat{1}, \quad [\hat{a}_j, \hat{a}_{j'}] = [\hat{a}_j^\dagger, \hat{a}_{j'}^\dagger] = \hat{0}. \tag{6.89}$$

Further

$$\mathbf{A}_j^{(\mathrm{phot})}(\mathbf{x}, t) = \sqrt{\frac{\hbar}{2\epsilon_0\omega_j}}\,\mathbf{u}_j(\mathbf{x})\exp(-\mathrm{i}\omega_j t), \tag{6.90}$$

with $\hbar$ the reduced Planck constant, ϵ_0 the vacuum (free-space) electric permittivity, ω_j and $\mathbf{u}_j(\mathbf{x})$ satisfying the Helmholtz equation

$$\nabla^2\mathbf{u}_j(\mathbf{x}) + \frac{\omega_j^2}{c^2}\mathbf{u}_j(\mathbf{x}) = \mathbf{0}, \tag{6.91}$$

the transversality condition

$$\nabla\cdot\mathbf{u}_j(\mathbf{x}) = 0, \tag{6.92}$$

and the boundary conditions

$$\mathbf{n_x} \times \mathbf{u}_j(\mathbf{x}) \,\big|_{\partial G} = \mathbf{0}, \tag{6.93}$$

$$\mathbf{n_x} \cdot \left[\nabla \times \mathbf{u}_j(\mathbf{x})\right]\big|_{\partial G} = (\mathbf{n_x} \times \nabla) \cdot \mathbf{u}_j(\mathbf{x})\big|_{\partial G} = 0, \tag{6.94}$$

where $\mathbf{n_x}$ is the normal vector at the point $\mathbf{x}$. It is required that the modal functions $\mathbf{u}_j(\mathbf{x})$ be orthogonal and normalized as expressed by the relation

$$\int_G \mathbf{u}_j(\mathbf{x}) \cdot \mathbf{u}_{j'}(\mathbf{x})\, \mathrm{d}^3\mathbf{x} = \delta_{jj'}. \tag{6.95}$$

From relation (6.90), it is seen that the harmonic time dependence has the form $\exp(-\mathrm{i}\omega t)$, not $\exp(\mathrm{i}\omega t)$ as in relation (6.1). We have adopted this change although it may be a source of obscurity.

(i) Empty rectangular cavity

For illustration, we will assume that

$$G = \{\mathbf{x} : -a_x < x < a_x,\ -a_y < y < a_y,\ -a_z < z < a_z\}, \tag{6.96}$$

where $a_x,\ a_y,\ a_z$ are positive. It can be proved that the index set J is a collection of $j = (n_x, n_y, n_z, s)$, where $n_r \in 0, 1, \ldots, \infty$, $r = x,\ y,\ z$, $s =$ TE, TM, $n_x > 0$ or $s =$ TE, $n_y > 0$ or $s =$ TE, and $n_z > 0$ or $s =$ TM, $n_x + n_y + n_z \geq 2$.

The solutions ω_j have the form

$$\omega_j = c\sqrt{k_x^2 + k_y^2 + k_z^2}, \tag{6.97}$$

with

$$c = \frac{1}{\sqrt{\epsilon_0 \mu_0}}, \quad k_r = \frac{n_r \pi}{2a_r}, \tag{6.98}$$

and the solutions $\mathbf{u}_j(\mathbf{x})$ are connected with the classical solutions

$$E_{jx}(\mathbf{x}) = \alpha_j \cos\left(\frac{n_x\pi}{2a_x}(x + a_x)\right) \sin\left(\frac{n_y\pi}{2a_y}(y + a_y)\right) \times \sin\left(\frac{n_z\pi}{2a_z}(z + a_z)\right),$$

$$E_{jy}(\mathbf{x}) = \beta_j \sin\left(\frac{n_x\pi}{2a_x}(x + a_x)\right) \cos\left(\frac{n_y\pi}{2a_y}(y + a_y)\right) \times \sin\left(\frac{n_z\pi}{2a_z}(z + a_z)\right),$$

$$E_{jz}(\mathbf{x}) = \gamma_j \sin\left(\frac{n_x\pi}{2a_x}(x+a_x)\right)\sin\left(\frac{n_y\pi}{2a_y}(y+a_y)\right)$$
$$\times \cos\left(\frac{n_z\pi}{2a_z}(z+a_z)\right), \tag{6.99}$$

to the equivalent boundary-value problem

$$\nabla\cdot\mathbf{E}_j = 0, \quad \nabla\times\mathbf{B}_j + \mathrm{i}\frac{\omega_j}{c^2}\mathbf{E}_j = \mathbf{0}, \tag{6.100}$$

$$\nabla\cdot\mathbf{B}_j = 0, \quad \nabla\times\mathbf{E}_j - \mathrm{i}\omega_j\mathbf{B}_j = \mathbf{0}, \tag{6.101}$$

$$\mathbf{n}_\mathbf{x}\cdot\mathbf{B}_j(\mathbf{x},\omega_j)\big|_{\partial G} = 0, \quad \mathbf{n}_\mathbf{x}\times\mathbf{E}_j(\mathbf{x},\omega_j)\big|_{\partial G} = \mathbf{0}. \tag{6.102}$$

Here

$$\alpha_j = -\mathrm{i}\omega_j\frac{k_y}{k_x^2+k_y^2}\gamma_j', \quad \beta_j = \mathrm{i}\omega_j\frac{k_x}{k_x^2+k_y^2}\gamma_j',$$
$$\gamma_j = 0, \quad \text{for } s = \mathrm{TE}, \tag{6.103}$$

$$\alpha_j = -\frac{k_xk_z}{k_x^2+k_y^2}\gamma_j, \quad \beta_j = -\frac{k_yk_z}{k_x^2+k_y^2}\gamma_j, \quad \text{for } s = \mathrm{TM}, \tag{6.104}$$

with γ_j', γ_j complex parameters.

The connecting relation is

$$\mathbf{u}_j(\mathbf{x}) = -\mathrm{i}\sqrt{\frac{2\epsilon_0}{\hbar\omega_j}}\mathbf{E}_j^{(\mathrm{phot})}(\mathbf{x}), \tag{6.105}$$

where

$$\mathbf{E}_j^{(\mathrm{phot})}(\mathbf{x}) = \sum_{r=x,y,z} E_{jr}^{(\mathrm{phot})}(\mathbf{x})\mathbf{e}_r, \tag{6.106}$$

with $E_{jr}^{(\mathrm{phot})}(\mathbf{x})$ given by the formulae (6.99), (6.103), (6.104), in which

$$\gamma_j' = 2\sqrt{\frac{\hbar}{\epsilon_0\omega_j}}\sqrt{\frac{k_x^2+k_y^2}{(1+\delta_{k_x0})(1+\delta_{k_y0})(1-\delta_{k_z0})V}}\,\zeta_j', \tag{6.107}$$

$$\gamma_j = 2c\sqrt{\frac{\hbar}{\epsilon_0\omega_j}}\sqrt{\frac{k_x^2+k_y^2}{(1-\delta_{k_x0})(1-\delta_{k_y0})(1+\delta_{k_z0})V}}\,\zeta_j \tag{6.108}$$

are substituted, $V = 8a_xa_ya_z$, $|\zeta_j'| = |\zeta_j| = 1$.

It can be easily derived that the vector-valued functions $\mathbf{u}_j(\mathbf{x})$, $j \in J$, satisfy a completeness relation,

$$\sum_{j\in J} \mathbf{u}_j(\mathbf{x})\mathbf{u}_j^*(\mathbf{x}') = \delta(\mathbf{x}-\mathbf{x}')\mathbf{1} - \nabla_{\mathbf{x}}\nabla_{\mathbf{x}'}\mathcal{G}(\mathbf{x},\mathbf{x}'), \tag{6.109}$$

where $\mathcal{G}(\mathbf{x},\mathbf{x}')$ is a Green's function for the Laplace operator (the Dirichlet problem).

(ii) Rectangular cavity filled with a refractive medium

In this case, the quantization can be performed according to the relations (6.88), (6.89), (6.90), with ω_j and $\mathbf{u}_j(\mathbf{x})$ satisfying the Helmholtz equation

$$\nabla^2\mathbf{u}_j(\mathbf{x}) + \epsilon_{\mathrm{r}0}\frac{\omega_j^2}{c^2}\mathbf{u}_j(\mathbf{x}) = \mathbf{0}, \tag{6.110}$$

where $\epsilon_{\mathrm{r}0}$ is the relative electric permittivity of the medium, the transversality condition (6.92), and the boundary conditions (6.93) and (6.94). It is required that the modal functions $\mathbf{u}_j(\mathbf{x})$ be orthogonal and normalized as expressed by the relation

$$\int_G \epsilon_{\mathrm{r}0}\mathbf{u}_j^*(\mathbf{x})\cdot\mathbf{u}_{j'}(\mathbf{x})\,\mathrm{d}^3\mathbf{x} = \delta_{jj'}. \tag{6.111}$$

For illustration, we will assume that G is defined by (6.96). The solutions ω_j have the form

$$\omega_j = v\sqrt{k_x^2+k_y^2+k_z^2}, \tag{6.112}$$

with

$$v = \frac{1}{\sqrt{\epsilon_0\epsilon_{\mathrm{r}0}\mu_0}},\quad k_r = \frac{n_r\pi}{2a_r},\ r = x,\ y,\ z, \tag{6.113}$$

and the solutions $\mathbf{u}_j(\mathbf{x})$ are connected with the solutions of the form (6.99) to the equivalent boundary-value problem

$$\nabla\cdot\mathbf{E}_j = 0,\quad \nabla\times\mathbf{B}_j + \mathrm{i}\epsilon_{\mathrm{r}0}\frac{\omega_j}{c^2}\mathbf{E}_j = \mathbf{0}, \tag{6.114}$$

$$\nabla\cdot\mathbf{B}_j = 0,\quad \nabla\times\mathbf{E}_j - \mathrm{i}\omega_j\mathbf{B}_j = \mathbf{0}, \tag{6.115}$$

$$\mathbf{n}_{\mathbf{x}}\cdot\mathbf{B}_j(\mathbf{x},\omega_j)\big|_{\partial G} = 0,\quad \mathbf{n}_{\mathbf{x}}\times\mathbf{E}_j(\mathbf{x},\omega_j)\big|_{\partial G} = \mathbf{0}. \tag{6.116}$$

The connecting relation is (6.105) with (6.106), where $E_{jr}^{(\mathrm{phot})}(\mathbf{x})$ are given by the formulae (6.99), (6.103), (6.104), in which

$$\gamma'_j = 2\sqrt{\frac{\hbar}{\epsilon_0\epsilon_{r0}\omega_j}}\sqrt{\frac{k_x^2+k_y^2}{(1+\delta_{k_x0})(1+\delta_{k_y0})(1-\delta_{k_z0})V}}\,\zeta'_j, \tag{6.117}$$

$$\gamma_j = 2c\sqrt{\frac{\hbar}{\epsilon_0\epsilon_{r0}\omega_j}}\sqrt{\frac{k_x^2+k_y^2}{(1-\delta_{k_x0})(1-\delta_{k_y0})(1+\delta_{k_z0})V}}\,\zeta_j \tag{6.118}$$

are substituted.

It can be easily derived that the vector-valued functions $\mathbf{u}_j(\mathbf{x})$, $j \in J$, satisfy a completeness relation

$$\sum_{j\in J} \epsilon_{r0}\mathbf{u}_j(\mathbf{x})\mathbf{u}_j^*(\mathbf{x}') = \delta(\mathbf{x}-\mathbf{x}')\mathbf{1} - \nabla_{\mathbf{x}}\nabla_{\mathbf{x}'}\mathcal{G}(\mathbf{x},\mathbf{x}'), \tag{6.119}$$

where $\mathcal{G}(\mathbf{x},\mathbf{x}')$ is a Green's function for the Laplace operator (the Dirichlet problem).

(iii) Rectangular waveguide filled with a refractive medium and located in a flat space

We will consider a subset $G = G_\perp \times \mathrm{S}^1(-a_z \le z < a_z)$, with the boundary $\partial G = \partial G_\perp \times \mathrm{S}^1(-a_z \le z < a_z)$, of a flat non-Euclidean space $\mathrm{R}^2 \times \mathrm{S}^1(-a_z \le z < a_z)$, where $\mathrm{S}^1(-a_z \le z < a_z)$ means a topological circle of the length $2a_z$. In this case, the quantization can be performed according to the relations (6.88), (6.89), (6.90), with ω_j and $\mathbf{u}_j(\mathbf{x})$ satisfying the Helmholtz equation of the form (6.110), the transversality condition (6.92), and the boundary conditions (6.93) and (6.94). It is required that the modal functions $\mathbf{u}_j(\mathbf{x})$ be orthogonal and normalized as expressed by the relation (6.111).

For illustration, we will assume that

$$G = \{\mathbf{x} : -a_x < x < a_x, -a_y < y < a_y, -a_z \le z < a_z\}, \tag{6.120}$$

where $a_x,\ a_y,\ a_z$ are positive. It can be proved that the index set J is a collection of $j = (n_x, n_y, n_z, s)$, where $n_r \in \{0\} \cup \mathrm{N}$, $r = x,\ y$, $n_z \in \mathrm{Z}$, $s = \mathrm{TE}, \mathrm{TM}$, $n_x > 0$ or $s = \mathrm{TE}$, $n_y > 0$ or $s = \mathrm{TE}$, and $n_x + n_y \ge 1$.

The solutions ω_j have the form

$$\omega_j = v\sqrt{k_x^2+k_y^2+k_z^2}, \tag{6.121}$$

with

$$v = \frac{1}{\sqrt{\epsilon_0\epsilon_{r0}\mu_0}},\quad k_r = \frac{n_r\pi}{2a_r},\quad r = x,\ y,\quad k_z = \frac{n_z\pi}{a_z}, \tag{6.122}$$

and the solutions $\mathbf{u}_j(\mathbf{x})$ are connected with the classical solutions

$$E_{jx}(\mathbf{x}) = \alpha_j \cos\left(\frac{n_x\pi}{2a_x}(x+a_x)\right)\sin\left(\frac{n_y\pi}{2a_y}(y+a_y)\right) \times \exp\left(\mathrm{i}\frac{n_z\pi}{a_z}(z+a_z)\right),$$

$$E_{jy}(\mathbf{x}) = \beta_j \sin\left(\frac{n_x\pi}{2a_x}(x+a_x)\right)\cos\left(\frac{n_y\pi}{2a_y}(y+a_y)\right)$$
$$\times \exp\left(\mathrm{i}\frac{n_z\pi}{a_z}(z+a_z)\right),$$
$$E_{jz}(\mathbf{x}) = \mathrm{i}\gamma_j \sin\left(\frac{n_x\pi}{2a_x}(x+a_x)\right)\sin\left(\frac{n_y\pi}{2a_y}(y+a_y)\right)$$
$$\times \exp\left(\mathrm{i}\frac{n_z\pi}{a_z}(z+a_z)\right), \qquad (6.123)$$

to the equivalent boundary-value problem of the forms (6.114), (6.115), (6.116).

The connecting relation is (6.105) with (6.106), where $E_{jr}^{(\mathrm{phot})}(\mathbf{x})$ are given by the formulae (6.123), (6.103), (6.104), in which

$$\gamma_j' = \sqrt{\frac{2\hbar}{\epsilon_0\epsilon_{\mathrm{r}0}\omega_j}}\sqrt{\frac{k_x^2+k_y^2}{(1+\delta_{k_x0})(1+\delta_{k_y0})V}}\,\zeta_j', \qquad (6.124)$$

$$\gamma_j = c\sqrt{\frac{2\hbar}{\epsilon_0\epsilon_{\mathrm{r}0}\omega_j}}\sqrt{\frac{k_x^2+k_y^2}{(1-\delta_{k_x0})(1-\delta_{k_y0})V}}\,\zeta_j \qquad (6.125)$$

are substituted. It can be easily derived that the vector-valued functions $\mathbf{u}_j(\mathbf{x})$, $j \in J$, satisfy a completeness relation (6.119). Usually the scalar product is considered

$$\int_G \epsilon_{\mathrm{r}}(z)\mathbf{u}^*(\mathbf{x})\cdot\mathbf{u}(\mathbf{x})\,\mathrm{d}^3\mathbf{x}. \qquad (6.126)$$

With these solutions of the problem defined, they have finite norms.

(iv) Rectangular waveguide filled with a homogeneous refractive medium

We will consider a subset $G = G_\perp \times \mathrm{R}^1$, with the boundary $\partial G = \partial G_\perp \times \mathrm{R}^1$, of the usual Euclidean space R^3. In optics, the quantization may be a definition of the vector potential operator $\hat{\mathbf{A}}(\mathbf{x}, t)$ by the relation

$$\hat{\mathbf{A}}(\mathbf{x}, t) = \sum_{j_\perp \in J_\perp}\int_{-\infty}^{\infty}\left[\mathbf{A}_{j_\perp}^{(\mathrm{phot})}(\mathbf{x}, k_z, t)\hat{a}_{j_\perp}(k_z)\right.$$
$$\left. + \mathbf{A}_{j_\perp}^{(\mathrm{phot})*}(\mathbf{x}, k_z, t)\hat{a}_{j_\perp}^\dagger(k_z)\right]\mathrm{d}k_z, \qquad (6.127)$$

where $J_\perp$ is an index set and the photon annihilation, and creation operators $\hat{a}_{j_\perp}(k_z)$ and $\hat{a}_{j_\perp}^\dagger(k_z)$ in the mode $(j_\perp, k_z)$ fulfil the commutation relations

$$[\hat{a}_{j_\perp}(k_z), \hat{a}_{j_\perp'}^\dagger(k_z')] = \delta_{j_\perp j_\perp'}\delta(k_z - k_z')\hat{1},$$
$$[\hat{a}_{j_\perp}(k_z), \hat{a}_{j_\perp'}(k_z')] = [\hat{a}_{j_\perp}^\dagger(k_z), \hat{a}_{j_\perp'}^\dagger(k_z')] = \hat{0}. \qquad (6.128)$$

Further

$$\mathbf{A}_{j_\perp}^{(\mathrm{phot})}(\mathbf{x}, k_z, t) = \sqrt{\frac{\hbar}{2\epsilon_0 \omega_{j_\perp}(k_z)}}\, \mathbf{u}_{j_\perp}(\mathbf{x}, k_z) \exp[-\mathrm{i}\omega_{j_\perp}(k_z)t], \tag{6.129}$$

with ϵ_0 the vacuum (free-space) electric permittivity, $\omega_{j_\perp}(k_z)$ and $\mathbf{u}_{j_\perp}(\mathbf{x}, k_z)$ satisfying the Helmholtz equation

$$\nabla^2 \mathbf{u}_{j_\perp}(\mathbf{x}, k_z) + \epsilon_{\mathrm{r}0} \frac{\omega_{j_\perp}^2(k_z)}{c^2} \mathbf{u}_{j_\perp}(\mathbf{x}, k_z) = \mathbf{0}, \tag{6.130}$$

the transversality condition

$$\nabla \cdot \mathbf{u}_{j_\perp}(\mathbf{x}, k_z) = 0, \tag{6.131}$$

and the boundary conditions

$$\mathbf{n}_\mathbf{x} \times \mathbf{u}_{j_\perp}(\mathbf{x}, k_z) \big|_{\partial G} = \mathbf{0}, \tag{6.132}$$

$$\mathbf{n}_\mathbf{x} \cdot \left[\nabla \times \mathbf{u}_{j_\perp}(\mathbf{x}, k_z)\right]\big|_{\partial G} = (\mathbf{n}_\mathbf{x} \times \nabla) \cdot \mathbf{u}_{j_\perp}(\mathbf{x}, k_z)\big|_{\partial G} = 0, \tag{6.133}$$

where $\mathbf{n}_\mathbf{x}$ is the normal vector at the point $\mathbf{x}$. It is required that the modal functions $\mathbf{u}_{j_\perp}(\mathbf{x}, k_z)$ be orthogonal and normalized as expressed by the relation

$$\int_G \epsilon_{\mathrm{r}0} \mathbf{u}_{j_\perp k_z}^*(\mathbf{x}, k_z) \cdot \mathbf{u}_{j'_\perp k'_z}(\mathbf{x}, k_z)\, \mathrm{d}^3\mathbf{x} = \delta_{j_\perp j'_\perp} \delta(k_z - k'_z). \tag{6.134}$$

For illustration, we will assume that

$$G = \{\mathbf{x} : -a_x < x < a_x, -a_y < y < a_y, -\infty < z < \infty\}, \tag{6.135}$$

where a_x, a_y are positive. It can be proved that the index set $J_\perp$ is a collection of $j_\perp = (n_x, n_y, s)$, where $n_r \in \{0\} \cup \mathrm{N}$, $r = x,\ y$, $s = \mathrm{TE}$, TM, $n_x > 0$ or $s = \mathrm{TE}$, $n_y > 0$ or $s = \mathrm{TE}$, and $n_x + n_y \geq 1$.

The solutions $\omega_{j_\perp}(k_z)$ have the form

$$\omega_{j_\perp}(k_z) = v\sqrt{k_x^2 + k_y^2 + k_z^2}, \tag{6.136}$$

with

$$v = \frac{1}{\sqrt{\epsilon_0 \epsilon_{\mathrm{r}0} \mu_0}}, \quad k_r = \frac{n_r \pi}{2a_r}, \quad r = x,\ y, \tag{6.137}$$

and the solutions $\mathbf{u}_{j_\perp}(\mathbf{x}, k_z)$ are connected with the classical solutions

$$E_{j_\perp x}(\mathbf{x}, k_z) = \alpha_{j_\perp}(k_z) \cos\left(\frac{n_x \pi}{2a_x}(x + a_x)\right) \times \sin\left(\frac{n_y \pi}{2a_y}(y + a_y)\right) \exp\left(\mathrm{i}k_z z\right),$$

$$E_{j_\perp y}(\mathbf{x}, k_z) = \beta_{j_\perp}(k_z) \sin\left(\frac{n_x \pi}{2a_x}(x + a_x)\right)$$
$$\times \cos\left(\frac{n_y \pi}{2a_y}(y + a_y)\right) \exp(\mathrm{i}k_z z),$$
$$E_{j_\perp z}(\mathbf{x}, k_z) = \mathrm{i}\gamma_{j_\perp}(k_z) \sin\left(\frac{n_x \pi}{2a_x}(x + a_x)\right)$$
$$\times \sin\left(\frac{n_y \pi}{2a_y}(y + a_y)\right) \exp(\mathrm{i}k_z z), \tag{6.138}$$

to the equivalent boundary-value problem

$$\nabla \cdot \mathbf{E}_{j_\perp} = 0, \quad \nabla \times \mathbf{B}_{j_\perp} + \mathrm{i}\epsilon_{\mathrm{r}0} \frac{\omega_{j_\perp}(k_z)}{c^2} \mathbf{E}_{j_\perp} = \mathbf{0}, \tag{6.139}$$
$$\nabla \cdot \mathbf{B}_{j_\perp} = 0, \quad \nabla \times \mathbf{E}_{j_\perp} - \mathrm{i}\omega_{j_\perp}(k_z) \mathbf{B}_{j_\perp} = \mathbf{0}, \tag{6.140}$$
$$\mathbf{n_x} \cdot \mathbf{B}_{j_\perp}\big|_{\partial G} = 0, \quad \mathbf{n_x} \times \mathbf{E}_{j_\perp}\big|_{\partial G} = \mathbf{0}. \tag{6.141}$$

Here $\mathbf{E}_{j_\perp} \equiv \mathbf{E}_{j_\perp}(\mathbf{x}, k_z, \omega_{j_\perp}(k_z))$, $\mathbf{B}_{j_\perp} \equiv \mathbf{B}_{j_\perp}(\mathbf{x}, k_z, \omega_{j_\perp}(k_z))$,

$$\alpha_{j_\perp}(k_z) = -\mathrm{i}\omega_{j_\perp}(k_z) \frac{k_y}{k_x^2 + k_y^2} \gamma'_{j_\perp}, \quad \beta_{j_\perp}(k_z) = \mathrm{i}\omega_{j_\perp}(k_z) \frac{k_x}{k_x^2 + k_y^2} \gamma'_{j_\perp},$$
$$\gamma_{j_\perp}(k_z) = 0, \quad \text{for } s = \mathrm{TE}, \tag{6.142}$$
$$\alpha_{j_\perp}(k_z) = -\frac{k_x k_z}{k_x^2 + k_y^2} \gamma_{j_\perp}, \quad \beta_{j_\perp}(k_z) = -\frac{k_y k_z}{k_x^2 + k_y^2} \gamma_{j_\perp},$$
$$\gamma_{j_\perp}(k_z) = \gamma_{j_\perp}, \quad \text{for } s = \mathrm{TM}, \tag{6.143}$$

with $\gamma'_{j_\perp}$, $\gamma_{j_\perp}$ complex parameters.

The connecting relation is

$$\mathbf{u}_{j_\perp}(\mathbf{x}, k_z) = -\mathrm{i}\sqrt{\frac{2\epsilon_0}{\hbar\omega_{j_\perp}}} \mathbf{E}^{(\mathrm{phot})}_{j_\perp}(\mathbf{x}, k_z), \tag{6.144}$$

where

$$\mathbf{E}^{(\mathrm{phot})}_{j_\perp}(\mathbf{x}, k_z) = \sum_{r=x,y,z} E^{(\mathrm{phot})}_{j_\perp r}(\mathbf{x}, k_z)\mathbf{e}_r, \tag{6.145}$$

with $E^{(\mathrm{phot})}_{j_\perp r}(\mathbf{x}, k_z)$ given by the formulae (6.138), (6.142), (6.143), in which

$$\gamma'_{j_\perp}(k_z) = \sqrt{\frac{\hbar}{\epsilon_0\epsilon_{\mathrm{r}0}\omega_{j_\perp}(k_z)}} \sqrt{\frac{k_x^2 + k_y^2}{(1 + \delta_{k_x 0})(1 + \delta_{k_y 0})\pi V_\perp}} \zeta'_{j_\perp}, \tag{6.146}$$

$$\gamma_{j_\perp}(k_z) = c\sqrt{\frac{\hbar}{\epsilon_0\epsilon_{\mathrm{r}0}\omega_{j_\perp}(k_z)}} \sqrt{\frac{k_x^2 + k_y^2}{(1 - \delta_{k_x 0})(1 - \delta_{k_y 0})\pi V_\perp}} \zeta_{j_\perp} \tag{6.147}$$

are substituted, $V_\perp = a_x a_y$, $|\zeta'_{j_\perp}| = |\zeta_{j_\perp}| = 1$.

The vector-valued functions $\mathbf{u}_{j_\perp}(\mathbf{x}, k_z)$, $j_\perp \in J_\perp$, satisfy a completeness relation,

$$\sum_{j_\perp \in J_\perp} \int_{-\infty}^{\infty} \epsilon_{r0} \mathbf{u}_{j_\perp}(\mathbf{x}, k_z) \mathbf{u}^*_{j_\perp}(\mathbf{x}', k_z)\, \mathrm{d}k_z$$
$$= \delta(\mathbf{x} - \mathbf{x}')\mathbf{1} - \nabla_{\mathbf{x}} \nabla_{\mathbf{x}'} \mathcal{G}(\mathbf{x}, \mathbf{x}'), \tag{6.148}$$

where $\mathcal{G}(\mathbf{x}, \mathbf{x}')$ is a Green's function for the Laplace operator (the Dirichlet problem).

(v) Rectangular waveguide filled with a nonhomogeneous refractive medium

Quantum electrodynamics in periodic dielectric media has been treated several times (Caticha and Caticha 1992, Kweon and Lawandy 1995, Tip 1997). For the quantization true modal functions may be useful which we have found in the Section 6.1.2(ii). They have not been presented just as needed, but complex conjugated. We must consider another difficulty that they are not orthogonal.

6.1.5 Quantization in Linear Nonhomogeneous Nonconducting Medium

Tip (1997) reminds one of the fact that a quantization of the electromagnetic field is made for a complete quantum description of interaction of radiation with atoms or molecules. These however may be placed in a "photonic material", and so a quantization of the electromagnetic field in the material is purposeful.

Tip (1997) points out the fact that photonic crystals are classical dielectric media with a periodic dielectric permittivity. The periodicity may give rise to a band structure. If a band gap is present and an embedded atom has a transition frequency in the gap, a single-photon emission is inhibited. However, even if no gap develops, decay rates may differ much from their free values. Free decay rates can be obtained through an application of Fermi's golden rule. Generalization for any electric permittivity and magnetic permeability uses the so-called local density of states related to the classical electromagnetic field again.

The author expounds a general approach to the quantization of linear evolution equations. Such an equation for $F(t)$ from a separable, real Hilbert space has the form

$$\frac{\partial F(t)}{\partial t} = NF(t) - G(t),\ N^\dagger = -N. \tag{6.149}$$

The matter is a quantization of the quantity $F(t)$.

Tip (1997) assumes a linear, nonhomogeneous, nonconducting material medium including external currents or a Schrödinger quantum particle system. We have mostly adopted the notation and have changed it in part only. In particular,

$\epsilon_0 = \mu_0 = 1$ upon the choice of units. The case $\epsilon = \mu = 1$ is called the vacuum case and, with arbitrary ϵ and μ, one may still speak of a free electromagnetic field. We will still use the asterisk for complex conjugates, the centred dot in multiplication of matrices, which indicates that they are treated as tensors.

In general, an n-component field over the space $\mathrm{R}^d \times \mathrm{R}$ is considered, which satisfies the equation

$$\frac{\partial}{\partial t}\mathbf{f}(\mathbf{x}, t) = \mathbf{M}(\mathbf{x}, \nabla) \cdot \mathbf{f}(\mathbf{x}, t), \tag{6.150}$$

where $\mathbf{x} \in \mathrm{R}^d$, and $\mathbf{M} \equiv \mathbf{M}(\mathbf{x}, \nabla)$ is an $n \times n$ matrix, whose entries are real partial differential operators with, in general, variable coefficients. There exists a bounded, real, invertible matrix $\boldsymbol{\rho}(\mathbf{x})$ with bounded inverse such that

$$\mathbf{M}^\dagger \cdot \boldsymbol{\rho}^2(\mathbf{x}) + \boldsymbol{\rho}^2(\mathbf{x}) \cdot \mathbf{M} = \mathbf{0}, \tag{6.151}$$

or the "energy"

$$\mathcal{E} = \frac{1}{2} \int [\boldsymbol{\rho}(\mathbf{x}) \cdot \mathbf{f}(\mathbf{x}, t)]^2 \, \mathrm{d}^3\mathbf{x} \tag{6.152}$$

can be introduced. Considering

$$\mathbf{F}(\mathbf{x}, t) = \boldsymbol{\rho}(\mathbf{x}) \cdot \mathbf{f}(\mathbf{x}, t), \tag{6.153}$$

$$\mathbf{N} = \boldsymbol{\rho} \cdot \mathbf{M} \cdot \boldsymbol{\rho}^{-1}, \tag{6.154}$$

where $\mathbf{N} \equiv \mathbf{N}(\mathbf{x}, \nabla)$, we obtain the equations

$$\frac{\partial}{\partial t}\mathbf{F}(\mathbf{x}, t) = \mathbf{N}(\mathbf{x}, \nabla) \cdot \mathbf{F}(\mathbf{x}, t), \tag{6.155}$$

$$\mathbf{N}^\dagger = -\mathbf{N}. \tag{6.156}$$

On introducing, in addition, a norm

$$\|\mathbf{F}(\mathbf{x}, t)\|^2 = 2\mathcal{E}, \tag{6.157}$$

equation (6.155) describes a unitary time evolution on the real space $\mathcal{H}_\mathrm{r} = L^2(\mathrm{R}^d, \, \mathrm{d}\mathbf{x}; \mathrm{R}^n)$

$$\mathbf{F}(\mathbf{x}, t) = \exp(\mathbf{N}t) \cdot \mathbf{F}(\mathbf{x}, 0). \tag{6.158}$$

Tip (1997) intends to work with a real Hilbert space as long as possible. He begins with two examples of field equations with a conservation law. Proceeding with Maxwell's equations for a nonconducting material medium, he writes the Maxwell equations, which are equations for real waves originally. He assumes that

$\epsilon(\mathbf{x})$ and $\mu(\mathbf{x})$ are real, smooth, bounded from below and above by positive constants.

For $J = 0$, the energy is conserved,

$$\begin{aligned}\mathcal{E} &= \frac{1}{2}\int \left\{\epsilon(\mathbf{x})[\mathbf{E}(\mathbf{x}, t)]^2 + \frac{1}{\mu(\mathbf{x})}[\mathbf{B}(\mathbf{x}, t)]^2\right\} \mathrm{d}^3\mathbf{x} \\ &= \frac{1}{2}\int |\mathbf{F}(\mathbf{x}, t)|^2 \,\mathrm{d}^3\mathbf{x},\end{aligned} \tag{6.159}$$

where

$$\mathbf{F}(\mathbf{x}, t) = \begin{pmatrix} \sqrt{\epsilon}\,\mathbf{E}(\mathbf{x}, t) \\ \frac{1}{\sqrt{\mu}}\,\mathbf{B}(\mathbf{x}, t) \end{pmatrix} = \begin{pmatrix} \mathbf{F}_1(\mathbf{x}, t) \\ \mathbf{F}_2(\mathbf{x}, t) \end{pmatrix}. \tag{6.160}$$

Relation (6.154) becomes

$$\mathbf{N} = \begin{pmatrix} \mathbf{0} & \mathbf{N}_{12} \\ \mathbf{N}_{21} & \mathbf{0} \end{pmatrix} = \begin{pmatrix} \mathbf{0} & -\frac{1}{\sqrt{\epsilon}}(\boldsymbol{\varepsilon}\cdot\nabla)\frac{1}{\sqrt{\mu}} \\ \frac{1}{\sqrt{\epsilon}}(\boldsymbol{\varepsilon}\cdot\nabla)\frac{1}{\sqrt{\mu}} & \mathbf{0} \end{pmatrix}, \tag{6.161}$$

where $\boldsymbol{\varepsilon}$ is the Levi-Cività pseudotensor, or

$$\mathbf{N} = \mathbf{W}\cdot\mathbf{N}_0\cdot\mathbf{W}, \tag{6.162}$$

where

$$\mathbf{N}_0 = \begin{pmatrix} \mathbf{0} & -\boldsymbol{\varepsilon}\cdot\nabla \\ \boldsymbol{\varepsilon}\cdot\nabla & \mathbf{0} \end{pmatrix}, \quad \mathbf{W} = \begin{pmatrix} \frac{1}{\sqrt{\epsilon}}\mathbf{1} & \mathbf{0} \\ \mathbf{0} & \frac{1}{\sqrt{\mu}}\mathbf{1} \end{pmatrix}. \tag{6.163}$$

The orthogonal eigenprojector of the matrix $\mathbf{N}$ at the eigenvalue 0 is

$$\mathbf{P}_0 = \begin{pmatrix} \mathbf{P}_1 & \mathbf{0} \\ \mathbf{0} & \mathbf{P}_2 \end{pmatrix} = \begin{pmatrix} \sqrt{\epsilon}\nabla[\nabla\cdot\epsilon\nabla]^{-1}\nabla\sqrt{\epsilon} & \mathbf{0} \\ \mathbf{0} & \sqrt{\mu}\nabla[\nabla\cdot\mu\nabla]^{-1}\nabla\sqrt{\mu} \end{pmatrix}. \tag{6.164}$$

Here $[\nabla\cdot\epsilon\nabla]^{-1}$ and $[\nabla\cdot\mu\nabla]^{-1}$ can be expressed by an integral transform in the vacuum case.

Tip (1997) pays attention to the Helmholtz operators and to the scattering theory, which is not reproduced here. The Helmholtz operators are

$$\begin{aligned}\mathbf{H}_1 &= -\mathbf{N}_{12}\mathbf{N}_{21} = \frac{1}{\sqrt{\epsilon}}(\boldsymbol{\varepsilon}\cdot\nabla)\cdot\frac{1}{\mu}(\boldsymbol{\varepsilon}\cdot\nabla)\frac{1}{\sqrt{\epsilon}}, \\ \mathbf{H}_2 &= -\mathbf{N}_{21}\mathbf{N}_{12} = \frac{1}{\sqrt{\mu}}(\boldsymbol{\varepsilon}\cdot\nabla)\cdot\frac{1}{\epsilon}(\boldsymbol{\varepsilon}\cdot\nabla)\frac{1}{\sqrt{\mu}}.\end{aligned} \tag{6.165}$$

Let $\mathbf{u}_{\lambda\alpha}$ denote eigenvectors of $\mathbf{H}_1$,

$$\mathbf{H}_1\cdot\mathbf{u}_{\lambda\alpha} = \lambda^2\mathbf{u}_{\lambda\alpha}, \tag{6.166}$$

where $\lambda \geq 0$ and α distinguish eigenvectors at the same eigenvalue. As $\mathbf{H}_1$ is a real operator, $\mathbf{u}^*_{\lambda\alpha}$ differs from some $\mathbf{u}_{\lambda\beta}$ by a constant factor only. We can always use real eigenvectors. As $\mathbf{H}_2 = \mathbf{N}_{12}^{-1} \cdot \mathbf{H}_1 \cdot \mathbf{N}_{12} = \mathbf{N}_{21} \cdot \mathbf{H}_1 \cdot \mathbf{N}_{21}^{-1}$, the eigenvector $\mathbf{H}_2$ can be obtained for instance as $\mathbf{N}_{21} \cdot \mathbf{u}_{\lambda\alpha}$.

The exposition of the Lagrange formalism is introduced by a warning that in the case

$$\mathcal{H} = \mathcal{H}_1 \oplus \mathcal{H}_2,\, N = \begin{pmatrix} 0 & N_{12} \\ N_{21} & 0 \end{pmatrix},\, N_{21} = -N_{12}^{\dagger}, \tag{6.167}$$

equation (6.149) for $G = 0$ is not obtained using the Lagrange formalism, if we take F for the coordinate field. For the two components of the vector $F = \begin{pmatrix} F_1 \\ F_2 \end{pmatrix}$, separate equations are obtained and their connection is lost.

It is recommended to use $\xi = N^{-1}F$ as coordinate field. In the scalar wave case, this means to define the coordinate field using solution of a generalization of the Poisson equation. In the Maxwell case, $\mathbf{N}$ cannot be inverted due to the zero eigenvalue. Tip (1997) reminds one of projection upon the propagating modes. In the vacuum situation

$$\begin{pmatrix} \xi_1 \\ \xi_2 \end{pmatrix} = \mathbf{N}_0^{-1} \begin{pmatrix} \mathbf{E}^{\perp} \\ \mathbf{B} \end{pmatrix} \tag{6.168}$$

is introduced. The quantity $-\boldsymbol{\xi}_1$ is the vector potential $\mathbf{A}$ in the Coulomb gauge. The theory comprises the relations

$$\begin{gathered} \mathbf{E}^{\perp} = -\frac{\partial \mathbf{A}}{\partial t}, \quad \mathbf{B} = \nabla \times \mathbf{A}, \\ \mathbf{E}^{\parallel} = -\nabla\Phi, \quad \Phi(\mathbf{x}) = \int \frac{\rho(\mathbf{x}')}{4\pi\,|\mathbf{x}-\mathbf{x}'|}\,\mathrm{d}^3\mathbf{x}', \quad \rho(\mathbf{x}) = \nabla\cdot\mathbf{E}(\mathbf{x}), \end{gathered} \tag{6.169}$$

where $\rho(\mathbf{x})$ is the external charge density.

In general case, we let P mean the projector upon the null space $\mathcal{N} = \mathcal{N}(N)$ of N and $Q = 1 - P$. It holds that

$$P = \begin{pmatrix} P_1 & 0 \\ 0 & P_2 \end{pmatrix},\, Q = \begin{pmatrix} Q_1 & 0 \\ 0 & Q_2 \end{pmatrix},\, Q_j = 1 - P_j, \tag{6.170}$$

where P_j acts in $\mathcal{H}_j$, $j = 1, 2$. We choose $G = \begin{pmatrix} G_1 \\ G_2 \end{pmatrix}$. As

$$\begin{pmatrix} 0 & N_{12} \\ N_{21} & 0 \end{pmatrix}^{-1} = \begin{pmatrix} 0 & -N_{21}^{-1} \\ -N_{12}^{-1} & 0 \end{pmatrix}, \tag{6.171}$$

a quantity

$$\tilde{\xi} = N_{21}^{-1} Q_2 F_2 \tag{6.172}$$

is introduced satisfying the generalization of the Coulomb gauge condition

$$P_1\tilde{\xi} = 0. \tag{6.173}$$

In terms of this quantity, propagating components can be expressed,

$$\begin{aligned} Q_1 F_1 &= -\frac{\partial \tilde{\xi}}{\partial t} - N_{21}^{-1} Q_2 G_2, \\ Q_2 F_2 &= -N_{21}\tilde{\xi}. \end{aligned} \tag{6.174}$$

Tip (1997) also considers a generalization of a gauge transformation

$$\xi = \tilde{\xi} + P_1\eta,\ \eta \in \mathcal{H}_1. \tag{6.175}$$

We make a replacement $N_{21}^{-1} Q_2 G_2 \to N_{21}^{-1} Q_2 G_2 - P_1 \frac{\partial \eta}{\partial t}$,

$$Q_1 F_1 = -\frac{\partial \xi}{\partial t} - N_{21}^{-1} Q_2 G_2 + P_1 \frac{\partial \eta}{\partial t} \tag{6.176}$$

and

$$Q_2 F_2 = -N_{21}\xi. \tag{6.177}$$

With respect to the application to the Maxwell equations, we assume that

$$G = \begin{pmatrix} G_1 \\ 0 \end{pmatrix},\ \ P_2 F_2|_{t=0} = 0. \tag{6.178}$$

Equations (6.176) and (6.177) simplify,

$$Q_1 F_1 = -\frac{\partial \xi}{\partial t} + P_1 \frac{\partial \eta}{\partial t}, \tag{6.179}$$

$$F_2 = -N_{21}\xi. \tag{6.180}$$

Tip (1997) introduces a generalization ξ_0 of the scalar potential Φ of the Maxwell theory. He considers another real Hilbert space $\mathcal{H}_3$ and an invertible operator M from $\mathcal{H}_3$ into $P_1\mathcal{H}_1$. Then

$$\xi_0 = -M^{-1} P_1 \left(F_1 + \frac{\partial \eta}{\partial t} \right). \tag{6.181}$$

He introduces a generalization ρ of the charge density,

$$\rho = -M^{\dagger} P_1 F_1. \tag{6.182}$$

He lets $(.,.)_j$ mean the inner product in $\mathcal{H}_j$ and presents the Lagrangian

$$L = \frac{1}{2}\left(\frac{\partial \xi}{\partial t} + M\xi_0, \frac{\partial \xi}{\partial t} + M\xi_0\right)_1 - \frac{1}{2}(N_{21}\xi, N_{21}\xi)_2$$
$$+ (G_1, \xi)_1 - (\rho, \xi_0)_3 . \tag{6.183}$$

Tip (1997) considers generalizations of the Coulomb, Lorentz, and temporal gauges. The field η has specific properties in each particular gauge. Expounding the C gauge, he writes the condition

$$P_1\xi = 0, \tag{6.184}$$

which leads to equations

$$\frac{\partial^2 \xi}{\partial t^2} - N_{12}N_{21}\xi = Q_1 G_1,$$
$$M^\dagger M\xi_0 = \rho. \tag{6.185}$$

He eliminates ξ_0 by expressing it in terms of ρ and presents a Lagrangian, the canonical momentum field associated with ξ, $\pi = \dot{\xi}$, and a Hamiltonian.

Expounding the L gauge, he needs the condition

$$\frac{\partial \xi_0}{\partial t} - M^\dagger P_1 \xi = 0, \tag{6.186}$$

which leads to equations

$$\frac{\partial^2 \xi_0}{\partial t^2} + M^\dagger M\xi_0 = \rho, \tag{6.187}$$

$$\frac{\partial^2 \xi}{\partial t^2} - N_{12}N_{21}\xi + MM^\dagger P_1\xi = G_1. \tag{6.188}$$

He writes a Lagrangian, the momentum field $\pi_0 = \dot{\xi}_0$, and a Hamiltonian.

Expounding the T gauge, the author writes the condition (and the equation)

$$\xi_0 = 0, \tag{6.189}$$

which leads to an equation

$$\frac{\partial^2 \xi}{\partial t^2} - N_{12}N_{21}\xi = G_1. \tag{6.190}$$

He presents a Lagrangian and a Hamiltonian.

At this level, Tip (1997) reminds one of the familiar method of canonical quantization and concentrates himself on the C gauge case and the L gauge case. The C

gauge case is discussed in full detail. He chooses an orthonormal basis $\{u_j\}$ in the subspace $Q_1\mathcal{H}_1 \subset \mathcal{H}_1$ and decomposes

$$\xi = \sum_j \xi_j u_j, \quad \pi = \sum_j \pi_j u_j, \tag{6.191}$$

where $\xi_j = \xi(u_j) = (\xi, u_j)_1, \pi_j = \pi(u_j) = (\pi, u_j)_1$. He expresses the Hamiltonian in terms of ξ_j, π_j, which obey the Poisson brackets

$$\{\xi_j, \pi_k\} = \delta_{jk}. \tag{6.192}$$

A quantization is accomplished by replacing the Poisson brackets by the commutators

$$[\hat{\xi}(u_j), \hat{\pi}(u_k)] = \mathrm{i}\delta_{jk}\hat{1}. \tag{6.193}$$

We utilize hats here, although Tip (1997) does not write them, or lets them mean something else. He defines the complexifications of the Hilbert spaces and the Fock space $\mathcal{F}(\mathcal{H})$ over any Hilbert space $\mathcal{H}$. He introduces the annihilation (creation) operator $\hat{a}(\varphi)$ ($\hat{a}^\dagger(\varphi)$), which acts in $\mathcal{F}(\mathcal{H})$, where φ is the wave function of an annihilated (created) boson. In a rather complicated manner, it must be shown that the Hamiltonian can be expressed in terms of $\hat{a}(u_j)$, $\hat{a}^\dagger(u_j)$, which act in $\mathcal{F}(\mathcal{H})$, where $\mathcal{H}$ is the complexification of $\mathcal{H}_1$.

In the comment on the L gauge case, Tip (1997) chooses, in addition, an orthonormal basis $\{v_j\}$ in the subspace $P_1\mathcal{H}$ and such a basis $\{w_j\}$ in the complexification $\mathcal{H}'$ of the real space $\mathcal{H}_3$. He expounds that the Hamiltonian can be expressed in terms of $\hat{a}(u_j)$, $\hat{a}(v_j)$, $\hat{b}(w_j)$ and their Hermitian conjugates. Here $\hat{b}(w_j)$ is the annihilation operator, which acts in $\mathcal{F}(\mathcal{H}')$.

Applying this formalism to the Maxwell equations, he determines that

$$\boldsymbol{\xi} = -\sqrt{\epsilon}\mathbf{A}, \boldsymbol{\pi} = -\sqrt{\epsilon}\frac{\partial \mathbf{A}}{\partial t}, \xi_0 = \Phi, \mathbf{M} = \sqrt{\epsilon}\nabla. \tag{6.194}$$

Moreover, it holds that $\mathbf{M}^\dagger = -\nabla\sqrt{\epsilon}, \mathbf{M}^\dagger\cdot\mathbf{M} = -\nabla\sqrt{\epsilon}\cdot\sqrt{\epsilon}\nabla, \mathbf{M}\mathbf{M}^\dagger = -\sqrt{\epsilon}\nabla\nabla\sqrt{\epsilon}$.

In the C gauge, the condition (6.184) becomes

$$\nabla \cdot \epsilon\mathbf{A} = 0, \tag{6.195}$$

leading to the equations

$$\frac{\partial^2(\epsilon\mathbf{A})}{\partial t^2} + \nabla \times \frac{1}{\mu}(\nabla \times \mathbf{A}) = \mathbf{Q}_1 \cdot \mathbf{J}, \tag{6.196}$$

$$\nabla\sqrt{\epsilon} \cdot \sqrt{\epsilon}\nabla\Phi = -\rho, \tag{6.197}$$

where $\mathbf{J}$ is the external current density. Tip (1997) also presents a Lagrangian and a Hamiltonian. The Poisson brackets have the form

$$\{\boldsymbol{\xi}(\mathbf{x}), \boldsymbol{\pi}(\mathbf{y})\} = \mathbf{Q}_1(\mathbf{x}, \mathbf{y}), \tag{6.198}$$

where $\mathbf{Q}_1(\mathbf{x}, \mathbf{y}) = \delta(\mathbf{x} - \mathbf{y}) - \mathbf{P}_1(\mathbf{x}, \mathbf{y})$, $\mathbf{P}_1(\mathbf{x}, \mathbf{y})$ are kernels associated with the projectors $\mathbf{Q}_1$ and $\mathbf{P}_1$, respectively.

In the L gauge, the condition (6.186) becomes

$$\frac{\partial \Phi}{\partial t} + \nabla \cdot \epsilon \mathbf{A} = 0, \tag{6.199}$$

leading to the equations of motion

$$\frac{\partial^2 \Phi}{\partial t^2} - \nabla \cdot \epsilon \nabla \Phi = \rho, \tag{6.200}$$

$$\frac{\partial^2 \mathbf{A}}{\partial t^2} + \frac{1}{\epsilon} \nabla \times \frac{1}{\mu} (\nabla \times \mathbf{A}) - \nabla^2 \epsilon \mathbf{A} = \frac{1}{\epsilon} \mathbf{J}. \tag{6.201}$$

A Lagrangian, the momentum field $\boldsymbol{\pi}_0 = \frac{\partial \Phi}{\partial t}$, and a Hamiltonian are also presented. The Poisson brackets have the form

$$\begin{aligned} \{\Phi(\mathbf{x}), \boldsymbol{\pi}_0(\mathbf{y})\} &= \delta(\mathbf{x}, \mathbf{y}), \\ \{\boldsymbol{\xi}(\mathbf{x}), \boldsymbol{\pi}(\mathbf{y})\} &= \mathbf{1}\delta(\mathbf{x}, \mathbf{y}). \end{aligned} \tag{6.202}$$

Tip (1997) devotes an appropriate amount of place to the application to the atomic radiative decay in dielectrics.

Under the usual assumptions on the dielectric permittivity, a quantization of the Hamiltonian formalism of the electromagnetic field using a method close to the microscopic approach was performed by Tip (1998). A proper definition of band gaps in the periodic case and a new continuity equation for energy flow was obtained, and an S-matrix formalism for scattering from absorbing objects was worked out. In this way, the generation of Čerenkov and transition radiation have been investigated.

6.2 Corrugated Waveguides

The use of dielectric optical waveguides (Marcuse 1974, Yariv and Yeh 1984) and the coupled-mode theory appropriate in the case of corrugated waveguides lead naturally to the task of describing the propagation of a general quantum state in these devices. The question of possibility and impossibility of quantizing the classical description is not posed usually. The apology for it is that the copropagation does not make difficulties. The description can be analyzed in the framework of the theoretical mechanics with the time variable replaced by the propagation distance. We work easily with the Heisenberg and Schrödinger pictures and formulate respective equations. The quantum momentum operator can be derived with respect to the modal orthonormalization property on a cross section of an optical waveguide (Liñares and Nistal 2003). The difficulties due to counterpropagation are regularly disregarded. In fact, the classical description is free of difficulties, and the theoretical mechanics that has been already modified by the replacement of the time variable with the space variable can be extended to involve the counterpropagation (Luis and Peřina 1996b). It has been concluded that the quantization will not be successful in the case of counterpropagation in a nonlinear medium (except optical parametric processes).

The propagation in a linear dielectric, and in the devices based on such materials, can be quantized in time (Dalton et al. 1996, Dalton et al. 1999b, Dalton and Knight 1999a,b). Here we shall not criticize the use of operators in situations which are classical essentially, since the literature abounds with this (Janszky et al. 1988, Peřina 1995a,b, Peřina and Peřina, Jr. 1995b,c, Korolkova and Peřina 1997c). The dependence of outputs on inputs can be formulated in the spatial Heisenberg and Schrödinger pictures without respective differential equations. This restriction is due to the peculiar nature of quantization. In the copropagation, the spatio-temporal description (cf. (Lukš and Peřinová 2002)) has not been used and therefore the unusual description has not been used by anybody even in the counterpropagation, where it is hardly dispensable.

The simplified quantization enables one to employ knowledge of quantum mechanical descriptions as follows. An amplifier should be described in the Schrödinger picture, if we use quasidistributions and the antinormal ordering (the Husimi functions) for the expression of the input–output dependence. An attenuator ought to be described in the Schrödinger picture, if we employ the quasidistributions and the normal ordering (the Glauber diagonal representation) to a similar goal. To show this, we present a formal definition of an amplifier and that of an attenuator. These definitions operate with the integrated quantum-noise terms which are needed for the input–output relations to preserve the commutators. The integrated quantum-noise terms can be formally decomposed into creation operators in amplifier case and into annihilation operators in the attenuator case. We believe that we can provide more than such a formal expansion considering fields of modes or mode densities coupled to the counterpropagating modes. These fields are well-known quantum reservoirs, e.g., (Louisell 1973), whose frequency dependence has been replaced by the position dependence.

In a more complicated case when in the mode a (let us say) an attenuation proceeds and in the mode b an amplification occurs (Severini et al. 2004), a more complicated behaviour in the Schrödinger picture can be expected, there is not a guide to choose the ordering. To describe the amplification and attenuation, one needs a full Heisenberg–Langevin approach.

Distributed feedback laser has attracted attention as a device, which in the framework of coupled-wave theory, deserves quantization (Toren and Ben Aryeh 1994). The quantization produced by the authors seems to be very complicated. Unfortunately, Toren and Ben Aryeh (1994) have not developed an overarching quantization of the analysis of amplification and that of the contradirectional coupling.

6.2.1 Lossless Propagation in a Waveguide Structure

Peřina Jr., et al. (2004) study an optical parametric process, namely a second-order process. They had studied photonic band-gap structures and continue the work (Tricca et al. 2004) for the second-harmonic generation in a planar nonlinear corrugated waveguide. They consider both the influence of the corrugation of the waveguide on the longitudinal confinement of the signal and idler modes (a modification of (Tricca 2004)) and this influence on the phase matching of the process.

They present the decomposition of the electric-field amplitude related to photons of the respective modes

$$\mathbf{E}(x, y, z, t) = \mathrm{i} \sum_m \sqrt{\frac{\hbar\omega_m}{2\epsilon_0 \bar{\epsilon}_\mathrm{r} L}} \mathbf{e}_m \times \{A_m(z) f_m(x, y) \exp[\mathrm{i}(k_{mz} z - \omega_m t)] - \mathrm{c.\,c.}\}, \tag{6.203}$$

where A_m is the amplitude of the mth mode, f_m means the transverse eigenfunction of the mth mode, $\mathbf{e}_m$ stands for the polarization vector, ω_m denotes the frequency, and $\mathbf{k}_m$ is the wave vector of the mth mode. The mean permittivity of the waveguide is denoted as $\bar{\epsilon}_\mathrm{r}$, ϵ_0 stands for the vacuum permittivity, $\hbar$ is the reduced Planck constant, and L is the length of the structure. The abbreviation c. c. stands for complex-conjugated terms.

The function $f_m(x, y)$ is a solution of the equation

$$\nabla_\mathrm{T}^2 f_m(x, y) - k_{mz}^2 f_m(x, y) + \mu\epsilon_0\bar{\epsilon}_\mathrm{r}\omega_m^2 f_m(x, y) = 0. \tag{6.204}$$

At present, one has not yet realized a perfect (or an imperfect) quantization. The function $f_m(x, y)$ is normalized, has the property

$$\int\int |f_m(x, y)|^2 \,\mathrm{d}x\,\mathrm{d}y = 1. \tag{6.205}$$

The electric-field amplitude $\mathbf{E} \equiv \mathbf{E}(\mathbf{r}, t)$ satisfies the wave equation inside the waveguide

$$\nabla^2\mathbf{E} - \mu\epsilon_0\epsilon_\mathrm{r}\frac{\partial^2\mathbf{E}}{\partial t^2} = \mu\frac{\partial^2\mathbf{P}_\mathrm{nl}}{\partial t^2}, \tag{6.206}$$

where $\epsilon_0\epsilon_\mathrm{r}$ stands for the permittivity of the waveguide, μ denotes the vacuum permeability, and $\mathbf{P}_\mathrm{nl}$ describes the nonlinear polarization of the medium. The relative permittivity $\epsilon_\mathrm{r}(\mathbf{r})$ can be written as follows

$$\epsilon_\mathrm{r}(x, y, z) = \bar{\epsilon}_\mathrm{r}(x, y) + \Delta\varepsilon_\mathrm{r}(x, y, z). \tag{6.207}$$

Here $\Delta\varepsilon_r(x, y, z)$ are small variations of the permittivity related to the corrugation. These variations decompose into harmonic functions

$$\Delta\varepsilon_r(x, y, z) = \sum_{q=-\infty}^{\infty} \varepsilon_q(x, y) \exp\left(\mathrm{i}q\frac{2\pi}{\Lambda_l}z\right), \ \varepsilon_0(x, y) = 0, \tag{6.208}$$

$\varepsilon_q(x, y)$ are coefficients of the decomposition and Λ_l is the spatial period of the grating. The polarization $\mathbf{P}_{nl}$ of the medium is determined using the second-order susceptibility tensor $\boldsymbol{\chi}$,

$$\mathbf{P}_{nl} = \epsilon_0 \boldsymbol{\chi} : \mathbf{EE}, \tag{6.209}$$

where : denotes a contraction, i.e. a double sum to be carried out after the tensors are replaced by their components and products of the corresponding components are formed. On the assumption of three monochromatic components, a substitution of the amplitude (6.203) into the wave equation (6.206) provides three coupled Helmholtz equations for these components. We consider two directions of propagation for each of the monochromatic components. We have six modes: the signal forward-propagating mode (with amplitude A_{s_F}), the signal backward-propagating mode (A_{s_B}), the idler forward-propagating mode (A_{i_F}), the idler backward-propagating mode (A_{i_B}), the pump forward-propagating mode (A_{p_F}) and, finally, the pump backward-propagating mode (A_{p_B}). In the framework of the coupled-mode theory, we represent each of the three Helmholtz equations with two ordinary differential equations for the amplitudes

$$\begin{aligned}
\frac{\mathrm{d}A_{s_F}}{\mathrm{d}z} &= \mathrm{i}K_s \exp(-\mathrm{i}\delta_s z)A_{s_B} + K_F \exp(\mathrm{i}\delta_F z)A_{p_F}A_{i_F}^*,\\
\frac{\mathrm{d}A_{i_F}}{\mathrm{d}z} &= \mathrm{i}K_i \exp(-\mathrm{i}\delta_i z)A_{i_B} + K_F \exp(\mathrm{i}\delta_F z)A_{p_F}A_{s_F}^*,\\
\frac{\mathrm{d}A_{s_B}}{\mathrm{d}z} &= -\mathrm{i}K_s^* \exp(\mathrm{i}\delta_s z)A_{s_F} - K_B \exp(-\mathrm{i}\delta_B z)A_{p_B}A_{i_B}^*,\\
\frac{\mathrm{d}A_{i_B}}{\mathrm{d}z} &= -\mathrm{i}K_i^* \exp(\mathrm{i}\delta_i z)A_{i_F} - K_B \exp(-\mathrm{i}\delta_B z)A_{p_B}A_{s_B}^*,\\
\frac{\mathrm{d}A_{p_F}}{\mathrm{d}z} &= \mathrm{i}K_p \exp(-\mathrm{i}\delta_p z)A_{p_B} - K_F^* \exp(-\mathrm{i}\delta_F z)A_{s_F}A_{i_F},\\
\frac{\mathrm{d}A_{p_B}}{\mathrm{d}z} &= -\mathrm{i}K_p^* \exp(\mathrm{i}\delta_p z)A_{p_F} + K_B^* \exp(\mathrm{i}\delta_B z)A_{s_B}A_{i_B},
\end{aligned} \tag{6.210}$$

where $K_p = 0$ and

$$\begin{aligned}
\delta_a &= |k_{a_F z}| + |k_{a_B z}| - \delta_l, \ a = \mathrm{s, i},\\
\delta_l &= \frac{2\pi}{\Lambda_l},\\
\delta_b &= |k_{p_b z}| - |k_{s_b z}| - |k_{i_b z}|, \ b = \mathrm{F, B}.
\end{aligned} \tag{6.211}$$

The linear coupling constants K_s and K_i are given as

$$K_a = \frac{\mu\epsilon_0\omega_a^2}{2|k_{a_\mathrm{F}z}|}\int\int \varepsilon_1(x,y) f_{a_\mathrm{F}}^*(x,y) f_{a_\mathrm{B}}(x,y)\,\mathrm{d}x\,\mathrm{d}y,\ a = \mathrm{s, i}, \tag{6.212}$$

where we have assumed that $|k_{a_\mathrm{F}z}| = |k_{a_\mathrm{B}z}|$. The expressions for the nonlinear coupling constants K_F and K_B are

$$\begin{aligned} K_b &= \frac{\mu\epsilon_0\omega_\mathrm{s}}{|k_{\mathrm{s}_b z}|}\sqrt{\frac{\hbar\omega_\mathrm{p}\omega_\mathrm{i}\omega_\mathrm{s}}{2\epsilon_0\bar{\epsilon}_\mathrm{r}L}}\,\boldsymbol{\chi}\vdots\mathbf{e}_\mathrm{p}\mathbf{e}_\mathrm{i}\mathbf{e}_\mathrm{s}\int\int f_{\mathrm{p}_b}(x,y) f_{\mathrm{i}_b}^*(x,y) f_{\mathrm{s}_b}^*(x,y)\,\mathrm{d}x\,\mathrm{d}y \\ &= \frac{\mu\epsilon_0\omega_\mathrm{i}}{|k_{\mathrm{i}_b z}|}\sqrt{\frac{\hbar\omega_\mathrm{p}\omega_\mathrm{s}\omega_\mathrm{i}}{2\epsilon_0\bar{\epsilon}_\mathrm{r}L}}\,\boldsymbol{\chi}\vdots\mathbf{e}_\mathrm{p}\mathbf{e}_\mathrm{s}\mathbf{e}_\mathrm{i}\int\int f_{\mathrm{p}_b}(x,y) f_{\mathrm{s}_b}^*(x,y) f_{\mathrm{i}_b}^*(x,y)\,\mathrm{d}x\,\mathrm{d}y \\ &= \frac{\mu\epsilon_0\omega_\mathrm{p}}{|k_{\mathrm{p}_b z}|}\sqrt{\frac{\hbar\omega_\mathrm{s}\omega_\mathrm{i}\omega_\mathrm{p}}{2\epsilon_0\bar{\epsilon}_\mathrm{r}L}}\,\boldsymbol{\chi}\vdots\mathbf{e}_\mathrm{s}\mathbf{e}_\mathrm{i}\mathbf{e}_\mathrm{p}\int\int f_{\mathrm{s}_b}^*(x,y) f_{\mathrm{i}_b}^*(x,y) f_{\mathrm{p}_b}(x,y)\,\mathrm{d}x\,\mathrm{d}y, \end{aligned} \tag{6.213}$$

where $\vdots$ denotes a contraction, i.e. a treble sum to be carried out after the tensors are replaced by their components and products of the corresponding components are formed. We have assumed that $\frac{\omega_\mathrm{s}}{|k_{\mathrm{s}_b z}|} \approx \frac{\omega_\mathrm{i}}{|k_{\mathrm{i}_b z}|} \approx \frac{\omega_\mathrm{p}}{|k_{\mathrm{p}_b z}|}$.

The dependencies of the solutions to equations (6.210) on the boundary data $A_{a_\mathrm{F}}(0)$, $A_{a_\mathrm{B}}(L)$, $a = \mathrm{s, i, p}$ (of the solutions to the boundary-value problem) can be considered as classical input–output relations. These are significant for investigation of the effect of stochastic boundary data.

Approximate results can be obtained by considering variations of the classical solutions δA_{a_b}, $a = \mathrm{s, i, p}$, $b = \mathrm{F, B}$. The variations verify linear equations

$$\begin{aligned} \frac{\mathrm{d}\delta A_{\mathrm{s_F}}}{\mathrm{d}z} &= \mathrm{i}K_\mathrm{s}\exp(-\mathrm{i}\delta_\mathrm{s}z)\delta A_{\mathrm{s_B}} + K_\mathrm{F}\exp(\mathrm{i}\delta_\mathrm{F}z)\delta(A_{\mathrm{p_F}}A_{\mathrm{i_F}}^*), \\ \frac{\mathrm{d}\delta A_{\mathrm{i_F}}}{\mathrm{d}z} &= \mathrm{i}K_\mathrm{i}\exp(-\mathrm{i}\delta_\mathrm{i}z)\delta A_{\mathrm{i_B}} + K_\mathrm{F}\exp(\mathrm{i}\delta_\mathrm{F}z)\delta(A_{\mathrm{p_F}}A_{\mathrm{s_F}}^*), \\ \frac{\mathrm{d}\delta A_{\mathrm{s_B}}}{\mathrm{d}z} &= -\mathrm{i}K_\mathrm{s}^*\exp(\mathrm{i}\delta_\mathrm{s}z)\delta A_{\mathrm{s_F}} - K_\mathrm{B}\exp(-\mathrm{i}\delta_\mathrm{B}z)\delta(A_{\mathrm{p_B}}A_{\mathrm{i_B}}^*), \\ \frac{\mathrm{d}\delta A_{\mathrm{i_B}}}{\mathrm{d}z} &= -\mathrm{i}K_\mathrm{i}^*\exp(\mathrm{i}\delta_\mathrm{i}z)\delta A_{\mathrm{i_F}} - K_\mathrm{B}\exp(-\mathrm{i}\delta_\mathrm{B}z)\delta(A_{\mathrm{p_B}}A_{\mathrm{s_B}}^*), \\ \frac{\mathrm{d}\delta A_{\mathrm{p_F}}}{\mathrm{d}z} &= \mathrm{i}K_\mathrm{p}\exp(-\mathrm{i}\delta_\mathrm{p}z)\delta A_{\mathrm{p_B}} - K_\mathrm{F}^*\exp(-\mathrm{i}\delta_\mathrm{F}z)\delta(A_{\mathrm{s_F}}A_{\mathrm{i_F}}), \\ \frac{\mathrm{d}\delta A_{\mathrm{p_B}}}{\mathrm{d}z} &= -\mathrm{i}K_\mathrm{p}^*\exp(\mathrm{i}\delta_\mathrm{p}z)\delta A_{\mathrm{p_F}} + K_\mathrm{B}^*\exp(\mathrm{i}\delta_\mathrm{B}z)\delta(A_{\mathrm{s_B}}A_{\mathrm{i_B}}), \end{aligned} \tag{6.214}$$

where the variations $\delta(XY)$ are to be further transformed using the Leibniz formula $\delta(XY) = Y\delta X + X\delta Y$. These or the resulting equations do not depend on whether or not solutions to the boundary-value problem for (6.210) or those to the problem with the initial data $A_{a_\mathrm{F}}(0)$, $A_{a_\mathrm{B}}(0)$, which seems to be easy, are the case.

The consideration of variations leads to an approximate quantization. The input–output relations for variations are linear and, on certain conditions, they may be interpreted even as input–output relations for quantum corrections $\delta\hat{A}_{a_b}$, $a = \mathrm{s, i, p}$, $b = \mathrm{F, B}$.

Let us suppose that we first express a solution of the initial-value problem for (6.214). On introducing the notation

$$\delta\mathcal{A}_b(z) = \begin{pmatrix} \delta A_{\mathrm{s}_b}(z) \\ \delta A^*_{\mathrm{s}_b}(z) \\ \delta A_{\mathrm{i}_b}(z) \\ \delta A^*_{\mathrm{i}_b}(z) \\ \delta A_{\mathrm{p}_b}(z) \\ \delta A^*_{\mathrm{p}_b}(z) \end{pmatrix},\ b = \mathrm{F, B}, \tag{6.215}$$

this solution is

$$\begin{pmatrix} \delta\mathcal{A}_{\mathrm{F}}(L) \\ \delta\mathcal{A}_{\mathrm{B}}(L) \end{pmatrix} = \begin{pmatrix} \mathcal{V}_{\mathrm{FF}}(L) & \mathcal{V}_{\mathrm{FB}}(L) \\ \mathcal{V}_{\mathrm{BF}}(L) & \mathcal{V}_{\mathrm{BB}}(L) \end{pmatrix} \begin{pmatrix} \delta\mathcal{A}_{\mathrm{F}}(0) \\ \delta\mathcal{A}_{\mathrm{B}}(0) \end{pmatrix}. \tag{6.216}$$

Then only we determine a solution of the boundary-value problem as

$$\begin{pmatrix} \delta\mathcal{A}_{\mathrm{F}}(L) \\ \delta\mathcal{A}_{\mathrm{B}}(0) \end{pmatrix} = \mathcal{U} \begin{pmatrix} \delta\mathcal{A}_{\mathrm{F}}(0) \\ \delta\mathcal{A}_{\mathrm{B}}(L) \end{pmatrix}, \tag{6.217}$$

where

$$\mathcal{U} = \begin{pmatrix} \mathcal{V}_{\mathrm{FF}}(L) - \mathcal{V}_{\mathrm{FB}}(L)\mathcal{V}^{-1}_{\mathrm{BB}}(L)\mathcal{V}_{\mathrm{BF}}(L) & \mathcal{V}_{\mathrm{FB}}(L)\mathcal{V}^{-1}_{\mathrm{BB}}(L) \\ -\mathcal{V}^{-1}_{\mathrm{BB}}(L)\mathcal{V}_{\mathrm{BF}}(L) & \mathcal{V}^{-1}_{\mathrm{BB}}(L) \end{pmatrix}. \tag{6.218}$$

The input–output relations for quantum corrections are

$$\begin{pmatrix} \delta\hat{\mathcal{A}}_{\mathrm{F}}(L) \\ \delta\hat{\mathcal{A}}_{\mathrm{B}}(0) \end{pmatrix} = \mathcal{U} \begin{pmatrix} \delta\hat{\mathcal{A}}_{\mathrm{F}}(0) \\ \delta\hat{\mathcal{A}}_{\mathrm{B}}(L) \end{pmatrix}, \tag{6.219}$$

where $\delta\hat{\mathcal{A}}_b(z)$, $b = \mathrm{F, B}$, are given by relation (6.215), but with operators instead of the classical variables. The complex conjugates are replaced with the Hermitian ones. Nonclassical properties of the device are assessed by the operators

$$\hat{A}_{a_b}(z) = A_{a_b}(z)\hat{1} + \delta\hat{A}_{a_b}(z),\ \ z = 0, L. \tag{6.220}$$

Let the input operators, the components of the vectors $\delta\hat{\mathcal{A}}_{\mathrm{F}}(0)$ and $\delta\hat{\mathcal{A}}_{\mathrm{B}}(L)$, fulfil the usual boson commutation relations. In other words, the first, the third, and the fifth component are annihilation operators, the second, the fourth, and the sixth one are creation operators. Then also the output operators, the components of the vectors $\delta\hat{\mathcal{A}}_{\mathrm{F}}(L)$ and $\delta\hat{\mathcal{A}}_{\mathrm{B}}(0)$, obey the boson commutation relations.

A sufficient condition for it to be possible to carry out this approximate quantization is the existence of a suitable momentum function $G_{\mathrm{int}}(z)$ and the expression of the equations (6.210) in the form

$$\frac{\mathrm{d}X}{\mathrm{d}z} = \frac{\mathrm{i}}{\hbar}[X, G_{\mathrm{int}}]. \tag{6.221}$$

Here X (as well as Y in what follows) is any function of the variables A_{a_b}, $A^*_{a_b}$, and $G_{\mathrm{int}}(z)$ is a real function of these variables and of the coordinate z. With respect to a complicated expression of the bracket $[X, Y]$, operators are written about and the bracket has the usual meaning of a commutator in the paper (Peřina, Jr., et al. 2004). We will prefer the classical variables and

$$\begin{aligned}[X, Y] = \sum_{a=\mathrm{s,i,p}} &\left(\frac{\partial X}{\partial A_{a_\mathrm{F}}} \frac{\partial Y}{\partial A^*_{a_\mathrm{F}}} - \frac{\partial Y}{\partial A_{a_\mathrm{F}}} \frac{\partial X}{\partial A^*_{a_\mathrm{F}}} \right) \\ - \sum_{a=\mathrm{s,i,p}} &\left(\frac{\partial X}{\partial A_{a_\mathrm{B}}} \frac{\partial Y}{\partial A^*_{a_\mathrm{B}}} - \frac{\partial Y}{\partial A_{a_\mathrm{B}}} \frac{\partial X}{\partial A^*_{a_\mathrm{B}}} \right).\end{aligned} \tag{6.222}$$

The appropriate momentum function $G_{\mathrm{int}}(z)$ is

$$\begin{aligned} G_{\mathrm{int}}(z) = \hbar\big[&K_\mathrm{s} \exp(\mathrm{i}\delta_\mathrm{s} z) A^*_{\mathrm{s_F}} A_{\mathrm{s_B}} + K_\mathrm{i} \exp(\mathrm{i}\delta_\mathrm{i} z) A^*_{\mathrm{i_F}} A_{\mathrm{i_B}} \\ &+ K_\mathrm{p} \exp(\mathrm{i}\delta_\mathrm{p} z) A^*_{\mathrm{p_F}} A_{\mathrm{p_B}} + \mathrm{c.\ c.}\big] \\ &- \hbar\big[\mathrm{i} K_\mathrm{F} \exp(\mathrm{i}\delta_\mathrm{F} z) A_{\mathrm{p_F}} A^*_{\mathrm{s_F}} A^*_{\mathrm{i_F}} + \mathrm{i} K_\mathrm{B} \exp(-\mathrm{i}\delta_\mathrm{B} z) A_{\mathrm{p_B}} A^*_{\mathrm{s_B}} A^*_{\mathrm{i_B}} + \mathrm{c.\ c.}\big]. \end{aligned} \tag{6.223}$$

In such a case, a search for conservation laws is facilitated. Equations (6.210) have the properties

$$\begin{aligned} \frac{\mathrm{d}}{\mathrm{d}z} \left(|A_{\mathrm{s_F}}|^2 + |A_{\mathrm{p_F}}|^2 - |A_{\mathrm{s_B}}|^2 - |A_{\mathrm{p_B}}|^2 \right) &= 0, \\ \frac{\mathrm{d}}{\mathrm{d}z} \left(|A_{\mathrm{i_F}}|^2 + |A_{\mathrm{p_F}}|^2 - |A_{\mathrm{i_B}}|^2 - |A_{\mathrm{p_B}}|^2 \right) &= 0. \end{aligned} \tag{6.224}$$

Peřina, Jr., et al. (2005) pay attention also to the production of the longitudinal confinement of the pump mode(s) through the corrugation. The equations (6.210) are utilized in full generality; the restriction $K_\mathrm{p} = 0$ has been lifted. The explanations (6.211) are completed with another one,

$$\delta_\mathrm{p} = |k_{\mathrm{p_F}z}| + |k_{\mathrm{p_B}z}| - 2\delta_1. \tag{6.225}$$

The linear coupling constant K_p is expressed according to (6.212), which is completed with $a = \mathrm{p}$, and we have assumed that $|k_{\mathrm{p_F}z}| = |k_{\mathrm{p_B}z}|$.

Peřina, Jr., et al. (2007) study degenerate optical parametric processes, namely second-harmonic and second-subharmonic generation. They have considered anisotropy of the waveguide.

We present the decomposition of the electric-field amplitude related to photons of the respective modes

$$\mathbf{E}(x, y, z, t) = \mathrm{i} \sum_m \sqrt{\frac{\hbar\omega_m}{2\epsilon_0 L}} \times \left\{ A_m(z) \bar{\boldsymbol{\epsilon}}^{-\frac{1}{2}} \cdot \mathbf{f}_m(x, y) \exp[\mathrm{i}(k_{mz} z - \omega_m t)] - \text{c. c.} \right\}, \tag{6.226}$$

where $\mathbf{f}_m$ is the vector-valued transverse eigenfunction of the mth mode. The mean permittivity tensor of the waveguide is denoted as $\bar{\boldsymbol{\epsilon}}$.

The vector-valued function $\mathbf{f}_m(x, y)$ is a solution of the equation

$$(\mathbf{I}\nabla_{\mathrm{T}}^2 - \nabla_{\mathrm{T}}\nabla_{\mathrm{T}}) \cdot \mathbf{f}_m(x, y) - \mathrm{i}\beta_m \nabla_{\mathrm{T}} \mathbf{e}_z \cdot \mathbf{f}_m(x, y) - \mathrm{i}\beta_m \mathbf{e}_z \nabla_{\mathrm{T}} \cdot \mathbf{f}_m(x, y) - \beta_m^2 (\mathbf{I} - \mathbf{e}_z \mathbf{e}_z) \cdot \mathbf{f}_m(x, y) + \mu\epsilon_0 \bar{\boldsymbol{\epsilon}} \omega_m^2 \cdot \mathbf{f}_m(x, y) = \mathbf{0}. \tag{6.227}$$

The function $\mathbf{f}_m(x, y)$ is normalized; it has the property

$$\int\int \mathbf{f}_m^*(x, y) \cdot \mathbf{f}_m(x, y) \, \mathrm{d}x \, \mathrm{d}y = 1. \tag{6.228}$$

The electric-field amplitude $\mathbf{E}(\mathbf{r}, t)$ satisfies the wave equation inside the waveguide

$$-\nabla \times (\nabla \times \mathbf{E}) - \mu\epsilon_0 \boldsymbol{\epsilon} \cdot \frac{\partial^2 \mathbf{E}}{\partial t^2} = \mu \frac{\partial^2 \mathbf{P}_{\mathrm{nl}}}{\partial t^2}, \tag{6.229}$$

where $\boldsymbol{\epsilon}$ stands for the permittivity tensor of the waveguide. Every spectral component of the permittivity tensor $\boldsymbol{\epsilon}(\mathbf{r}, \omega)$ can be written as follows

$$\boldsymbol{\epsilon}(x, y, z, \omega) = \bar{\boldsymbol{\epsilon}}(x, y, \omega) + \Delta\boldsymbol{\varepsilon}(x, y, z, \omega). \tag{6.230}$$

Here $\Delta\boldsymbol{\varepsilon}(x, y, z, \omega)$ are small variations of the permittivity tensor related to the corrugation. These variations decompose into tensor-valued harmonic functions

$$\Delta\boldsymbol{\varepsilon}(x, y, z, \omega) = \sum_{q=-\infty}^{\infty} \boldsymbol{\varepsilon}_q(x, y, \omega) \exp\left(\mathrm{i}q \frac{2\pi}{\Lambda_1} z\right), \ \boldsymbol{\varepsilon}_0(x, y, \omega) = \mathbf{0}^{(1)}, \tag{6.231}$$

where $\boldsymbol{\varepsilon}_q(x, y, \omega)$ are coefficients of the decomposition, and $\mathbf{0}^{(1)}$ is the second-rank zero tensor. The polarization $\mathbf{P}_{\mathrm{nl}}(\mathbf{r}, t)$ of the medium is determined using the second-order susceptibility tensor $\boldsymbol{\chi}(\mathbf{r})$,

$$\mathbf{P}_{\mathrm{nl}}(\mathbf{r}, t) = \epsilon_0 \boldsymbol{\chi}(\mathbf{r}) : \mathbf{E}(\mathbf{r}, t)\mathbf{E}(\mathbf{r}, t). \tag{6.232}$$

A spectral component $\tilde{\chi}(\mathbf{r}, -2\omega; \omega, \omega)$ of the second-order susceptibility can be expressed as

$$\tilde{\chi}(x, y, z, -2\omega; \omega, \omega) = \sum_{q=-\infty}^{\infty} \tilde{\chi}_q(x, y, -2\omega; \omega, \omega) \exp\left(\mathrm{i}q\frac{2\pi}{\Lambda_{\mathrm{nl}}}z\right), \tag{6.233}$$

where Λ_{nl} describes the period of a possible periodical poling of the nonlinear material. On the assumption of two monochromatic components, a substitution of the amplitude (6.226) into the wave equation (6.229) provides two coupled Helmholtz equations for these components. We consider two directions of propagation for each of the monochromatic components. We have four modes: the signal forward-propagating mode (with amplitude $A_{\mathrm{s_F}}$), the signal backward-propagating mode ($A_{\mathrm{s_B}}$), the pump forward-propagating mode ($A_{\mathrm{p_F}}$), and, finally, the pump backward-propagating mode ($A_{\mathrm{p_B}}$). In the framework of the coupled-mode theory, we represent each of the two Helmholtz equations with two ordinary differential equations for the amplitudes

$$\begin{aligned}
\frac{\mathrm{d}A_{\mathrm{s_F}}}{\mathrm{d}z} &= \mathrm{i}K_{\mathrm{s}} \exp(-\mathrm{i}\delta_{\mathrm{s}}z)A_{\mathrm{s_B}} + 2K_{\mathrm{F},q} \exp(\mathrm{i}\delta_{\mathrm{F}}z)A_{\mathrm{p_F}}A^*_{\mathrm{s_F}}, \\
\frac{\mathrm{d}A_{\mathrm{s_B}}}{\mathrm{d}z} &= -\mathrm{i}K^*_{\mathrm{s}} \exp(\mathrm{i}\delta_{\mathrm{s}}z)A_{\mathrm{s_F}} - 2K_{\mathrm{B},q} \exp(-\mathrm{i}\delta_{\mathrm{B}}z)A_{\mathrm{p_B}}A^*_{\mathrm{s_B}}, \\
\frac{\mathrm{d}A_{\mathrm{p_F}}}{\mathrm{d}z} &= \mathrm{i}K_{\mathrm{p}} \exp(-\mathrm{i}\delta_{\mathrm{p}}z)A_{\mathrm{p_B}} - K^*_{\mathrm{F},q} \exp(-\mathrm{i}\delta_{\mathrm{F}}z)A^2_{\mathrm{s_F}}, \\
\frac{\mathrm{d}A_{\mathrm{p_B}}}{\mathrm{d}z} &= -\mathrm{i}K^*_{\mathrm{p}} \exp(\mathrm{i}\delta_{\mathrm{p}}z)A_{\mathrm{p_F}} + K^*_{\mathrm{B},q} \exp(\mathrm{i}\delta_{\mathrm{B}}z)A^2_{\mathrm{s_B}},
\end{aligned} \tag{6.234}$$

where

$$\begin{aligned}
\delta_a &= |\beta_{a_{\mathrm{F}}}| + |\beta_{a_{\mathrm{B}}}| - \delta_1, \, a = \mathrm{p}, \mathrm{s}, \\
\delta_{b,q} &= |\beta_{\mathrm{p}_b}| - 2|\beta_{\mathrm{s}_b}| + q\frac{2\pi}{\Lambda_{\mathrm{nl}}}, \, b = \mathrm{F}, \mathrm{B}.
\end{aligned} \tag{6.235}$$

The linear coupling constants K_{p} and K_{s} are given as

$$K_a = \frac{\mu\epsilon_0\omega_a^2}{2|\beta'_{a_{\mathrm{F}}}|} \int\int \mathbf{f}^*_{a_{\mathrm{F}}}(x, y) \cdot \boldsymbol{\varepsilon}_1(x, y, \omega_a) \cdot \mathbf{f}_{a_{\mathrm{B}}}(x, y)\, \mathrm{d}x\, \mathrm{d}y, \, a = \mathrm{p}, \mathrm{s}, \tag{6.236}$$

where

$$\begin{aligned}
\beta'_{a_{\mathrm{F}}} = \beta_{a_{\mathrm{F}}} - \beta_{a_{\mathrm{F}}} \int\int |\mathbf{e}_z \cdot \mathbf{f}_{a_{\mathrm{F}}}(x, y)|^2\, \mathrm{d}x\, \mathrm{d}y \\
+ \mathrm{Im} \int\int [\nabla_{\mathrm{T}} \cdot \mathbf{f}^*_{a_{\mathrm{F}}}(x, y)][\mathbf{e}_z \cdot \mathbf{f}_{a_{\mathrm{F}}}(x, y)]\, \mathrm{d}x\, \mathrm{d}y,
\end{aligned} \tag{6.237}$$

and we have assumed that $|\beta'_{a_F z}| = |\beta'_{a_B z}|$. The expressions for the nonlinear coupling constants K_F and K_B are

$$K_{b,q} = \frac{\mu\epsilon_0\omega_s}{2|\beta'_{s_b}|}\sqrt{\frac{\hbar\omega_p\omega_s\omega_s}{2\epsilon_0 L}} \int\int [\bar{\boldsymbol{\epsilon}}^{-\frac{1}{2}}(\omega_p)\cdot \mathbf{f}_{p_b}(x,y)]\cdot\tilde{\boldsymbol{\chi}}_q(x,y,-\omega_p;\omega_s,\omega_s)$$
$$:[\bar{\boldsymbol{\epsilon}}^{\frac{1}{2}}(\omega_s)\cdot \mathbf{f}^*_{s_b}(x,y)][\bar{\boldsymbol{\epsilon}}^{-\frac{1}{2}}(\omega_s)\cdot \mathbf{f}^*_{s_b}(x,y)]\,dx\,dy$$
$$= \frac{\mu\epsilon_0\omega_p}{2|\beta'_{p_b}|}\sqrt{\frac{\hbar\omega_s\omega_s\omega_p}{2\epsilon_0 L}} \int\int [\bar{\boldsymbol{\epsilon}}^{\frac{1}{2}}(\omega_p)\cdot \mathbf{f}_{p_b}(x,y)]\cdot\tilde{\boldsymbol{\chi}}_q(x,y,-\omega_p;\omega_s,\omega_s)$$
$$:[\bar{\boldsymbol{\epsilon}}^{-\frac{1}{2}}(\omega_s)\cdot \mathbf{f}^*_{s_b}(x,y)][\bar{\boldsymbol{\epsilon}}^{-\frac{1}{2}}(\omega_s)\cdot \mathbf{f}^*_{s_b}(x,y)]\,dx\,dy, \qquad (6.238)$$

where we have assumed that $\frac{\omega_s}{|\beta'_{s_b}|} \approx \frac{\omega_p}{|\beta'_{p_b}|}$.

The dependencies of the solutions to the equations (6.234) on the boundary data $A_{a_F}(0)$, $A_{a_B}(L)$, $a = s, p$ (of the solutions to the boundary-value problem) can be considered as classical input–output relations. These are significant for the study of stochastic boundary data.

Approximate results can be obtained by considering variations of the classical solutions δA_{a_b}, $a = s, p$, $b = F, B$. The variations satisfy linear equations

$$\frac{d\delta A_{s_F}}{dz} = iK_s \exp(-i\delta_s z)\delta A_{s_B} + 2K_{F,q}\exp(i\delta_F z)\delta(A_{p_F}A^*_{s_F}),$$
$$\frac{d\delta A_{s_B}}{dz} = -iK^*_s \exp(i\delta_s z)\delta A_{s_F} - 2K_{B,q}\exp(-i\delta_B z)\delta(A_{p_B}A^*_{s_B}),$$
$$\frac{d\delta A_{p_F}}{dz} = iK_p \exp(-i\delta_p z)\delta A_{p_B} - K^*_{F,q}\exp(-i\delta_F z)\delta(A^2_{s_F}),$$
$$\frac{d\delta A_{p_B}}{dz} = -iK^*_p \exp(i\delta_p z)\delta A_{p_F} + K^*_{B,q}\exp(i\delta_B z)\delta(A^2_{s_B}). \qquad (6.239)$$

The consideration of variations leads to an approximate quantization. The input–output relations for variations are linear and, as in the previous case, they may lead to input–output relations for quantum corrections $\delta\hat{A}_{a_b}$, $a = s, p$, $b = F, B$.

Let us suppose that we first express a solution of the initial-value problem for (6.239). On introducing notation

$$\delta\mathcal{A}_b(z) = \begin{pmatrix} \delta A_{s_b}(z) \\ \delta A^*_{s_b}(z) \\ \delta A_{p_b}(z) \\ \delta A^*_{p_b}(z) \end{pmatrix},\ b = F, B, \qquad (6.240)$$

this solution becomes (6.216). Thereafter we determine a solution of the boundary-value problem as (6.217), with (6.218). The input–output relations for quantum corrections are (6.219), where $\delta\hat{\mathcal{A}}_b(z)$, $b = F, B$, are given by relation (6.240), but with operators instead of the classical variables. The complex conjugates are replaced with the Hermitian ones. Nonclassical behaviour of a process is assessed by the operators (6.220).

In this case, the equations (6.234) have the form (6.221), with

$$[X, Y] = \sum_{a=\mathrm{s,p}} \left(\frac{\partial X}{\partial A_{a_\mathrm{F}}} \frac{\partial Y}{\partial A^*_{a_\mathrm{F}}} - \frac{\partial Y}{\partial A_{a_\mathrm{F}}} \frac{\partial X}{\partial A^*_{a_\mathrm{F}}} \right) - \sum_{a=\mathrm{s,p}} \left(\frac{\partial X}{\partial A_{a_\mathrm{B}}} \frac{\partial Y}{\partial A^*_{a_\mathrm{B}}} - \frac{\partial Y}{\partial A_{a_\mathrm{B}}} \frac{\partial X}{\partial A^*_{a_\mathrm{B}}} \right). \tag{6.241}$$

The appropriate momentum function $G_{\mathrm{int}}(z)$ is

$$G_{\mathrm{int}}(z) = \hbar \left[K_\mathrm{s} \exp(\mathrm{i}\delta_\mathrm{s} z) A^*_{\mathrm{s_F}} A_{\mathrm{s_B}} + K_\mathrm{p} \exp(\mathrm{i}\delta_\mathrm{p} z) A^*_{\mathrm{p_F}} A_{\mathrm{p_B}} + \mathrm{c.\,c.} \right] - \hbar \left[\mathrm{i} K_\mathrm{F} \exp(\mathrm{i}\delta_\mathrm{F} z) A_{\mathrm{p_F}} A^{*2}_{\mathrm{s_F}} + \mathrm{i} K_\mathrm{B} \exp(-\mathrm{i}\delta_\mathrm{B} z) A_{\mathrm{p_B}} A^{*2}_{\mathrm{s_B}} + \mathrm{c.\,c.} \right]. \tag{6.242}$$

Equations (6.234) have the property

$$\frac{\mathrm{d}}{\mathrm{d}z} \left(|A_{\mathrm{s_F}}|^2 + 2|A_{\mathrm{p_F}}|^2 - |A_{\mathrm{s_B}}|^2 - 2|A_{\mathrm{p_B}}|^2 \right) = 0. \tag{6.243}$$

6.2.2 Coupled-Mode Theory Including Gain or Losses

We assume a monochromatic wave propagating in a waveguide in the form

$$\mathbf{E}(x, y, z, t) = \sum_m A_m(z) \boldsymbol{\mathcal{E}}_m(x, y) \exp[\mathrm{i}(\omega t - k_{mz} z)], \tag{6.244}$$

where $A_m(z)$, $\frac{\mathrm{d}}{\mathrm{d}z} A_m(z) = 0$, is the amplitude of the mth mode, ω is a frequency, and k_{mz} are propagation constants (components along the direction of propagation of the wave vector of each mode). Let us note the difference from relation (6.203), where the wave is real. These eigenmodes have the electric vectors and the magnetic vectors of the form

$$\mathbf{E}_m(x, y, z, t) = \boldsymbol{\mathcal{E}}_m(x, y) \exp[\mathrm{i}(\omega t - k_{mz} z)], \quad \mathbf{H}_m(x, y, z, t) = \boldsymbol{\mathcal{H}}_m(x, y) \exp[\mathrm{i}(\omega t - k_{mz} z)], \tag{6.245}$$

respectively. In fact, the fields are real, and they must be recovered as $\frac{1}{2}[\mathbf{E}_m(x, y, z, t) + \mathbf{E}^*_m(x, y, z, t)]$, etc.

It is understood that counterpropagating modes are orthogonal. The normalization and the orthogonality property of copropagating modes are expressed by the relation

$$\frac{1}{2} \int\int (\mathbf{E}_k \times \mathbf{H}^*_m) \cdot \mathbf{e}_z \, \mathrm{d}x \, \mathrm{d}y = \frac{v_{gm}}{\mathcal{L}} \hbar\omega\delta_{km}, \tag{6.246}$$

where $\mathbf{e}_z$ is the unit vector in the direction of the z-axis, v_{gm} is the group velocity, $v_{gm} = \frac{\partial\omega}{\partial k_z}|_{k_z = k_{mz}}$, $\mathcal{L}$ is a quantization length, and $\hbar$ is the reduced Planck constant. The arguments of $\mathbf{E}_k$ and $\mathbf{H}^*_m$ have been omitted for convenience. The treatment will be restricted to dielectric structures, which consist of pieces of homogeneous and isotropic materials, or those with a small gradient of the refractive index. Then in (6.245), $\mathcal{E}_m(x, y)$ and $\mathcal{H}_m(x, y)$ obey the vectorial wave equation

$$\left[\frac{\partial^2}{\partial x^2} + \frac{\partial^2}{\partial y^2} + \omega^2 \mu \bar{\epsilon}(x, y) - k_{mz}^2\right] \mathcal{U}_m(x, y) = 0, \tag{6.247}$$

where μ is the magnetic permeability and $\bar{\epsilon}(x, y) = \epsilon_0 \bar{\epsilon}_r(x, y)$. In the case of a piecewise homogeneous dielectric structure, equation (6.247) is valid separately in each of homogeneous domains. So the field must be determined separately in every domain and, then, the tangential components of the fields must be joined at each of the interfaces. Another important boundary condition for waveguide modes is the vanishing of the field amplitudes at infinity. For the boundary conditions to be satisfied at all the points of the interfaces between homogeneous media, the paraxial propagation constant k_{mz} must be the same in the whole waveguide structure (Yariv and Yeh 1984). For definiteness, we will describe a planar waveguide and put $\mathcal{U}_m = \mathcal{E}_m$.

The solutions should be determined in the core (guiding region), which we designate as D. We let C denote the boundary of D, which are two straight lines parallel to the y-axis. The quantity $\bar{\epsilon}_r(x, y)$ is independent of y, and the solutions that are independent of y are looked for. The transverse electric (TE) modes and the transverse magnetic (TM) modes are distinguished. The TE modes have

$$\mathcal{E}_{mx}(x, y) = \mathcal{E}_{mz}(x, y) = \mathcal{H}_{my}(x, y) = 0, \tag{6.248}$$

where the first component is included by the definition of these modes, the second one is present here due to the condition $\nabla \cdot \mathbf{E} = 0$, and the third one follows from a Maxwell equation. The TM modes have

$$\mathcal{H}_{mx}(x, y) = \mathcal{H}_{mz}(x, y) = \mathcal{E}_{my}(x, y) = 0, \tag{6.249}$$

where the first component is included by the definition of these modes, the second one is present here due to the condition $\nabla \cdot \mathbf{H} = 0$, and the third one follows from a Maxwell equation. For suitable k_{mz}, the TE mode boundary conditions are the continuity requirements on $\mathcal{E}_{my}(x, y)_{|C}$ and $\mathcal{H}_{mz}(x, y)_{|C}$. The TM mode boundary conditions are related to $\mathcal{H}_{my}(x, y)_{|C}$ and $\mathcal{E}_{mz}(x, y)_{|C}$. Equation (6.247) is solved in the whole x–y plane excepting the point set C perhaps. We have neglected a term $\nabla(\nabla \cdot \mathbf{E})$, which is justified if the change of the quantity $\bar{\epsilon}_0(x, y)$ over a wavelength is small. In the case of TE modes of planar dielectric waveguides $\nabla \cdot \mathbf{E} = 0$, since $\nabla \cdot (\epsilon \mathbf{E}) = \epsilon \nabla \cdot \mathbf{E} + \mathbf{E} \cdot \nabla \epsilon = 0$, where $\mathbf{E} \cdot \nabla \epsilon = 0$. In fact, ϵ has a jump in the x-direction, whereas E_x vanishes. This implies that equation (6.247) is exact at C. In the case of TM modes of a planar waveguide, equation (6.247) does not hold at the interface C.

In some cases, it is useful to consider complex dielectric permittivity. In this generalization, the definition of unperturbed modes is based on $\mathrm{Re}\{\bar{\epsilon}_r(x, y)\}$. Respecting it, the replacement of $\bar{\epsilon}(x, y)$ by $\mathrm{Re}\{\bar{\epsilon}_{\mathrm{r}}(x, y)\}$ should be made, where it is appropriate. Particularly, relation (6.207) becomes

$$\epsilon_{\mathrm{r}}(x, y, z) = \mathrm{Re}\{\bar{\epsilon}_{\mathrm{r}}(x, y)\} + \Delta\varepsilon_{\mathrm{r}}(x, y, z). \tag{6.250}$$

Now

$$\mathrm{Re}\{\Delta\varepsilon_{\mathrm{r}}(x, y, z)\} = \sum_{q=-\infty}^{\infty} \varepsilon_q^{(\mathrm{Re})}(x, y) \exp\left(\mathrm{i}q\frac{2\pi}{\Lambda_1}z\right), \; \varepsilon_0^{(\mathrm{Re})}(x, y) = 0, \tag{6.251}$$

$$\mathrm{Im}\{\Delta\varepsilon_{\mathrm{r}}(x, y, z)\} = \sum_{q=-\infty}^{\infty} \varepsilon_q^{(\mathrm{Im})}(x, y) \exp\left(\mathrm{i}q\frac{2\pi}{\Lambda_1}z\right),$$
$$\varepsilon_0^{(\mathrm{Im})}(x, y) = \mathrm{Im}\{\bar{\epsilon}_{\mathrm{r}}(x, y)\}. \tag{6.252}$$

In the following, we restrict ourselves to the TE modes. Here $E(x, y, z, t)$ $(\mathcal{E}_m(x, y), \mathcal{H}_m(x, y))$ will be a shorthand for the component $E_y(x, y, z, t)$ $(\mathcal{E}_{my}(x, y),$ $\mathcal{H}_{my}(x, y))$ along the direction of y. The orthogonality condition of the modes reads

$$\langle \mathcal{E}_m^* | \mathcal{E}_k^* \rangle = \int\int \mathcal{E}_m^*(x, y) \cdot \mathcal{E}_k(x, y)\, \mathrm{d}x\, \mathrm{d}y = \frac{|v_{gm}|}{\mathcal{L}} \frac{2\hbar\omega^2\mu}{|k_{mz}|} \delta_{km}. \tag{6.253}$$

With respect to the quantum treatment, we introduce the negative- and positive-frequency parts

$$E^{(-)}(x, y, z, t) = \frac{1}{2} E(x, y, z, t),$$
$$E^{(+)}(x, y, z, t) = [E^{(-)}(x, y, z, t)]^*, \tag{6.254}$$

the corresponding envelope

$$\mathrm{A}_m(z) = \frac{1}{2} A_m^*(z), \tag{6.255}$$

and we note that relation (6.244) can be rewritten as

$$E^{(+)}(x, y, z, t) = \sum_m \mathrm{A}_m(z) \mathcal{E}_m^*(x, y) \exp[-\mathrm{i}(\omega t - k_{mz} z)]. \tag{6.256}$$

This classical field is an eigenfunction (rather, eigenvalue that depends on parameters),

$$E^{(+)}(x, y, z, t) \left| \left\{ \mathrm{A}_m \begin{pmatrix} 0 \\ L \end{pmatrix} \right\} \right\rangle = \hat{E}^{(+)}(x, y, z, t) \left| \left\{ \mathrm{A}_m \begin{pmatrix} 0 \\ L \end{pmatrix} \right\} \right\rangle, \tag{6.257}$$

where $\hat{E}^{(+)}(x, y, z, t)$ is the positive-frequency part of the electric strength operator, $\left| \left\{ \mathrm{A}_m \binom{0}{L} \right\} \right\rangle$ is a coherent state, 0 (L) corresponds to $k_{mz} > 0$ ($k_{mz} < 0$), and L is the length of the optical device. We consider expansion (6.256) with $E^{(+)}(x, y, z, t)$ replaced by $\hat{E}^{(+)}(x, y, z, t)$ and $\mathrm{A}_m(z)$ replaced by $\hat{A}_m(z)$.

From the viewpoint of the integrated optics, the interaction of modes or coupling of modes has been interesting and also the possibility of quantization is attractive. Here we have touched even the impossibility of quantization. Whenever a linear relation between pairs of complex amplitudes is appropriate (a linear canonical transformation), quantization is possible and the complex amplitudes can be interpreted as operators. In the literature (Milburn et al. 1984) concerning the nonlinear processes, it has been shown that such operators do not obey the usual commutation relations. In this case, we assert that the quantization has not been accomplished. Nevertheless, in this study, we interpret the input–output relations as quantal.

A special use of quantum-mechanical descriptions has been mentioned in Section 3.1.4 already. The full Heisenberg–Langevin approach has been utilized, which is based on the concept of a quantum reservoir.

We classify reservoirs as forward- and backward-propagating ones, even though it is not quite usual to combine these terms. But we do not utilize the concept of a quantum reservoir. We rely on the concept of quasi-continuous measurements (Ban 1994). It is sufficient to know that the quasi-continuous measurement is realized using a system of lossless beam splitters or using a system of parametric amplifiers. A transition to a continuous limit is possible, and the differential equation of its description coincides with the master equation for the description of a single mode obtained on an elimination of the quantum reservoir (Peřinová and Lukš 2000). Independent of this, a return to the classical description is possible provided that the quantum measurements are not studied.

The system of lossless beam splitters represents an "attenuator", since the energy of the light mode is by parts reflected by the beam splitters to detectors. The system of parametric amplifiers represents an "amplifier", since photons of pump beam are converted to photon-twin pairs. One of each pair is directed to the detector, and the other supplies energy to the light mode going through aligned axes of nonlinear crystals. Similarly, repeated nondemolition measurements can be implemented.

Another idea, which however requires the replacement of the time variable by the space variable, is the measurement of an observable of a cavity mode using Rydberg atoms. Literature is devoted to the useful case of repeated nondemolition measurements, but we image easily also an attenuator similar to the system of lossless beam splitters and an amplifier similar to the system of parametric amplifiers. This idea allows the transition to a continuous limit as well. Different frequencies of a quantum reservoir are replaced with different times when the atoms interact with the cavity mode. On the change of the time variable by the space variable, we describe the physical fact that the guided mode is in short but densely distributed segments of a waveguide coupled to the sources or sinks of the energy.

We will approach the distributed feedback laser (Yariv and Yeh 1984, Toren and Ben Aryeh 1994) as a quantum amplifier. In the usual coupled-mode theory, it is assumed that the perturbation $\Delta\varepsilon_{\mathrm{r}}(x, y, z)$ of the dielectric permittivity is real, but the presence of a small gain can be also considered a perturbation and then $\Delta\varepsilon_{\mathrm{r}}(x, y, z)$ is to be held for a complex quantity. Describing a lossy medium, one assumes that the imaginary part of $\Delta\varepsilon_{\mathrm{r}}(x, y, z)$ is negative, but the gain medium exhibits a positive imaginary part of $\Delta\varepsilon_{\mathrm{r}}(x, y, z)$. We assume that modes 1 and 2 are coupled, and we let k_{1z}, k_{2z} denote the z-components of the respective wave vectors. We will treat the particular case when $k_{1z} \approx \frac{l\pi}{\lambda}$, i.e., mode 1 is strongly coupled with the backward-propagating mode 2, $k_{2z} = -k_{1z}$. The classical description is based on the differential equations. We complex conjugate the differential equations of classical description (Yariv and Yeh 1984) and replace $A_j^*(z)$ by $\mathrm{A}_j(z)$. This results in the differential equations

$$\frac{\mathrm{d}}{\mathrm{d}z}\mathrm{A}_1(z) = \mathrm{i}\kappa^*\mathrm{A}_2(z)\exp(2\mathrm{i}\delta z) + \frac{\gamma}{2}\mathrm{A}_1(z),$$
$$\frac{\mathrm{d}}{\mathrm{d}z}\mathrm{A}_2(z) = -\mathrm{i}\kappa\mathrm{A}_1(z)\exp(-2\mathrm{i}\delta z) - \frac{\gamma}{2}\mathrm{A}_2(z), \tag{6.258}$$

where

$$\kappa = \frac{\epsilon_0\mathcal{L}}{4\hbar|v_\mathrm{g}|}\int\int \boldsymbol{\mathcal{E}}_1^*(x,y)\cdot\varepsilon_{-l}^{(\mathrm{Re})}(x,y)\boldsymbol{\mathcal{E}}_1(x,y)\,\mathrm{d}x\,\mathrm{d}y,$$
$$\delta = \frac{1}{2}(k_{1z} - k_{2z}) - \frac{l\pi}{\lambda},$$
$$\gamma = \frac{\epsilon_0\mathcal{L}}{2\hbar|v_\mathrm{g}|}\int\int \boldsymbol{\mathcal{E}}_1^*(x,y)\cdot\mathrm{Im}\{\bar{\epsilon}_\mathrm{r}(x,y)\}\boldsymbol{\mathcal{E}}_1(x,y)\,\mathrm{d}x\,\mathrm{d}y, \tag{6.259}$$

with v_g the group velocity of light. The input–output relations are given as

$$\mathrm{A}_1(L) = u_{11}(L)\mathrm{A}_1(0) + u_{12}(L)\mathrm{A}_2(L),$$
$$\mathrm{A}_2(0) = u_{21}(L)\mathrm{A}_1(0) + u_{22}(L)\mathrm{A}_2(L). \tag{6.260}$$

Here $u_{jk}(L)$, $j, k = 1, 2$, are given by generalized relations from (Peřinová et al. 1991),

$$u_{11}(L) = \exp(\mathrm{i}\delta L)\left[\cosh(DL) + \mathrm{i}\frac{\Delta}{D}\sinh(DL)\right]^{-1},$$
$$u_{12}(L) = \mathrm{i}\frac{\kappa^*}{D}\exp(\mathrm{i}2\delta L)\sinh(DL)$$
$$\times\left[\cosh(DL) + \mathrm{i}\frac{\Delta}{D}\sinh(DL)\right]^{-1},$$
$$u_{21}(L) = \mathrm{i}\frac{\kappa}{D}\sinh(DL)\left[\cosh(DL) + \mathrm{i}\frac{\Delta}{D}\sinh(DL)\right]^{-1},$$
$$u_{22}(L) = u_{11}(L), \tag{6.261}$$

with

$$D = \sqrt{|\kappa|^2 - \Delta^2}, \quad \Delta = \delta + \mathrm{i}\frac{\gamma}{2}. \tag{6.262}$$

The quantization of the classical description in the case of no gain can be accomplished in the spatial Heisenberg picture using equations (6.260). The complex amplitudes $\mathrm{A}_j(z)$ are replaced by the annihilation operators $\hat{A}_j(z)$, $j = 1, 2$, $z = 0, L$. In general, the spatial Schrödinger picture is appropriate. Any input statistical operator $\hat{\rho}(0)$ evolves to the output operator $\hat{\rho}(L)$. In contrast, the operators $\hat{A}_1(0)$ and $\hat{A}_2(L)$ do not change and that is why we abbreviate $\hat{A}_1 \equiv \hat{A}_1(0)$, $\hat{A}_2 \equiv \hat{A}_2(L)$ here.

The statistical properties of the input and output fields are described equivalently by the characteristic functions. The definition of these functions depends on the ordering of field operators. We choose the antinormal ordering, which is suitable for the amplifier. The antinormal characteristic functions of the input and output state, respectively, are

$$\begin{aligned} &C_{\mathcal{A}}(\beta_1, \beta_2, 0) \\ &= \mathrm{Tr}\left\{\hat{\rho}(0)\exp(-\beta_1^*\hat{A}_1 - \beta_2^*\hat{A}_2)\exp(\beta_1\hat{A}_1^\dagger + \beta_2\hat{A}_2^\dagger)\right\}, \\ &C_{\mathcal{A}}(\beta_1, \beta_2, L) \\ &= \mathrm{Tr}\left\{\hat{\rho}(L)\exp(-\beta_1^*\hat{A}_1 - \beta_2^*\hat{A}_2)\exp(\beta_1\hat{A}_1^\dagger + \beta_2\hat{A}_2^\dagger)\right\}. \end{aligned} \tag{6.263}$$

The antinormal characteristic function for the output can be defined taking into account the coefficients $u_{jk}(L)$, $j, k = 1, 2$, in (6.261),

$$\begin{aligned} C_{\mathcal{A}}(\beta_1, \beta_2, L) &= \mathrm{Tr}\Big\{\hat{\rho}(0) \\ &\times \exp\Big[-\Big(\beta_1^* u_{11}(L) + \beta_2^* u_{21}(L)\Big)\hat{A}_1 - \Big(\beta_1^* u_{12}(L) + \beta_2^* u_{22}(L)\Big)\hat{A}_2\Big] \\ &\times \exp\Big[\Big(\beta_1 u_{11}^*(L) + \beta_2 u_{21}^*(L)\Big)\hat{A}_1^\dagger + \Big(\beta_1 u_{12}^*(L) + \beta_2 u_{22}^*(L)\Big)\hat{A}_2^\dagger\Big]\Big\} \\ &= C_{\mathcal{A}}\Big(\beta_1 u_{11}^*(L) + \beta_2 u_{21}^*(L), \beta_1 u_{12}^*(L) + \beta_2 u_{22}^*(L), 0\Big). \end{aligned} \tag{6.264}$$

Here we have reduced an alternative definition to a substitution into the characteristic function for the input. We may conclude with the inversion of the second equation in (6.263),

$$\begin{aligned} \hat{\rho}(L) = \frac{1}{\pi^2}\int\int & C_{\mathcal{A}}(\beta_1, \beta_2, L)\exp(\beta_1\hat{A}_1^\dagger + \beta_2\hat{A}_2^\dagger) \\ & \times \exp(-\beta_1^*\hat{A}_1 - \beta_2^*\hat{A}_2)\,\mathrm{d}^2\beta_1\,\mathrm{d}^2\beta_2, \end{aligned} \tag{6.265}$$

where $\mathrm{d}^2\beta_j$=d(Re $\{\beta_j\}$)d(Im $\{\beta_j\}$), $j = 1, 2$. The above procedure is a completely positive map. We will not present a proof, which can be established similarly as below in the case of attenuator. With the characteristic functions, the quasidistributions are associated related to the same ordering

$$\begin{aligned} \Phi_{\mathcal{A}}(\mathrm{A}_1, \mathrm{A}_2, 0) &= \frac{1}{\pi^4}\int\int C_{\mathcal{A}}(\beta_1, \beta_2, 0)\exp(\mathrm{A}_1\beta_1^* + \mathrm{A}_2\beta_2^* - \mathrm{c.\,c.})\,\mathrm{d}^2\beta_1\,\mathrm{d}^2\beta_2, \\ \Phi_{\mathcal{A}}(\mathrm{A}_1, \mathrm{A}_2, L) &= \frac{1}{\pi^4}\int\int C_{\mathcal{A}}(\beta_1, \beta_2, L)\exp(\mathrm{A}_1\beta_1^* + \mathrm{A}_2\beta_2^* - \mathrm{c.\,c.})\,\mathrm{d}^2\beta_1\,\mathrm{d}^2\beta_2. \end{aligned} \tag{6.266}$$

The relationships between the characteristic functions become the relationships between the quasidistributions. The latter are more complicated (one needs elements of the inverse to the matrix $U(L)$ consisting of the elements $u_{jk}(L)$, $j, k = 1, 2$, and its determinant $|U(L)|$).

According to our statement below the relation (6.262), when $\gamma = 0$, the quantization can be accomplished in the Heisenberg picture. Relationships between statistical operators may be intricate sometimes. Then we can adopt the Heisenberg–Langevin approach. We can formally define an amplifier to be a device which can be described with input–output relations

$$\begin{aligned} \hat{A}_1(L) &= u_{11}(L)\hat{A}_1(0) + u_{12}(L)\hat{A}_2(L) + \hat{M}_1(L), \\ \hat{A}_2(0) &= u_{21}(L)\hat{A}_1(0) + u_{22}(L)\hat{A}_2(L) + \hat{M}_2(L), \end{aligned} \tag{6.267}$$

where $\hat{M}_j(L)$, $j = 1, 2$, are integrated quantum-noise terms. A characteristic property of the amplifier by the definition is that the commutator matrix

$$\begin{pmatrix} [\hat{M}_1(L), \hat{M}_1^\dagger(L)] & [\hat{M}_1(L), \hat{M}_2^\dagger(L)] \\ [\hat{M}_2(L), \hat{M}_1^\dagger(L)] & [\hat{M}_2(L), \hat{M}_2^\dagger(L)] \end{pmatrix} \leq \begin{pmatrix} 0 & 0 \\ 0 & 0 \end{pmatrix} \hat{1}, \tag{6.268}$$

i.e., both its eigenvalues, when $\hat{1}$ is factored out, are nonpositive. It means that we can find numbers $\overline{u}_{jk}(L)$ and creation operators $\hat{A}_k^\dagger(L)$, $j = 1, 2$, $k = 3, 4$, such that

$$\begin{aligned} \hat{M}_1(L) &= \overline{u}_{13}(L)\hat{A}_3^\dagger(0) + \overline{u}_{14}(L)\hat{A}_4^\dagger(L), \\ \hat{M}_2(L) &= \overline{u}_{23}(L)\hat{A}_3^\dagger(0) + \overline{u}_{24}(L)\hat{A}_4^\dagger(L). \end{aligned} \tag{6.269}$$

It is required that the commutation relations between input annihilation and creation operators (6.305) below and those between such operators related to the expression (6.269),

$$[\hat{A}_3(0), \hat{A}_3^\dagger(0)] = [\hat{A}_4(L), \hat{A}_4^\dagger(L)] = \hat{1}, \tag{6.270}$$

could be rewritten as those between annihilation and creation output operators in (6.305). It is assumed that the operators of distinct modes mutually commute.

Instead of proving that the procedure for the derivation of the output statistical operator from the input one is a completely positive map, we can find expressions for integrated quantum-noise terms and prove the characteristic property of the amplifier (6.268). In Yariv and Yeh (1984), much attention is paid to the regime of light generation, defined by the condition

$$\cosh(DL) + \mathrm{i}\frac{\Delta}{D}\sinh(DL) = 0. \tag{6.271}$$

In the explicit expression of the coefficients $u_{jk}(L)$, $j = 1, 2$, in (6.261), the division by zero occurs. It is obvious that the model is only tentative; for instance, the effect of saturation can be described by a nonlinear model, while the model under discussion is linear.

As the reciprocity theorem is derived on the assumption of the lossless medium, the attenuator has not been elaborated in the coupled-mode theory, contrary to the amplifier. But it is rather reasonable to assume that a perturbation $\Delta\varepsilon_r(x, y, z)$ is a complex quantity with a negative imaginary part. Using the same procedure as in the case of amplifier, we arrive at the differential equations

$$\frac{d}{dz}A_1(z) = i\kappa^* A_2(z)\exp(2i\delta z) - \frac{\gamma}{2}A_1(z),$$
$$\frac{d}{dz}A_2(z) = -i\kappa A_1(z)\exp(-2i\delta z) + \frac{\gamma}{2}A_2(z), \qquad (6.272)$$

where κ is defined in (6.259) and

$$\gamma = -\frac{\epsilon_0 \mathcal{L}}{2\hbar|v_g|}\int\int \boldsymbol{\mathcal{E}}_1^*(x, y)\cdot \mathrm{Im}\{\bar{\epsilon}_r(x, y)\}\boldsymbol{\mathcal{E}}_1(x, y)\,dx\,dy. \qquad (6.273)$$

Here we expound the usual approach (the coherent-state technique), which has been implicitly used also in the previous exposition. The solution of an initial problem for (6.272) can be obtained in the form

$$A_1(z) = v_{11}(z)A_1(0) + v_{12}(z)A_2(0),$$
$$A_2(z) = v_{21}(z)A_1(0) + v_{22}(z)A_2(0), \qquad (6.274)$$

where $v_{jk}(L)$, $j, k = 1, 2$, are given by generalized relations from Peřinová et al. (1991),

$$v_{11}(z) = \exp(i\delta z)\left[\cosh(Dz) - i\frac{\Delta}{D}\sinh(Dz)\right],$$
$$v_{12}(z) = i\frac{\kappa^*}{D}\exp(i\delta z)\sinh(Dz),$$
$$v_{21}(z) = -i\frac{\kappa}{D}\exp(-i\delta z)\sinh(Dz),$$
$$v_{22}(z) = \exp(-i\delta z)\left[\cosh(Dz) + i\frac{\Delta}{D}\sinh(Dz)\right], \qquad (6.275)$$

with

$$D = \sqrt{|\kappa|^2 - \Delta^2}, \quad \Delta = \delta - i\frac{\gamma}{2}. \qquad (6.276)$$

The input–output relations have the form (6.260), where $u_{jk}(L)$, $j, k = 1, 2$, are given in (6.261) except Δ, $\Delta = \delta - i\frac{\gamma}{2}$.

The case of no losses coincides with the case of no gain, when the quantization is possible as mentioned in the previous exposition. In general, the Schrödinger picture can be recommended. With respect to (6.263), the input and output operators are denoted as $\hat{\rho}(0)$ and $\hat{\rho}(L)$, respectively. In the description, the normal ordering is used to simplify the form. The normal characteristic functions of the input and output states, respectively, are

$$C_{\mathcal{N}}(\beta_1, \beta_2, 0) = \mathrm{Tr}\left\{\hat{\rho}(0)\exp(\beta_1\hat{A}_1^\dagger + \beta_2\hat{A}_2^\dagger)\exp(-\beta_1^*\hat{A}_1 - \beta_2^*\hat{A}_2)\right\},$$
$$C_{\mathcal{N}}(\beta_1, \beta_2, L) = \mathrm{Tr}\left\{\hat{\rho}(L)\exp(\beta_1\hat{A}_1^\dagger + \beta_2\hat{A}_2^\dagger)\exp(-\beta_1^*\hat{A}_1 - \beta_2^*\hat{A}_2)\right\}. \quad (6.277)$$

Intuitively, the normal characteristic function is

$$\begin{aligned} C_{\mathcal{N}}(\beta_1, \beta_2, L) &= \mathrm{Tr}\Big\{\hat{\rho}(0) \\ &\times \exp\left[\left(\beta_1 u_{11}^*(L) + \beta_2 u_{21}^*(L)\right)\hat{A}_1^\dagger + \left(\beta_1 u_{12}^*(L) + \beta_2 u_{22}^*(L)\right)\hat{A}_2^\dagger\right] \\ &\times \exp\left[-\left(\beta_1^* u_{11}(L) + \beta_2^* u_{21}(L)\right)\hat{A}_1 - \left(\beta_1^* u_{12}(L) + \beta_2^* u_{22}(L)\right)\hat{A}_2\right]\Big\} \\ &= C_{\mathcal{N}}\left(\beta_1 u_{11}^*(L) + \beta_2 u_{21}^*(L), \beta_1 u_{12}^*(L) + \beta_2 u_{22}^*(L), 0\right). \end{aligned} \quad (6.278)$$

Here we have expressed the statistical properties of the output through the normal characteristic function for the inputs. The procedure concludes with the inversion of the second equation in (6.277),

$$\begin{aligned} \hat{\rho}(L) = \frac{1}{\pi^2}\int\int & C_{\mathcal{N}}(\beta_1, \beta_2, L)\exp(-\beta_1^*\hat{A}_1 - \beta_2^*\hat{A}_2) \\ & \times \exp(\beta_1\hat{A}_1^\dagger + \beta_2\hat{A}_2^\dagger)\,\mathrm{d}^2\beta_1\,\mathrm{d}^2\beta_2. \end{aligned} \quad (6.279)$$

We assert that we have defined a completely positive map. We will provide a proof of this proposition in what follows. Meanwhile, we remark that for some states, there also exist quasidistributions related to the normal ordering as ordinary functions

$$\begin{aligned} \Phi_{\mathcal{N}}(\mathrm{A}_1, \mathrm{A}_2, 0) &= \frac{1}{\pi^4} \\ &\times \int\int C_{\mathcal{N}}(\beta_1, \beta_2, 0)\exp(\mathrm{A}_1\beta_1^* + \mathrm{A}_2\beta_2^* - \text{c. c.})\,\mathrm{d}^2\beta_1\,\mathrm{d}^2\beta_2, \\ \Phi_{\mathcal{N}}(\mathrm{A}_1, \mathrm{A}_2, L) &= \frac{1}{\pi^4} \\ &\times \int\int C_{\mathcal{N}}(\beta_1, \beta_2, L)\exp(\mathrm{A}_1\beta_1^* + \mathrm{A}_2\beta_2^* - \text{c. c.})\,\mathrm{d}^2\beta_1\,\mathrm{d}^2\beta_2. \end{aligned} \quad (6.280)$$

The relationships between the statistical operators may be involved sometimes. In response, we can adopt the Heisenberg–Langevin approach. We can formally define an attenuator to be a device which can be described with the input–output relations (6.267). A characteristic property of the attenuator by the definition is that the commutator matrix

$$\begin{pmatrix} [\hat{M}_1(L), \hat{M}_1^\dagger(L)] & [\hat{M}_1(L), \hat{M}_2^\dagger(L)] \\ [\hat{M}_2(L), \hat{M}_1^\dagger(L)] & [\hat{M}_2(L), \hat{M}_2^\dagger(L)] \end{pmatrix} \geq \begin{pmatrix} 0 & 0 \\ 0 & 0 \end{pmatrix}\hat{1}, \quad (6.281)$$

i.e., both its eigenvalues (let $\hat{1}$ be factored out) are nonnegative. It means that we can find numbers $u_{jk}(L)$ and annihilation operators $\hat{A}_k(L)$, $j = 1, 2$, $k = 3, 4$, such that

$$\begin{aligned}\hat{M}_1(L) &= u_{13}(L)\hat{A}_3(0) + u_{14}(L)\hat{A}_4(L),\\ \hat{M}_2(L) &= u_{23}(L)\hat{A}_3(0) + u_{24}(L)\hat{A}_4(L).\end{aligned} \tag{6.282}$$

It is required that the commutation relations (6.305) between input annihilation and creation operators and the relations (6.270) related to the expression (6.282) could be rewritten as those between annihilation and creation output operators in (6.311). The operators of different modes mutually commute, the input operators with the input ones and the output operators with the output ones.

In place of proving that the procedure for derivation of the output statistical operator from the input one is a completely positive map, we can find expressions for integrated quantum noise terms and demonstrate the characteristic property of the attenuator (6.281).

(i) Expressions for the integrated quantum-noise terms

In order to take into account losses, we use the extended differential equations

$$\begin{aligned}\frac{\mathrm{d}}{\mathrm{d}z}\mathrm{A}_1(z) &= \mathrm{i}\kappa^*\mathrm{A}_2(z)\exp(2\mathrm{i}\delta z) - \mathrm{i}G_1(z, z),\\ \frac{\mathrm{d}}{\mathrm{d}z}\mathrm{A}_2(z) &= -\mathrm{i}\kappa\mathrm{A}_1(z)\exp(-2\mathrm{i}\delta z) + \mathrm{i}G_2(z, z),\end{aligned} \tag{6.283}$$

$$\begin{aligned}\frac{\mathrm{d}}{\mathrm{d}z}G_1(\zeta, z) &= -\mathrm{i}\gamma\delta(\zeta - z)\mathrm{A}_1(z),\\ \frac{\mathrm{d}}{\mathrm{d}z}G_2(\zeta, z) &= \mathrm{i}\gamma\delta(\zeta - z)\mathrm{A}_2(z),\end{aligned} \tag{6.284}$$

where $G_j(\zeta, z)$, $j = 1, 2$, is a continuum of modes, which are right- and left-going in dependence on $j = 1$ and $j = 2$, respectively. In terms of these modes, losses are modelled. The detail that the losses of each mode $\mathrm{A}_j(z)$, $j = 1, 2$, are modelled by coupling with modes propagating in the same direction is not essential, but it has been chosen for simplicity. Quite a novel thing in this description is that the coupling of the mode $G_j(\zeta, z)$ is concentrated into the position $z = \zeta$. Solving (6.284) for complex amplitudes $G_j(\zeta, z)$, $j = 1, 2$, we obtain that

$$\begin{aligned}G_1(\zeta, z) &= \begin{cases} G_1(\zeta, 0) & \text{for } z < \zeta,\\ G_1(\zeta, 0) - \mathrm{i}\frac{\gamma}{2}\mathrm{A}_1(\zeta) & \text{for } z = \zeta,\\ G_1(\zeta, 0) - \mathrm{i}\gamma\mathrm{A}_1(\zeta) & \text{for } z > \zeta,\end{cases}\\ G_2(\zeta, z) &= \begin{cases} G_2(\zeta, 0) + \mathrm{i}\gamma\mathrm{A}_2(\zeta) & \text{for } z > \zeta,\\ G_2(\zeta, 0) + \mathrm{i}\frac{\gamma}{2}\mathrm{A}_2(\zeta) & \text{for } z = \zeta,\\ G_2(\zeta, 0) & \text{for } z < \zeta.\end{cases}\end{aligned} \tag{6.285}$$

As the contradirectional coupling presents more difficulties than the codirectional coupling, we may use the solutions of (6.285) to create apparent paradoxes. On substituting relation (6.285) into (6.283), we obtain differential equations

$$\begin{aligned}\frac{\mathrm{d}}{\mathrm{d}z}\mathrm{A}_1(z) &= \mathrm{i}\kappa^*\mathrm{A}_2(z)\exp(2\mathrm{i}\delta z) - \frac{\gamma}{2}\mathrm{A}_1(z) - \mathrm{i}G_1(z,0),\\ \frac{\mathrm{d}}{\mathrm{d}z}\mathrm{A}_2(z) &= -\mathrm{i}\kappa\mathrm{A}_1(z)\exp(-2\mathrm{i}\delta z) - \frac{\gamma}{2}\mathrm{A}_2(z) + \mathrm{i}G_2(z,0).\end{aligned} \tag{6.286}$$

These equations will contradict Equations (6.272) after one omits the Langevin terms as is usual in the treatment of the co-propagation, where such a manipulation is correct. The solutions of the initial-value problem for equations (6.286) and (6.284) are of the form

$$\begin{aligned}\mathrm{A}_1(z) &= v_{11\mathrm{f}}(z)\mathrm{A}_1(0) + v_{12\mathrm{f}}(z)\mathrm{A}_2(0)\\ &\quad - \mathrm{i}\int_0^z v_{11\mathrm{f}}(z|\zeta')G_1(\zeta',0)\,\mathrm{d}\zeta' + \mathrm{i}\int_0^z v_{12\mathrm{f}}(z|\zeta')G_2(\zeta',0)\,\mathrm{d}\zeta',\end{aligned} \tag{6.287}$$

$$\begin{aligned}\mathrm{A}_2(z) &= v_{21\mathrm{f}}(z)\mathrm{A}_1(0) + v_{22\mathrm{f}}(z)\mathrm{A}_2(0)\\ &\quad - \mathrm{i}\int_0^z v_{21\mathrm{f}}(z|\zeta')G_1(\zeta',0)\,\mathrm{d}\zeta' + \mathrm{i}\int_0^z v_{22\mathrm{f}}(z|\zeta')G_2(\zeta',0)\,\mathrm{d}\zeta',\end{aligned} \tag{6.288}$$

$$\begin{aligned}G_1(\zeta,z) &= -\gamma\theta(z-\zeta)\Big[\mathrm{i}v_{11\mathrm{f}}(\zeta)\mathrm{A}_1(0) + \mathrm{i}v_{12\mathrm{f}}(\zeta)\mathrm{A}_2(0)\\ &\quad + \int_0^\zeta v_{11\mathrm{f}}(\zeta|\zeta')G_1(\zeta',0)\,\mathrm{d}\zeta'\\ &\quad - \int_0^\zeta v_{12\mathrm{f}}(\zeta|\zeta')G_2(\zeta',0)\,\mathrm{d}\zeta'\Big] + G_1(\zeta,0),\end{aligned} \tag{6.289}$$

$$\begin{aligned}G_2(\zeta,z) &= \gamma\theta(z-\zeta)\Big[\mathrm{i}v_{21\mathrm{f}}(\zeta)\mathrm{A}_1(0) + \mathrm{i}v_{22\mathrm{f}}(\zeta)\mathrm{A}_2(0)\\ &\quad + \int_0^\zeta v_{21\mathrm{f}}(\zeta|\zeta')G_1(\zeta',0)\,\mathrm{d}\zeta'\\ &\quad - \int_0^\zeta v_{22\mathrm{f}}(\zeta|\zeta')G_2(\zeta',0)\,\mathrm{d}\zeta'\Big] + G_2(\zeta,0),\end{aligned} \tag{6.290}$$

where $\theta(z-\zeta)$ is the Heaviside (unit-step) function, $v_{jk\mathrm{f}}(z) = v_{jk\mathrm{f}}(z|0)$, $j,k = 1,2$, with

$$v_{jk\mathrm{f}}(z|z') = \exp\left[-\frac{\gamma}{2}(z-z')\right] v_{jk}(z|z')\big|_{\gamma=0}, \tag{6.291}$$

where

$$\begin{aligned}v_{11}(z|z')\big|_{\gamma=0} &= \exp[\mathrm{i}\delta(z-z')]\\ &\quad\times\left\{\cosh\Big[D_0(z-z')\Big] - \mathrm{i}\frac{\delta}{D_0}\sinh\Big[D_0(z-z')\Big]\right\},\end{aligned}$$

$$\begin{aligned}
v_{12}(z|z')\big|_{\gamma=0} &= \mathrm{i}\frac{\kappa^*}{D_0}\exp[\mathrm{i}\delta(z+z')]\sinh\Big[D_0(z-z')\Big],\\
v_{21}(z|z')\big|_{\gamma=0} &= -\mathrm{i}\frac{\kappa}{D_0}\exp[-\mathrm{i}\delta(z+z')]\sinh\Big[D_0(z-z')\Big],\\
v_{22}(z|z')\big|_{\gamma=0} &= \exp[-\mathrm{i}\delta(z-z')]\\
&\quad\times\left\{\cosh\Big[D_0(z-z')\Big]+\mathrm{i}\frac{\delta}{D_0}\sinh\Big[D_0(z-z')\Big]\right\},
\end{aligned}\tag{6.292}$$

with

$$D_0 = \sqrt{|\kappa|^2-\delta^2}\,. \tag{6.293}$$

Taking into account the relation

$$G_2(\zeta,z)=\begin{cases} G_2(\zeta,L) & \text{for } z>\zeta,\\ G_2(\zeta,L)-\mathrm{i}\frac{\gamma}{2}\mathrm{A}_2(\zeta) & \text{for } z=\zeta,\\ G_2(\zeta,L)-\mathrm{i}\gamma\mathrm{A}_2(\zeta) & \text{for } z<\zeta,\end{cases}\tag{6.294}$$

we replace equations (6.286) by the following equations

$$\begin{aligned}
\frac{\mathrm{d}}{\mathrm{d}z}\mathrm{A}_1(z) &= \mathrm{i}\kappa^*\mathrm{A}_2(z)\exp(2\mathrm{i}\delta z)-\frac{\gamma}{2}\mathrm{A}_1(z)-\mathrm{i}G_1(z,0),\\
\frac{\mathrm{d}}{\mathrm{d}z}\mathrm{A}_2(z) &= -\mathrm{i}\kappa\mathrm{A}_1(z)\exp(-2\mathrm{i}\delta z)+\frac{\gamma}{2}\mathrm{A}_2(z)+\mathrm{i}G_2(z,L).
\end{aligned}\tag{6.295}$$

The solutions of the boundary-value problem

$$\begin{aligned}
\hat{A}_j(z)\big|_{z=0} &= \hat{A}_j(0),\quad j=1,2,\\
G_1(\zeta,z)|_{z=0} = G_1(\zeta,0),&\quad G_2(\zeta,z)|_{z=L}=G_2(\zeta,L),
\end{aligned}\tag{6.296}$$

for equations (6.283) and (6.284) transformed to the form (6.295) and (6.296) are for $z=L$ as follows

$$\begin{aligned}
\mathrm{A}_1(L) &= v_{11\mathrm{g}}(L)\mathrm{A}_1(0)+v_{12\mathrm{g}}(L)\mathrm{A}_2(0)-\mathrm{i}\int_0^L v_{11\mathrm{g}}(L|\zeta')G_1(\zeta',0)\,\mathrm{d}\zeta'\\
&\quad+\mathrm{i}\int_0^L v_{12\mathrm{g}}(L|\zeta')G_2(\zeta',L)\,\mathrm{d}\zeta',
\end{aligned}\tag{6.297}$$

$$\begin{aligned}
\mathrm{A}_2(L) &= v_{21\mathrm{g}}(L)\mathrm{A}_1(0)+v_{22\mathrm{g}}(L)\mathrm{A}_2(0)\\
&\quad-\mathrm{i}\int_0^L v_{21\mathrm{g}}(L|\zeta')G_1(\zeta',0)\,\mathrm{d}\zeta'+\mathrm{i}\int_0^L v_{22\mathrm{g}}(L|\zeta')G_2(\zeta',L)\,\mathrm{d}\zeta',
\end{aligned}\tag{6.298}$$

$$\begin{aligned}
G_1(\zeta,L) &= -\mathrm{i}\gamma v_{11\mathrm{g}}(\zeta)\mathrm{A}_1(0)-\mathrm{i}\gamma v_{12\mathrm{g}}(\zeta)\mathrm{A}_2(0)\\
&\quad-\gamma\int_0^\zeta v_{11\mathrm{g}}(\zeta|\zeta')G_1(\zeta',0)\,\mathrm{d}\zeta'+G_1(\zeta,0)\\
&\quad+\gamma\int_0^\zeta v_{12\mathrm{g}}(\zeta|\zeta')G_2(\zeta',L)\,\mathrm{d}\zeta'+G_2(\zeta,L),
\end{aligned}\tag{6.299}$$

$$
\begin{aligned}
G_2(\zeta, 0) = & -\mathrm{i}\gamma v_{21\mathrm{g}}(\zeta)\mathrm{A}_1(0) - \mathrm{i}\gamma v_{22\mathrm{g}}(\zeta)\mathrm{A}_2(0) \\
& - \gamma \int_0^{\zeta} v_{21\mathrm{g}}(\zeta|\zeta')G_1(\zeta', 0)\,\mathrm{d}\zeta' + G_1(\zeta, 0) \\
& + \gamma \int_0^{\zeta} v_{22\mathrm{g}}(\zeta|\zeta')G_2(\zeta', L)\,\mathrm{d}\zeta' + G_2(\zeta, L),
\end{aligned} \tag{6.300}
$$

where $v_{jk\mathrm{g}}(z) = v_{jk\mathrm{g}}(z|0) = v_{jk}(z)$, $j, k = 1, 2$, with

$$
\begin{aligned}
v_{11\mathrm{g}}(z|z') &= \exp[\mathrm{i}\delta(z - z')] \left\{\cosh\left[D(z - z')\right] - \mathrm{i}\frac{\Delta}{D}\sinh\left[D(z - z')\right]\right\}, \\
v_{12\mathrm{g}}(z|z') &= \mathrm{i}\frac{\kappa^*}{D}\exp[\mathrm{i}\delta(z + z')]\sinh\left[D(z - z')\right], \\
v_{21\mathrm{g}}(z|z') &= -\mathrm{i}\frac{\kappa}{D}\exp[-\mathrm{i}\delta(z + z')]\sinh\left[D(z - z')\right], \\
v_{22\mathrm{g}}(z|z') &= \exp[-\mathrm{i}\delta(z - z')] \\
&\quad \times \left\{\cosh\left[D(z - z')\right] + \mathrm{i}\frac{\Delta}{D}\sinh\left[D(z - z')\right]\right\}.
\end{aligned} \tag{6.301}
$$

The solutions of differential equations (6.286) and (6.284) conserve pseudonorms (Peřinová et al. 2006)

$$
\begin{aligned}
& |A_1(L)|^2 - |A_2(L)|^2 + \frac{1}{\gamma}\int_0^L |G_1(\zeta, L)|^2\,\mathrm{d}\zeta - \frac{1}{\gamma}\int_0^L |G_2(\zeta, L)|^2\,\mathrm{d}\zeta \\
& = |A_1(0)|^2 - |A_2(0)|^2 + \frac{1}{\gamma}\int_0^L |G_1(\zeta, 0)|^2\,\mathrm{d}\zeta - \frac{1}{\gamma}\int_0^L |G_2(\zeta, 0)|^2\,\mathrm{d}\zeta,
\end{aligned} \tag{6.302}
$$

$$
\begin{aligned}
& |A_1(L)|^2 - |A_2(L)|^2 + \frac{1}{\gamma}\int_0^L |G_1(\zeta, L)|^2\,\mathrm{d}\zeta + \frac{1}{\gamma}\int_0^L |G_2(\zeta, 0)|^2\,\mathrm{d}\zeta \\
& = |A_1(0)|^2 - |A_2(0)|^2 + \frac{1}{\gamma}\int_0^L |G_1(\zeta, 0)|^2\,\mathrm{d}\zeta + \frac{1}{\gamma}\int_0^L |G_2(\zeta, L)|^2\,\mathrm{d}\zeta.
\end{aligned} \tag{6.303}
$$

On the inversion of equations (6.297) through (6.300), we see that the input and output complex amplitudes conserve the norm

$$
\begin{aligned}
& |A_1(L)|^2 + |A_2(0)|^2 + \frac{1}{\gamma}\int_0^L |G_1(\zeta, L)|^2\,\mathrm{d}\zeta + \frac{1}{\gamma}\int_0^L |G_2(\zeta, 0)|^2\,\mathrm{d}\zeta \\
& = |A_1(0)|^2 + |A_2(L)|^2 + \frac{1}{\gamma}\int_0^L |G_1(\zeta, 0)|^2\,\mathrm{d}\zeta + \frac{1}{\gamma}\int_0^L |G_2(\zeta, L)|^2\,\mathrm{d}\zeta
\end{aligned} \tag{6.304}
$$

and the corresponding scalar product. We can establish the Heisenberg picture in the following sense. We assume that the input annihilation and creation operators obey the usual commutation relations

$$
[\hat{A}_1(0), \hat{A}_1^{\dagger}(0)] = [\hat{A}_2(L), \hat{A}_2^{\dagger}(L)] = \hat{1}, \tag{6.305}
$$

where the operators in different modes commute. With respect to the noise operators, we make similar assumptions

$$[\hat{G}_1(\zeta, 0), \hat{G}_1^\dagger(\zeta', 0)] = [\hat{G}_2(\zeta, L), \hat{G}_2^\dagger(\zeta', L)] = \gamma\delta(\zeta - \zeta')\hat{1}. \tag{6.306}$$

The output operators are

$$\begin{aligned}\hat{A}_1(L) &= u_{11\mathrm{g}}(L)\hat{A}_1(0) + u_{12\mathrm{g}}(L)\hat{A}_2(L)\\ &\quad - \mathrm{i}\int_0^L w_{22\mathrm{g}}(L|\zeta')\hat{G}_1(\zeta', 0)\,\mathrm{d}\zeta' - \mathrm{i}\int_0^L w_{12\mathrm{g}}(L|\zeta')\hat{G}_2(\zeta', L)\,\mathrm{d}\zeta',\end{aligned} \tag{6.307}$$

$$\begin{aligned}\hat{A}_2(0) &= u_{21\mathrm{g}}(L)\hat{A}_1(0) + u_{22\mathrm{g}}(L)\hat{A}_2(L)\\ &\quad - \mathrm{i}\int_0^L w_{21\mathrm{g}}(L|\zeta')\hat{G}_1(\zeta', 0)\,\mathrm{d}\zeta' - \mathrm{i}\int_0^L w_{11\mathrm{g}}(L|\zeta')\hat{G}_2(\zeta', L)\,\mathrm{d}\zeta',\end{aligned} \tag{6.308}$$

where

$$u_{jk\mathrm{g}}(L) = u_{jk}(L), \quad j, k = 1, 2, \tag{6.309}$$

$$\begin{aligned}w_{22\mathrm{g}}(L|\zeta') &= \exp[\mathrm{i}\delta(L - \zeta')]\left[\cosh(D\zeta') + \mathrm{i}\frac{\Delta}{D}\sinh(D\zeta')\right]\\ &\quad\times\left[\cosh(DL) + \mathrm{i}\frac{\Delta}{D}\sinh(DL)\right]^{-1},\\ w_{21\mathrm{g}}(L|\zeta') &= \mathrm{i}\frac{\kappa}{D}\exp(-\mathrm{i}\delta\zeta')\sinh[D(L - \zeta')]\\ &\quad\times\left[\cosh(DL) + \mathrm{i}\frac{\Delta}{D}\sinh(DL)\right]^{-1},\\ w_{12\mathrm{g}}(L|\zeta) &= \mathrm{i}\frac{\kappa^*}{D}\exp[\mathrm{i}\delta(L + \zeta')]\sinh(D\zeta')\\ &\quad\times\left[\cosh(DL) + \mathrm{i}\frac{\Delta}{D}\sinh(DL)\right]^{-1},\\ w_{11\mathrm{g}}(L|\zeta') &= \exp(\mathrm{i}\delta\zeta')\left\{\cosh[D(L - \zeta')] + \mathrm{i}\frac{\Delta}{D}\sinh[D(L - \zeta')]\right\}\\ &\quad\times\left[\cosh(DL) + \mathrm{i}\frac{\Delta}{D}\sinh(DL)\right]^{-1}.\end{aligned} \tag{6.310}$$

The output operators obey the same commutation relations as (6.305)

$$[\hat{A}_1(L), \hat{A}_1^\dagger(L)] = [\hat{A}_2(0), \hat{A}_2^\dagger(0)] = \hat{1}. \tag{6.311}$$

Now we see that the relations (6.307) and (6.308) have the form (6.267), where the integrated quantum noise terms are

$$\hat{M}_1(L) = -\mathrm{i}\int_0^L w_{22\mathrm{g}}(L|\zeta')\hat{G}_1(\zeta'|0)\,\mathrm{d}\zeta' \\ -\mathrm{i}\int_0^L w_{12\mathrm{g}}(L|\zeta')\hat{G}_2(\zeta'|L)\,\mathrm{d}\zeta', \tag{6.312}$$

$$\hat{M}_2(L) = -\mathrm{i}\int_0^L w_{21\mathrm{g}}(L|\zeta')\hat{G}_1(\zeta'|0)\,\mathrm{d}\zeta' \\ -\mathrm{i}\int_0^L w_{11\mathrm{g}}(L|\zeta')\hat{G}_2(\zeta'|L)\,\mathrm{d}\zeta'. \tag{6.313}$$

Their commutators are

$$[\hat{M}_1(L), \hat{M}_1^\dagger(L)] = \int_0^L \left[|w_{22\mathrm{g}}(L|\zeta')|^2 + |w_{12\mathrm{g}}(L|\zeta')|^2\right] \mathrm{d}\zeta'\,\hat{1},$$
$$[\hat{M}_1(L), \hat{M}_2^\dagger(L)]$$
$$= \int_0^L \left[w_{22\mathrm{g}}(L|\zeta')w_{21\mathrm{g}}^*(L|\zeta') + w_{12\mathrm{g}}(L|\zeta')w_{11\mathrm{g}}^*(L|\zeta')\right] \mathrm{d}\zeta'\,\hat{1},$$
$$[\hat{M}_2(L), \hat{M}_1^\dagger(L)]$$
$$= \int_0^L \left[w_{21\mathrm{g}}(L|\zeta')w_{22\mathrm{g}}^*(L|\zeta') + w_{11\mathrm{g}}(L|\zeta')w_{12\mathrm{g}}^*(L|\zeta')\right] \mathrm{d}\zeta'\,\hat{1},$$
$$[\hat{M}_2(L), \hat{M}_2^\dagger(L)] = \int_0^L \left[|w_{21\mathrm{g}}(L|\zeta')|^2 + |w_{11\mathrm{g}}(L|\zeta')|^2\right] \mathrm{d}\zeta'\,\hat{1}. \tag{6.314}$$

The commutator matrix is

$$\begin{pmatrix} [\hat{M}_1(L), \hat{M}_1^\dagger(L)] & [\hat{M}_1(L), \hat{M}_2^\dagger(L)] \\ [\hat{M}_2(L), \hat{M}_1^\dagger(L)] & [\hat{M}_2(L), \hat{M}_2^\dagger(L)] \end{pmatrix}$$
$$= \int_0^L \begin{pmatrix} |w_{22\mathrm{g}}(L|\zeta')|^2 & w_{22\mathrm{g}}(L|\zeta')w_{21\mathrm{g}}^*(L|\zeta') \\ w_{21\mathrm{g}}(L|\zeta')w_{22\mathrm{g}}^*(L|\zeta') & |w_{21\mathrm{g}}(L|\zeta')|^2 \end{pmatrix} \mathrm{d}\zeta'\,\hat{1}$$
$$+ \int_0^L \begin{pmatrix} |w_{12\mathrm{g}}(L|\zeta')|^2 & w_{12\mathrm{g}}(L|\zeta')w_{11\mathrm{g}}^*(L|\zeta') \\ w_{11\mathrm{g}}(L|\zeta')w_{12\mathrm{g}}^*(L|\zeta') & |w_{11\mathrm{g}}(L|\zeta')|^2 \end{pmatrix} \mathrm{d}\zeta'\,\hat{1}$$
$$\geq \begin{pmatrix} 0 & 0 \\ 0 & 0 \end{pmatrix} \hat{1}. \tag{6.315}$$

We have shown that the device under investigation is an attenuator according to the formal definition. It has the attenuator characteristic property (6.281).

(ii) Derivation of normal characteristic function

We can pass from the Heisenberg picture to the Schrödinger picture using characteristic functionals

$$C_{\mathcal{N}}(\beta_1, \beta_2, \beta_1(z), \beta_2(z), 0) = \mathrm{Tr}\Bigg\{\hat{\rho}_{\mathrm{e}}(0)$$
$$\times \exp\left[\beta_1 \hat{A}_1^\dagger + \beta_2 \hat{A}_2^\dagger + \int_0^L \beta_1(z')\hat{G}_1^\dagger(z')\,\mathrm{d}z' + \int_0^L \beta_2(z')\hat{G}_2^\dagger(z')\,\mathrm{d}z'\right]$$
$$\times \exp\left[-\beta_1^* \hat{A}_1 - \beta_2^* \hat{A}_2 - \int_0^L \beta_1^*(z')\hat{G}_1(z')\,\mathrm{d}z' - \int_0^L \beta_2^*(z')\hat{G}_2(z')\,\mathrm{d}z'\right]\Bigg\}, \quad (6.316)$$

$$C_{\mathcal{N}}(\beta_1, \beta_2, \beta_1(z), \beta_2(z), L) = \mathrm{Tr}\Bigg\{\hat{\rho}_{\mathrm{e}}(L)$$
$$\times \exp\left[\beta_1 \hat{A}_1^\dagger + \beta_2 \hat{A}_2^\dagger + \int_0^L \beta_1(z')\hat{G}_1^\dagger(z')\,\mathrm{d}z' + \int_0^L \beta_2(z')\hat{G}_2^\dagger(z')\,\mathrm{d}z'\right]$$
$$\times \exp\left[-\beta_1^* \hat{A}_1 - \beta_2^* \hat{A}_2 - \int_0^L \beta_1^*(z')\hat{G}_1(z')\,\mathrm{d}z' - \int_0^L \beta_2^*(z')\hat{G}_2(z')\,\mathrm{d}z'\right]\Bigg\}. \quad (6.317)$$

Here we abbreviate $\hat{G}_1(z) \equiv \hat{G}_1(z, 0)$, $\hat{G}_2(z) \equiv \hat{G}_2(z, L)$. Choosing the input quantum noise fields in the vacuum state, we have

$$C_{\mathcal{N}}(\beta_1, \beta_2, \beta_1(z), \beta_2(z), 0) = C_{\mathcal{N}}(\beta_1, \beta_2, 0). \quad (6.318)$$

We do not describe the output quantum noise fields and so we are interested in the normal characteristic function

$$\begin{aligned} C_{\mathcal{N}}(\beta_1, \beta_2, L) &= C_{\mathcal{N}}(\beta_1, \beta_2, 0, 0, L) \\ &= C_{\mathcal{N}}\Big(\beta_1 u_{11}^*(L) + \beta_2 u_{21}^*(L), \beta_1 u_{12}^*(L) + \beta_2 u_{22}^*(L), \\ &\qquad \mathrm{i}\beta_1 w_{22\mathrm{g}}^*(L, z) + \mathrm{i}\beta_2 w_{21\mathrm{g}}^*(L, z), \\ &\qquad \mathrm{i}\beta_1 w_{12\mathrm{g}}^*(L, z) + \mathrm{i}\beta_2 w_{11\mathrm{g}}^*(L, z), 0\Big) \\ &= C_{\mathcal{N}}\Big(\beta_1 u_{11}^*(L) + \beta_2 u_{21}^*(L), \beta_1 u_{12}^*(L) + \beta_2 u_{22}^*(L), 0\Big). \end{aligned} \quad (6.319)$$

This already suffices for the proof that the procedure yields a completely positive map.

In Nielsen and Chuang (2000), the notion of quantum fidelity of two states $\hat{\rho}$, $\hat{\rho}'$ is introduced,

$$F = \mathrm{Tr}\left\{\sqrt{\hat{\rho}^{\frac{1}{2}}\hat{\rho}'\hat{\rho}^{\frac{1}{2}}}\right\}. \quad (6.320)$$

Here we put $\hat{\rho} = \hat{\rho}(0)$, $\hat{\rho}' = \hat{\rho}(L)$. For pure states, $\hat{\rho}(0) = |\psi(0)\rangle\langle\psi(0)|$, $\hat{\rho}(L) = |\psi(L)\rangle\langle\psi(L)|$, and the formula for the fidelity simplifies

$$F = |\langle\psi(L)|\psi(0)\rangle|. \quad (6.321)$$

Attenuated coherent states are coherent again,

$$|\psi(0)\rangle = |A_1(0), A_2(L)\rangle = |A_{1\text{in}}, A_{2\text{in}}\rangle,$$
$$|\psi(L)\rangle = |A_1(L), A_2(0)\rangle = |A_{1\text{out}}, A_{2\text{out}}\rangle, \tag{6.322}$$

where (see (6.260)) $A_{1\text{in}} = A_1(0)$, $A_{2\text{in}} = A_2(L)$, $A_{1\text{out}}=A_1(L)$, $A_{2\text{out}} = A_2(0)$. It is known that for the coherent states (Peřina 1991)

$$|\langle \mathrm{A}_{1\text{out}}, \mathrm{A}_{2\text{out}}|\mathrm{A}_{1\text{in}}, \mathrm{A}_{2\text{in}}\rangle|$$
$$= \exp\left(-\frac{1}{2}|\mathrm{A}_{1\text{out}} - \mathrm{A}_{1\text{in}}|^2 - \frac{1}{2}|\mathrm{A}_{2\text{out}} - \mathrm{A}_{2\text{in}}|^2\right). \tag{6.323}$$

The quantum fidelity should be applied in the time or space Schrödinger picture.

In Severini et al. (2004), with focusing on mode 1, a transmission coefficient has been introduced

$$\mathcal{T} = \frac{\langle \hat{A}_1^\dagger(L)\hat{A}_1(L)\rangle}{\langle \hat{A}_1^\dagger(0)\hat{A}_1(0)\rangle}. \tag{6.324}$$

The quantum averages are calculated in the state $|\psi(0)\rangle$. The transmission coefficient ought to be utilized in the space Heisenberg picture. Both in this and in the Schrödinger picture the numerical analysis simplifies for $|\psi(0)\rangle = |\mathrm{A}_{\text{in}}, 0\rangle$. The transmission spectrum is a function of a mismatch coefficient δ (cf. Figs. 6.1, 6.2),

$$\mathcal{T}(\delta) = |u_{11}(L)|^2. \tag{6.325}$$

The quantum fidelity spectrum is a function of the same coefficient

$$F(\delta) = \exp\left\{-\frac{1}{2}\left[|u_{11}(L) - 1|^2 + |u_{21}(L)|^2\right]|\mathrm{A}_{\text{in}}|^2\right\}, \tag{6.326}$$

but we restrict ourselves to $|\mathrm{A}_{\text{in}}|^2 = 1$ in Fig. 6.3.

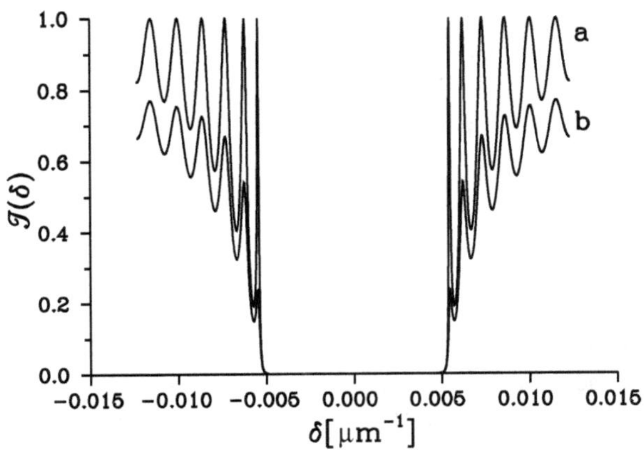

Fig. 6.1 Transmission spectrum for a corrugated $LiNbO_3$ planar waveguide as a function of the mismatch coefficient δ, for $\kappa = \frac{3\pi}{L}$

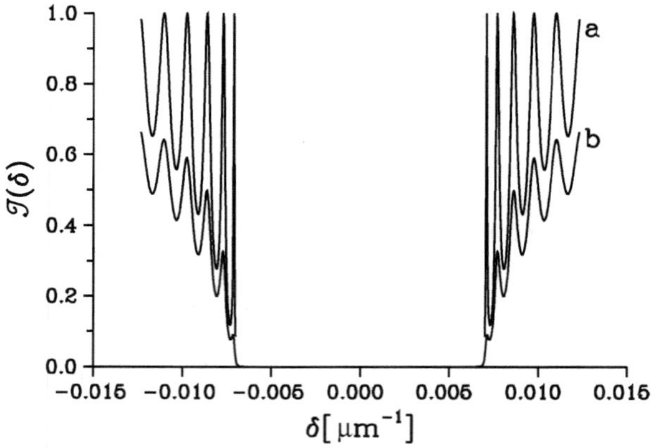

Fig. 6.2 The same as in Fig. 6.1, but for $\kappa = \frac{4\pi}{L}$

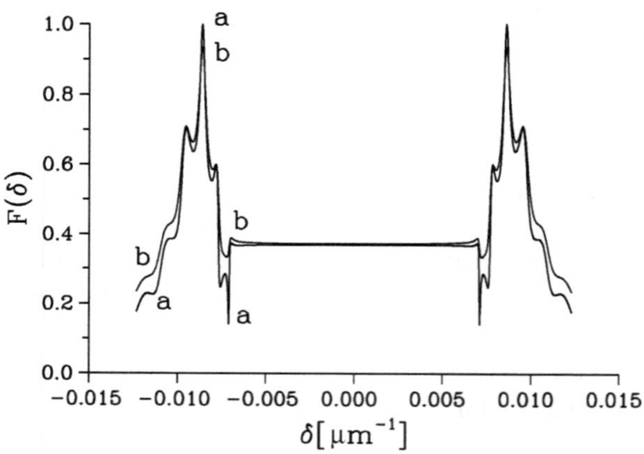

Fig. 6.3 Fidelity spectrum for a corrugated $LiNbO_3$ planar waveguide as a function of the mismatch coefficient δ. Here $\kappa = \frac{4\pi}{L}$, but other parameters are the same as in Fig. 6.1. For $|\kappa| = \frac{3\pi}{L}$, $F(\delta) = 1$ is not obtained

We have calculated the semiclassical and quantum transmission and fidelity spectra for a corrugated $LiNbO_3$ planar waveguide as functions of the mismatch coefficient δ between the spatial corrugation of the refractive index of the guide and the wavenumber of the propagating mode. We have used units $[\delta] = \mu m^{-1}$. The spatial corrugation has caused a coupling, which is characterized by the coupling constant κ, $\kappa = \frac{3\pi}{L}$ (Fig. 6.1) or $\kappa = \frac{4\pi}{L}$ (Figs. 6.2, 6.3). The waveguide length is $L = 1824.598\ \mu m$. We assume that

$$-\frac{\sqrt{9+6.5^2}}{L}\pi \leq \delta \leq \frac{\sqrt{9+6.5^2}}{L}\pi. \tag{6.327}$$

Here 6 is the number of maxima of the spectrum plotted in Fig. 6.1, and 6.5 is a value which makes the curves end with the next minimum approximately. The losses are characterized by the damping constant $\gamma \geq 0$, which is chosen as 0 (in the case without losses) and $0.01\kappa\sqrt{5}$. Line a (b) corresponds to the damping coefficient $\gamma = 0$ ($\gamma = 0.01\kappa\sqrt{5}$).

The output two-mode state differs from the input state only by an inessential phase factor when either

$$\delta \equiv 0 \bmod \frac{2\pi}{L}, \quad \sqrt{\delta^2 - |\kappa|^2} \equiv 0 \bmod \frac{2\pi}{L}, \tag{6.328}$$

or

$$\delta \equiv \frac{\pi}{L} \bmod \frac{2\pi}{L}, \quad \sqrt{\delta^2 - |\kappa|^2} \equiv \frac{\pi}{L} \bmod \frac{2\pi}{L}. \tag{6.329}$$

These conditions cannot be fulfilled for $|\kappa| = \frac{3\pi}{L}$, more generally, for $|\kappa| = \frac{\pi}{L} \bmod \frac{2\pi}{L}$. It can be satisfied for $|\kappa| = \frac{4\pi}{L}$, more generally, $|\kappa| = 0 \bmod \frac{2\pi}{L}$. In Fig. 6.3 this condition is met for $\delta = \frac{5\pi}{L}$, $|\kappa| = \frac{4\pi}{L}$, $\sqrt{\delta^2 - |\kappa|^2} = \frac{3\pi}{L}$ (one of the Pythagorean triangles).

In (Severini et al. 2004), a contradirectional coupler has been described by the following equations

$$\begin{aligned}\frac{\mathrm{d}}{\mathrm{d}z}\mathrm{A}_1(z) &= \mathrm{i}K_L\mathrm{A}_2(z)\exp(2\mathrm{i}\delta z) - \frac{\gamma}{2}\mathrm{A}_1(z),\\ \frac{\mathrm{d}}{\mathrm{d}z}\mathrm{A}_2(z) &= -\mathrm{i}K_L\mathrm{A}_1(z)\exp(-2\mathrm{i}\delta z) - \frac{\gamma}{2}\mathrm{A}_2(z),\end{aligned} \tag{6.330}$$

where $K_L \equiv \kappa \geq 0$. Although the authors specify that $\alpha \equiv \frac{\gamma}{2}$ characterizes leakage phenomena, they do not write signs conformable to the attenuator case, cf. equations (6.272). The second equation describes rather amplification in mode 2. The amplification has perhaps led to the derivation of "a state whose quantum properties are preserved".

Let us assume that in a medium, an attenuator in mode 1 and an amplifier in mode 2 occur, although we do not know of such a case. We could work with semiclassical and quantum noises as in the previous sections, but neither the normal ordering nor the antinormal one lead to a simple Schrödinger picture. With respect to the quantum noise of the amplifiers, one can assert that there is no state in which all the quantum properties are preserved.

We have dealt with the problem of quantum description of light modes which propagate in a periodic medium in opposite directions (Peřinová et al. 2006). Although we believe that partial differential equations comprising both time and space derivatives would be appropriate for the description, we have neglected the time ones and retained the space ones. As the coupled-mode theory is a classical description by means of ordinary differential equations involving a space derivative which leads to a quantum description of copropagating modes, it is also widely used for such a simple quantum description of counterpropagating modes. We have included also the gain and losses which have not been described to our knowledge yet. As application we have treated the conditions that can be imposed on a waveguide for the output state to be the same as the incoming one.

Bozhevolnyi et al. (2005) study propagation of long-range surface plasmon polaritons along periodically modulated medium both theoretically and experimentally. Surface plasmon polaritons are quasi-two-dimensional electromagnetic excitations that propagate along a dielectric–metal interface. Their application prospects are narrow. More complex excitations are an exception that are created in the configuration of two similar and very close metal–dielectric interfaces, such as surfaces of a thin metal film embedded in a dielectric. Then it is appropriate to speak of long-range surface plasmon polaritons.

Similarity to dielectric symmetric waveguides suggests to realize the band-gap effect for the long-range surface plasmon polaritons. The metal films are periodically thickness modulated. This is achieved with a periodic array of metal ridges. Provided that we know the electric field $\mathbf{E}^0(\mathbf{r})$ propagating along the metal film and the electric field Green tensor $\mathbf{G}(\mathbf{r}, \mathbf{r}')$ for the same structure, we can obtain the total electric field $\mathbf{E}(\mathbf{r})$ resulting in the process of multiple scattering by the ridges by solving the equation

$$\mathbf{E}(\mathbf{r}) = \mathbf{E}^0(\mathbf{r}) + k_0^2 \int \mathbf{G}(\mathbf{r}, \mathbf{r}') \left[\epsilon(\mathbf{r}') - \epsilon_{\text{ref}}(\mathbf{r}')\right] \mathbf{E}(\mathbf{r}')\, \mathrm{d}^2\mathbf{r}'. \tag{6.331}$$

Here k_0 is the free-space wave number, ϵ is the dielectric constant of the total structure inclusive of the metal ridges, and ϵ_{ref} is the dielectric constant of the reference structure (only a metal film embedded in a dielectric).

The gap in transmission and the peak in reflection are centred at $\lambda_g \approx 2n\Lambda$, where n is the refractive index of the dielectric and Λ is the grating period. For low ridges, the gap and the peak improve with the increase of the ridge height. For larger heights, the band-gap effect was not achieved. For $n = 1.543$ and $\Lambda = 500$ nm, we obtain $\lambda_g = 1543$ nm. Band gaps centred at 1550 nm and 20 nm wide have been simulated and experimentally investigated. The lengths of the structures were $L = 20, 40, 80, 160$ μm. The band-gap effect has been utilized for design and fabrication of a compact wavelength add-drop filter.

Deng et al. (2006) have reported second-harmonic generation in a sample made of lithium niobate. Near the surface, a waveguide was fabricated applying the proton-exchange technique. Ultraviolet laser lithography was applied to make photonic band-gap gratings on the sample. On the sample, two different gratings are inscribed. The first one couples the pump into the waveguide and the pump wave may come at an angle of around 45°. The second one is the photonic band-gap grating. A numerical model utilizes the coupled-mode theory. The corrugation couples TM to TM modes. The authors have measured that the second-harmonic generation in a waveguide mode is very weak compared with the second harmonic radiated into the substrate from the Cherenkov condition.

6.3 Photonic Crystals

The integral equation for quantum mechanical Green operator is a pattern for other integral equations of the field theory, in particular for the relation comprising the input and retarded electromagnetic fields (Białynicki-Birula and Białynicka-Birula

1975). A treatment of two- and three-dimensional photonic crystals may use a similar approach.

(i) Quadrature-phase squeezing in photonic crystals

The Green function method has been used for classical fields in Sakoda and Ohtaka (1996a,b). Sakoda (2002) has obtained the enhancement of a quantum optical process by use of a perturbation theory based on a Green function formalism. The results are related to degenerate optical parametric amplification, but the perturbation theory is not limited to this process and can be applied to other quantum-optical processes in the photonic crystals.

The quantization proceeds according to Glauber and Lewenstein (1991). The eigenmode of the electric field is denoted as $\mathbf{E}_{\mathbf{k}n}(\mathbf{r})$, where $\mathbf{k}$ is a wave vector in the first Brillouin zone and n is a band index. The eigenmodes are normalized with the condition

$$\int_V \epsilon(\mathbf{r})\mathbf{E}^*_{\mathbf{k}n}(\mathbf{r})\mathbf{E}_{\mathbf{k}'n'}(\mathbf{r})\,\mathrm{d}^3\mathbf{r} = V\delta_{\mathbf{k}\mathbf{k}'}\delta_{nn'}, \tag{6.332}$$

where V means the volume of the photonic crystal, and ϵ is a spatially periodic dielectric constant. The volume of unit cell is denoted as V_0.

On expressing the electric-field operator in the form

$$\hat{\mathbf{E}}(\mathbf{r},t) = \sum_{\mathbf{k},n} \mathrm{i}\sqrt{\frac{\hbar\omega_{\mathbf{k}n}}{2\epsilon_0 V}}[\hat{a}_{\mathbf{k}n}(t)\mathbf{E}_{\mathbf{k}n}(\mathbf{r}) - \hat{a}^\dagger_{\mathbf{k}n}(t)\mathbf{E}^*_{\mathbf{k}n}(\mathbf{r})], \tag{6.333}$$

where $\hat{a}_{\mathbf{k}n}$ and $\hat{a}^\dagger_{\mathbf{k}n}$ are the usual photon annihilation and creation operators, respectively, and writing the magnetic-induction operator similarly, the total electromagnetic energy (in the volume) is reduced to a quantum-mechanical Hamiltonian

$$\hat{H} = \sum_{\mathbf{k},n} \hbar\omega_{\mathbf{k}n}\left[\hat{a}^\dagger_{\mathbf{k}n}(t)\hat{a}_{\mathbf{k}n}(t) + \frac{1}{2}\hat{1}\right]. \tag{6.334}$$

The nonlinear medium is described by a second-order susceptibility tensor $\boldsymbol{\chi}^{(2)}(\mathbf{r})$ that has the same spatial periodicity as $\epsilon(\mathbf{r})$. But $\boldsymbol{\chi}^{(2)}(\mathbf{r})$ is nonzero only in the region $0 \le z \le l = an_z$, where a is the lattice constant of the crystal and n_z is a positive integer.

It is assumed that a pump wave denoted $\mathbf{E}_\mathrm{p}(\mathbf{r},t)$ and a signal wave denoted $\mathbf{E}_\mathrm{s}(\mathbf{r},t)$ propagate along the z-axis. Both waves are single mode. We let $k_{\mathrm{p}z}$ and $k_{\mathrm{s}z}$ denote the z-components of their wave vectors, and we introduce a phase mismatch Δk_z, $\Delta k_z = k_{\mathrm{p}z} - 2k_{\mathrm{s}z}$. The frequency of the pump wave, ω_p, is twice that of the signal wave, ω_s. We let v_g denote the group velocity of the signal wave.

Since

$$\mathbf{E}_\mathrm{p}(\mathbf{r},t) = \mathrm{i}A[\mathbf{E}_\mathrm{p}(\mathbf{r})\exp(-\mathrm{i}\omega_\mathrm{p}t + \mathrm{i}\theta) - \mathbf{E}^*_\mathrm{p}(\mathbf{r})\exp(\mathrm{i}\omega_\mathrm{s}t - \mathrm{i}\theta)] \tag{6.335}$$

is a classical quantity and

$$\hat{\mathbf{E}}_{\mathrm{s}}(\mathbf{r},t)=\mathrm{i}\sqrt{\frac{\hbar\omega_{\mathrm{s}}}{2\epsilon_0 V}}[\hat{a}_{\mathrm{s}}(0)\mathbf{E}_{\mathrm{s}}(\mathbf{r})\exp(-\mathrm{i}\omega_{\mathrm{s}}t)-\hat{a}_{\mathrm{s}}^{\dagger}(0)\mathbf{E}_{\mathrm{s}}^{*}(\mathbf{r})\exp(\mathrm{i}\omega_{\mathrm{s}}t)], \tag{6.336}$$

where $\mathbf{E}_{\mathrm{p}}(\mathbf{r})$ and $\mathbf{E}_{\mathrm{s}}(\mathbf{r})$ are eigenmodes of the electric field, A is the amplitude of the pump wave, θ the shift of its phase, and $\hat{a}_{\mathrm{s}}(0)$ and $\hat{a}_{\mathrm{s}}^{\dagger}(0)$ are photon annihilation and creation operators, respectively, can be considered to be an output electric-field operator, the integral equation for this operator is formulated,

$$\epsilon_0\epsilon(\mathbf{r})\hat{\mathbf{E}}(\mathbf{r},t)+\hat{\mathbf{P}}(\mathbf{r},t)=\epsilon_0\epsilon(\mathbf{r})\hat{\mathbf{E}}_{\mathrm{s}}(\mathbf{r},t)+\frac{\epsilon(\mathbf{r})}{V}$$

$$\times\sum_{\mathbf{k},n}\omega_{\mathbf{k}n}\mathbf{E}_{\mathbf{k}n}(\mathbf{r})\int_V\int_{-\infty}^{t}\mathbf{E}_{\mathbf{k}n}^{*}(\mathbf{r}')\cdot\hat{\mathbf{P}}(\mathbf{r}',t')\sin\left[\omega_{\mathbf{k}n}(t-t')\right]\,\mathrm{d}^3\mathbf{r}'\,\mathrm{d}t', \tag{6.337}$$

where $\hat{\mathbf{P}}(\mathbf{r},t)$ is the output nonlinear polarization operator,

$$\hat{\mathbf{P}}(\mathbf{r},t)\approx 2\boldsymbol{\chi}^{(2)}(\mathbf{r}):\mathbf{E}_{\mathrm{p}}(\mathbf{r},t)\hat{\mathbf{E}}(\mathbf{r},t). \tag{6.338}$$

The solution of the integral equation (6.337) is of the form

$$\hat{\mathbf{E}}(\mathbf{r},t)\approx\mathrm{i}\sqrt{\frac{\hbar\omega_{\mathrm{s}}}{2\epsilon_0 V}}[\hat{b}\mathbf{E}_{\mathrm{s}}(\mathbf{r})\exp(-\mathrm{i}\omega_{\mathrm{s}}t)-\hat{b}^{\dagger}\mathbf{E}_{\mathrm{s}}^{*}(\mathbf{r})\exp(\mathrm{i}\omega_{\mathrm{s}}t)], \tag{6.339}$$

where

$$\begin{aligned}\hat{b}&=\hat{a}_{\mathrm{s}}(0)\cosh(|\beta'|l)+\exp[\mathrm{i}(\theta+\phi')]\hat{a}_{\mathrm{s}}^{\dagger}(0)\sinh(|\beta'|l),\\ \hat{b}^{\dagger}&=\hat{a}_{\mathrm{s}}^{\dagger}(0)\cosh(|\beta'|l)+\exp[-\mathrm{i}(\theta+\phi')]\hat{a}_{\mathrm{s}}(0)\sinh(|\beta'|l),\end{aligned} \tag{6.340}$$

with ϕ' being the phase of the effective coupling constant (inclusive of the amplitude A),

$$\begin{aligned}\beta'&=\beta\eta\xi,\quad \beta=\frac{\omega_{\mathrm{s}}AF_{\mathrm{s,p,s}}}{\epsilon_0 v_{\mathrm{g}}},\\ \eta&=\frac{\sin\left(\frac{l\Delta k_z}{2}\right)}{n_z\sin\left(\frac{a\Delta k_z}{2}\right)},\quad \xi=\exp\left[\mathrm{i}\frac{a(n_z-1)\Delta k_z}{2}\right],\\ F_{\mathrm{s,p,s}}&=\frac{1}{V_0}\int_{V_0}\mathbf{E}_{\mathrm{s}}^{*}(\mathbf{r})\cdot\boldsymbol{\chi}^{(2)}(\mathbf{r}):\mathbf{E}_{\mathrm{p}}(\mathbf{r})\mathbf{E}_{\mathrm{s}}^{*}(\mathbf{r})\,\mathrm{d}^3\mathbf{r}.\end{aligned} \tag{6.341}$$

The nonlinear properties of photonic crystals were reviewed in Slusher and Eggleton (2003).

(ii) Parametric down conversion in a multilayered structure simply described

A description of spontaneous parametric down conversion in finite-length multilayer structure has been developed using semiclassical and quantum approaches (Centini et al. 2005). The semiclassical model has allowed one to find the criterion for designing and optimizing the structure. The quantum model is related to the properties of emitted entangled photon pairs.

One considers a one-dimensional dispersive lossless inhomogeneous medium, where both the dielectric constant, $\epsilon(z, \omega)$, and the nonlinear susceptibility, $d^{(2)}(z)$, are functions of a single spatial coordinate z. The study is limited to s-polarized plane monochromatic waves that fall onto the interfaces in the normal direction. The planes $z = 0$ and $z = L$ are the first and last interfaces of the structure embedded in air. The classical treatment begins with the following nonlinear, coupled Helmholtz equations:

$$\begin{aligned}\frac{\mathrm{d}^2 E_\mathrm{s}}{\mathrm{d}z^2} + \frac{\omega_\mathrm{s}^2 \epsilon_\mathrm{s}(z)}{c^2} E_\mathrm{s} &= -2\frac{\omega_\mathrm{s}^2}{c^2} d^{(2)}(z) E_\mathrm{i}^* E_\mathrm{p},\\ \frac{\mathrm{d}^2 E_\mathrm{i}}{\mathrm{d}z^2} + \frac{\omega_\mathrm{i}^2 \epsilon_\mathrm{i}(z)}{c^2} E_\mathrm{i} &= -2\frac{\omega_\mathrm{i}^2}{c^2} d^{(2)}(z) E_\mathrm{s}^* E_\mathrm{p},\\ \frac{\mathrm{d}^2 E_\mathrm{p}}{\mathrm{d}z^2} + \frac{\omega_\mathrm{p}^2 \epsilon_\mathrm{p}(z)}{c^2} E_\mathrm{p} &= -2\frac{\omega_\mathrm{p}^2}{c^2} d^{(2)}(z) E_\mathrm{i}^* E_\mathrm{s},\end{aligned} \tag{6.342}$$

where $\epsilon_n(z) \equiv \epsilon(z, \omega_n)$, ω_n is the angular frequency of the field n, $n = \mathrm{s, i, p}$, and s, i, and p stand for signal, idler, and pump, respectively. It is assumed that $\omega_\mathrm{p} = \omega_\mathrm{s} + \omega_\mathrm{i}$. The treatment has followed (D'Aguanno et al. 2002).

The solutions of the corresponding linear equations E can easily be decomposed into forward- and backward-propagating waves $E = E_\mathrm{F} + E_\mathrm{B}$, where F (B) means the forward (backward) propagation. The solutions to these equations are introduced, $\Theta_n^{(+)}$ and $\Theta_n^{(-)}$, which fulfil the following boundary conditions,

$$\begin{aligned}\Theta_{n\mathrm{F}}^{(+)}(-0) = 1,\ \Theta_{n\mathrm{B}}^{(+)}(L+0) = 0,\\ \Theta_{n\mathrm{F}}^{(-)}(-0) = 0,\ \Theta_{n\mathrm{B}}^{(-)}(L+0) = 1.\end{aligned} \tag{6.343}$$

These solutions have the familiar properties

$$\begin{aligned}\Theta_{n\mathrm{F}}^{(+)}(L+0) = \mathtt{t}_a^{(+)},\ \Theta_{n\mathrm{B}}^{(+)}(-0) = \mathtt{r}_a^{(+)},\\ \Theta_{n\mathrm{B}}^{(-)}(-0) = \mathtt{t}_n^{(-)},\ \Theta_{n\mathrm{F}}^{(-)}(L+0) = \mathtt{r}_n^{(-)},\end{aligned} \tag{6.344}$$

where $\mathtt{r}_n^{(+)}$ and $\mathtt{t}_n^{(+)}$ ($\mathtt{r}_n^{(-)}$ and $\mathtt{t}_n^{(-)}$) are the linear reflection and transmission complex coefficients for left-to-right (right-to-left) propagation, cf. (Born and Wolf 1999, Yeh 1988). The solutions of the nonlinear equations are decomposed as

$$E_n = A_n^{(+)}(z)\Theta_n^{(+)}(z) + A_n^{(-)}(z)\Theta_n^{(-)}(z), \tag{6.345}$$

where $A_n^{(+)}(z)$ and $A_n^{(-)}(z)$ are slowly varying complex envelope functions. Especially,

$$E_{n\mathrm{B}}(z) = \left[A_n^{(+)}(0)\mathtt{r}_n^{(+)} + A_n^{(-)}(0)\mathtt{t}_n^{(-)}\right]\exp\left(-\mathrm{i}\frac{\omega_n z}{c}\right) \text{ for } z \leq 0,$$
$$E_{n\mathrm{F}}(z) = \left[A_n^{(-)}(L)\mathtt{r}_n^{(-)} + A_n^{(+)}(L)\mathtt{t}_n^{(+)}\right] \times \exp\left(\mathrm{i}\frac{\omega_n(z-L)}{c}\right) \text{ for } z \geq L. \quad (6.346)$$

As usual with the semiclassical approach, a scalar product is introduced,

$$\langle f|g\rangle = \frac{1}{L}\int_0^L f^*(z)g(z)\,\mathrm{d}z. \quad (6.347)$$

A description has been developed using, e.g., the integrals

$$\Gamma_{(\mathrm{s},j)}^{(k,l)} = \left\langle \Theta_\mathrm{s}^{(j)}|d^{(2)}\Theta_\mathrm{p}^{(k)}\Theta_\mathrm{i}^{(l)*}\right\rangle, \quad (6.348)$$

where $j,k,l = +,-$.

In contrast, we present a quantum description using the solutions of linear equations $\tilde{\Theta}_a^{(+)}$ and $\tilde{\Theta}_a^{(-)}$, $a = \mathrm{s, i}$, which fulfil the boundary conditions

$$\tilde{\Theta}_{a\mathrm{B}}^{(+)}(-0) = 0,\ \tilde{\Theta}_{a\mathrm{F}}^{(+)}(L+0) = 1,$$
$$\tilde{\Theta}_{a\mathrm{B}}^{(-)}(-0) = 1,\ \tilde{\Theta}_{a\mathrm{F}}^{(-)}(L+0) = 0. \quad (6.349)$$

The connecting relations are

$$\tilde{\Theta}_a^{(+)} = \Theta_a^{(+)}\mathtt{t}_a^{(+)*} + \Theta_a^{(-)}\mathtt{r}_a^{(-)*},$$
$$\tilde{\Theta}_a^{(-)} = \Theta_a^{(+)}\mathtt{r}_a^{(+)*} + \Theta_a^{(-)}\mathtt{t}_a^{(-)*}. \quad (6.350)$$

We introduce the overlap integrals

$$\tilde{\Gamma}_{(\mathrm{s},j)}^{(k,l)} = \left\langle \tilde{\Theta}_\mathrm{s}^{(j)}|d^{(2)}\Theta_\mathrm{p}^{(k)}\tilde{\Theta}_\mathrm{i}^{(l)*}\right\rangle. \quad (6.351)$$

The nonlinear interaction in the entire photonic band-gap structure is described by an interaction Hamiltonian $\hat{H}(t)$ given as a sum of operators $\hat{H}^{(l)}(t)$ ($l = 1,\ldots,N$) that characterize every layer of the structure,

$$\hat{H}(t) = \sum_{l=1}^{N} \hat{H}^{(l)}(t), \quad (6.352)$$

where N is the number of layers of the structure. Strong pump positive-frequency electric-field amplitudes $E_\mathrm{p}^{(l,+)}(z,t)$, weak signal and idler positive-frequency electric-field operator amplitudes $\hat{E}_a^{(l,+)}(z,t)$, $a = \mathrm{s, i}$, and their Hermitian-conjugated expressions $\hat{E}_a^{(l,-)}(z,t)$ are introduced. Then

$$\hat{H}^{(l)}(t) = \tilde{d}^{(l)}\int_{z_{l-1}}^{z_l}\left[E_\mathrm{p}^{(l,+)}(z,t)\hat{E}_\mathrm{s}^{(l,-)}(z,t)\hat{E}_\mathrm{i}^{(l,-)}(z,t) + \mathrm{H.c.}\right]\mathrm{d}z. \quad (6.353)$$

Here $\tilde{d}^{(l)} = \frac{\hbar}{c}\sqrt{\omega_s\omega_i}d^{(l)}$, and $d^{(l)}$ means the second-order susceptibility of the lth layer.

A monochromatic pump electric field has the positive-frequency amplitude

$$E_p^{(l,+)}(z,t) = \left\{B_{pF}^{(l)} \exp\left[ik_p^{(l)}(z - z_{l-1})\right] + B_{pB}^{(l)} \exp\left[-ik_p^{(l)}(z - z_{l-1})\right]\right\} \exp(-i\omega_p t); \quad (6.354)$$

$k_p^{(l)}$ means the wave vector of the pump field in the lth layer and $B_{pF}^{(l)}$ ($B_{pB}^{(l)}$) is a complex coefficient. Down-converted fields are polychromatic, and it is convenient to express their amplitudes in units of $\sqrt{\frac{2\hbar\omega_a}{\epsilon_0 V}}$, where V is the quantization volume. Then the positive-frequency operator amplitudes are

$$\hat{E}_a^{(l,+)}(z,t) = \int b_a^{(l)} \left\{\hat{b}_{aF}^{(l)}(\omega_a) \exp\left[ik_a^{(l)}(\omega_a)(z - z_{l-1})\right] + \hat{b}_{aB}^{(l)}(\omega_a) \exp\left[-ik_a^{(l)}(\omega_a)(z - z_{l-1})\right]\right\} \exp(-i\omega_a t)\, d\omega_a. \quad (6.355)$$

Here $b_a^{(l)} = \frac{1}{\sqrt{\epsilon_a^{(l)}}}$, $\epsilon_a^{(l)}$ stands for the relative permittivity in the lth layer for the field a.

It holds that

$$\begin{aligned} B_{pF}^{(l)} &= A_p^{(+)}(0)\Theta_{pF}^{(+)}(z_{l-1}) + A_p^{(-)}(L)\Theta_{pF}^{(-)}(z_{l-1}), \\ B_{pB}^{(l)} &= A_p^{(+)}(0)\Theta_{pB}^{(+)}(z_{l-1}) + A_p^{(-)}(L)\Theta_{pB}^{(-)}(z_{l-1}),\ l = 1, \ldots, N+1; \end{aligned} \quad (6.356)$$

particularly, $B_{pB}^{(N+1)} = A_p^{(-)}(L)$, but $B_{pF}^{(0)} = A_p^{(+)}(0)$. Similarly,

$$\begin{aligned} b_a^{(l)}\hat{b}_{aF}^{(l)}(\omega_a) &= \hat{b}_{aF}^{(N+1)}(\omega_a)\tilde{\Theta}_{aF}^{(+)}(z_{l-1}) + \hat{b}_{aB}^{(0)}(\omega_a)\tilde{\Theta}_{aF}^{(-)}(z_{l-1}), \\ b_a^{(l)}\hat{b}_{aB}^{(l)}(\omega_a) &= \hat{b}_{aF}^{(N+1)}(\omega_a)\tilde{\Theta}_{aB}^{(+)}(z_{l-1}) + \hat{b}_{aB}^{(0)}(\omega_a)\tilde{\Theta}_{aB}^{(-)}(z_{l-1}),\ l = 1, \ldots, N. \end{aligned} \quad (6.357)$$

Here $\hat{b}_{aF}^{(N+1)}(\omega_a)$ and $\hat{b}_{aB}^{(0)}(\omega_a)$ are output operators.

The solution $|\psi\rangle_{s,i}$ of the Schrödinger equation correct up to first order on the assumption of the initial vacuum state $|\text{vac}\rangle_{s,i}$ for the down-converted fields reads as

$$|\psi\rangle_{s,i} = |\text{vac}\rangle_{s,i} - \frac{i}{\hbar} \lim_{T\to\infty} \int_{-T}^{T} \hat{H}(t)|\text{vac}\rangle_{s,i}\, dt. \quad (6.358)$$

It can be written in the form

$$\begin{aligned} |\psi\rangle_{s,i} = |\text{vac}\rangle_{s,i} + \int\!\!\int \Big[&\Phi_{FF}(\omega_s,\omega_i)\hat{b}_{sF}^{(N+1)\dagger}\hat{b}_{iF}^{(N+1)\dagger}|\text{vac}\rangle_{s,i} \\ &+ \Phi_{FB}(\omega_s,\omega_i)\hat{b}_{sF}^{(N+1)\dagger}\hat{b}_{iB}^{(0)\dagger}|\text{vac}\rangle_{s,i} + \Phi_{BF}(\omega_s,\omega_i)\hat{b}_{sB}^{(0)\dagger}\hat{b}_{iF}^{(N+1)\dagger}|\text{vac}\rangle_{s,i} \\ &+ \Phi_{BB}(\omega_s,\omega_i)\hat{b}_{sB}^{(0)\dagger}\hat{b}_{iB}^{(0)\dagger}|\text{vac}\rangle_{s,i}\, d\omega_s\, d\omega_i \Big]. \end{aligned} \quad (6.359)$$

Here

$$\Phi_{\rm FF}(\omega_{\rm s},\omega_{\rm i}) = \frac{2\pi\sqrt{\omega_{\rm s}\omega_{\rm i}}L}{{\rm i}c}\delta(\omega_{\rm p}-\omega_{\rm s}-\omega_{\rm i})\left[A_{\rm p}^{(+)}\tilde{\Gamma}_{({\rm s},+)}^{(+,+)} + A_{\rm p}^{(-)}\tilde{\Gamma}_{({\rm s},+)}^{(-,+)}\right],$$
$$\Phi_{\rm FB}(\omega_{\rm s},\omega_{\rm i}) = \frac{2\pi\sqrt{\omega_{\rm s}\omega_{\rm i}}L}{{\rm i}c}\delta(\omega_{\rm p}-\omega_{\rm s}-\omega_{\rm i})\left[A_{\rm p}^{(+)}\tilde{\Gamma}_{({\rm s},+)}^{(+,-)} + A_{\rm p}^{(-)}\tilde{\Gamma}_{({\rm s},+)}^{(-,-)}\right],$$
$$\Phi_{\rm BF}(\omega_{\rm s},\omega_{\rm i}) = \frac{2\pi\sqrt{\omega_{\rm s}\omega_{\rm i}}L}{{\rm i}c}\delta(\omega_{\rm p}-\omega_{\rm s}-\omega_{\rm i})\left[A_{\rm p}^{(+)}\tilde{\Gamma}_{({\rm s},-)}^{(+,+)} + A_{\rm p}^{(-)}\tilde{\Gamma}_{({\rm s},-)}^{(-,+)}\right],$$
$$\Phi_{\rm BB}(\omega_{\rm s},\omega_{\rm i}) = \frac{2\pi\sqrt{\omega_{\rm s}\omega_{\rm i}}L}{{\rm i}c}\delta(\omega_{\rm p}-\omega_{\rm s}-\omega_{\rm i})\left[A_{\rm p}^{(+)}\tilde{\Gamma}_{({\rm s},-)}^{(+,-)} + A_{\rm p}^{(-)}\tilde{\Gamma}_{({\rm s},-)}^{(-,-)}\right]. \quad (6.360)$$

Even though these functions are not probability amplitudes, they can be combined, e.g., with parameters of a finite spatial region to give such amplitudes.

(iii) Parametric down conversion in a multilayered structure including polarization

The foregoing description has been generalized and revised in part in (Peřina Jr., et al. 2006). For instance, the structure may be embedded in a medium with the relative permittivity $\epsilon_n^{(0)}$ ($\epsilon_n^{(N+1)}$) in front of (beyond) the sample. The linear indices of refraction are introduced, $n_m^{(l)}=\sqrt{\epsilon_m^{(l)}}$, $l=0,\ldots,N+1$, $m={\rm s,i,p}$.

The treatment has been restricted to plane waves with wave vectors parallel to the yz-plane. The forward-propagating fields have the wave vectors $\mathbf{k}_{m{\rm F}}^{(l)} = \mathbf{e}_y k_m^{(l)}\sin(\vartheta_m^{(l)}) + \mathbf{e}_z k_m^{(l)}\cos(\vartheta_m^{(l)})$, where

$$k_m^{(l)} = \frac{\omega_m}{c}n_m^{(l)}. \quad (6.361)$$

The angles $\vartheta_m^{(l)}$ fulfil the Snell law:

$$n_m^{(l)}\sin(\vartheta_m^{(l)}) = \text{constant},\ l=0,\ldots,N+1. \quad (6.362)$$

The wave-vectors $\mathbf{k}_{m{\rm B}}^{(l)} = \mathbf{e}_y k_m^{(l)}\sin(\vartheta_m^{(l)}) - \mathbf{e}_z k_m^{(l)}\cos(\vartheta_m^{(l)})$ characterize the backward-propagating fields. For simplicity, $\mathbf{k}_m^{(l)} \equiv \mathbf{k}_{m{\rm F}}^{(l)}$, $k_{m,x}^{(l)}=0$, $k_{m,y}^{(l)} = k_m^{(l)}\sin(\vartheta_m^{(l)})$, $k_{m,z}^{(l)} = k_m^{(l)}\cos(\vartheta_m^{(l)})$.

At this moment, we may still use classical concepts and restrict ourselves to monochromatic waves. We distinguish the TE- and TM-waves. These have the electric fields of the forms

$$\mathbf{E}_{m,{\rm TE}} = E_{m,{\rm TE}}\mathbf{e}_x,$$
$$\mathbf{E}_{m,{\rm TM}} = E_{m,{\rm TM}}\mathbf{e}_y\cos\left[\vartheta_m(z)\right] - \bar{E}_{m,{\rm TM}}\mathbf{e}_z\sin\left[\vartheta_m(z)\right], \quad (6.363)$$

where $\vartheta_m(z) = \vartheta_m^{(l)}$ for z in the lth layer and

$$\bar{E}_{m,{\rm TM}} = \frac{1}{{\rm i}k_{m,y}(z)}\frac{\partial}{\partial y}E_{m,{\rm TM}} = \frac{1}{{\rm i}k_{m,z}(z)}\frac{\partial}{\partial z}E_{m,{\rm TM}}, \quad (6.364)$$

with $k_{m,y}(z) = k_{m,y}^{(l)}$ and $k_{m,z}(z) = k_{m,z}^{(l)}$ (in the lth layer).

The linear Helmholtz equations read

$$\frac{\partial^2 E_{m,\alpha}}{\partial z^2} + k_{m,z}^2 E_{m,\alpha} = 0, m = \text{s, i, p}, \quad \alpha = \text{TE, TM}. \tag{6.365}$$

The prolongation conditions depend on the polarization. The case $\alpha = \text{TE}$ is very attractive, since the conditions require that $E_{m,\alpha}$ and $\frac{\partial}{\partial z} E_{m,\alpha}$ be continuous at the points $z = z_l$. Let x and y be zero for definiteness. The case $\alpha = \text{TM}$ includes the conditions that $E_{m,\alpha} \cos[\vartheta_m(z)]$ and $\frac{1}{\cos[\vartheta_m(z)]} \frac{\partial}{\partial z} E_{m,\alpha}$ are continuous at $z = z_l$.

We introduce solutions to these equations, $\Theta_{m,\alpha}^{(+)}$ and $\Theta_{m,\alpha}^{(-)}$, which satisfy the following boundary conditions:

$$\begin{aligned} &\Theta_{m\text{F},\alpha}^{(+)}(-0, \omega_m) = 1, \Theta_{m\text{B},\alpha}^{(+)}(L+0, \omega_m) = 0, \\ &\Theta_{m\text{F},\alpha}^{(-)}(-0, \omega_m) = 0, \Theta_{m\text{B},\alpha}^{(-)}(L+0, \omega_m) = 1. \end{aligned} \tag{6.366}$$

Similarly, we will need some of the solutions $\tilde{\Theta}_{m,\alpha}^{(+)}$ and $\tilde{\Theta}_{m,\alpha}^{(-)}$, which fulfil the boundary conditions

$$\begin{aligned} &\tilde{\Theta}_{m\text{B},\alpha}^{(+)}(-0, \omega_m) = 0, \tilde{\Theta}_{m\text{F},\alpha}^{(+)}(L+0, \omega_m) = 1, \\ &\tilde{\Theta}_{m\text{B},\alpha}^{(-)}(-0, \omega_m) = 1, \tilde{\Theta}_{m\text{F},\alpha}^{(-)}(L+0, \omega_m) = 0. \end{aligned} \tag{6.367}$$

In quantum physics, we will use $\mathbf{E}_{\text{p},\alpha}^{(+)}(z, \omega_\text{p})$ instead of $\mathbf{E}_{\text{p},\alpha}(z)$ for the positive-frequency electric-field amplitude of a monochromatic component at frequency ω_p with polarization α. The positive-frequency electric-field amplitude $\mathbf{E}_\text{p}^{(+)}(z,t)$ is decomposed into the TE- and TM-wave contributions $\mathbf{E}_{\text{p,TE}}^{(+)}(z,t)$ and $\mathbf{E}_{\text{p,TM}}^{(+)}(z,t)$ and expressed in the forms

$$\begin{aligned} \mathbf{E}_\text{p}^{(+)}(z,t) &= \mathbf{E}_{\text{p,TE}}^{(+)}(z,t) + \mathbf{E}_{\text{p,TM}}^{(+)}(z,t) \\ &= \frac{1}{\sqrt{2\pi}} \int_0^\infty \mathbf{E}_\text{p}^{(+)}(z, \omega_\text{p})\, \text{d}\omega_\text{p} \\ &= \frac{1}{\sqrt{2\pi}} \int_0^\infty \left[\mathbf{E}_{\text{p,TE}}^{(+)}(z, \omega_\text{p}) + \mathbf{E}_{\text{p,TM}}^{(+)}(z, \omega_\text{p}) \right] \text{d}\omega_\text{p}. \end{aligned} \tag{6.368}$$

Here

$$\begin{aligned} \mathbf{E}_{\text{p},\alpha}^{(+)}(z, \omega_\text{p}) = {} & A_{\text{pF},\alpha}^{(l)}(\omega_\text{p}) \mathbf{e}_{\text{pF},\alpha}^{(l)}(\omega_\text{p}) \exp[\text{i}k_{\text{p},z}^{(l)}(z - z_{l-1})] \\ & + A_{\text{pB},\alpha}^{(l)}(\omega_\text{p}) \mathbf{e}_{\text{pB},\alpha}^{(l)}(\omega_\text{p}) \exp[-\text{i}k_{\text{p},z}^{(l)}(z - z_{l-1})], \end{aligned} \tag{6.369}$$

with

$$\begin{aligned} A_{\text{pF},\alpha}^{(l)}(\omega_\text{p}) = {} & A_{\text{pF},\alpha}^{(0)}(\omega_\text{p}) \Theta_{\text{pF},\alpha}^{(+)}(z_{l-1}, \omega_\text{p}) + A_{\text{pB},\alpha}^{(N+1)}(\omega_\text{p}) \Theta_{\text{pF},\alpha}^{(-)}(z_{l-1}, \omega_\text{p}), \\ A_{\text{pB},\alpha}^{(l)}(\omega_\text{p}) = {} & A_{\text{pF},\alpha}^{(0)}(\omega_\text{p}) \Theta_{\text{pB},\alpha}^{(+)}(z_{l-1}, \omega_\text{p}) \\ & + A_{\text{pB},\alpha}^{(N+1)}(\omega_\text{p}) \Theta_{\text{pB},\alpha}^{(-)}(z_{l-1}, \omega_\text{p}), \, l = 1, \ldots, N+1. \end{aligned} \tag{6.370}$$

For $l = 0$, we use this formula on replacement of z_{l-1} by z_0. Further,

$$\begin{aligned} \mathbf{e}^{(l)}_{m\mathrm{F,TE}}(\omega_m) &= \mathbf{e}^{(l)}_{m\mathrm{B,TE}}(\omega_m) = \mathbf{e}_x, \\ \mathbf{e}^{(l)}_{m\mathrm{F,TM}}(\omega_m) &= \mathbf{e}_y \cos(\vartheta^{(l)}_m) - \mathbf{e}_z \sin(\vartheta^{(l)}_m), \\ \mathbf{e}^{(l)}_{m\mathrm{B,TM}}(\omega_m) &= \mathbf{e}_y \cos(\vartheta^{(l)}_m) + \mathbf{e}_z \sin(\vartheta^{(l)}_m), \end{aligned} \tag{6.371}$$

where $m = \mathrm{p}$. The positive-frequency electric-field operators $\hat{\mathbf{E}}^{(+)}_{\mathrm{s}}(z,t)$ and $\hat{\mathbf{E}}^{(+)}_{\mathrm{i}}(z,t)$ for the signal and idler fields can be decomposed into the TE- and TM-wave contributions $\hat{\mathbf{E}}^{(+)}_{a,\mathrm{TE}}(z,t)$ and $\hat{\mathbf{E}}^{(+)}_{a,\mathrm{TM}}(z,t)$ and expressed as follows (Vogel et al. 2001)

$$\begin{aligned} \hat{\mathbf{E}}^{(+)}_a(z,t) &= \hat{\mathbf{E}}^{(+)}_{a,\mathrm{TE}}(z,t) + \hat{\mathbf{E}}^{(+)}_{a,\mathrm{TM}}(z,t) \\ &= \frac{1}{\sqrt{2\pi}} \int_0^\infty \hat{\mathbf{E}}^{(+)}_a(z,\omega_a)\,\mathrm{d}\omega_a \\ &= \frac{1}{\sqrt{2\pi}} \int_0^\infty \left[\hat{\mathbf{E}}^{(+)}_{a,\mathrm{TE}}(z,\omega_a) + \hat{\mathbf{E}}^{(+)}_{a,\mathrm{TM}}(z,\omega_a)\right] \mathrm{d}\omega_a,\ a = \mathrm{s}, \mathrm{i}. \end{aligned} \tag{6.372}$$

Here

$$\begin{aligned} \hat{\mathbf{E}}^{(+)}_{a,\alpha}(z,\omega_a) = \sqrt{\frac{\hbar\omega_a}{2\epsilon_0 c\mathcal{B}}} \Big\{ & \hat{a}^{(l)}_{a\mathrm{F},\alpha}(\omega_a)\mathbf{e}^{(l)}_{a\mathrm{F},\alpha}(\omega_a) \exp[\mathrm{i}k^{(l)}_{a,z}(z - z_{l-1})] \\ & + \hat{a}^{(l)}_{a\mathrm{B},\alpha}(\omega_a)\mathbf{e}^{(l)}_{a\mathrm{B},\alpha}(\omega_a) \exp[-\mathrm{i}k^{(l)}_{a,z}(z - z_{l-1})] \Big\}, \end{aligned} \tag{6.373}$$

with $\mathcal{B}$ the area of the transverse profile of a beam and

$$\begin{aligned} b^{(l)}_a \hat{a}^{(l)}_{a\mathrm{F},\alpha}(\omega_a) &= b^{(N+1)}_a \hat{a}^{(N+1)}_{a\mathrm{F},\alpha}(\omega_a) \tilde{\Theta}^{(+)}_{a\mathrm{F},\alpha}(z_{l-1},\omega_a) \\ &\quad + b^{(0)}_a \hat{a}^{(0)}_{a\mathrm{B},\alpha}(\omega_a) \tilde{\Theta}^{(-)}_{a\mathrm{F},\alpha}(z_{l-1},\omega_a), \\ b^{(l)}_a \hat{a}^{(l)}_{a\mathrm{B},\alpha}(\omega_a) &= b^{(N+1)}_a \hat{a}^{(N+1)}_{a\mathrm{F},\alpha}(\omega_a) \tilde{\Theta}^{(+)}_{a\mathrm{B},\alpha}(z_{l-1},\omega_a) \\ &\quad + b^{(0)}_a \hat{a}^{(N+1)}_{a\mathrm{B},\alpha}(\omega_a) \tilde{\Theta}^{(-)}_{a\mathrm{B},\alpha}(z_{l-1},\omega_a),\ l = 1,\ldots,N. \end{aligned} \tag{6.374}$$

Further, $\mathbf{e}^{(l)}_{a\mathrm{F},\alpha}(\omega_a)$ and $\mathbf{e}^{(l)}_{a\mathrm{B},\alpha}(\omega_a)$ are defined by relation (6.371), where $m = a$.

The operators $\hat{a}^{(N+1)}_{a\mathrm{F},\alpha}(\omega_a)$ and $\hat{a}^{(0)}_{a\mathrm{B},\alpha}(\omega_a)$ obey the following commutation relations

$$\begin{aligned} \left[\hat{a}^{(N+1)}_{a\mathrm{F},\alpha}(\omega_a), \hat{a}^{(N+1)\dagger}_{a'\mathrm{F},\alpha'}(\omega'_a)\right] &= \delta_{\alpha,\alpha'}\delta_{a,a'}\delta(\omega_a - \omega'_a)\hat{1}, \\ \left[\hat{a}^{(N+1)}_{a\mathrm{F},\alpha}(\omega_a), \hat{a}^{(N+1)}_{a'\mathrm{F},\alpha'}(\omega'_a)\right] &= \hat{0}, \\ \left[\hat{a}^{(0)}_{a\mathrm{B},\alpha}(\omega_a), \hat{a}^{(0)\dagger}_{a'\mathrm{B},\alpha'}(\omega'_a)\right] &= \delta_{\alpha,\alpha'}\delta_{a,a'}\delta(\omega_a - \omega'_a)\hat{1}, \\ \left[\hat{a}^{(0)}_{a\mathrm{B},\alpha}(\omega_a), \hat{a}^{(0)}_{a'\mathrm{B},\alpha'}(\omega'_a)\right] &= \hat{0}, \end{aligned}$$

$$\left[\hat{a}^{(N+1)}_{a\mathrm{F},\alpha}(\omega_a), \hat{a}^{(0)\dagger}_{a'\mathrm{B},\alpha'}(\omega'_a)\right] = \hat{0},$$
$$\left[\hat{a}^{(N+1)}_{a\mathrm{F},\alpha}(\omega_a), \hat{a}^{(0)}_{a'\mathrm{B},\alpha'}(\omega'_a)\right] = \hat{0}. \tag{6.375}$$

The interaction Hamiltonian $\hat{H}(t)$ describing spontaneous parametric down-conversion can be written as

$$\hat{H}(t) = \frac{\epsilon_0 \mathcal{B}}{(2\pi)^{\frac{3}{2}}} \int_{z_0}^{z_N} \int_0^\infty \int_0^\infty \int_0^\infty \sum_{\alpha,\beta,\gamma=\mathrm{TE,TM}} \mathbf{d}(z)$$
$$\vdots \left[\mathbf{E}^{(+)}_{\mathrm{p},\alpha}(z,\omega_\mathrm{p})\hat{\mathbf{E}}^{(-)}_{\mathrm{s},\beta}(z,\omega_\mathrm{s})\hat{\mathbf{E}}^{(-)}_{\mathrm{i},\gamma}(z,\omega_\mathrm{i}) + \mathrm{H.\ c.}\right] \mathrm{d}z\, \mathrm{d}\omega_\mathrm{p}\, \mathrm{d}\omega_\mathrm{s}\, \mathrm{d}\omega_\mathrm{i}, \tag{6.376}$$

where $\mathbf{d}(z)$ means a third-order tensor of nonlinear susceptibility and $\vdots$ denotes a contraction, i.e., treble sum after the tensors are replaced by their components, and products of the corresponding components are formed.

The solution $|\psi\rangle_{\mathrm{s,i}}$ of the Schrödinger equation correct up to first order on the assumption of the initial vacuum state $|\mathrm{vac}\rangle_{\mathrm{s,i}}$ for the down-converted fields is given by the relation

$$|\psi\rangle_{\mathrm{s,i}} = |\mathrm{vac}\rangle_{\mathrm{s,i}} + \int_0^\infty \int_0^\infty \sum_{\beta,\gamma=\mathrm{TE,TM}} \Big[\Phi_{\mathrm{F}\beta\mathrm{F}\gamma}(\omega_\mathrm{s},\omega_\mathrm{i})$$
$$\times b^{(N+1)}_\mathrm{s} \hat{a}^{(N+1)\dagger}_{\mathrm{sF},\beta}(\omega_\mathrm{s}) b^{(N+1)}_\mathrm{i} \hat{a}^{(N+1)\dagger}_{\mathrm{iF},\gamma}(\omega_\mathrm{i})|\mathrm{vac}\rangle_{\mathrm{s,i}}$$
$$+ \Phi_{\mathrm{F}\beta\mathrm{B}\gamma}(\omega_\mathrm{s},\omega_\mathrm{i}) b^{(N+1)}_\mathrm{s} \hat{a}^{(N+1)\dagger}_{\mathrm{sF},\beta}(\omega_\mathrm{s}) b^{(0)}_\mathrm{i} \hat{a}^{(0)\dagger}_{\mathrm{iB},\gamma}(\omega_\mathrm{i})|\mathrm{vac}\rangle_{\mathrm{s,i}}$$
$$+ \Phi_{\mathrm{B}\beta\mathrm{F}\gamma}(\omega_\mathrm{s},\omega_\mathrm{i}) b^{(0)}_\mathrm{s} \hat{a}^{(0)\dagger}_{\mathrm{sB},\beta}(\omega_\mathrm{s}) b^{(N+1)}_\mathrm{i} \hat{a}^{(N+1)\dagger}_{\mathrm{iF},\gamma}(\omega_\mathrm{i})|\mathrm{vac}\rangle_{\mathrm{s,i}}$$
$$+ \Phi_{\mathrm{B}\beta\mathrm{B}\gamma}(\omega_\mathrm{s},\omega_\mathrm{i}) b^{(0)}_\mathrm{s} \hat{a}^{(0)\dagger}_{\mathrm{sB},\beta}(\omega_\mathrm{s}) b^{(0)}_\mathrm{i} \hat{a}^{(0)\dagger}_{\mathrm{iB},\gamma}(\omega_\mathrm{i})|\mathrm{vac}\rangle_{\mathrm{s,i}}\Big] \mathrm{d}\omega_\mathrm{s}\, \mathrm{d}\omega_\mathrm{i}. \tag{6.377}$$

Here

$$\Phi_{\mathrm{F}\beta\mathrm{F}\gamma}(\omega_\mathrm{s},\omega_\mathrm{i}) = -\frac{\mathrm{i}}{4\pi} \int_0^\infty \sum_{\alpha=\mathrm{TE,TM}} \sqrt{\omega_\mathrm{s}\omega_\mathrm{i}}\,\delta(\omega_\mathrm{p}-\omega_\mathrm{s}-\omega_\mathrm{i})$$
$$\times A^{(0)}_{\mathrm{pF},\alpha}(\omega_\mathrm{p}) \sum_{m,n,o=\mathrm{F,B}} \int_{z_0}^{z_N} \Theta^{(+)}_{\mathrm{p}m,\alpha}(z,\omega_\mathrm{p}) \tilde{\Theta}^{(+)*}_{\mathrm{s}n,\beta}(z,\omega_\mathrm{s}) \tilde{\Theta}^{(+)*}_{\mathrm{i}o,\gamma}(z,\omega_\mathrm{i})$$
$$\times \mathbf{d}(z) \vdots \mathbf{e}_{\mathrm{p}m,\alpha}(z,\omega_\mathrm{p}) \mathbf{e}_{\mathrm{s}n,\beta}(z,\omega_\mathrm{s}) \mathbf{e}_{\mathrm{i}o,\gamma}(z,\omega_\mathrm{i})\, \mathrm{d}z\, \mathrm{d}\omega_\mathrm{p}, \tag{6.378}$$

$$\Phi_{\mathrm{F}\beta\mathrm{B}\gamma}(\omega_\mathrm{s},\omega_\mathrm{i}) = -\frac{\mathrm{i}}{4\pi} \int_0^\infty \sum_{\alpha=\mathrm{TE,TM}} \sqrt{\omega_\mathrm{s}\omega_\mathrm{i}}\,\delta(\omega_\mathrm{p}-\omega_\mathrm{s}-\omega_\mathrm{i})$$
$$\times A^{(0)}_{\mathrm{pF},\alpha}(\omega_\mathrm{p}) \sum_{m,n,o=\mathrm{F,B}} \int_{z_0}^{z_N} \Theta^{(+)}_{\mathrm{p}m,\alpha}(z,\omega_\mathrm{p}) \tilde{\Theta}^{(+)*}_{\mathrm{s}n,\beta}(z,\omega_\mathrm{s}) \tilde{\Theta}^{(-)*}_{\mathrm{i}o,\gamma}(z,\omega_\mathrm{i})$$
$$\times \mathbf{d}(z) \vdots \mathbf{e}_{\mathrm{p}m,\alpha}(z,\omega_\mathrm{p}) \mathbf{e}_{\mathrm{s}n,\beta}(z,\omega_\mathrm{s}) \mathbf{e}_{\mathrm{i}o,\gamma}(z,\omega_\mathrm{i})\, \mathrm{d}z\, \mathrm{d}\omega_\mathrm{p}, \tag{6.379}$$

$$\Phi_{\mathrm{B}\beta\mathrm{F}\gamma}(\omega_{\mathrm{s}},\omega_{\mathrm{i}}) = -\frac{\mathrm{i}}{4\pi}\int_0^\infty \sum_{\alpha=\mathrm{TE,TM}} \sqrt{\omega_{\mathrm{s}}\omega_{\mathrm{i}}}\delta(\omega_{\mathrm{p}}-\omega_{\mathrm{s}}-\omega_{\mathrm{i}})$$
$$\times A^{(0)}_{\mathrm{pF},\alpha}(\omega_{\mathrm{p}}) \sum_{m,n,o=\mathrm{F,B}} \int_{z_0}^{z_N} \Theta^{(+)}_{pm,\alpha}(z,\omega_{\mathrm{p}})\tilde{\Theta}^{(-)*}_{sn,\beta}(z,\omega_{\mathrm{s}})\tilde{\Theta}^{(+)*}_{io,\gamma}(z,\omega_{\mathrm{i}})$$
$$\times \mathbf{d}(z) \vdots \mathbf{e}_{pm,\alpha}(z,\omega_{\mathrm{p}})\mathbf{e}_{sn,\beta}(z,\omega_{\mathrm{s}})\mathbf{e}_{io,\gamma}(z,\omega_{\mathrm{i}})\,\mathrm{d}z\,\mathrm{d}\omega_{\mathrm{p}}, \tag{6.380}$$

$$\Phi_{\mathrm{B}\beta\mathrm{B}\gamma}(\omega_{\mathrm{s}},\omega_{\mathrm{i}}) = -\frac{\mathrm{i}}{4\pi}\int_0^\infty \sum_{\alpha=\mathrm{TE,TM}} \sqrt{\omega_{\mathrm{s}}\omega_{\mathrm{i}}}\delta(\omega_{\mathrm{p}}-\omega_{\mathrm{s}}-\omega_{\mathrm{i}})$$
$$\times A^{(0)}_{\mathrm{pF},\alpha}(\omega_{\mathrm{p}}) \sum_{m,n,o=\mathrm{F,B}} \int_{z_0}^{z_N} \Theta^{(+)}_{pm,\alpha}(z,\omega_{\mathrm{p}})\tilde{\Theta}^{(-)*}_{sn,\beta}(z,\omega_{\mathrm{s}})\tilde{\Theta}^{(-)*}_{io,\gamma}(z,\omega_{\mathrm{i}})\mathbf{d}(z)$$
$$\vdots\, \mathbf{e}_{pm,\alpha}(z,\omega_{\mathrm{p}})\mathbf{e}_{sn,\beta}(z,\omega_{\mathrm{s}})\mathbf{e}_{io,\gamma}(z,\omega_{\mathrm{i}})\,\mathrm{d}z\,\mathrm{d}\omega_{\mathrm{p}}, \tag{6.381}$$

with $\mathbf{e}_{pm,\alpha}(z,\omega_{\mathrm{p}}) = \mathbf{e}^{(l)}_{pm,\alpha}(\omega_{\mathrm{p}})$, $\mathbf{e}_{sn,\beta}(z,\omega_{\mathrm{s}}) = \mathbf{e}^{(l)}_{sn,\beta}(\omega_{\mathrm{s}})$, $\mathbf{e}_{io,\gamma}(z,\omega_{\mathrm{i}}) = \mathbf{e}^{(l)}_{io,\gamma}(\omega_{\mathrm{i}})$ (in the lth layer) when we restrict ourselves to the case $A^{(N+1)}_{\mathrm{pB},\alpha}(\omega_{\mathrm{p}}) = 0$.

Corona and U'Ren (2007) study type-II, frequency degenerate, collinear parametric down-conversion in a $\chi^{(2)}$ material with uniaxial birefringence. The type-II interaction means that the pump is an extraordinary wave, and the signal and idler are extraordinary and ordinary. For this operation, signal and idler photons are orthogonally polarized. The material is characterized by a spatial periodicity in its linear optical properties. Introducing $\mu = \mathrm{o}$ for the ordinary ray and $\mu = \mathrm{e}$ for the extraordinary ray, the index of refraction will be

$$n_\mu(\omega,z) = \begin{cases} n_{\mu 1}(\omega), & 0 < z < a, \\ n_{\mu 2}(\omega), & a < z < \Lambda. \end{cases} \tag{6.382}$$

The authors assume $a = \frac{\Lambda}{2}$ in a numerical analysis. The Bloch waves are written in the form

$$E(z,t) = E_K(z,\omega)\exp\{\mathrm{i}[K(\omega)z - \omega t]\}. \tag{6.383}$$

Here $E_K(z,\omega)$ is the Bloch envelope, which has the same period, Λ, as the material, and $K(\omega)$ stands for the Bloch wave number. The Bloch waves are described in the vicinity of $K = \frac{m\pi}{\Lambda}$. Further $m = 1$.

A standard perturbative approach to the quantum description has been adopted by Corona and U'Ren (2007). There $\hat{E}_\mu(\mathbf{r},t)$ ($\mu = \mathrm{p, s, i}$) represents the electric-field operators related to each of the interacting fields. They assume that the pump field is classical or that the replacement

$$\hat{E}^{(+)}_{\mathrm{p}}(\mathbf{r},t) \to \int \alpha_{\mathrm{p}}(\omega)E_{K_{\mathrm{p}}}(z,\omega)\exp\{\mathrm{i}[K_{\mathrm{p}}(\omega)z - \omega t]\}\,\mathrm{d}\omega, \tag{6.384}$$

where $K_{\mathrm{p}}(\omega)$ is the Bloch wave number, $E_{K_{\mathrm{p}}}(z,\omega)$ is the Bloch envelope, and $\alpha_{\mathrm{p}}(\omega)$ is the spectral amplitude, may be done in the interaction Hamiltonian. The Bloch envelope may be expressed as a Fourier series,

$$E_{K_p}(z,\omega) = \sum_l \varepsilon_{pl}(\omega) e^{iG_l z}, \tag{6.385}$$

in terms of the spatial harmonics $G_l = \frac{2\pi l}{\Lambda}$. The authors touch the quantization of the signal and idler fields when they present the operator

$$\hat{E}_\mu^{(+)}(\mathbf{r},t) = i \int \sum_l \varepsilon_{\mu l}(\omega) l_\mu(\omega) \hat{a}_\mu \left[K_\mu(\omega) + G_l\right] \times \exp\left(i\left\{\left[K_\mu(\omega) + G_l\right] z - \omega t\right\}\right) d\omega, \tag{6.386}$$

where $K_\mu(\omega)$ is the Bloch wave number, $\varepsilon_{\mu l}$ are the Bloch envelope Fourier series coefficients, and $\hat{a}_\mu\,(K(\omega))$ is the annihilation operator for the signal (s) or idler (i) mode. The normalization constant is

$$l_\mu(\omega) = \sqrt{\frac{\hbar\omega K'_\mu(\omega)}{2\epsilon_\mu(\omega) S}}, \tag{6.387}$$

where $K'_\mu(\omega)$ is the first frequency derivative of K_μ, $\varepsilon_\mu(\omega)$ is the permittivity in the nonlinear medium, and S is the transverse beam area.

The approach to quantization is macroscopic. The joint spectral amplitude has been defined, which depends on the length of the crystal, L. The authors prove with a numerical calculation that each of the three interacting fields propagates essentially as a plane wave. They may utilize this approximation.

To obtain conditions for factorizability, the authors expand the mismatch $L\Delta K(\omega)$, where $\Delta K(\omega) = K_p(\omega) - K_s(\omega) - K_i(\omega)$. They let ω_o denote the degenerate frequency and introduce the mismatch $\tau_\mu^{(j)}$, $\mu = s, i$, $j = 2, 3, 4$, in the jth frequency derivatives between the wave numbers of the pump and the signal (idler) wave packets

$$\tau_\mu^{(j)} = L(K_p^{(j)} - K_\mu^{(j)}), \tag{6.388}$$

where

$$K_\mu^{(j)} = \frac{1}{j!} \left.\frac{d^j K_\mu(\omega)}{d\omega^j}\right|_{\omega=\omega_o}, \quad K_p^{(j)} = \frac{1}{j!} \left.\frac{d^j K_p(\omega)}{d\omega^j}\right|_{\omega=2\omega_o}, \tag{6.389}$$

and the frequency detunings $\nu_{s,i} = \omega_{s,i} - \omega_o$. Whereas the zeroth-order approximation is

$$L\Delta K(\omega) \approx L\Delta K^{(0)} + O(1), \tag{6.390}$$

where

$$\Delta K^{(0)} = K_p(2\omega_o) - K_s(\omega_o) - K_i(\omega_o), \tag{6.391}$$

and $O(J+1)$ represents $(J+1)$th- and higher-order terms in the detunings; the authors present a fourth-order approximation, $J=4$.

They then choose $\alpha_{\rm p}(\omega_{\rm s}+\omega_{\rm i})$ in the relation $f_{000}(\omega_{\rm s},\omega_{\rm i}) = \omega_{\rm p}(\omega_{\rm s}+\omega_{\rm i})\,\phi_{000}(\omega_{\rm s},\omega_{\rm i})$, where $\phi_{000}(\omega_{\rm s},\omega_{\rm i})$ is given by relation (15) in Corona and U'Ren (2007), as a Gaussian function with the parameter σ^2 and, in that defining relation, they replace $\mathrm{sinc}(\frac{L\Delta K(\omega)}{2})$ by another Gaussian (with the parameter $\frac{1}{\gamma}$). The approximation of the function $f(\omega_{\rm s},\omega_{\rm i}) \equiv f_{000}(\omega_{\rm s},\omega_{\rm i})$ comprises the quantity $\Phi_{\rm si}(\nu_{\rm s},\nu_{\rm i})$ given in (25) in the cited paper.

The expression is complicated. The factorability occurs if $\Phi_{\rm si}(\nu_{\rm s},\nu_{\rm i})=0$. Provided that $\tau_{\rm s}^{(1)}=\tau_{\rm i}^{(1)}=0$, the expression simplifies. Then relations for $|f(\nu_{\rm s},\nu_{\rm i})|$ and $\arg[f(\nu_{\rm s},\nu_{\rm i})]$ can be written. Further the authors assume that the signal and idler photons undergo much stronger dispersion than the pump. Particularly, they assume that $|\tau_{\rm p}^{(j)}| \ll |\tau_{\rm s}^{(j)}|, |\tau_{\rm i}^{(j)}|$, $j=2,3,4$, $\tau_{\rm p}^{(j)}=LK_{\rm p}^{(j)}$. At last, they assume that the pump is broad-band,

$$\sigma \gg 2\sqrt[4]{\frac{4}{\gamma}}\frac{1}{\sqrt{\tau_{\rm s}^{(2)}+\tau_{\rm i}^{(2)}}}. \tag{6.392}$$

While the phase of the joint spectral amplitude has been factorable without the condition (6.392), only now the modulus of the joint spectral amplitude reduces to

$$|f(\nu_{\rm s},\nu_{\rm i})| \approx \exp\left[\frac{\gamma}{4}\left(\tau_{\rm s}^{(2)}\nu_{\rm s}^2+\tau_{\rm i}^{(2)}\nu_{\rm i}^2\right)^2\right]. \tag{6.393}$$

In Corona and U'Ren (2007), it has been shown also that this modulus can describe a nearly factorable two-photon state.

The authors first evidence that in nonlinear photonic crystals, complete group velocity matching, $K_{\rm p}'=K_{\rm s}'=K_{\rm i}'$, or $\tau_{\rm s}^{(1)}=\tau_{\rm i}^{(1)}=0$, can be achieved. Here it is assumed that $a=\frac{\Lambda}{2}$.

Then one may search for the lattice period Λ, the permittivity contrast

$$\alpha = \frac{2\left[\epsilon_{\mu 1}(\omega)-\epsilon_{\mu 2}(\omega)\right]}{\epsilon_{\mu 1}(\omega)+\epsilon_{\mu 2}(\omega)}, \tag{6.394}$$

in which an independence of the frequency and of $\mu = {\rm e},{\rm o}$ is actually assumed, and the crystal propagation angle θ_{pm} such that $\Delta K^{(0)}=0$, $K_{\rm p}'=K_{\rm s}'$, $K_{\rm p}'=K_{\rm i}'$. Here $\omega_{\rm p}=2\omega_{\rm o}$, $\omega_{\rm s}=\omega_{\rm i}=\omega_{\rm o}$ as in relation (6.389). The fact that the produced two-photon state is nearly factorable has been shown in Law et al. (2000).

(iv) Further principles and effects

Diao and Blair (2007) have paid attention to the use of multilayer thin film structures for optical bistability and multistability. They have considered single-cavity and coupled-cavity structures. In the case of a single-cavity structure, mirrors consist of M quarter-wave low-index layers and $M-1$ quarter-wave high-index layers. The cavity is based on L quarter-wave high-index layers.

The coupled-cavity structures have N cavities and $N + 1$ mirrors. It is assumed that the low-index layers are constructed from silicon dioxide. They have a nonlinear coefficient of the refractive index change $n_{2,\text{silica}} \approx 3 \times 10^{-16}$ cm^{2}W^{-1}. The high-index layers have a coefficient n_2.

For these structures, linear transmission (in magnitude and phase) and group delay may be calculated. The optical bistability and multistability are analyzed mainly by dependence of nonlinear transmission (in magnitude and phase) on normalized input intensity. As it is a multivalued function of the input, also the nonlinear transmission (in magnitude only) is plotted versus the intensity within the cavity.

Photonic crystal fibres (PCFs) are dielectric optical fibres with an array of airholes running along the fibre. Usually, the fibres employ a single dielectric material. Mortensen (2005) notes that other base materials have been studied besides the silica. Typically, the airholes are arranged in a triangular lattice with a pitch Λ. A waveguide is formed as a cladding and a core using the core defect, i.e., by removal of a single airhole. From this, the author has realized a call for a theory of photonic crystal fibres with an arbitrary base material. Solving a scalar two-dimensional Schrödinger-like equation, geometrical eigenvalues γ^2 have been calculated, dependent on the normalized airhole diameter $\frac{d}{\Lambda}$. If $\frac{d}{\Lambda}$ is below a critical value, only the eigenvalue for the fundamental core mode $\gamma_{\text{c},1}^2$ and that for the fundamental cladding mode γ_{cl}^2 are seen. If $\frac{d}{\Lambda}$ is above the critical value, an eigenvalue for the second-order core mode $\gamma_{\text{c},2}^2$ diverges from that for the fundamental cladding mode. Then an abbreviation, $\gamma_\text{c}^2 \equiv \gamma_{\text{c},1}^2$, is used and, for $\frac{d}{\Lambda} \leq 0.8$, a third-order polynomial is presented, which fits the eigenvalues for the fundamental core mode well. One considers the V parameter of the form

$$V_{\text{PCF}} = \sqrt{\gamma_{\text{cl}}^2 - \gamma_\text{c}^2}, \tag{6.395}$$

and the endlessly single-mode regime is associated with the condition $V_{\text{PCF}} < \pi$ (Mortensen et al. 2003).

The dependence of the effective index $n_{\text{eff}} = \frac{c\beta}{\omega}$ on the normalized free-space wavelength $\frac{\lambda}{\Lambda}$ is expressed with a second-order polynomial, which agrees with fully vectorial plane-wave simulations in the short-wavelength limit $\lambda \ll \Lambda$. The comparison has been performed for a single slightly subcritical value of the normalized airhole diameter $\frac{d}{\Lambda}$ and both for the fundamental core mode and for the fundamental cladding mode.

Della Villa et al. (2005) study formation of band gaps in photonic quasicrystals. They have considered a photonic quasicrystal with a Penrose-type lattice. They have inferred a band gap from the normalized local density of states

$$\rho(\mathbf{r}_0, \omega) = \text{Im}\left\{G(\mathbf{r}_0, \mathbf{r}_0, \omega)\right\}, \tag{6.396}$$

where $G(\mathbf{r}, \mathbf{r}_0, \omega)$ is the Green function. Numerical calculations were performed for finite-size quasicrystals made of hundreds of rods. The normalized local density was determined at the centre of the quasicrystal, and the choice of the central point did not affect results. In comparison with the photonic crystal with square lattice, the quasicrystal exhibits small additional band gaps. In the central band gap of the photonic quasicrystal, the normalized local density of states exhibits the exponential decay similar to that of the photonic crystal. The central band gap seems to stem from relatively short-distance interactions. Lateral band gaps stem from long-range interactions.

The Fourier spectrum of the permittivity profile for the quasicrystal, together with the usual Bragg condition, predicts the central and upper band gap, and a lower contrast of the permittivities is more advantageous. The frequency, at which the lower band gap occurs, may not be explained using single scattering and should so include multiple scattering.

The use of two-dimensional photonic crystals instead of the conventional one-dimensional feedback grating can lower the lasing threshold of distributed feedback lasers. Such photonic-crystal-based organic lasers have been studied (Harbers et al. 2005).

The photonic crystals can change the spontaneous emission dynamics of excited atoms (Yablonovich 1987, John 1987). The photonic crystals modify also spontaneous emission of quantum dots (Lodahl et al. 2004, Yoshie et al. 2004).

Hughes (2005a) introduces a scheme that enables one to study quantum correlations between two quantum dots in a planar-photonic-crystal nanocavity. He considers the fundamental cavity mode $\mathbf{e}_{\mathrm{c}}(\mathbf{r})$, which fulfils the normalization condition

$$\int_{\text{all space}} \epsilon_{\mathrm{c}}(\mathbf{r})|\mathbf{e}_{\mathrm{c}}(\mathbf{r})|^2 \,\mathrm{d}^3\mathbf{r} = 1, \tag{6.397}$$

where $\epsilon_{\mathrm{c}}(\mathbf{r})$ is the permittivity of the nanocavity structure. He considers a tensor-valued Green function

$$\mathbf{G}^{\mathrm{b}}(\mathbf{r}, \mathbf{r}'; \omega) = \mathbf{G}^{\mathrm{t}}(\mathbf{r}, \mathbf{r}'; \omega) - \left(\frac{c}{\omega}\right)^2 \frac{\delta(\mathbf{r} - \mathbf{r}')\mathbf{1}}{|\mathbf{e}_{\mathrm{c}}(\mathbf{r})|}, \tag{6.398}$$

where for simplicity

$$\mathbf{G}^{\mathrm{t}}(\mathbf{r}, \mathbf{r}'; \omega) = \left(\frac{c}{\omega}\right)^2 \frac{\omega_{\mathrm{c}}^2 \mathbf{e}_{\mathrm{c}}(\mathbf{r})\mathbf{e}_{\mathrm{c}}^*(\mathbf{r})}{\omega_{\mathrm{c}}^2 - \omega^2 - \mathrm{i}\omega\Gamma_{\mathrm{c}}}, \tag{6.399}$$

where ω_{c} is the cavity resonance frequency and $\Gamma_{\mathrm{c}} = \frac{\omega_{\mathrm{c}}}{Q}$ is the cavity linewidth. Here we have digressed from Hughes (2005b), who has made some modification of the function.

The quantum dots are modelled as two-level atoms (Dung et al. 2002b). We introduce a quantum mechanical basis for the quantum dots a and b and the cavity mode as $|A\rangle \otimes |B\rangle \otimes |k\rangle = |ABk\rangle$, where each variable can assume a value of 0 or 1. We concentrate on wave functions of the form

$$|\psi(t)\rangle_{\mathrm{e}} = C_a(t)|100\rangle + C_b(t)|010\rangle + C_{\mathrm{p}}(t)|001\rangle, \tag{6.400}$$

where $C_a(t)$, $C_b(t)$, and $C_{\mathrm{p}}(t)$ are complex amplitudes. We consider the reduced wave function of the form

$$|\psi(t)\rangle = \frac{1}{\sqrt{|C_a(t)|^2 + |C_b(t)|^2}}[C_a(t)|10\rangle + C_b(t)|01\rangle], \tag{6.401}$$

where $C_a(t)$ and $C_b(t)$ obey integro-differential equations (Hughes 2005c). The time dependence of the entanglement ($E(t)$) is calculated for simple initial conditions from the concurrence ($C(t)$) using (Hughes 2005c)

$$E(t) = -x\log_2(x) - (1-x)\log_2(1-x), \tag{6.402}$$

where

$$x = \frac{1}{2} + \frac{1}{2}\sqrt{1-C(t)}, \quad C(t) = 4|C_a(t)|^2|C_b(t)|^2. \tag{6.403}$$

As a continuation of the paper Sakoda and Haus (2003), Sibilia et al. (2005) have studied the properties of super-radiant emission from a two-level atomic system embedded in a one-dimensional photonic band gap structure. The description by a reduced system of equations has further been reduced. Attention has been paid to the Rabi splitting. The effect of the location of the atoms has been obtained using a classical model of an amplifier.

In rotating Bose–Einstein condensates vortices develop. Müstecaplıoğlu and Öktel (2005) have shown that a vortex lattice can act as a photonic crystal and generate photonic band gaps. They have considered a two-dimensional triangular lattice. A numerical simulation of the propagation of an electromagnetic wave in a finite lattice has indicated that tens of vortices are enough for the infinite lattice properties to occur. Those authors have proposed a method to measure the rotation frequency of the condensate using a directional band gap.

6.4 Quantization in Disordered Media

Quantization of the electromagnetic field in disordered media may be realized by any of the expounded or mentioned approaches. In particular it is likely that an approach as that in Section 2.2.4 could be chosen. Many explanations from Section 6.1 remain valid even on the assumption of a disordered medium.

With respect to application to random lasers, a great emphasis is laid on the notion of a mode. As the electric permittivity is a scalar random field on the usual assumption of an isotropic dielectric, eigenfrequencies and modal functions of a device are random. We form an idea on properties of the eigenenergies and modal functions by the condensed-matter theory and even by nuclear physics as well. It may lead to the restriction that the vectorial character is neglected in the modal functions. Certain models are mathematically very demanding to the contrary. Many studies encounter the laser dynamics, which forces one to determine pseudomodes (after Garraway and Knight 1996), lossy modes, quasimodes otherwise).

Patra (2002), e.g., assumes a random medium closed in a cavity to be allowed to suppose orthogonal modal functions. He neglects the vectorial character of modal functions. Whereas he may assume that, in an opening from the cavity, the value of a modal function has a Gaussian distribution, he encounters a difficulty in the whole cavity. He mentions that the modal functions are Gaussian random fields according to the literature, but he does not use, in fact, this assumption. Patra (2002) restricts himself to values of a finite set of modal functions in a finite number of points and implements orthogonality using columns of a random unitary matrix.

Loudon (1999) has expounded the polariton dispersion relation in the framework of the classical Lorentz theory. The linear response theory for a perfect crystal has been generalized to include a randomly diluted crystal. The dependence of polariton radiative damping rates on the occupation probability with admixture atoms has been expressed.

The quantum, semiclassical, and classical theories of spontaneous emission have been characterized. The Glauber–Lewenstein model and more quantum theory have been applied to the radiative decay of dilute active atoms in lossy homogeneous and inhomogeneous dielectrics. The phase, group, and energy velocities for optical pulse propagation through a lossy dielectric have been defined and their physical meanings have been illustrated.

6.4.1 Quantization in Chaotic Cavity

Recently (Patra 2002) a model of laser has been adopted that ignores the phase of the field and, on including suitable Langevin terms like in Mishchenko and Beenakker (1999), provides the photon statistics. Peřinová et al. (2004) hope that the calculations can be refined, when the open systems theory is invoked (see, for example Peřinová and Lukš (2000) and references therein).

An optical cavity is considered which is coupled to the environment by a small opening of a diameter d. We concentrate on $N_{\rm p}$ cavity eigenmodes described by the chaotic cavity modal functions $\Theta_i(\mathbf{r})$, each with an eigenfrequency ω_i. The quantum description is reduced to only considering the number $n_i(t)$ of photons in each mode i. Photons in mode i escape through the opening with rate γ_i.

The cavity is filled with an amplifying medium. The medium can be a four-level laser dye, in which the lasing transition is directed from the third to the second level, the transition's resonance frequency being Ω. In Patra (2002), the density of excited atoms has been considered at every point $\mathbf{r}$ in the cavity. The coupling of mode i to the medium at the point $\mathbf{r}$ has been given by $K_i(\mathbf{r}) = w(\omega_i)|\Theta_i(\mathbf{r})|^2$, $i = 1, \ldots, N_{\rm p}$, where $w(\omega_i)$ has been the transition matrix element of the atomic transition $3 \to 2$ (de-excitation). At the level of a numerical solution, a linearization and a discretization have been performed. We discretize the space by introducing the Borel measurable neighbourhoods $U(\mathbf{0}_j)$ of "uniformly" located centres $\mathbf{0}_j$, $j = 1, \ldots, N_{\rm s}$, which exhaust all the space and have the same volume ΔV each, $N_{\rm s}\Delta V = V$, the cavity volume. The description is reduced to only considering the density of excited atoms N_j in each neighbourhood $U(\mathbf{0}_j)$. Excitations are created by pumping with a rate P_j and are lost nonradiatively with a rate a_j. The coupling of mode i to the medium in the neighbourhood $U(\mathbf{0}_j)$ is given by K_{ij}, $K_{ij} \equiv K_i(\mathbf{0}_j)$. The original notation due to (Patra 2002) has been changed as follows: $g_i \to \gamma_i$, $K_{ij} \to K_{ij}\Delta V$, $N_j \to N_j \Delta V$, $P_j \to P_j \Delta V$, $a_j \to a_j \Delta V$.

We have utilized the annihilation (creation) operators $\hat{a}_i(t)$ ($\hat{a}_i^\dagger(t)$), $i = 1, \ldots, N_\mathrm{p}$, which are assigned to the cavity modes and obey the commutation relations

$$[\hat{a}_i(t), \hat{a}_{i'}^\dagger(t)] = \delta_{ii'}\hat{1}, \quad [\hat{a}_i(t), \hat{a}_{i'}(t)] = \hat{0}. \tag{6.404}$$

Using them, we can reinterpret the photon numbers $n_i(t)$ as the operators $\hat{n}_i(t) = \hat{a}_i^\dagger(t)\hat{a}_i(t)$.

In fact, we still have to consider the matter field operators $\hat{A}_j(t)$, $\hat{A}_j^\dagger(t)$, which obey the commutation relations

$$[\hat{A}_j(t), \hat{A}_{j'}^\dagger(t)] = \frac{1}{\Delta V}\delta_{jj'}\hat{1}, \quad [\hat{A}_j(t), \hat{A}_{j'}(t)] = \hat{0}. \tag{6.405}$$

Using these operators, we can reinterpret the densities of excited atoms $N_j(t)$ as the operators $\hat{N}_j(t) = \hat{A}_j^\dagger(t)\hat{A}_j(t)$.

In analogy with exponential phase operators (Peřinová et al. 1998)

$$\begin{aligned} \widehat{\exp}[-\mathrm{i}\varphi_i(t)] &= \hat{a}_i^\dagger(t)\left[\hat{n}_i(t) + \hat{1}\right]^{-\frac{1}{2}}, \\ \widehat{\exp}[\mathrm{i}\varphi_i(t)] &= \left[\hat{n}_i(t) + \hat{1}\right]^{-\frac{1}{2}}\hat{a}_i(t), \end{aligned} \tag{6.406}$$

we introduce the quantum phase operators

$$\begin{aligned} \widehat{\exp}[-\mathrm{i}\Phi_j(t)] &= \hat{A}_j^\dagger(t)\left[\hat{N}_j(t) + \frac{\hat{1}}{\Delta V}\right]^{-\frac{1}{2}}, \\ \widehat{\exp}[\mathrm{i}\Phi_j(t)] &= \left[\hat{N}_j(t) + \frac{\hat{1}}{\Delta V}\right]^{-\frac{1}{2}}\hat{A}_j(t). \end{aligned} \tag{6.407}$$

In this exposition, we restrict ourselves to the quantum description in the framework of the Schrödinger picture. The time dependence of the operators was related to the Heisenberg picture, is not present in the Schrödinger picture, and is dropped. We adopt the master equation approach, which concerns the temporal evolution of the statistical operator $\hat{\rho}(t)$ normalized such that

$$\mathrm{Tr}\{\hat{\rho}(t)\} = 1 \tag{6.408}$$

and leads to the rate equations (6.418) straightforwardly. We propose that the master equation has the form

$$\frac{\partial}{\partial t}\hat{\rho}(t) = \hat{\hat{L}}_{\mathrm{att}}\hat{\rho}(t) + \hat{\hat{L}}_{\mathrm{amp}}\hat{\rho}(t) + \hat{\hat{L}}_{\mathrm{Att}}\hat{\rho}(t) + \hat{\hat{L}}_{\mathrm{nln}}\hat{\rho}(t), \tag{6.409}$$

where $\hat{\hat{L}}_{\mathrm{att}}$, $\hat{\hat{L}}_{\mathrm{amp}}$, $\hat{\hat{L}}_{\mathrm{Att}}$, $\hat{\hat{L}}_{\mathrm{nln}}$ are the Liouvillian superoperators with respective properties

$$\hat{\hat{L}}_{\mathrm{att}}\hat{\rho}(t) = \sum_{i=1}^{N_{\mathrm{p}}} \gamma_i \left[\hat{a}_i\hat{\rho}(t)\hat{a}_i^\dagger - \frac{1}{2}\hat{n}_i\hat{\rho}(t) - \frac{1}{2}\hat{\rho}(t)\hat{n}_i \right], \tag{6.410}$$

$$\hat{\hat{L}}_{\mathrm{amp}}\hat{\rho}(t) = \Delta V \sum_{j=1}^{N_{\mathrm{s}}} P_j \left[\widehat{\exp}(-\mathrm{i}\Phi_j)\hat{\rho}(t)\widehat{\exp}(\mathrm{i}\Phi_j) - \hat{\rho}(t) \right], \tag{6.411}$$

$$\hat{\hat{L}}_{\mathrm{Att}}\hat{\rho}(t) = \Delta V \sum_{j=1}^{N_{\mathrm{s}}} a_j \left[\hat{A}_j\hat{\rho}(t)\hat{A}_j^\dagger - \frac{1}{2}\hat{N}_j\hat{\rho}(t) - \frac{1}{2}\hat{\rho}(t)\hat{N}_j \right], \tag{6.412}$$

$$\hat{\hat{L}}_{\mathrm{nln}}\hat{\rho}(t) = \Delta V \sum_{i=1}^{N_{\mathrm{p}}} \sum_{j=1}^{N_{\mathrm{s}}} K_{ij} \left[\hat{a}_i^\dagger\hat{A}_j\hat{\rho}(t)\hat{A}_j^\dagger\hat{a}_i - \frac{1}{2}(\hat{n} + \hat{1})\hat{N}_j\hat{\rho}(t) \right.$$
$$\left. - \frac{1}{2}\hat{\rho}(t)\hat{N}_j(\hat{n} + \hat{1}) \right]. \tag{6.413}$$

The subscript "att" denotes the attenuation (escape of photons), the subscript "amp" denotes the amplification (pumping), the subscript "Att" another attenuation (relaxation of the medium), and the subscript "nln" denotes a nonlinear process. The term $\hat{\hat{L}}_{\mathrm{att}}\hat{\rho}(t)$ is well known from the quantum theory of damping at zero temperature (Haken 1970). In analogy with it, the term $\hat{\hat{L}}_{\mathrm{Att}}\hat{\rho}(t)$ has been proposed. This is equivalent to treating the excited atoms as bosons. All the terms have the Lindblad form Lindblad (1976). For instance, the term $\hat{\hat{L}}_{\mathrm{nln}}\hat{\rho}(t)$ has such a form with the Lindblad operators $\hat{O}_{ij\mathrm{nln}} = \hat{a}_i^\dagger\hat{A}_j$. The Lindblad form of the term $\hat{\hat{L}}_{\mathrm{amp}}\hat{\rho}(t)$ has been gained at the cost of considering the Lindblad operators $\hat{O}_{j\mathrm{amp}} = \widehat{\exp}(-\mathrm{i}\Phi_j)$. The unusual operators may be related either to the discretization of the space or to a treatment of the excitations as bosons. To our knowledge, the phenomenology added has not been proved to be wrong. The master equation must be completed with an initial condition that gives the statistical operator $\hat{\rho}(t_0)$ which commutes with all the operators $\hat{n}_i$, $i = 1, \ldots, N_{\mathrm{p}}$, $\hat{N}_j$, $j = 1, \ldots, N_{\mathrm{s}}$.

It is natural to ask whether a limit of $N_{\mathrm{s}} \to \infty$ may be taken to absolve the quantum description from the parameters N_{s} and ΔV. This is not excluded, but the emerging model has too much in common with a quantum field theory. A need for a renormalization would bring us farther than an appropriate choice of ΔV.

We let $|n_1, \ldots, n_{N_{\mathrm{p}}}, N_1, \ldots, N_{N_{\mathrm{s}}}\rangle$ denote the (normalized) simultaneous eigenkets of the operators $\hat{n}_i$, $i = 1, \ldots, N_{\mathrm{p}}$, $\hat{N}_j$, $j = 1, \ldots, N_{\mathrm{s}}$. Using properties of the operators which underlie to averaging, we obtain the rate equations for probabilities

$$p(n_1, \ldots, n_{N_{\mathrm{p}}}, N_1, \ldots, N_{N_{\mathrm{s}}}, t)$$
$$= \langle n_1, \ldots, n_{N_{\mathrm{p}}}, N_1, \ldots, N_{N_{\mathrm{s}}}|\hat{\rho}(t)|n_1, \ldots, n_{N_{\mathrm{p}}}, N_1, \ldots, N_{N_{\mathrm{s}}}\rangle, \tag{6.414}$$

normalized such that

$$\sum_{n_1=0}^{\infty} \cdots \sum_{n_{N_{\mathrm{p}}}=0}^{\infty} \sum_{N_1=0}^{\infty}{}' \cdots \sum_{N_{N_{\mathrm{s}}}=0}^{\infty}{}' p(n_1, \ldots, n_{N_{\mathrm{p}}}, N_1, \ldots, N_{N_{\mathrm{s}}}, t) = 1, \tag{6.415}$$

where each prime means that the summation proceeds with the step $\frac{1}{\Delta V}$. We get also an initial condition related to the time t_0.

A derivation of rate equations is made easier when we write

$$\hat{a}_i(t) = \widehat{\exp}[\mathrm{i}\varphi_i(t)][\hat{n}_i(t)]^{\frac{1}{2}}, \quad \hat{A}_j(t) = \widehat{\exp}[\mathrm{i}\Phi_j(t)][\hat{N}_j(t)]^{\frac{1}{2}}, \tag{6.416}$$

$$\hat{a}_i^\dagger(t)\hat{A}_j(t) = \widehat{\exp}[-\mathrm{i}\varphi_i(t)]\widehat{\exp}[\mathrm{i}\Phi_j(t)]\left[\hat{n}_i(t) + \hat{1}\right]^{\frac{1}{2}} \hat{N}_j(t)^{\frac{1}{2}}. \tag{6.417}$$

They have the form

$$\begin{aligned}\frac{\partial}{\partial t}p(n_1, \ldots, n_{N_\mathrm{p}}, N_1, \ldots, N_{N_\mathrm{s}}, t) = {} & L_\mathrm{att}p(n_1, \ldots, n_{N_\mathrm{p}}, N_1, \ldots, N_{N_\mathrm{s}}, t)\\ & + L_\mathrm{amp}p(n_1, \ldots, n_{N_\mathrm{p}}, N_1, \ldots, N_{N_\mathrm{s}}, t)\\ & + L_\mathrm{Att}p(n_1, \ldots, n_{N_\mathrm{p}}, N_1, \ldots, N_{N_\mathrm{s}}, t)\\ & + L_\mathrm{nln}p(n_1, \ldots, n_{N_\mathrm{p}}, N_1, \ldots, N_{N_\mathrm{s}}, t),\end{aligned} \tag{6.418}$$

where L_att, L_amp, L_Att, L_nln are operators with the respective properties

$$\begin{aligned}& L_\mathrm{att}p(n_1, \ldots, n_{N_\mathrm{p}}, N_1, \ldots, N_{N_\mathrm{s}}, t)\\ & = \sum_{i=1}^{N_\mathrm{p}} \gamma_i[(n_i + 1)p(n_1, \ldots, n_i + 1, \ldots, n_{N_\mathrm{p}}, N_1, \ldots, N_{N_\mathrm{s}}, t)\\ & \quad - n_i p(n_1, \ldots, n_{N_\mathrm{p}}, N_1, \ldots, N_{N_\mathrm{s}}, t)],\end{aligned} \tag{6.419}$$

$$\begin{aligned}& L_\mathrm{amp}p(n_1, \ldots, n_{N_\mathrm{p}}, N_1, \ldots, N_{N_\mathrm{s}}, t)\\ & = \Delta V \sum_{j=1}^{N_\mathrm{s}} P_j\left[p\left(n_1, \ldots, n_{N_\mathrm{p}}, N_1, \ldots, N_j - \frac{1}{\Delta V}, \ldots, N_{N_\mathrm{s}}, t\right)\right.\\ & \quad \left. - p(n_1, \ldots, n_{N_\mathrm{p}}, N_1, \ldots, N_{N_\mathrm{s}}, t)\right],\end{aligned} \tag{6.420}$$

$$\begin{aligned}& L_\mathrm{Att}p(n_1, \ldots, n_{N_\mathrm{p}}, N_1, \ldots, N_{N_\mathrm{s}}, t)\\ & = \Delta V \sum_{j=1}^{N_\mathrm{s}} a_j\left[\left(N_j + \frac{1}{\Delta V}\right)\right.\\ & \quad \times p\left(n_1, \ldots, n_{N_\mathrm{p}}, N_1, \ldots, N_j + \frac{1}{\Delta V}, \ldots, N_{N_\mathrm{s}}, t\right)\\ & \quad \left. - N_j p(n_1, \ldots, n_{N_\mathrm{p}}, N_1, \ldots, N_{N_\mathrm{s}}, t)\right],\end{aligned} \tag{6.421}$$

$$
\begin{aligned}
&L_{\text{nln}}p(n_1,\ldots,n_{N_\text{p}},N_1,\ldots,N_{N_\text{s}},t)\\
&=\Delta V\sum_{i=1}^{N_\text{p}}\sum_{j=1}^{N_\text{s}}K_{ij}\left[n_i\left(N_j+\frac{1}{\Delta V}\right)\right.\\
&\quad\times p\left(n_1,\ldots,n_i-1,\ldots,n_{N_\text{p}},N_1,\ldots,N_j+\frac{1}{\Delta V},\ldots,N_{N_\text{s}},t\right)\\
&\quad\left.-(n_i+1)N_j p(n_1,\ldots,n_{N_\text{p}},N_1,\ldots,N_{N_\text{s}},t)\right].
\end{aligned}
\tag{6.422}
$$

The relation (6.419) indicates that a photon escapes from the ith mode with the probability $\gamma_i n_i \Delta t$ within a period of duration Δt. Similarly, the term (6.421) suggests that an excited atom near $\mathbf{0}_j$ decays with the probability $a_j N_j \Delta V \Delta t$ within a period of Δt. But the term (6.420) expresses that an excited atom near $\mathbf{0}_j$ emerges with probability $P_j \Delta V\, \Delta t$ within such a period. Finally, the term (6.422) tells that an excited atom near $\mathbf{0}_j$ emits a photon into the ith mode with the probability $K_{ij}(n_i+1)N_j\Delta V\Delta t$ within such a period. The initial condition related to the time t_0 gives the probabilities $p(n_1, \ldots, n_{N_\text{p}}, N_1, \ldots, N_{N_\text{s}}, t_0)$.

6.4.2 Open Systems Approach

The open systems approach enables us to "unravel" the master equation (6.409) with respect to one, two, three, or all of the Liouvillian superoperators $\hat{\hat{L}}_{\text{att}}$, $\hat{\hat{L}}_{\text{amp}}$, $\hat{\hat{L}}_{\text{Att}}$, $\hat{\hat{L}}_{\text{nln}}$. The unravelling is interesting in the steady state too. This state evolves from an old initial datum after a long time period since $t_0 \le 0$ and it is considered to be part of a new initial condition at the time $t = 0$.

The new description presents a stochastic process $\underline{\hat{\rho}}_{\text{c}}(t)$ whose values are statistical operators $\hat{\rho}_\text{c}(t)$. Here the subscript c stands for condition, and it will be explained in the following. We suppose that such a process has a Markovian property. The event of a continuous change or maybe conservation and the event of an instantaneous discontinuous change of the statistical operator whose probability is asymptotically proportional to Δt for $\Delta t \to 0$ are statistically independent of such past events. The continuous change could be described with a master equation

$$
\frac{\partial}{\partial t}\underline{\hat{\rho}}_{\text{c}}(t)=\hat{\hat{L}}_{\mp,\text{att}}\underline{\hat{\rho}}_{\text{c}}(t)+\hat{\hat{L}}_{\mp,\text{amp}}\underline{\hat{\rho}}_{\text{c}}(t)+\hat{\hat{L}}_{\mp,\text{Att}}\underline{\hat{\rho}}_{\text{c}}(t)+\hat{\hat{L}}_{\mp,\text{nln}}\underline{\hat{\rho}}_{\text{c}}(t),
\tag{6.423}
$$

where

$$
\begin{aligned}
\hat{\hat{L}}_{-,\text{att}}&=\hat{\hat{L}}_{\text{att}},\\
\hat{\hat{L}}_{+,\text{att}}\underline{\hat{\rho}}_{\text{c}}(t)&=\sum_{i=1}^{N_\text{p}}\gamma_i\left[\langle\hat{n}_i\rangle_\text{c}(t)\underline{\hat{\rho}}_{\text{c}}(t)-\frac{1}{2}\hat{n}_i\underline{\hat{\rho}}_{\text{c}}(t)-\frac{1}{2}\underline{\hat{\rho}}_{\text{c}}(t)\hat{n}_i\right],
\end{aligned}
\tag{6.424}
$$

$$\hat{\hat{L}}_{-,\mathrm{amp}} = \hat{\hat{L}}_{\mathrm{att}},$$

$$\hat{\hat{L}}_{+,\mathrm{amp}}\underline{\hat{\rho}}_{\mathrm{c}}(t) = \Delta V \sum_{j=1}^{N_{\mathrm{s}}} P_j \left[\underline{\hat{\rho}}_{\mathrm{c}}(t) - \underline{\hat{\rho}}_{\mathrm{c}}(t)\right] = \hat{0}, \tag{6.425}$$

$$\hat{\hat{L}}_{-,\mathrm{Att}} = \hat{\hat{L}}_{\mathrm{Att}},$$

$$\hat{\hat{L}}_{+,\mathrm{Att}}\underline{\hat{\rho}}_{\mathrm{c}}(t) = \Delta V \sum_{j=1}^{N_{\mathrm{s}}} a_j \left[\langle \hat{N}_j \rangle_{\mathrm{c}}(t)\underline{\hat{\rho}}_{\mathrm{c}}(t) - \frac{1}{2}\hat{N}_j\underline{\hat{\rho}}_{\mathrm{c}}(t) - \frac{1}{2}\underline{\hat{\rho}}_{\mathrm{c}}(t)\hat{N}_j\right], \tag{6.426}$$

$$\hat{\hat{L}}_{-,\mathrm{nln}} = \hat{\hat{L}}_{\mathrm{nln}},$$

$$\hat{\hat{L}}_{+,\mathrm{nln}}\underline{\hat{\rho}}_{\mathrm{c}}(t) = \Delta V \sum_{i=1}^{N_{\mathrm{p}}} \sum_{j=1}^{N_{\mathrm{s}}} K_{ij} \Big[\langle (\hat{n} + \hat{1})\hat{N}_j \rangle_{\mathrm{c}}(t) \hat{a}_i^{\dagger} \hat{A}_j \underline{\hat{\rho}}_{\mathrm{c}}(t)$$

$$- \frac{1}{2}(\hat{n} + \hat{1})\hat{N}_j\underline{\hat{\rho}}_{\mathrm{c}}(t) - \frac{1}{2}\underline{\hat{\rho}}_{\mathrm{c}}(t)\hat{N}_j(\hat{n} + \hat{1})\Big]. \tag{6.427}$$

Having introduced the subscripts $\mp$, we have alluded to the choice of $2^4 - 1$ unravellings, $+$ means unravelled and $-$ means unmodified.

After the discontinuous change, the new statistical operator

$$\underline{\hat{\rho}}_{\mathrm{c}}'(t) = \frac{\hat{a}_i \underline{\hat{\rho}}_{\mathrm{c}}(t) \hat{a}_i^{\dagger}}{\langle \hat{n}_i \rangle_{\mathrm{c}}(t)}, \quad \frac{\widehat{\exp}(-\mathrm{i}\Phi_j)\underline{\hat{\rho}}_{\mathrm{c}}(t)\widehat{\exp}(\mathrm{i}\Phi_j)}{1}, \quad \frac{\hat{A}_j \underline{\hat{\rho}}_{\mathrm{c}}(t) \hat{A}_j^{\dagger}}{\langle \hat{N}_j \rangle_{\mathrm{c}}(t)}, \quad \frac{\hat{a}_i^{\dagger} \hat{A}_j \underline{\hat{\rho}}_{\mathrm{c}}(t) \hat{A}_j^{\dagger} \hat{a}_i}{\langle (\hat{n}_i + 1)\hat{N}_j \rangle_{\mathrm{c}}(t)} \tag{6.428}$$

and the factors of asymptotical proportionality are $\gamma_i \langle \hat{n}_i \rangle_{\mathrm{c}}(t)$, P_j, $a_j \langle \hat{N}_j \rangle_{\mathrm{c}}(t) \Delta V$, $K_{ij} \langle (\hat{n}_i + \hat{1})\hat{N}_j \rangle_{\mathrm{c}}(t) \Delta V$. We call them intensities. The discontinuous changes and intensities are related to the components of superoperators $\hat{\hat{L}}_{\mathrm{att}}$ (N_{p} components), $\hat{\hat{L}}_{\mathrm{amp}}$ (N_{s} components), $\hat{\hat{L}}_{\mathrm{Att}}$ (N_{s} components), $\hat{\hat{L}}_{\mathrm{nln}}$ ($N_{\mathrm{p}}N_{\mathrm{s}}$).

The application of the new description is perspicuous and it does not contradict a physical intuition when it is related only to components of the superoperator $\hat{\hat{L}}_{\mathrm{att}}$. The new description becomes more transparent after introducing variables $\underline{m}_1(t), \ldots, \underline{m}_{N_{\mathrm{p}}}(t)$, which obey the initial condition $\underline{m}_1(0) = \ldots = \underline{m}_{N_{\mathrm{p}}}(0) = 0$. On the continuous change of the statistical operator $\underline{\hat{\rho}}_{\mathrm{c}}(t)$, all of these variables are conserved. In the discontinuous change of the statistical operator related to the ith component of $\hat{\hat{L}}_{\mathrm{att}}$, $\underline{m}_i(t)$ is increased by unity. Even if from this it follows only that the variables $\underline{m}_1(t), \ldots, \underline{m}_{N_{\mathrm{p}}}(t)$ do not burden the description, we state that the expression of $\underline{\hat{\rho}}_{\mathrm{c}}(t)$ in dependence on $\underline{m}_1(t'), \ldots, \underline{m}_{N_{\mathrm{p}}}(t')$, $0 \le t' \le t$, and the treatment of $\underline{m}_1(t), \ldots, \underline{m}_{N_{\mathrm{p}}}(t)$ as classical stochastic processes with the intensities, which are given as quantum averages, is lucid. Moreover, it is appropriate to the physical intuition, when we identify $\underline{m}_i(t)$ with numbers of photons (or photocounts), which have been registered with a detector since the time $t = 0$.

The unravelling can be most easily understood as the equality

$$\hat{\rho}(t) = E[\underline{\hat{\rho}}_{\mathrm{c}}(t)], \tag{6.429}$$

where E stands for the expectation value. We introduce the notation

$$\hat{\rho}_{m_1,\ldots,m_{N_\mathrm{p}}}(t) = E[\underline{\hat{\rho}}_{\mathrm{c}}(t)|\underline{m}_1(t) = m_1, \ldots, \underline{m}_{N_\mathrm{p}}(t) = m_{N_\mathrm{p}}] \times p(m_1, \ldots, m_{N_\mathrm{p}}, t), \tag{6.430}$$

where $E[\underline{\hat{\rho}}_{\mathrm{c}}(t)|A]$ is the conditioned expectation value of the random operator $\underline{\hat{\rho}}_{\mathrm{c}}(t)$ conditioned on the event A and

$$p(m_1, \ldots, m_{N_\mathrm{p}}, t) = \Pr[\underline{m}_1(t) = m_1, \ldots, \underline{m}_{N_\mathrm{p}}(t) = m_{N_\mathrm{p}}], \tag{6.431}$$

with Pr denoting the probability. Invoking the probability theory, we note that

$$\hat{\rho}(t) = \sum_{m_1=0}^{\infty} \cdots \sum_{m_{N_\mathrm{p}}=0}^{\infty} \hat{\rho}_{m_1,\ldots,m_{N_\mathrm{p}}}(t). \tag{6.432}$$

On introducing the Hilbert space with a complete orthonormal basis $|m_1, \ldots, m_{N_\mathrm{p}}\rangle_{\mathrm{det}}$ and considering this space in a tensor product with the original Hilbert space, we can define a statistical operator

$$\hat{\rho}_{\mathrm{e}}(t) = \sum_{m_1=0}^{\infty} \cdots \sum_{m_{N_\mathrm{p}}=0}^{\infty} \hat{\rho}_{m_1,\ldots,m_{N_\mathrm{p}}}(t) \otimes |m_1, \ldots, m_{N_\mathrm{p}}\rangle_{\mathrm{det}\;\mathrm{det}}\langle m_1, \ldots, m_{N_\mathrm{p}}|, \tag{6.433}$$

where "e" means extended. Letting $\mathrm{Tr}_{\mathrm{det}}$ denote the partial trace over the extending Hilbert-space factor, $\mathrm{Tr}_{\mathrm{det}} \equiv \mathrm{Tr}_{\mathrm{d}_1} \ldots \mathrm{Tr}_{\mathrm{d}_{N_\mathrm{p}}}$, we see easily that

$$\hat{\rho}(t) = \mathrm{Tr}_{\mathrm{det}}\, \hat{\rho}_{\mathrm{e}}(t). \tag{6.434}$$

The master equation has the form

$$\frac{\partial}{\partial t}\hat{\rho}_{\mathrm{e}}(t) = \hat{\hat{L}}_{\mathrm{e,att}}\hat{\rho}_{\mathrm{e}}(t) + \hat{\hat{L}}_{\mathrm{e,amp}}\hat{\rho}_{\mathrm{e}}(t) + \hat{\hat{L}}_{\mathrm{e,Att}}\hat{\rho}_{\mathrm{e}}(t) + \hat{\hat{L}}_{\mathrm{e,nln}}\hat{\rho}_{\mathrm{e}}(t), \tag{6.435}$$

where

$$\hat{\hat{L}}_{\mathrm{e,att}}\hat{\rho}_{\mathrm{e}}(t) = \sum_{i=1}^{N_\mathrm{p}} \gamma_i \left[\hat{a}_i \widehat{\exp}(-\mathrm{i}\theta_i)\hat{\rho}_{\mathrm{e}}(t)\widehat{\exp}(\mathrm{i}\theta_i)\hat{a}_i^{\dagger} - \frac{1}{2}\hat{n}_i\hat{\rho}_{\mathrm{e}}(t) - \frac{1}{2}\hat{\rho}_{\mathrm{e}}(t)\hat{n}_i\right], \tag{6.436}$$

$$\begin{aligned}\hat{\hat{L}}_{\mathrm{e,amp}}\hat{\rho}_{\mathrm{e}}(t) &= \hat{\hat{L}}_{\mathrm{amp}}\hat{\rho}_{\mathrm{e}}(t),\\ \hat{\hat{L}}_{\mathrm{e,Att}}\hat{\rho}_{\mathrm{e}}(t) &= \hat{\hat{L}}_{\mathrm{Att}}\hat{\rho}_{\mathrm{e}}(t),\\ \hat{\hat{L}}_{\mathrm{e,nln}}\hat{\rho}_{\mathrm{e}}(t) &= \hat{\hat{L}}_{\mathrm{nln}}\hat{\rho}_{\mathrm{e}}(t),\end{aligned} \tag{6.437}$$

$$\widehat{\exp}(\mathrm{i}\theta_i) = \hat{1} \otimes \sum_{m_1=0}^{\infty} \cdots \sum_{m_i=0}^{\infty} \cdots \sum_{m_{N_\mathrm{p}}=0}^{\infty} 1 \\ \times |m_1, \ldots, m_i, \ldots, m_{N_\mathrm{p}}\rangle_{\mathrm{det}\ \mathrm{det}}\langle m_1, \ldots, m_i + 1, \ldots, m_{N_\mathrm{p}}|, \tag{6.438}$$

$$\widehat{\exp}(-\mathrm{i}\theta_i) = \left[\widehat{\exp}(\mathrm{i}\theta_i)\right]^\dagger, \\ \hat{\rho}_\mathrm{e}(0) = \hat{\rho}(0) \otimes |0, \ldots, 0\rangle_{\mathrm{det}\ \mathrm{det}}\langle 0, \ldots, 0|. \tag{6.439}$$

Now we extend the joint probability distribution of photon numbers and the densities of excited atoms by numbers of emitted photons and introduce the probabilities

$$p_\mathrm{e}(n_1, \ldots, n_{N_\mathrm{p}}, N_1, \ldots, N_{N_\mathrm{s}}, m_1, \ldots, m_{N_\mathrm{p}}, t) \\ = \langle n_1, \ldots, n_{N_\mathrm{p}}, N_1, \ldots, N_{N_\mathrm{s}}|\hat{\rho}_{m_1,\ldots,m_{N_\mathrm{p}}}(t) \\ \times |n_1, \ldots, n_{N_\mathrm{p}}, N_1, \ldots, N_{N_\mathrm{s}}\rangle. \tag{6.440}$$

The rate equations for these probabilities have the form

$$\frac{\partial}{\partial t} p_\mathrm{e}(n_1, \ldots, n_{N_\mathrm{p}}, N_1, \ldots, N_{N_\mathrm{s}}, m_1, \ldots, m_{N_\mathrm{p}}, t) \\ = L_{\mathrm{e,att}} p_\mathrm{e}(n_1, \ldots, n_{N_\mathrm{p}}, N_1, \ldots, N_{N_\mathrm{s}}, m_1, \ldots, m_{N_\mathrm{p}}, t) \\ + L_{\mathrm{e,amp}} p_\mathrm{e}(n_1, \ldots, n_{N_\mathrm{p}}, N_1, \ldots, N_{N_\mathrm{s}}, m_1, \ldots, m_{N_\mathrm{p}}, t) \\ + L_{\mathrm{e,Att}} p_\mathrm{e}(n_1, \ldots, n_{N_\mathrm{p}}, N_1, \ldots, N_{N_\mathrm{s}}, m_1, \ldots, m_{N_\mathrm{p}}, t) \\ + L_{\mathrm{e,nln}} p_\mathrm{e}(n_1, \ldots, n_{N_\mathrm{p}}, N_1, \ldots, N_{N_\mathrm{s}}, m_1, \ldots, m_{N_\mathrm{p}}, t), \tag{6.441}$$

where

$$L_{\mathrm{e,att}} p_\mathrm{e}(n_1, \ldots, n_{N_\mathrm{p}}, N_1, \ldots, N_{N_\mathrm{s}}, m_1, \ldots, m_{N_\mathrm{p}}, t) \\ = \sum_{i=1}^{N_\mathrm{p}} \gamma_i[(n_i + 1) p_\mathrm{e}(n_1, \ldots, n_i + 1, \ldots, n_{N_\mathrm{p}}, N_1, \ldots, N_{N_\mathrm{s}}, \\ m_1, \ldots, m_i - 1, \ldots, m_{N_\mathrm{p}}, t) \\ - n_i p_\mathrm{e}(n_1, \ldots, n_{N_\mathrm{p}}, N_1, \ldots, N_{N_\mathrm{s}}, m_1, \ldots, m_{N_\mathrm{p}}, t)], \tag{6.442}$$

$$L_{\mathrm{e,amp}} p_\mathrm{e}(n_1, \ldots, n_{N_\mathrm{p}}, N_1, \ldots, N_{N_\mathrm{s}}, m_1, \ldots, m_{N_\mathrm{p}}, t) = \Delta V \sum_{j=1}^{N_\mathrm{s}} P_j \\ \times \left[p_\mathrm{e}\left(n_1, \ldots, n_{N_\mathrm{p}}, N_1, \ldots, N_j - \frac{1}{\Delta V}, \ldots, N_{N_\mathrm{s}}, m_1, \ldots, m_{N_\mathrm{p}}, t \right) \right. \\ \left. - p_\mathrm{e}(n_1, \ldots, n_{N_\mathrm{p}}, N_1, \ldots, N_{N_\mathrm{s}}, m_1, \ldots, m_{N_\mathrm{p}}, t) \right], \tag{6.443}$$

$$L_{\mathrm{e,Att}} p_{\mathrm{e}}(n_1, \ldots, n_{N_{\mathrm{p}}}, N_1, \ldots, N_{N_{\mathrm{s}}}, m_1, \ldots, m_{N_{\mathrm{p}}}, t)$$
$$= \Delta V \sum_{j=1}^{N_{\mathrm{s}}} a_j \left[\left(N_j + \frac{1}{\Delta V} \right) \right.$$
$$\times p_{\mathrm{e}}\left(n_1, \ldots, n_{N_{\mathrm{p}}}, N_1, \ldots, N_j + \frac{1}{\Delta V}, \ldots, N_{N_{\mathrm{s}}}, m_1, \ldots, m_{N_{\mathrm{p}}}, t \right)$$
$$\left. - N_j p_{\mathrm{e}}(n_1, \ldots, n_{N_{\mathrm{p}}}, N_1, \ldots, N_{N_{\mathrm{s}}}, m_1, \ldots, m_{N_{\mathrm{p}}}, t) \right], \tag{6.444}$$

$$L_{\mathrm{e,nln}} p_{\mathrm{e}}(n_1, \ldots, n_{N_{\mathrm{p}}}, N_1, \ldots, N_{N_{\mathrm{s}}}, m_1, \ldots, m_{N_{\mathrm{p}}}, t)$$
$$= \Delta V \sum_{i=1}^{N_{\mathrm{p}}} \sum_{j=1}^{N_{\mathrm{s}}} K_{ij} \left[n_i \left(N_j + \frac{1}{\Delta V} \right) p_{\mathrm{e}}\left(n_1, \ldots, n_i - 1, \ldots, n_{N_{\mathrm{p}}}, N_1, \ldots, N_j \right. \right.$$
$$\left. + \frac{1}{\Delta V}, \ldots, N_{N_{\mathrm{s}}}, m_1, \ldots, m_{N_{\mathrm{p}}}, t \right)$$
$$\left. - (n_i + 1) N_j p_{\mathrm{e}}(n_1, \ldots, n_{N_{\mathrm{p}}}, N_1, \ldots, N_{N_{\mathrm{s}}}, m_1, \ldots, m_{N_{\mathrm{p}}}, t) \right]. \tag{6.445}$$

The initial condition related to the time origin is

$$p_{\mathrm{e}}(n_1, \ldots, n_{N_{\mathrm{p}}}, N_1, \ldots, N_{N_{\mathrm{s}}}, m_1, \ldots, m_{N_{\mathrm{p}}}, t)\Big|_{t=0}$$
$$= p(n_1, \ldots, n_{N_{\mathrm{p}}}, N_1, \ldots, N_{N_{\mathrm{s}}}, 0) \delta_{m_1 0} \ldots \delta_{m_{N_{\mathrm{p}}} 0}. \tag{6.446}$$

Only the term (6.442) means some added phenomenology. It is assumed that an array of ideal detectors is available and, whenever a photon escapes from the ith mode, it is absorbed by the ith detector.

By analogy with (6.438), we introduce the operators

$$\hat{m}_i = \hat{1}_{\mathrm{e}} \otimes \sum_{m_1=0}^{\infty} \ldots \sum_{m_i=0}^{\infty} \ldots \sum_{m_{N_{\mathrm{p}}}=0}^{\infty} |m_1, \ldots, m_{N_{\mathrm{p}}}\rangle_{\mathrm{det}\ \mathrm{det}}\langle m_1, \ldots, m_{N_{\mathrm{p}}}|. \tag{6.447}$$

We will introduce a shorthand notation

$$\hat{n}^k \equiv \hat{n}_1^{k_1} \ldots \hat{n}_{N_{\mathrm{p}}}^{k_{N_{\mathrm{p}}}},$$
$$\hat{N}^l \equiv \hat{N}_1^{l_1} \ldots \hat{N}_{N_{\mathrm{s}}}^{l_{N_{\mathrm{s}}}},$$
$$\hat{m}^r \equiv \hat{m}_1^{r_1} \ldots \hat{m}_{N_{\mathrm{p}}}^{r_{N_{\mathrm{p}}}}. \tag{6.448}$$

Considering the moments $\langle \hat{n}^k \hat{N}^l \hat{m}^r \rangle(t)$, we can rewrite equation (6.441) in the form of a hierarchy of equations

$$\frac{\mathrm{d}}{\mathrm{d}t} \langle \hat{n}^k \hat{N}^l \hat{m}^r \rangle(t) = \langle \hat{\hat{L}}_{\mathrm{e,att}}^{\dagger} \hat{n}^k \hat{N}^l \hat{m}^r \rangle(t) + \langle \hat{\hat{L}}_{\mathrm{e,amp}}^{\dagger} \hat{n}^k \hat{N}^l \hat{m}^r \rangle(t)$$
$$+ \langle \hat{\hat{L}}_{\mathrm{e,Att}}^{\dagger} \hat{n}^k \hat{N}^l \hat{m}^r \rangle(t) + \langle \hat{\hat{L}}_{\mathrm{e,nln}}^{\dagger} \hat{n}^k \hat{N}^l \hat{m}^r \rangle(t), \tag{6.449}$$

where

$$\hat{\hat{L}}^{\dagger}_{\mathrm{e,att}}\hat{n}^k\hat{N}^l\hat{m}^r = -\sum_{i=1}^{N_\mathrm{p}}\gamma_i\hat{n}_i\Big[\hat{n}_1^{k_1}\dots\hat{n}_{N_\mathrm{p}}^{k_{N_\mathrm{p}}}\hat{N}^l\hat{m}_1^{r_1}\dots\hat{m}_{N_\mathrm{p}}^{r_{N_\mathrm{p}}}$$
$$-\hat{n}_1^{k_1}\dots(\hat{n}_i-\hat{1}_\mathrm{e})^{k_i}\dots\hat{n}_{N_\mathrm{p}}^{k_{N_\mathrm{p}}}\hat{N}^l\hat{m}_1^{r_1}\dots(\hat{m}_i+\hat{1}_\mathrm{e})^{r_i}\dots\hat{m}_{N_\mathrm{p}}^{r_{N_\mathrm{p}}}\Big], \tag{6.450}$$

$$\hat{\hat{L}}^{\dagger}_{\mathrm{e,amp}}\hat{n}^k\hat{N}^l\hat{m}^r = \Delta V\sum_{j=1}^{N_\mathrm{s}}P_j\left[\hat{n}^k\hat{N}_1^{l_1}\dots\left(\hat{N}_j+\frac{\hat{1}_\mathrm{e}}{\Delta V}\right)^{l_j}\dots\hat{N}_{N_\mathrm{s}}^{l_{N_\mathrm{s}}}\hat{m}^r\right.$$
$$\left.-\hat{n}^k\hat{N}_1^{l_1}\dots\hat{N}_{N_\mathrm{s}}^{l_{N_\mathrm{s}}}\hat{m}^r\right], \tag{6.451}$$

$$\hat{\hat{L}}^{\dagger}_{\mathrm{e,Att}}\hat{n}^k\hat{N}^l\hat{m}^r = -\Delta V\sum_{j=1}^{N_\mathrm{s}}a_j\hat{N}_j\left[\hat{n}^k\hat{N}_1^{l_1}\dots\hat{N}_{N_\mathrm{s}}^{l_{N_\mathrm{s}}}\hat{m}^r\right.$$
$$\left.-\hat{n}^k\hat{N}_1^{l_1}\dots\left(\hat{N}_j-\frac{\hat{1}_\mathrm{e}}{\Delta V}\right)^{l_j}\dots\hat{N}_{N_\mathrm{s}}^{l_{N_\mathrm{s}}}\hat{m}^r\right], \tag{6.452}$$

$$\hat{\hat{L}}^{\dagger}_{\mathrm{e,nln}}\hat{n}^k\hat{N}^l\hat{m}^r = \pm\Delta V\sum_{i=1}^{N_\mathrm{p}}\sum_{j=1}^{N_\mathrm{s}}K_{ij}(\hat{n}_i+\hat{1}_\mathrm{e})\hat{N}_j\Big[\pm\hat{n}_1^{k_1}$$
$$\dots(\hat{n}_i+\hat{1}_\mathrm{e})^{k_i}\dots\hat{n}_{N_\mathrm{p}}^{k_{N_\mathrm{p}}}\hat{N}_1^{l_1}\dots\left(\hat{N}_j-\frac{\hat{1}_\mathrm{e}}{\Delta V}\right)^{l_j}\dots\hat{N}_{N_\mathrm{s}}^{l_{N_\mathrm{s}}}\hat{m}^r$$
$$\mp\hat{n}_1^{k_1}\dots\hat{n}_{N_\mathrm{p}}^{k_{N_\mathrm{p}}}\hat{N}_1^{l_1}\dots\hat{N}_{N_\mathrm{s}}^{l_{N_\mathrm{s}}}\hat{m}^r\Big], \tag{6.453}$$

with $\dagger$ denoting the Hermitian conjugation and $\hat{1}_\mathrm{e}$ denoting the identity operator. The moment equations (6.449) can be decoupled by various assumptions and approximations. First, we observe that the equations for the means $\langle\hat{n}_i\rangle(t)$, $\langle\hat{N}_j\rangle(t)$ need those for the second moments $\langle\hat{n}_i\hat{N}_j\rangle(t)$

$$\frac{\mathrm{d}}{\mathrm{d}t}\langle\hat{n}_i\rangle(t) = -\gamma_i\langle\hat{n}_i\rangle(t)+\Delta V\sum_{j=1}^{N_\mathrm{s}}K_{ij}\langle(\hat{n}_i+\hat{1})\hat{N}_j\rangle(t), \tag{6.454}$$

$$\frac{\mathrm{d}}{\mathrm{d}t}\langle\hat{N}_j\rangle(t) = P_j - a_j\langle\hat{N}_j\rangle(t)-\sum_{i=1}^{N_\mathrm{p}}K_{ij}\langle(\hat{n}_i+\hat{1})\hat{N}_j\rangle(t). \tag{6.455}$$

We neglect such a coupling by the factorizing approximation (Patra 2002, Rice and Carmichael 1994)

$$\langle\hat{n}_i\hat{N}_j\rangle(t)\approx\langle\hat{n}_i\rangle(t)\langle\hat{N}_j\rangle(t). \tag{6.456}$$

We introduce the shorthand $\overline{n}_i(t) \equiv \langle \hat{n}_i \rangle(t)$, $\overline{N}_j(t) \equiv \langle \hat{N}_j \rangle(t)$, and $\overline{m}_i(t) \equiv \langle \hat{m}_i \rangle(t)$, and, from now on, we consider the differential equations

$$\frac{\mathrm{d}}{\mathrm{d}t}\overline{n}_i(t) = -\gamma_i \overline{n}_i(t) + \Delta V \sum_{j=1}^{N_\mathrm{s}} [\overline{n}_i(t) + 1] K_{ij} \overline{N}_j(t), \tag{6.457}$$

$$\frac{\mathrm{d}}{\mathrm{d}t}\overline{N}_j(t) = P_j - a_j \overline{N}_j(t) - \sum_{i=1}^{N_\mathrm{p}} [\overline{n}_i(t) + 1] K_{ij} \overline{N}_j(t), \tag{6.458}$$

$$\frac{\mathrm{d}}{\mathrm{d}t}\overline{m}_i(t) = \gamma_i \overline{n}_i(t), \tag{6.459}$$

and the appropriate initial conditions related to the time t_0 and to the time origin, which give averages of the initial data of the unlinearized model if possible.

Further, we introduce the variations $\delta\hat{n}_i(t) = \hat{n}_i - \overline{n}_i(t)\hat{1}_\mathrm{e}$, $\delta\hat{N}_j(t) = \hat{N}_j - \overline{N}_j(t)\hat{1}_\mathrm{e}$, $\delta\hat{m}_i(t) = \hat{m}_i - \overline{m}_i(t)\hat{1}_\mathrm{e}$, $i = 1, \ldots, N_\mathrm{p}$, $j = 1, \ldots, N_\mathrm{s}$, of zero expectation values. Assuming that the equations (6.457), (6.458), (6.459) have been solved for the means $\overline{n}_i(t)$, $\overline{N}_j(t)$, $\overline{m}_i(t)$, $i = 1, \ldots, N_\mathrm{p}$, $j = 1, \ldots, N_\mathrm{s}$, and neglecting higher (third) moments, we derive approximate at most linear equations for the variances

$$\langle [\delta\hat{n}_i(t)]^2 \rangle(t), \quad \langle [\delta\hat{N}_j(t)]^2 \rangle(t), \quad \langle [\delta\hat{m}_i(t)]^2 \rangle(t) \tag{6.460}$$

and the covariances

$$\begin{aligned} &\langle \delta\hat{n}_i(t)\delta\hat{n}_j(t)\rangle(t), \quad \langle \delta\hat{n}_i(t)\delta\hat{N}_j(t)\rangle(t), \quad \langle \delta\hat{N}_j(t)\delta\hat{N}_{j'}(t)\rangle(t), \\ &\langle \delta\hat{n}_i(t)\delta\hat{m}_j(t)\rangle(t), \quad \langle \delta\hat{N}_j(t)\delta\hat{m}_i(t)\rangle(t), \quad \langle \delta\hat{m}_i(t)\delta\hat{m}_{i'}(t)\rangle(t) \end{aligned} \tag{6.461}$$

from the hierarchy of moment equations. Also in the relations (6.460) and (6.461), we have written the argument t to the right of the angular brackets to indicate that the quantum description is still being carried out in the Schrödinger picture, even though some time dependence has been created by the subtracted means.

6.4.3 Semiclassical Approach

Semiclassical approach consists in the equations of motion for the photon numbers in the eigenmodes $1, \ldots, N_\mathrm{p}$ and the densities of excited atoms in the neighbourhoods $U(\mathbf{0}_j)$, $j = 1, \ldots, N_\mathrm{s}$. These equations may be supplemented with those for the number of counts taken by the detectors.

Stationary values of the means $\overline{n}_i(t)$, $\overline{N}_j(t)$ can be characteristic of stationary processes $n_i(t)$, $N_j(t)$, which can be obtained in the limit $t_0 \to -\infty$. We also obtain the time independence of the variances $\langle [\delta n_i(t)]^2 \rangle(t)$, $\langle [\delta N_j(t)]^2 \rangle(t)$ and the covariances $\langle \delta n_i(t)\delta n_j(t)\rangle$, $\langle \delta n_i(t)\delta N_j(t)\rangle$, $\langle \delta N_j(t)\delta N_{j'}(t)\rangle$, and $\overline{m}_i(t) = \gamma_i t \overline{n}_i(0)$. In this case, the number of modes above the threshold N_l is interesting, mode i being above the threshold if and only if $\overline{n}_i(t) \geq 2$.

We will restrict ourselves to the Fano factor (Peřina 1991) which can be calculated as

$$\mathcal{F}_{\text{sugg}} = \frac{\sum_{i=1}^{N_p}\sum_{j=1}^{N_p} \langle \delta m_i(T)\delta m_j(T)\rangle}{\sum_{i=1}^{N_p} \overline{m}_i(T)}, \tag{6.462}$$

where the subscript sugg stands for suggested, T is the time needed to explore the entire space inside the cavity, $T = \frac{\Omega^2 V^2}{\pi^2 c^3}$ is chosen as a detection time, and c is the speed of light. As a short-T approximation, we obtain the usual formula for the Fano factor, in fact

$$\left.\frac{\mathrm{d}}{\mathrm{d}T}(\mathcal{F}_{\text{sugg}} - 1)\right|_{T=0} = \frac{1}{T}(\mathcal{F}_{\text{corr}} - 1), \tag{6.463}$$

where $\mathcal{F}_{\text{corr}}$ means the Fano factor, the subscript corr stands for correlated, which is given as

$$\mathcal{F}_{\text{corr}} = \frac{\sum_{i=1}^{N_p}\sum_{j=1}^{N_p} \mathcal{T}_i\mathcal{T}_j \langle \delta n_i(0)\delta n_j(0)\rangle}{\sum_{i=1}^{N_p} \mathcal{T}_i\overline{n}_i(0)}, \tag{6.464}$$

with the transmission probability $\mathcal{T}_i = \gamma_i T$, $0 \leq \mathcal{T}_i \leq 1$.

With respect to reduced information on an individual random laser, the problem becomes a stochastic problem, an ensemble of cavities with small variations in a shape or scatterer positions being considered. The coefficients γ_i and K_{ij} thus become random quantities. The coefficients γ_i are independent and identically distributed, the probability density is

$$P(\{\gamma_i\}) = \prod_{i=1}^{N_p} P(\gamma_i), \tag{6.465}$$

where

$$P(\gamma_i) = \frac{1}{\sqrt{2\pi\gamma_i\overline{\gamma}}} e^{-\frac{\gamma_i}{2\overline{\gamma}}}, \tag{6.466}$$

which is the gamma or the Porter–Thomas distribution (Porter 1965) with the mean loss rate of a cavity $\overline{\gamma} = \frac{\overline{\mathcal{T}}}{T}$, $\overline{\mathcal{T}}$ being the mean transmissivity of a cavity, $\overline{\mathcal{T}} = \frac{16\pi^2 d^6\Omega^6}{c^6}$. The expectation values $\overline{\gamma}_i \equiv \langle\gamma_i\rangle$, $i = 1, \ldots, N_p$, are equal to the mean loss rate of the cavity, $\overline{\gamma}_i = \overline{\gamma}$. K_{ij}, $i = 1, \ldots, N_p$, $j = 1, \ldots, N_s$, are independent of γ_i, and they are distributed as squared moduli of elements of a random unitary matrix. We hope to be consistent with Patra (2002), even after the correction of its formula (16).

As an illustration, a laser with a cavity supporting ten modes has been considered, where one mode is coupled out much less than the others, i.e., $\gamma_1 = 0.01 = g$, $\gamma_2 = \cdots = \gamma_{10} = 0.1$, $K_{ij} = 0.1$ for all i, $a_1 = \cdots = a_{10} = 1$.

It can be accepted that some characteristics of a random laser fluctuate not very much about the ensemble means. Such characteristics are the scaled Fano factor $\frac{\mathcal{F}-1}{g}$ of the lasing mode ($\gamma_i = g$)

$$\frac{\mathcal{F}-1}{g} = T\left(\frac{\langle(\delta n_i)^2\rangle}{\overline{n}_i} - 1\right) \tag{6.467}$$

and the number of modes above threshold N_l. In computer simulations, the scaled Fano factor may and may not fluctuate very much about the conditional mean $\langle\frac{\mathcal{F}-1}{g}\rangle|_{N_l}$ when the probability of N_l, $p(N_l)$, is small or not as can be seen from Fig. 6.4.

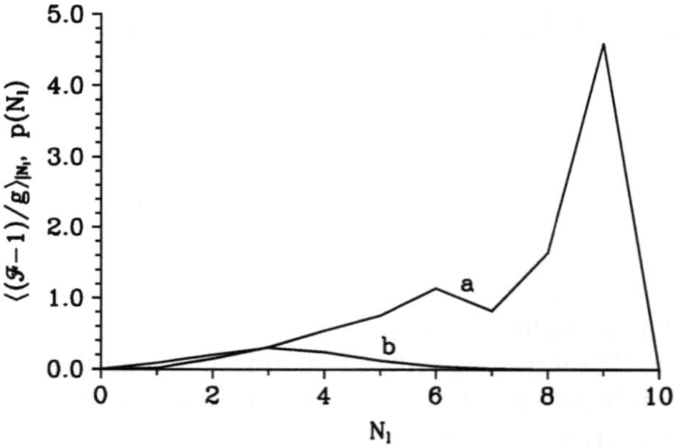

Fig. 6.4 The conditional mean of the scaled Fano factor in the lasing mode in the dependence on the number of modes above the threshold (*curve* a) and the probability of the number of modes above the threshold (*curve* b) for cavities with a fixed number $N_p = N_s$ modes

In this figure, the conditional mean increases in dependence on N_l (curve a). The irregularity at 7–10 is likely due to the simulation. The expectation value of N_l is about 3 (curve b). In the simulation $N_p = N_s = 10$, $\overline{\gamma} = 0.125$, and $P_1 = \cdots = P_{N_s} = P$, $P = 30$.

The simulation study of the conditional mean $\langle\frac{\mathcal{F}-1}{g}\rangle|_{\frac{g}{\overline{\gamma}}}$ for the lasing mode is more difficult than the previous one, because $\frac{g}{\overline{\gamma}}$ is a continuous variable. Denoting the probability density of g with $P_{(1)}(g)$, we note that $\frac{g}{\overline{\gamma}}$ has the probability density $\overline{\gamma}P_{(1)}(g)$. The calculated values of the conditional means are plotted in Fig. 6.5 as the curve a and the probabilities

$$\overline{P}_{(1)}\left(\frac{g}{\overline{\gamma}}\right) = \int_g^{g+0.025\overline{\gamma}} P_{(1)}(g')\,\mathrm{d}g' \tag{6.468}$$

are depicted as the curve b. The distribution of $\frac{g}{\overline{\gamma}}$ indicates that the conditional mean may be calculated well only for the smallest values of $\frac{g}{\overline{\gamma}}$, and it is estimated worse otherwise.

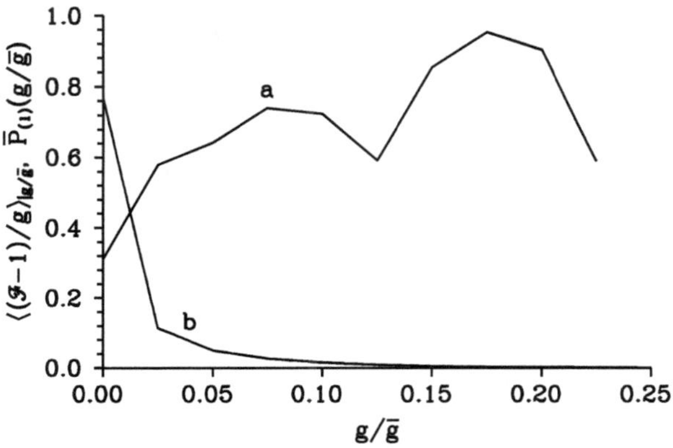

Fig. 6.5 The conditional mean of the scaled Fano factor in the lasing mode in the dependence on $\frac{g}{\gamma}$ (*curve* a) and the probability of $\frac{g}{\gamma}$ in bins of length of 0.025 (*curve* b) for cavities with a fixed number $N_{\mathrm{p}} = N_{\mathrm{s}}$ modes

In Fig. 6.6 the scaled Fano factors $\frac{\mathcal{F}_{\mathrm{corr}}-1}{g}$ and $\frac{\mathcal{F}_{\mathrm{sugg}}-1}{g}$ are plotted (curves a, c, e and b, d, f, respectively) in dependence on $T \in [0, 5]$ for $P = g, 10^{0.2}g, 10^{0.4}g$ (pairs (a, b), (c, d), (e, f)). The straight lines depicting $\frac{\mathcal{F}_{\mathrm{corr}}-1}{g}$ are tangent to the respective curves $\frac{\mathcal{F}_{\mathrm{sugg}}-1}{g}$ at the origin.

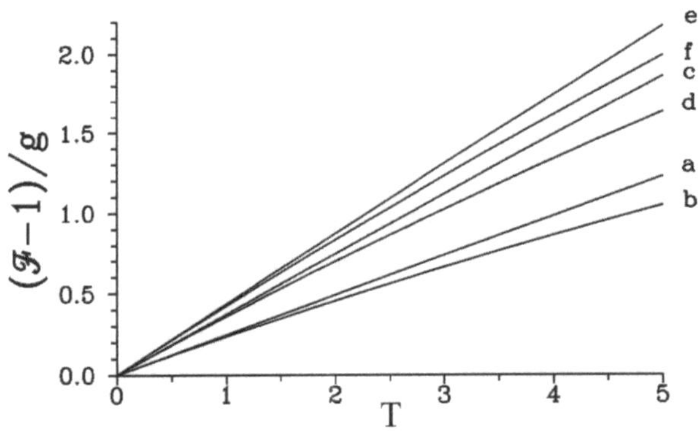

Fig. 6.6 The scaled Fano factors $\frac{\mathcal{F}_{\mathrm{corr}}-1}{g}$ and $\frac{\mathcal{F}_{\mathrm{sugg}}-1}{g}$ (*curves* a, c, e and b, d, f, respectively) in the dependence on $T \in [0, 5]$ for $P = g, 10^{0.2}g, 10^{0.4}g$ (pairs (a, b), (c, d), (e, f)), respectively

Considering the master equation in the Lindblad form, we have derived rate equations for the probability distributions describing "classical" state of the random laser. From this, using standard approximations, we have rederived well-known equations for the means and linear equations for the correlators. Both for traditional and random lasers, the Fano factor has been proposed based on open systems theory. Comparison of this proposal with the usual Fano factor has been made in the traditional laser. In the random laser, the scaled Fano factor in the lasing mode has been averaged in dependence on the number of modes above threshold and, alternatively, in dependence on the scaled loss rate.

Fu and Berman (2005) have complemented the Green function approach to the spontaneous emission from an atom embedded inside a disordered dielectric (Fleischhauer 1999). The virtual cavity result has been reproduced using an amplitude approach which has been extended to second order.

6.5 Propagation in Amplifying Random Media

The light propagation in disordered media is contiguous with the concept of light transport analogous to the electron transport in the condensed-matter physics, and its description is of importance for the experimental study of random lasers.

Quantum statistical properties of light are determined in disordered media. Models of a random laser with incoherent and coherent feedback are mentioned and it is stated that, in the framework of such models, photon statistics of optical modes were determined.

Both localization and laser theories which were developed in the 1960s have been jointly utilized in the study of a random laser. They have been used in studies of strongly scattering gain media. Lasing in disordered media has been a subject of intense theoretical and experimental studies. Random lasers have been classified into incoherent and coherent random lasers. Research works on both types of random lasers have been surveyed in the monographic chapter (Cao 2003). In order to explain quantum-statistical properties of random lasers, quantum theory is needed. Standard quantum theory for lasers applies only to quasidiscrete modes and cannot account for lasing in the presence of overlapping modes. In a random medium, the character of lasing modes depends on the amount of disorder. Weak disorder leads to a poor confinement of light and to strongly overlapping modes. Statistics naturally belongs to the theory of amplifying random media (Beenakker 1998, Patra and Beenakker 1999, 2000, Mishchenko et al. 2001), which is restricted to linear media and has not been used for the description of random lasers above lasing threshold (Cao 2003). The random laser model of Patra (2002, 2003), who calculated more than only statistics of the photon number, has been completed (Lukš and Peřinová 2003). In the framework of the open systems theory, the equations of motion involve those for numbers of photons absorbed by detector. This extension corrects the photon-number statistics. Hackenbroich et al. (2002) have developed a quantization scheme for optical resonators with overlapping (nonorthogonal) modes. Cheng and Siegman (2003) have derived a generalized formalism of radiation-field quantization which need not rely on a set of orthogonal eigenmodes. True eigenmodes of a system will be nonorthogonal and the method is intended for quantization of an open system which contains a gain or loss medium.

6.5.1 Strongly Scattering Media

John (1984) has proposed a range of wavelengths or frequencies, in which electromagnetic waves in a strongly scattering disordered medium undergo the Anderson localization (Anderson 1958, Abrahams et al. 1979). Although the derivation is conducted in $d = 2 + \varepsilon$ dimensions, in consequence it holds that the photon mobility edge ω^* is as $(\omega^* l)^2 \simeq \frac{1}{2\pi}$ $(d = 3)$, where l is the photon elastic mean free path. The range of wavelengths has the property $\lambda \sim l$.

In the case where $\lambda \ll l$, early experiments showed that the intensity of light scattered from a concentrated suspension of latex microspheres in water presented a sharp peak centred at the backscattering direction (Kuga and Ishimaru 1984, van Albada and Lagendijk 1985, Wolf and Maret 1985). This peak is a coherence effect, which is present in disordered media. Akkermans et al. (1986) have analyzed the multiple scattering of light to explain the peak line shape. The explanation is based on the constructive interferences between time-reversed paths of light in a semi-infinite medium. The intensity reflected in directions distant by more than one degree is almost constant and is incoherent. The analysis is not restricted to scalar waves. It respects that polarization was analyzed parallel and perpendicular to the incident one.

John et al. (1996) have contributed to the field of optical tomography (see, e.g., Huang et al. (1991) and references in John et al. (1996)). They assume a wave of frequency ω and velocity c. They recall that a simplified view of a photon as a quantum mechanical particle leads to the use of the Wigner coherence function, which is

$$I(\mathbf{R},\mathbf{k}) = \int \exp(\mathrm{i}\mathbf{k}\cdot\mathbf{r})\left\langle E^*\left(\mathbf{R}+\frac{\mathbf{r}}{2}\right)E\left(\mathbf{R}-\frac{\mathbf{r}}{2}\right)\right\rangle_{\text{ensemble}} \mathrm{d}^3\mathbf{r}. \tag{6.469}$$

The authors let $I_0(\mathbf{R},\mathbf{k})$ denote the source coherence function. On choosing $\mathbf{R}'$, $\mathbf{k}'$, the source $I_0(\mathbf{R},\mathbf{k}) = \delta(\mathbf{R}-\mathbf{R}')\delta(\mathbf{k}-\mathbf{k}')$ "radiates" a coherence function $\Gamma(\mathbf{R}-\mathbf{R}';\mathbf{k},\mathbf{k}')$, which is called a propagator. The Wigner coherence function can be measured (Raymer et al. 1994).

Here we must also refer to John and Stephen (1983) and McKintosh and John (1989) for reviews of the theory. In what follows we mention only the description of a homogeneously disordered dielectric material. Its statistical properties are given by the ensemble-averaged autocorrelation function

$$B_\mathrm{h}(\mathbf{r}-\mathbf{r}') = k_0^4\langle\epsilon_\mathrm{h}^*(\mathbf{r})\epsilon_\mathrm{h}(\mathbf{r}')\rangle_{\text{ensemble}}, \tag{6.470}$$

where h stands for homogeneous and $k_0 = \frac{\omega}{c}$. The Fourier transform is

$$\tilde{B}_\mathrm{h}(\mathbf{q}) = \int \exp(-\mathrm{i}\mathbf{q}\cdot\mathbf{r})B_\mathrm{h}(\mathbf{r})\,\mathrm{d}^3\mathbf{r}. \tag{6.471}$$

They define

$$\tilde{\Gamma}_\mathrm{h}(\mathbf{Q};\mathbf{k}',\mathbf{k}) = \frac{1}{k_0^4}\int \exp(-\mathrm{i}\mathbf{Q}\cdot\mathbf{R})\Gamma_\mathrm{h}(\mathbf{R};\mathbf{k}',\mathbf{k})\,\mathrm{d}^3\mathbf{R}. \tag{6.472}$$

In the field of optical tomography, conventional radiative transfer theory has been applied (Case and Zweifel 1976). In this theory, the (time-independent) specific light intensity $I^\mathrm{c}(\mathbf{R},\mathbf{k})$ (c means conventional) of a homogeneous medium without absorption obeys the following phenomenological Boltzmann transport equation (Case and Zweifel 1976)

$$\begin{aligned}&(\mathbf{k}\cdot\nabla_\mathbf{R})I^\mathrm{c}(\mathbf{R},\mathbf{k}) + k\sigma(k)I^\mathrm{c}(\mathbf{R},\mathbf{k})\\ &= I_0^\mathrm{c}(\mathbf{R},\mathbf{k}) + \int k'\sigma(\mathbf{k}'\to\mathbf{k})I^\mathrm{c}(\mathbf{R},\mathbf{k}')\frac{\mathrm{d}^3\mathbf{k}'}{(2\pi)^3},\end{aligned} \tag{6.473}$$

where $\sigma(k)$ and $\sigma(\mathbf{k}' \to \mathbf{k})$ are the total and angular scattering coefficients, respectively, that satisfy the relation

$$\int \sigma(\mathbf{k}' \to \mathbf{k}) \frac{\mathrm{d}^3\mathbf{k}'}{(2\pi)^3} = \sigma(k). \tag{6.474}$$

In (6.473), $I_0^{\mathrm{c}}(\mathbf{R}, \mathbf{k})$ is the source specific intensity. Replacing $\mathbf{k}'$ by $\mathbf{k}''$, $I_0^{\mathrm{c}}(\mathbf{R}, \mathbf{k})$ by $(2\pi)^3\delta(\mathbf{R}-\mathbf{R}')\delta(\mathbf{k}-\mathbf{k}')$, and $I^{\mathrm{c}}(\mathbf{R}, \mathbf{k})$ by $\Gamma_{\mathrm{h}}^{\mathrm{c}}(\mathbf{R}-\mathbf{R}'; \mathbf{k}, \mathbf{k}')$, we obtain the equation for the Green function for the specific light intensity. Its Fourier transform $\tilde{\Gamma}_{\mathrm{h}}^{\mathrm{c}}(\mathbf{Q}; \mathbf{k}, \mathbf{k}')$ satisfies the equation

$$\begin{aligned} \mathrm{i}\mathbf{k} \cdot \mathbf{Q}\tilde{\Gamma}_{\mathrm{h}}^{\mathrm{c}}(\mathbf{Q}; \mathbf{k}, \mathbf{k}') &= (2\pi)^3\delta(\mathbf{k} - \mathbf{k}') \\ &+ \int k_1\sigma(\mathbf{k}_1 \to \mathbf{k})\tilde{\Gamma}_{\mathrm{h}}^{\mathrm{c}}(\mathbf{Q}; \mathbf{k}_1, \mathbf{k}') \frac{\mathrm{d}^3\mathbf{k}_1}{(2\pi)^3} - k\sigma(k)\tilde{\Gamma}_{\mathrm{h}}^{\mathrm{c}}(\mathbf{Q}; \mathbf{k}, \mathbf{k}'). \end{aligned} \tag{6.475}$$

From the optical coherence theory, it follows that

$$\begin{aligned} 2\mathbf{k} \cdot \mathbf{Q}\tilde{\Gamma}_{\mathrm{h}}(\mathbf{Q}; \mathbf{k}, \mathbf{k}') &= \Delta G_{\mathbf{k}}(\mathbf{Q})(2\pi)^3\delta(\mathbf{k} - \mathbf{k}') \\ &+ \int \Delta G_{\mathbf{k}}(\mathbf{Q})\tilde{B}_{\mathrm{h}}(\mathbf{k} - \mathbf{k}_1)\tilde{\Gamma}_{\mathrm{h}}(\mathbf{Q}; \mathbf{k}_1, \mathbf{k}') \frac{\mathrm{d}^3\mathbf{k}_1}{(2\pi)^3} \\ &- \int \Delta G_{\mathbf{k}_1}(\mathbf{Q})\tilde{B}_{\mathrm{h}}(\mathbf{k} - \mathbf{k}_1)\frac{\mathrm{d}^3\mathbf{k}_1}{(2\pi)^3}\tilde{\Gamma}_{\mathrm{h}}(\mathbf{Q}; \mathbf{k}, \mathbf{k}'), \end{aligned} \tag{6.476}$$

where

$$\begin{aligned} \Delta G_{\mathbf{k}}(\mathbf{Q}) &= G^{+}\left(\mathbf{k} + \frac{\mathbf{Q}}{2}\right) - G^{-}\left(\mathbf{k} - \frac{\mathbf{Q}}{2}\right), \\ G^{\pm}(\mathbf{k}) &= \frac{1}{k_0^2 - k^2 - \Sigma^{\pm}(\mathbf{k})}, \end{aligned} \tag{6.477}$$

with

$$\Sigma^{\pm}(\mathbf{k}) = \int \frac{\tilde{B}_{\mathrm{h}}(\mathbf{k} - \mathbf{q})}{k_0^2 - q^2 - \Sigma^{\pm}(\mathbf{q})} \frac{\mathrm{d}^3\mathbf{q}}{(2\pi)^3}. \tag{6.478}$$

A comparison of (6.475) with (6.476) may be made in a thorough and expert manner. Even though the coherent backscattering effects have not been included in John et al. (1996), they may be described using the results of MacKintosh and John (1989).

On considering multiple light scattering near an inhomogeneity, the analysis of the propagation and measurement of the Wigner distribution function may enhance the resolving power of optical tomography. A formal analogy between wave propagation in a multiple scattering medium with a statistical inhomogeneity and the quantum mechanical scattering of a particle by a localized potential has been explored.

Bruce and Chalker (1996) have generalized the treatment of quasi-one-dimensional systems due to Dorokhov (1982) and Mello et al. (1988b) for it to include absorption. Interestingly, they recall that it is not easy to obtain transmission properties. They consider a waveguide or optical fibre, along which N modes can propagate in each direction.

Slightly deviating from them, we let $\mathsf{a}(L)$ and $\mathsf{b}(L)$ mean vectors of wave amplitudes ($N \times 1$ matrices) for the right-hand propagation and the left-hand propagation, respectively. Here L stands for a propagation distance. With respect to the disordered medium, the coupled modes are described by stochastic differential equations, which we write in the matrix form

$$\begin{aligned} &\begin{pmatrix} \mathsf{a}(L+\delta l) \\ \mathsf{b}(L+\delta l) \end{pmatrix} - \begin{pmatrix} \mathsf{a}(L) \\ \mathsf{b}(L) \end{pmatrix} \\ &= \left[\mathrm{i}\mu \begin{pmatrix} \mathsf{x} & \mathsf{y} \\ -\mathsf{y}^* & -\mathsf{x}^* \end{pmatrix} - \mu^2 a \begin{pmatrix} 1 & 0 \\ 0 & -1 \end{pmatrix} \right. \\ &\quad \left. - \frac{1}{2}\mu^2 \begin{pmatrix} \mathsf{x} & \mathsf{y} \\ -\mathsf{y}^* & -\mathsf{x}^* \end{pmatrix}^2 \right] \begin{pmatrix} \mathsf{a}(L) \\ \mathsf{b}(L) \end{pmatrix}. \end{aligned} \tag{6.479}$$

Here $\mu = \sqrt{\delta l}$, with $\delta l > 0$ an infinitesimal length, a parametrizes the strength of absorption, 1 is an $N \times N$ unit matrix, $\mathsf{x} \equiv \mathsf{x}_{[L,L+\delta l]}$ is a random $N \times N$ Hermitian matrix, and $\mathsf{y} \equiv \mathsf{y}_{[L,L+\delta l]}$ is a random $N \times N$ symmetric matrix. The two matrices $\mathsf{x}_{[L_1,L_2]}$, $\mathsf{y}_{[L_1,L_2]}$ are statistically independent of the two matrices $\mathsf{x}_{[L_3,L_4]}$, $\mathsf{y}_{[L_3,L_4]}$, when the intervals $[L_1, L_2]$ and $[L_3, L_4]$ do not overlap. Considering again $\mathsf{x} \equiv \mathsf{x}_{[L,L+\delta l]}$ and $\mathsf{y} \equiv \mathsf{y}_{[L,L+\delta l]}$, the elements $x_{\alpha\beta}$, $y_{\alpha\beta}$ are random variables, which are specified in terms of the first and second moments,

$$\begin{aligned} &\langle x_{\alpha\beta} \rangle = \langle y_{\alpha\beta} \rangle = 0, \quad \langle x_{\alpha\beta} y_{\gamma\delta} \rangle = 0, \quad \langle y_{\alpha\beta} y_{\gamma\delta} \rangle = 0, \\ &\langle x_{\alpha\beta} x_{\gamma\delta}^* \rangle = \frac{\delta_{\alpha\gamma}\delta_{\beta\delta}}{N}, \quad \langle y_{\alpha\beta} y_{\gamma\delta}^* \rangle = \frac{\delta_{\alpha\gamma}\delta_{\beta\delta} + \delta_{\alpha\delta}\delta_{\beta\gamma}}{N+1}. \end{aligned} \tag{6.480}$$

A general solution of equations (6.479) has the form

$$\begin{pmatrix} \mathsf{a}(L) \\ \mathsf{b}(L) \end{pmatrix} = \begin{pmatrix} \mathsf{v}_{\mathrm{FFf}}(L|L_0) & \mathsf{v}_{\mathrm{FBf}}(L|L_0) \\ \mathsf{v}_{\mathrm{BFf}}(L|L_0) & \mathsf{v}_{\mathrm{BBf}}(L|L_0) \end{pmatrix} \begin{pmatrix} \mathsf{a}(L_0) \\ \mathsf{b}(L_0) \end{pmatrix}, \tag{6.481}$$

where L_0, $L_0 \leq L$, is another propagation length and subscripts F and B mean forward (to the right) and backward (to the left), cf. Section 6.2.

Now we introduce the matrix

$$\begin{pmatrix} \tau & \rho' \\ \rho & \tau' \end{pmatrix}, \tag{6.482}$$

with

$$\begin{aligned} \tau &= \mathsf{v}_{\mathrm{FFf}}(L_1|L) - \mathsf{v}_{\mathrm{FBf}}(L_1|L)\mathsf{v}_{\mathrm{BBf}}^{-1}(L_1|L)\mathsf{v}_{\mathrm{BFf}}(L_1|L), \\ \rho' &= \mathsf{v}_{\mathrm{FBf}}(L_1|L)\mathsf{v}_{\mathrm{BBf}}^{-1}(L_1|L), \end{aligned}$$

$$\rho = -\mathsf{v}_{\mathrm{BBf}}^{-1}(L_1|L)\mathsf{v}_{\mathrm{BFf}}(L_1|L),$$
$$\tau' = \mathsf{v}_{\mathrm{BBf}}^{-1}(L_1|L), \tag{6.483}$$

where $L_1 \equiv L + \delta l$ and a reflection matrix for waves incident from the right

$$\mathsf{r}'(L) \equiv \mathsf{v}_{\mathrm{FBf}}(L|0)\mathsf{v}_{\mathrm{BBf}}^{-1}(L|0). \tag{6.484}$$

From the stochastic differential equation (6.479), it follows that

$$\begin{pmatrix} \tau & \rho' \\ \rho & \tau' \end{pmatrix} = \begin{pmatrix} 1 & 0 \\ 0 & 1 \end{pmatrix} + \mathrm{i}\mu \begin{pmatrix} \mathsf{x} & \mathsf{y} \\ \mathsf{y}^* & \mathsf{x}^* \end{pmatrix}$$
$$- \mu^2 a \begin{pmatrix} 1 & 0 \\ 0 & 1 \end{pmatrix} - \frac{1}{2}\mu^2 \begin{pmatrix} \mathsf{x} & \mathsf{y} \\ \mathsf{y}^* & \mathsf{x}^* \end{pmatrix}^2 . \tag{6.485}$$

The authors assume the increase of the system length L by δl. As the new reflection matrix r_1' is given by a relatively simple relation

$$\mathsf{r}_1' = \rho' + \tau(\mathsf{r}' + \mathsf{r}'\rho\mathsf{r}' + \ldots)\tau', \tag{6.486}$$

where $\mathsf{r}_1' \equiv \mathsf{r}'(L_1)$, $\mathsf{r}' \equiv \mathsf{r}'(L)$, the authors derive that

$$\mathsf{r}_1' = \mathsf{r}' + \mathrm{i}\mu(\mathsf{y} + \mathsf{x}\mathsf{r}' + \mathsf{r}'\mathsf{x}^* + \mathsf{r}'\mathsf{y}^*\mathsf{r}')$$
$$+ \mu^2 \left[-2a\mathsf{r}' - \frac{1}{2}(\mathsf{y}\mathsf{y}^* + \mathsf{x}^2)\mathsf{r}' - \frac{1}{2}\mathsf{r}'(\mathsf{y}^*\mathsf{y} + \mathsf{x}^{*2}) - \mathsf{x}\mathsf{r}'\mathsf{x}^* \right]. \tag{6.487}$$

On subtracting r' from each side and dividing by μ^2, we obtain a usual idea of a stochastic differential equation. The mathematical notation does not require even the division by μ^2.

It is relatively easy to derive the stochastic differential equation for Λ, the diagonal matrix of eigenvalues Λ_α of the matrix $\mathsf{r}'^\dagger\mathsf{r}'$. The authors introduce λ_α as $\lambda_\alpha = \frac{\Lambda_\alpha}{1-\Lambda_\alpha}$. The joint probability distribution of the set $\{\lambda_\alpha\}$, $W(\lambda_1, \lambda_2, \ldots, \lambda_N, L)$, evolves with the system length L according to the Fokker–Planck equation

$$\frac{\partial W}{\partial L} = \frac{2}{N+1} \sum_{\alpha=1}^{N} \frac{\partial}{\partial \lambda_\alpha} \Biggl\{ \lambda_\alpha(1+\lambda_\alpha)$$
$$\times \left[\sum_{\beta,\beta\neq\alpha} \frac{1}{\lambda_\beta - \lambda_\alpha} W + 2a(N+1)W + \frac{\partial W}{\partial \lambda_\alpha} \right] \Biggr\}. \tag{6.488}$$

This equation has a stationary solution in the limit of long samples ($L \to \infty$). Without absorption, this limit is trivial:

$$r_{mn}' = \sum_k U_{mk} U_{nk}, \tag{6.489}$$

where U_{mn} are elements of the matrix U that has a uniform distribution on the unitary group U(N). They present a result equivalent to relation (6.503) by Beenakker et al. (1996).

Brouwer and Beenakker (1996) have expounded a diagrammatic method for averaging over the circular ensemble of random-matrix theory. The role of the circular ensemble of unitary matrices in the scattering matrix approach has been respected. The method has been modified to the ensemble of uniformly distributed unitary symmetric matrices, which is referred to as the circular orthogonal ensemble. Even though these matrices are of the form $\mathsf{U} = \mathsf{V}\mathsf{V}^T$, with the matrix V uniformly distributed over the unitary group, this efficient method is available. The results have been extended to unitary matrices of quaternions. Brouwer and Beenakker (1996) have applied the method to two types of mesoscopic systems in the condensed-matter physics.

In the article (Beenakker 1997), the author concentrates himself on two types of mesoscopic systems in the condensed-matter theory. In conclusions, he mentions that the propagation of electromagnetic waves through a waveguide is the optical analogue of conduction through a wire. In the problem of localization by disorder, the analogy is incomplete. The (relative) dielectric constant $\epsilon(x, y)$ not only fluctuates around unity but is always positive. Potentials V greater than the Fermi energy have no optical analogue.

One new aspect of the optical problem is the behaviour in the case that the dielectric constant has a nonzero imaginary part. The intensity of radiation which has propagated over a distance L is multiplied by a factor $\exp(\sigma L)$, with σ negative (positive) for absorption (amplification). Here the growth rate σ is related to the dielectric constant by the relation $\sigma = -2k \operatorname{Im} \sqrt{\epsilon}$. The Dorokhov–Mello–Pereyra–Kumar equation, which applies to the conduction through a wire, is accordingly generalized. Another new aspect of the optical problem is the frequency dependence of the term $\left(\frac{\omega}{c}\right)^2 \epsilon$ ($k_0^2\epsilon$) in the Helmholtz equation, whereas an energy-dependent potential does not occur in the mesoscopic systems.

van Rossum and Nieuwenhuizen (1999) have provided a discussion of the propagation of waves in random media. The description of radiation transport respects three length scales: macroscopic, mesoscopic, and microscopic. The diffusion theory presents the diffusion approximation at the macroscopic level. Important corrections are calculated with the radiative transfer equation, which describes intensity transport at the mesoscopic level and is derived from the microscopic wave equation. A precise treatment of the diffuse intensity includes the effects of boundary layers. Situations such as the enhanced backscattering cone and imaging of objects in opaque media are also discussed in the cited work. Mesoscopic correlations between multiple scattered intensities are introduced. The correlation functions and intensity distribution functions are derived.

Guo (2002) has modelled the propagation of a plane scalar wave through a uniform dielectric slab using the multiple scattering method. This approach can be used to model light propagation in stratified media, which represents also one-dimensional photonic crystals. The multiple scattering method can easily be generalized to treat light propagation in nonuniform media, such as light propagation in random media.

The scalar approximation is based on the assumption that the transverse electric waves are propagated. An incident plane wave, $E_{\mathrm{inc}}(\mathbf{r}) = e^{i\mathbf{k}_0\cdot\mathbf{r}}$, with a wave vector $\mathbf{k}_0$, may be a component of an incident pulse. The corresponding total field $E(\mathbf{r})$ may be a component of the corresponding diffracted, transmitted, and reflected pulses. The total field $E(\mathbf{r})$ obeys the following integral equation (in Gaussian units),

$$E(\mathbf{r}) = E_{\mathrm{inc}}(\mathbf{r}) + 4\pi k_0^2 \int_\sigma G_0(\mathbf{r}, \mathbf{r}_1)\chi(\mathbf{r}_1)E(\mathbf{r}_1)\,\mathrm{d}^3\mathbf{r}_1, \tag{6.490}$$

where

$$G_0(\mathbf{r}, \mathbf{r}_1) = \frac{1}{4\pi|\mathbf{r}-\mathbf{r}_1|}e^{ik_0|\mathbf{r}-\mathbf{r}_1|}. \tag{6.491}$$

The subscript σ indicates exclusion of a small volume surrounding the position at $\mathbf{r}$.

Guo (2002) has assumed that the slab is formed by uniformly distributed dipoles. He has concentrated on the resonant scattering case, where the complex electric permittivity is pure imaginary and energy is lost.

In mathematics, random motions in random media are treated (Bolthausen and Sznitman 2002). Lectures are restricted to discrete models. Then a one-dimensional model of diffusion in a constant medium is the nearest neighbour random walk x_n, x_n assuming integer values. Here are two ways to introduce the medium randomness in a simple random walk.

(i) The probabilities of jumping to the right neighbour are chosen as independent identically distributed random variables $p(x)$, $0 \le p(x) \le 1$, x being an integer.

(ii) The probabilities of jumping to the right neighbour are given by the relations

$$p(x) = \frac{c_{x,x+1}}{c_{x-1,x} + c_{x,x+1}}, \tag{6.492}$$

where $c_{x,x+1}$ are independent identically distributed random variables, $c_{x,x+1} > 0$. In disordered media physics, rather, this model occurs (Bolthausen and Sznitman 2002, Hughes 1996).

6.5.2 Incoherent and Coherent Random Lasers

Gouedard et al. (1993) have developed mirrorless light sources based on heavily doped neodymium materials pumped by nanosecond laser pulses. These sources generate quasimonochromatic short pulses and present characteristics of spatial and temporal incoherence. Such devices may find applications in domains such as holography, transport of energy in fibres for medical applications, and laser inertial confinement fusion. A result of their study is the existence of a threshold below which amplified spontaneous emission occurred in one of two compounds mentioned by the authors. Above the threshold the specific behaviour has been found. This behaviour was reported by Ter-Gabrielyan et al. (1991).

The authors have considered two schemes of the origin:

1. Many different microcrystallites are lasing sequentially in very short pulses.
2. The grains of the powder emit collectively due to distributed feedback provided by multiple scattering.

Scheme 2 may be related to the photon localization effect, but this has not been proved or disproved in Gouedard et al. (1993). Second-harmonic generation in strongly scattering media has been investigated by Kravtsov et al. (1991).

Lawandy et al. (1994) have opined that composite systems may have spectral and temporal properties characteristic of a multimode laser oscillator, even though these systems do not comprise an external cavity. They have investigated a laser dye dispersed in a strongly scattering medium. This medium was a colloidal suspension of titanium dioxide particles. Colloidal solutions were optically pumped by linearly polarized 532-nm radiation. Either single 7-ns-long pulses or a 125-ns-long train of nine 80-ps-long pulses were used.

Most experiments were performed using the long pulses. The presence of the TiO_2 nanoparticles led first to a larger emission linewidth, but when the energy of the excitation pulses was increased, the emission at 617 nm grew rapidly and the line narrowed. The emission at the peak wavelength was studied as a function of the pump pulse energy for four different TiO_2 nanoparticle densities. A threshold of the pump energy has been observed. This threshold is obvious in plots of emission linewidth as a function of the pump pulse energy when the TiO_2 nanoparticle concentration is varied. Excitation with a train of 80-ps pulses demonstrated a threshold behaviour in the temporal characteristics of the colloid. When the pump pulse energy was increased beyond the threshold, the response was a sharp peak.

The authors also concluded that a theory for this process did not exist. The literature cited by them required that every dimension of the sample be large compared to the optical scattering length.

Sha et al. (1994) performed similar experiments. Single 3-ns-long laser pulses were used. A series of spectral experiments were carried out with a density of 5×10^{11} cm^{-3} of TiO_2 nanoparticles. When the pump pulse energy was varied, a threshold at 620 nm and a possible one at 650 nm were demonstrated.

The highest dye concentration exhibited a reduction of the lasing threshold, when the density of scattering particles was increased. For high particle density from 5×10^{11} to 2.5×10^{12} cm^{-3} the lowest dye concentration revealed the threshold of 0.17 mJ, higher than 0.07 mJ for this concentration in the neat solvent. Temporal characteristics of the response have been investigated as well. Above the threshold, the bandwidth of emission and the temporal width of the emitted pulses are narrowed. Also these authors stated that the exact mechanism for this process had not yet been explained.

Wiersma et al. (1995a) declared that Lawandy et al. (1994) had prepared only amplified spontaneous emission. For the pump geometry used the amplified spontaneous emission proceeds perpendicular to the propagation direction of the pump pulse (side emission). It can be achieved that the amplification is negligible parallel to this direction. The paper under criticism does not record the side emission.

Addition of the TiO_2 particles brings about scattering or the detection of some amplified spontaneous emission light.

Lawandy and Balachandran (1995) have mainly presented data that clearly show that for a fixed pump pulse energy there exists a threshold scatterer concentration for both side and front emissions. For the front emission it can be demonstrated that the threshold pump pulse energy decreases with increasing particle density.

Wiersma et al. (1995b) performed coherent backscattering measurements from amplifying random media. They used optically pumped Ti:sapphire powders. They have found that the light intensity as a function of the angular distance from the exact backscattering direction exhibits a top, which sharpens with increasing gain. They have solved the stationary diffusion equation with gain (Davison and Sykes 1958, Letokhov 1968) and have used an approach by Akkermans et al. (1986) to obtain the coherent component of the backscattered intensity.

Beenakker et al. (1996) have recognized the notion of a random laser. Letokhov (1968) called it a "laser with incoherent feedback", but randomness admits the "last coherence effect that survives" (Akkermans et al. 1986). It is appropriate when the model includes an illuminated area, S. The authors associate the number of modes $N \simeq \frac{S}{\lambda^2}$ with it, where λ is a wavelength of the incident light. The authors were in fact motivated by a paper of Pradhan and Kumar (1994) on the case $N = 1$ and have generalized it.

The reflection of a monochromatic plane wave (frequency ω, wavelength λ) by a slab (thickness L, transverse dimension W) is considered, which represents a disordered medium (mean free path l). This medium either amplifies or absorbs the radiation. For a statistical description an ensemble of slabs with different configurations of scatterers is considered. The authors let σ mean the amplification per unit length. A negative value of σ indicates absorption. The parameter $\gamma = \sigma l$ is the amplification per mean free path. The treatment is limited to scalar waves.

It is assumed that the slab is embedded in an optically passive waveguide without disorder. We let N be the number of modes which can propagate in the waveguide at frequency ω. The modal functions are normalized such that each mode carries unit power. The $N \times N$ reflection matrix r has the elements r_{mn}, r_{mn} means the amplitude of a wave reflected into mode m from an incident mode n.

The matrix r is symmetric, $\mathsf{r}^T = \mathsf{r}$, and its singular-value decomposition is a product of U, the diagonal form of the matrix r, and U^T. Here U is a unitary matrix, which has elements U_{mn} (Mello et al. 1988a). It is specific that they assume the reflection eigenvalues to be nonnegative. They let $\sqrt{R_n}$, $n = 1, 2, \ldots, N$ mean the singular values. Then

$$r_{mn} = \sum_k U_{mk} U_{nk}^{(*)} \sqrt{R_k}. \tag{6.493}$$

The difference between the symmetric matrix r and the Hermitian one r is signed by the omission and the use of an asterisk, respectively.

The unit power of the mode n is amplified (or reduced) to the value

$$a_n = \sum_m |r_{mn}|^2. \tag{6.494}$$

The statistical calculation uses the assumption that $W \ll L$ and U is uniformly distributed in the unitary group. As a consequence, a_n has a distribution independent of n. Further it is assumed that $\lambda \ll l$, $\lambda \ll \frac{1}{\sigma}$. On introducing

$$\mu_n = \frac{1}{R_n - 1}, \tag{6.495}$$

the distribution $P(\mu_1, \mu_2, \ldots, \mu_N, L)$ obeys the Fokker–Planck equation

$$\begin{aligned} l\frac{\partial P}{\partial L} &= \frac{2}{N+1}\sum_{i=1}^{N}\frac{\partial}{\partial \mu_i}\left\{\mu_i(1+\mu_i)\left[\sum_{j,j\neq i}\frac{1}{\mu_j - \mu_i} + \gamma(N+1)\right]P\right\} \\ &\quad + \frac{2}{N+1}\sum_{i=1}^{N}\frac{\partial}{\partial \mu_i}\left[\mu_i(1+\mu_i)\frac{\partial P}{\partial \mu_i}\right], \end{aligned} \tag{6.496}$$

with the initial condition $P|_{L=0} = N\prod_i \delta(\mu_i + 1)$.

On introducing the notation $\bar{a} \equiv \langle a_n \rangle$, $\operatorname{var} a \equiv \langle (a_n - \bar{a})^2 \rangle$, it can be derived that

$$l\frac{\mathrm{d}}{\mathrm{d}L}\bar{a} = (\bar{a} - 1)^2 + 2\gamma\bar{a}, \tag{6.497}$$

$$l\frac{\mathrm{d}}{\mathrm{d}L}\operatorname{var} a = 4(\bar{a} - 1 + \gamma)\operatorname{var} a + \frac{2}{N}\bar{a}(\bar{a} - 1)^2. \tag{6.498}$$

Equation (6.498) has been derived for large N. Equation (6.497) is in agreement with Selden (1974). The initial conditions for Equations (6.497) and (6.498) are $\bar{a}(0)=0$, $\operatorname{var} a(0)=0$, respectively. In the case of absorption ($\gamma < 0$) and in the limit $L \to \infty$ the solution, which we do not quote here, yields

$$\bar{a}_\infty = 1 - \gamma - \sqrt{\gamma^2 - 2\gamma}, \tag{6.499}$$

$$\operatorname{var} a_\infty = \frac{1}{2N}\frac{\bar{a}_\infty(1 - \bar{a}_\infty)^2}{1 - \gamma - \bar{a}_\infty}, \tag{6.500}$$

which is a stationary solution of equations (6.497) and (6.498) obviously.

In the case of amplification ($\gamma > 0$), the condition $L < L_c$ must be fulfilled, where L_c is a critical length,

$$L_c = l\frac{\arccos(\gamma - 1)}{\sqrt{2\gamma - \gamma^2}}. \tag{6.501}$$

At this length $\bar{a}$ and $\operatorname{var} a$ diverge. It does not imply that a probability distribution of a_n independent of n, which has the density

$$P(a, L) = \left\langle \delta\left(a - 1 - \sum_k U_{nk}U_{nk}^*\frac{1}{\mu_k}\right)\right\rangle, \tag{6.502}$$

does not exist for $L \geq L_c$. It can be characterized by a modal value $a_{\max}$.

The stationary solution of equation (6.496) is the Laguerre ensemble (Bronk 1965)

$$P(\mu_1, \mu_2, \ldots, \mu_N, \infty) = C \prod_i \exp[-\gamma(N+1)\mu_i] \prod_{i<j} |\mu_j - \mu_i|, \tag{6.503}$$

where C is a normalization constant. The density

$$\rho(\mu) = \left\langle \sum_i \delta(\mu - \mu_i) \right\rangle \tag{6.504}$$

is introduced, which in the large-N limit yields

$$\rho(\mu) = \frac{N}{\pi}\sqrt{\frac{2\gamma}{\mu} - \gamma^2}, \quad 0 < \mu < \frac{2}{\gamma}. \tag{6.505}$$

The average in (6.502) consists in the average of U_{nk}'s over the unitary group followed by the average of the μ_k's over the Laguerre ensemble (6.503). The pertinent result is known (Dyson 1962), even though only for large N, and it is an inverse Laplace transform. It has been found that the modal value of the distribution of a_n independent of n is $a_{\max}$=1 + $0.8\gamma N$. Predictions of random-matrix theory have been compared with numerical simulations of the analogous electronic Anderson model with a complex scattering potential.

John and Pang (1996) have determined the emission intensity properties of a model dye system, which is immersed in a multiply scattering medium with transport mean free path l^*. Since they considered a rhodamine 640 dye solution based on the literature, they respected the emission at 620 and 650 nm. They assumed that the dye solution with scattering titanium particles fills the whole sample region between the two planes $z = 0$ and $z = L$. They have used a generic dye laser scheme (Svelto and Hanna 1977, 1989) that explains bichromatic emission from the singlet states and the triplet states. The description comprises laser rate equations for the singlet states and those for the triplet states. The intensity of the pumping beam is of the form $I_{\text{inc}} \exp(\frac{-z}{l_z})$, where I_{inc} is the intensity at $z = 0$ and $l_z = \sqrt{l^* \frac{l_a}{3}}$, with l_a the absorption length. Those equations are completed with a diffusion equation for the photon flux. A nonlinear diffusion equation for a dimensionless intensity is presented, from which populations of dye molecules in singlet and triplet states are eliminated.

The emission spectra at nine different pump intensities agree with experiments (Sha et al. 1994). The linewidth and emission intensity at 620 nm as a function of pump intensity for different values of transport mean free path l^* are consistent with observations (Balachandran and Lawandy 1995, Lawandy et al. 1994).

Wiersma and Lagendijk (1996) have confirmed that they set high standards upon random lasers. In their paper they have presented calculations on light diffusion with amplification and have reported on experiments.

In a random medium light is multiply scattered. The relevant length scales that describe the scattering process are the scattering mean free path l_s defined as the average distance between two scattering events and transport mean free path l defined as the average distance a wave travels before its propagation direction randomizes. In an amplifying random medium it is necessary to define two more length scales: the gain length l_g and amplification length l_{amp}. The gain length is defined as the path length over which the intensity is multiplied by a factor $e = \exp(1)$. The amplification length is defined as the (rms) distance between the beginning and ending points for paths of length l_g, $l_{amp} = \sqrt{\frac{ll_g}{3}}$. A sample of an amplifying random medium in the form of a slab has been studied. Light and the amplifying medium become unstable if the thickness L is above the critical thickness L_{cr}, $L_{cr} = \pi l_{amp}$.

Wiersma and Lagendijk (1996) have assumed the laser material to be a four-level system. They have considered an incident pump and probe pulse and spontaneous emission. They have described their system with coupled differential equations. The set is formed by three diffusion equations and the rate equation for the concentration $N_1(\mathbf{r}, t)$ of laser particles in a metastable state. The first two diffusion equations are linear relative to the energy densities of the pump light and the probe light. External fields in those equations are the intensities of the incoming pump and probe pulses, respectively.

The boundary condition is the vanishing of the energy densities outside the slab at the distance $z_0 \approx 0.7l$ from the surfaces of the slab. The authors have been able to calculate the outgoing flux at either the front or rear surface of the slab, which is determined by the gradient of the energy density at the sample surface. Experiments have demonstrated that it is possible to realize an amplifying random medium.

Wiersma and Lagendijk (1996) have distinguished three regimes depending on the amount of scattering.

1. *Weak scattering and gain*. If l is of the order of the sample size, one says that the scattering is very weak. Addition of scatterers lifts a directionality in the output of the system, cf. (Wiersma et al. 1995a). It is assumed that l continues to be of the order of the sample size.
2. *Modest scattering and gain*. If $\lambda \ll l \ll L$, one says that the scattering is temperate. The calculations have shown that modest scattering with gain can lead to a pulsed output, cf. (Gouedard et al. 1993).
3. *Strong scattering and gain*. If l is smaller than or equal to the wavelength λ, one says that the scattering is strong. In this regime the Anderson localization of light is expected to occur. Around 1996 there was no experimental evidence for the photonic Anderson localization.

Paasschens et al. (1996) have studied the propagation of radiation through a disordered waveguide with a complex dielectric constant ϵ. They have called the systems dual, which differ only in the sign of the imaginary part of ϵ. In the case of the scattering matrix

$$\mathsf{S} = \begin{pmatrix} \mathsf{r} & \mathsf{t}' \\ \mathsf{t} & \mathsf{r}' \end{pmatrix}, \tag{6.506}$$

they have introduced transmittances T, T' and reflectances R, R'

$$T = \frac{1}{N}\,\mathrm{Tr}\{\mathsf{t}\mathsf{t}^\dagger\},\quad R = \frac{1}{N}\,\mathrm{Tr}\{\mathsf{r}\mathsf{r}^\dagger\},$$
$$T' = \frac{1}{N}\,\mathrm{Tr}\{\mathsf{t}'\mathsf{t}'^\dagger\},\quad R' = \frac{1}{N}\,\mathrm{Tr}\{\mathsf{r}'\mathsf{r}'^\dagger\}. \tag{6.507}$$

They have let $\mathcal{T}_n$ mean the eigenvalues of the matrix $\mathsf{t}\mathsf{t}^\dagger$ (all of them depend on L) and have introduced their localization lengths ξ_n with the properties

$$\frac{1}{\xi_n} = -\lim_{L\to\infty}\frac{1}{L}\ln[\mathcal{T}_n(L)]. \tag{6.508}$$

The decay length ξ is defined as $\xi = \max(\xi_1, \xi_2, \ldots, \xi_N)$. For $L \to \infty$ it holds that $\xi(\sigma) = \xi(-\sigma)$, where a dependence on σ is indicated, $\sigma = -2k\,\mathrm{Im}\sqrt{1+\mathrm{i}\,\mathrm{Im}\,\epsilon}$, with k the free-space wave number of the radiation.

In the case $N = 1$, the authors have presented also the Fokker–Planck equation for the joint probability distribution $P(R, T, L)$ of the reflectance R and transmittance T. They have included the familiar relation for $P(R, L = \infty)$ and have introduced $\gamma = \sigma l$.

Since $\langle T\rangle$ diverges for $\gamma > 0$ at the lasing threshold $L_{\mathrm{c}} \approx \frac{l_{\mathrm{c}}(\gamma)}{|\gamma|}$, where

$$c(\gamma) = \mathbf{C} + (\ln 2)\gamma - e^{2\gamma}\,\mathrm{Ei}(-2\gamma), \tag{6.509}$$

where $\mathbf{C}$ is Euler's constant and $\mathrm{Ei}(x) = \int_{-\infty}^{x}\frac{e^t}{t}\,\mathrm{d}t$ is the exponential integral, $\langle\ln T\rangle$ $(N \geq 1)$ and $\mathrm{var}\ln T(N = 1)$ are studied. The sum $\langle\ln T\rangle + \frac{L}{\xi_0}$, where $\xi_0 = (N+1)\frac{l}{2}$ is the localization length for $\sigma = 0$, is mostly negative in consequence of the approximate equality $\langle\ln T\rangle + \frac{L}{\xi} \approx 0$ and the observation that $\xi \leq \xi_0$.

Beenakker (1998) has generalized results concerning amplified spontaneous emission from a random medium for them to include thermal radiation. He has applied the method of random-matrix theory (Mehta 1991) to quantum optics. He thinks of a linear amplifier as a system in thermal equilibrium at a negative temperature (Jeffers 1993, Matloob et al. 1997).

It is assumed that radiation, maybe no photons, comes into a random medium via an N-mode waveguide. The radiation is transformed in the random medium and goes out from it via the same waveguide. Annihilation operators $\tilde{\hat{a}}_n^{\mathrm{in}}(\omega)$, $\tilde{\hat{a}}_n^{\mathrm{out}}(\omega)$, $\tilde{\hat{b}}_n(\omega)$, $\tilde{\hat{c}}_n(\omega)$, $n = 1, 2, \ldots, N(\omega)$, are introduced. They satisfy the commutation relations

$$[\tilde{\hat{a}}_n(\omega), \tilde{\hat{a}}_m^\dagger(\omega')] = \delta_{nm}\delta(\omega-\omega')\hat{1},\quad [\tilde{\hat{a}}_n(\omega), \tilde{\hat{a}}_m(\omega')] = \hat{0}, \tag{6.510}$$

for $\tilde{\hat{a}} = \tilde{\hat{a}}^{\mathrm{in}}, \tilde{\hat{a}}^{\mathrm{out}}, \tilde{\hat{b}}, \tilde{\hat{c}}$. The operators $\tilde{\hat{a}}_n^{\mathrm{in}}(\omega)$ are related to the incoming modes and the operators $\tilde{\hat{a}}_n^{\mathrm{out}}(\omega)$ describe the outgoing modes, $\tilde{\hat{b}}_n(\omega)$ and $\tilde{\hat{c}}_n(\omega)$ are quantum noises for absorption and amplification, respectively.

On introducing

$$\tilde{\mathsf{a}}^{\mathrm{in}}(\omega) = \begin{pmatrix} \tilde{a}_1^{\mathrm{in}}(\omega) \\ \vdots \\ \tilde{a}_N^{\mathrm{in}}(\omega) \end{pmatrix}, \; \tilde{\mathsf{a}}^{\mathrm{out}}(\omega) = \begin{pmatrix} \tilde{a}_1^{\mathrm{out}}(\omega) \\ \vdots \\ \tilde{a}_N^{\mathrm{out}}(\omega) \end{pmatrix},$$
$$\tilde{\mathsf{b}}(\omega) = \begin{pmatrix} \tilde{b}_1(\omega) \\ \vdots \\ \tilde{b}_N(\omega) \end{pmatrix}, \; \tilde{\mathsf{c}}^{\dagger}(\omega) = \begin{pmatrix} \tilde{c}_1^{\dagger}(\omega) \\ \vdots \\ \tilde{c}_N^{\dagger}(\omega) \end{pmatrix}, \tag{6.511}$$

the input–output relations take the form

$$\tilde{\mathsf{a}}^{\mathrm{out}}(\omega) = \mathsf{S}\tilde{\mathsf{a}}^{\mathrm{in}}(\omega) + \mathsf{Q}\tilde{\mathsf{b}}(\omega) + \mathsf{V}\tilde{\mathsf{c}}^{\dagger}(\omega), \tag{6.512}$$

where S, Q, V are matrices, which satisfy the conditions

$$\mathsf{Q}\mathsf{Q}^{\dagger} - \mathsf{V}\mathsf{V}^{\dagger} = \mathsf{1} - \mathsf{S}\mathsf{S}^{\dagger}, \tag{6.513}$$

where $\mathsf{1}$ is the unit matrix. Besides zero mean values, the quantum noises have the properties

$$\langle \tilde{b}_n^{\dagger}(\omega)\tilde{b}_m(\omega')\rangle = \delta_{nm}\delta(\omega-\omega')f(\omega,T),$$
$$\langle \tilde{c}_n^{\dagger}(\omega)\tilde{c}_m(\omega')\rangle = \delta_{nm}\delta(\omega-\omega')f(\omega,T), \tag{6.514}$$

where T is the temperature and

$$f(\omega,T) = \frac{1}{\exp\left(\frac{\hbar\omega}{k_{\mathrm{B}}T}\right) - 1}. \tag{6.515}$$

The negative temperature is obtained according to the relation

$$-[f(\omega,T)+1] = f(\omega,-T). \tag{6.516}$$

The author also mentions the photodetection theory: The probability that n photons are counted in a time t is given by the relation

$$p(n) = \frac{1}{n!}\langle : \hat{W}^n \exp(-\hat{W}) : \rangle, \tag{6.517}$$

where : : means the normal ordering of the operators inside and

$$\hat{W}(t) = \int_0^t \hat{\mathsf{a}}^{\mathrm{out}\dagger}(t')\hat{\mathsf{a}}^{\mathrm{out}}(t')\,\mathrm{d}t', \tag{6.518}$$

with

$$\hat{\mathsf{a}}^{\mathrm{out}}(t) = \frac{1}{\sqrt{2\pi}}\int_0^{\infty} e^{-\mathrm{i}\omega t}\tilde{\mathsf{a}}^{\mathrm{out}}(\omega)\,\mathrm{d}\omega. \tag{6.519}$$

The generating function $F(\xi)$ of factorial cumulants

$$F(\xi) = \ln\left[\sum_{n=0}^{\infty}(1+\xi)^n p(n)\right] = \ln\langle: \exp[\xi \hat{W}(t)] :\rangle \tag{6.520}$$

is introduced. Here factorial means that the connection of usual cumulants with moments of the photon–number distribution is applied to the factorial moments of the distribution $p(n)$ and $\hat{W}(t)$ stands for the integrated intensity. The factorial cumulants are

$$\kappa_p = \left.\frac{\mathrm{d}^p F(\xi)}{\mathrm{d}\xi^p}\right|_{\xi=0}, \quad p = 1, 2, \ldots, \infty. \tag{6.521}$$

If $\omega_c t \gg 1$, where ω_c is the frequency interval within which $\mathrm{SS}^\dagger$ does not vary significantly, then

$$F(\xi) = -t\int_0^{\infty} \ln\left\{\det[1-(1-\mathrm{SS}^\dagger)\xi f(\omega, T)]\right\}\frac{\mathrm{d}\omega}{2\pi}, \tag{6.522}$$

where det indicates the determinant. If $\Omega_c t \ll 1$, where Ω_c is the frequency range over which $\mathrm{SS}^\dagger$ differs appreciably from the unit matrix, then

$$F(\xi) = -\ln\left\{\det\left[1 - t\int_0^{\infty}(1-\mathrm{SS}^\dagger)\xi f(\omega, T)\right]\frac{\mathrm{d}\omega}{2\pi}\right\}. \tag{6.523}$$

The long-time limit depends only on the set of eigenvalues $\sigma_1, \sigma_2, \ldots, \sigma_N$ of $\mathrm{SS}^\dagger$ (Beenakker 1998). These eigenvalues are called scattering strengths. As a frequency-resolved measurement leads to

$$\begin{aligned} F(\xi) &= -\frac{t\delta\omega}{2\pi}\sum_{n=1}^{\infty}\ln[1-(1-\sigma_n)\xi f(\omega, T)] \\ &= -\frac{t\delta\omega}{2\pi}\ln\left\{\det\left[1-(1-\mathrm{SS}^\dagger)\xi f(\omega, T)\right]\right\}, \end{aligned} \tag{6.524}$$

where $\delta\omega$ is a frequency interval, the factorial cumulants are

$$\kappa_p = (p-1)!\frac{t\delta\omega}{2\pi}[f(\omega, T)]^p \sum_{n=1}^{N}(1-\sigma_n)^p, \quad p = 1, 2, \ldots, \infty. \tag{6.525}$$

Particularly,

$$\begin{aligned} \bar{n} = \kappa_1 &= \frac{t\delta\omega}{2\pi} f(\omega, T)\sum_{n=1}^{N}(1-\sigma_n), \\ \operatorname{var} n = \kappa_2 &= \frac{t\delta\omega}{2\pi}[f(\omega, T)]^2\sum_{n=1}^{N}(1-\sigma_n)^2. \end{aligned} \tag{6.526}$$

But the blackbody radiation has

$$\bar{n} = \nu f, \quad \operatorname{var} n = \nu f(\omega, T)[1 + f(\omega, T)], \tag{6.527}$$

where

$$\nu = N \frac{t \delta\omega}{2\pi}. \tag{6.528}$$

It has the property

$$\operatorname{var} n = \bar{n} + \frac{1}{\nu}\bar{n}^2, \tag{6.529}$$

whence it is advantageous to introduce

$$\nu_{\mathrm{eff}} = \frac{\bar{n}^2}{\operatorname{var} n - \bar{n}}, \tag{6.530}$$

an effective number of degrees of freedom. Hence,

$$\nu_{\mathrm{eff}} = \nu \frac{\left[\sum_{n=1}^{N}(1-\sigma_n)\right]^2}{N \sum_{n=1}^{N}\left[(1-\sigma_n)^2 - \frac{1}{f(\omega,T)}\sum_{n=1}^{N}(1-\sigma_n)\right]}. \tag{6.531}$$

As $f(\omega, T) \to \infty$ for $T \to \infty$, relation

$$\nu_{\mathrm{eff}} = \nu \frac{\left[\sum_{n=1}^{N}(1-\sigma_n)\right]^2}{N \sum_{n=1}^{N}(1-\sigma_n)^2} \tag{6.532}$$

is obtained for $T \to \infty$. Turning to applications, Beenakker (1998) concentrates on the long-time regime with $N \gg 1$. For random media, a scattering-strength density $\rho(\sigma)$ is considered. Relations (6.525) and (6.532) are written using

$$\sum_{n=1}^{N} M(\sigma_n) \to \int \rho(\sigma) M(\sigma)\, \mathrm{d}\sigma, \tag{6.533}$$

where $M(\sigma_n)$=$(1-\sigma_n)^p$.

Beenakker (1998) presents further results for a semi-infinite random medium. Let τ_{s} denote the inverse scattering rate and τ_{a} the inverse absorption or amplification rate. Let us introduce $\gamma = \frac{16}{3}\frac{\tau_{\mathrm{s}}}{\tau_{\mathrm{a}}}$. In the regime $\gamma N^2 \gg 1$,

$$\rho(\sigma) = \frac{N}{\pi}\sqrt{\gamma}\frac{1}{(1-\sigma)^2}\sqrt{\frac{1}{\sigma} - 1 - \frac{\gamma}{4}}, \quad \text{for} \quad 0 < \sigma < \frac{1}{1+\frac{\gamma}{2}}, \tag{6.534}$$

$\rho(\sigma) = 0$ elsewhere. From this,

$$\bar{n} = \frac{1}{2}\nu f(\omega, T)\gamma\left(\sqrt{1 + \frac{4}{\gamma}} - 1\right). \tag{6.535}$$

Beenakker (1998) has arrived at the following conclusions. For strong absorption, $\gamma \gg 1$, $\nu_{\mathrm{eff}} = \nu$ is obtained as for the blackbody radiation. For weak absorption,

$\gamma \ll 1$, it is found that $\nu_{\text{eff}} = 2\nu\sqrt{\gamma}$. It has been recognized that it is Glauber's result for the Lorentzian spectrum (Glauber 1963).

Specific results are added for an optical cavity coupled to a photodetector via an N-mode waveguide. Here $N \gg 1$ is assumed and the modes overlap. Let us introduce $\gamma = \frac{\tau_{\text{dwell}}}{\tau_{\text{a}}}$, where τ_{dwell} is the mean dwell time of a photon in the cavity without absorption. In the limit of weak absorption, $\gamma \ll 1$,

$$\rho(\sigma) = \frac{N}{2\pi}\frac{1}{(1-\sigma)^2}\sqrt{(\sigma-\sigma_-)(\sigma_+-\sigma)} \quad \text{for} \quad \sigma_- < \sigma < \sigma_+, \tag{6.536}$$

where $\sigma_\pm = 1 - 3\gamma \pm 2\gamma\sqrt{2}$. In the limit of strong absorption, $\gamma \gg 1$, relation (6.534) holds. From this,

$$\bar{n} = \nu f \frac{\gamma}{1+\gamma}. \tag{6.537}$$

It holds that $\nu_{\text{eff}} = \nu$ for $\gamma \gg 1$. For $\gamma \ll 1$, we find that $\nu_{\text{eff}} = \frac{\nu}{2}$, which is finite (nonvanishing) for $\gamma \to 0$.

The general formulae can also be applied to amplified spontaneous emission. Beenakker (1998) investigates only the random laser below the laser threshold. The semi-infinite medium is above the laser threshold for an arbitrarily small amplification, but the cavity is below the threshold, as long as $\gamma < 1$.

Cao et al. (1999) observed random laser action with coherent feedback in semiconductor powder. They found that the scattering mean free path is less than the emission wavelength. A comparison with the random laser theory (Wiersma and Lagendijk 1996, Zhang 1995) was realized. The laser emission from the powder could be observed in all directions. Their work is very different from the work on powder laser (Markushev et al. 1986, Ter-Gabriėlyan et al. 1991). When the particle size was much larger than the wavelength, a single particle could serve as a resonator. When the particle size was less than this wavelength, a single particle was too small to serve as a laser resonator. Laser resonators were formed by recurrent light scattering.

Patra and Beenakker (1999) have continued the study of an amplifying disordered cavity (Beenakker 1998). Besides the cavity they have investigated an amplifying disordered waveguide. The disordered medium is illuminated by monochromatic radiation of a single propagating mode in a coherent state.

First they consider an amplifying disordered medium embedded in a waveguide that supports $N(\omega)$ propagating modes at frequency ω. The incoming radiation in mode n is described by an annihilation operator $\tilde{\hat{a}}_n^{\text{in}}(\omega)$, where $n = 1, 2, \ldots, N$ for a mode on the left-hand side of the medium and $n = N+1, N+2, \ldots, 2N$ for a mode on the right-hand side. The outgoing radiation in mode n is described by an annihilation operator $\tilde{\hat{a}}_n^{\text{out}}(\omega)$, where again $n = 1, 2, \ldots, N$ for a mode on the left-hand side of the medium and $n = N+1, N+2, \ldots, 2N$ for a mode on the right-hand side. These two sets of operators are connected by the input–output relations

$$\tilde{\hat{\mathsf{a}}}^{\text{out}}(\omega) = \mathsf{S}(\omega)\tilde{\hat{\mathsf{a}}}^{\text{in}}(\omega) + \mathsf{V}(\omega)\tilde{\hat{\mathsf{c}}}^{\dagger}(\omega). \tag{6.538}$$

Here $\mathsf{S}(\omega)$ is a $2N \times 2N$ scattering matrix, $\mathsf{V}(\omega)$ is a $2N \times 2N$ matrix, and $\tilde{\hat{\mathsf{c}}}^\dagger(\omega)$ is a vector of $2N$ creation operators $\tilde{\hat{c}}_1^\dagger(\omega)$, $\tilde{\hat{c}}_2^\dagger(\omega)$,..., $\tilde{\hat{c}}_{2N}^\dagger(\omega)$. The scattering matrix S has the form

$$\mathsf{S}(\omega) = \begin{pmatrix} \mathsf{r}' & \mathsf{t}' \\ \mathsf{t} & \mathsf{r} \end{pmatrix}, \tag{6.539}$$

where $\mathsf{r} \equiv \mathsf{r}(\omega)$ and $\mathsf{r}' \equiv \mathsf{r}'(\omega)$ are $N \times N$ reflection matrices and $\mathsf{t} \equiv \mathsf{t}(\omega)$ and $\mathsf{t}' \equiv \mathsf{t}'(\omega)$ are $N \times N$ transmission matrices. In the case under consideration $\mathsf{t}' = \mathsf{t}^T$, $\mathsf{r} = \mathsf{r}^T$, and $\mathsf{r}' = \mathsf{r}'^T$.

It is assumed that the operators $\tilde{\hat{c}}_n(\omega)$ and $\tilde{\hat{c}}_n^\dagger(\omega)$ commute with the operators $\tilde{\hat{a}}_m(\omega)$ and $\tilde{\hat{a}}_m^\dagger(\omega)$. This implies that

$$\mathsf{V}(\omega)\mathsf{V}^\dagger(\omega) = \mathsf{S}(\omega)\mathsf{S}^\dagger(\omega) - 1. \tag{6.540}$$

The state of the incoming radiation is

$$|\psi\rangle = \bigotimes_{\substack{m=1 \\ m \neq m_0}}^{2N} |\mathrm{vac}\rangle_m \otimes |\psi(\omega)\rangle_{m_0}, \tag{6.541}$$

where $|\mathrm{vac}\rangle_m$ is a vacuum state of mode m of the incoming radiation and $|\psi(\omega)\rangle_m$ is a coherent state related to an unnormalized wave function $\psi(\omega)$, $|\psi(\omega)|^2 \to I_0\delta(\omega - \omega_0)$. The coherent state is defined in terms of the usual unnormalized single-photon states $|\omega\rangle_n$ with the property

$${}_n\langle\omega|\omega'\rangle_m = \delta_{nm}\delta(\omega - \omega'). \tag{6.542}$$

The counting of photons is studied using the integrated intensity

$$\hat{W}(\tau) = \int_0^\tau \hat{I}(t)\,\mathrm{d}t, \tag{6.543}$$

where

$$\hat{I}(t) = \eta\hat{\mathsf{a}}^{\mathrm{out}\dagger}(t)\mathsf{P}\hat{\mathsf{a}}^{\mathrm{out}}(t), \tag{6.544}$$

with η the efficiency of the photodetector,

$$\mathsf{P} = \begin{pmatrix} 0 & 0 \\ 0 & 1 \end{pmatrix}, \tag{6.545}$$

being a $2N \times 2N$ matrix divided into four $N \times N$ matrices, and

$$\hat{a}_n^{\mathrm{out}}(t) = \frac{1}{\sqrt{2\pi}} \int_0^\infty e^{-i\omega t}\tilde{\hat{a}}_n^{\mathrm{out}}(\omega)\,\mathrm{d}\omega. \tag{6.546}$$

The generating function is

$$F(\xi) = \ln\langle: \exp(\xi\hat{W}) :\rangle. \tag{6.547}$$

The result for the detection in the long-time regime $\omega_c \tau \gg 1$ is relatively simple. It is found that

$$F(\xi) = F_{\mathrm{ex}}(\xi) - \frac{\tau}{2\pi} \int_0^\infty \ln \left\{ \det[1 - \eta \xi f(\omega, T)(1 - \mathsf{r}\mathsf{r}^\dagger - \mathsf{t}\mathsf{t}^\dagger)] \right\} \mathrm{d}\omega, \tag{6.548}$$

where

$$F_{\mathrm{ex}}(\xi) = \eta \xi \tau \left\{ \mathsf{t}_0^\dagger \left[1 - \eta \xi f(\omega_0, T)(1 - \mathsf{r}_0 \mathsf{r}_0^\dagger - \mathsf{t}_0 \mathsf{t}_0^\dagger) \right]^{-1} \mathsf{t}_0 \right\}_{m_0 m_0}, \tag{6.549}$$

with $\{\ldots\}_{m_0 m_0}$ denoting the matrix element located in m_0th row and m_0th column. In relation (6.549) $\mathsf{t}_0 \equiv \mathsf{t}(\omega_0)$, $\mathsf{r}_0 \equiv \mathsf{r}(\omega_0)$.

The general description may be applied also to an optical cavity filled with an amplifying random medium. It holds that $\mathsf{t} = 0$ because there is no transmission. Patra and Beenakker (1999) have studied how the noise figure $\mathcal{F}$ increases on approaching the laser threshold. Near the laser threshold the noise figure has a divergent ensemble average. Its modal value is of the order of the number N of propagating modes in the medium.

The noise power is increased by

$$P_{\mathrm{ex}} = 2\eta^2 f I_0 \left[\mathsf{t}_0^\dagger (1 - \mathsf{r}_0 \mathsf{r}_0^\dagger - \mathsf{t}_0 \mathsf{t}_0^\dagger) \mathsf{t}_0 \right]_{m_0 m_0}. \tag{6.550}$$

It has been found that P_{ex} increases monotonically with increasing amplification rate, but it has a maximum as a function of absorption rate for certain geometries.

Mishchenko and Beenakker (1999) say clearly that they borrow from the field of electronic conduction in disordered metals. Besides, they take into account the absorption and emission of photons. They let $f_{\mathbf{k}}(\mathbf{r}, t)$ denote the density of photon number at the position $\mathbf{r}$ and such that $\int f_{\mathbf{k}}(\mathbf{r}, t)\, \mathrm{d}^3\mathbf{r}$ is the total photon number in the mode $\mathbf{k}$. The authors consider the random field $f_{\mathbf{k}}(\mathbf{r}, t)$ and its mean $\bar{f}_{\mathbf{k}}(\mathbf{r})$. The system of wave vectors $\mathbf{k}$ is discrete. This has been adopted for ease of notation.

It is appropriate that this notion is independent of the time. But this does not facilitate reading of a Boltzmann equation

$$c\mathbf{s} \cdot \nabla_{\mathbf{r}} \bar{f}_{\mathbf{k}}(\mathbf{r}, t) = \bar{I}_{\mathbf{k}}(\mathbf{r}, t), \tag{6.551}$$

where $\mathbf{s} = \frac{\mathbf{k}}{|\mathbf{k}|}$ and

$$\bar{I}_{\mathbf{k}}(\mathbf{r}, t) = \sum_{\mathbf{k}'} \left[\bar{J}_{\mathbf{k}\mathbf{k}'}(\mathbf{r}, t) - \bar{J}_{\mathbf{k}'\mathbf{k}}(\mathbf{r}, t) \right] + \bar{I}_{\mathbf{k}}^{+}(\mathbf{r}, t), \tag{6.552}$$

where

$$\begin{aligned} \bar{J}_{\mathbf{k}\mathbf{k}'}(\mathbf{r}, t) &\equiv J_{\mathbf{k}\mathbf{k}'}\left(\bar{f}_{\mathbf{k}}(\mathbf{r}, t), \bar{f}_{\mathbf{k}'}(\mathbf{r}, t)\right), \\ \bar{J}_{\mathbf{k}'\mathbf{k}}(\mathbf{r}, t) &\equiv J_{\mathbf{k}'\mathbf{k}}\left(\bar{f}_{\mathbf{k}'}(\mathbf{r}, t), \bar{f}_{\mathbf{k}}(\mathbf{r}, t)\right). \end{aligned} \tag{6.553}$$

On writing it in the form

$$0 = -c\mathbf{s} \cdot \nabla_{\mathbf{r}} \bar{f}_{\mathbf{k}}(\mathbf{r}, t) + \bar{I}_{\mathbf{k}}(\mathbf{r}, t), \tag{6.554}$$

we see that a continuity equation was first generalized

$$\frac{\partial \bar{f}_{\mathbf{k}}(\mathbf{r}, t)}{\partial t} = -c\mathbf{s} \cdot \nabla_{\mathbf{r}} \bar{f}_{\mathbf{k}}(\mathbf{r}, t) + \tilde{I}_{\mathbf{k}}(\mathbf{r}, t). \tag{6.555}$$

Here $\tilde{I}_{\mathbf{k}}(\mathbf{r}, t)$ has to mean gain and loss terms.

Let us note the form of the gain term due to amplification

$$\tilde{I}_{\mathbf{k}}^{+}(\mathbf{r}, t) = w_{\mathbf{k}}^{+} \left[1 + \bar{f}_{\mathbf{k}}(\mathbf{r}, t)\right], \tag{6.556}$$

with $w_{\mathbf{k}}^{+}$ the amplification rate. The unity enables a zero density to be amplified. The gain term due to scattering from the mode with the wave vector $\mathbf{k}'$,

$$\tilde{J}_{\mathbf{k}\mathbf{k}'}(\mathbf{r}, t) = w_{\mathbf{k}\mathbf{k}'} \bar{f}_{\mathbf{k}'}(\mathbf{r}, t)[1 + \bar{f}_{\mathbf{k}}(\mathbf{r}, t)], \tag{6.557}$$

is similar. Obviously, it is nonlinear in the described field. This does not mean that Equation (6.551) is not linear. It holds that

$$\begin{aligned} \tilde{J}_{\mathbf{k}\mathbf{k}'}(\mathbf{r}, t) - \tilde{J}_{\mathbf{k}'\mathbf{k}}(\mathbf{r}, t) &= w_{\mathbf{k}\mathbf{k}'}[\bar{f}_{\mathbf{k}'}(\mathbf{r}, t) - \bar{f}_{\mathbf{k}}(\mathbf{r}, t)] \\ &= w_{\mathbf{k}'\mathbf{k}}[\bar{f}_{\mathbf{k}'}(\mathbf{r}, t) - \bar{f}_{\mathbf{k}}(\mathbf{r}, t)]. \end{aligned} \tag{6.558}$$

Mishchenko and Beenakker (1999) continue the results such as those in Kogan (1996). Consistently with (6.551), the Boltzmann–Langevin equation for the random field itself is presented

$$\begin{aligned} c\mathbf{s} \cdot \nabla_{\mathbf{r}} f_{\mathbf{k}}(\mathbf{r}, t) + \frac{\partial f_{\mathbf{k}}(\mathbf{r}, t)}{\partial t} &= \sum_{\mathbf{k}'} [J_{\mathbf{k}\mathbf{k}'}(\mathbf{r}, t) - J_{\mathbf{k}'\mathbf{k}}(\mathbf{r}, t)] \\ &\quad + I_{\mathbf{k}}^{+}(\mathbf{r}, t) - I_{\mathbf{k}}^{-}(\mathbf{r}, t) + \mathcal{L}_{\mathbf{k}}(\mathbf{r}, t), \end{aligned} \tag{6.559}$$

where

$$\begin{aligned} J_{\mathbf{k}\mathbf{k}'}(\mathbf{r}, t) &\equiv J_{\mathbf{k}\mathbf{k}'}\big(f_{\mathbf{k}}(\mathbf{r}, t), f_{\mathbf{k}'}(\mathbf{r}, t)\big), \\ J_{\mathbf{k}'\mathbf{k}}(\mathbf{r}, t) &\equiv J_{\mathbf{k}'\mathbf{k}}\big(f_{\mathbf{k}'}(\mathbf{r}, t), f_{\mathbf{k}}(\mathbf{r}, t)\big). \end{aligned} \tag{6.560}$$

Here $I_{\mathbf{k}}^{+}(\mathbf{r}, t)$ is a gain term due to amplification,

$$I_{\mathbf{k}}^{+}(\mathbf{r}, t) = w_{\mathbf{k}}^{+} [1 + I_{\mathbf{k}}(\mathbf{r}, t)] \tag{6.561}$$

and $I_{\mathbf{k}}^{-}(\mathbf{r}, t)$ is a loss term due to absorption,

$$I_{\mathbf{k}}^{-}(\mathbf{r}, t) = w_{\mathbf{k}}^{-} f_{\mathbf{k}}(\mathbf{r}, t), \tag{6.562}$$

with $w_{\mathbf{k}}^{-}$ the absorption rate. In (6.559), $\mathcal{L}_{\mathbf{k}}(\mathbf{r}, t)$ is a Langevin term,

$$\mathcal{L}_{\mathbf{k}}(\mathbf{r}, t) = \sum_{\mathbf{k}'} [\delta J_{\mathbf{k}\mathbf{k}'}(\mathbf{r}, t) - \delta J_{\mathbf{k}'\mathbf{k}}(\mathbf{r}, t)] + \delta I_{\mathbf{k}}^{+}(\mathbf{r}, t) - \delta I_{\mathbf{k}}^{-}(\mathbf{r}, t), \tag{6.563}$$

which is remarkable for copying the form of the previous terms. The elementary stochastic processes $\delta J_{\mathbf{k}\mathbf{k}'}(\mathbf{r}, t)$, $\delta I_{\mathbf{k}}^{\pm}(\mathbf{r}, t)$ have zero means,

$$\overline{\delta J_{\mathbf{k}\mathbf{k}'}(\mathbf{r}, t)} = 0, \quad \overline{\delta I_{\mathbf{k}}^{+}(\mathbf{r}, t)} = 0, \quad \overline{\delta I_{\mathbf{k}}^{-}(\mathbf{r}, t)} = 0, \tag{6.564}$$

and they have properties

$$\overline{\delta J_{\mathbf{kk}'}(\mathbf{r},t)\delta J_{\mathbf{qq}'}(\mathbf{r}',t')} = \Delta(\mathbf{r},t,\mathbf{r}',t')\delta_{\mathbf{kq}}\delta_{\mathbf{k}'\mathbf{q}'}J_{\mathbf{kk}'}(\mathbf{r},t),$$
$$\overline{\delta I_{\mathbf{k}}^{\pm}(\mathbf{r},t)\delta I_{\mathbf{k}'}^{\pm}(\mathbf{r}',t')} = \Delta(\mathbf{r},t,\mathbf{r}',t')\delta_{\mathbf{kk}'}I_{\mathbf{k}}^{\pm}(\mathbf{r},t), \tag{6.565}$$

where

$$\Delta(\mathbf{r},t,\mathbf{r}',t') = \delta(\mathbf{r}-\mathbf{r}')\delta(t-t'). \tag{6.566}$$

They are also characterized by the stochastic independence between $\delta J_{\mathbf{kk}'}(\mathbf{r},t)$ and $\delta I_{\mathbf{q}}^{\pm}(\mathbf{r}',t')$ and by the same relationship between $\delta I_{\mathbf{k}}^{+}(\mathbf{r},t)$ and $\delta I_{\mathbf{q}}^{-}(\mathbf{r}',t')$.

The authors make a diffusion approximation, which is related to the expansion of the random field with respect to **s**. Then they consider the propagation through an absorbing or amplifying disordered waveguide (of length L). The noise power is decomposed into the fluctuations in the transmitted radiation, those in the thermal radiation, and the excess noise, which is characterized in Henry and Kazarinov (1996). The expressions for the thermal fluctuations and the excess noise agree with Beenakker (1999).

As a contribution the noise power of the thermal radiation emitted by a sphere is given (per unit surface area). Numerical results both for the waveguide geometry and for the sphere geometry are presented.

Beenakker (1999) has presented the statistics of thermal radiation in dependence on the deviation $1 - \mathsf{SS}^{\dagger}$ from the unitarity of the scattering matrix S of the system. He has recovered the familiar results for black-body radiation in the limit $\mathsf{S} \to 0$. A simple expression for the mean photocount has been identified as Kirchhoff's law. A generalization of the Kirchhoff law has been derived. For the extension of the Kirchhoff law to the statistics of quanta, which exists in the case of single-mode detection, reference is made to (Bekenstein and Schiffer 1994). Due to a similarity, the theory has been easily applied to a random amplifying medium (or a "random laser") below the laser threshold.

Zacharakis et al. (2000) measured photon–number distributions of fluorescence of an organic dye. The dye was mixed with poly(methyl methacrylate), which fixed scatterers. They were able to measure the photon–number distributions for different time delays and at different wavelengths. The source of excitations was a frequency-doubled 200-fs pulsed laser emitting at 400 nm.

The photon-number distribution from the sample when it was pumped above threshold had different character for different time delays. For a small time delay this distribution was Poisson-shaped with an imperfect vacuum value. When the time delay increased, a Bose–Einstein distribution was appropriate. In contrast to the high-energy case, when the excitation energy is below threshold, the photon number has the Bose–Einstein distribution, which is independent of the time delay.

In Patra and Beenakker (2000), a continuation of Patra and Beenakker (1999) and Beenakker (1998) is contained. The treatment includes a noise characteristic averaged over an ensemble of random media with different positions of the scatterers. The authors assume a waveguide with $N(\omega)$ propagating modes at frequency ω. Modes $1, 2, \ldots, N$ are on the left-hand side of the medium and modes

$N+1, N+2, \ldots, 2N$ are on the right-hand side. The outgoing radiation in mode n is described by an annihilation operator $\tilde{\hat{a}}_n^{\mathrm{out}}(\omega)$. A vector $\tilde{\hat{\mathsf{a}}}^{\mathrm{out}}(\omega)$ is defined. Similar notation is introduced for incoming radiation.

The operators, which belong to the same stage, fulfil the commutation relations between the annihilation and creation ones,

$$[\tilde{\hat{a}}_n(\omega), \tilde{\hat{a}}_m^\dagger(\omega')] = \delta_{nm}\delta(\omega-\omega')\hat{1}, \quad [\tilde{\hat{a}}_n(\omega), \tilde{\hat{a}}_m(\omega')] = \hat{0}, \tag{6.567}$$

where $\tilde{\hat{a}}$=$\tilde{\hat{a}}^{\mathrm{in}}$, $\tilde{\hat{a}}^{\mathrm{out}}$.

In the input–output relations, also the vectors $\hat{\mathsf{b}}$ and $\hat{\mathsf{c}}$ occur, each of which has $2N$ annihilation operators for its elements. Their correlation functions are determined dependent on the temperature of the medium.

The usual quantization of discrete frequencies is obtained by considering those in the relations for a frequency step Δ, the frequencies $\omega_p = p\Delta$, and subscripts as those in the relations

$$\begin{aligned} \tilde{\hat{a}}_{np}^{\mathrm{out}} &= \frac{1}{\sqrt{\Delta}} \int_{\omega_p}^{\omega_{p+1}} \tilde{\hat{a}}_n^{\mathrm{out}}(\omega)\,\mathrm{d}\omega, \\ S_{np,n'p'} &= S_{nn'}(\omega_p)\delta_{pp'}. \end{aligned} \tag{6.568}$$

Patra and Beenakker (2000) have considered a useful modification

$$\tilde{\hat{a}}_{np}^{\prime\mathrm{out}} = \frac{1}{\sqrt{\Delta}} \tilde{\hat{a}}_{np}^{\mathrm{out}}. \tag{6.569}$$

We have used the prime to distinguish this modification from the usual annihilation operator.

A characteristic function is

$$\chi(\beta, \Delta) = \left\langle : \exp\left[\Delta^{\frac{1}{2}}\left(\tilde{\hat{\mathsf{a}}}'^\dagger\beta - \beta^\dagger\tilde{\hat{\mathsf{a}}}'\right)\right] : \right\rangle. \tag{6.570}$$

The vector β has the elements $\beta_{np} = \beta_n(\omega_p)$.

The statistical properties of the bath are

$$\chi_{\mathrm{abs}}(\beta, \Delta) = \exp(-\beta^\dagger \mathsf{f} \beta) \tag{6.571}$$

for an absorbing medium and

$$\chi_{\mathrm{amp}}(\beta, \Delta) = \exp(\beta^\dagger \mathsf{f} \beta) \tag{6.572}$$

for an amplifying medium. In these relations f means a matrix with the elements $f_{np,n'p'} = \delta_{nn'}\delta_{pp'} f(\omega_p, T)$, where $f(\omega_p, T)$ is given in (6.515). The characteristic function of the outgoing state is

$$\chi_{\mathrm{out}}(\beta, \Delta) = \exp\left[-\beta^\dagger(1-\mathsf{S}\mathsf{S}^\dagger)\mathsf{f}\beta\right] \chi_{\mathrm{in}}(\mathsf{S}^\dagger\beta, \Delta). \tag{6.573}$$

The photocount distribution is the probability $P(n, \tau)$ that n photons are absorbed by a photodetector within a time τ. The appropriate generating function is denoted by $F(\xi, \tau)$. The integrated intensity $\hat{W}(\tau)$ is defined by the relation

$$\hat{W}(\tau) = \int_0^\tau \sum_{n=1}^{2N} \eta_n \hat{a}_n^{\text{out}\dagger}(t)\hat{a}_n^{\text{out}}(t)\,\mathrm{d}t. \tag{6.574}$$

Here $\eta_n \in [0, 1]$ is the detection efficiency of the nth mode. We let η denote a $2N \times 2N$ diagonal matrix containing the detection efficiencies η_n on the diagonal ($\eta_{nm} = \eta_n \delta_{nm}$). The discretization of frequencies will lead to the integrated intensity $\hat{W}(\tau, \Delta)$. It can be written using the matrix $\tilde{\eta} = \eta \otimes 1_{\text{frequencies}}$, with the elements $\tilde{\eta}_{np,n'p'} = \eta_{nn'}\delta_{pp'}$. Here $1_{\text{frequencies}}$ is the unit matrix with the elements $\delta_{pp'}$. The respective generating function is denoted by $F(\xi, \tau, \Delta)$.

The generating function $F(\xi, \tau, \Delta)$ may be determined from $\exp[F(\xi, \tau, \Delta)]$, which in turn is a linear integral transform of the characteristic function $\chi_{\text{out}}(\beta, \Delta)$. On respecting the input–output relation (6.573), we obtain the relation

$$\begin{aligned}\exp[F(\xi, \tau, \Delta)] = {} & \frac{1}{\det(-\xi\pi\tilde{\eta})} \\ & \times \int \chi_{\text{in}}(\mathsf{S}^\dagger\beta, \Delta) \exp\left[\frac{1}{\xi}\beta^\dagger(\tilde{\eta})^{-1}\beta - \beta^\dagger(1 - \mathsf{S}\mathsf{S}^\dagger)\mathsf{f}\beta\right] \mathrm{d}\beta,\end{aligned} \tag{6.575}$$

where $\Delta = \frac{2\pi}{\tau}$ is chosen. The thermal fluctuations can be separated and we obtain the relations

$$\begin{aligned}F(\xi, \tau, \Delta) = {} & F_{\text{th}}(\xi, \tau, \Delta) \\ & + \ln\left[\frac{1}{\det(\pi\mathsf{M})}\int \chi_{\text{in}}(\beta, \Delta) \exp(-\beta^\dagger \mathsf{M}^{-1}\beta)\,\mathrm{d}\beta\right],\end{aligned} \tag{6.576}$$

$$F_{\text{th}}(\xi, \tau, \Delta) = -\ln\{\det[1 - \xi\tilde{\eta}(1 - \mathsf{S}\mathsf{S}^\dagger)\mathsf{f}]\}, \tag{6.577}$$

and

$$\mathsf{M} = -\xi\mathsf{S}^\dagger[1 - \xi\tilde{\eta}(1 - \mathsf{S}\mathsf{S}^\dagger)\mathsf{f}]^{-1}\tilde{\eta}\mathsf{S}. \tag{6.578}$$

Returning to the continuous frequency, relation (6.577) can be written as

$$F_{\text{th}}(\xi, \tau) = -\frac{\tau}{2\pi}\int_0^\infty \ln\left\{\det\left(1 - \xi\eta[1 - \mathsf{S}(\omega)\mathsf{S}^\dagger(\omega)]f(\omega, T)\right)\right\} \mathrm{d}\omega, \tag{6.579}$$

where $f(\omega, T)$ is given in (6.515). It is stated that all factorial cumulants depend linearly on the detection time in the long-time limit. The notion of detection efficiency may be utilized also to treatment of particular cases, such as the detection at one side of the waveguide (Beenakker 1999).

Patra and Beenakker (2000) assume that the incident radiation is in the ideal squeezed state $|\zeta, \alpha\rangle = \hat{C}\hat{S}|0\rangle$, where

$$\hat{S} = \exp\left[\frac{1}{2}\Delta(\hat{\mathsf{a}}'^{\text{in}\,T}\zeta^\dagger\tilde{\hat{\mathsf{a}}}'^{\text{in}} - \tilde{\hat{\mathsf{a}}}'^{\text{in}\dagger}\zeta\tilde{\hat{\mathsf{a}}}'^{\text{in}*})\right], \tag{6.580}$$

is the squeezing operator and

$$\hat{C} = \exp\left[\Delta^{\frac{1}{2}}(\tilde{\mathsf{a}}^{\prime\mathrm{in}\dagger}\alpha - \alpha^{\dagger}\tilde{\mathsf{a}}^{\prime\mathrm{in}})\right] \tag{6.581}$$

is the displacement operator.

In relation (6.580), T means the transposition, ζ is the diagonal matrix with the elements $\zeta_{np,n'p'} = \zeta_n(\omega_p)\delta_{nn'}\delta_{pp'}$, and α is the vector with the elements $\alpha_{np} = \alpha_n(\omega_p)$. Useful are the real parameters $\rho_n(\omega_p)$, $\phi_n(\omega_p)$ such that $\zeta_n(\omega) = \rho_n(\omega)\exp(\mathrm{i}\phi_n(\omega))$. Here $\tilde{\mathsf{a}}^{\prime\mathrm{in}\dagger}_{np}$ is the column with the elements $\tilde{a}^{\prime\mathrm{in}}_{np}$.

The characteristic function of the incident radiation in the case where this radiation is in the ideal squeezed state is

$$\begin{aligned}\chi_{\mathrm{in}}(\beta,\Delta) = \exp\Big\{&\alpha^{\dagger}\beta - \beta^{\dagger}\alpha - \frac{1}{4}\beta^{T}[e^{-\mathrm{i}\phi}\sinh(2\rho)]\beta \\ &- \frac{1}{4}\beta^{\dagger}[e^{\mathrm{i}\phi}\sinh(2\rho)]\beta^{*} - \beta^{\dagger}(\sinh\rho)^{2}\beta\Big\}.\end{aligned} \tag{6.582}$$

On substitution into relation (6.573), we obtain that the characteristic function of the outgoing radiation is

$$\begin{aligned}\chi_{\mathrm{out}}(\beta,\Delta) = \exp\Big\{&\alpha^{\dagger}\mathsf{S}^{\dagger}\beta - \beta^{\dagger}S\alpha - \frac{1}{4}\beta^{T}\mathsf{S}^{*}[e^{-\mathrm{i}\phi}\sinh(2\rho)]\mathsf{S}^{\dagger}\beta \\ &- \frac{1}{4}\beta^{\dagger}\mathsf{S}[e^{\mathrm{i}\phi}\sinh(2\rho)]\mathsf{S}^{T}\beta^{*} \\ &- \beta^{\dagger}[\mathsf{f} - \mathsf{S}(\mathsf{f} - (\sinh\rho)^{2})\mathsf{S}^{\dagger}]\beta\Big\}.\end{aligned} \tag{6.583}$$

Using an integral transformation, we obtain the generating function $F(\xi,\tau,\Delta)$,

$$F(\xi,\tau,\Delta) = F_{\mathrm{th}}(\xi,\tau,\Delta) - \frac{1}{2}\ln(\det\mathsf{X}) - \frac{1}{2}\begin{pmatrix}\alpha^{*}\\ \alpha\end{pmatrix}^{T}\mathsf{X}^{-1}\begin{pmatrix}\mathsf{M}\alpha\\ \mathsf{M}^{*}\alpha^{*}\end{pmatrix}, \tag{6.584}$$

where the matrix X is defined in terms of the matrix M,

$$\mathsf{X} = 1 + \begin{pmatrix}\mathsf{M}\sinh\rho & -\mathsf{M}e^{\mathrm{i}\phi}\cosh\rho\\ -\mathsf{M}^{*}e^{-\mathrm{i}\phi}\cosh\rho & \mathsf{M}^{*}\sinh\rho\end{pmatrix}\begin{pmatrix}\sinh\rho & 0\\ 0 & \sinh\rho\end{pmatrix}. \tag{6.585}$$

Further Patra and Beenakker (2000) consider the case, where only the mode m_0 is squeezed.

The Fano factor is the ratio

$$\mathcal{F} = \frac{P}{\bar{I}}, \tag{6.586}$$

where $P = \frac{1}{\tau}(\kappa_2 + \kappa_1)$ is the noise power and $\bar{I} = \frac{1}{\tau\kappa_1}$ is the mean current. For simplicity this characteristic is considered in the limit of $\tau \to \infty$. The detection efficiency does not depend on the mode. The thermal contributions may be neglected in the considered case, since they are spread out over a wide range of frequencies.

In the exposition, η ceases to be a matrix and it means the common value of the detector efficiencies η_n. For (6.582), one has

$$\mathcal{F}_{\text{in}} = 1 + \frac{|\alpha \cosh \rho - \alpha^* e^{i\phi} \sinh \rho|^2 - |\alpha|^2 + \frac{1}{2}(\sinh \rho)^2 \cosh(2\rho)}{|\alpha|^2 + (\sinh \rho)^2}. \tag{6.587}$$

Patra and Beenakker (2000) consider the Fano factor both for the direct detection ($\mathcal{F}_{\text{direct}}$) and for the homodyne detection ($\mathcal{F}_{\text{homo}}$). In the case of direct detection, the authors find that

$$\begin{aligned}\mathcal{F}_{\text{direct}} - 1 &= \eta(\mathsf{t}_0^\dagger \mathsf{t}_0)_{m_0 m_0}(\mathcal{F}_{\text{in}} - 1) \\ &\quad + 2\eta f(\omega_0, T)\frac{[\mathsf{t}_0^\dagger(1 - \mathsf{r}_0\mathsf{r}_0^\dagger - \mathsf{t}_0\mathsf{t}_0^\dagger)\mathsf{t}_0]_{m_0 m_0}}{(\mathsf{t}_0^\dagger \mathsf{t}_0)_{m_0 m_0}}.\end{aligned} \tag{6.588}$$

The Fano factors for the direct and homodyne detections depend on the reflection and transmission matrices of the waveguide. These matrices depend on the positions of the scatterers inside the waveguide. The distribution of these matrices can be chosen by random-matrix theory (Beenakker 1997). The details may be found in Brouwer (1998). It is examined how the distribution depends on the mean free path l and the amplification (absorption) length $\xi_a = \sqrt{D\tau_a}$, where $D = \frac{cl}{3}$ is the diffusion constant and $\frac{1}{\tau_a}$ is the amplification (absorption) rate.

On neglect of the correlation between numerator and denominator for an absorbing disordered waveguide it is obtained that

$$\begin{aligned}\mathcal{F}_{\text{direct}} &= 1 + \frac{4l\eta}{3\xi_a \sinh s}(\mathcal{F}_{\text{in}} - 1) \\ &\quad + \frac{\eta}{2} f(\omega_0, T)\left[3 - \frac{2s + \coth s}{\sinh s} - \frac{s \coth s - 1}{(\sinh s)^2} + \frac{s}{(\sinh s)^3}\right].\end{aligned} \tag{6.589}$$

Here $s \equiv \frac{L}{\xi_a}$. In the limit of strong absorption, the Fano factor $\mathcal{F}_{\text{direct}} = 1 + \frac{3}{2\eta} f(\omega_0, T)$. The Fano factor $\mathcal{F}_{\text{in}}$ may be given by equation (6.587), but Equation (6.589) is valid even for any state of the incident radiation.

For an amplifying disordered waveguide it is found that

$$\begin{aligned}\mathcal{F}_{\text{direct}} &= 1 + \frac{4l\eta}{3\xi_a \sin s}(\mathcal{F}_{\text{in}} - 1) \\ &\quad + \frac{\eta}{2} f(\omega_0, T)\left[3 - \frac{2s - \cot s}{\sin s} + \frac{s \cot s - 1}{(\sin s)^2} - \frac{s}{(\sin s)^3}\right].\end{aligned} \tag{6.590}$$

The laser threshold is $s = \pi$. In Patra and Beenakker (2000) also the average Fano factors for the homodyne detection are presented. The effect of absorption on quasi-modes of a random waveguide has been studied in Sebbah et al. (2007).

Cao et al. (2000) have mentioned the concept of the Anderson localization. Optical absorption counteracts photon localization. Also optical gain reduces photon localization length in a one-dimensional random medium. After the previous experiment on coherent feedback for lasing, the authors have arrived at enhancement of the scattering strength and at spatial confinement of laser light in the disordered medium. They opine that the optical gain enhances the photon localization at least in a three-dimensional medium. The coherent amplification of the scattered light enhances the interference effect and helps the spatial confinement.

ZnO particles of the average size about 50 nm and a ZnO powder film of thickness about 30 μm were prepared. The scattering mean free path l in the ZnO powder has been estimated $l \sim 0.5\lambda$. The ZnO powder film is photoluminescent, when it is pumped at 266 nm. The pump beam falls in the normal direction on an about 20 μm spot of the film. The spectrum of emission from the powder film is measured. At the same time, the spatial distribution of the emitted light intensity is imaged in the ultraviolet.

Figure 1, which we do not reproduce here, presents the measured spectra and spatial distribution of emission in a ZnO powder film at two different pump powers. For the lower pump level, the spectrum consists of a single broad spontaneous emission peak. The spatial distribution of the spontaneous emission intensity is almost uniform across the excitation area. It depends on the pump intensity distribution.

For the higher pump level, when the pump intensity exceeds a threshold, sharp peaks emerge in the emission spectrum. Bright tiny spots appear in the spatial distribution of the emission. When the pump intensity increases further, additional sharp peaks emerge in the emission spectrum. Also more bright spots appear in the emitted light pattern. Above the threshold, the total emission intensity begins to approach the pump power.

The fact that the bright spots in the emission pattern and the lasing modes in the emission spectrum always occur simultaneously could mean that the bright spots are efficient scatterers. Then the bright spots should scatter the spontaneously emitted light below the lasing threshold. As the bright spots do not exist below the lasing threshold, the laser light intensity is high at their locations.

The authors have measured the short scattering mean free path, i.e. very strong light scattering on average. They expect small regions of higher disorder and stronger scattering. In other words, they assume many resonant cavities formed by multiple scattering and interference. Every cavity has its lasing threshold. The lasing peaks in the emission spectrum illustrate cavity resonant frequencies and the bright spots in the spatial light pattern give positions and shapes of the cavities.

To verify this hypothesis, the authors have reduced the size of the random medium to a cluster of ZnO nanoparticles. The existence of the lasing threshold is related to nonradiative and radiative recombination of the excited carriers. The authors have calculated the electromagnetic-field distribution in a random medium using the finite-difference time-domain method according to the book (Taflove 1995). In the calculations the assumption that the ZnO particles are surrounded by air has been used. Optical gain has been introduced to the Maxwell equations by the negative conductance σ.

The simulation has shown that, when the optical gain is just above the lasing threshold, the emission spectrum consists of a single peak. When the optical gain increases further, additional lasing modes appear. The authors have concluded that optical gain helps spatial confinement of light in a random medium.

The authors have paid more attention to the Anderson localization. In spite of the fact that there is no criterion for the Anderson localization in an active random medium, they have determined the Thouless number $\delta = 0.75 < 1$ in favour of the Anderson localization in the lasing mode.

Thareja and Mitra (2000) have reported on an experiment on optically pumped ZnO powder. In this medium a random laser has been demonstrated. The theoretic explanation of this effect has been based on the paper (Cao et al. 1999).

Jiang and Soukoulis (2000) emphasize that, in contrast to the paper (Lawandy et al. 1994), where discrete lasing peaks were not observed, a new interesting property was reported, e.g. in Cao et al. (1999).

The authors provide references, e.g. in (Zyuzin 1995, John and Pang 1996), but they have pointed out a limitation of the diffusion approach. They see a limitation, though mild, also in the approach as in (Paasschens et al. 1996, Jiang and Soukoulis 1999, Jiang et al. 1999). Essentially, they return to the semiclassical laser theory, e.g., (Siegman 1986).

The authors have combined the equations for electron densities with Maxwell's equations and have used the finite-difference time-domain method according to the book (Taflove 1995). After a long relaxation time stationary solutions can be obtained. The time dependence of the electric field inside the system and in its vicinity is examined. The emission spectra and the modes inside the system can be obtained after the Fourier transformation.

The system is a one-dimensional simplification of the reported experiments. It consists of many dielectric layers of fixed thickness bestowed between two surfaces, with the space among the dielectric layers filled with a gain medium. The distance between the neighbouring dielectric layers is assumed to be a random variable. The total length of the system is L.

The numerical simulations have shown the following:

(i) In periodic and short ($L < \xi$) random systems an extended mode dominates in the field and the spectrum.
(ii) For either strong disorder or a long ($L \gg \xi$) system a low threshold value for lasing is obtained. By increasing the length or the gain more peaks appear in the spectrum. The peaks are coming from localized modes.
(iii) The saturation can be observed. The number of the peaks is proportional to the length of the system.
(iv) The emission spectra are not the same for various output directions. In the three-dimensional case lasing peaks need not be so sharp as in the one-dimensional case.

The alternating layers are made of dielectric materials with dielectric constants $\epsilon_1 = \epsilon_0$ and $\epsilon_2 = 4\epsilon_0$. The thickness of the first layer, which simulates the gain medium, is a random variable $a_n = a_0(1 + W\gamma)$, where $a_0 = 300$ nm, W is the strength of randomness, and γ is a random value in the range $[-0.5, 0.5]$. The thickness of the second layer, which simulates the scatterers, is a constant $b = 180$ nm. In the layers, which represent the gain medium, a four-level electronic material is admixed. The electron densities at a ground, first, second, and third levels are $N_0(x,t)$,...,$N_3(x,t)$, respectively. An external mechanism pumps electrons from the ground level to the third one at a certain pumping rate P_r. Nonradiative transitions occur from the higher level to the lower one with the lifetimes of the upper levels τ_{32}, τ_{21}, τ_{10}. The radiative transition from the second level to the first one, or back has the centre frequency ω_a.

According to the monograph (Siegman 1986), the polarization density $P(x,t)$ depends nonlinearly on the population inversion $\Delta N(x,t) = N_1(x,t) - N_2(x,t)$ and on the electric field $E(x,t)$. An equation for the polarization density $P(x,t)$ can be written. Equations for electron densities at every level can be utilized.

One must introduce sources into the system. The distance between the two sources L_s must be smaller than the localization length ξ. The sources simulate the spontaneous emission. They have a Lorentzian spectrum centred around ω_a and their amplitudes depend on N_2. Two leads are assumed at the left-hand and right-hand sides of the system. A numerical method for solving the mentioned equations with an absorbing-boundary condition is described.

Jiang and Soukoulis (2000) have performed the numerical simulations for periodic and random systems. They associate a lasing threshold with each of the systems. With the increase of the randomness, the threshold intensity decreases. It has been found that, in the case of a periodic system and a short ($L < \xi$) random system, one mode dominates, even if the gain increases far above the threshold. In the case of a long ($L \gg \xi$) random system, the stationary behaviour is marked with beats. There are more than one localized mode and each one has its specific frequency. In a figure, which is not reproduced here, it is shown how, for the pumping rates $P_r = 10^4, 10^6, 10^{10}\ \mathrm{s}^{-1}$, one lasing mode appears and then more lasing modes. So more than one mode can exist together and each mode seems to repel others to reserve itself some space. There exists a saturated number of lasing modes N_m, which is proportional to the length of the system L. There exists an average mode length $L_m = \frac{L}{N_m}$, which is proportional to the localization length. The emission spectra at the right-hand and left-hand sides of the system are different. In the real three-dimensional experiments, Jiang and Soukoulis (2000) assume that every localized mode has its direction, strength, and position.

Cao et al. (2001) have measured the photon statistics of random lasers with resonant feedback. They have found that, when the pump intensity increases, the photon–number distribution changes continuously from the Bose–Einstein distribution at the threshold to the Poisson distribution well above the threshold. The normalized second factorial moment $G_2 = \frac{\langle n(n-1)\rangle}{\langle n\rangle^2}$ decreases correspondingly from 2 to 1. By comparing the photon statistics of a random laser with resonant feedback and this statistics of a random laser with nonresonant feedback, the authors have formed the idea about two lasing mechanisms.

For a random laser with nonresonant feedback, the fluctuation of the total number of photons in all modes of laser emission is smaller than the fluctuation of this number in blackbody radiation with the same number of modes (Zacharakis et al. 2000). However, the photon–number distribution in a single mode remains the Bose–Einstein distribution even well above the threshold. The quasimodes correspond to eigenfrequencies, whose imaginary parts represent the decay rates, in fact they are pseudomodes. When $kl > 1$ (k is a wave number and l is the transport mean free path), the quasimodes overlap spectrally and the emission spectrum is continuous.

In other words, in the case of weak scattering, the quasimodes decay fast and they are strongly coupled. So the loss of quasimodes is much lower than the loss of a single quasimode. In an active random medium, when the optical gain for interacting quasimodes reaches the loss of these quasimodes, lasing with nonresonant feedback emerges. A significant spectral narrowing is observed. Well above the threshold, the total photon–number fluctuation decreases due to gain saturation. However, strong coupling of quasimodes excludes stabilization of the lasing in a single quasimode.

When the amount of optical scattering increases, the decay rates of the quasimodes decrease and the mixing of the quasimodes weakens. When the optical gain increases, lasing with nonresonant feedback occurs first. As the optical gain increases further, it exceeds the loss of a quasimode that has a long lifetime. This resembles a traditional laser. A further increase of optical gain leads to lasing in more low-loss quasimodes. Laser emission manifests itself by discrete peaks. This process is lasing with resonant feedback.

When the scattering strength increases further, the lasing threshold in individual low-loss quasimodes drops below the threshold for lasing in coupled quasimodes. So the mentioned stage of lasing with nonresonant feedback is absent. Well above the threshold, the fluctuations of individual photon numbers decrease due to gain saturation.

Vanneste and Sebbah (2001) also perceived that the studies of the dependence of laser action on the strength of the disorder (the randomness) and, for highly scattering media, of the Anderson localization on laser gain were not finished. The Anderson localization of electronic waves has later been extended to electromagnetic waves (John 1984). The average localization length ξ characterizes an exponential decrease of the envelope of a localized mode. Also properties of the electronic or photonic transport depend on this parameter. The localized eigenmodes are microcavities in fact and they can serve as the feedback cavities of the laser. The reports of experiments and theoretic interpretations admit only nonresonant feedback of spontaneous emission amplified along open scattering paths (Vanneste and Sebbah 2001).

Only recently laser action in a random medium with resonant feedback has been reported, e.g. (Cao et al. 1999). In the experiment, a semiconductor powder was used, which played simultaneously the roles of the random and active media. Possible connection with the Anderson localization was merely mentioned (Cao et al. 2000). In theory it is assumed that the localized modes of the passive random system are preserved in the presence of gain, but one may ask, how much they are modified by the gain. The doubt whether localization is enhanced or inhibited by the gain ended in a low quotation index in one of the papers (Zhang 1995, Paasschens et al. 1996, Jiang 1999).

Vanneste and Sebbah (2001) examine the role of strong localization in the lasing action process. The numerical model describes the full dynamics of the field and the levels' populations in a two-dimensional active random medium. First they choose a window of modes strongly localized in the spectrum of the passive medium and examine the spatial and spectral characteristics of these modes. Next the gain is activated and the passive modes are compared with the laser modes. It results that the active medium is described by the modes of the passive system. The amplifying medium has only a small effect on the frequencies. The two-dimensional spatial profile of the localized wave functions is reproduced without distortion.

They consider two-dimensional disordered medium of size L^2 made of circular scatterers with radius r, refractive index n_2, and surface filling fraction ϕ, imbedded in a matrix of index n_1. This system is equivalent to an array of dielectric cylinders parallel to the z-axis. The matrix also plays the role of an active medium. They utilize the rate equations of a four-level atomic system and Maxwell's equations with a polarization term including atomic population inversion. It is a generalization of the paper (Jiang and Soukoulis 2000). A TM field defined by the components E_z, H_x, and H_y is considered.

The modes of the passive system have been studied as follows. The time response to a short pulse is recorded and Fourier transformed. It is damped. The first and second half of the time record can be Fourier transformed. It leads to a conclusion that the leaky modes (quasimodes) with shorter lifetimes have not survived in the second half of the time record. The modes with longer lifetimes are examined by a monochromatic source on their spatial pattern and time evolution. It can be concluded that the regime of the Anderson localization has been attained. The investigation has continued with introducing gain by uniform pumping of the atoms in the whole system. Above threshold a stationary regime is attained after a transient exponential growth of the field amplitude. The structure of mode has been preserved.

At higher pump levels, the laser emission is multimode. After a transient regime the field becomes stationary in beats between several excited modes. The choice of individual localized modes is possible by pumping locally. So it is meaningful to consider local thresholds for lasing. Vanneste and Sebbah (2001) have concluded with a question, if the introduction of gain contributes somehow to the discrimination between the diffusive and localized regimes in actual experiments.

Burin et al. (2001) have based their model for a random laser on a planar system of resonant scatterers pumped by an external laser. At the beginning they expound also the following classification: Random lasing with nonresonant feedback appears as the remarkable narrowing of the luminescence spectrum to a single peak of width about several nanometres. The coherent feedback lasing is identified as the series of high and narrow peaks having the width decreasing with the increase of pump power to at least the tenth nanometre scale. They distinguish two different theoretical approaches to the description of random lasing:

(i) The diffusion model with coherent backscattering corrections, e.g. (Wiersma and Lagendijk 1996) appropriate for describing the regime of nonresonant feedback, but which fails to predict the lasing threshold behaviour for the laser operation. The criticism is directed to the disability of prediction of the formation of high-quality random cavities. It seems to comprise even the rejection of the Anderson localization (considered in the paper (Jiang and Soukoulis 1999)).

(ii) In another approach, the criticism is contrasted with the intended comparison of the model with the random matrix approach (Frahm et al. 2000, Misirpashaev and Beenakker 1998).

The authors have analysed an experiment on ZnO disk-shaped powder samples using a classical microscopic model. The medium is represented by a set of random scatterers. The number of these particles is denoted by N. Each of them is considered as an electric dipole oscillator. The resonant frequency of the kth particle is denoted by ω_k and the transition dipole moment of the kth particle is denoted by $\mathbf{d}_k$, its length is $|\mathbf{d}_k|=d_k$. The position vector of a particle k relative to a centre j is denoted by $\mathbf{R}_{kj}$. It is assumed that the polarization component $\mathbf{p}_k$, $|\mathbf{p}_k|=p_k$, is parallel to the transition dipole moment,

$$\mathbf{p}_k = p_k \frac{\mathbf{d}_k}{d_k}. \tag{6.591}$$

We suppose that

$$\mathbf{p}_k = \tilde{\mathbf{p}}_k e^{\mathrm{i}zt},\ \mathbf{E}_k = \tilde{\mathbf{E}}_k e^{\mathrm{i}zt},\ \mathbf{E}_{kj} = \tilde{\mathbf{E}}_{kj} e^{\mathrm{i}zt}. \tag{6.592}$$

If it may be put $\hbar = 1$, then the equations for the collective eigenfrequencies z and the collective eigenvectors $\{\tilde{\mathbf{p}}_k\}$ of the system have the form

$$-z^2\tilde{\mathbf{p}}_k = -(\omega_k - \mathrm{i}g)^2\tilde{\mathbf{p}}_k + 2\omega_k \mathbf{d}_k(\mathbf{d}_k \cdot \tilde{\mathbf{E}}_k), \tag{6.593}$$

where g is a gain rate and

$$\tilde{\mathbf{E}}_k = \sum_{\substack{j \\ j \neq k}} \tilde{\mathbf{E}}_{kj} + \mathrm{i}\frac{2}{3}q^3 \tilde{\mathbf{p}}_k \tag{6.594}$$

are the electric fields of other particles and the damping term, with

$$\begin{aligned} \tilde{\mathbf{E}}_{kj} &= e^{\mathrm{i}qR_{kj}} \frac{\tilde{\mathbf{p}}_j - 3\mathbf{n}(\mathbf{n}\cdot\tilde{\mathbf{p}}_j)}{R_{kj}^3} \left(1 - \mathrm{i}qR_{kj}\right) \\ &\quad + q^2 e^{\mathrm{i}qR_{kj}} \frac{\tilde{\mathbf{p}}_j - \mathbf{n}(\mathbf{n}\cdot\tilde{\mathbf{p}}_j)}{R_{kj}}, \end{aligned} \tag{6.595}$$

where

$$\mathbf{n} = \frac{\mathbf{R}_{kj}}{R_{kj}},\quad q = \frac{z}{c}. \tag{6.596}$$

The equations are linear, but z enters in a relatively complicated manner. It is a complex number, the imaginary part of which is the decay rate.

The iteration method they have used calculates the lasing threshold. They could not treat a very large system with the number of particles exceeding $N = 1000$.

They have restricted the analysis to a two-dimensional system of scatterers. They have ignored the difference of dielectric constants of substrate and air. They report that they have studied the case $\omega_k = \omega_0$. The positions of all N particles were generated randomly within the circle of the radius $R = \sqrt{N}\frac{\eta c}{\omega_0}$. Three different values of the parameter η, $\eta = 0.3, 1, 3$, have been probed. In comparison with the random matrix approach, they have seen that the high-quality collective modes, in fact, occupy a few scatterers (e. g., from 5 to 10 for $N = 100$).

Ling et al. (2001) have presented a detailed experimental study of random lasers with resonant feedback. One of two materials was a poly(methyl methacrylate) film that contained dye and titanium dioxide particles. The other was zinc oxide polycrystalline film on a sapphire substrate. The dependences of the incident pump-pulse energy at the lasing threshold and the number of lasing modes for a fixed pump intensity on the transport mean free path have been measured. The effects of the pump area and the sample size have been determined too. The idea published in Cao et al. (2001) has been developed to an analytical model and the theoretical predictions have agreed with the experimental results.

Burin et al. (2002) have announced an analytical approach to random lasing in a one-dimensional medium. They have dealt with the lasing threshold. They have discussed application to the regime of strong three-dimensional localization of light. They have derived that the lasing threshold has strong fluctuations from sample to sample.

The original approach (Letokhov 1968) is based on a diffusion formalism. It predicts the lasing instability, when the length of the diffusion path $\frac{L^2}{l_t}$, with l_t being the mean-free path length of light, attains the gain length l_g. Not even John (1984) modifies this criterion. But experimental and numerical studies have shown a different value of gain rate at which lasing sets in.

The authors consider a different physical mechanism from diffusive motion. They study a one-dimensional medium as studied by (Jiang and Soukoulis 1999). They identify relevant channels responsible for lasing with the quasimodes. The results may be applied to higher dimension in the strong localization regime. This phenomenon has been reported in the studies (Chabanov et al. 2000, Wiersma et al. 1997).

It is assumed that a one-dimensional scattering medium is situated between the planes $x = 0$ and $x = L$. A gain medium is assumed. The description is based on an imaginary correction of the frequency $\omega \rightarrow \omega + \frac{ig}{2}$, where g is the gain rate. The lasing threshold is associated with the singularity in the transmission through the sample ($\frac{0}{0}$) as in papers (Jiang and Soukoulis 1999, Beenakker 1998). The authors relate the threshold with the intensity of the field near the source point. They show the equivalence convincingly. In the strong localization regime the lasing threshold is very small.

We will consider a source in the middle of the structure. We let r_l, r_r denote the reflection coefficients from the left-hand and right-hand halves, respectively. Provided that $l_t \ll L$, $|r_l| \approx 1$, $|r_r| \approx 1$, it is essential that the reflected waves interfere constructively and the sum of reflection phases ($r(\omega) = |r(\omega)|e^{i\Phi(\omega)}$) is a multiple of 2π. The authors call this resonance. The resonances approach eigenmodes of the whole system closed between mirrors.

But in the passive medium the equality is not reached, $|r_l| < 1$, $|r_r| < 1$. The equation

$$r_l(\omega)r_r(\omega) = 1 \tag{6.597}$$

is rewritten as

$$|r_l r_r| \exp\left\{i\left[\Phi_l\left(\omega + i\frac{g}{2}\right) + \Phi_r\left(\omega + i\frac{g}{2}\right)\right]\right\} = 1, \tag{6.598}$$

or

$$|r_l r_r| \exp\left[-\frac{g}{2}\left(\frac{d\Phi_l}{d\omega} + \frac{d\Phi_r}{d\omega}\right)\right] = 1. \tag{6.599}$$

Letting g_c denote the solution of equation (6.599), we can express it, approximately, as

$$g_c \approx -\frac{|t_l|^2 + |t_r|^2}{\frac{d\Phi_l}{d\omega} + \frac{d\Phi_r}{d\omega}}. \tag{6.600}$$

We differ in the sign. The validity or invalidity of the sign could be ultimately determined according to examples of $\Phi_l(\omega)$, $\Phi_r(\omega)$. The exposition comprises other minor errors. For example, the Green function

$$G(x, x_s) = \begin{cases} c_r\left[e^{ik(x-x_s)} + r_r e^{-ik(x-x_s)}\right], & x \geq x_s, \\ c_l\left[e^{-ik(x-x_s)} + r_l e^{ik(x-x_s)}\right], & x \leq x_s, \end{cases} \tag{6.601}$$

is written without brackets, but also without c_r, c_l, which would have been obtained on a possible removal of the brackets. Here $k = \frac{\omega}{c}$ and x_s is the source point. The Green function should be continuous at x_s, but $\frac{\partial}{\partial x}G(x, x_s)$ should have a jump $\frac{2}{c}$ at $x = x_s$. The authors set the jump to unity dubiously.

In fact, we solve the following equations:

$$\begin{aligned} c_r(1 + r_r) - c_l(1 + r_l) &= 0, \\ c_r(1 - r_r) - c_l(-1 + r_l) &= \frac{2}{i\omega}. \end{aligned} \tag{6.602}$$

We find

$$c_r = \frac{1}{i\omega}\frac{1 + r_l}{1 - r_l r_r}, \quad c_l = \frac{1}{i\omega}\frac{1 + r_r}{1 - r_l r_r}. \tag{6.603}$$

A consideration of gain rate may be restricted to the range $(\omega_0, \omega_0 + \delta\omega)$. The density of modes $\rho(\omega)$ per unit energy and length can be introduced. In the regime of the strong localization, the transmission t decreases exponentially with the size of the sample,

$$t \approx \exp\left(-\frac{L}{l_1}\right), \tag{6.604}$$

where l_1 is the localization length. The fluctuations are described by the logarithmically normal distribution, with the variance of the transmission logarithm $\frac{L}{l_2}$, $l_1, l_2 \approx l_t$.

A resonance state is described by the coordinate of its centre and two transmission coefficients

$$t_l(x) \approx e^{-\frac{x}{l_1}}, \quad t_r(x) \approx e^{-\frac{L-x}{l_1}}. \tag{6.605}$$

The nominator of (6.600) is in fact denoted by τ_0. A study of dwell times has been mentioned in Burin et al. (2002). But the fact that the nominator comprises a sum has not been touched on

$$g(x) \approx \frac{1}{\tau_0}\left\{\exp\left(-\frac{2x}{l_1}\right) + \exp\left[-\frac{2(L-x)}{l_1}\right]\right\}, \tag{6.606}$$

$$g \equiv g\left(\frac{L}{2}\right) \approx \frac{2}{\tau_0}\exp\left(-\frac{L}{l_1}\right). \tag{6.607}$$

Here we have completed a coefficient which is not present in the cited paper, however.

The authors remark that the exponential dependence has not been obtained in the literature as, e.g. in (Jiang and Soukoulis 1999). The lasing threshold strongly fluctuates from mode to mode depending on the fluctuation of the centre x, transmission, and dwell time. Strong fluctuations of lasing threshold are also demonstrated by the numerical simulations.

Patra (2002) considers an optical cavity, which is coupled to the outside by an opening that is small compared to the wavelength of the emitted radiation. As the opening is small, it is possible to work with the true modes of the cavity. These modes are distinguished by their eigenfrequencies ω_i, $i = 1, 2, \ldots, N_p$, as is asserted with respect to the irregular shape of the cavity. A finite number of modes can be connected with the size of the cavity, but the dependence has not been expressed. Each mode i can be described by the number n_i of photons in it. Photons in mode i can escape through the opening with the rate g_i.

The cavity is filled with an amplifying medium. This medium is modelled as a four-level laser dye. The density of excited atoms at the point $\mathbf{r}$ in the cavity is $N(\mathbf{r})$. Excitations are created by pumping with the rate $P(\mathbf{r})$ and are lost nonradiatively with the rate $a(\mathbf{r})$. The coupling of mode i to the medium at the point $\mathbf{r}$ is quantified by a coefficient $K_i(\mathbf{r}) \equiv w(\omega_i)|\Theta_i(\mathbf{r})|^2$, where $w(\omega)$ is the transition matrix element and $\Theta_i(\mathbf{r})$ is an eigenfunction of the mode i.

The semiclassical equations of motion for n_i and $N(\mathbf{r})$ respect the quantum physics of the matter only. Here n_i cannot mean the photon number. The Langevin terms in the equations are in fact to correct the "continuous variation" of the photon numbers, which is not appropriate, to jumps of the photon numbers. Similarly also the medium can be treated, when the cavity volume is divided into N_s cells (Lukš and Peřinová 2003).

To linearize the description, we write $n_i = \bar{n}_i + \delta n_i$ and $N(\mathbf{r}) = \bar{N}(\mathbf{r}) + \delta N(\mathbf{r})$, where $\bar{n}_i$ and $\bar{N}(\mathbf{r})$ are the average solutions, $\bar{n}_i \equiv \langle n_i \rangle$, $\bar{N}(\mathbf{r}) \equiv \langle N(\mathbf{r}) \rangle$, and δn_i and $\delta N(\mathbf{r})$ are new variables. Patra (2002) discretizes the equations in space as follows. In the cavity, he picks N_s points $\mathbf{r}_j$, $j = 1, 2, \ldots, N_s$. Then $K_{ij} \equiv K_i(\mathbf{r}_j)$, $N_j \equiv N(\mathbf{r}_j)$, and analogues for other quantities are introduced. For simplicity, P_j does not depend on j, $P_j \equiv P$. The stationary mean numbers $\bar{n}_i$ and mean densities $\bar{N}_j$ are the solution of equations

$$0 = -g_i \bar{n}_i + \frac{V}{N_s} \sum_{j=1}^{N_s} (\bar{n}_i + 1) K_{ij} \bar{N}_j, \quad i = 1, 2, \ldots, N_p, \tag{6.608}$$

$$0 = P_j - a_j \bar{N}_j - \sum_{i=1}^{N_p} (\bar{n}_i + 1) K_{ij} \bar{N}_j, \quad j = 1, 2, \ldots, N_s. \tag{6.609}$$

The factor $\frac{V}{N_s}$ is set equal to unity. On the stationary average variables the correlations $\langle \delta n_i(t) \delta n_{i'}(t) \rangle$, $\langle \delta n_i(t) \delta N_{j'}(t) \rangle$, $\langle \delta N_j(t) \delta n_{i'}(t) \rangle$, and $\langle \delta N_j(t) \delta N_{j'}(t) \rangle$ are determined.

The author defines the Fano factor $\mathcal{F}_i$ of the radiation emitted from the cavity in the mode i and the Fano factor $\mathcal{F}$ for a measurement where the photons emitted from the cavity in all modes are detected simultaneously. He discusses three simple cases of lasing regimes. He characterizes the single-mode laser. He treats the dependence of the Fano factor of the total radiation $\mathcal{F}$ and the dependence of the Fano factor of the radiation of one of lasing modes $\mathcal{F}_i$ on the pumping rate P. On several simplifying assumptions he compares the laser with $N_s = N_p = 10$ and $g_1 = \frac{1}{100}$, $g_i = \frac{1}{10}$ for $i = 2, \ldots, 10$ with the laser with $N_s = N_p = 10$ and all $g_i = \frac{1}{100}$. In the second case, the mode with the smallest g_i becomes the lasing mode and the behaviour is similar to a single-mode laser. In the third case, the mode competition occurs. Even though the total radiation behaves similarly as in the case of a single-mode laser, the Fano factor of any of lasing modes grows excessively.

Patra (2002) intends to study the mode-competition noise. He holds a spatial overlap of the modes for the necessary condition of the mode competition. He recalls the ensemble of cavities with small variations in shape or scatterer positions. Neither the volume V nor the hole diameter d should vary. Then, for instance, in the descriptions (6.608) and (6.609) oriented to the average variables, the coefficients begin to be interpreted as random variables. A parameter of their distribution is, for instance, the mean loss rate $\bar{g}$ that depends on the frequency ω. The expression we do not reproduce here is related to the assumption that neither the volume V nor the hole diameter d changes.

It is assumed that the loss rates g_i are independent and identically distributed. For simplicity, $K_{ij} = |\Theta_{ij}|^2$ is computed, where Θ_{ij} are elements of an $N_s \times N_s$ random unitary matrix. The author, for instance, has calculated the Fano factor $\mathcal{F}$ of the radiation emitted from the primary lasing mode depending on the pumping rate P. Along with the quantity $\mathcal{F}$ he has presented the number of modes above the lasing threshold N_1. Actually, the axes are denoted as $\frac{P}{g}$ and $\frac{\mathcal{F}-1}{g}$. Whereas these quantities have been determined for a particular sample, a number of further quantities have been found using 9×10^5 samples, e.g. the average of the Fano factor depending on the number of modes above the threshold N_1.

Trying to interpret experiments, the author recalls the spatial confinement of light, which is assumed in the experiments with the random lasers and has been confirmed, as we know, by simulations. The author holds these modes somewhat contradictorily for virtual multimode chaotic cavities. The deliberateness of this interpretation is underlined by considering the difference between these totally open chaotic cavities and the chaotic cavities with a small opening, which are studied in Patra (2002). Again he reminds of the fact that his paper does not include the modes which overlap. Therefore a comparison between the papers by (Zacharakis 2000, Cao 2001), in which the author is interested, cannot be based on the model. Also the concept of a "more or less" resonant feedback only glints in the exposition.

Cao (2003) has contributed a chapter on lasing in disordered media. Lasing is light amplification by stimulated emission with feedback. The author has distinguished two kinds of feedback: intensity or energy feedback and field or amplitude feedback. These mechanisms are also called incoherent (or nonresonant) feedback and coherent (or resonant) feedback, respectively. She has let l_t denote the transport mean free path.

Many experiments have been classified as random lasers with incoherent feedback. Some experiments have been reviewed as random lasers with coherent feedback. Transition between the two types of random lasers has been characterized as, e.g. in Cao et al. (2001). Classical and quantum theories have been reviewed in short. It has been admitted that the characteristics of lasing modes in three-dimensional random media are not fully understood. A previous theory has predicted that in a three-dimensional random medium with $l_{\mathrm{t}} > \lambda$, the quasimodes extend over the entire sample. The experiments have shown that the size of individual lasing modes is much smaller than that of the entire sample even when $l_{\mathrm{t}} \gg \lambda$, cf. (Apalkov et al. 2002).

Patra (2003) has noted that, in the theoretical treatment of disordered materials, two particular geometries are of great importance, the disordered slab and the chaotic cavity. Random-matrix theory began at the moment, when Wigner (1956) formulated the eigenvalue distribution for a closed chaotic cavity. For chaotic cavities with broken time-reversal symmetry, the distribution of imaginary part of eigenvalue (the decay rate distribution) is known analytically for an opening of any size (Fyodorov and Sommers 1997). The open systems in optics are not attributed broken time-reversal symmetry, but only "approximability" by a cavity with broken symmetry and an opening of half the real size.

Patra (2003) has studied a disordered slab of the length L and the width N. He has added an argument that N can be identified with the number of propagating modes for an ideal coupling at the centre of the conduction band. From the results of the numerical simulations, he has found mainly that the mean free path l is in very good approximation given by the relation

$$l = \frac{6}{w^{\frac{3}{2}}}, \tag{6.610}$$

where w is an amount of randomness, $0.3 \leq w \leq 1.0$. He has distinguished the diffusive and localized regimes. In both cases, he has also been able to find simple analytical formulae, which describe the decay rate distribution well. It has been proposed that these distributions depend on a scaling factor γ_0,

$$\gamma_0 = \frac{2l}{L^2} \tag{6.611}$$

in the diffusive regime and

$$\gamma_0 = \frac{a}{N^2} \exp\left(-\frac{L}{a\xi}\right), \tag{6.612}$$

with $a = 1.12$, $\xi = \frac{N+1}{2}l$, in the localized regime. The results can be applied to both electronic and photonic systems. It has been shown that, on certain conditions, the lasing threshold of a random laser is close to γ_0 in both the diffusive and the localized regime.

Florescu and John (2004) have considered the probability distribution for photons in the vicinity of position **r**, travelling with a wave vector **k**, and a master equation for the diffusion model with the gain describing this distribution. The steady-state photon distribution function has been characterized. The statistical properties of the emitted laser light have been studied. The dependence of the photon statistics on scatterer density, gain concentration, and position within a sample has been discussed.

van der Molen et al. (2007) have listed five optical and material properties of the sample, seven experimental details, and seven experimental data of the random laser needed for a quantitative comparison between different experiments. Comparisons between different theories and those between experiments and theory are foreseen too. Their recent experiment has been advertised, in fact, by a comparison with 8 perfectly reported ones from among 160 papers. An analysis of the model in (Mujumdar et al. 2004) has been added.

6.5.3 Modal Decomposition in Optical Resonators

Cheng and Siegman (2003) have reapproached the quantum description of a laser. They have mentioned the paper (Deutsch 1991), where the quantization of quasimodes (let us call them according to other work) is expounded, and the paper (Ho 1998), where rather the pseudomodes (let us call them with respect to other literature) are paid attention to. Cheng and Siegman (2003) contribute with a kind of generalized quasimode approach.

The authors introduce nonorthogonal modes u_n and nonstandard annihilation (creation) operators $\hat{a}_{u_n}$ ($\hat{a}_{u_n}^\dagger$). On using these quantities, the electric-field operator is expressed. A kind of normalization is represented just by a formal identity of the proposed expansion with that of the electric-field operator in terms of orthogonal modes.

Next, adjoint modes ϕ_n are introduced, which satisfy the biorthogonal relation

$$(\phi_n|u_m) \equiv \int \phi_n^* u_m \, \mathrm{d}^3\mathbf{x} = \delta_{nm}. \tag{6.613}$$

The authors recall that $(\phi_n|\phi_n) > 1$ (Siegman 1989). A description of the laser using quantum Langevin equations is provided, where the correlation relations of the atomic noise operators can be obtained from the Einstein relation (Sargent III 1977). We would remark that these correlation relations involve reservoir averages on the right-hand sides, which may suggest other commutation relations. The authors assume that the laser is in a single mode oscillation and label the oscillation mode as u_0. Finally, they derive the well-known quantum laser line-width equations

$$\Delta\omega_\mathrm{l} \approx (\phi_0|\phi_0)\frac{\hbar\omega_\mathrm{l}\gamma_\mathrm{c}^2}{2P_\mathrm{l}}, \tag{6.614}$$

where ω_l is the centre frequency, γ_l is the cavity decay rate, and $P_\mathrm{l} = |A_0|^2\hbar\omega_\mathrm{l}\gamma_\mathrm{c}$, with $A_0 = \langle\hat{a}_{u_0}\rangle_\mathrm{R}$ being the reservoir averaged steady-state solutions, which are quantum corrected.

6.5.4 Chaotic Resonators

Gräf et al. (1992) studied the distribution of eigenfrequencies in a microwave stadium billiard with chaotic dynamics. Alt et al. (1995) paid attention to wave functions or decay amplitudes (widths). The cavity was coupled to three antennas. The widths of the resonances are proportional to the squares of the eigenfunction at the respective locations of the antennas. Tests were in perfect agreement with predictions of Gaussian orthogonal ensemble for the statistics of eigenfunctions.

A resonance labelled μ has a frequency ω_μ and decay widths $\Gamma_{\mu j}$, $j = 1, 2, 3$, where j labels an antenna. The total width, Γ_μ, is

$$\Gamma_\mu = \sum_{j=1}^{3} \Gamma_{\mu j} + \Gamma_{\mu,\mathrm{wall}}, \tag{6.615}$$

where $\Gamma_{\mu,\mathrm{wall}}$ represents dissipation in the walls of a superconducting cavity. The Gaussian orthogonal ensemble predicts a Gaussian distribution for the decay amplitudes $\pm\sqrt{\Gamma_{\mu j}}$, or a gamma distribution for $\Gamma_{\mu j}$ with fixed j. The prediction can be tested that the decay amplitudes for different channels are uncorrelated. This holds provided that the distances between neighbouring antennas are larger than the maximum wavelength used in the experiment. The Gaussian orthogonal ensemble predicts that the resonance widths are statistically independent of the resonance energies.

Nöckel and Stone (1997) have focused on the effectively two-dimensional case of a deformed cylindrical resonator. Any resonator is characterized by a set of modes, each of which has a resonant frequency ω and resonance width $\delta\omega = \frac{1}{\tau}$, where τ

is the lifetime of a photon in the mode. A cylindrical dielectric resonator has long-lived resonances due to "whispering gallery" modes in which light is trapped by total internal reflection. The lifetime is finite in consequence of evanescent leakage, the optical analogue of quantum tunnelling. Asymmetric resonant cavities exhibit partially chaotic ray dynamics. A deformation of the cavity gives rise to a broadening of the whispering-gallery resonances. For small deformations the evanescent leakage increases. For large deformations the lifetime is shortened by a process called "chaos-assisted tunnelling" (Doron and Frischat 1995).

The predictions of ray optics are compared with wave solutions. Rays escaping near the critical angle are emitted tangent to the surface. In the framework of the theory of two-dimensional "billiards" (Gutzwiller 1990), it can be derived that the points of escape are the points of maximum curvature of the asymmetric resonant cavities. This implies strong emission maxima in the far field in directions tangent to the points of maximum curvature for appropriate values of the refractive index. For smaller indices the points of highest curvature may be screened by stable islands.

Gornik (1998) reminds of the interest in lasing not only at extremely low currents but also in directionality control expected from micrometre-size lasers. A circular symmetric cylinder or disk laser may be based on a whispering-gallery resonance. Light only weakly leaks out from the cavity. A power increase and output directionality is achieved by deformation, at which a circular cross section becomes quadrupolar. At small deformations chaotic whispering-gallery resonances occur. At larger deformations the lasers operate on bow-tie-shaped modes.

Gmachl et al. (1998) have reported on semiconductor microcylinder lasers, which have a high power and high directionality due to the deformation of cross section of an optical resonator. They have focused on semiconductor lasers that have an effective index of $n \approx 3.3$. The deformation has been called a flattened quadrupole. The quadrupolar shape has been derived from a resist pattern, which has a stadium shape. They chose quantum cascade lasers to emit light of wavelength $\lambda = 5.2\ \mu\mathrm{m}$ (mid-infrared).

While in circular and weakly deformed lasers whispering-gallery type modes occur, in the higher range of deformations a bow-tie shaped resonance is born, which is related to the improved performance of the lasers. Figures in the cited work have been obtained by the numerical solution of the Helmholtz equation. This analysis has been supported by a study of the short-wavelength limit of the problem by (Gutzwiller 1990, Reichl 1992). Poincaré's surface of section has been completed by the Husimi function for a particular deformation (LeBoeuf and Saraceno 1990).

Patra et al. (2000) have presented a random-matrix theory for the linewidth of a laser cavity in which the radiation propagates chaotically. They continue the paper (Misirpashaev and Beenakker 1998), which contains results on the lasing threshold and mode competition. They investigate the Petermann factor K, or excess noise factor. The factor K is expressed in terms of a non-Hermitian matrix H, $\mathsf{H} = \mathsf{H}_0 - \mathrm{i}\pi \mathsf{W}\mathsf{W}^T$. Here H_0 is an $M \times M$ Hermitian random matrix and W is a real, nonrandom $M \times N$ coupling matrix, M is the number of considered modes, and N is the number of the output channels. At the end of the calculation the limit

$M \to \infty$ is taken. The distribution of the factor K is calculated exactly on the assumption that $N = 1$. It is surprising that the mean Petermann factor converges nonanalytically to unity for the transmission probability T of the output channel going to zero.

Frahm et al. (2000) relate the Petermann factor with the general theory of resonant scattering. They concentrate themselves on the Petermann factor in chaotic cavities in the case of broken time-reversal symmetry. They have calculated that the average $\langle K \rangle$ scales with the square root of the number of scattering channels N. The exact result is compared with the result of a numerical simulation. Only from a numerical simulation they have found that the sublinear increase holds also in the case of preserved time-reversal symmetry.

Hackenbroich et al. (2001) recall the facts known since the development of the quantum theory of lasing in the 1960s. Below the laser threshold, the photon statistics is well represented by a thermal distribution. Above the threshold, nonlinear interactions stabilize the field intensity, and the photon statistics approaches the Poissonian distribution. The authors mention the renewal of interest in the photon statistics of amplifying media. This interest goes hand in hand with the attention paid to random media. While linear random amplifiers were intensively studied, in about the same time, there were few results for an amplifying random medium above the lasing threshold. The authors concentrate themselves on a chaotic laser resonator, which is considered to be a random medium. They restrict themselves to the single-mode lasing.

The randomness means that an experiment is performed on an ensemble of modes. The variation of parameters of a single resonator and the variation of shape of different resonators are typical, manners how to reach the randomness.

A chaotic resonator is filled with an active medium. The radiation is allowed to go out through a single opening, a partially transmitting mirror. This mirror reflects back a photon with the mean probability R. It transmits a photon with the mean probability T. The cavity is connected to a waveguide and a frequency selective photodetector. The waveguide supports M modes. It is assumed that $MT \ll 1$. The mean cavity escape rate $\bar{\Gamma}$ and the counting time t are introduced. The frequency resolution requires a longer counting time t. The active medium is described by three parameters $\mathcal{A}$, $\mathcal{B}$, and $\mathcal{C}$, which characterize the linear gain, the nonlinear saturation, and the total loss, respectively. The coefficients $\mathcal{A}$ and $\mathcal{B}$ are considered to be nonrandom, but $\mathcal{C}$ is a random quantity, since $\mathcal{C} = \Gamma + \kappa$, where Γ is the photon escape rate and κ is the absorption rate, which is nonrandom, but Γ is a random variable.

The authors apply the input–output theory (Gardiner and Collet 1985), mainly a linear coupling between the waveguide and the cavity field, which is described in terms of constants γ_p, $p = 1, 2, \ldots, M$. In the linear regime below the threshold, the input–output theory yields the theory according to (Beenakker 1998).

Then the authors consider the nonlinear regime near and above the threshold. The photocount statistics is given in dependence on statistical properties of the integrated cavity field intensity. Considering an ensemble of modes, Γ has the χ_ν^2 distribution

$$P(\Gamma) = A_\nu \Gamma^{\frac{\nu}{2}-1} \exp\left(-\frac{\nu\Gamma}{2\bar{\Gamma}}\right), \tag{6.616}$$

where $\nu = \beta M$, with $\beta = 1, 2$, and A_ν is a normalization constant. Here $\beta = 1$, if the system is time-reversal invariant and $\beta = 2$ otherwise. The special case $M = \beta = 1$ is known as the Porter–Thomas distribution. The factorial moments

$$\mu_r = \sum_{m=0}^{\infty} \frac{m!}{(m-r)!} p(m) \tag{6.617}$$

in the dependence on Γ become random quantities $\mu_r|_\Gamma$ and their probability density is

$$\mathcal{P}(\mu_r) = \int P(\Gamma)\delta\left(\mu_r - \mu_r|_\Gamma\right) \mathrm{d}\Gamma, \tag{6.618}$$

where

$$\mu_r|_\Gamma = \left\langle : W^r : \right\rangle = \Gamma^r \left\langle : I^r : \right\rangle. \tag{6.619}$$

The case $M \gg 1$ can be analyzed from the viewpoint of small relative fluctuations $\frac{\Delta\Gamma}{\bar{\Gamma}} \to 0$ for $M \to 0$. In the weak-coupling limit $\Gamma \to 0$, the moments have a power-law behaviour.

For short counting intervals, the photocount distribution is connected simply with the stationary photon-number distribution P_n in the lasing mode. This distribution depends on the parameters of lasing $\mathcal{A}$, $\mathcal{B}$, and $\mathcal{C}$. The authors have found the probability density $\mathcal{P}(\mu_1)$. The existence of $\mu_{\max} \equiv \sup_\Gamma \{\mu_1|_\Gamma\}$ may be utilized. They discuss the dependence of μ_1 on Γ, which in turn depends on $\mathcal{A}$, $\mathcal{B}$, κ. Above the threshold, it means that $\kappa < \mathcal{A}$; below the threshold, it means that $\kappa > \mathcal{A}$. For example, (a) $\kappa = 0.7 < 1 = \mathcal{A}$ is above the threshold and (b) $\kappa = 2.0 > 1 = \mathcal{A}$ is below the threshold. The mean photocount μ_1 in dependence on Γ, above the threshold, has a maximum, which can be approximated by a quadratic polynomial, and so its probability density has a peak in this maximum. Below the threshold, a supremum is rather the right term, and the density vanishes therein.

In the limit $\Gamma \to \infty$, $\mu_1 > 0$, which is an amplified spontaneous emission. Monotonic decrease (above the threshold) and monotonic increase (below the threshold) contribute with a broad peak in the right-hand neighbourhood $\mu_1(\infty)$ (above the threshold) and in the left-hand neighbourhood $\mu_1(\infty)$ (below the threshold), if $\bar{\Gamma}$ is not small. The mean photocount can be expressed in terms of parameters of the stationary distribution of the laser, but μ_r, with $r \geq 2$, already depends on other parameters.

Related fluctuation phenomena occur in neutron waves and in semiconductor quantum dots. Chaotic lasers feature not only wave chaos but also the nonlinear dynamics of the laser. The distribution of Γ may not be the Porter–Thomas distribution then.

Yang and Kellman (2000, 2002) have reported on a method of calculation of the semiclassical wave function for an Einstein–Brillouin–Keller quantizing torus. The semiclassical wave function of the "primitive type", which diverges at the classical caustics, is computed directly from running a classical trajectory.

Hackenbroich et al. (2002) recall the laser action in amplifying random media. While (Beenakker 1998) is devoted to the subthreshold radiation, the nonlinear lasing regime has been treated only modestly. Random laser modes have unusual properties. The modal functions and modal frequencies in the random lasers are analyzed in a statistical manner. The character of modes depends on the amount of disorder. For strong disorder, the localization of light may occur, i.e., the rise of well-separated modes in different regions of space. Weak disorder leads to an imperfect confinement of light and to strongly overlapping modes.

The authors state the lack of satisfactory scheme for the field quantization in random media. They pay attention to optical resonators with overlapping modes and develop a quantization scheme for such resonators. This scheme does not neglect the case of a chaotic resonator. The frequency spectrum of the resonator modes with appropriate correlations may be utilized. The mode coupling by dissipation has a stochastic character as well and its description has been conceived.

The authors consider a two-dimensional optical resonator for the sake of simplicity. It is actually a waveguide, which does not guide any fields. It is assumed that the electric field contained in the resonator is polarized in the direction of the translational symmetry of the resonator, which is the direction of the z-axis. The authors call these fields TM fields in conformity with the waveguide terminology. The terminology of the reflection and refraction at a planar interface, which is still a little related, would be "TE fields" on the contrary.

It is proper to perform an exact quantum description of the total system comprising the resonator and the external radiation field, since there are openings of the total width $\mathcal{W}$ in the boundary of the resonator. Hackenbroich et al. (2002) have provided an outline of a two-dimensional generalization of the approach to which we allude in Section 3.3. By the authors' exposition, this approach consists in the finding of the exact eigenfunctions $\psi_m(\omega, \mathbf{r})$ of the Helmholtz equation. This equation describes a harmonic wave in the free space along with boundary conditions of significance. The eigenfunctions are labelled by the continuous frequency ω and the integer m. The latter can be interpreted as a channel number. It is important that the conjugate of the field $\mathbf{A}(\mathbf{r}, t) = A(\mathbf{r}, t)\mathbf{e}_z$ is the field $\mathbf{\Pi}(\mathbf{r}, t) = \Pi(\mathbf{r}, t)\mathbf{e}_z$, where

$$A(\mathbf{r}, t) = c \int \psi^T(\omega, \mathbf{r})\mathsf{q}(\omega, t)\,\mathrm{d}\omega,$$

$$\Pi(\mathbf{r}, t) = \frac{1}{c} \int \psi^T(\omega, \mathbf{r})\mathsf{p}(\omega, t)\,\mathrm{d}\omega, \tag{6.620}$$

with $\psi(\omega, \mathbf{r})$ being an M-component vector with components $\psi_m(\omega, \mathbf{r})$, $\mathsf{q}(\omega, t)$ being an M-component vector with components $q_m(\omega, t)$, and $\mathsf{p}(\omega, t)$ is an M-component vector with components $p_m(\omega, t)$. To quantize the fields, we impose the canonical commutation relations $[\hat{q}_m(\omega, t), \hat{p}_n(\omega', t)] = \mathrm{i}\hbar\delta_{mn}\delta(\omega - \omega')\hat{1}$.

The authors opine that the modes-of-the-universe approach does not provide any information about the resonator itself. Evoking Feshbach's projector formalism, the authors assume the single-photon Hamiltonian of the form

$$\mathcal{H} = \sum_{\lambda} E_{\lambda}|\phi_{\lambda}\rangle\langle\phi_{\lambda}| + \sum_{m}\int E|\chi_m(E)\rangle\langle\chi_m(E)|\,\mathrm{d}E$$
$$+\sum_{\lambda,m}\int [W_{\lambda m}(E)|\phi_{\lambda}\rangle\langle\chi_m(E)| + \mathrm{H.\,c.}]\,\mathrm{d}E, \tag{6.621}$$

where $|\phi_{\lambda}\rangle$ are photon states with the energy E_{λ} in the decoupled resonator and $|\chi_m(E)\rangle$ are photon states with the energy E in the decoupled channel m.

The resonator wave functions $\phi_{\lambda}(\mathbf{r})$ are nonzero only within the resonator, and the channel wave functions $\chi(E,\mathbf{r})$ are nonzero only outside. These functions must satisfy the Dirichlet conditions along the boundary of the total system except the openings. The boundary condition along the ideal surface bridging the openings is arbitrary with the exception that the total Hamiltonian must be self-adjoint. Then the coupling amplitudes $W_{\lambda m}(E)$ are given by surface integrals of products of resonator and channel wave functions. The diagonalization of the Hamiltonian is performed according to the papers by Viviescas and Hackenbroich (2003) and Dittes (2000). It leads to a transformation of the relations (6.620). Particularly, it can be derived that the intracavity fields have the same representation as in the case of the closed cavity (and the Gaussian system of units). The field Hamiltonian is rather similar to the well-known system-bath Hamiltonian of quantum optics. As the nonresonant terms can be neglected, provided that the broadening of the resonator modes is much smaller than their frequency, this similarity is remarkable.

The equations for mode operators have the form

$$\dot{\hat{a}}_{\lambda}(t) = -\mathrm{i}\omega_{\lambda}\hat{a}_{\lambda}(t) - \pi\sum_{\lambda'}(\mathsf{W}\mathsf{W}^{\dagger})_{\lambda\lambda'}\hat{a}_{\lambda'}(t) + \hat{F}_{\lambda}(t), \tag{6.622}$$

where $\hat{F}_{\lambda}(t)$ is a well-defined noise operator and W is the coupling matrix with the elements $W_{\lambda m}$. The mode operators $\hat{a}_{\lambda}$ are coupled by the damping matrix $\mathsf{W}\mathsf{W}^{\dagger}$. The noise operators $\hat{F}_{\lambda}$ are appropriately correlated, $\langle\hat{F}_{\lambda}(t)\hat{F}_{\lambda'}^{\dagger}(t')\rangle \propto (\mathsf{W}\mathsf{W}^{\dagger})_{\lambda\lambda'}\delta(t-t')$. Hackenbroich et al. (2002) formulate the condition, on which equation (6.622) simplifies to the standard equation of motion for nonoverlapping modes. From the relation (6.622), they conclude that for the interesting case of wave chaos, the mode dynamics will be governed by a non-Hermitian random matrix (Fyodorov and Sommers 1997, Chalker and Mehlig 1998), eventually. They study also an open resonator above the laser threshold. They announce that they have derived the linewidth

$$\delta\omega = K\delta\omega_{\mathrm{ST}}, \tag{6.623}$$

where $\delta\omega_{\mathrm{ST}}$ is the fundamental (Shawlow–Townes) linewidth and K is the Petermann factor

$$K = \langle l|l\rangle\langle r|r\rangle. \tag{6.624}$$

Here $|r\rangle$ and $\langle l|$ are, respectively, the right and left eigenvectors of the non-Hermitian matrix $\mathcal{H}$, whose anti-Hermitian part expresses the damping and gain.

An experimental evidence that random-matrix theory is helpful has been provided in Méndez-Sánchez et al. (2003). In many microwave experiments with chaotic and disordered billiards, the distribution of reflection coefficients was investigated. For all regimes, the agreement between experimental and theoretical predictions has been almost perfect.

Chapter 7
Conclusions

In this book we have mainly reviewed canonical quantum descriptions of light propagation in a nonlinear dispersionless dielectric medium and in linear and nonlinear dispersive dielectric media. These descriptions have regularly been simplified by a transition to one-dimensional propagation, which has been illustrated also by some original simple descriptions. Besides this we have reported criticisms of description of light propagation in a nonlinear medium using a spatial variable instead of and similarly as the quantum-mechanical time parameter.

We have adopted the standpoint that macroscopic quantum electrodynamics arises both through reduction of a microscopic description and through immediate application of a quantization scheme to macroscopic fields. The origins of immediate macroscopic theories are connected with the possibility of determining the electrical displacement field as the canonical momentum (up to the sign) to the vector potential. One may also consider as a member of this class the possibility of starting from the assumption that the canonical momentum is the magnetic-induction field with the dual vector potential whose curl is the electric displacement field.

Due to a relatively large number of various fields and their components and the relative complexity of commutators, discrete and continuous expansions in terms of annihilation and creation operators must be presented. In the papers reviewed here, the ground (vacuum) state of the electromagnetic field has mostly been assumed. The Heisenberg picture of time evolution has been preferred. In one of these papers the opinion has been formulated that the quantum description of sources is to be treated separately, but a detailed treatment cannot be found there. As a criterion for the correctness of the theories, equivalence between the Heisenberg equations of motion for the electric displacement and magnetic-induction fields and the Maxwell equations is required.

Not always it is necessary to utilize the formalism of the electromagnetic field in the matter. For description of experiments with correlated photons it suffices to describe the electromagnetic field between optical devices and to know the input–output relations for the optical elements, both passive and active, with which the radiation is transformed.

The role of the momentum operator has been analyzed: its limited domain of application, one-dimensional propagation, is still appealing for its simplicity. The integral expression for the momentum operator has been given. It has been stressed

A. Lukš, V. Peřinová, *Quantum Aspects of Light Propagation*,
DOI 10.1007/b101766_7,

that the use of this operator as the generator of spatial progression depends on the knowledge of appropriate equal-space commutators between field operators. Results on the connection with the slowly-varying-envelope approximation have been presented, including some important applications.

The possibility has been examined of determining a phenomenological Hamiltonian, and the quantization scheme with dual potential has been expounded and appropriately modified, so that the real relative permittivity of the lossless medium is expressed at least locally (in terms of a quadratic Taylor polynomial). In simplifying to the one-dimensional case, an application to a nonlinear dispersive (Kerr-like) medium and quantum solitons has been presented. Various notions of the slowly-varying-envelope approximation and the quantum-paraxial approximation have been presented, along with applications of these concepts in both linear and nonlinear cases.

Multimode consideration of nonclassical effects is needed when one wants to improve the processing of images and of parallel signals. The quantum fluctuations of light at different spatial points in the plane perpendicular to the propagation direction of the light beam have been investigated.

A new application to a nonlinear dispersionless (Kerr-like) medium has been concerned with the propagation of arbitrary pulses. The need for renormalization has been declared even in this, one-dimensional case. As an example, the nonlinear process has been presented, whose description includes the renormalization. Finally, the application of one of the macroscopic approaches has lead to the description of several linear optical devices and to the study of radiating atoms in a linear medium, which is a recurrent theme by the way.

Derivation of a macroscopic description from a microscopic model has been performed; it is not possible – in the prevailing Heisenberg picture – to eliminate the matter fields completely. The argument has been provided not only with the description of the dispersion up to any order or accuracy, but also with losses, as follows from the Kramers–Kronig relations.

The exposition of this microscopic model has been delivered from the viewpoint of the macroscopic variables and the operator-valued noise current, which is a form of the matter field. A nonlinear modification of this description of the linear dispersive absorbing medium has been performed for the Kerr medium. The application to a two-level atom positioned in the centre of a spherical cavity has been mentioned.

The magnetic properties are usually neglected, but they must be included in the phenomenological quantum description of negative-index materials. Even though the Casimir effect is not regularly connected with the propagation, an expression of the noise which is quantal in essence fits in the framework of the electromagnetic-field quantization.

A different notion of the macroscopic fields has been mentioned as well: These are not comprised in a microscopic model in advance, but approximately equal operators of measurable modes of the dressed matter fields.

The periodic and disordered media have been paid attention to, respecting their importance in photonics. The macroscopic approach to quantization of the electromagnetic field in a periodic medium has mainly been spoken of, but also the papers have been mentioned, which have utilized specific approaches. Finite periodic media can couple free-space modes. In applications, the corrugated waveguides are important. Photonic crystals are infinite or finite periodic media. The literature on one-dimensional periodic media dominates, since both theory and fabrication of such media are simpler.

We have mentioned the quantization of the electromagnetic field in a disordered medium mainly in connection with various physical studies. The macroscopic approach to the quantization of the electromagnetic field suffices usually, but a description of the disordered or random medium is not easy. We have cited the papers, whose authors have restricted themselves to the quantum input–output relations and the detection theory. This is also related to application of results of further fields of the quantum physics. Many papers have reported the random lasers. Even though we have intended to review rather the theory, we see that we have been able to only provide a partial review. In the essence of the matter is that much theoretical work is not published as the optical physics.

References

Abraham, M. (1909). *Rendiconti del Circolo Matematico di Palermo*, 28:1.
Abraham, M. (1910). *Rendiconti del Circolo Matematico di Palermo*, 30:33.
Abrahams, E., Anderson, P. W., Licciardello, D. C., and Ramakrishnan, T. V. (1979). *Phys. Rev. Lett.*, 42:673.
Abram, I. (1987). *Phys. Rev. A*, 35:4661.
Abram, I. and Cohen, E. (1991). *Phys. Rev. A*, 44:500.
Abram, I. and Cohen, E. (1994). *J. Mod. Opt.*, 41:847.
Agarwal, G. S. (1975). *Phys. Rev. A*, 11:230.
Agarwal, G. S. and Wolf, E. (1970). *Phys. Rev. D*, 2:2206.
Agrawal, G. P. (1989). *Nonlinear Fiber Optics*. Academic Press, New York.
Agrawal, G. P. (1995). *Nonlinear Fiber Optics*. Academic Press, San Diego, CA.
Ahmedov, H. and Duru, I. H. (2003). *J. Math. Phys.*, 44:5487.
Aiello, A. (2000). *Phys. Rev. A*, 62:063813.
Aiello, A. and Woerdman, J. P. (2005). *Phys. Rev. A*, 72:060101.
Akkermans, E., Wolf, P. E., and Maynard, R. (1986). *Phys. Rev. Lett.*, 56:1471.
Allen, L., Barnett, S. M., and Padgett, M. J. (2003). *Optical Angular Momentum*. Institute of Physics Publishing (IOP), Bristol.
Allen, L. and Stenholm, S. (1992). *Opt. Commun.*, 93:253.
Alt, H., Gräf, H.-D., Harney, H. L., Hofferbert, R., Lengeler, H., Richter, A., Schardt, P., and Weidenmüller, H. A. (1995). *Phys. Rev. Lett.*, 74:62.
Altman, A. R., Köprülü, K. G., Corndorf, E., Kumar, P., and Barbosa, G. A. (2005). *Phys. Rev. Lett.*, 94:123601.
Anderson, P. W. (1958). *Phys. Rev.*, 109:1492.
Antezza, M., Pitaevskii, L. P., Stringari, S., and Svetovoy, V. B. (2006). *Phys. Rev. Lett.*, 97:223203.
Apalkov, V. M., Raikh, M. E., and Shapiro, B. (2002). *Phys. Rev. Lett.*, 89:016802.
Artoni, M. and Loudon, R. (1997). *Phys. Rev. A*, 55:1347.
Artoni, M. and Loudon, R. (1998). *Phys. Rev. A*, 57:622.
Artoni, M. and Loudon, R. (1999). *Phys. Rev. A*, 59:2279.
Assanto, G., Laureti-Palma, A., Sibilia, C., and Bertolotti, M. (1994). *Opt. Commun.*, 110:599.
Babiker, M. and Loudon, R. (1983). *Philos. Trans. R. Soc. Lond. Ser. A*, 385:439.
Bache, M., Brambilla, E., Gatti, A., Lugiato, L. A. (2004). *Phys. Rev. A*, 70:023823.
Balachandran, R. M. and Lawandy, N. M. (1995). *Opt. Lett.*, 20:1271.
Ban, M. (1994). *Phys. Rev. A*, 49:5078.
Barash, Yu. S. and Ginzburg, V. L. (1975). *Sov. Phys. Usp.*, 18:305.
Barnett, S. M., Gilson, C. R., Huttner, B., and Imoto, N. (1996). *Phys. Rev. Lett.*, 77:1739.
Barnett, S. M., Jeffers, J., Gatti, A., and Loudon, R. (1998). *Phys. Rev. A*, 57:2134.
Barnett, S. M., Matloob, R., and Loudon, R. (1995). *J. Mod. Opt.*, 42:1165.
Barnett, S. M. and Radmore, P. M. (1988). *Opt. Commun.*, 68:364.
Barton, G. (1987). *Proc. R. Soc. London, Ser. A*, 410:175.

A. Lukš, V. Peřinová, *Quantum Aspects of Light Propagation*,
DOI 10.1007/b101766_BM2,

Barton, G. (1991a). *J. Phys. A*, 24:991.
Barton, G. (1991b). *J. Phys. A*, 24:5533.
Barton, G. (2001). *J. Phys. A: Math. Gen.*, 34:4083.
Bechler, A. (1999). *J. Mod. Opt.*, 46:901.
Beenakker, C. W. J. (1997). *Rev. Mod. Phys.*, 69:731.
Beenakker, C. W. J. (1998). *Phys. Rev. Lett.*, 81:1829.
Beenakker, C. W. J. (1999). In Fouque, J.-P., editor, *Diffuse Waves in Complex Media*, NATO Science Series C531, pp. 137–164. Kluwer Academic Publishers, Dordrecht.
Beenakker, C. W. J., Paasschens, J. C. J., and Brouwer, P. W. (1996). *Phys. Rev. Lett.*, 76:1368.
Bekenstein, J. D. and Schiffer, M. (1994). *Phys. Rev. Lett.*, 72:2512.
Ben-Aryeh, Y., Lukš, A., and Peřinová, V. (1992). *Phys. Lett. A*, 165:19.
Ben-Aryeh, Y. and Serulnik, S. (1991). *Phys. Lett. A*, 155:473.
Bennink, R. S., Bentley, S. J., Boyd, R. W., and Howell, J. C. (2004). *Phys. Rev. Lett.*, 94:033601.
Bergman, K. and Haus, H. A. (1991). *Optics Lett.*, 16:663.
Berman, P. R. (2004). *Phys. Rev. A*, 69:022101.
Berman, P. R. (2007). *Phys. Rev. A*, 76:042106.
Bespalov, V. G., Kozlov, S. A., Shpolyanskiy, Yu. A., and Walmsley, I. A. (2002). *Phys. Rev. A*, 66:013811.
Białynicka-Birula, Z. and Białynicki-Birula, I. (1987). *J. Opt. Soc. Am.*, 4:1621.
Białynicki-Birula, I. (1996a). In Wolf, E., editor, *Progress in Optics*, Vol. 36, p. 245. North-Holland, Amsterdam.
Białynicki-Birula, I. (1996b). In Eberly, J. H., Mandel, L., and Wolf, E., editors, *Coherence and Quantum Optics VII*, p. 313. Plenum Press, New York.
Białynicki-Birula, I. (1998). *Phys. Rev. Lett.*, 80:5247.
Białynicki-Birula, I. (2000). *Opt. Commun.*, 179:237.
Białynicki-Birula, I. and Białynicka-Birula, Z. (1975). *Quantum Electrodynamics*. Pergamon, Oxford.
Björk, G., Söderholm, J., and Sánchez-Soto, L. L. (2004). *J. Opt. B: Quantum Semiclass. Opt.*, 6:S478.
Bjorken, J. D. and Drell, S. D. (1965). *Relativistic Quantum Fields*. McGraw-Hill, New York.
Bleany, B. I. and Bleany, B. (1985). *Electricity and Magnetism*. Oxford University Press, Oxford.
Bloembergen, N. (1965). *Nonlinear Optics*. Benjamin, New York.
Bloembergen, N. and Pershan, P. S. (1962). *Phys. Rev.*, 128:606.
Blow, K. J., Loudon, R., and Phoenix, S. J. D. (1991). *J. Opt. Soc. Am. B*, 8:1750.
Blow, K. J., Loudon, R., and Phoenix, S. J. D. (1992). *Phys. Rev. A*, 45:8064.
Blow, K. J., Loudon, R., Phoenix, S. J. D., and Sheperd, T. J. (1990). *Phys. Rev. A*, 42:4102.
Bolthausen, E. and Sznitman, A.-S. (2002). *Ten Lectures on Random Media*. Birkhäuser, Basel.
Bordag, M., Geyer, B., Klimchitskaya, G. L., and Mostepanenko, V. M. (2000). *Phys. Rev. Lett.*, 85:503.
Bordag, M., Mohideen, U., and Mostepanenko, V. M. (2001). *Phys. Rep.*, 353:1.
Born, M. and Infeld, L. (1934). *Proc. Roy. Soc. Lond. A*, 147:522.
Born, M. and Infeld, L. (1935). *Proc. Roy. Soc. Lond. A*, 150:141.
Born, M. and Wolf, E. (1968). *Principles of Optics*. Pergamon Press, New York.
Born, M. and Wolf, E. (1999). *Principles of Optics*, 7th ed. Cambridge University Press, Cambridge, England.
Boyd, R. W. (2003). *Nonlinear Optics*, 2nd ed. Academic Press, New York.
Boyer, T. H. (1974). *Phys. Rev. A*, 9:2078.
Boyer, T. H. (2003). *Am. J. Phys.*, 71:990.
Bowmeester, D., Ekert, A., and Zeilinger, A., editors (2000). *The Physics of Quantum Information*. Springer-Verlag, Berlin.
Bozhevolnyi, S. I., Boltasseva, A., Søndergaard, T., Nikolajsen, T., and Leosson, K. (2005). *Opt. Commun.*, 250:328.

Bressi, G., Carugno, G., Onofrio, R., and Ruoso, G. (2002). *Phys. Rev. Lett.*, 88:041804.
Brevik, I., Aarseth, J. B., and Høye, J. S. (2002). *Phys. Rev. E*, 66:026119.
Bronk, B. V. (1965). *J. Math. Phys. (NY)*, 6:228.
Brouwer, P. W. (1998). *Phys. Rev. B*, 57:10526.
Brouwer, P. W. and Beenakker, C. W. J. (1996). *J. Math. Phys.*, 37:4904.
Brown, S. A. and Dalton, B. J. (2001a). *J. Mod. Opt.*, 48:597.
Brown, S. A. and Dalton, B. J. (2001b). *J. Mod. Opt.*, 48:639.
Brown, S. A. and Dalton, B. J. (2002). *J. Mod. Opt.*, 49:1009.
Brown, L. S. and Maclay, G. J. (1969). *Phys. Rev.*, 184:1272.
Bruce, N. A. and Chalker, J. T. (1996). *J. Phys. A: Math. Gen.*, 29:3761.
Brun, T. A. and Barnett, S. M. (1998). *J. Mod. Opt.*, 45:777.
Bruno, P. (2002). *Phys. Rev. Lett.*, 88:240401.
Burin, A. L., Ratner, M. A., Cao, H., and Chang, R. P. H. (2001). *Phys. Rev. Lett.*, 87:215503.
Burin, A. L., Ratner, M. A., Cao, H., and Chang, S. H. (2002). *Phys. Rev. Lett.*, 88:093904.
Burnham, D. C. and Weinberg, D. L. (1970). *Phys. Rev. Lett.*, 25:84.
Caetano, D. P. and Souto Ribeiro, P. H. (2004). *Opt. Commun.*, 239:121.
Cahill, K. E. and Glauber, R. J. (1999). *Phys. Rev. A*, 59:1538.
Campos, R. A., Saleh, B. E. A., and Teich, M. A. (1990). *Phys. Rev. A*, 42:4127.
Cao, H. (2003). In Wolf, E., editor, *Progress in Optics*, Vol. 45, pp. 317–370. Elsevier Science (North-Holland), Amsterdam.
Cao, H., Ling, Y., Xu, J. Y., Cao, C. Q., and Kumar, P. (2001). *Phys. Rev. Lett.*, 86:4524.
Cao, H., Xu, J. Y., Zhang, D. Z., Chang, S.-H., Ho, S. T., Seelig, E. W., Liu, S., and Chang, R. P. H. (2000). *Phys. Rev. Lett.*, 84:5584.
Cao, H., Zhao, Y. G., Ho, S. T., Seelig, E. W., Wang, Q. H., and Chang, R. P. (1999). *Phys. Rev. Lett.*, 82:2278.
Carmichael, H. J. (1987). *J. Opt. Soc. Am. B*, 4:1565.
Carniglia, C. K. and Mandel, L. (1971). *Phys. Rev. D*, 3:280.
Carter, S. J. and Drummond, P. D. (1991). *Phys. Rev. Lett.*, 67:3757.
Carter, S. J., Drummond, P. D., Reid, M. D., and Shelby, R. M. (1987). *Phys. Rev. Lett.*, 58:1841.
Casado, A., Fernández-Rueda, A., Marshall, T., Risco-Delgado, R., and Santos, E. (1997a). *Phys. Rev. A*, 55:3879.
Casado, A., Marshall, T. W., and Santos, E. (1997b). *J. Opt. Soc. Am. B*, 14:494.
Case, K. M. and Zweifel, P. F. (1976). *Linear Transport Theory*. Addison-Wesley, Reading, MA.
Casimir, H. B. G. and Polder, D. (1948). *Phys. Rev.*, 73:360.
Caticha, A. and Caticha, N. (1992). *Phys. Rev. B*, 46:479.
Caves, C. M. (1981). *Phys. Rev. D*, 23:1693.
Caves, C. M. (1982). *Phys. Rev. D*, 26:1817.
Caves, C. M. and Crouch, D. D. (1987). *J. Opt. Soc. Am. B*, 4:1535.
Caves, C. M. and Schumaker, B. L. (1985). *Phys. Rev. A*, 31:3068.
Centini, M., Peřina, Jr., J., Sciscione, L., Sibilia, C., Scalora, M., Bloemer, M. J., and Bertolotti, M. (2005). *Phys. Rev. A*, 72:033806.
Cerboneschi, E., Renzoni, F., and Arimondo, E. (2002). *J. Opt. B: Quantum Semiclass. Opt.*, 4:S267.
Chabanov, A. A., Stoytchev, M., and Genack, A. Z. (2000). *Nature*, 404:850.
Chalker, J. T. and Mehlig, B. (1998). *Phys. Rev. Lett.*, 81:3367.
Chan, H. B., Aksyuk, V. A., Kleiman, R. N., Bishop, D. J., and Capasso, F. (2001a). *Science*, 291:1941.
Chan, H. B., Aksyuk, V. A., Kleiman, R. N., Bishop, D. J., and Capasso, F. (2001b). *Phys. Rev. Lett.*, 87:211801.
Chefles, A. and Barnett, S. M. (1996). *J. Mod. Opt.*, 43:709.
Chen, F., Klimchitskaya, G. L., Mohideen, U., and Mostepanenko, V. M. (2003). *Phys. Rev. Lett.*, 90:160404.

Chen, F., Klimchitskaya, G. L., Mostepanenko, V. M., and Mohideen, U. (2006). *Phys. Rev. Lett.*, 97:170402.
Chen, F., Mohideen, U., Klimchitskaya, G. L., and Mostepanenko, V. M. (2002a). *Phys. Rev. Lett.*, 88:101801.
Chen, F., Mohideen, U., Klimchitskaya, G. L., and Mostepanenko, V. M. (2002b). *Phys. Rev. A*, 66:032113.
Chen, F., Mohideen, U., Klimtchitskaya, G. L., and Mostepanenko, V. M. (2005). *Phys. Rev. A*, 72:020101.
Chen, F., Mohideen, U., Klimtchitskaya, G. L., and Mostepanenko, V. M. (2006). *Phys. Rev. A*, 74:022103.
Chen, F., Mohideen, U., Klimchitskaya, G. L., and Mostepanenko, V. M. (2007). *Phys. Rev. Lett.*, 98:068901.
Cheng, Y.-J. and Siegman, A. E. (2003). *Phys. Rev. A*, 68:043808.
Chew, W. C. (1995). *Waves and Fields in Inhomogeneous Media*. IEEE Press, New York.
Chizhov, A. V., Nazmitdinov, R. G., and Shumovsky, A. S. (1991). *Quantum Opt.*, 3:1.
Chui, S. T. and Hu, L. B. (2002). *Phys. Rev. B*, 65:144407.
Clemmow, P. C. (1996). *The Plane Wave Spectrum Representation of Electromagnetic Fields*. Oxford University Press/IEEE Press, Oxford, UK.
Cohen-Tannoudji, C., Dupont-Roc, J., and Grynberg, G. (1989). *Photons and Atoms: Introduction to Quantum Electrodynamics*. Wiley, New York.
Collett, M. J. and Gardiner, C. W. (1984). *Phys. Rev. A*, 30:1386.
Collett, M. J., Loudon, R., and Gardiner, C. W. (1987). *J. Mod. Opt.*, 34:881.
Collett, M. J. and Walls, D. F. (1985). *Phys. Rev. A*, 35:2887.
Corona, M. and U'Ren, A. B. (2007). *Phys. Rev. A*, 76:043829.
Coulson, C. A. (1952). *Valence*. Clarendon Press, Oxford.
Crenshaw, M. E. (2003). *Phys. Rev. A*, 67:033805.
Crenshaw, M. E. (2004). *Opt. Commun.*, 235:153.
Crenshaw, M. E. and Bowden, C. M. (2000a). *Phys. Rev. Lett.*, 85:1851.
Crenshaw, M. E. and Bowden, C. M. (2000b). *Phys. Rev. A*, 63:013801.
Crenshaw, M. E. and Bowden, C. M. (2002). *J. Mod. Opt.*, 49:511.
Crouch, D. D. (1988). *Phys. Rev. A*, 38:508.
Cubukcu, E., Aydin, K., Ozbay, E., Foteinopoulou, S., and Soukoulis, C. M. (2003). *Phys. Rev. Lett.*, 91:207401.
Cui, T. J., Hao, Z.-C., Yin, X. X., Hong, W., and Kong, J. A. (2004). *Phys. Lett. A*, 323:484.
Dalton, B. J. and Babiker, M. (1997). *Phys. Rev. A*, 56:905.
Dalton, B. J., Barnett, S. M., and Garraway, B. M. (2003). In Bigelow, N. P., Eberly, J. H., Stroud, Jr., C. R., Walmsley, I. A., editors, *Coherence and Quantum Optics VIII, Proceedings of the Eighth Rochester Conference on Coherence and Quantum Optics*, p. 495. Kluwer Academic/Plenum Publishers, New York.
Dalton, B. J., Barnett, S. M., and Knight, P. L. (1999a). *J. Mod. Opt.*, 46:1107.
Dalton, B. J., Barnett, S. M., and Knight, P. L. (1999b). *J. Mod. Opt.*, 46:1315.
Dalton, B. J., Barnett, S. M., and Knight, P. L. (1999c). *J. Mod. Opt.*, 46:1495.
Dalton, B. J., Barnett, S. M., and Knight, P. L. (1999d). *J. Mod. Opt.*, 46:1559.
Dalton, B. J., Guerra, E. S., and Knight, P. L. (1996). *Phys. Rev. A*, 54:2292.
Dalton, B. J. and Knight, P. L. (1999a). *J. Mod. Opt.*, 46:1817.
Dalton, B. J. and Knight, P. L. (1999b). *J. Mod. Opt.*, 46:1839.
D'Aguanno, G., Akozbek, N., Mattiucci, N., Scalora, M., Bloemer, M. J., and Zheltikov, A. M. (2005). *Opt. Lett.*, 30:1998.
D'Aguanno, G., Centini, M., Scalora, M., Sibilia, C., Bertolotti, M., Bloemer, M. J., and Bowden, C. M. (2002). *J. Opt. Soc. Am. B*, 19:211.
da Silva, J. C., Khanna, F. C., Matos Neto, A., and Santana, A. E. (2002). *Phys. Rev. A*, 66:052101.
Davison, B. and Sykes, J. B. (1958). *Neutron Transport Theory*. Oxford University Press, New York.

Dechoum, K., Marshall, T. W., and Santos, E. (2000). *J. Mod. Opt.*, 47:1273.
de Groot, S. R. (1969). In de Boer, G. E. U. J., editor, *Studies in Statistical Mechanics*, Vol. 4, p. 11. North-Holland, Amsterdam.
de Lange, O. L. and Raab, R. E. (2006). *Am. J. Phys.*, 74:301.
Della Villa, A., Enoch, S., Tayeb, G., Pierro, V., Galdi, V., and Capolino, F. (2005). *Phys. Rev. Lett.*, 94:183903.
den Dekker, A. J. and van den Bos, A. (1997). *J. Opt. Soc. Am. B*, 14:547.
Deng, C., Haus, J. W., Sarangan, A., Mahfoud, A., Sibilia, C., Scalora, M., and Zheltikov, A. (2006). *Laser Physics*, 16:927.
Derjaguin, B. V. (1960). *Sci. Am.*, 203:47.
Derjaguin, B. V. and Abrikosova, I. I., (1957). *Sov. Phys. JETP*, 3:819.
Deutsch, I. H. (1991). *Am. J. Phys.*, 59:834.
Deutsch, D. and Candelas, P. (1979). *Phys. Rev. D*, 20:3063.
Deutsch, I. H. and Garrison, J. C. (1991a). *Opt. Commun.*, 86:311.
Deutsch, I. H. and Garrison, J. C. (1991b). *Phys. Rev. A*, 43:2498.
Diao, L. and Blair, S. (2007). *J. Opt. A: Pure Appl. Opt.*, 9:972.
Di Stefano, O., Savasta, S., and Girlanda, R. (1999). *Phys. Rev. A*, 60:1614.
Di Stefano, O., Savasta, S., and Girlanda, R. (2000). *Phys. Rev. A*, 61:023803.
Di Stefano, O., Savasta, S., and Girlanda, R. (2001a). *J. Mod. Opt.*, 48:67.
Di Stefano, O., Savasta, S., and Girlanda, R. (2001b). *J. Opt. B: Quantum Semiclass. Opt.*, 3:288.
Dirac, P. A. M. (1927). *Proc. Roy. Soc. Lond. A*, 11:4243.
Dirac, P. A. M. (1964). *Lectures on Quantum Mechanics*. Belfer Graduate School of Science, New York.
Dittes, F. M. (2000). *Phys. Rep.*, 339:215.
Dolling, G., Wegener, M., Soukoulis, C. M., and Linden, S. (2007). *Opt. Lett.*, 32:53.
Dorokhov, O. (1982). *Pis. Zh. Ėksp. Teor. Fiz.*, 36:259.
Doron, E. and Frischat, S. D. (1995). *Phys. Rev. Lett.*, 75:3661.
Drummond, P. D. (1990). *Phys. Rev. A*, 42:6845.
Drummond, P. D. (1994). In Evans, M. V. and Kielich, S., editors, *Modern Nonlinear Optics*, Part 3, p. 379, J. Wiley, New York.
Drummond, P. D. and Carter, S. J. (1987). *J. Opt. Soc. Am. B*, 4:1565.
Drummond, P. D., Carter, S. J., and Shelby, R. M. (1989). *Opt. Lett.*, 14:373.
Drummond, P. D. and Hardman, A. D. (1993). *Europhys. Lett.*, 21:279.
Drummond, P. D. and He, H. (1997). *Phys. Rev. A*, 56:R1107.
Drummond, P. D. and Hillery, M. (1999). *Phys. Rev. A*, 59:691.
Dung, H.-T., Buhmann, S. Y., Knöll, L., Welsch, D.-G., Scheel, S., and Kästel, J. (2003). *Phys. Rev. A*, 68:043816.
Dung, H.-T., Knöll, L., and Welsch, D.-G. (1998). *Phys. Rev. A*, 57:3931.
Dung, H.-T., Knöll, L., and Welsch, D.-G. (2000). *Phys. Rev. A*, 62:053804.
Dung H.-T., Knöll, L., and Welsch, D.-G. (2001). *Phys. Rev. A*, 64:013804.
Dung, H.-T., Knöll, L., and Welsch, D.-G. (2002a). *Phys. Rev. A*, 65:043813.
Dung, H.-T., Knöll, L., and Welsch, D.-G. (2002b). *Phys. Rev. A*, 66:063810.
Dutra, S. M. and Furuya, K. (1997). *Phys. Rev. A*, 55:3832.
Dutra, S. M. and Furuya, K. (1998a). *Europhys. Lett.*, 43:13.
Dutra, S. M. and Furuya, K. (1998b). *Phys. Rev. A*, 57:3050.
Dutra, S. M. and Nienhuis, G. (2000). *Phys. Rev. A*, 62:063805.
Dyson, F. J. (1962). *J. Math. Phys. (NY)*, 3:157.
Dzyaloshinskii, I. E., Lifshitz, E. M., and Pitaevskii, L. P. (1960). *Sov. Phys. JETP*, 9:1282.
Dzyaloshinskii, I. E., Lifshitz, E. M., and Pitaevskii, L. P. (1961). *Adv. Phys.*, 10:165.
Dzyaloshinskii, I. E. and Pitaevskii, L. P. (1959). *Sov. Phys. JETP*, 9:1282.
Emig, T. (2003). *Europhys. Lett.*, 62:466.
Emig, R. (2007). *Phys. Rev. Lett.*, 98:060801.
Emig, R., Graham, N., Jaffe, R. L., and Kardar, M. (2007). *Phys. Rev. Lett.*, 99:170403.

Engheta, N. (1998). *Microwave Opt. Technolog. Lett.*, 17:86.
Fano, U. (1956). *Phys. Rev.*, 103:1202.
Fano, U. (1961). *Phys. Rev.*, 124:1866.
Felbacq, D. and Bouchitté, G. (2005). *Phys. Rev. Lett.*, 94:183902.
Fermani, R., Scheel, S., and Knight, P. L. (2007). *Phys. Rev. A*, 75:062905.
Finlayson, N., Banyai, W. C., Wright, E. M., Seaton, C. T., Stegeman, G. I., Cullen, T. J., and Ironside, C. N. (1988). *Appl. Phys. Lett.*, 53:1144.
Fisher, R. A. (1983). *Optical Phase Conjugation*. Academic Press, New York.
Fiurášek, J., Křepelka, J., and Peřina, J. (1999a). *Opt. Commun.*, 167:115.
Fiurášek, J., Křepelka, J., and Peřina, J. (1999b). *Acta Phys. Slovaca*, 49:689.
Fiurášek, J. and Peřina, J. (1998). *Acta Phys. Slovaca*, 48:361.
Fiurášek, J. and Peřina, J. (1999). Quantum statistics of light in Raman and Brillouin couplers with losses. In Hrabovský, M., Štrba, A., and Urbańczyk, W., editors, *SPIE*, Vol. 3820, pp. 23–33. SPIE, Bellingham.
Fiurášek, J. and Peřina, J. (2000a). *J. Opt. B: Quantum Semiclass. Opt.*, 2:10.
Fiurášek, J. and Peřina, J. (2000b). *J. Mod. Opt.*, 47:1399.
Fleischhauer, M. (1999). *Phys. Rev. A*, 60:2534.
Florescu, L. and John, S. (2004). *Phys. Rev. E*, 69:046603.
Foteinopoulou, S., Economou, E. N., and Soukoulis, C. M. (2003). *Phys. Rev. Lett.*, 90:107402.
Foteinopoulou, S. and Soukoulis, C. M. (2003). *Phys. Rev. B*, 67:235107.
Fox, A. G. and Li, T. (1961). *Bell Sys. Tech. J.*, 40:453.
Frahm, K. M., Schomerus, H., Patra, M., and Beenakker, C. W. J. (2000). *Europhys. Lett.*, 49:48.
Franson, J. D. (1989). *Phys. Rev. Lett.*, 62:2205.
Friberg, S. R., Machida, S., and Yamamoto, Y. (1992). *Phys. Rev. Lett.*, 69:3165.
Fu, H. and Berman, P. R. (2005). *Phys. Rev. A*, 72:022104.
Fulling, S. A., Kaplan, L., and Wilson, J. H. (2007). *Phys. Rev. A*, 76:012118.
Fyodorov, Y. V. and Sommers, H.-J. (1997). *J. Math. Phys.*, 38:1918.
Galvez, E. J., Holbrow, G. H., Pysher, M. J., Martin, J. W., Cortemanche, N., Heilig, L., and Spencer, J. (2005). *Am. J. Phys.*, 73:127.
Garabedian, P. A. (1964). *Partial Differential Equations*. Chelsea, New York, Chap. 10.
Gardiner, C. W. and Collett, M. J. (1985). *Phys. Rev. A*, 31:3761.
Garraway, B. M. (1997a). *Phys. Rev. A*, 55:2290.
Garraway, B. M. (1997b). *Phys. Rev. A*, 55:4636.
Garraway, B. M. and Knight, P. L. (1996). *Phys. Rev. A*, 54:3592.
Gatti, A. (2003). In Bigelow, N. P., Eberly, J. H., Stroud, Jr., C. R., Walmsley, I. A., editors, *Coherence and Quantum Optics VIII, Proceedings of the Eighth Rochester Conference on Coherence and Quantum Optics*, p. 129. Kluwer Academic/Plenum Publishers, New York.
Gatti, A., Brambilla, E., Bache, M., Lugiato, L. A. (2004). *Phys. Rev. A*, 70:013802.
Gatti, A., Brambilla, E., and Lugiato, L. (2003). *Phys. Rev. Lett.*, 90:133603.
Gatti, A. and Lugiato, L. A. (1995). *Phys. Rev. A*, 52:1675.
Gea-Banacloche, J., Lu, N., Pedrotti, L. M., Prasad, S., Scully, M. O., and Wódkiewicz, K. (1990a). *Phys. Rev. A*, 41:369.
Gea-Banacloche, J., Lu, N., Pedrotti, L. M., Prasad, S., Scully, M. O., and Wódkiewicz, K. (1990b). *Phys. Rev. A*, 41:381.
Genet, C., Lambrecht, A., Maia Neto, P., and Reynaud, S. (2003). *Europhys. Lett.*, 62:484.
Geyer, B., Klimchitskaya, G. L., and Mostepanenko, V. M. (2003). *Phys. Rev. A*, 67:062102.
Ghosh, R., Hong, C. K., Ou, Z. Y., and Mandel, L. (1986). *Phys. Rev. A*, 34:3962.
Ghosh, R. and Mandel, L. (1987). *Phys. Rev. Lett.*, 59:1903.
Ginzburg, V. L. (1940). *J. Phys. USSR*, 2:441.
Glauber, R. J. (1963). *Phys. Rev.*, 130:2529.
Glauber, R. J. (1965). In Blandin, A., DeWitt, C., and Cohen-Tannoudji, C., editors, *Quantum Optics and Electronics*, p. 144. Gordon and Breach, London.

Glauber, R. J. and Lewenstein, M. (1989). In Tombesi, P. and Pike, E. R., editors, *Squeezed and Nonclassical Light*, p. 203. Plenum Press, New York.

Glauber, R. J. and Lewenstein, M. (1991). *Phys. Rev. A*, 43:467.

Gmachl, C., Capasso, F., Narimanov, E. E., Nöckel, J. U., Stone, A. D., Faist, J., Sivco, D. L., and Cho, A. Y. (1998). *Science*, 280:1556.

Gordon, J. P. (1986). *Opt. Lett.*, 11:662.

Gornik, E. (1998). *Science*, 280:1544.

Gouedard, C., Husson, D., Sauteret, C., Auzel, F., and Migus, A. (1993). *J. Opt. Soc. Am. B*, 10:2358.

Gräf, H.-D., Harney, H. L., Lengeler, H., Lewenkopf, C. H., Rangacharyulu, C., Richter, A., Schardt, P., and Weidenmüller, H. A. (1992). *Phys. Rev. Lett.*, 69:1296.

Graham, R. and Haken, H. (1968). *Z. Phys.*, 213:420.

Greenberger, D. M., Horne, M. A., and Zeilinger, A. (1993). *Phys. Today*, 46(8):22.

Greiner, W. (1998). *Classical Electrodynamics*. Springer, New York.

Grimes, C. A. and Grimes, D. M. (1991). *Phys. Rev. B*, 43:10780.

Gruner, T. and Welsch, D.-G. (1995). *Phys. Rev. A*, 51:3246.

Gruner, T. and Welsch, D.-G. (1996a). *Phys. Rev. A*, 53:1818.

Gruner, T. and Welsch, D.-G. (1996b). *Phys. Rev. A*, 54:1661.

Guo, W. (2002). *Am. J. Phys.*, 70:1039.

Guo, W. (2005). *Opt. Commun.*, 250:137.

Guo, W. (2007). *Phys. Rev. A*, 76:023834.

Gutzwiller, M. C. (1990). *Chaos in Classical and Quantum Mechanics*. Springer, New York.

Hackenbroich, G., Viviescas, C., Elattari, B., and Haake, F. (2001). *Phys. Rev. Lett.*, 86:5262.

Hackenbroich, G., Viviescas, C., and Haake, F. (2002). *Phys. Rev. Lett.*, 89:083902.

Haken, H. (1970). *Laser Theory*. Springer-Verlag, Berlin.

Hansen, P. B., Kloch, A., Aakjer, T., and Rasmussen, T. (1995). *Opt. Commun.*, 119:178.

Harbers, R., Moll, N., Mahrt, R. F., Erni, D., and Bächtold, W. (2005). *J. Opt. A: Pure Appl. Opt.*, 7:S230.

Hatami-Hanza, H. and Chu, P. L. (1995). *Opt. Commun.*, 119:347.

Haus, H. A. and Kärtner, F. X. (1992). *Phys. Rev. A*, 46:R1175.

Haus, H. A. and Lai, Y. (1990). *J. Opt. Soc. Am. B*, 7:386.

Haus, H. A., Watanabe, K., and Yamamoto, Y. (1989) *J. Opt. Soc. Am. B*, 6:1138.

Hawton, M. (1999). *Phys. Rev. A*, 59:3223.

Healey, W. P. (1982). *Nonrelativistic Quantum Electrodynamics*. Academic Press, London.

Heitler, W. (1954). *The Quantum Theory of Radiation*, 3rd edition. Oxford University Press/Clarendon Press, Oxford.

Henry, C. H. and Kazarinov, R. F. (1996). *Rev. Mod. Phys.*, 68:801.

Herec, J. (1999). *Acta Phys. Slovaca*, 49:731.

Herzog, T. J., Rarity, J. G., Weinfurter, H., and Zeilinger, A. (1994). *Phys. Rev. Lett.*, 72:629.

Hillery, M. and Drummond, P. D. (2001). *Phys. Rev. A*, 64:013815.

Hillery, M. and Mlodinow, L. D. (1984). *Phys. Rev. A*, 30:1860.

Hillery, M. and Mlodinow, L. D. (1997). *Phys. Rev. A*, 55:678.

Hinds, E. A. (1990). In Bates, D. and Bederson, B., editors, *Advances in Atomic, Molecular and Optical Physics*, Vol. 28, p. 237. Academic Press, New York.

Ho, K. C., Leung, P. T., Maassen van den Brink, A., and Young, K. (1998). *Phys. Rev. E*, 58:2965.

Ho, S.-T. and Kumar, P. (1993). *J. Opt. Soc. Am. B*, 10:1620.

Hopfield, J. J. (1958). *Phys. Rev.*, 112:1555.

Horne, M., Shimony, A., and Zeilinger, A. (1990). In Anandan, J. S., editor, *Quantum Coherence*, pp. 356–372. World Scientific, Singapore.

Høye, J. S., Brevik, I., and Aarseth, J. B. (2001). *Phys. Rev. E*, 63:051101.

Hradil, Z. (1996). *Phys. Rev. A*, 53:3687.

Huang, Y. and Gao, L. (2003). *Phys. Lett. A*, 318:592.

Huang, X. and Schaich, W. L. (2004). *Am. J. Phys.*, 72:1232.

Huang, D., Swanson, E. A., Lin, C. P., Schuman, J. S., Stinson, W. G., Chang, W., Hee, M. R., Flotte, T., Gregory, K., Puliafito, C. A., and Fujimoto, J. G. (1991). *Science*, 254:1178.
Hughes, B. D. (1996). *Random Walks and Random Environments*, Vol. 2. Clarendon Press, Oxford.
Hughes, S. (2005a). *Opt. Lett.*, 30:1393.
Hughes, S. (2005b). *Phys. Rev. Lett.*, 94:033903.
Hughes, S. (2005c). *Phys. Rev. Lett.*, 94:227402.
Huttner, B. and Barnett, S. M. (1992a). *Europhys. Lett.*, 18:487.
Huttner, B. and Barnett, S. M. (1992b). *Phys. Rev. A*, 46:4306.
Huttner, B., Baumberg, J. J., and Barnett, S. M. (1991). *Europhys. Lett.*, 16:177.
Huttner, B., Serulnik, S., and Ben-Aryeh, Y. (1990). *Phys. Rev. A*, 42:5594.
Hynne, F. and Bullough, R. K. (1990). *Phil. Trans. Roy. Soc. A* (London), 330:253.
Iannuzzi, D. and Capasso, F. (2003). *Phys. Rev. Lett.*, 91:029101.
Imoto, N. (1989). In Kobayashi, S., Egawa, H., Murayama, Y., and Nomura, S., editors, *Proc. 3rd Int. Symp. on Foundations of Quantum Mechanics (Tokyo)*, p. 306. Physical Society of Tokyo, Tokyo.
Inagaki, T. (1998). *Phys. Rev. A*, 57:2204.
Inoue, T. and Hori, H. (2001). *Phys. Rev. A*, 63:063805.
Intravaia, F., Henkel, C., and Lambrecht, A. (2007). *Phys. Rev. A*, 76:033820.
Itzykson, C. and Zuber, J. B. (1980). *Quantum Field Theory*. McGraw-Hill, New York.
Jaekel, M.-T. and Reynaud, S. (1991). *J. Phys. I*, 1:1395.
Janowicz, M., Reddig, D., and Holthaus, M. (2003). *Phys. Rev. A*, 8:043823.
Janszky, J., Sibilia, C., Bertolotti, M., Adam, P., and Petak, A. (1995). *J. Opt. B: Quantum Semiclass. Opt.*, 7:509.
Janszky, J., Sibilia, C., Bertolotti, M., and Yushin, Y. (1988). *J. Mod. Opt.*, 35:1757.
Jauch, J. M. and Watson, K. M. (1948). *Phys. Rev.*, 74:950.
Javanainen, J., Ruostekoski, J., Vestergaard, B., and Francis, M. R. (1999). *Phys. Rev. A*, 64:1959.
Jeffers, J. and Barnett, S. M. (1994). *J. Mod. Opt.*, 41:1121.
Jeffers, J., Barnett, S. M., Loudon, R., Matloob, R., and Artoni, M. (1996). *Opt. Commun.*, 131:66.
Jeffers, J. R., Imoto, N., and Loudon, R. (1993). *Phys. Rev. A*, 47:3346.
Jensen, S. M. (1982). *IEEE J. Quantum Electron.*, QE-18:1580.
Jiang, X., Li, Q., and Soukoulis, C. M. (1999). *Phys. Rev. B*, 59:R9007.
Jiang, X. and Soukoulis, C. M. (1999). *Phys. Rev. B*, 59:6159.
Jiang, X. and Soukoulis, C. M. (2000). *Phys. Rev. Lett.*, 85:70.
Joannapoulos, J. D., Meade, R. D., and Winn, J. N. (1995). *Photonic Crystals*. Princeton University, Princeton, NJ.
John, S. (1984). *Phys. Rev. Lett.*, 53:2169.
John, S. (1987). *Phys. Rev. Lett.*, 58:2486.
John, S. and Pang, G. (1996). *Phys. Rev. A*, 54:3642.
John, S., Pang, G., and Yang, Y. (1996). *J. Biomed. Opt.*, 1:180.
John, S. and Stephen, M. J. (1983). *Phys. Rev. B*, 28:6358.
Joobeur, A., Saleh, B. E. A., and Teich, M. C. (1994). *Phys. Rev. A*, 50:3349.
Karpati, A., Adam, P., Janszky, J., Bertolotti, M., and Sibilia, C. (2000). *J. Opt. B: Quantum Semiclass. Opt.*, 2:133.
Kelley, P. L. and Kleiner, W. H. (1964). *Phys. Rev.*, 136:316.
Kennedy, T. A. B. and Wright, E. M. (1988). *Phys. Rev. A*, 38:212.
Kenneth, O., Klich, I., Mann, A., and Revzen, M. (2002). *Phys. Rev. Lett.*, 89:033001.
Kenneth, O., Klich, I., Mann, A., and Revzen, M. (2003). *Phys. Rev. Lett.*, 91:029102.
Khanbekyan, M., Knöll, L., and Welsch, D.-G. (2003). *Phys. Rev. A*, 67:063812.
Khosravi, H. and Loudon, R. (1991). *Proc. Roy. Soc. Lond. A*, 433:337.
Khosravi, H. and Loudon, R. (1992). *Proc. Roy. Soc. Lond. A*, 436:373.
Kitagawa, M. and Yamamoto, Y. (1986). *Phys. Rev. A*, 34:3974.
Kittel, C. (1987). *Quantum Theory of Solids*, 2nd ed. J. Wiley, New York.
Klimchitskaya, G. L. and Mostepanenko, V. M. (2001). *Phys. Rev. A*, 63:062108.

Klingshirn, C. F. (1995). *Semiconductor Optics*. Springer-Verlag, Berlin, p. 57.
Klyshko, D. N. (1988). *Photons and Nonlinear Optics*. Gordon and Breach, New York.
Knöll, L. and Leonhardt, U. (1992). *J. Mod. Opt.*, 39:1253.
Knöll, L., Scheel, S., Schmidt, E., Welsch, D.-G., and Chizhov, A. V. (1999). *Phys. Rev. A*, 59:4716.
Knöll, L., Scheel, S., and Welsch, D.-G. (2001). In Peřina, J., editor, *Coherence and Statistics of Photons and Atoms*, p. 1. J. Wiley, New York.
Knöll, L., Vogel, W., and Welsch, D.-G. (1986). *J. Opt. Soc. Am. B*, 3:1315.
Knöll, L., Vogel, W., and Welsch, D.-G. (1987). *Phys. Rev. A*, 36:3803.
Knöll, L., Vogel, W., and Welsch, D.-G. (1990). *Phys. Rev. A*, 42:503.
Knöll, L., Vogel, W., and Welsch, D.-G. (1991). *Phys. Rev. A*, 43:543.
Knöll, L. and Welsch, D.-G. (1992). *Prog. Quantum Electron.*, 16:135.
Kobe, D. H. (1999). *Phys. Lett. A*, 253:7.
Kogan, Sh. (1996). *Electronic Noise and Fluctuations in Solids*. Cambridge University Press, Cambridge.
Kolobov, M. I. (1999). *Rev. Mod. Phys.*, 71:1539.
Kolobov, M. I. (2007). In Kolobov, M. I., editor, *Quantum Imaging*, p. 316. Springer-Verlag, New York.
Kolobov, M. I. and Kumar, P. (1993). *Opt. Lett.*, 18:849.
Kolobov, M. I. and Lugiato, L. A. (1995). *Phys. Rev. A*, 52:4930.
Kolobov, M. I. and Sokolov, I. V. (1989a). *Zh. Eksp. Teor. Fiz.*, 96:1945.
Kolobov, M. I. and Sokolov, I. V. (1989b). *Sov. Phys. JETP*, 69:1097.
Kondrat'ev, I. G. and Smirnov, A. I. (2003). *Phys. Rev. Lett.*, 91:249401.
Korolkova, N., Loudon, R., Gardavsky, G., Hamilton, M. W., and Leuchs, G. (2001). *J. Mod. Opt.*, 48:1339.
Korolkova, N. and Peřina, J. (1997a). *Opt. Commun.*, 137:263.
Korolkova, N. and Peřina, J. (1997b). *Opt. Commun.*, 136:135.
Korolkova, N. and Peřina, J. (1997c). *J. Mod. Opt.*, 44:1525.
Kovalev, V. F., Bychenkov, V. Yu., and Tikhonchuk, V. T. (2000). *Phys. Rev. A*, 61:033809.
Krause, D. E., Decca, R. S., López, D., and Fischbach, E. (2007). *Phys. Rev. Lett.*, 98:050403.
Kravtsov, V. E., Agranovich, V. M., and Grigorishin, K. I. (1991). *Phys. Rev. B*, 44:4931.
Kubo, R. (1962). *J. Phys. Soc. Jpn.*, 17:1100.
Kuchment, P. (2004). *Waves in Random Media*, 14:S107.
Kuga, Y. and Ishimaru, A. (1984). *J. Opt. Soc. Am. A*, 8:831.
Kumar, P. and Shapiro, J. H. (1984). *Phys. Rev. A*, 30:1568.
Kupiszewska, D. (1992). *Phys. Rev. A*, 46:2286.
Kupiszewska, D. and Mostowski, J. (1990). *Phys. Rev. A*, 41:4636.
Kurizki, G. and Haus, J. W., editors. (1994). *Journal of Modern Optics*, Special Issue Photonic Band Structures, Vol. **41**, pp. 171–408. Taylor and Francis, London.
Kurokawa, N. and Wakayama, M. (2002). *Indagationes Mathematicae, Nova Series*, 13:63.
Kweon, G. and Lawandy, N. M. (1995). *Opt. Commun.*, 118:388.
Lai, W. K., Bužek, V., and Knight, P. L. (1991). *Phys. Rev. A*, 43:6323.
Lai, Y. and Haus H. A. (1989). *Phys. Rev. A*, 40:844.
Lamoreaux, S. K. (1997). *Phys. Rev. Lett.*, 78:5.
Lamoreaux, S. K. (1999). *Am. J. Phys.*, 67:850.
Lamoreaux, S. K. (2005). *Rep. Prog. Phys.*, 68:201.
Landau, L. D. and Lifshitz, E. M. (1960). *Electrodynamics of Continuous Media*. Pergamon Press, Oxford.
Landau, L. D. and Lifshitz, E. M. (1963). *Electrodynamics of Continuous Media*. Pergamon Press, Oxford.
Landau, L. and Lifshitz, E. (1980). *Statistical Physics, Part II*. Pergamon Press, Oxford.
Landau, L. D., Lifshitz, E. M., and Pitaevskii, L. P. (1984). *Electrodynamics of Continuous Media*, Pergamon Press, Oxford.
Lang, R. and Scully, M. O. (1973). *Opt. Commun.*, 9:331.

Lang, R., Scully, M. O., and Lamb, Jr, W. E. (1973). *Phys. Rev. A*, 7:1788.
Law, C. K., Walmsley, I. A., and Eberly, J. H. (2000). *Phys. Rev. Lett.*, 84:5304.
Lawandy, N. M. and Balachandran, R. M. (1995). *Nature*, 373:204.
Lawandy, N. M., Balachandran, R. M., Gomes, A. S. L., and Sauvain, E. (1994). *Nature*, 368:436.
Lax, M., Louisell, W. H., and McKnight, W. B. (1974). *Phys. Rev. A*, 11:1365.
LeBoeuf, P. and Saraceno, M. (1990). *J. Phys. A: Math. Gen.*, 23:1745.
Ledinegg, E. (1966). *Zeit. Phys.*, 191:177.
Lenac, Z. and Tomaš, M. S. (2007). *Phys. Rev. A*, 75:042101.
Leonhardt, U. (2000). *Phys. Rev. A*, 62:012111.
Leonhardt, U. and Philbin, T. G. (2007a). *New J. Phys.*, 9:254.
Leonhardt, U. and Philbin, T. G. (2007b). *J. Opt. A: Pure Appl. Opt.*, 9:S289.
Leonhardt, U. and Piwnicki, P. (2001). *J. Mod. Opt.*, 48:977.
Letokhov, V. S. (1968). *Sov. Phys. JETP*, 26:835.
Leutheuser, V., Langbein, U., and Lederer, F. (1990). *Opt. Commun.*, 75:251.
Levenson, M. D., Shelby, R. M., Aspect, A., Reid, M., and Walls, D. F. (1985). *Phys. Rev. A*, 32:1550.
Li, L. W., Kooi, P. S., Leong, M. S., and Yeo, T. S. (1994). *IEEE Transactions Microwave Theory Tech.*, 42:2302.
Li, Y., Mikami, H., Wang, H., and Kobayashi, T. (2005). *Phys. Rev. A*, 72:063801.
Lifshitz, E. M. (1955). *Zh. Eksp. Teor. Fiz.*, 29:94.
Lifshitz, E. M. (1956). *Sov. Phys. JETP*, 2:73.
Lifshitz, E. M. and Pitaevskii, L. P. (1980). *Statistical Physics: Part 2*. Pergamon, Oxford.
Liñares, J. and Nistal, N. C. (2003). *J. Mod. Opt.*, 50:781.
Lindblad, G. (1976). *Commun. Math. Phys.*, 48:119.
Ling, Y., Cao, H., Burin, A. L., Ratner, M. A., Liu, X., and Chang, R. P. (2001). *Phys. Rev. A*, 64:063808.
Liu, G. J., Liu, J., Liang, B. M., Li, Q., and Jin, G. L. (2003). *Opt. Lett.*, 28:1347.
Lodahl, P., van Driel, A. F., Nikolaev, I. S., Irman, A., Overgaag, K., Vanmaekelbergh, D., and Vos, W. L. (2004). *Nature* (London), 430:654.
Lombardi, A. M. (2002). *J. Opt. B: Quantum Semiclass. Opt.*, 4:S473.
London, F. (1930). *Z. Phys.*, 63:245.
Loudon, R. (1963). *Proc. Phys. Soc.*, 83:393.
Loudon, R. (1970). *J. Phys. A*, 3:233.
Loudon, R. (1998). *Phys. Rev. A*, 57:4904.
Loudon, R. (1999). In Fouque, J.-P., editor, *Diffuse Waves in Complex Media*, NATO Science Series C531, pp. 213–247. Kluwer Academic Publishers, Dordrecht.
Loudon, R. (2000). *The Quantum Theory of Light*, 3rd ed. Oxford University Press, New York.
Loudon, R. (2002). *J. Mod. Opt.*, 49:821.
Loudon, R. (2003). *Phys. Rev. A*, 68:013806.
Loudon, R., Barnett, S. M., and Baxter, C. (2005). *Phys. Rev. A*, 71:063802.
Loudon, R. and Knight, P. L. (1987). *J. Mod. Opt.*, 34:709.
Louisell, W. H. (1973). *Quantum Statistical Properties of Radiation*. J. Wiley, New York.
Lu, D., Hu, W., Zheng, Y., and Yang, Z. (2003). *Opt. Commun.*, 228:217.
Lugiato, L. A., Brambilla, M., and Gatti, A. (1999). In Bederson, B. and Walther, H., editors, *Advances in Atomic, Molecular, and Optical Physics*, p. 229. Academic Press, Boston.
Lugiato, L. A. and Gatti, A. (1993). *Phys. Rev. Lett.*, 70:3868.
Lugiato, L. A. and Marzoli, I. (1995). *Phys. Rev. A*, 52:4886.
Luis, A. and Peřina, J. (1996a). *Phys. Rev. A*, 53:1886.
Luis, A. and Peřina, J. (1996b). *Quantum Semiclass. Opt.*, 8:39.
Lukosz, W. (1971). *Physica* (Amsterdam), 56:109.
Lukš, A. and Peřinová, V. (2002). In Wolf, E., editor, *Progress in Optics*, Vol. 45, pp. 295–431. Elsevier Science, Amsterdam.

Lukš, A. and Peřinová, V. (2003). In Nowak, J., Zajac, M., and Masajada, J., editors, *Proceedings of SPIE*, Vol. 5259, pp. 22–29. SPIE, Bellingham, WA.
Maclay, J. (2000). *Phys. Rev. A*, 61:052110.
Mandel, L. (1964). *Phys. Rev.*, 136:B1221.
Mandel, L. (1966). *Phys. Rev.*, 144:1071.
Mandel, L. and Wolf, E. (1995). *Optical Coherence and Quantum Optics*. Cambridge University Press, New York.
Marcuse, D. (1974). *Theory of Dielectric Optical Waveguides*. Academic Press, New York.
Marinescu, N. (1992). *Prog. Quantum Electron.*, 16:183.
Markoš, P. (2005). *Čs. čas. fyz.*, 55:531 (in Slovak).
Markoš, P., Rousochatzakis, I., and Soukoulis, C. M. (2002). *Phys. Rev. E*, 66:045601.
Markoš, P. and Soukoulis, C. M. (2002a). *Phys. Rev. B*, 65:033401.
Markoš, P. and Soukoulis, C. M. (2002b). *Phys. Rev. E*, 65:036622.
Markushev, V. M., Zolin, V. F., and Briskina, Ch. M. (1986). *Sov. J. Quant. Electron.*, 16:281.
Marqués, R., Martel, J., Mesa, F., and Medina, F. (2002). *Phys. Rev. Lett.*, 89:183901.
Marqués, R., Martel, J., Mesa, F., and Medina, F. (2003). *Phys. Rev. Lett.*, 91:249402.
Matloob, R. (1999a). *Phys. Rev. A*, 60:50.
Matloob, R. (1999b). *Phys. Rev. A*, 60:3421.
Matloob, R. (2001). *Opt. Commun.*, 192:287.
Matloob, R. (2004a). *Phys. Rev. A*, 69:052110.
Matloob, R. (2004b). *Phys. Rev. A*, 70:022108.
Matloob, R. (2005). *Phys. Rev. A*, 71:062105.
Matloob, R. and Falinejad, H. (2001). *Phys. Rev. A*, 64:042102.
Matloob, R., Keshavarz, A., and Sedighi, D. (1999). *Phys. Rev. A*, 60:3410.
Matloob, R., Loudon, R., Artoni, M., Barnett, S. M., and Jeffers, J. (1997). *Phys. Rev. A*, 55:1623.
Matloob, R., Loudon, R., Barnett, S. M., and Jeffers, J. (1995). *Phys. Rev. A*, 52:4823.
Matloob, R. and Pooseh, G. (2000). *Opt. Commun.*, 181:109.
Matsko, A. B. and Kozlov, V. V. (2000). *Phys. Rev. A*, 62:033811.
Mazzitelli, F. D., Sánchez, M. J., Scoccola, N. N., and von Stecher, J. (2003). *Phys. Rev. A*, 76:013807.
McDonald, K. T. (2001). *Am. J. Phys.*, 69:607.
McKintosh, F. C. and John, S. (1989). *Phys. Rev. B*, 40:2383.
Mehra, J. (1967). *Physica*, 37:145.
Mehta, M. L. (1991). *Random Matrices*. Academic Press, New York.
Mello, P. A., Akkermans, E., and Shapiro, B. (1988a). *Phys. Rev. Lett.*, 61:459.
Mello, P. A., Pereyra, P., and Kumar, N. (1988b). *Ann. Phys.*, 181:290.
Méndez-Sánchez, R. A., Kuhl, U., Barth, M., Lewenkopf, C. H., and Stöckmann, H.-J. (2003). *Phys. Rev. Lett.*, 91:174102.
Mendonça, J. T., Martins, A. M., and Guerreiro, A. (2000). *Phys. Rev. E*, 62:2989.
Messina, R. and Passante, R. (2007a). *Phys. Rev. A*, 75:042113.
Messina, R. and Passante, R. (2007b). *Phys. Rev. A*, 76:032107.
Milburn, G. J., and Walls, D. F. (1981). *Opt. Commun.*, 39:40.
Milburn, G. J., Walls, D. F., and Levenson, M. D. (1984). *J. Opt. Soc. Am. B*, 1:390.
Milonni, P. W. (1994). *The Quantum Vacuum: An Introduction to Quantum Electrodynamics*. Academic Press, San Diego.
Milonni, P. W. (1995). *J. Mod. Opt.*, 42:1991.
Milonni, P. W., Cook, R. J., and Goggin, M. E. (1988). *Phys. Rev. A*, 38:1621.
Milonni, P. W. and Maclay, G. J. (2003). *Opt. Commun.*, 228:161.
Milton, K. A. (2004). *J. Phys. A: Math. Gen.*, 37:R209.
Mishchenko, E. G. and Beenakker, C. W. J. (1999). *Phys. Rev. Lett.*, 83:5475.
Mishchenko, E. G., Patra, M., and Beenakker, C. W. J. (2001). *Europhys. J. D*, 13:289.
Mishra, S. and Satpathy, S. (2003). *Phys. Rev. B*, 68:045121.

Misirpashaev, T. Sh. and Beenakker, C. W. J. (1998). *Phys. Rev. A*, 57:2041.

Misner, C. W., Thorne, K. C., and Wheeler, J. A. (1970). *Gravitation*. Freeman, San Francisco.

Mišta, Jr., L. (1999). *Acta Phys. Slovaca*, 49:737.

Mišta, Jr., L. and Peřina, J. (1997). *Czech. J. Phys.*, 47:629.

Mogilevtsev, D. N., Korolkova, N., and Peřina, J. (1996). *J. Opt. B: Quantum Semiclass. Opt.*, 8:1169.

Mogilevtsev, D. N., Korolkova, N., and Peřina, J. (1997). *J. Mod. Opt.*, 44:1293.

Mohideen, U. and Roy, A. (1998). *Phys. Rev. Lett.*, 81:4549.

Monken, C. H., Ribeiro, P. H. S., and Pádua, S. (1998). *Phys. Rev. A*, 57:3123.

Morris, G. M. (1989). In Arsenault, H. H., Szoplik, T., and Macukow, B., editors, *Optical Processing and Computing*, p. 343. Academic Press, New York.

Mortensen, N. A. (2005). *Opt. Lett.*, 30:1455.

Mortensen, N. A., Folkenberg, J. R., Nielsen, M. D., and Hansen, K. P. (2003). *Opt. Lett.*, 28:1879.

Mosset, A., Devaux, F., and Lantz, E. (2005). *Phys. Rev. Lett.*, 94:223603.

Mrówczyński, S. and Müller, B. (1994). *Phys. Rev. D*, 50:7542.

Mujumdar, S., Ricci, M., Torre, R., and Wiersma, D. S. (2004). *Phys. Rev. Lett.*, 93:53903.

Munday, J. N. and Capasso, F. (2007). *Phys. Rev. A*, 75:060102.

Müstecaplıoğlu, Ö. E. and Öktel, M. Ö. (2005). *Phys. Rev. Lett.*, 94:220404.

Nagasako, E. M., Boyd, R. W., and Agarwal, G. S. (1997). *Phys. Rev. A*, 55:1412.

Naqvi, Q. A. and Abbas, M. (2003). *Opt. Commun.*, 227:143.

Nesterenko, V. V. and Pirozhenko, I. G. (1997). *Phys. Rev. D*, 57:1284.

Newton, R. G. (1966). *Scattering Theory of Waves and Particles*. McGraw-Hill, New York.

Nielsen, M. A. and Chuang, I. L. (2000). *Quantum Computation and Quantum Information*. Cambridge University Press, Cambridge.

Nilsson, O., Yamamoto, Y., and Machida, S. (1986). *IEEE J. Quant. Electr.*, QE-2:2043.

Nöckel, J. U. and Stone, D. (1997). *Nature*, 385:45.

Notomi, M. (2000). *Phys. Rev. B*, 62:10696.

Oughstun, K. E. and Cartwright, N. A. (2005). *J. Mod. Opt.*, 52:1089.

Ozbay, E., Guven, K., and Aydin, K. (2007). *J. Opt. A: Pure Appl. Opt.*, 9:S301.

Paasschens, J. C. J., Misirpashaev, T. S., and Beenakker, C. W. J. (1996). *Phys. Rev. B*, 54:11887.

Padgett, M., Barnett, S. M., and Loudon, R. (2003). *J. Mod. Opt.*, 50:1555.

Paspalakis, E. and Kis, Z. (2002). *Phys. Rev. A*, 66:025802.

Passante, R., Power, E. A., and Thirunamachandran, T. (1998). *Phys. Lett. A*, 249:77.

Passante, R. and Spagnolo, S. (2007). *Phys. Rev. A*, 76:042112.

Patra, M. (2002). *Phys. Rev. A*, 65:043809.

Patra, M. (2003). *Phys. Rev. E*, 67:016603.

Patra, M. and Beenakker, C. W. J. (1999). *Phys. Rev. A*, 60:4059.

Patra, M. and Beenakker, C. W. J. (2000). *Phys. Rev. A*, 61:063805.

Patra, M., Schomerus, H., and Beenakker, C. W. J. (2000). *Phys. Rev. A*, 61:023810.

Pedrotti, F. L. and Pedrotti, L. S. (1993). *Introduction to Optics*. Prentice-Hall, Englewood Cliffs, NJ.

Peierls, R. (1976). *Proc. R. Soc. A* (London), 347:75.

Peierls, R. (1985). In Bassani, F., Fumi, F., Loudon, R., and Tosi, M. P., editors, *Highlights of Condensed Matter Theory*. North-Holland, Amsterdam.

Pendry, J. B. (2000). *Phys. Rev. Lett.*, 85:3966.

Pendry, J. B. (2003). *Nature*, 423:22.

Pendry, J. B., Holden, A. J., Robbins, D. J., and Stewart, W. J. (1999). *IEEE Trans. Microwave Theory Tech.*, 47:2075.

Pendry, J. B., Holden, A. J., Stewart, W. J., and Youngs, I. (1996). *Phys. Rev. Lett.*, 76:4773.

Peřina, J. (1981a). *Optica Acta*, 28:325.

Peřina, J. (1981b). *Optica Acta*, 28:1529.

Peřina, J. (1991). *Quantum Statistics of Linear and Nonlinear Optical Phenomena*, 2nd edn. Kluwer Academic Publishers, Dordrecht.

Peřina, J. (1995a). *J. Mod. Opt.*, 42:1517.
Peřina, J. (1995b). *Acta Physica Slovaca*, 45:279.
Peřina, J. and Bajer, J. (1995). *J. Mod. Opt.*, 42:2337.
Peřina, J. and Křepelka, J. (2005). *J. Opt. B: Quantum Semiclass. Opt.*, 7:246.
Peřina, J. and Křepelka, J. (2006). *Opt. Commun.*, 265:632.
Peřina, J., Křepelka, J., Peřina, Jr., J., Bondani, M., Allevi, A., and Andreoni, A. (2007). *Phys. Rev. A*, 76:043806.
Peřina, J. and Peřina, Jr., J. (1995a). *J. Opt. B: Quantum Semiclass. Opt.*, 7:541.
Peřina, J. and Peřina, Jr., J. (1995b). *J. Opt. B: Quantum Semiclass. Opt.*, 7:849.
Peřina, J. and Peřina, Jr., J. (1995c). *J. Opt. B: Quantum Semiclass. Opt.*, 7:863.
Peřina, J. and Peřina, Jr., J. (1996). *J. Mod. Opt.*, 43:1951.
Peřina, Jr., J. (2008). *Phys. Rev. A*, 77:013803.
Peřina, Jr., J., Centini, M., Sibilia, C., Bertolotti, M., and Scalora, M. (2006). *Phys. Rev. A*, 73:033823.
Peřina, Jr., J., Haderka, O., Sibilia, C., Bertolotti, M., and Scalora, M. (2007). *Phys. Rev. A*, 76:033813.
Peřina, Jr., J. and Peřina, J. (1997). *J. Opt. B: Quantum Semiclass. Opt.*, 9:443.
Peřina, Jr., J. and Peřina, J. (2000). In Wolf, E., editor, *Progress in Optics*, Vol. 41, pp. 361–419. North-Holland, Amsterdam.
Peřina, Jr., J., Sibilia, C., Tricca, D., and Bertolotti, M. (2004). *Phys. Rev. A*, 70:043816.
Peřina, Jr., J., Sibilia, C., Tricca, D., and Bertolotti, M. (2005). *Phys. Rev. A*, 71:043813.
Peřinová, V. and Lukš, A. (2000). In Wolf, E., editor, *Progress in Optics*, Vol. XL, pp. 115–269. North-Holland, Amsterdam.
Peřinová, V. and Lukš, A. (2003). *Fort. Phys.*, 51:210.
Peřinová, V., Lukš, A., and Křepelka, J. (2003). *J. Opt. B: Quantum Semiclass. Opt.*, 5:254.
Peřinová, V., Lukš, A., and Křepelka, J. (2004). *J. Opt. B: Quantum Semiclass. Opt.*, 6:S104.
Peřinová, V., Lukš, A., and Křepelka, J. (2006). *J. Phys. B: At. Mol. Opt. Phys.*, 39:2267.
Peřinová, V., Lukš, A., Křepelka, J., and Jelínek, V. (2000). *J. Opt. B: Quantum Semiclass. Opt.*, 2:746.
Peřinová, V., Lukš, A., Křepelka, J., Sibilia, C., and Bertolotti, M. (1991). *J. Mod. Opt.*, 38:2429.
Peřinová, V., Lukš, A., and Peřina, J. (1998). *Phase in Optics*. World Scientific, Singapore.
Petersson, L. E. R. and Smith, G. S. (2003). *J. Opt. Soc. Am. A*, 20:2378.
Pfeifer, R. N. C., Nieminen, T. A., Heckenberg, N. R., and Rubinsztein-Dunlop, H. (2007). *Rev. Mod. Phys.*, 79:1197.
Pitaevskii, L. P. (2006). *Phys. Rev. A*, 73:047801.
Plunien, G., Muller, B., and Greiner, W. (1986). *Phys. Rep.*, 134:87.
Pokrovsky, A. L. and Efros, A. L. (2002). *Phys. Rev. Lett.*, 89:093901.
Porter, C. E. (1965). *Statistical Theory of Spectra: Fluctuations*. Academic Press, New York.
Potasek, M. J. and Yurke, B. (1987). *Phys. Rev. A*, 35:3974.
Power, E. A. and Thirunamachandran, T. (1993). *Phys. Rev. A*, 48:4761.
Power, E. A. and Zienau, S. (1959). *Philos. Trans. R. Soc. Lond.* Series A, 251:427.
Pradhan, P. and Kumar, N. (1994). *Phys. Rev. B*, 50:9644.
Raabe, C., Knöll, L., and Welsch, D.-G. (2003). *Phys. Rev. A*, 68:033810.
Raabe, C., Knöll, L., and Welsch, D.-G. (2004). *Phys. Rev. A*, 69:019901.
Raabe, C. and Welsch, D.-G. (2005). *Phys. Rev. A*, 71:013814.
Raabe, C. and Welsch, D.-G. (2006). *Phys. Rev. A*, 73:047802.
Rarity, J. G. and Tapster, P. R. (1990). *Phys. Rev. Lett.*, 64:2495.
Raymer, M. G., Beck, M., and McAlister, D. F. (1994). *Phys. Rev. Lett.*, 72:1137.
Raymond Ooi, C. H. and Scully, M. O. (2007). *Phys. Rev. A*, 76:043822.
Reed, M. and Simon, B. (1979). *Methods of Modern Mathematical Physics: III, Scattering Theory*. Academic Press, New York.
Řeháček, J. and Peřina, J. (1996). *Opt. Commun.*, 132:549.

Reichl, L. (1992). *The Transition to Chaos in Conservative Classical Systems: Quantum Manifestations*. Springer-Verlag, New York.
Reimann, P. (2002). *Phys. Rep.*, 361:57.
Rice, P. R. and Carmichael, H. J. (1994). *Phys. Rev. A*, 50:4318.
Rizzuto, L. (2007). *Phys. Rev. A*, 76:062114.
Robinson, F. N. H. (1971). *Physica* (Utrecht), 54:329.
Robinson, F. N. H. (1973). *Macroscopic Electromagnetism*, 1st ed., Pergamon Press, Oxford.
Rodriguez, A., Ibanescu, M., Iannuzzi, D., Capasso, F., Joannopoulos, J. D., and Johnson, S. G. (2007a). *Phys. Rev. Lett.*, 99:080401.
Rodriguez, A., Ibanescu, M., Iannuzzi, D., Joannopoulos, J. D., and Johnson, S. G. (2007b). *Phys. Rev. A*, 76:032106.
Rodrigues, R. B., Maia Neto, P. A., Lambrecht, A., and Reynaud, S. (2006a). *Europhys. Lett.*, 76:822.
Rodrigues, R. B., Maia Neto, P. A., Lambrecht, A., and Reynaud, S. (2006b). *Phys. Rev. Lett.*, 96:100402.
Rodrigues, R. B., Maia Neto, P. A., Lambrecht, A., and Reynaud, S. (2007a). *Phys. Rev. A*, 75:062108.
Rodrigues, R. B., Maia Neto, P. A., Lambrecht, A., and Reynaud, S. (2007b). *Phys. Rev. Lett.*, 98:068902.
Roman, P. (1969). *Introduction to Quantum Field Theory*. J. Wiley, New York.
Roppo, V., Centini, M., Sibilia, C., Bertolotti, M., de Ceglia, D., Scalora, M., Akozbek, N., Bloemer, M. J., Haus, J. W., Kosareva, O. G., and Kandidov, V. P. (2007). *Phys. Rev. A*, 76:033829.
Rosenau da Costa, M., Caldeira, A. O., Dutra, S. M., and Westfahl, Jr., H. (2000). *Phys. Rev. A*, 61:022107.
Rosenbluh, M. and Shelby, R. M. (1991). *Phys. Rev. Lett.*, 66:153.
Rosewarne, D. M. (1991). *Quantum Opt.*, 3:193.
Roth, J.-P. (1985). In Mokobodzki, G. and Pinchon, D., editors, *Théorie du Potentiel, Proceedings of the Colloquia Jacques Deny*, Lecture Notes in Mathematics, Vol. 1096, pp. 521–539. Springer-Verlag, Berlin.
Ruostekoski, J. (2000). *Phys. Rev. A*, 61:033605.
Ruppin, R. (2000). *Solid State Commun.*, 116:411.
Ruppin, R. (2002). *Phys. Lett. A*, 299:309.
Sakoda, K. (2002). *J. Opt. Soc. Am. B*, 19:2060.
Sakoda, K. and Haus, J. W. (2003). *Phys. Rev. A*, 68:053809.
Sakoda, K. and Ohtaka, K. (1996a). *Phys. Rev. B*, 54:5732.
Sakoda, K. and Ohtaka, K. (1996b). *Phys. Rev. B*, 54:5742.
Saleh, B. E. A. and Teich, M. C. (1991). *Fundamentals of Photonics*. J. Wiley, New York.
Sanz, M., Papageorgopoulos, A. C., Egelhoff, W. F., Nieto-Vesperinas, M., and García, N. (2003). *Phys. Rev. E*, 67:067601.
Sargent III, M., Scully, M. O., and Lamb Jr., W. E. (1977). *Laser Physics*. Addison-Wesley, Reading, MA.
Savage, G. M. and Walls, D. F. (1987). J. Opt. Soc. Am. B, 4:1514.
Scardicchio, A. and Jaffe, R. L. (2006). *Nucl. Phys.*, B743:249.
Schaden, M. and Spruch, L. (1998). *Phys. Rev. A*, 58:935.
Schaden, M. and Spruch, L. (2000). *Phys. Rev. Lett.*, 84:459.
Scheel, S., Knöll, L., and Welsch, D.-G. (1998). *Phys. Rev. A*, 58:700.
Scheel, S., Knöll, L., and Welsch, D.-G. (1999a). *Phys. Rev. A*, 60:4094.
Scheel, S., Knöll, L., and Welsch, D.-G. (2000). *Phys. Rev. A*, 61:069901.
Scheel, S., Knöll, L., Welsch, D.-G., and Barnett, S. M. (1999b). *Phys. Rev. A*, 60:1590.
Schmidt, E., Jeffers, J., Barnett, S. M., Knöll, L., and Welsch, D.-G. (1998). *J. Mod. Opt.*, 45:377.
Schmidt, E., Knöll, L., and Welsch, D.-G. (1996). *Phys. Rev. A*, 54:843.
Schram, K. (1960). *Physica* (Utrecht), 26:1080.
Schram, K. (1973). *Phys. Lett.*, 43A:282.

Schubert, M. and Wilhelmi, B. (1986). *Nonlinear Optics and Quantum Electronics*. J. Wiley, New York.
Schwinger, J., DeRaad, Jr., L. L., and Milton, K. A. (1978). *Ann. Phys.* (New York), 115:1.
Scully, M. O. and Zubairy, M. S. (1997). *Quantum Optics*. Cambridge University Press, Cambridge.
Sczaniecki, L. (1983). *Phys. Rev. A*, 28:3493.
Sebbah, P., Hu, B., Kopp, V. I., and Genack, A. Z. (2007). *J. Opt. Soc. Am. B*, 24:A77.
Selden, A. C. (1974). *Opt. Commun.*, 10:1.
Serulnik, S. and Ben-Aryeh, Y. (1991). *Quantum Opt.*, 3:63.
Severini, S., Tricca, D., Sibilia, C., Bertolotti, M., and Peřina, J. (2004). *J. Opt. B: Quantum Semiclass. Opt.*, 6:110.
Sha, W. L., Liu, C.-H., and Alfano, R. R. (1994). *Opt. Lett.*, 19:1922.
Shelby, R. M., Drummond, P. D., and Carter, S. J. (1990). *Phys. Rev. A*, 2:2966.
Shelby, R. A., Smith, D. R., and Schultz, S. (2001). *Science*, 292:77.
Shen, Y. R. (1967). *Phys. Rev.*, 155:921.
Shen, Y. R. (1969). In Glauber, R. J., editor, *Quantum Optics*, p. 473. Academic Press, New York.
Shen, Y. R., (1984). *The Principles of Nonlinear Optics*. J. Wiley, New York.
Shen, J.-Q. (2004). *J. Opt. A: Pure Appl. Opt.*, 6:239.
Shih, Y. (2003). *Rep. Prog. Phys.*, 66:1009.
Shirasaki, M. and Haus, H. A. (1990). *J. Opt. Soc. Am. B*, 7:30.
Shirasaki, M., Haus, H. A., and Liu Wong, D. (1989). *J. Opt. Soc. Am. B*, 6:82.
Shirkov, D. V. and Kovalev, V. F. (2001). *Phys. Rep.*, 352:219.
Shukla, P. K., Marklund, M., Brodin, G., and Stenflo, L. (2004). *Phys. Lett. A*, 330:131.
Sibilia, C., Centini, M., Sakoda, K., Haus, J. W., and Bertolotti, M. (2005). *J. Opt. A: Pure Appl. Opt.*, 7:S198.
Slusher, R. E. and Eggleton, B. J. (2003). *Nonlinear Photonic Crystals*. Springer-Verlag, Berlin.
Siegman, A. E. (1986). *Lasers*. Mill Valley, California.
Siegman, A. E. (1989). *Phys. Rev. A*, 39:1253.
Slusher, R. E., Grangier, P., La Porta, A., Yurke, B., and Potasek, M. J. (1987). *Phys. Rev. Lett.*, 59:2566.
Smith, G. S. (1997). *An Introduction to Classical Electromagnetic Radiation*. Cambridge University Press, Cambridge, UK.
Smith, D. R., Schultz, S., Markoš, P., and Soukoulis, C. M. (2002). *Phys. Rev. B*, 65:195104.
Sokolov, I. V. (1991a). *Zh. Eksp. Teor. Fiz.*, 100:780.
Sokolov, I. V. (1991b). *Sov. Phys. JETP*, 73:431.
Solimeno, S., Crosignani, B., and Di Porto, P. (1986). *Guiding, Diffraction, and Confinement of Optical Radiation*. Academic Press, Orlando.
Sparnaay, M. J. (1989). In Sarlemijn, A. and Sparnaay, M. J., editors, *Physics in the Making*, pp. 235–246. Elsevier, Amsterdam.
Stegeman, G. I., Bowden, C. M., and Zheltikov, A. M., editors. (2002). *Journal of the Optical Society of America B*, Special Issue Nonlinear Optics of Photonic Crystals, Vol. **19**, pp. 2046–2296. Optical Society of America, Washington, DC.
Steinberg, S. (1985). In Sanchez Mondragon, J. and Wolf, K. B., editors, *Lie Methods in Optics*. Springer-Verlag, Berlin.
Stenholm, S. (2000). *Opt. Commun.*, 179:247.
Strekalov, D. V., Sergienko, A. V., Klyshko, D. N., and Shih, Y. H. (1995). *Phys. Rev. Lett.*, 74:3600.
Sudarshan, E. C. G. (1963). *Phys. Rev. Lett.*, 10:277.
Suttorp, L. G. and Wubs, M. (2004). *Phys. Rev. A*, 70:013816.
Svelto, O. and Hanna, D. C. (1977). In Schäfer, F. P., editor, *Dye Lasers*, p. 78. Springer-Verlag, Berlin.
Svelto, O. and Hanna, D. C. (1989). *Principles of Lasers*. Plenum Press, New York.

Svetovoy, V. B. (2007). *Phys. Rev. A*, 76:062102.
Taflove, A. (1995). *Computational Electrodynamics: The Finite-Difference Time-Domain Method*. Artech House, London.
Tai, C.-T. (1994). *Dyadic Green Functions in Electromagnetic Theory*. IEEE Press, New York.
Tanaka, H., Niwa, H., Hayami, K., Furue, S., Nakayama, K., Kohmoto, T., Kunitomo, M., and Fukuda, Y. (2003). *Phys. Rev. A*, 68:053801.
Tapster, P. R., Rarity, J. G., and Owens, P. C. M. (1994). *Phys. Rev. Lett.*, 73:1923.
Tarasov, V. E. (2001). *Phys. Lett. A*, 288:173.
Tatarinova, L. L. and Garcia, M. E. (2007). *Phys. Rev. A*, 76:043824.
Taylor, J. R. (1972). *Scattering Theory: The Quantum Theory of Nonrelativistic Collisions*. Krieger, Malabar, Florida.
Ter-Gabriėlyan, N. Ė., Markushev, V. M., Belan, V. R., Briskina, C. M., Dimitrova, O. V., Zolin, V. F., and Lavrov, A. V. (1991a). *Sov. J. Quantum Electron.*, 21:281.
Ter-Gabriėlyan, N. Ė., Markushev, V. M., Belan, V. R., Briskina, Ch. M., and Zolin, V. F. (1991b). *Sov. J. Quantum Electron.*, 21:32.
Ter Haar, D. (1961). *Elements of Hamiltonian Mechanics*. Interscience, New York.
Thareja, R. K. and Mitra, A. (2000). *Appl. Phys. B*, 71:181.
Tip, A. (1997). *Phys. Rev. A*, 56:5022.
Tip, A. (1998). *Phys. Rev. A*, 57:4818.
Tip, A. (2004). *Phys. Rev. A*, 69:013804.
Tip, A. (2007). *Phys. Rev. A*, 76:043835.
Tip, A., Knöll, L., Scheel, S., and Welsch, D.-G. (2001). *Phys. Rev. A*, 63:043806.
Toda, M., Adachi, S., and Ikeda, K. (1989). *Prog. Theor. Phys. Suppl.*, 98:323.
Tomaš, M. S. (1995). *Phys. Rev. A*, 51:2545.
Tomaš, M. S. (2002). *Phys. Rev. A*, 66:052103.
Toren, M. and Ben-Aryeh, Y. (1994). *Quantum Opt.*, 6:425.
Townsend, P. D., Baker, G. L., Jackel, J. L., Shelburne III, J. A., and Etemad, S. (1989). *SPIE*, 1147:256.
Tricca, D., Sibilia, C., Severini, S., Bertolotti, M., Scalora, M., Bowden, M., and Sakoda, K. (2004). *J. Opt. Soc. Am. B*, 21:671.
Tucker, J. and Walls, D. F. (1969). *Phys. Rev.*, 178:2036.
Ujihara, K. (1975). *Phys. Rev. A*, 12:148.
Ujihara, K. (1976). *Jpn. J. Appl. Phys.*, 15:1529.
Ujihara, K. (1977). *Phys. Rev. A*, 16:562.
Valanju, P. M., Walser, R. M., and Valanju, A. P. (2002). *Phys. Rev. Lett.*, 88:187401.
Valencia, A., Scarcelli, G., D'Angelo, M., and Shih, Y. (2005). *Phys. Rev. Lett.*, 94:063601.
van Albada, M. P. and Lagendijk, A. (1985). *Phys. Rev. Lett.*, 55:2692.
van der Molen, K. L., Mosk, A. P., and Lagendijk, A. (2007). *Opt. Commun.*, 278:110.
van Kampen, N. G., Nijboer, B. R. A., and Schram, K. (1968). *Phys. Lett. A*, 26:307.
van Kranendonk, J. and Sipe, J. E. (1977). In Wolf, E., editor, *Progress in Optics*, Vol. 15, p. 245. North-Holland, Amsterdam.
Vanneste, C. and Sebbah, P. (2001). *Phys. Rev. Lett.*, 87:183903.
van Rossum, M. C. W. and Nieuwenhuizen, T. M. (1999). *Rev. Mod. Phys.*, 71:313.
Veselago, V. G. (1967). *Sov. Phys. Solid State*, 8:2854.
Veselago, V. G. (1968). *Sov. Phys. Uspechi*, 10:509.
Veselago, V. G. (2002). *Uspekhi Fiz. Nauk*, 172:1215 (in Russian).
Viviescas, C. and Hackenbroich, G. (2003). *Phys. Rev. A*, 67:013805.
Vogel, W. and Welsch, D.-G. (1994). *Lectures on Quantum Optics*. Akademie Verlag, Berlin.
Vogel, W., Welsch, D.-G., and Wallentowitz, S. (2001). *Quantum Optics*. Wiley-VCH, Weinheim.
Walls, D. F. (1983). *Nature* (London), 306:141.
Walls, D. F. and Milburn, G. J. (1994). *Quantum Optics*. Springer-Verlag, Berlin, p. 127.
Wang, L. J., Zou, X. Y., and Mandel, L. (1991a). *Phys. Rev. A*, 44:4614.

Wang, L. J., Zou, X. Y., and Mandel, L. J. (1991b). *J. Opt. Soc. Am. B*, 8:978.

Watanabe, K., Nakano, H., Honold, A., and Yamamoto, Y. (1989). *Phys. Rev. Lett.*, 62:2257.

Weigert, S. (1996). *Phys. Lett. A*, 214:215.

Weinert-Raczka, E. (1996). *Pure Appl. Opt.*, 5:251.

Weinert-Raczka, E. and Lederer, F. (1993). *Opt. Commun.*, 102:478.

Wennerström, H., Daicic, J., and Ninham, B. W. (1999). *Phys. Rev. A*, 60:2581.

Wiersma, D. S., Bartolini, P., Lagendijk, A., and Righini, R. (1997). *Nature*, 390:671.

Wiersma, D. S. and Lagendijk, A. (1996). *Phys. Rev. E*, 54:4256.

Wiersma, D. S., van Albada, M. P., and Lagendijk, A. (1995a). *Nature*, 373:203.

Wiersma, D. S., van Albada, M. P., and Lagendijk, A. (1995b). *Phys. Rev. Lett.*, 75:1739.

Wigner, E. P. (1932). *Phys. Rev.*, 40:749.

Wigner, E. P. (1956). In *Proceedings of the Conference on Neutron Physics by Time-of-Flight*, pp. 59–70. Gatlinburg, Tennessee. (Oak Ridge National Laboratory Report No. ORNL 2309, unpublished).

Williamson, J. (1936). *Am. J. Math.*, 58:141.

Wiseman, H. M. and Mølmer, K. (2000). *Phys. Lett. A*, 270:245.

Wolf, P. E. and Maret, G. (1985). *Phys. Rev. Lett.*, 55:2696.

Woolley, R. G. (1971). *Philos. Trans. R. Soc. Lond.* Series A, 321:557.

Wright, E. M. (1990). *J. Opt. Soc. Am. B*, 7:1141.

Wubs, M. and Suttorp, L. G. (2001). *Phys. Rev. A*, 63:043809.

Yablonovich, E. (1987). *Phys. Rev. Lett.*, 58:2059.

Yablonovitch, E. and Gmitter, T. J. (1987). *Phys. Rev. Lett.*, 58:2486.

Yablonovitch, E. and Gmitter, T. J. (1989). *Phys. Rev. Lett.*, 63:1950.

Yamamoto, Y. and Imoto, N. (1986). *IEEE J. Quant. Electr.*, QE–22:2032.

Yang, C. N. (1967). *Phys. Rev. Lett.*, 19:1312.

Yang, C. N. (1968). *Phys. Rev.*, 168:1920.

Yang, S. and Kellman, M. E. (2000). *Phys. Rev. A*, 62:022105.

Yang, S. and Kellman, M. E. (2002). *Phys. Rev. A*, 65:034103.

Yanik, M. F. and Fan, S. (2005). *Phys. Rev. A*, 71:013803.

Yariv, A. (1985). *Optical Electronics*, 3rd ed. J. Wiley, New York.

Yariv, A. (1989). *Quantum Electronics*, 3rd ed. J. Wiley, New York, p. 136.

Yariv, A. and Yeh, P. (1984). *Optical Waves in Crystals*. J. Wiley, New York.

Yeh, P. (1988). *Optical Waves in Layered Media*. J. Wiley, New York.

Yoshie, T., Scherer, A., Hendrickson, J., Khitrova, G., Gibbs, H. M., Rupper, G., Ell, C., Shchekin, O. B., and Deppe, D. G. (2004). *Nature* (London), 432:200.

Yuen, H. P. and Shapiro, J. H. (1979). Opt. Lett., 4:334.

Yurke, B. (1984). *Phys. Rev. A*, 29:408.

Yurke, B. (1985). *Phys. Rev. A*, 32:300.

Yurke, B., Grangier, P., Slusher, R. F., and Potasek, M. J. (1987). *Phys. Rev. A*, 35:3586.

Zacharakis, G., Papadogiannis, N. A., Filippidis, G., and Papazoglou, T. G. (2000). *Opt. Lett.*, 25:923.

Zaleśny, J. (2001). *Phys. Rev. E*, 63:026603.

Zhang, Z.-Q. (1995). *Phys. Rev. B*, 52:7960.

Zhang, J., Peng, K., and Braunstein, S. L. (2003). *Phys. Rev. A*, 68:013808.

Zharov, A. A., Shadrivov, I. V., and Kivshar, Y. S. (2003). *Phys. Rev. Lett.*, 91:037401.

Zhou, F. and Spruch, L. (1995). *Phys. Rev. A*, 52:297.

Zinn-Justin, J. (1989). *Quantum Field Theory and Critical Phenomena*. Oxford University Press, Oxford.

Ziolkowski, R. W. (2001). *Phys. Rev. E*, 63:046604.

Ziolkowski, R. W. and Heyman, E. (2001). *Phys. Rev. E*, 64:056625.

Zou, X. Y., Wang, L. J., and Mandel, L. (1991). *Phys. Rev. Lett.*, 67:318.

Zyuzin, A. Yu. (1995). *Phys. Rev. E*, 51:5274.

Index

MIX
Papier aus verantwortungsvollen Quellen
Paper from responsible sources
FSC® C105338

If you have any concerns about our products,
you can contact us on
ProductSafety@springernature.com

In case Publisher is established outside the EU,
the EU authorized representative is:
Springer Nature Customer Service Center GmbH
Europaplatz 3, 69115 Heidelberg, Germany

Printed by Libri Plureos GmbH
in Hamburg, Germany